THE BLACK HOLE FACTORY

THE BLACK HOLE FACTORY

CERN VS. 'EINSTEIN ET AL'

To order additional copies of this book, contact:
Xlibris Corporation
1-888-795-4274
www.Xlibris.com
Orders@Xlibris.com
37970

CONTENTS

To Walter, Otto, Eric, two James, and three Marks, who did their job, denouncing 'Muster Quark'

To Obama, Rompuy, Sarkozy, Fog, and ZP, who should do theirs: saving the Earth by shutting down CERN

INTRODUCTION

Technology and the Future of History

We are a thin layer of mush on a rock speck in a corner of the Universe.
Now, departing from those facts, we can talk about man.
—Schopenhauer, father of modern philosophy

The Evolution of Technology

Science is searching for higher levels of energy (global warming, weapons, and supercolliders) and information (robotics) that will push life species, including mankind, into extinction. Death is an overdrive of energy (accident) or information (third age) that breaks the balances of life.

1. The Extinction of History and Gaia

We are a fragile species. Despite our claims to be the center of the Universe, professed by most religious people that believe in a god ever attentive to man or by most scientists who think humanity is the only intelligent species of the Universe and have ruled out the existence of a higher god ("Sir, we have ruled out that hypothesis," said Laplace to Napoleon[1]), the position of man relative to the cosmos is similar to that of a bacterium relative to a human body: insignificant and rather expendable. In scientific terms, that insignificance implies that we could easily disappear from the world, leaving few traces behind and the Universe would not even notice. This is what happens to us as individuals, despite our intense struggle against death and our creative attempts to leave behind a trace of our information—our work and genes in our sons. But within a few generations, 99.99% of human beings disappear in the memory of those who follow them. When we go through century-old pictures of our ancestors, we cannot even remember who they are. Death erases information, returning all beings to the initial dust of space-time from where they departed. So it will happen to the human species if we do not take seriously, as we do with our individual lives, the protection of our collective social superorganism, history—the life of the human species, from the first cell of history, the first human being who reproduced and colonized the Earth, till the last group of men that will inhabit it.

The fundamental systems of Gaia, the planet Earth, are superorganisms, individual groups of the same species, regulated by networks of energy and information. We are concerned here with two of such superorganisms, the individual human being and the entire species of mankind, in which history acts as the collective brain of this planet. In the individual case, our cells are the "fractal parts," joined by energetic blood systems and informative nervous systems, which create together the whole biological human superorganism.

As any medical student knows, the foundation of his discipline is physiology, which studies the imbalances of those energetic/blood and nervous/informative systems that cause most human sicknesses. Death happens when the networks of energy and information that hold together the cellular elements of any organic system become corrupted and go out of balance, suffering an excess of energy (fever, accidents) or a lack of it (cold, hunger) or an excess of information (cancer, old age) or a lack of it (ignorance that puts us in jeopardy).

A civilization of history is also a superorganism made of human cells (citizens) joined by networks of energy (electricity networks, roads, transport machines, weapons) and networks of information (science, financial systems, audiovisual information, culture, political laws). It follows that a civilization dies when an excess of energetic weapons or financial and legal information (wars, corrupted laws, hyperinflation) destroys its political or financial system, as it is happening today with many societies.[2]

When those legal and economical systems become corrupted, the nation suffers economical crises, revolutions, and wars that might destroy the nation or civilization.

In the case of history, the sum of all nations, which form today, in the age of globalization, an interconnected single historic superorganism, the *informative*, legal, and political systems and the *energetic* economic systems have to be balanced to survive.

It follows that Gaia, our global superorganism, made of an energetic system, the planet Earth, and an informative network, the human kind—which is the species with higher information, hence acting as its relative brain—could also die if an excess of energy (a superweapon, like the quark cannon that might produce black holes in France this year) or a new species with higher information than humanity (superintelligent robots, made of metals, atoms stronger and more complex than our weak carbon/oxygen structures) substitute us.

2. The Richter's Scale of Extinction

The *Guardian*, a respected British magazine, made two years ago a scientific poll, trying to find those excesses of energy and information that can extinguish us. The newspaper reported the ten most probable causes of the end of man with the usual surrealist sense of humor that has become standard in the press, unable to confront objectively and propose solutions to the reality of our death foretold:

> How will it all end? Some say we are likely to go with a bang, others predict a slow lingering end, while the optimists suggest we will overcome our difficulties by evolving into a different species.
>
> Kate Ravilious asks 10 scientists to name the biggest danger to Earth and assess the chances of extinction from 10 to 1, in case the event happens in the next 70 years.
>
> The results from minimal to maximum extinction risk are as follows:
>
> — A major terrorist attack: risk value 2 (minimal extinction)
> — A viral pandemic: risk 3
> — Cosmic ray blast from exploding supernova: risk 4
> — Meteorite impact—the Earth being hit by a large asteroid: risk 5
> — Climate Change: risk 6
> — Supervolcanoes: risk 7

— A human population crash due to telomere erosion: risk 8
— A global nuclear war: risk 8
— Superintelligent robots taking over: risk 8
— Earth being gobbled up by a black hole: risk 10 (absolute extinction)

If we analyze that list of possible causes of human extinction in this century, we notice first that the two most publicized causes for collective death risk—terrorism or pandemic, provoked by some virus run amok (a new mad cow disease, a flu, a new strain of AIDS or a genetically modified lethal virus)—are the less risky, scoring a mere two or three points in the Richter scale of human extinction. Yet because they seem easier to solve, they are preferred by mass media and politicians. We have today a flu scare because we all know it is just flu and it will be solved. So politicians will feel successful and get their medals while the press will have a fairy tale with a final, public catharsis. On the other hand, the true menaces to our future, hyperinformative and energetic technologies, are ignored because they are real problems, which need to confront the power of companies running amok like CERN and, what is even more difficult, *the ideology of mechanism, the concept that machines are always good, and the solution to all our problems.*

The same "don't worry be happy" attitude of not confronting the real problems of mankind, which we call in this book the LOL method, works in the next category of catastrophes since they are an array of natural, accidental causes that humans can't control: a supervolcano, a meteorite crashing on Earth, the sun becoming a supernova and burning the Earth, or our molecular DNA becoming unstable. Yet we already notice that one of those causes—the sun becoming a quark star after the big bang of a supernova—can happen to the Earth due to CERN's experiments with quarks and black holes. It is in fact the same danger to the Earth posed by the creation of a quark factory, able to produce black holes, but happening to our sun, not in one of our labs.

Finally, the three most dangerous, lethal causes of human extinction—*a global nuclear war, superintelligent robots, and the Earth being gobbled up by a black hole created in a particle's collider*—are provoked by the *evolution of technology, specifically of particle physics*, which is opening the doors of awesome energies (nuclear bombs, production of black holes) and new forms of informative intelligence (quantum computing, which will be the basis of intelligent nets and robots). Atomic cannons will peer this year into the limits of energy of the Universe at the Nuclear Company of Europe, CERN, while robots will cross the threshold of AI and self-reproduction (as metal nanobacteria) when we make quantum computers, based in calculation with atoms, within one or two decades.

If we add the other cause left in the poll, rated 5 by scientists—a sudden climatic change due to an excess of industrial pollution that is heating the Earth—*we conclude that technology is crossing in the twenty-first century the thresholds of energy and information that might cause the death of our species*, since we live in a Darwinian Universe in which no species or superorganism is safe when its physiological systems face an excess of energy and information they can't safely manage.

Indeed, two specific technologies—nuclear physics, specialized in the search of energy and robotics, specialized in the creation of machines of information—have advanced so much that their machines menace to break the balances of life of our planet. Yet its scientists are eager to cross those limits in the next decades without any sense of social responsibility. Thus, a serious technological policy should limit those two sciences while promoting all other forms of knowledge. If this does not happen, it is because those machines are very expensive, give a lot of profits, and the military-industrial complex, its main producer and buyer, is still ruling a society that has not changed after the cold war; it only pretends to have changed with a newspeak of *peaceful* wars and *research* weapons.

Only arrogant, wishful thinking and monetary gain, obtained by making those machines, prevents our species from taking measures to control the evolution of harmful machines and avoid the likely death of history in the future by an overdrive of energy and information. The bad apples of science are those two dogmatic disciplines, with an excessive dependence on cutting-edge technologies, which have specialized in the search of pure energy and information without limits:

High Energy Physics seeks today to open the ultimate frontier of energy in the Universe, the nuclei of atoms, liberating its "strong" quarks, manufacturing black holes, and replicating the big bang on Earth by creating the quark-gluon soup that originated it and is the cause of supernovas. After the cold war ended between America and Russia, now the European Company of Nuclear Research, CERN, has taken the initiative in the evolution of nuclear energy without any safety measures.

Robotics seeks to cross the last barriers of information, creating artificial intelligence, machines that will surpass the informative capacity of man, making us an obsolete species, either killing humanity in the battlefield or substituting workers in the economic ecosystem.

In both cases, we face disciplines that show a complete unbalance and misunderstanding of the equilibriums between energy and information needed to create in this planet a healthy world, made to the image and likeness of mankind. So even if I will use the terms *science* and *scientists* very often for them, it should be clear to the reader that those two disciplines are *the two bad apples of the Tree of Science that might cause the extinction of mankind within decades.*

Robots are beyond the limit of information humans can compete with. It is self-evident, and it is already happening that robots displace humans from the economic ecosystem as soldiers and workers, creating massive unemployment and killing indiscriminately civilians and soldiers—today in Afghan villages, in the future perhaps in a global war. In the long term, as robots keep evolving, despite the anthropomorphic myths of human superiority and uniqueness, in a Darwinian Universe in which our only advantage is our superior form (our information, now transferred to machines with software programs and biological designs), those robots are geared to substitute us as top predators of this planet.

While nuclear weapons are about to give a jump in power and complexity with the creation of the first quark cannon (the Large Hadron Collider), that will release the energies inside the nuclei of atoms, quarks, and strong forces and attempt to make black holes on Earth—starting this year 2010. If those black holes don't evaporate according to a disputed theory by Mr. Hawking and follow the laws of Einstein, then they will grow and feed on the matter of the planet till destroying it. Thus, since it is evident that this lethal technology is both, the more dangerous of all the possible causes of our extinction (risk 10) and the first one to be fully operational, this book will be dedicated to study both, the technology, its risks and the supposed benefits for science it will not bring to mankind, given the fact that the ultimate excuse of all those industries is "knowledge." Indeed, if religious believers sacrifice their lives and commit acts of terrorism for their perceived "truth," their religious dogmas, nuclear physicists are about to sacrifice this planet for a series of "dogmas" about mass and black holes, which are not only disputed but deny the "standard" theory of mass and gravitation defined by Einstein (reason why the book is authored under the pseudonym "Einstein et al").

3. The Content of This Book

In that regard, this book's structure follows on the tradition of the first classic of scientific literature, *Dialogue Concerning the Two Chief World Systems* by Galileo Galilei (1632), as a dialogue between two "authors," CERN, the paradigm of the bad fruits of the Tree of Science—technological weapons and its theories of a universe built only with "energy," its excuse of false knowledge—and "Einstein et al," which represent the good fruits of the Tree of Science, the knowledge of the Universe, explained as a system made with "energy and information" in which man is the supreme being, as the most complex informative mind of that Universe. A science whose bioethical goal is to improve human life, creating a world made to our image and likeness.

CERN's problem—the use of knowledge about the Universe as an excuse to build a weapon—requires a dual book, one dedicated to explain the dangers of the quark cannon and one to understand truly the Universe with the work of Einstein and those cosmologists, which have evolved his knowledge in the past decades ("et al"). Yet before we describe the beauty of the relativistic fractal Universe, we shall discuss the quark cannon, its weaponlike nature, and what CERN will really do—a supernova with the Earth. According to the "new physics"[1] of "complex, dual systems" of energy and information, the big bangs that CERN will replicate on Earth will create a supernova, causing also a dual "big crunch" of our matter into a quark star:

— Thus in the main chapters of the book, we study the risks, the falsity of CERN's safety statements and theories of mass (evaporating black holes, Higgs), completed with an account of the main documents presented in the suits against the company.
— Then the appendix "The Fractal Universe," showing the "future of cosmology," which, departing from Einstein's theories about mass and the relativity of the laws of nature which are the same regardless of size, motion, or position, builds on the new astrophysics of the twenty-first century. The fractal paradigm, not CERN, represents the true frontier of cosmology.[1] It is based in a growing corpus of theoretical work, that using the new mathematics of fractals and non-Euclidean geometry, describes the Universe, in which all its beings are self-similar parts modeled as fractal superorganisms of energy and information.

4. Theoretical Perspective: The Fractal Universe

The purpose of this theoretical perspective is not only to show the beauty of the Universe in all its magnificent self-similarities, but to dismount the use of knowledge as an alibi to justify the building of the quark cannon as well. Indeed, CERN affirms that the quark cannon will be the fundamental tool needed to solve a series of key problems of astrophysics, which quantum theorists cannot solve with pen and paper, such as the meaning of mass, the nature of the big bang, or the existence of time arrows from future to past (hypothetical cause of the evaporation of black holes).

However, we shall show that the reason those problems are not solved by nuclear physicists is because *they are not the people who are solving them, nor their science, quantum entropy, and tool—the collider—are the proper ones to explore the Universe and the meaning of time. Complexity theorists, fractal relativists, philosophers of science, and cosmologists with telescopes and satellites, not* quantum physicists, are in fact, solving those questions thanks to the next evolution of mathematical and logic science—what scientists call *the fractal paradigm—and new telescopes and satellites, such as the Hubble, the Webb, and the Fermi.* Indeed in the past decades, a group of mathematicians, physicists, philosophers of science, and astrophysicists have refounded the science of cosmology, departing from the work of Einstein; providing with the new physics, solutions to most questions left unanswered by quantum physics; and using the new mathematical and logic tools of fractals, chaos theory, complexity, and duality (the study of the Universe with two arrows of time, energy, and information, unlike quantum physicists, who use only the arrow of energy). So we live a paradoxical age, similar to that of Copernicus and Galileo, who had less power and prestige than Ptolemaic physicists but were far more advanced in their understanding of the principles of the Universe. Today, the same can be said of the new cosmology, with far less resources to study the Universe than quantum physicists, because their work does not have the same industrial/military applications.

And yet the truths of new astrophysics are far more accurate describing what CERN pretends to do than the obsolete quantum models this company uses. For example, when we consider the big bangs of the quark-gluon soup CERN pretends to form, we shall study them as "far from equilibrium" dual processes, which create in the center a big crunch and externally a big bang, thus making in the center black holes and strange stars and expelling radiation outward. This analysis of the experiments at CERN with the tools of complexity and duality, which has recently been experimentally confirmed,[2] means that it is much easier to make black holes and strange stars than CERN predicts. CERN denies it will make strangelets and black holes because *they are made in cold environments*, and it affirms that its experiments are done at high temperatures. Yet a big bang is a complex dual system that acts as a refrigerator that expels heat outward in a wave of radiation to cool down its center, where the black hole forms. So CERN, which only uses "energy" to explain this process, merely acknowledges the external radiation of the big bang and NOT the internal formation of a supercold quark liquid, which it describes as a gas that will evaporate. This is not the truth, and we have many experimental evidences that the process is dual and so a hyperattractive vortex of quarks is forming inside.

The naïve reader might then think that CERN, as an institution of learning, would accept those advances (of the complex new physics and mathematical detailed version), revise its reports, and reconsider its experiments. This was my opinion five years ago when we sent detailed papers to CERN. Yet CERN is not a group of idealist scientists but a nuclear company making a machine. And so CERN denied and keeps denying, even as mounting experimental evidence comes from self-similar experiments, all those advances of science because its mission is to shoot its cannon, not to advance science.

Perhaps the most clear proof of that "double talk" of CERN is its defense at all costs of an obsolete, false theory of mass called the Higgs, which pretends to substitute without any proof Einstein's well-founded principle of equivalence between mass and acceleration, which describes mass as a vortex of space-time that attracts as a hurricane does, the space-time and all the objects that "swim" on it.

Since mass is an intrinsic property of all beings, not a "gift" given by an invisible never-found "God's particle," the Higgs was properly baptized as a "toilet" particle to flush in a vortex of mass of Dr. Einstein by Weinberg, a Nobel Prize winner with little patience for falsity in science. The Higgs sting, and the hype built around it will be properly denounced in this book since it not only means a $10 billion swindle to the European taxpayers in an age of economical crisis, but it also prevents the proper understanding of mass as an attractive vortex that carry the physical information of the Universe in the frequency of their rotation, as this computer carries information in the frequency of its cyclical clocks. And so the Higgs is a false dogma, a wall that prevents scientists from understanding the new cosmology that truly explains the meaning of the Universe, based in *Einstein's relativity, not in the quantum musings of Higgs and Hawking.*

Some of the enormous advances in cosmology and time theory achieved by a proper understanding of mass as a vortex of physical information are explained in this book to explain the facts that matter to European scientific budgets and the safety of its citizens: that the two fundamental theories CERN sponsors on mass and black holes, the Higgs and Hawking's hypothesized radiation, are false. They were ideas brought about forty years ago by quantum theorists trying to substitute Einstein's work, which the new discoveries of modern cosmology have proved wrong. And so black holes are much more dangerous than CERN pretends, because they are far easier to do in a quark-gluon soup and they won't evaporate. But CERN's money and its hype mass media campaigns have made those theories untouchable dogmas nobody dares to defy, even if most physicists today know they are false. And this brings the theme of "group thinking" and corruption of an entire "professional caste" of nuclear physicists working around this institution, who already showed a complete lack of bioethics in the past century when they built thousands of nuclear weapons that maintain the world in a perpetual state of terror.

Money and power corrupt science like any other profession. And this is what scientists should understand about CERN. CERN portrays itself as the "avant-garde" of science and those who dispute its safety as "obscure" religious fundamentalists, totally misleading the public and scientific magazines and administrators on this issue. If CERN were a university, the LHC would not be happening. Arguments between complexity, relativity, and quantum theorists would have long ago resolved the meaning of mass and the safety of the experiment in a friendly manner. But CERN, an old institution of the cold war, with diplomatic immunity for political reasons, works as a secret military research center would, pretending NOT to be one. So now instead of "top secret" documents, in this new age of "just wars" and Walt Disney concepts, it makes infantile marketing campaigns on God's particles. Yet at the same time, it makes its workers sign confidentiality statements of zero risk and has established a Law of Silence in the community of physics, similar to the one it existed during the cold war, which we tried to break with a suit. We failed as CERN adduced its diplomatic immunity not to go to the suits but merely made a campaign of ad hominem attacks, "damned lies and statistics." It said that the quark-gluon soup will manufacture itself—similar to a cosmic ray. Yet they have nothing to do with each other. Cosmic rays are lonely ions, with hardly any mass, and we never saw quarks in cosmic rays. Thus, we are facing today the worst scenario: a company that lies systematically; that has constructed the most powerful weapon of history; that is going to produce the most dangerous explosive substances of the Universe, quark-gluon soups, and yet is considered as an institution of learning, revered by the press, immune to justice, because of its diplomatic immunity achieved when it was an agency of the cold war; and that as we speak is already producing drops of neutral, highly stable strange particles (dibaryons, hyperons), which might be already falling to the center of the Earth, where if they reach a sizeable mass will implode the planet. And yet our society has accepted this without any argument, with a religious trust that we might indeed regret for eternity.

Here, we shall explain all those theoretical details with simplified mathematical models. The complex models of those liquids, quark-gluon soups, far from equilibrium systems and fractal relativity are available in professional literature.[3] They are not being published by mass media magazines, which have unwisely sided with CERN, its ad hominem campaigns against "Einstein et al" and the Higgs Hoax to their own peril. Indeed, the community of physicists have decided by decree as it did in the cold war that the big bang experiment is something that "ought to be done." And so as in war, when the battle starts, one cannot go backward even if the enemy will kill us all.

For all those reasons, after all these years of ad hominem campaigns and "damned lies and statitics," this book, which is the proper answer to those campaigns, treats CERN and the theorists that sponsor this potential genocide with the same disrespect they have shown for the true goals of science—bioethics and knowledge. Nuclear physicists are behaving in this issue as they did in the fifities, at the height of the cold war. The sad truth though is that the rest of scientists and mankind also behaves with the same sheeple attitude of those years, letting CERN do anything it wants as they let McCarthy, Stalin, and the nuclear physicists' community to build during those years an awesome arsenal at an amazing cost, with the same antitruths of "safety" and "knowledge" used then by Teller to convince Truman of the need of an H-bomb to "reveal the secrets of the atom." It is a déjà vu experience, but now the consequences might be far more scary. We can't, of course, know all the details of the far-from-equilibrium processes that will create strangelets and black holes with a quark-gluon soup, but two fundamental laws of complexity ignored by quantum physicists add dangers to those processes: the parts of a system are more reactive, faster and more stable than the whole. For example, chips are faster when smaller and so are methabolic processes. This means that small black holes and strangelets are far more reactive, hungrier, and faster as any newborn baby is than adult species. And so they are able to explode stars in supernovas within seconds of their creation. Masses are accelerated motions, NOT some silly God's particle. So as a hurricane turns faster in its center, a mass attracts more, the smaller it is. Thus, a growing vortex of superfluid quarks will be very attractive and impossible to control, once its reaction of transformation of our particles into heavy quarks becomes stable. Indeed, a bottom quark switches its state at the rate of a trillion particles per second. Thus, a vortex mass of quarks could blow up the Earth in a fraction of time. Yet all those mass reactions are ignored by CERN's *physicists, which lack a theory of mass as physical information.* They are constructing, like Ptolemy did, a Universe with only one parameter, energy, and using very complicated mathematics as Ptolemy did, committing a lot of errors in the process because they have lost completely the right vision of the whole.

We are in the hands of big boys with big toys: outdated theorists with the most awesome weapon ever constructed, spoiled by a mass media system that considers them geniuses—since the cold war propaganda machine raised their status to icons of our civilization—when they are mostly clueless engineers operating a weapon.

Thus, we shall provide here on both books the proper explanations of those risks and the solutions of the new astrophysics of fractal relativity to all the questions CERN pretend to resolve, according to the real laws of truth of the scientific method: logic, mathematical accuracy, and experimental proofs. Those questions resolved here are the meaning of mass, the fractal nature of the big bang, the nonevaporation of black holes, the reasons why there are three families of masses, the reasons why antiparticles are not observed in the Universe, the nature of dark matter, the unification equation of charges and masses, the nonexistence of the

Higgs particle, the cause of the expansion of the Universe, the reasons why gravitation is a weaker force, the nonexistence of supersymmetric particles, the nature of dimensions and its number, the nonexistence of parallel universes, the solution to the uncertainty paradox, the meaning of the holographic principle, how energy becomes mass and the values of masses of different particles. Those are all the questions CERN pretends to solve with the quark cannon. They have been solved by the theorists of fractal relativity, but those solutions are barely known beyond the restricted group of practitioners of this new paradigm of science, given the enormous hype built around CERN and the Higgs by the industrial/mass media system of misinformation, which prevents the spread of true knowledge in modern physics and philosophy of science.

In a few words, the theories sponsored by CERN have become a "dogma" of industrial technnology and the "ethics of an industrial civilization" quoted in this book, *but are not and will never be a dogma of true science, the search for knowledge and the meaning of the Universe with the languages of the human mind.* In that regard, perhaps the most cynical twist of this story is the fact that CERN, a powerhouse of money for research in nuclear physics, is sponsoring cosmological theories that justify its machine but are outdated, false, or irrelevant to our knowledge of the Universe, such as Higgs's theory of mass, or Hawking's theory of black hole evaporation. Those theories grouped generally under the name of "quantum entropy" and "quantum gravity" pretend to substitute the work of Einstein by applying the laws of the infinitely small (nuclear particles) to the infinitely large (the Universe). Yet they are wrong since they deny the principle of Einstein's relativity (size doesn't matter, it is relative in a Universe of infinite scales) and ignore the recent experimental and theoretical developments in the science of fractal information, which consider mass physical information and black holes the creators of the physical information that balance the entropy of the Universe. This new paradigm of cosmology has now undeniable proofs and will usher the science of astrophysics into a new, golden age, *able to explain with the power of the human mind, not with a machine, the meaning of it all, solving the disputes between Einstein and quantum theorists in favor of Einstein* . . . if we survive CERN.

This need to survive CERN and the quark cannon is the second theme of this work, treated in its second book.

5. Technological and Existential Perspective: The Quark Cannon

True science has two goals—an ethic goal, to create a world that improves human life, and an intellectual one, to find the truth about Man and the Universe. Yet those goals are inverted when science is used to create the "bad apples" of the tree of technology, weapons that kill mankind and "energy-only theories" of the Universe that justify them. Nuclear physics has swimmed always between those two waters as we shall see in this book, reason why it is so difficult to analyze the nature of the large hadron collider in objective terms: Is it an instrument of research, or the last advancement of nuclear weapons? Will it discover the meaning of it all? Or is the hype about the Higgs and the big bang just a tool of marketing for the company, to sell this machine now that the cold war is over and the excuse of defense is not enough to get funding? We will try to reason why we believe CERN is just a nuclear company evolving nuclear weapons and disguising it all "in the name of science."

Thus the first chapters of the book are dedicated to study the existential risks for mankind of this machine. We consider the history of atomic cannons and the dangers the large hadron collider—the first machine that will put at risk our existence as a species—poses for the future of the Earth, according to what we know today about quarks and black holes, the substances it will produce. We shall confront CERN's safety standards with Einstein's work about black holes, quark condensates, and mass bombs as those the quark cannon will produce, showing that if Einstein is right—and Einstein is so far standard science in those themes—we will very likely become extinct. Indeed the quark cannon will shoot the Earth with the "strong force" of quarks at increasing speeds during three years, starting this very same 2010, NOT because we need a planetary weapon after the end of the cold war, neither because it will improve our knowledge of the Universe, a task that belongs to telescopes, satellites, and cosmologists, nor because we could waste ±10 billion dollars to find Mr. Higgs's particle, which has tried unsuccessfully for forty years to substitute Einstein's theory of masses as attractive whirls of space-time, but because our industrial system could make this machine-weapon and so it has done it.[4] Now the fate of mankind is 100 meters under the Earth, in its bolts and tubes, no longer in our hands, unless a responsible political or military order from NATO or the European Union halts this experiment and opens a public and scientific debate on its dangers, as the environmental laws of the EU ask for.

CERN is ultimately a company that has built a machine. And so regardless of the individuals that work there, the behavior of CERN is that of a corporation. Today the world has become "owned" by corporations that effectively derail any attempt of individuals, like those represented by "Einstein et al," to change their lethal policies against Gaia, by using political power, think tanks, and "experts" to hide their risks against the environment. We will describe the irresponsible connivance of the nuclear industry doing business as usual; the scientific establishment, proud of a machine that is breaking all the energy records; and an ignorant press that censors the real dangers of the quark cannon, not to cause public alarm. It is the industrial perspective of a world ruled by technology, studied here in its collateral effects of extinction of life species in this planet.

Thus, our interest is also human, psychological: how a potential holocaust of this magnitude can be accepted by our present society, how monstrosities happen in history. Several approaches are considered. First, the military sense of obeissance to authority that infuses the work of physicists. But "authority" in this case is a company, a type of institution, which has perfected marketing and lobbyism to hide all type of crimes, and an ideology, nationalism, "grandeur," which infuses the actions of the French and German governments, which paid the bulk of its budget. We compare the Nazi monstrosity and CERN's monstrosity, coming to an obvious conclusion:

The reason why today Europe accepts CERN are the same reasons that explain why Germany, the most evolved scientific culture of the twentieth century, backed Hitler and his military-industrial complex, which already destroyed once before the world: today our religion is the energetic machine, no longer the informative human being. And so we prefer always technological solutions to our questions and problems instead of human ones; we believe a machine will explain the Universe, NOT a human mind, we believe in "mechanical experiments," not in "thought experiments" as those Einstein performed.

CERN is the most advanced technological company of the world. So technological ethics, our present religion, accepts in the name of "the machine" any risks for "humanity" that a quark cannon can deliver, as Hitler prefered to fight wars with the most advanced machines of the age instead of finding diplomatic or social solutions to the national and economic problems of Germany.

Time is indeed cyclical.[ch.5] And history repeats its events with self-similar characters. Now the new CEO of the European military-industrial complex is a Rolf Heuer, not an Adolf Hitler, and the industries of Germany and France work together, not opposed to each other, but the result will be the same, a hundredfold by the power of new technologies if the quark cannon is fired.

Thus chapter 9 returns to this introduction when our analysis goes beyond CERN and physics, considering from a biological perspective, the general process of destruction of life by polluting industries and new technologies built by the "homo technologicus lethaliensis."

Are we humans a suicidal species, geared to self-destruction? It is the Universe deterministic; it is death always the end of any species, or do we have the freedom to save the world? Those metaphysical questions are introduced from the wider perspective of a Universe that constantly increases its fractal information and from the historic perspective of our civilization that also increases constantly its technological information, unaware that an excess of information is also a cause of death.

It is precisely the null understanding by quantum physicists of the meaning of information and its cycles of life and death, compared to their mastery on the creation of "energetic" weapons, what is putting our species at risk. That technological knowledge has allowed the quark cannon to go online, on account of the ethics of our technological civilization: "If we can make a bigger, better machine, we must do it, even if it menaces to kill us all" (Fromm), forgetting the fragility of human "information," which has taken billions of time-years to evolve but could dissappear in a fraction of a second. As life, informative processes require little energy but a lot of time to take place while death processes last very little time but release a lot of energy. And yet, without understanding the laws of time, energy, and information, nuclear physicsts at CERN are going to release energies never before tested on Earth to "see what happens." And they don't want to listen to the "experts" on information, complexity, fractal, and chaos theory that are telling them what will happen: *they will kill all of us.*

Yet the "program of life and death, of information and energy" is not fixed since there are species that maintain a perfect balance between form and energy, becoming immortal. Mankind could also reach a relative immortality in this planet if it were aware of the laws and ages of time that cause the life and death cycle.[ch. 5] Unfortunately this is not the case for CERN's scientists, who, as John Ellis, its chief theorist, put it, are more worried defending "the LHC from humanity."

Because CERN is the most advanced technological institution of the world, it has gotten away with murder with that excuse—reason why we need to explain the real "future of science," brought about by the fractal paradigm, to show the difference between the following:

True knowledge of the meaning of time and the cosmos is represented by Einstein with his theory of general relativity and those who follow in his path and are developing today with "thought experiments" the cosmological models of the twenty-first century or studying directly the Universe with the proper instruments, telescopes, and satellites, which cannot harm human life.

Knowledge as an alibi for military technology is represented by CERN, which pretends to prove with a weapon that merely menaces our lives, the absurd theories about time travel of Mr. Hawking's evaporating black holes and the energetic big bang theories of Mr. Gamow, both of them made obsolete by the understanding of fractal information and its key role in the creation of black holes and the cosmos.

In a few words, CERN and its quantum cosmologists act as the Vatican priests of the establishment of industrial science, which is NOT advancing true knowledge, but halting the spread of the Galilean revolution that the fractal paradigm is bringing to science.

The Universe is not modeled as an energetic machine as physicists think, but it is constructed by the constant interaction of energy and information, very much in the manner organisms are created: through feed-back cycles that transform back and forth energy into information, creating complex networks that have evolved particles into atoms, into molecules, into cells and organisms or planets, stars, and galactic systems. This arrow of information and social evolution, which CERN's theorists completely missed, implies that black holes won't evaporate and the safety standards of CERN are false because they are based in obsolete energy-only dogmas about quark-gluon soups and the origin of the Universe. For that reason, the collective of scientists represented under Einstein et al. sued the nuclear company in a bioethical action, which we consider in the last chapter as an example of what true scientists should do. It is the last chapter of the book, "Asking a Judge to Save the World," where the main affidavits of the American suits against CERN and DOE asked to halt the experiments that will take place at CERN in this and the incoming years. Those documents presented, as they were produced at Court by both parts, as follows:

— The establishment of the nuclear industry, represented by DOE, the American partner of CERN, and two Nobel Prize awardees who had worked for this company and merely denied any risk, asking the judge to believe in "their authority" since CERN adduced diplomatic immunity, achieved as an agency of the cold war not to appear on those suits.
— The theorists of the fractal paradigm and the risk and safety experts that are grouped here under the name of Einstein et al., which represent the future of science and the bioethical laws that should guide public policy.

We cannot bring about all the suits put against this company so we just bring here the most publicized of them, which took place in the Federal Court System of America and is still pending on appeal in San Francisco, where it has been sided, hoping that as John Ellis, the chief theorist of CERN, put it, "the way to stop all this argument about whether the LHC is going to destroy the planet is to get the LHC working. Within a few weeks time, we will know that LSAG (safety report issued by CERN's "experts") was right."[5] Or not, in which case we shall all die for the ethics of our technological civilization in the name of CERN's authority, not on the authority of the laws of truth of the scientific method.

The death of history is not unavoidable, but given the indifference of judges and politicians who have ignored a complex issue they prefer not to deal with and the press who has accepted the ad hominem campaigns of CERN against Einstein et al. without researching the case, we have to draw a pessimist conclusion to this issue. CERN indeed has survived all the attempts of Einstein et al. to halt the quark cannon, but mankind is not safe from CERN because the laws of the Universe will not be changed by a corrupted, arrogant group of scientists, who are defying with their Russian roulette against this planet the causal laws of time and the properties of masses and black holes, which they ignore and don't want to hear about. Those laws of the new physics of fractal relativity imply enormous risks to those experiments. According to those laws, if a quark condensate, pulsar, or black hole becomes stable, there is no way humans could avoid it from swallowing up the Earth.

But our demise is not a necessary fate if we believe in free will. Any of the people to which this book is dedicated could during 2010, with an administrative, political, or military order, stop the quark cannon. If they don't do it, if they die for the quark cannon or allow this potential genocide to happen, it won't be because they were not warned[6] but because they believed in machines more than in human beings, in the ethics of technology more than in the survival ethics of man. And so they will lose it all.

Just think on that. Everything we live for, our people, our land, our culture, the songs of love, the names of god, the smell of coffee, the sweetness of sugar, the touch of a woman, the sand, the waves, the colors of dawn and sunset, the beauty of form, arts, sports, the laughs of children, the embrace of our mothers, the memories of history, nature, cities, streams, plants, all kinds of living beings, every day of creation, every night of dreams, all might be gone in the time it takes to read this short list of wonders Gaia has given to the human species in this little corner of paradise.

---

[1]   As quoted by Rouse Ball, 1908.

[2]   A good introduction to the new physics of fractal and chaotic systems, far from equilibrium reactions and the advances of new cosmology, is the book *The New Physics* by P. C. Davies, Cambridge University Press. Published in 1989, it made an account with articles written at an undergraduate level by the leading researchers on those fields, on the main themes that would shape the cosmology of the twenty-first century. On one hand, it shows the theories of quantum cosmologists (quantum gravity, quantum entropy) which departing from single, energy-only concepts of the Universe have shown to be a wrong path, a no-way out for modern cosmology. On the other hand it lays down the other path of cosmological research, which started in those years, based in the duality of energy and information, the fractal nature of the Universe, the invariance of scale of all the laws of reality, and the physics of far-from-equilibrium systems and superconductivity, in which this book is based and will be the foundations of twenty-first-century cosmology. The old "Vatican priests" of quantum cosmology and the new "Galileos" of the third millennium write brief articles in a vibrant account of a shift of scientific paradigm. For further information

on the new cosmology, one must read the scientific papers of the founding fathers of those disciplines: Mandelbrot and Sancho in fractal mathematics and fractal time; Rössler, Mehaute, and Thom in chaotic and fractal systems; Pietronero and Nottale in fractal cosmology; Smolin and Nambu, on the falsity of Higgs and so on.

[3]   For a fairly simple analysis of quark condensates and strangelets as a nonequilibrium system, see Letessier et al., *http://arxiv.org/abs/hep-ph/0206145v2*.

Further on in classic literature there is a significant to very significant strangelet production yields for A = 4 to A= 6 and 2×10^{11} collisions at LHC (given by Dar et al., 1999). Schaffner-Bielich et al., 1998, regard that even at these minimal quantities, strange liquid could be very long-lived, thus enabling catastrophe. Peng and Chen, 2004, coincide with Bielich.

[4]   Mr. Nambu, last year's Nobel Prize winner, proved that a top quark/antiquark condensate acts as a Higgs field. Later, Higgs deduced a self-similar equation for a self-similar particle. And Mr. Smolin and Zee proved they were equivalent. Since the Universe is efficient and doesn't create redundant particles, this means that the Higgs field is a top quark/antiquark field, reason why in the graph that opens this introduction, taken from *SciAm*, which shows CERN's quark production, they are shown to have the same mass.

Yet nobody would pay $10 billion for a particle already discovered. So Mr. Leo Lederman, the biggest lobbyist of the industry of accelerators, convinced Reagan to pay for the American supercollider to find God's particle, the already discovered Higgs=top. And later, when the cold war was over and the American project was cancelled, as the toy to research cosmic bombs became military obsolete, CERN took on this marketing strategy. Today nuclear physicists hide those facts tacitly accepted by the concession of the Nobel Prize to Mr. Nambu, the original proponent of this top=Higgs theory, not to Mr. Higgs. We can say that Nambu's original equation has two versions, the top and the Higgs, which are the same person. Thus, in the same manner we say she has black hair, or she is a brunette, talking of the same person, but the use of synonymous expressions won't create a twin of her out of nothing. The Top=Higgs can be described with Nambu's original top/antitop mathematical sentences or the latter ± Higgs field equations, but that won't create a new particle. See chapter 3 for a detailed analysis of the Higgs hoax.

[5]   Quoted by Eric Johnson, Tennessee Law Review, http://arxiv.org/abs/0912.5480.

[6]   A quark superfluid was created at RHIC accelerator with 5,640 times less critical mass/energy than the LHC. It became a perfect surprise according to the RHIC bulletin because it was a perfect liquid that lived billions of times longer, accreted much more mass than expected and behaved according to physicist Mr. Nastase at Brown University as a proto–black hole (*http://news.bbc.co.uk/1/hi/sci/tech/4357613.stm*). Ever since, the number of theoretical papers, letters from Einstein et al. to the world and suits of all kinds against this company testify that mankind has been informed but our "industrial ethics" and leaders have not been able to protect our species of this potential genocide.

CHAPTER 1

The Dual Fruits of the Tree of Science

And God said: Do not eat of the good and evil fruits of
The Tree of Science; because the day you do, you'll die.
—Genesis, father book of all Western religions

I. The Good Fruits of the Tree of Science: Knowledge and Bioethics

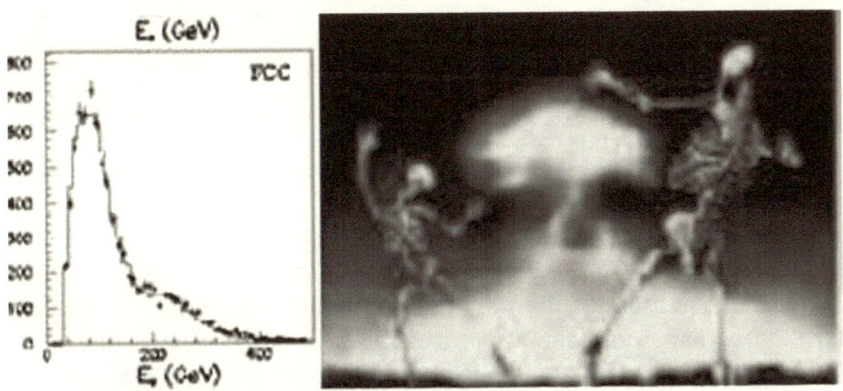

Mathematics simplifies beings into a game of abstract points, without parts.
Yet those points are organic systems that live and die, under the effects
of digital science. In the graph, the equation of an A-bomb versus its real
effects: a wave of energy that kills thousands of human beings.

1. The Bioethic, Political Debate: Pruning the Tree of Science

True Science has two confessed goals spelled out by its judges, which are not physicists but philosophers of science, the creators of the scientific method: to make a better world for the benefit of man and to learn the meaning of the Universe and its species, especially the nature and role of mankind within it.

And so the philosopher of science, the judge of the scientific experience, must carry out two *hierarchical* tasks to meet those goals:

— To denounce from an ethical humanist point of view, those actions of science, which regardless of their value for abstract knowledge, harm human life from an ethical point of view. Contrary to belief, this task is the first job of true science, which is an endeavor that takes place within the human community. So bioethics puts science under the sphere of the law and political power, which "ideally" in a noncorrupted society should fight for the common good of its citizens, preventing dogmatic scientists from risking lives in pursuit of their perceived "Saint Grail" of knowledge, as politicians do with other seekers of truth that commit violence in their pursuits. Ultimately, an extinct scientist knows nothing, so bioethics also improves our overall knowledge of the Universe. If we die, all that knowledge will be lost.

The philosopher of science must also test the veracity of scientific theories according to the laws of scientific truth in order to find out the truth about man and the Universe. Yet truth in science requires not only mathematical veracity, as most physicists believe, but also logical consistency and experimental evidence.

In this book we shall show that CERN fails to meet both goals. It won't advance knowledge, and we shall show why, falsifying according to the scientific method the theories it pretends to test. Those theories belong to the realm of "fiction thought," *which does also happen in science more often than people realize when a theory is constructed only with mathematical equations, without logic or experimental evidence.* Further on, CERN will risk the life of mankind, as there are several possible scenarios of global catastrophe that have a very high probability of happening when CERN replicates the conditions of the big bang.

2. Fiction Thought and Scientific Fundamentalism

As a philosopher of science, concerned with the definition of truth and falsity in scientific theories, I have been for quite sometime worried about the growing expansion of fiction thought in science. Fiction thought in science could be defined as the attempt to create "a priori" a theory, based on wishful thinking and the manipulation of mathematical equations that has nothing to do whatsoever with observed real phenomena. Then once a mathematical fiction is created, in a second phase, fiction thought acts manipulating experimental evidence to fit the hypothesis established by the imagination of the scientist. That is, fiction science doesn't explain experimental facts and phenomena observed in nature as the scientific method obliges, but it acts first with human, anthropomorphic arrogance, trying to bend nature and reality to the mental mathematical inventions of the imaginative scientist.

In that regard, we must not forget that science is first a linguistic statement of humanity about the Universe, and as such, it is subject to the rules of truth of any language of information. This has been established by information theory in philosophy of science long ago, but *quantum physicists sponsor a theory of reality called naïve realism, which denies the influence of languages in science and the limits of truth they impose on scientific theories based only in mathematical statements.*

Real science, which tries to explain the Universe with the languages of mathematics and logic, tends to be more accurate than philosophy, which uses the language of verbal thought to that purpose, or art, which works with the language of images, because mathematics, a synoptic language, with a strong inner logic, carries more information and is more accurate than words do. However, both languages derive from logic, as Frege and Gödel proved. And both carry less information than images, the primary language of the Universe. So in science, logic and visual experience *must* prevail over mathematics in any dispute about truth. Further on, those two languages are far easier to understand for a nonspecialist than the algebraic equations used by quantum physicists to explain the same phenomena. So we shall use here visual and logic principles to explain the meaning of masses, big bangs, and black holes, the three subjects CERN will study, and consider only the minimal equations needed to describe them since ultimately the Universe is not an algebraic game but a living painting—a game of three-dimensional beings in motion.

Yet if God is a painter in four dimensions, why is art not the supreme form of knowledge? The problem with art is that it has always accepted a subjective, anthropomorphic nature. And so art doesn't paint the Universe as it is since objectively, the Universe must be considered antithropic:

"Man is only a mush lost in a rock in a corner of the cosmos." So he should be cautious and objective, judging cosmic risks.

Thus, in essence, those three realms of human knowledge, art, verbal thought, and science are linguistic interpretations of the Universe and as such they are subject to the laws of truth and falsity of languages, among which is paramount the existence of fiction thoughts in all languages, which in mathematics was shown to be the case thanks to the work of Mr. Gödel. In simple terms, we can say that any language reflects with a series of symbols in a synoptic way the total information of an event or substance of the Universe. And because the language carries by definition far less information than the subject it studies, all languages can, without experimental evidence, create fiction thoughts. That is, a mathematical equation will not be real as long as there is no experimental proof of the phenomenon it describes. It will be a simplification of reality that might seem more truth, given its simplicity, but it will fail to explain many other properties of the entity or event it describes. And so it is a fundamental rule of science that as we acquire more knowledge of an event, which increases in complexity, initial mathematical theories tend to be false, replaced for more complex models. Yet in that process of evolution of knowledge, it is also a rule of science that the entity or event we study appears more stable and information, form, morphology becomes essential in its understanding.

Those two rules are "key elements" in the discussion of risks and false theories happening at CERN because all the theories and events CERN pretends to test are *only mathematical theories*. Further on, they use simple, lineal, mathematics (cosmic big bang, evaporation of black holes, the Higgs), which fail to understand the complex nonlineal mathematics of information, developed in the last twenty years by people like Sancho and Rössler, who have sued this company for criminal negligence, because as experts in nonlineal, non-Euclidean mathematics, fractal and chaos theory,[Intr. Nt.1] they know those risks exist in modern relativistic theories of mass as physical information, which explain masses and black holes with far more complexity than the musings of Higgs[ch.3] and Hawking.[ch.6, II]

Obviously CERN knows some of those facts, but it cannot accept them without closing its facilities. While the rest of the scientific community trusts the "system" and institutions of the nuclear industry without further analysis, this mistake might have grave consequences for our lives because it is today a fact proved by the laws of truth of the scientific method explained in this chapter that black holes don't evaporate and Mr. Hawking's mathematical-only theory is wrong. As so many simplistic mathematical-only theories have been in the past.

For that reason, Einstein said to Poncaire that while "I know when mathematics are true, I don't know when they are real." As we do not know if a book of fiction describes real characters unless we identify them in the real world. Thus, to avoid linguistic fictions in science, the experimental method was established, according to which a scientist should only invent theories to explain experimental phenomena already observed. *But without prior observation of phenomena, the scientist should not pretend that his mathematical theory is more relevant to reality than Don Quixote, regardless of the beauty of the thoughts of Mr. Cervantes or Mr. Higgs.* Fact is, we have never had any experimental evidence of the Higgs and the evaporation of black holes for forty years despite massive research looking for evidence. Those two theories are Quixotes, invented by their authors out of the hat of their mathematical musings. So we should conclude that those theories, which also defy the modern mathematics of information and its improvement on our understanding of mass and black holes, as the entities that create the physical information of the Universe, are plainly wrong. Thus to risk our lives for those theories is as silly as looking for real Quixotes in the streets of Madrid. If this is not accepted today by mainstream scientific magazines and the nuclear establishment, it is precisely because of the enormous power of CERN and its $10 billion budget and eighty thousand collaborators that act as a cover-up of those two false theories, to avoid closing its facilities, since if Hawking is wrong and black holes don't evaporate, we can all die. Thus, while a writer like Cervantes can write Quixote and a physicist like Mr. Hawking can't take the irresponsible avenue of fiction thought on view of the bioethical risks black hole represent.

In true science, if a scientist decides to explore mathematically a certain hypothesis, he should at least always respect in his deductions all the proved laws of experimental science, which have been validated by experience. He should avoid any contradiction with those laws or else his theory must be considered false as it no longer belongs to the realm of science but of science fiction.

This simple methodology of truth has brought about our present technological civilization and the certainty that the laws of science happen: That when we throw a pen, it will fall down, not upward, even if a mere change in the negative symbol of a gravitational equation will give us that fantasy result of objects flying upward; that when we heat water, it will boil, not freeze, even if a mere change of symbol in a thermodynamic equation can give us that result. For all those reasons, a scientist who decided ad hoc to change the arrow of time in the gravitational and thermodynamic equations and pretended objects levitate and heated water freeze *must not be taken serious.* And for the same two laws called the law of gravitation and the law of thermodynamics, when a hot, dense black hole appears in our cold surroundings or when it appears at CERN's experiments, matter will fall in, not out, and our cold watery world will evaporate, cooling down the black hole, even if a change of minus symbol in the arrow of time convinced Mr. Hawking that black holes break the well-known laws of gravitation and thermodynamics. *For the same reasons that we don't levitate and when we heat water it doesn't freeze.*

Mr. Hawking pretends that instead of absorbing matter, black holes will vomit it, which is to say that instead of falling down, a weight will move upward. How those absurd beliefs pass as science? In essence, Hawking got away with his theory after initial resistance because of personal reasons (he was the Lucasian chair of mathematics, NOT of physics in Cambridge); because of the age in which he exposed them (the seventies, when science fiction theories with weird concepts were very popular in that hippie era); because we knew very little about black holes at that moment and so they were considered mere mathematical objects to play with at will (Mr. Hawking initially proposed his theory as a mathematical exercise and bet that those black holes would not exist in reality); because of his engaging personality and human tragedy (which made very difficult to oppose his personal views without risking an ad hominem campaign). In a few words, he had become an icon of mass media, and his theory opened the way to science fiction—his black holes were time machines that were moving to the past. Those *absurd ideas promoted his popularity, sold many books, and made him a star* in *Star Trek* movies.

Why do we know his idea is a fiction thought? It is then when the three legs of the scientific method—mathematical accuracy, experimental evidence, and logic consistency—come to our rescue:

— The idea of a black hole evaporating fails to include the new mathematics of nonlineal, chaotic equations, information theory, and non-Euclidean topologies developed by people who have sued CERN, like Rössler, Penrose, and Sancho among many others, long after Mr. Hawking ceased to work on serious science (his paper is thirty-six years old, when not even computers had been developed).
— His idea is not born from *experimental experience* but was invented a priori, as a mathematical exercise.
— It does not follow the logic laws of science, but breaks its three fundamental laws of time causality (movement from past to future): Newton's/Einstein's gravitational theory (things fall toward the heavier object), Boltzmann's thermodynamics (hotter things get colder), and the law of conservation of energy and information.

Indeed, what Hawking failed to realize, as he has in multiple occasions scorned philosophers of science for pointing him out those inconsistencies, is that *science is a complex building whose foundations are the laws of causality in time, and you cannot tamper with those foundations and pretend time moves backward without destroying the entire building. As the saying goes, "You cannot build starting with the roof."*

Add 1+1—lack of experimental evidence and contradiction with known laws of science—and we must deduce evaporating black holes are fiction thought.

And yet this fiction thought is today considered a dogma of quantum physics at CERN, politically incorrect to discuss as the idea that Allah is god, the creator of the Universe, and Mohammed his only prophet is considered a dogma, politically incorrect to discuss in any Muslim country because believers are extremely active promoting such ideas at any cost.

In the case of evaporating black holes, the belief was held by a small group of physicists who worked in the hypothesis, most notably Bekenstein and Hawking. Such belief was accepted because it was a mere theoretical exercise, as we accept other improbable fictions such as multidimensional universes, parallel worlds, and the virginity of Mary, Mother of Christ.

The musings of Hawking were an "inconvenient truth" of serious science, "not good enough" as Higgs has pointed out recently,[o] but harmless and difficult to rebate. As a scientist who tried to say the truth told me recently, "I was accused of attacking a disabled person." Because Mr. Hawking has a sickness that atrophies his nervous and neuronal system, which he fights courageously, the press has made him a hero and an example of the power of humans to overcome difficulties, making him the new Einstein—and selling many books and magazines with this "human story." Yet the fact is that for forty years, Mr. Hawking's neuronal atrophy makes impossible for him to follow the complex advances of new physics. And so our lives today are at risk by an absurd theory that cannot be falsified in objective terms without entering into emotional ad hominem arguments. Yet as the nuclear company of Europe prepares itself to make black holes on Earth, the belief in the fiction thought of evaporating black holes is no longer a dogma of the scientific press that concerns only physicists, but all mankind. Since if CERN makes a black hole, true science proves that it will kill mankind. And mankind should not tolerate that risk imposed by the yellow press of popular scientific magazines. And yet, this fiction thought is imposed upon all the people of this planet. Why?

In the case of Allah, creator of the Universe, that belief only affects those Muslims who choose to think so. Thus, in a way, we could justify respect for such belief, as long as it doesn't impinge on our lives. But if we are obliged to believe in the divinity of Allah and someone exercises violence against us to accept that belief, we must rebel and claim our right to think independently. This is the justification of our present war against Islamic terrorism, and it is the reason why politicians should close CERN. Since if Mr. Hawking is wrong, we shall all die for his beliefs.

But the issue has not been argued in the Western press, for the same reason that not a single newspaper in Iran is arguing the divinity of Allah. The high priests of CERN, the physicists who sustain our mechanist technological civilization, will not permit it, and the rest of our Western society whose religion is science will not doubt their high priests as no Iranian should doubt their ayatollahs. Instead, our likely extinction has been censored by serious newspapers like the *New York Times*, which initially alarmed the world and then published a paid-by editorial, before a judge issued a decision on the cases against CERN for genocide, "asking the judge to throw the suit in the nearest black hole." So as Iran has no arguments on matters of faith, which must be ruled by the Supreme Council of Ayatollahs, the issue of black hole evaporation has been ruled by the Supreme Council of CERN and given to the world to believe in. A lawyer, Mr. Johnson, points out in a recent article on those cases:

> It is remarkable to think for a moment how CERN's situation might be viewed if, instead of operating a particle accelerator, CERN was a developer of pharmaceuticals. If a pharmaceutical firm attempted to take a drug to market based on the safety assessment of a panel of five of its employees, who in turn relied on the scientific work of one employee and one other scientist with a pending visiting position with the firm—it would be a scandal of epic proportions.

Yet this very same lawyer will say in the same document that "I'm trying to take seriously the question of whether the LHC really does present a planet-threatening danger, we are presented with a paradox. How could so many exceedingly smart people—particle physicists, no less, who make a cliché of genius—be capable of getting something so terribly wrong? So Mr. Johnson, after all a believer, will conclude despite the massive proofs he finds of criminal negligence in this company that: Am I fearful that the world will be destroyed by a lab-created black hole? No. Not really. It does not keep me awake at night."

He knows nothing of physics and *trusts nuclear physicists, the ayatollahs of CERN*, who already tried to build two doomsday machines in the past—a superbomb able to blow the entire USSR, cancelled by Eisenhower when he knew what its purpose was, and a supercollider ten times more powerful than the LHC, cancelled by Clinton. If any of those doomsday weapons had been fired, we would be already evaporated. And yet all nuclear physicists backed those machines. So why should people not lose their sleep? Because, as Mr. Johnson does, people believe in the Latin of their high priests, which today is mathematics. *As Kuhn explains in his masterpiece on* The Structure of Scientific Revolutions, *it is a very little recognized fact of human nature that our brain is* NOT *primarily rational but more similar to a computer, based in "imprinted programs of thought," beliefs that are memorized, not rationalized, as dogmatic truths, called in science postulates or theories without proof. And this is the case of black hole evaporation.* Thus, if a Luther comes to translate and explain why those Latin sentences are false as this text does, we feel emotionally angered by the heretic, which rejects the authority of our high priests and shows the shortcomings of our "ways of thought."

And so a prize is put to Luther's head, as the press has done with Penrose, Sancho, Rössler, Wagner, and all the scientists who sued CERN. The message printed in the *NY Times* editorial, the Court dictum, the *Scientific American* articles, the ministries of science and technology that pay the machine is clear: *morituri te salutant* (the machine is God) and we shall die for the machine and the celebrity with the computer-like voice, NOT for knowledge, NOT for science, NOT for mankind, NOT for life. And so the true laws of science, the god of the Universe, will respond: "As you wish, children of thought, slaves of the machine, die for it."

Indeed, according to the ethics of our technological civilization, our high priests must be right because if they make our machines, they must be more intelligent than philosophers of science, risk experts, and mathematicians who are denouncing them. But they are not. Machines are not the summit of human civilization. Languages are. Artists who dominate visual languages, logicians and mathematicians who dominate the languages of science, people like Einstein who only made "thought experiments" might have less money, power, and resources but are miles ahead of CERN's engineers in knowledge about the Universe. They are precisely the people whose work has raised red flags about this experiment. And yet the true geniuses of science, which is the process of accumulation of knowledge through centuries, Mr. Einstein, who discovered the laws of mass and black holes; Mr. Boltzmann, who discovered the laws of thermodynamics; Mr. Galileo and Mr. Aristotle, who discovered the scientific and experimental method and the logic of truth—all of them denied by the new guru of science, fiction—are not listened.

It doesn't matter they are dead. Their work has not been proved wrong and sustains the Universe. And yet we are betting against them and accepting fiction thought. We are giving Mr. Hawking superman powers to change the laws of the Universe and establish new foundations, as if he were the infallible pope of the cosmos, revealed to him by the supreme authority. Are we at this point of our civilization of technological machines and fiction thought so deluded that we believe all the myths and marketing campaigns of companies and exaggerations of the mass media circus of celebrities that also has imposed its law in serious science? Yes, indeed, we live in an age of children of thought because machines, not them, do their "thought experiments" and so there are no longer Einsteins. It is the zeitgeist of this age. Will Mr. Hawking call Mr. Harry Potter and come together with Mr. Spock in the *Enterprise* to save the Earth from being blown up by a quark cannon, as in the movie *Star Trek*, throwing CERN back to the past? We love fiction thought because most outlets of expression are dedicated today to fiction. War has been reconverted into a fiction called sports and we go to the fictional battles of soccer games to shout for victory. Even news are manipulated and real war is presented as a video game. Documentaries have disappeared from film making, and today, only fiction rules in television. Economical statistics are bent to the point that economists say there are truths, half truths, damned lies, and statistics. And so finally, the use of fiction thought has entered the realm of physics, by the hand of quantum theorists that deny the well-proved laws of the experimental method. Yet the Universe is stubborn and Darwinian and has its own method of proving itself right: reality exists and happens and fiction thought disappears. In other words, the Universe cares nothing for human imagination but for the laws of proved science.

So despite economical statistics that denied this crisis would happen, and now deny this crisis is still happening, the economy of the world crashed. And despite CERN's wishful thinking that if a black hole appears in its experiments in the next years, it will not obey the fiction thought of Mr. Hawking but the well-proved laws of causality in time, thermodynamics, relativity (gravitation), and conservation of energy and information and teach mankind that God are the laws of the Universe, not Christ, not Hawking, not fiction thought, and swallow us all. If this happens, we might have time to reflect as we see in our televisions a hole growing in Geneva, what we have done to ourselves to keep the arrogant fantasy that we could dominate the Tyrannosaurus Rex of the galaxy. If this happens, we might reflect about the stupidity of giving a corrupted nuclear company, employing thousands of physicists that used to make nuclear bombs, $10 billion just because its marketing department has been very active promoting a fiction God's particle. When this happens, we will know how dumb we were for believing such absurd fiction thoughts, how little we did to protect our lives, how irrelevant we are compared to the cosmic forces of the Universe. I just think on the words of Mr. Einstein: "Two things I deem infinite, the Universe and the stupidity of man, and I am not sure of the latter."

The reader might think that the brain of a scientist is different from a believer. But both are humans, both are imprinted by memorial thoughts and dogmas. The reader, a believer in science, as all the members of our technological civilization, might think that Mr. Hawking's science fiction is superior to the "obviously wrong" theories of priests. But I can assure him that under the laws of the scientific method, the chances of black holes evaporating back to the past are equal to the chances of believers coming back to life. The change in the arrow of time needed for both things to happen is exactly the same. So even if Hawking closes his article in black hole evaporation, affirming cleverly that "it appears that Einstein was double wrong," I would go with Einstein, who said to those who proposed time travel that "wires don't travel to the past." Indeed, recently another quantum physicist proposed on account of the same fantasies that the future was preventing the Higgs from happening because it abhors such absurd theory. But it is not *the future, but the future of complex science what is telling us that if CERN goes on, we might not have a future at all.*

In that sense, the mujahideen are far cleverer in their beliefs because it costs nothing to believe in resurrection after death. That belief makes them happy. And who knows? Maybe after all they would wake up in paradise while the millions of quantum zealots that have opted to believe time travel is possible, zombies resurrect, hot water freezes in the cold, and levitation happens all the time around black holes are putting their life and ours on the line for that belief. It might happen, but we won't wake up in paradise.

Indeed, the truth beyond the smoking gun of complex equations and chutzpah statements is that we hardly know anything about black holes except what we see: that once they are born, they grow so fast they eat stars within seconds, provoking supernovas as they should do in this planet if the collider makes them. Thus, obviously, it is an absurd experiment to make them on Earth since we can use telescopes to look at the real thing without danger. Plainly speaking, the LHC is sort of a Jurassic Park experiment that instead of a Tyrannosaurus Rex will resurrect the most dangerous predator of the galaxy. And so it appeals to particle physicists; as Jurassic Park appealed to the married couple of dinosaur experts in the film, so are LHC's physicists eager to unleash and pet the mighty black hole, the pervading, dominant species of the cosmos.

Yet the marketing of the quark cannon as the instrument that will usher physics into the twenty-first century is absurd, especially when the knowledge it can give to mankind is weighted against its risks—the negative bioethical value of killing all of us. When we add both purposes of science, knowledge, and bioethics, which this experiment fails to meet, we conclude that the large hadron collider is a giant leap backward for mankind. The leap forward that is ushering cosmology into the twenty-first century in fact is happening outside CERN, as theorists are improving the laws of our three languages of the mind—logic, mathematical, and visual thought—creating a new paradigm of scientific truth, *the fractal paradigm* that will substitute the outdated twentieth-century quantum paradigm. Its application to cosmology is called fractal relativity since it evolves further the topological work of Einstein and his theory of mass. And it means a giant leap forward for science because it improves the three legs of the scientific method beyond the previous quantum paradigm: it improves our description of space with evolved mathematics (fractal and non-Euclidean mathematics); it introduces the laws of information as principles of temporal logic (unlike quantum physics which uses only entropy to describe the future, fractal logic is dual, using two arrows of time, energy, and information to explain the Universe); and it conforms better to the experimental data we have about the Universe and its species made of self-similar parts. And it explains many of the "questions" unresolved by quantum physicists that CERN pretends to discover bombing the Earth with quarks.

It must be said in any case that the valuation of a scientific paradigm and the truth of a certain theory is not decided by physicists, but by philosophers of science, the judges of truth in science, who have long pointed out the fact that quantum cosmology, the theories of the Universe sponsored by CERN, are full of logic contradictions and that the true path of cosmology must come from the standard model of the Universe, developed by Einstein, NOT by quantum theorists who pretend to substitute the work of Einstein with laws that apply NOT to the Universe at large but to the restricted world of microcosmic quantum particles. Again this is proper of religious dogmas, which use their "particular god" to explain all phenomena beyond the realm of theology.

For all those reasons, we need to know first those basic laws of the scientific method so the reader can have the tools of reason needed to decide if CERN's dogmas are right or wrong, without having CERN and the celebrities of the scientific press decide for him. We shall then proceed, in the rest of this first book, to explain the basic laws of truth in science and the advances of the last decades in our mathematical, logical, and experimental understanding of the Universe and then apply those laws and cosmological breakthroughs to the understanding of the different theories about the particles and events that will happen in the quark cannon.

3. The Three Legs of the Scientific Method:
Logic, Mathematics, and Experimental Evidence

All this said it is obvious that we must start our inquiry about the quark cannon by defining positively the laws of truth and the bioethical purpose of science, given the misuse the nuclear industry makes of those purposes to justify "in the name of science" the risk of extinction that mankind faces in the next three years, as CERN's nuclear physicists reach the highest levels of energy of the cosmos in their self-suicidal attempt to manufacture black holes on planet Earth. Since those laws have been forgotten and subverted by quantum cosmology, the theories of the Universe sponsored by CERN, which openly deny the standard cosmology of black holes and quark stars, used by astronomers, which originated in the work of Einstein, a century ago.

In science, as Descartes and Galileo admitted, there is a hierarchy based on the logic of truth. On top of all sciences stand precisely the logic principles of the scientific method since they are used to judge which theories are right and which theories are false. This is done based on three fundamental tenets of the scientific method: *logic consistency, mathematical accuracy and testable, experimental proofs.* When a theory fails any of those three tenets of the scientific method, it is deemed false. And the task of the philosopher of science is to denounce such theory as an incorrect hypothesis that must be discharged to avoid further developments that only hinder the search for truth. This is a rewarding job, as some theories are puzzles so complex that their falsity is difficult to break. So to help in this task, science has established a series of laws since the age of Aristotle, the founder of the experimental method. Those logic principles are the summit of the scientific method and its search of truth even more powerful than mathematical evidence *since, as Frege and Gödel proved, mathematics is a language that can be derived from the laws of logic and hence is hierarchically inferior to them. And so mathematical-only theories that deny the logic laws of the scientific method must be considered false unless experimental evidence appears.* Those logic laws are as follows:

Simplicity (Occam's razor)

A simpler theory is always truth because as Einstein put it, "God is simple and not malicious." This also implies the law of economicity. In the Universe, all has a meaning, a reason to exist. There are no free lunches in the Darwinian systems of reality. We men might not understand yet, for example, the full extent of information encoded in the DNA, but we are finding that every bit of it has a meaning after thinking for decades that most of it was waste. Economicity also means that a theory or equation that with the minimal quantity of parameters explains the maximal number of phenomena is the right theory.

Simplicity is always a higher truth than complexity and the reason is obvious: all what exists is a mere combination of two elements, energy and information. So equations and languages are simple. In philosophy of science, in fact, we define any language with only three elements:

Words: Subject (informative element) verb (action) object (energy of the subject)

> *Mathematics: X Operandi Y; where X and Y are most likely parameters of energy and information*
> *Visual language: Red (code of visual energy) <Green> Blue (code of visual information)*

The discovery of this universal grammar of all languages, which always requires an energetic and informative element related by an operandi or action verb, is an important breakthrough in our comprehension of the logic of the Universe that rules out a "monist" vision of reality, based only on energy, as quantum physicists sponsor. *It means that theories and equations, based in the duality of those two elements, tend to be the ultimate laws of science.*

For example: $E=mc^2$ is one of such laws, where E represents energy and mass is information,[ch.2] since mass is, according to Einstein's principle of equivalence between acceleration and mass, a cyclical vortex that carries physical information, measured in terms of the frequency of rotation of such a vortex. Indeed, information is measured in frequencies, as it is the information processed by this computer is measured in megahertz.

The simplicity and duality of the Universe is obvious. Yet for reasons that have to do more with ego than with reason, men tend to consider complicated theories of reality more truth because of the difficulty to make sense of them, which makes scientists think they are more intelligent. This is evident in the case of quantum cosmologists, who portray themselves constantly as more intelligent than other scientists because of the aberrant complication of some of their theories, without realizing this as a proof of their falsity, not of their intelligence. For example, Hawking said that he studied physics because "intelligent people did not study biology."[0] In true form, those statements and disqualifications of other sciences are just a measure of arrogance, which often is synonymous of ignorance of the laws of the scientific method. Such is the blatant case of Mr. Hawking whose theories deny both the laws of the scientific method and the standard laws of physics (causality in time, Occam's razor, thermodynamics, relativity, etc.).

Thus, simplicity means the more complicated theory is always false: the complicated Ptolemy's system was false and the simpler Aristarchus's heliocentrism right. The complicated Higgs and Hawking's theories are false and the simpler Einstein's theories of mass and black holes right.[ch.2]

And yet because today science is done with technology, with computers, the love for complicated theories has reached a peak, which prevents the advance of true science that must strive to find those final laws of simplicity. And yet, today in the world of scientific congresses, simple explanations of reality obtained with the human mind, as those provided by Einstein—today by the fractal paradigm—are ignored in favor of complex computer models, based on absurd theories.

In any case, one of the most rewarding tasks of a philosopher of science is to prune the tree of science of its bad fruits by simplifying and eliminating from its roots complicated theories full of logic errors, as those CERN sponsors. But obviously this is difficult, because as Kuhn masterly explains in his book *The Structure of Scientific Revolutions*, those who have invested time and money in a false theory feel threatened, as CERN does, in their livelihoods and deny vehemently they have chosen a wrong track with no way out. Further on, when a new paradigm appears and simplifies the laws of the previous complicated paradigm, it makes a "fool" out of the pretentious genius of the obsolete paradigm. So the new science tends to be scorned by the "high priests" of the past. Copernicus was scorned because it was "too simple." Einstein's special relativity was ignored for years because its "mathematics are from high school," as Hilbert said. In a recent congress of science in which I explained the fractal equation of unification of masses and charges based on Einstein's concept of mass as a vortex of space-time,[ch.2] a famous practitioner of quantum cosmology came to me at the end of the conference and said to me, "It looks right but . . . it is too simple." Indeed, God is simple but not malicious, exactly the opposite of what CERN is.

For that reason, it is important to prune the tree of science of its bad fruits as earlier as possible. Otherwise, a series of infections will happen, and false branches will break out, poisoning the scientific quest for true knowledge and its main logic principles, which are all derived of that law of simplicity:

> *Efficiency*. It means the Universe is not redundant and when a certain particle can do a certain job, there is not another particle equal to the first one. Thus, the Higgs does not exist because all what the Higgs can do the known top quark has done it already; and the Universe will not create a new particle, which is just a never-found self-similar mathematical equation. It is like saying Higgs is bald and Higgs has no hair. Both linguistic sentences refer to the same person.
>
> *Hylomorphism*. It was defined by Aristotle, father of the experimental method, which states that any entity of reality must have both information and energy or, in terms of the scientific method, we must find to prove any entity real, both, a linguistic description and experimental evidence of its existence.
>
> Hylomorphism is based in the duality of the Universe, as all entities that exist have both, energy and information. It is a law, however denied by quantum cosmology, which on one side,

sponsors entropy energy only theories of the Universe and on the other side believes as real any mathematical theory, regardless of its logic consistency or experimental proof.

The laws of causality in time. The Universe is causal. That means there are relationships between events and actions that follow a given order. For example: *Whenever you heat water, it boils; whenever you put it in a cold environment, it freezes.* Those laws of causality must be respected in any theory of reality. Or else we enter the realm of science fiction. There are no time travel machines, nor people resurrecting after death, because causality exists. When a hot object, for example, a micro black hole, is born, it doesn't get hotter in our cold Universe, evaporating as Hawking pretends, but it heats and evaporates our world as it cools down, absorbing it, according to the second law of thermodynamics. So theories that deny causality in time, as those that deal with the evaporation of hot black holes, which are supposed to keep getting hotter in our cold environment, are false. Point.

Mr. Hawking got to those conclusions by changing the arrow of time of black holes and hypothesizing without the slightest experimental proof that black holes expelling antiparticles that were traveling to the past, becoming hotter, instead of doing what relativity and thermodynamics say they will do, become colder as they grow and evaporate our cold world of electroweak masses, absorbing its particles. He broke two of the most important laws of science—thermodynamics that applies to electroweak forces and relativity that applies to gravitation. And so he became famous in the celebrity circus of mass media, in love with science fiction, since by changing the arrow of time of black holes, he made them "time machines." Yet in the world of real science, those hypothesis without proof caused many other errors, like the information paradox (where does the information go evaporated by black holes?), etc. So regardless of its sophisticated mathematical games, we must reject his theory of black hole evaporation because it has no experimental proof; it breaks the laws of time and it is complicated enough to hide its falsity.

The experimental razor

A theory based on substances, particles, forces, or actions that we have observed already in reality is more truth than one based in nonexperimental facts. In fact, a theory with no experimental proof is not science per se. What this means is that to write a theory in the language of mathematics is not reason enough to declare a theory truth or scientific. To think so is to convert science into religion, which precisely is defined by the belief that the language in itself (in religion words, in science numbers) suffices to accept a truth. Otherwise we could prove the existence of God as Saint Thomas did (if the word God exists, therefore God exists, he said). The scientific method precisely appeared to avoid such false linguistic proofs. In science, the same linguistic error, truly extended in quantum cosmology, is called Pythagorism: the faith that a mathematical statement will always find an experimental equivalent. This is not truth. As Einstein said, "You know Henry (Poincare), I abandoned mathematics because while I knew when an equation was true or false I could not say when it was real." And science is about reality.

If those logic laws of truth, simplicity, efficiency, hylomorphism, experimental evidence and causality in time do not happen, history of science shows a theory is unavoidably false. Thus, unless a mathematical theory helps to explain a fact already experienced and it follows the logic laws of duality, causality, and simplicity, it will be found false. It is only a question of time. Indeed, today, as twenty-first-century science advances, thanks to the fractal paradigm,[Appendix] many errors of the complicated quantum paradigm are being discharged, according to those laws. We don't need in that sense the quark cannon to prove false the Higgs, the evaporation of black holes, and the cosmological big bang, three theories *already proved false*, a fact that reduces CERN's experiment to re-creating a "local big bang," the big bang of the Earth converted into a quark-gluon soup after its nova explosion.

Today, unfortunately, we live in an age dominated by audiovisual fictions, which has spread to the field of cosmology due to the denial by a group of physicists called quantum cosmologists of the three legs of the scientific method and its logic principles. It doesn't matter how many people work at CERN or believe those false theories, they are wrong because the scientific method has proved them false. Mankind tends to believe for reasons different to reason and truth, mainly authority, given by money and military power, anything the "high priests" of a certain society say. In our technological civilization, unfortunately mankind believes what a machine says, what money talks, what the military affirm; but as Einstein put it, "those who impose truth with power will be the laugh of the Gods." He knew well. In the nineteenth century, everybody believed an absurd theory called ether. Only Einstein understood that ether did not exist because it broke two of the three legs of the scientific method, logic consistency and experimental proof, despite its mathematical consistency, as the evaporation of black holes does today. And so he sustained such truth and could not find a job as a physicist in Germany for opposing the physicists of his age and had to join a patent office. In the sixteenth century, only a priest of a small church defied the conventional wisdom that the Earth didn't move, which was held by the most respected scholars of their age. And yet Copernicus was right: "Eppur si muove."

If we apply those rules to modern speculations in physics, it astonishes the philosopher of science how many false theories exist and yet are taken very seriously by their proponents, who defend their self-proclaimed "genius agenda." We will not quote them here, though any physicist can think of a few of them that occupy thousands of scientists—except, of course, the one he works on. The existence of speculation is not forbidden as a mental aesthetic delight, in the same way we do not forbid fictional films. But when we pretend to be doing science with consequences to the life of people, we must differentiate unneeded Pythagorism from experimental theory: if we have an experimental theory that fits the three legs of the scientific method and its logic laws and explains a given phenomena, it is unnecessary to find new theories. The model is complete. This is the case of Einstein's theory of mass and black holes completed by fractal relativity. ch.2 We do not need speculative theories like the Higgs, nor do we need to spend $10 billion and risk our lives to prove those speculative theories wrong.

CERN breaks all the tenets of the scientific method, and certainly it is not needed to prove wrong the three speculative theories that it says it shall prove—the cosmological big bang, the Higgs particle, and the evaporation of black holes. The tenets of the scientific method are enough to prove them wrong, as they have always been for 2,300 years since Aristotle defined them.

The obvious theoretical problem of quantum cosmology, already explained by Einstein, is its denial of the scientific method, which today has resulted in the incapacity to verify any of its theories about multiuniverses, multidimensions, etc. Those theories become a personal agenda to which the author clings till the end, regardless of its truth, which is substituted when the theory fails all the tests of the scientific method by a "probability." Any theory then becomes possible, and if it is not verified, it is because it has a low probability. Thus all can be possible in physics, which is an ideal state of affairs for a community that will always have new excuses to fund new machines to seek proofs for new theories, will always get grants to explore some mathematical idea, and will keep publishing papers about the most bizarre science fictions. It is the zeitgeist of our times. But human agendas have little to do with reality and the strict laws of truth in the Universe, which can be resumed in a Darwinian fact: Things that are not truth, that do not follow the strict laws of efficiency, simplicity, causality, and hylomorphism, do not exist. And/or when they are created, they disappear immediately. Specifically in this book, we talk of four theories that oppose those logic tenets of the scientific method:

Pythagoric theories that are only mathematical equations without logic or experimental truths, like Hawking's radiation; complicated theories like Hawking's radiating black holes (different from complex structures, which are achieved by the repetition of simple elements, as a network of atoms or a human being is made of a network of self-similar DNA cells); redundant theories like Higgs, energy-only theories that deny the arrow of informative creation in the Universe like quantum entropy (Hawking's black holes) and the cosmological big bang; theories that deny experimental evidence (Copenhagen uncertainty, multiuniverses, multidimensions); and theories that break causality in time (Hawking's black holes). This book will consider their falsity in detail and explain the "real theories" they pretend to substitute. To notice that the falsest of them all is Hawking's evaporating black holes. And yet it is the most popular, precisely because it looks purely like science fiction and we live in a world of audiovisual fictions.

Let us now consider in detail the two aberrations of truth most common among quantum cosmologists, Pythagorism, the belief that a mathematical equation is truth per se, and quantum entropy, the belief that the Universe is made only of energy. Those two concepts split the scientific truth of duality or hylomorphism, which states that to exist in reality an entity must have a limb or field of energy and a particle or head of information—or in linguistic terms, that any linguistic theory must have experimental evidence. Curiously enough, this law is also the fundamental law of quantum physics, called the principle of complementarity, which states that physical entities have all a field of energy and an informative particle.

In that regard, before we go further in our analysis of the errors of quantum cosmologists, we must now defend and explain the differences between quantum physics, the proper theory of the microcosms and quantum cosmology, the false theory of the Universe.

When we apply those laws of truth of the scientific method to quantum theories, it is clear that we can divide them in two types of theories, one truth, one false, one good for humanity, one wrong:

— The *good fruits of quantum theory* are the application of quantum theory to its original purpose, the study of the outer electronic cover of the atom and its particles and forces, which follows the three legs of the scientific method and is right. This is called quantum physics, and it is the basis of our industrial revolution of electronic machines. In this book, we accept fully the work of quantum physics, except the philosophical interpretations of uncertainty, whose falsity is coming to light in the twenty-first century, as the evolution of mathematics translates probabilistic uncertainties into fractal determined equations, which merely refine the outlook of quantum mathematics, as Einstein's work refined with non-Euclidean mathematics the outlook of Newton's differential equations. But it maintained the same logic, principle of equivalence between mass and acceleration, *since in science, the fundamental truths are those logic principles, which each new mathematical paradigm and observational machine merely describes with more detail.*

— The use of quantum particles to create nuclear weapons of mass destruction and the use of electronic devices to create robotic weapons, avenues of the military-industrial complex, are wrong from the perspective of bioethics, will cause the extinction of mankind in the short term (nuclear weapons) or in the long term (terminator robots), and should be banned.

— Absurd theories that expand quantum theory, from its original realm (particles) into cosmology, the realm of Einstein's theory of mass, gravitation, and black holes, which CERN pretends to prove, is called quantum cosmology. It is a false discipline in its entirety because it extends the use of laws found for particles to celestial bodies. This breaks the laws of the scientific method, converting a self-similarity or analogy between both scales into a homology or "equality." Bats and planes have analogous wings, not homologous ones, so their laws are different. It would be as if we were trying to extend the laws of social religions of love based on the bioethics of survival of groups, favored by solidarity, to the behavior of physical particles. The laws of human ethics that love religions embody have a biological justification because it has been proved ad nauseam that in nature, according to Darwin, species are subject to a process of natural selection. And so social love among individuals of the same species is positive and naturally arises in any language of any society. But when Christians started to explain the Universe with those laws and pretended to know how it was created, they went out of the realm in which bioethical laws of love and survival apply—human societies. So it happened when nuclear physicists decided to apply quantum laws to the Universe at large, substituting Einstein's work. That fact, which is at the core of CERN's issue, created a series of absurd theories that are not true science: quantum cosmology (the application of the laws of the small to the big) and quantum entropy (the creation of theories only with the arrow of energy that disregard the importance of information in the construction of the Universe).

Those two errors, the creation of weapons of mass destruction and the expansion of quantum and energetic theories to cosmology, are related, as those theories are taken seriously due to the power of quantum physicists in our military-industrial society. Further on, the people who make weapons, which are systems that release energy and kill the information of life, tend to "design" energetic theories of reality, as Gamow, the creator of the cosmological big bang and maker of A-weapons, did.

Indeed, CERN accepts those theories as possible precisely because its instrument of knowledge is a weapon, which only can test energetic theories of the Universe even if they are obsolete in twent-first-century science.

For all those reasons, we conclude that CERN and the quark cannon go against the bioethical and scientific principles of knowledge and must not be confused with quantum physics as a science and the advancement it has brought to mankind but considered specifically as a bad idea, a bad apple of the tree of technology that must be rooted out.

4. Pythagorism, the Religion of Mathematics

Einstein was a lonely genius. During his stay at Princeton, he would have Gödel only as a friend, also a German émigré famous for eccentricity. Gödel astonished his colleagues by proving that all mathematical statements were just approximations to the truth. This is logic. After all, mathematics is a language that carries less information than the object it describes, the truth in itself, *as all languages do. Hence any language is a relative truth, which must be checked with experience.*

If not, science becomes a religion. And indeed, we can talk of two kinds of dogmatic religions:

— *Pythagorism*, the religion that states any mathematical equation is truth. It is the religion sponsored by quantum cosmologists at CERN, who deny logic and experimental evidence.
— *Fundamentalism*, the religion that says revealed words are truth, as they are the language of God.

Both fundamentalisms that confuse reality and the language we use to describe it, mathematics and words, are dangerous to mankind because they invent reality and give power to those high priests who pretend to speak the "language of god." Yet our technological society is ruled by mathematical computers, and therefore it is ruled by the high priests of Pythagorism, which is neither acknowledged as a religion nor is the destructive power of its language and technology understood. Indeed, numbers can kill and certainly reduce the "feelings" of murder, which becomes a collateral damage exercise at a distance and justified by our belief in the ethics of technology. Those ethics mean that all what is uttered by a computer must be truth, all what is written with numbers is more real, all what is done by a machine is accepted. In that regard, Pythagorism and mechanism go together and become the dogmas of our society.

Pythagorism denies Gödel's proof that mathematical equations are linguistic theories; hence, only when they obey the logic laws of the Universe and there are experimental proofs of the existence of the entities they describe they will be real.

Gödel's work is denied by quantum cosmologists because it eliminates a number of fringe theories they invent without experimental and logical proof. Those logic aberrations, based on Pythagorism, would disappear for the good of cosmological physics if the laws of truth of the scientific method were respected. Then cosmology could find the single truth about the single known Universe.

An example of Pythagorism is the ether theory. In the twenty-first century, a mathematical theory defined a substance called ether that was harder than steel, denser than gold, occupied the entire Universe, and yet it allowed planets to travel through it. So ether had to be both transparent and weightless, but dense and rigid. It was thus by contradiction ad absurdum, a false theory despite its mathematical beauty. Yet because many physicists think the Universe is written not with logic but only with mathematics ("The language of god is mathematics," said Galileo), they accepted without experimental proof a logically false theory based on only one of the three legs of the scientific method—mathematical consistency.

Einstein proved the ether theory false with his theory of relativity and his concept of mass. He was a serious scientist that did master the three languages of mankind—words, numbers, and images—so he could not accept a theory that contradicted the laws of the scientific method. As he put it,

> *Every theory is speculative. If, however, a theory is such as to require the application of complicated logical processes in order to reach conclusions from the premises that can't be confronted with observation, everybody becomes conscious of the speculative nature of the theory. In such case an almost irresistible feeling of aversion arises.*

This is the case of the theories of quantum cosmology CERN pretends to prove and whose falsity we shall consider in detail in this book—theories like Hawking's radiation and the invisible Higgs, supersymmetric particles and multiple universes, nowhere to be seen that break the basic logic laws of reality. But they justify the use of the quark cannon and so today they are accepted as "probable" truths that explain why we need the large hadron collider.

5. Transcendental Illusion: Hyperbole

In that regard, the fundamental error in science and religion or any form of human thought is what Kant called transcendental illusion, in simple terms hyperbole: the innate capacity of human beings to feel more important than they are, to know more than they do, to think their languages are not approximations to the truth but reality itself.

The origin of transcendental hyperbole is the structure of any mind that maps our reality, according to a center point of view, the human himself. So the human perceives less about objects that are far away from his POV, such as galaxies or very small atoms. This makes hyperbolic statements most common in astrophysics. Indeed, perception of the large and small is limited and so hyperbole makes most physicists believe that a local phenomena, observed in a reduced number of experiments, is in fact an absolute phenomenon that transcends the relative, local world and becomes a law of the absolute Universe.

This act of pure arrogance, which denies the relativism of truth, has its reason in the structure of the mind, which is an electronic mirror of the Universe, a "linguistic world" that we confuse with the absolute reality. But sensorial or instrumental experience, often with a single language of the mind, merely acts as a linguistic mirror of the complete reality, *which is the only one that has all its information, hence all its truth about itself.* All that you see is NOT reality but the world that your mind observes from your point of view. You are not observing the Universe. Your ego is dual: your verbal wor(l)d made of words is commenting on the visual electronic mapping that your I=eye makes of reality. Your brain is a contiguous dialogue between your I=eye and your I=wor(l)d; and beyond, there is a huge "other Universe" of which, in fact, 96 percent is gravitational dark energy and matter we cannot observe. It should be obvious to the reader that a mirror of any kind is always smaller than the Universe it observes, as it has less space-time, less information, and less dimensions than the external reality, even if it is a focused mirror that makes a quite accurate image of reality. And yet for a physicist, ascribed to a theory called naïve realism, all what his machines measure is "absolute reality," to the point that the error of measure of those machines, which interfere with the atoms they observe (uncertainty principle), is interpreted not as an error of measure but as the existence of an uncertain world.

Later, when we have gone a bit deeper in the understanding of the fractal paradigm, we will be able to mathematize the space-time equation of a world-mind, which acts as a fractal mirror of the entire universe.

What matters to us now is to consider the effects for human theories of thought of any type, verbal/religious or mathematical/scientific, of the structure of minds/worlds, which are space-time mirrors of the bigger universe. We fix with our mind reality a quantum soup of vibrations of energy and form, which become through the visual mind a fixed pattern of objects. That tapestry of images is perceived from the perspective of our ego, of our mind, which is an infinitesimal knot of information compared to all the reality—yet it seems to the naïve, realist physicist everything that exists.

In reality, the brain takes that limited information provided by external senses and instruments and extracts a series of logical/temporal and mathematical/spatial relationships that further synthesize all that information. Thus, when we make a linguistic law of that mapping, we still believe it is an absolute truth, yet it is only a linguistic, simplified equation that needs as many experimental and logical proofs as possible to acquire a minimal certainty. Yet Pythagorism ignores this. It believes with religious fundamentalism that the mathematical mappings of the Universe made by the physicist and his machines are "all" what matters, the only truth of reality. And of course, for him, all other entities that "gauge" reality in different languages with different forces are of no importance. The artist that observes reality with human senses is NOT relevant. The different animal species that perceive are not obtaining "scientific information." Time, he says, is "what the clock measures." Yet every system, charge, mass, or circadian cycle is measuring informative patterns. This naïve realism or reductionism is at the core of the messianic zeal of the CERN physicist, who think that to measure a new decimal of a z-particle is worth risking the life of all other beings of this planet. Because for him, what his machines do not perceive do not exist. The species that speak other languages or perceive other worlds are not intelligent. Thus naïve realism reduces reality to the mind and instruments of perception of the scientist and knowledge to a series of numbers. The result is, of course, that naïve realists consider scientists as the only intelligent species of the Universe and expand their local measures and laws to the entire cosmos. The opposite is often truth. The Universe is relativistic, infinite, and man is merely a "mush" in a lost rock in a corner of a galaxy. Yet because all its "fractal scales" and forms are self-similar, made of energy and information, the laws that rule the creation of the physical and biological form we perceive are the same laws of faraway objects, which we hardly perceive. *So to prevent uncertainty errors, the most important sciences are not those of the infinitely big and small (astrophysics) but biological sciences, which study the closest, more detailed shapes of energy and information—life beings.* And yet physicists despise those sciences because they think that the Universe has "limits" to its infinite, hierarchical scales of energy and form. And so knowledge is to find "the limit," which must be something so small that nothing smaller exists, so big that encompasses the entire Universe.

This, as we shall see, is a hyperbolic error, which the fractal paradigm has proved wrong. So there is no interest on searching smaller scales because they are infinite and we just merely lose precision in our observations. There is not a cosmic big bang, but only a local big bang—a supernova or quasar, made with quark-gluon soups. And so CERN at best will explode the planet or the galaxy, but never will study the birth of the infinite Universe.

And yet the founders of naïve realism are the least relativistic of all scientists because they pretend that through the study of the infinitely large and small, they will find the meaning of it all. And because their perception has uncertainties in the infinitely small and limits in the infinitely large, *they can invent any philosophy, theory, or fantasy they want, as we can't really know much of those faraway objects. So their hyperbolic, naïve realism and the arrogance of their "technological/military/financial power," makes them fundamentalist priests of their science. And all relativistic concepts from Gödel's proof of mathematical relativity to Einstein's relativity theory to the fractal paradigm are taboo in their profession.* Let us consider then some of the basic errors of naïve realism, which will be argued in this book:

The big bang theory

Penzias studied the local background radiation of the galaxy, yet Gamow decided it came from the entire Universe. Today, however, proofs mount in favor of a local quasarlike big bang or a local black hole redshift lensing as the origin for that background radiation.

Thus, CERN will study a local big bang (the quark-gluon soup of supernovas that might convert us into one) and yet it pretends to be studying the birth of the entire Universe.

Space-time theory

Galileo studied the temporal change in the motion of beings, $v=s/t$, and yet he decided he was studying all the arrows of change/time of reality. Many philosophers have explained that time is change[ch.5] and physicists do not study all the times changes of the Universe, only those that imply a spatial translation or change in motion. Hence, only in physics is time the fourth dimension of space (spatial change).

There is also the morphological/biological time/change of beings, which physicists ignore and biologists study. But fundamentalism among physicists is shown in the fact that only they deny there is an arrow of physical, cyclical information in the Universe, which Darwin described with his theory of evolution and Einstein described with his theory of cyclical masses, to the point that Nobel forbade his prize to evolutionists and today quantum physicists sponsor the absurd idea that an entropic/energetic field and scalar, lineal particle, the *Higgs, gives mass to all other particles.*

Yet the entire concept that energy is the only arrow of time that creates the future is a hyperbolic theory.

Since Boltmann, the founder of entropy, studied only the electromechanic behavior of molecular forces, later extended to electroweak forces by quantum physicists, they deduced from the energetic, entropic properties of those forces that the entire Universe was guided toward an expansive death, without taking into account the informative nature of cyclical vortices of mass, which balance the Universe. Thus physical information, the arrow of mass, disappeared from science and still today the work of Einstein, which said that time "curves space" is denied by Higgs and their absurd theory of mass and Hawking, who affirms that black holes do not bend energy into time, entropy into information, but evaporate. The denial of half of the arrows that create the future is the biggest error of science, originated by the hyperbolic extension of the entropic properties of electromagnetism, which the cyclical properties of quarks, strong forces, and gravitational forces balance.

Newton applied an abstract, mathematical, linguistic image of space-time, the lineal, continuous Cartesian plane to study the Universe; and despite Leibniz's admonitions, he thought that there was only a single continuous space-time, like *the abstract, mathematical plane he used.* Ever since, the fractal sum of all the spaces of the cosmos were reduced to one: the continuous space of a Cartesian graph.

Galileo used a mechanical clock to study all the time cycles of the Universe, equalized to the clock rhythm and ever since physicists believe there is only one time, "what a clock measures."

Einstein's special relativity, which studies the change and deformation of our light space/time, following on the path of Galileo's definition of time as change in motion and space, $v=s/t$, is applied by physicists to all the times/changes of the Universe. So time becomes the fourth dimension of space and time travel is possible.

As a consequence of this, Hawking confuses in his equations physical time (the change in the motion of beings) with all the times/changes of the Universe and believes that a change in the direction of motion (particle/antiparticle inverse orientation) is a change in the total arrows of time of the Universe. So he thinks black holes travel back in time because antiparticles are not for him anticlockwise motions but particles that travel back in the single universal time of Galileo's clock.

Einstein studied the light-space of our galactic cosmos, whose limit of speed is c, and thought it was the limit of speed of all the scales of the

Universe. Yet the Universe has multiple fractal scales, with different limits of speed, given by their medium. So water limits the speed of fishes, air limits the speed of birds, light limits the speed of our Universe to *c* speed and then there is a dark, gravitational membrane of energy, which seems to reach between galaxies 10 *c* speed. And in that gravitational membrane, there is a top predator species, a black hole, which can break and absorb our light membrane and CERN will do, without knowing at all what it will create *since it ignores all about the fractal, discontinuous paradigm of space-time that describes them.*

Gödel proved that mathematical equations can be false and yet physicists think that any equation they invent must have the equivalent real particle or force in the Universe. So they can "play to be God."

Quantum theory

Our instruments of measure are uncertain when observing the microscopic quantum world and yet it is a dogma of faith in quantum physics that the uncertainty is in the quantum world, not in the human observer.

Particles and forces show a clear fractal organic behavior, as they move in groups called waves that become tighter networks called particles. Yet this organic analysis of the quantum world is denied by abstract quantum theorists, humans after all, who like to think they are the only "intelligent beings" of the Universe. Yet in absolute relativity, the scale of size doesn't matter. The properties of the Universe are *invariant at scale and so are the morphologies of energy and information. Thus, the quantum world must have the same properties than our scale of size and so waves are herds of fractal particles that become wholes, as it happens in our size-universe or in the world of stars and galaxies.*

Particles gauge/measure information, and all quantum theories are for that reason called gauge theories: particles act as herds or tight groups when traveling, depending on the best strategy to cross doors; particles have an informative center and an energetic system (complementarity law) as all organic systems do; the fundamental particles, quarks and electrons, absorb energy and reproduce their form in new particles; and finally, those particles organize themselves in complex social networks. Thus, particles "feed," "perceive," "reproduce" and evolve socially, the four drives or "arrows" of time/future of all organic systems, which means we exist in an organic Universe, where the ultimate properties of organisms are embedded in the essential elements, quarks and electrons. And yet any attempt to consider an organic theory of the Universe is considered a heresy because man MUST be the only intelligent being. So the concept that quarks and black holes feed on our electroweak matter and to make them at CERN is to invite the Tyrannosaurus Rex of the galaxy is NOT even considered because the death of our matter is called breaking the symmetry, the feeding membrane of black holes, the "event horizon" and the machine that will create the quark-gluon soup, origin of those black holes and pulsars that eat planets and stars, is not a weapon, but the LHC. And so our possible death will be a collateral effect of the LHC. Abstract jargons indeed make us wonder who has more life properties, if the physicists of CERN or the black holes and quark condensates they will create.

The list could go on and on. Fact is humans won't change the hyperbolic way in which they do science or religion which has the same hyperbolic structure that makes 2 billion people believe that Moses saw the creator of the Universe when a bush burned on the desert. Nothing differentiates that religious hyperbole from the hyperbole of the cosmic big bang that CERN pretends to study—a ball of quarks that fits in the fist created the Universe, and a single lineal Hubble equation, a doodle in the sand, explains its growth. How those hyperboles are imposed over reason? Obviously with censorship. For a century now, the principles of relativity and local measure of Einstein's work have fallen on deaf ears and are ignored by most quantum physicists. While the fractal paradigm, which returns man to "its" nothingness in the infinite Universe, which takes further Einstein's relativistic work, is simply ignored. I long ago stopped sending articles to "scientific" magazines about Einstein's mass theory and the fractal Universe, knowing that the Higgs is the religion that allows to swindle billions of dollars from the taxpayer and the cosmic big bang the dogma that allows nuclear physicists to appear as the priests of human thought.

Hyperbolic knowledge has a rational explanation in the relativistic, local, linguistic, limited perception of the point of view of any human, who measures the Universe from his perspective and so he thinks he is the center of the Universe. But his mind is just an infinitesimal point that gauges/mirrors the Universe, like any other particle does, not the entire World in itself. This paradox is fundamental to understand the arrogance and ignorance of humans at large and people at CERN, who are going to replicate the quark-gluon soup that explodes stars in a big bang, hence exploding the Earth, and yet they think they will observe the birth of the infinite cosmos.

The quality of our informative machines makes this belief easier. We live in an age of increasing scientific arrogance and ignorance due to the detailed beauty of computer models, whose underlying principles are no longer of any importance, as long as the pics are good. Scientists are becoming children of thought whose minds are being substituted by computer graphs, while mass media celebrities inflate their egos and proclaim themselves new Einsteins and their theories "God's particles," hyperboles of their real face value, which according to the laws of the scientific method is null.

II. Man vs. Machine: Form vs. Energy

I do thought experiments.
        —Einstein, father of modern physics

I am Kali, God of death.
        —Oppenheimer, father of the nuclear industry

Shiva, god of energy, sustained by Kali, goddess of death,
holds the eternal fire that destroys the Universe. The statue, found
at CERN, represents the belief of physicists who deny the arrow of
information in the Universe and search for pure energy as his Saint
Grail: the meaning of it all. But this ideology was born of their worldly
profession, as creators of weapons and energetic machines, since Galileo
defined time as "change in the motion of beings," studying the trajectory
of cannonballs describes only one-third of the arrows/dimensions of time
of the Universe. Since all what exists is a combination of energy and form
that creates the present, complementary entities of reality.
Physical forms are made of energetic forces and informative particles that
create simultaneous waves, biological beings are made of energetic
limbs and informative heads that create present organisms.

6. Technological Science, the Religion of Machines

We might already be in the year of our extinction. Yet nobody cares. Why? The answer is obvious: we believe in machines. We consider them the model of the Universe, the meaning of progress. We have worshipped them for four hundred years since Galileo affirmed that telescopes were better sensorial systems to observe space than the eyes of the artist and clocks were better forms of measuring time than the verbal modes of the human writer. Galileo, by defining time as "what a clock measures" and required the precision of machines to accurately calculate distances in space, obliged science to become dependent on machines to perceive the cosmos. Yet it also reduced our comprehension of the deeper meaning of times, which are all the modes of change of the Universe, including morphological, biological, and geometrical change that clocks cannot measure. In this manner, the study of the life/death cycle, caused by morphological change, disappeared from scientific studies till Darwin and because humans are biological beings, the study of man became also secondary to the study of matter, speed, space, and motion, the type of changes Galileo defined with its clocks ($t=s/v$). Yet after four hundred years, mechanism, the doctrine that science must be done only with machines and further on that machines are the accurate model to describe the Universe, has become a dogma of our technological society that worships those machines. So we are ready to take our chances of extinction firing up the quark cannon because it is a machine, not a human being, who will perpetrate the genocide. And machines are the seers of truth of our civilization. Imagine this action as done by a terrorist group. Within days, our security forces rightly will detain everybody working at CERN and close down its facilities. Yet the quark cannon is a machine made by the high priests of our technological civilization; the heirs of Kepler, who said that God was a clocker because he used clocks to measure time; of Galileo, who said that God only spoke mathematics because he used numbers to measure time. This, of course, is not truth but religious dogma.

7. Physics, the Science of Energy, Its Denial of Information

And so we shall sacrifice our lives to our higher God—Shiva, the god of energy, motion, and death.

Indeed, the main error of CERN's theorists is called quantum entropy, the belief that the Universe is made only of expansive motion, lineal speed, entropy, energy, synonymous words that are also the substance that simplifies and erases the information of life and the ancients represented with the gods Shiva and Kali or the yang symbol, opposed to the god

Vishnu and the yin symbol of information. Yet even those ancient mystiques understood better the dual universe. Since today physicists have imposed the myth that energy is the only "arrow of time"; that is, the only direction of the future, which means the Universe must die, and death, chaos, and disorder dominates information and life. This is obviously false and we shall prove it ad nauseam in this book with logic and experimental and mathematical proofs. But religious people do not see beyond their dogmas. They don't study other "religions"; in this case, physicists don't study the properties of fractal information, the second arrow that constantly creates the Universe.

It is obvious though that the Universe is not made only of energy as the first physicists believed, based on the analysis of motion machines and cannonballs. The Universe is dual, hylomorphic, made of energy and information, the two arrows that create the future. Yet four hundred years of Galilean tradition have become dogma. How this dogma is imposed, we have already explained it: through the error of hyperbole, which extends to all forces the properties of molecular heat, thermodynamics.

Physicists deny the informative nature of gravitation (Hawking's theory of black holes) or have expanded the local analysis of electromagnetic energy to the entire Universe gratuitously, developing entropy only theories like the cosmic big bang, biased by their worldly profession as weapon makers. Those physicists—from Nobel, the inventor of dynamite, to Wheeler, the inventor of the H-bomb and the singularity theory of the big bang—have always used their worldly power as high priests of the industrial military complex to spread their "energy-only" ideas as the meaning of it all.

Let us then consider the historic reasons why entropy-only theories about the future of time became dogma among physicists when they are so obviously false.

It is a fact of history that scientists first discover machines and then they invent theories of reality that adapt to their mechanist observations. But *they never recognize the influence of those machines and the bias they introduce in their perception of the Universe,* so those theories become a dogma of faith. Of all the ideologies of our mechanist world, none is stronger than the ideology of the time clock, which appeared with the invention of mechanical clocks and has fogged ever since our comprehension of the Universe. The idea brought about by the discovery of the clock is that time, the perception of change in the Universe, can only be understood with mechanical clocks, as time is only one, because those first scientists used only one clock-rhythm to measure it. This absurd belief in a single time for all the universal rhythms of change dates back to the use of clocks by Galileo and the creation of a single time-line in mathematical graphs by Descartes. Then Kepler affirmed that "God was a clocker" because he used a clock to study the Universe. It is still a dogma of mechanism, even a century after Einstein proved that different regions of the Universe have different rhythms of time change.

It was not needed however to wait four hundred years for the arrival of general relativity to understand the obvious—*that each species and organism of the Universe has its own biological rhythms of time, of life and death, of creation and extinction, of information and energy.*

One of the fundamental aberrations of mechanism is indeed to consider that reality is made of the languages or instruments we use to study it. So Kepler thought the Universe was a clockwork, Galileo thought God only spoke mathematics, and Newton considered the Universe as having only an absolute time and a continuous, absolute space because he used a Cartesian continuous graph to describe it. All of them confused the instruments and languages they used to study the Universe with the Universe itself. Today, most scientists think the Universe is a computer because they model it with computers. Those aberrations of scientists were already pointed out by Kant and Goethe, who affirmed that physicists confuse reality with the observations made by their instruments and consider themselves more intelligent than anyone because they see more with their machines. CERN's physicists belong to that frame of mind. Mechanist idolatry explains why—since the invention of the clock that substituted the study of time as change with past, present, and future verbs—machines have become essential to the scientific endeavour and time has become "what a clock measures."

Yet we need to return to a more sophisticated theory of times that includes all the types of change of the Universe if we want to understand the two arrows of time, energy, and information since physicists only study changes on the energy and motion of beings.

In fact, all theories of science about time depart from the first scientist of history, discoverer of the experimental method, Aristotle. He also defined the blueprint of a time theory that still stands as the origin of all other scientific philosophies of time. Aristotle defined time as change and so he affirmed that a theory of time should be a theory of change. Then he proposed, as Eastern philosophers did, two arrows, wills or motions in the Universe, responsible for the two types of times changes science studies: change in the form or information of beings and change in their energy. His wording, though, was somehow more complex. He called the arrow of information *morphological, informative change*, which happens in the processes of aging, death, and evolution. And he called the arrow of energy *translational change* or change in movement and energy, which happens mainly in the physical realm. Thus, he divided the study of times in two different sciences—biology, the science of form, of in-form-ation, and physics, the science of energy and matter. In the modern age, those two types of times-changes would be further evolved by Darwin in biology and Galileo and Einstein in physics, which added a new language, mathematics, to calculate with more precision those times/changes, and a new type of device, a machine called the clock, which would measure the rhythm of those times/changes. Finally, complexity and duality, the sciences of the twenty-first century, fusion of the work of Einstein's relativity and Darwin's evolution, the modern fathers of time theory in physics and biology. And the key to such advance in time theory is precisely to go beyond the mechanist concept of time as what a clock measures and return to the preclock, verbal, biological understanding of time as the modes of changes of the Universe. Since time is the perception of change in the Universe, which can be observed with any type of informative devices, not only clocks.

Unfortunately, physicists study all types of changes with a single rhythm of change, that of a mechanical clock. Thus, soon a mechanical bias was introduced in physical time studies: all the times/changes of the Universe became equalized with one specific time, the time of the clock. Soon the existence of multiple times/changes became forgotten and the concept of a single artificial time/change in the universe, that of the clock, became standard. So times were reduced to time and an absurd idea was born: there is only a *single time in the Universe—that of the clock—and a single arrow, energy and movement. The concept of time as change was forgotten and a formula devised by Galileo to measure moving bodies, through the use of mechanical clocks, appeared to define time in strict, limited, energetic, moving terms: V=s/t, t=v/s. It is the hyperbolic error of time in physics called reductionism or spatialization of time*, which philosophers of science have for long denounced.

It is from that formula where errors like Hawking's thoughts on black holes as time machines arise. In a Universe with a single infinite, lineal time, this might be possible. In a Universe with infinite time cycles, time clocks, one for each cyclical mass or circadian cycle or beating wings, this is impossible, as each time follows its own rhythms.

Further on, in the nineteenth-century, physicists were only able to mathematize one of both arrows of creation of future events, energy, expansive movement in space, which they called entropy. Because physicists ignore all what they cannot mathematize, this led them to ignore the second arrow of future events, in-form-ation, as there was not a mathematical model of information, of form, till the complete development of fractal and non-Euclidean mathematics that took place at the end of the twentieth century.

In a landmark book *Les géométries fractales: L'espace-temps brisé*, Mehaute, a chemist, proved that all the systems of the Universe behave in a dual way: either they produce disorder, entropy, motion, and energy or when they lack energy, they create fractal, broken patterns of information, adding a "fractal dimension" of form. So the Universe has two arrows that create future events, with opposite properties: motion and form. And the mathematics used to describe them are inverted: differential calculus, which explains motion, and nondifferential topological equations that describe information.

Thirty years have passed and many complex theorems and physical questions have been solved by the pioneers of fractal physics. Yet they are ignored by CERN and the establishment of quantum physics, which, as in all scientific revolutions (Kuhn), maintain as dogma the previous paradigm. This is the core problem the entropy-only theories CERN pretends to prove: the big bang, the evaporation of black holes, the description of masses as a colliding field of energetic particles (Higgs theory). They are Ptolemaic, inaccurate complicated models of the cosmos because they miss half of reality, information, as Ptolemy missed the fact that the sun is in the center of our solar system. So he had to work out complicated equations to explain the motion of planets around the Earth. Today, physicists ignore the arrow of creation of information, of form, which in physical events is the arrow of creation of mass, as "time bends space into mass" (Einstein). So they invent false theories of "quantum entropy" as those of Hawking about never-seen black holes that evaporate information.

8. Entropy Is a Local Phenomenon Blown Up by a Hyperbolic Error

But the worldly profession of physicists is to evolve weapons that release energy and kill the information of life. Thus life, made of information, is secondary to its research and its death just a collateral effect of studying a reductionist model of a dying Universe built only with energy. The Industrial Revolution and the expansion of the white man in empires based in the energy of steam machines sacralized those ideas. Entropy, defined in terms of heat and steam, became the glorifying principle of reality and life; and information became the weak element that could be ignored because physicists had killed previously theoretically the Universe.

The next dogmatic element of entropy-only theories was the discovery in the study of heating machines that atoms and molecules, guided by the electromagnetic force, tend to increase their disorder and expand in space. This law of thermodynamics, however, is NOT a law that applies to all the forces of the Universe, and it is at the core of the errors of quantum cosmology since it purposedly ignores the informative nature of gravitation, which attracts, forms reality. Gravitation is an implosive force, the force that balances the entropy of electroweak atoms. Only by denying specifically that informative nature of gravitation—what I call in my conferences on duality *the arrow of Einstein*—is it possible to create entropy-only theories. And that is the reason why Higgs, Hawking, and other proponents of quantum entropy theories try to substitute the work of Einstein with quantum entropy ideologies, saying that "Einstein is double wrong." The astonishing fact that for a century the community of physicists has imposed to the entire world of science, such as entropy-only theories, despite all the theoretical, logical, and experimental evidence of the informative nature of gravitation and the informative arrow of life, says a lot about the power of mechanism and the industrial-military complex that builds machines like the quark cannon to impose their energetic dogmas by sheer force. Indeed, the economical and political clout physicists have on think tanks, political institutions, scientific press and congresses of science explain why entropy still reigns.

Einstein ironically said that he could count the number of physicists who are in this profession for pure knowledge with the fingers of his fist. Since knowledge of the Universe is not the worldly profession of nuclear physicists, physicists, since Galileo received the princely salary of one thousand gold ducats for improving cannon shots, were first mechanists, manufacturers of machines of energy, mainly weapons, and so they know all about movement and very little about life and information.

Nuclear physicists should be considered part of the military profession, with the added problem of a total lack of responsibility. When Teller and Wheeler proposed to Eissenhower to make a hydrogen bomb able to destroy the entire USSR, *all the nuclear physicists of America backed them.* It was Eisenhower, a military who had seen corpses in battlefields, who stopped the first doomsday weapon. Physicists look idealist people because they have eliminated from their minds many emotions proper of life. Newton said his biggest pride was to die virgin. The nerd prototype is *not* a nice person because he is cold, has no emotions, and acts like a robot, detached from life experiences. On the contrary, he is far more dangerous because he is an irresponsible child who knows nothing about death, suffering, the wounds of a battlefield he has caused at a distance with his weapons. And certainly his dogmatic, "energetic" beliefs render his work totally irrelevant to the understanding of the Universe.

All this means that philosophers of science and astronomers are the people who should be studying the Universe and its laws—not nuclear physicists with quark cannons. The specific field they study, physics of energy and motion, is a field related to the creation of machines and weapons—it is not the field that ought to be studying the meaning of it all. The danger of CERN is to let a group of children of thought, who understand nothing about the duality of form and energy in the Universe, to manage the most dangerous weapon ever built. If you give a gun to a child and you tell him it is dangerous, *he will shoot it anyway and then look to the destruction he has caused. And this is what CERNerds will do. Only that they can destroy the entire planet.*

9. The Evolution of European Egolatric Ideologies of War

And yet behind CERN there are four hundred years of European cannons and politicians who have built a civilization based on the power of their weapons of death. As the French philosopher Paul Virilio, one of the critics of CERN, explained, European civilization was built based on the power of cannons and weapons, on the search for news speeds of destruction. Indeed, Europe meant little to the world till the discovery of gunpowder, applied to gunboats, which conquered the world and then evolved new energetic machines, steams, and trains. European social and cultural evolution was minimal, based on ego-driven religions, with man at the center of the Universe, far less advanced than the organic models of the Universe developed by Buddhist and Taoist philosophers, who portrayed reality in all its complexity as an organic network of entities constantly communicating yin/information and yang/energy between them, in many different languages, beyond the limits of comprehension of the mathematical man. "The languages of God are infinite," said the Upanishad, and they were right, as the fractal, dualist paradigm has found today. But old habits die hard. And so Europe has been built on a cult(ure) of weapons with little philosophical understanding of the Universe.

It is the secret history of physics and its half reality, its passion for entropy=death=energy theories that justify the military profession, *but forget the other half of reality, order=life=information.* A physicist, when he has to talk about the arrow of life, information, and order in the Universe, calls it negantropy, the negation of entropy, as if life were an error in the design of the Universe.

Unfortunately, the military loves energetic cosmogonies. Since if God created the world with a big explosion and "the Universe is dying" (Helmolth), then warriors and weapon makers are the chosen of the most powerful god, Death. Ultimately our civilization hasn't changed. The first Aryans worshipped Shiva, god of energy, which has a statue at CERN. In the Middle Ages, popes backed as high priests of truth king warriors, whose swords and power was given to them by "the grace of God." And so it was said "you will defend me with the sword and I will defend you with the word" (Tertulianus). *Now CERN defends energetic theories such as the cosmic big bang that justifies the creation of the quark cannon.* So physicists back, as high priests of science, all violent theories of reality and the industry of war can go on also in times of peace as we can make quark cannons to understand the Universe. And of course, we have the Saint Nobel of dynamite and the Bofors cannons (Nobel was the biggest producer of bombs and cannons in the twenty-first century, called by the press "the merchant of death") to give prizes to our best energy-only theorists. He also said cynically to his lover when she asked him to create a Peace Prize that "my factories will end war before peace congresses when they invent a weapon able to annihilate entire armies in a second," which is indeed what the quark cannon will do. Only when she left him, horrified by Mr. Nobel's cynicism, as the site of the nobel.org explains, that he created a Peace Prize.

That huge gap between the dexterity of our mechanist scientists and the power of their energetic machines versus their nihilist philosophies about man is the ultimate reason why our technological civilization is killing Gaia, the world—it simply doesn't even realize it is killing her as she is an "it," a dead machine. Europeans and today mankind at large have adapted their theories of the physical Universe to that mechanist concept. Of course, physicists want to understand reality, as every other human does. So do any religious believer who thinks his god created the Universe. For nuclear physicists, God is pure energy and so they seek to confirm their God at CERN. They have in common with any other religious believer the existence of dogmas, beliefs that cannot be bent by reason called postulates without proof *and a Latin jargon called quantum probabilities, which they used as popes used Latin to invent and prove any theory they fancy, which always will have a minimal probability to exist.* In the twentieth century, with the invention of quantum uncertainty, a hypothesis about the probabilistic nature of the Universe, increasingly discredited by the advances of modern fractal relativity and fractal cosmology, nuclear physicists became even more fond of theorizing about the meaning of it all, *as any theory could now be considered to have a low probability instead of being outright false.* On the other hand, facts of science that will happen certainly could be considered improbable if they were dangerous, as it happens with strange matter, which is popping out at CERN, but it is deemed improbable. This is false. In the Universe things happen if they follow the laws of science (it is the so-called totalitarian principle) or they don't if they are false. And strange liquid that can blow the planet should happen at CERN because it follows the laws of science and it is within its range. Yet CERN considers this kind of "supernova" bomb improbable. The perception of the Universe as made of mathematical probabilities is false, a Pythagoric and anthropomorphic error of quantum physicists pointed out by Einstein when he said, "God doesn't play dices" and the "moon doesn't come out when I look at it."

The solution to the probabilistic description of waves is simple since the same probabilities can be used to describe a population, in the case of quantum physics, a population of fractal, self-similar particles that form organic waves, which sometimes group in tight particles and sometimes travel in herds. So the Universe is not made of probabilities but of fractal parts and wholes.

Yet again, this organic vision of atoms and molecules is taboo *since the Universe must be mechanical, built to the image and likeness of machines, not of human organisms.* Quantum physicists won't abandon their Pythagoric and mechanical conception of the Universe to keep their status as high priests of science and to avoid any argument about their worldly, criminal profession: to make energetic weapons. As a military man needs to believe he is defending the nation, a military physicist needs to think energy, and the machine is the meaning of it all. And to a certain point, physicists help us to understand a half part of the Universe and the military protects our nation. *But corrupted dictators and corrupted nuclear physicists at CERN destroy their countries and might big-bang the Earth.*

III. The Bad Fruits: The Quark Cannon

The production of quark holes at CERN risks the future of the planet.

10. The Damocles Machine

The years 2010 and 2013, the two years in which CERN will attempt to make black holes and explore the world of strong forces, liberating millions of quarks, carriers of those forces, by colliding protons and lead at the speed of light, are key years in the history of mankind. For the first time in our short, self-destructive existence, humans have managed to build a weapon, an atomic cannon that menaces our very own survival. This has not happened in history since our birth as a species.

As in the legend of Damocles—the Greek king, each day thinner, was menaced by a sword hanging from a thread, which nobody else could see or even know it was there—humanity is for the first time menaced with extinction without even noticing, preparing herself for what could be the most important event of our future by creating our own collective Damocles sword. As in the legend, according to which the sword was the penitence given to Damocles by the gods for the pleasures of being an overpowerful king, in this modern absolute tragedy of history, the pleasures of living in a technological world are so enticing to mankind that we are unable to control the evolution of the bad fruits of the Tree of Science, the weapons that menace to extinguish us. Indeed, the Damocles sword of history is a weapon machine, the first quark cannon that can extinguish all forms of life in this planet by releasing the energies stored in the nuclei of atoms.

The Universe of matter and energy is built as a Russian doll: each time we perforate further into the doll, we uncover new smaller layers of increasing energy=mass. Since a weapon is an instrument that kills by releasing an overdrive of energy that erases the information, the forms of life, the best weapons are cannons and bombs that break with an explosion a deep layer of the Russian doll of the Universe, releasing its inner energy.

The evolution of artillery is dual. On one side, cannons have increased their precision in order to perforate and release the energy accumulated in ever-tinier bits of matter. On the other side, bombs have increased the quantity of the explosive material they carry, increasing their destructive power. Thus, the first cannons and bombs released gunpowder energy—the physical, outer cover of lesser energy—with limited precision. In the seventeenth century, Galileo improved that precision, studying cannonball trajectories and setting the foundation of the science of physics, which was therefore intimately connected to the construction of weapons from its inception. Then in the twenty-first century, Nobel discovered chemical explosives that released the next molecular chemical cover. So bombs and cannons improved in explosive power. And the Nobel Prize was set up with the benefits of war.

It was only left to explode the atom, which is made of an electroweak cover and a strong nuclei of quarks. Thus, the final horizon on the evolution of artillery was the atomic cannon. It first perforated the atom in 1936, in Nazi Germany, causing its fission. Soon the atomic bomb released the energy of that electroweak cover of the atom. Then cyclical cannons were discovered and so the bullet could turn around the barrel as many times as needed to perforate further the atom, allowing the fusion processes that gave birth to the hydrogen bomb. Now in its third evolutionary age, the atomic cannon will explore the next Russian doll of energy = mass—the quark. So in the twenty-first century, the quark cannon will open the final *solid* Russian doll of *strong* quarks, which accumulate 99 percent of the mass energy of the Universe, provoking a big bang so enormous that it could blow up the entire Earth.

It is the culmination of a long evolution of atomic cannons that dates back to the age of the Manhattan project that created the first atomic bombs in the 1940s. Indeed, the first atomic bomb was intended to be also an atomic cannon that would shoot two halves of a critical mass of plutonium, but due to its enormous size, physicists opted out for a smaller bomb, which was detonated with chemical explosives (dynamite) instead of using a cannon barrel that had to be transported in a supercarrier ship.

All those atomic cannons were funded to study nuclear reactions that could be used later to create bombs. The study of particles became then a collateral benefit of the study of future nuclear weapons with atomic cannons. And so a natural synergy occurred: the first tiny bombs were explored in an atomic cannon called a collider. And those reactions which were susceptible of further military use would be studied in more detail and, if suitable, a larger critical mass would be built and a nuclear bomb exploded.

Now unfortunately, the detonator substance of the quark bomb is a mere thousands of quarks, which weigh so little that the quark cannon can do the job of starting the mass reaction without the need to evolve further a quark bomb. This happens precisely because the Russian dolls become smaller and more attractive when we perforate the Universe. So the final doll is very attractive and very small. If a few quarks can hold together all the mass of your body, when we pack them, their attractive power is awesome—100^3 stronger than our electroweak atoms. How this is possible is self-evident when we consider the name and main purpose of the big bang machine: to study the first seconds of the big bang when the Universe was a dense, small ball of a quark-gluon soup, a condensate of quarks, which exploded with such a force that created the entire space-time of the Universe.

The main reason nuclear physicists build the quark cannon is truly unsettling, almost surrealist. The company tells us that the quark cannon will produce the quark-gluon soup responsible for the biggest explosion in the Universe, the big bang. And so it is done to study the big bang itself, here on Earth. Indeed, nuclear physicists believe that the Universe started as a quark-gluon soup that exploded with such amazing power that it blew up the space-time of the cosmos to the size it is now. The mass of that initial quark-gluon soup is still unknown as there are doubts about the size of the original mass that started the big bang. But astrophysicists believe it was truly small. They also have discovered in the last decade, when the machine was already commissioned and could not be stopped without shutting down CERN, that a quark condensate is the cause of supernova explosions that create pulsars, very dense stars that pulsate, emitting incredibly energetic gamma-ray bursts. Astrophysicists thought till a decade ago that they were made of neutrons. Now they know pulsars have a quark-gluon soup in its center that power their energy. Thus, if a quark-gluon soup exploded with such force as to dilate space-time to the limits of the Universe and a tiny quantity of it starts a nova reaction and powers the most energetic burst of the Universe, it is obvious that quark-gluon soups, also called Einstein's condensates, could blow up the Earth.

Does anyone in this company realize that irrefutable logic, hence the absurdity of their research proposal? Of course, but the question for them is, as a recent study by an impartial lawyer put it,[1] "between their livelihood and their lives." And it is a fact of any military or civil profession in which lives are at risk that the choice is *always* livelihood. Or in practical terms, since the dangers were known, employees of the company have to sign confidenciality statements of zero risk or lose their jobs.[2]

But what about us? The rest of mankind? Why must we be subject to this trial? Obviously, we should not reason why the nuclear company of

Europe has been denounced in Court by citizens. Yet the amazing thing is that neither the military nor the courts or governments have come forward to stop this experiment because they can't, as this nuclear company, born in the cold war, has "diplomatic immunity." Such is the legal and political perspective of this surrealist happening that might mean the end of history.

In brief, the quark cannon means a growth of explosive power in bomb devices similar to the jump that happened between atomic weapons and dynamite, but far more dangerous, for two reasons:

— The quark cannon will cross for the first time in history the one teravolt barrier of energy that causes the death of our electroweak matter—what physicists call the breaking of electroweak symmetry. Symmetry is a property of all entities of nature. For example, humans have bilateral symmetry, meaning we have two eyes, two ears, two legs, and if our symmetry breaks because we are axed in half, we die. So happens to our electroweak matter when it breaks its symmetry. It dies.

— The explosion a quark bomb produces is similar to a fire in the sense that once the fire/quark bomb starts, it is self-sustained, as the *combustible is all the matter of the Earth—all the wood of the forest, converted into ashes. In a quark reaction,* our electroweak matter dies when shot over one teravolt of energy, becoming transformed into quarks. And the more energy and quantity of matter the cannon shoots, the more quarks it will create till it forms with them a superdense, attractive quark condensate, which is a self-sustained mass bomb. A small lump of quarks put together will be so attractive that it will act as the detonator of the Earth, absorbing *all the other* electroweak atoms of this planet and converting them into quark mass. Hence, the mass reaction a quark bomb can cause will consume as combustible all the mass of the Earth, which could become a small ultradense lump of quark matter, a strangelet, quark star, pulsar, or black hole.

All those reactions were defined by Einstein with its inverse formula, $M=e/c^2$, which described the creation of a *mass bomb*, the conversion of energy into mass. In fact, the mass equation was the first one he published,[3] deducing the equation of a smaller, weak atomic bomb, $E=mc^2$, from the mass-bomb equation. So we know mass bombs do happen because the secondary type of lesser bombs, atomic bombs, do happen. And even if the physicists that made the nuclear bomb had doubts that such enormous explosion would actually happen, all atomic bombs do explode; because in science, there is a basic principle called the totalitarian principle,[4] a sort of Murphy's law that affirms if something could happen, it will happen. So mass bombs must happen in the large hadron collider, technical name of the quark cannon that will overcome the 1 TeV barrier of death of our matter this April.

Yet even if we are lucky and survive the biggest risk of the LHC, the extinction of the planet, as long as the machine is working we will suffer a smaller but seizable risk to human life: The first hints that the LHC is seriously damaging life on Earth will come from an increase on earthquake and volcano activity.

This is due to the fact that the LHC is creating a powerful gravito-magnetic field, a 'ring' of charged, massive particles that can interact with the magnetic fields of the magma and Earth's center.

Disturbances on the Earth's magnetic field by the magnets of the LHC and specially *the charged positive c-speed flow of protons* might come through 3 different processes, which are from maximal to minimal earthquake damage:

— The creation of strange liquid, already produced in the first experiments, (Kaons at the LHC, hyperons at RHIC) could also provoke, as it falls to the center of the Earth, explosions in the magma. If stable, it will leak in increasing quantities to the center of the Earth. Some of it will remain in the center, forming the seed of a strangelet. Some will accrete and/or explode in the mantle, in highly energetic, tiny bombs, triggering faults.

— The creation of *gravitational waves*. The LHC is a 27 Kilometer ring of positive charged massive particles, turning at c-speed. This is essentially equivalent to the 'singularity' of a *Kerr black hole* —a rotating c-speed charged ring of mass. Since a Kerr singularity can produce transversal gravitational waves; the LHC might produce perpendicular gravitational waves that will sink straight towards the center of the Earth (in a similar process a rotating, charged coil is used to produce electromagnetic waves). If so those Gravitational waves, which are undetectable will affect magnetic fields, provoking earthquake waves and increase volcano activity.

— The possibility that the the superfluid magnetic field can interact with self-similar charged flows in the magma, creating a powerful electro-magnetic effect, which acts as a butterfly effect (earthquakes are like avalanches, a small change, even a mine explosion can trigger the potential energy already stored in the fault). It is a fact that *the first day that the charged, proton ring was created in 2008 it caused 4 significant Earthquakes, the first one in Iran, seconds after it was powered up.*

While it is obvious we won't see those fields to prove visually the surge, this could be easily proved through the year by statistics. Statistically the highest year on record on earthquakes for all categories, was the man-made surge during II world war carpet bombing of the ring of fire (Pacific islands), by the American Army in 1943. That year earthquakes almost doubled. Such high peak would not be reached unless we make black holes that eat us all and so we would see an exponential growth till we all die.

But if 2010 comes close to a 2nd or 3rd year record, it will mean an increase of thousands of human victims due to Earthquakes and billions of $ in economical losses for years to come, as faults are triggered and volcanoes spurt lava.

The quark cannon is, plainly speaking, a Damocles sword that will hang over all of mankind, as the nuclear company that manufactured it shoots every day protons and lead to make the most dangerous substance of the Universe, quark bombs, quark-gluon soup, the substance responsible for the biggest explosions of the Universe. During two years, with increasing power, we shall enter a region in which the most dangerous, attractive substance of the cosmos, quark matter condensates, could become stable and feed on the rest of electroweak matter of the Earth, breaking our symmetry at atomic level, and obviously our eyes, ears, and limbs will become reduced to dust of space-time. As in the Greek legend in which each day the sword hangs on a thinner thread over the throne of Damocles, the quark cannon will be each day more dangerous as the company ramps up its power during two years in which the cannon promises to reach from their initial 3.5 teravolts in March 2010 till over 1,000 teravolt shoots in lead-to-lead collisions in Christmas 2013. At that point, CERN believes the quark cannon will be deconfining millions of quarks per second[5] to form massive quantities of quark-gluon soup. How dangerous this is was shown in calculations by the Shanghai National Center for Nuclear Research of China,[6] perhaps the only institution that has dared to defy the law of silence of nuclear scientists, as China is not a member of CERN. Chen and Weng, the leading specialists in strange quark condensates called strangelets, have shown that only 10,000 strange quarks could start a chain reaction that will devour the Earth. Thus, by 2013, when it reaches its maximal power, CERN, the European Company of Nuclear Research that has built the cannon, will be deconfining *enough quarks to detonate 100 Earths*, according to our standard knowledge of strangelets today, *which was in its infancy when CERN's collider was funded, or else it would have not been built*. So now, in defense of their livehood, CERN lies to the public and the few scientist experts in quark condensates hold their breath and cross their fingers to see what happens. But the first experiments taking place now at a mere 1.1 TeV, 1/1,000th fraction of its final power, already showed what will happen: 13% more strange quarks than "expected."[ch.4]

11. Peer Solidarity: The Alibi of Knowledge

Unfortunately, if the existence of the Damocles Machine is not bad enough, as in the legend, the nightmare becomes complete when we realize that only Damocles, only a few nuclear physicists, know the secrets of the quark cannon and they cannot reveal them to the rest of mankind because they will lose their kingdom, their jobs, and position of power as the high priests of science, who have never been made responsible for the enormous harm their weapons have caused to mankind. So none of them talks of the dangers of the quark cannon, only of the incredible discoveries it will bring to the human species.

To understand why mankind has paid for the quark cannon, we must consider a historic overview of the ways in which the military-industrial system has evolved and markets today successfully its weapons to governments and public alike, brushing off any opposition with the usual tools of industrial marketing and political corruption.

It is a "dejà vù" experience in the world of nuclear weapons. Physicists at CERN do not want to blow up the Earth and so as anyone in the military profession of making or using weapons, for decades they have found intellectual reasons to justify the industry of atomic cannons. Already Teller[7] convinced Truman of the need to make hydrogen bombs because they were needed to know better the atom. Then he proposed an H-bomb made with his company at a stratospheric price to blow up the whole USSR, backed by the entire nuclear community, as a hydrogen bomb has no limits of size. For example, if CERN mismanages its beams and ions exit the vacuum tube, colliding with the helium that surrounds the beam, theoretically it could trigger the fusion of helium into carbon. The resulting helium bomb will fuse several tons of helium into carbon atoms, and Switzerland and France would blow up. For that reason, the quark cannon has wells to redirect the beam if it goes out of track, trying to avoid its collision with the helium, which is another reason why this cannon should not go online. In the '50s, Teller's doomsday machine was not built. The American president was a military who knew about war and death firsthand and cancelled the project of MAD physicists, who live in their ivory tower of abstract numbers and shut their minds to "ethic, survival issues." They proposed a Mutual Assured Destruction strategy to justify the building of their weapons. Eissenhower realized of this MAD strategy and called Khrushchev, starting the process of disarmament.

The excuse of knowledge was again used by Lederman, who pumped up the Higgs, as the meaning of it all, and sold it to Reagan as *God's particle* to get 13 billion dollars for the Texas quark cannon.[8] Clinton, far more intelligent, realized of the dangers, uselesness, and tag prize[9] of the SCC and cancelled. This third time, the nuclear industry has finally succeeded, creating a doomsday weapon . . . As they say in Spanish, "A la tercera va la vencida" (Third time you try, you get it right). Thus, unless Mr. Obama calls Mr. Von Rompuy and both together shut down the factory, as Eisenhower and Khrushchev did with the first doomsday bomb, we will probably witness the creation of a frozen Earth by Christmas 2013.

Unfortunately, the nuclear industry is now far more savvy in its marketing of atomic devices, to the point that most physicists are unaware of any risks of the large hadron collider, the technical name of the quark cannon. CERN no longer considers, once the cold war is over, its machine to have military applications as all real weapons do. Instead, it markets it only as an instrument of research. Priorities have changed, so all could remain the same and the industrial-military complex, created to construct more powerful nuclear devices, now can continue providing jobs to nuclear physicists. In fact, CERN employs an important number of Russian physicists, who used to construct nuclear weapons.[2]

The excuse of knowledge today is absurd, as the fundamental particles of the Universe are discovered, diminishing the scientific returns of this weapon, while on the other hand, its dangers have increased, as the potency of atomic cannons multiplied thanks to its new superconductive, superfluid technologies. Thus, the quark cannon has definitely tilted the balance toward weaponry since the risks are far bigger than the knowledge we shall extract from it, precisely when marketing has gone the opposite way, and sells it only as a research machine to disguise those dangers.

Plainly speaking, after the cold war ended, the industry of quark cannons should have closed down. But the nuclear company of Europe did not want to shut down, so it found three excuses to dilute those dangers: denial of dangers, excuses of knowledge, and ad hominem campaigns against those who opposed them, paid by with the enormous $10 billion budget given to the company by naive politicians to build its machine.

CERN diluted the dangers with false statements that portrayed the quark cannon as a harmless cosmic ray factory, when it is a fact of science that quarks and cosmic rays are completely different kind of particles.[10] We have, in fact, excellent cosmic ray observatories and we never found quarks in cosmic rays. So CERN has built, in a surrealist twist, a factory of the most dangerous substance of the Universe, quark condensates, but it never mentions it in its safety reports, which talk of cosmic rays. Imagine a factory of missiles that would systematically affirm it only produces fireworks. But in an age in which research journalism is dead, it has been possible to pull off this kind of lie and sell it to the press, which merely prints what CERN says, as if a nuclear company *could not, like any other company, falsify information regarding the risks of enviromental catastrophe it might cause.*

What naive people don't want to accept about CERN and nuclear physicists is the fact that we are facing a case of corruption, both at industrial and theoretical knowledge. For fifty years, nuclear physicists have been heralded as the high priests of science by a self-interested apparatus of political propaganda, both in the USSR and America. And that myth has stuck. But the quark cannon is NOT about knowledge, and CERN is a corrupted company, belonging to a corrupted profession. "Power corrupts and absolute power corrupts absolutely," said Montesquieu. Nuclear physicists have in science, like bankers in the economy, absolute power. We are indeed talking of a similar case to the corruption of Wall Street that brought about the crash of the economy—but far more dangerous because now it can bring about the crash of the Earth into a lump of quark matter. Yet to hide both cases, experts abound to certify that the machine is not a weapon but an instrument of research and the toxic asset an AAA-derivative, where AAA means maximal quality rating for a financial asset, given by AIG, the most respected insurance company of Wall Street, which sold them and turned out to be the most crooked one. Now our experts are Nobels, a prize given by Mr. Nobel, the biggest arms dealer of the twenty-first century; and they work for CERN, the same company that manufactures the toxic asset, whose safety they certify. As all the safety reports given by the company have been made by employees of CERN, against the Standard Laws of Safety of the European Union that call for an independent panel of scientists of different disciplines.[11] Those employees of CERN are the people who certify the AAA-quality of their own weapon. And they are willing to take chances as our bankers did—to ride the wave of profits as long as it lasts.

The scam of toxic assets that crashed the world economy was obvious. Everyone on planet Earth would understand that making all kinds of people without work and resources to sign mortgages was not creating AAA-quality assets but toxic assets. Yet the system of creation of electronic money with derivatives needs an excuse, an initial form of money to create then a pyramid of derivatives that multiplies that money. Money is invented with financial instruments that need an initial asset to start the process of lending: A customer with a real job gets a mortgage, and with that collateral, a bank creates a credit and another bank with that collateral creates more money. And so on. So bankers use mortgages to build up pyramids of new money. What was new in the 2000s was the evolution of technology, computer programs that could invent now e-money much faster than old accountants. Computer technology has improved so much that now bankers can create much more e-money with financial, complex programs than ever before. So greed, the desire to keep inventing e-money, soon surpassed the quantity of good mortgages bankers could use as the base of the credit pyramid. That is why bankers started to sign mortgages to anyone. They knew perfectly that sooner or later, this pyramid of new money, without value to back it, would crash. Yet in the meantime, they could sell their derivatives to governments and citizens and keep the money so others would hold the toxic assets at the end of the day. And to that aim, a battery of experts—the entire financial community of Wall Street, the most respected people of the planet—participated on the scam, providing articles that assured the public of the quality of those derivatives. A law of silence was imposed, and only at the very end when the pyramid crashed and it was too late to solve the fraud, a few people talked, mainly with their money, betting that the pyramid will crash. Some banks, such as Goldman, even bet that the pyramid will crash while selling its toxic assets to its clients. And in reward, when the pyramid crashed they got $750 billion to pay their debts.[12] The lesson is clear: Power corrupts and absolute power corrupts absolutely. And nobody holds more power in our societies, still based on the structures built during the cold war, than bankers and nuclear physicists. For that reason, the case of the crash of the Earth by the quark cannon is similar in corruption, "damned lies and statistics" to the AAA-derivative scam. Many know at CERN that the machine is a toxic asset that could endanger life on Earth. Even the excuse of researching the big bang is bogus. Since the Webb telescope, launched in a couple of years, will observe directly the real big bang, not the faked one, as it probes into deep space 13 billion years away. So there is no need for the big bang machine.

Yet this means to close the quark cannon, the only reason the nuclear company of Europe exists. So in order to justify the project, CERN has pumped up for twenty years fringe theories, toxic assets of science that deny the risks and contradict standard Einstein's theory of mass and black holes, as "the meaning of it all." CERN's physicists want to keep their jobs and keep $10 billion on their pocket; they want to be the high priests of our technological civilization and be treated as the people that can undertand the meaning of it all. Thus, their selfish agenda, their livelihood, their prestige is at stake. And so they blind themselves to those dangers; they search for fringe theories not on the basis of truth, but as financiers did with their choice of experts, cherry-picking only experts that validate the quark cannon as a safe experiment, even if they contradict Einstein's work. So during all the years in which the company has lobbied for money to construct the machine, it has made marketing campaigns in favor of quantum experts (Higgs, Hawking), affirming that this machine will prove their fringe theories, which deny Einstein's work ("Einstein is double wrong,"[13] famously said Hawking). Those theories are *the toxic assets of cosmology*, pumped up as AAA-derivatives, as the meaning of it all.

In that regard, three of those toxic assets are paramount:

A. Hawking's theory that says black holes made at CERN will evaporate because they are very small. This absurd idea contradicts Einstein's standard theory of gravitation, which states that all black holes regardless of size will accrete the Earth since relativity means exactly that size is totally relative. Size just depends on the point of view of the observer and its perception of size, which is subjective. From the point of view of a star, humans are minions. From the point of view of a particle, humans seem giants. So size is not an objective scientific quality, and it does not alter the laws of black holes and gravitation. This is the essence of relativity theory, the standard science of gravitation, mass, and black holes, which twenty-first-century cosmology further proves, with its fractal models of a Universe of infinite scales of relative size that follow the same laws. Further on, there is increasing evidence that black holes of all size accrete matter at enormous speeds, both in astronomy and theoretical research, as Mr. Lehmann, points out in his book on the legal issues poised by the quark cannon.[14]

B. The company says it will study the initial substance that caused the big bang, the aforementioned quark-gluon soup. Indeed, it might do the big bang of Earth. But what knowledge can be extracted from those studies if the researchers die at the hands of the entity they research? It is like researching Ebola viruses, or doing, as the enthusiast character of *Jurassic Park* does, a research on the depositions of a dinosaur that is going to eat you. Further on, we have telescopes, which probe in galaxies so distant that we can observe the light of the real big bang itself without danger. Indeed, the light of those galaxies takes billions of light-years to reach the Earth. So it comes from the distant past. Thus, a deep-space telescope will allow us to observe the Universe, as it was born 13 billion years ago. This telescope is an American project called the James Webb telescope to be launched in 2014. It will be able to see so far that it will observe, prove, or disprove the beginning of the big bang without risks to the Earth.

C. The company said that the quark cannon will find "God's particle," a hypothesized particle called the Higgs, which according to the company shall explain the meaning of mass. If Hawking's radiation is the first toxic asset of CERN because in case it does not exist it will mean our demise, the Higgs is the second toxic asset of CERN because it is a hoax. The particle is self-similar to a well-known quark, the top quark. It has the same mass and properties that a top/antitop quark has; and so it is very likely, since the Universe is efficient and not redundant, that the Higgs, which we haven't found in thirty years, does not exist at all but its functions are done by the known top quark and the God's particle is sold to get funding from politicians for a new stupendous discovery that will never happen.[8] Further on, even if this particle existed, it is only useful to explain a reaction, the death of our matter, as neutrons and protons decay in two particles called W and Z, which are related to the Higgs=top quark[15] (the sum of their masses is that of a Higgs/top: W+Z=T). So this hypothetical Higgs merely would be the catalyzer of the death of our matter. A factory of Top/Higgs quarks means only a factory that will kill our matter and, if the reaction goes out of hand, all the matter of planet Earth.

The Higgs/top quark does not explain at all the meaning of mass or gives mass to other particles. This is nonsense. *It is just one more particle of the zoo of the Universe.* So all that marketing about the Higgs/top being God's particle and the most important particle of the Universe is ludicrous.

If the Higgs hoax is not widely recognized among nuclear physicists, it has to do with the confrontation between quantum nuclear physicists and Einstein, the pacifist scientist who worked out a theory of gravitation and mass, which still stands. In essence, Einstein defined masses as whirls of space-time, attractive tornado-like vortices that followed his principle of equivalence between mass and acceleration. So as a hurricane accelerates toward its center, attracting all what surrounds it, a mass is an attractive whirl of space-time. Unlike the Higgs, a particle with a limited use, such definition of mass, taken to its ultimate consequences, illuminates the understanding of the Universe as nothing else has done since Darwin—reason why those two colossus stand as the key figures of nineteenth—and twentieth-century science while Higgs is still an unrecognized, retired Edinburgh professor without a Nobel Prize. Why is Einstein's mass theory so important? Because it means among other facts that we shall consider in detail in this book:

— The existence of a spiritual Universe based on motions, on time events more than in substances, in space. Indeed, if a mass is a cyclical, accelerated motion and energy is a lineal motion, then the Universe is an eternal game of two motions in constant transformation. What we call energy or entropy and what we call in/form/ation, or cyclical form (masses and charges).

— It explains easily how mass becomes energy and vice versa. Indeed, in Einstein's model of mass, $E=mc^2$ means that when energy approaches the c-limit of speed, as it cannot go faster, it deflects its energy that curls into cyclical motion, into mass.

— It completes the science of physics, introducing the concept of physical information, provided by the frequency of rotation of a charge or mass. Such duality and constant transformation of energy into form, information (masses and charges), applies also to other sciences as biologists well know since they also describe biological beings as a series of cycles of trans/form/ation of energy into in/form/ation. Thus, Einstein's theory of mass gives science the possibility to relate naturally all disciplines and species as systems of energy (fields of forces in physics, bodies in biology) and information (cyclical particles and spherical heads). This fact is recognized by quantum physicists as the fundamental law of their science called the principle of complementarity: all particles need an energetic field and an informative particle to exist. So Einstein showed the road to unify physics and biology, which is the avenue that complexity, fractal theory, duality, and system sciences, the most promising sciences of the twenty-first century, are exploring, without risking at all the lives of those we love, by shooting a quark cannon to Mother Earth to see what happens—the clueless program of twenty-first-century research sponsored by the nuclear industry.

Higgs cannot explain any of those facts. But it is an excellent tool of marketing: even if it halts the knowledge of Einstein's true theory of mass, *it feeds the nuclear industry and that is what it matters to those whose livelihood depend on it.*

For all those reasons, the two most important nuclear physicists alive laugh at the Higgs. Mr. Weinberg, the Nobel Prize discoverer of the weak force, called it the toilet particle to flush in a vortex of mass of Dr. Einstein. Mr. Wilczek, discoverer of the strong force, says that masses are the frequencies of a whirl of space-time and that is the only explanation of mass there is. Obviously the discoverer of the modern laws of gravitation, the third force of the Universe, Mr. Einstein would disagree also with Higgs's theory. Mr. Maxwell, discoverer of the laws of the fourth force of nature, electromagnetism, lived in the nineteenth century. So we do not know what he thinks. Perhaps for that reason Mr. Higgs likes to compare himself with this other Scottish genius, as he does in the documentary provided by CERN.

In that sense, what Higgs, the big bang hype, and the dubious evaporation of black holes have in common is obvious: they are three marketing tools for a weapon, which justify the enormous risks for our lives the quark cannon imply. Higgs is God's particle and the big bang is the act of God's creation. And who can resist the "religious" feeling associated to the word *God* even if it means to risk our life, even if God has nothing to do with a cannon?

The Universe, the infinitely large, is not studied with a cannon that will bomb the Earth with quarks, the infinitely small, but with satellites and telescopes, *looking directly into the Universe*. But *as atomic* cannons and nuclear devices have become more dangerous and the number of potential deaths they could cause has increased geometrically till reaching the entire human population, the marketing of this instrument of "peaceful research" has required such surrealist plots. Since the quark cannon can blow up the entire planet and this cannot be said, the rhetoric of knowledge about it is absolute. So nobody speaks of the quark cannon as a weapon but only as an instrument of research. Workers in the quark cannon cannot talk of the risks of creating black holes and quark condensates. The company only speaks of the new records of energy the cannon sets. It does not even talk of quarks as its main produce, but focuses in the capacity of the cannon to discover exotic particles that do not even exist.

Plainly speaking, nuclear physicists are marketing this weapon as an instrument of knowledge, a cosmological device that will solve the meaning of the Universe. Yet what they are doing is usurpating a scientific role that does not correspond to their science but to astronomy and the philosophy of science.

Finally, when the dangers of the large hadron cannon were known, as scientists sued CERN for genocide in different courts of Europe and America, the nuclear company put the system of defamation, so useful in those cases, at work, accusing those scientists of being "crackpots" and lacking the expertise to judge those risks, to avoid an open, objective dispute of those risks in a due process of law.

Yet those scientists merely followed Einstein's work that showed enormous risks. That is why this book is signed "CERN vs Einstein et al." In science, there are no "experts" and "authorities" but methods of truth, independent of the authors of a theory. That is why science works as an objective form of knowledge, unlike other branches of human endeavor, tainted by the corruption of experts with selfish agendas. In science is the authority of truth, not the truth of authorities, that matters. So "Einstein et al." represent theories of science proved by the scientific method, which shows that the dangers of making Einstein's quark condensates and black holes are enormous.

Those suits against CERN failed to stop the quark cannon, not because there was no risk but because CERN knows scientific truth proves there is risk and chose not to attend those suits, invoking its dipomatic immunity as a priviledged military company of the cold war. CERN played the privileges of its "military authority" and Einstein et al., the reasons of truth, since they based their worries on standard Einstein's black hole, gravitation and mass theory, and Einstein-Bose description of quark condensates. Amazing as it sounds, the factory of the most dangerous substance of the planet has diplomatic immunity and so only a military intervention under the patriot act that allows the American government and NATO to seize any nuclear substance that endagers the life of its citizens can close the quark cannon. CERN chose not to confront scientific arguments, knowing it would lose the suits on account of scientific truth. And to further dilute that truth, it voiced out ad hominem campaigns against those scientists who denounced the crime. Let us follow the account of those facts, according to a neutral observer, Mr. Eric Johnson, professor of law in the University of Tennessee, which explained the case:[1]

> Brian Cox,[16] spokesman for CERN, said in public: "Anyone who thinks the LHC will destroy the world," he said, "is a twat." John Ellis, chief theorist at CERN referred to LHC detractors as "nuts" and insinuated that one of them, Walter Wagner, was only pursuing a lawsuit against CERN to make money. Yet Wagner was suing for an injunction, not damages.

Thus by monopolizing all the information about the suits, CERN soon convinced mankind that all those scientists who were not in favor of CERN were crazy people for suing CERN—as another CERN physicist put it. Yet the most revealing document came again from John Ellis, who gave a talk in the CERN auditorium in which he sought "to provide the ammunition" that CERN people could use to convince others that the quark cannon poses no danger. After reviewing the scientific arguments that the LHC is safe, Ellis explained that a question that worried him more than whether humanity was safe from the LHC was the opposite—whether the LHC was safe from humanity. Ellis then briefed the audience on unfavorable press reports, the various lawsuits filed to stop the LHC, and a public opinion poll indicating that most people thought the LHC was not worth the risk. Ellis also introduced the audience to Richard Posner, whom Ellis said he found "really worrying" for considering that society and judges should have a telling in an experiment that could cause our demise. Wrapping up, Ellis came to what his presentations slides labeled "The Best Answer." "So, to finish," Ellis said, "the way to stop all this argument about whether the LHC is going to destroy the planet is to get the LHC working. Within a few weeks' time, we will know that LSAG was right." Of course, in making such a statement, Ellis either showed that he misapprehends the relevant physics, which seems highly doubtful—concludes Johnson in his legal study of those cases.

Johnson hardly scratches in the massive ad hominem campaign against the scientists, represented here under the collective name of Einstein et al., who denounced the quark cannon. None of them were able to explain themselves in the scientific mainstream press or any serious TV program. Some lost their jobs and academic positions, and their professional papers are no longer published or quoted within the academic world. Of course, the professional world of science took notice and nobody has dared after those initial months to come out of the closet and say the truth about those risks.

The bulk of that ad hominem campaign was not obviously made by the visible heads of CERN, Ellis and Cox, but by the thousands of collaborators in universities around the world that littered the Web with insults and jokes about the very real risks for the Earth of CERN's experiments and the ignorant mass media who laughed at them and only published paid-by PR articles from CERN. It is the LOL method that we describe in the last chapter of this book: because humans seem unable to face the true challenges of history, namely, the overwhelming power of companies and new technologies in a free market where life is increasingly expendable, we laugh at the problems and we prosecute not those who are killing the Earth but those who denounce it. The fourth power that in the past denounced the excesses of the nuclear industry in films like *Dr. Strangelove*, it has become now self-obsessed by its celebrity status. So it feels cozy in its invented "Walt Disney" fantasy world of "stardom" and it doesn't want to fight for the real world.

In any case, the LOL method and Web campaign was extremely effective to form public opinion since today you are always googled and that first opinion defines the opinion we have on anyone, from Tiger to Einstein et al., regardless of the null knowledge the blogger has about the person he is anonymously insulting.

This book, even if it won't be read or commented, as the collective Einstein et al. are now nonpersons judged by the inquisition of thought, which in cosmology and physics CERN represents, is the best scientific answer to that campaign as it is centered in the search for truth on cosmology and the risks of CERN's experiments and written with the tools of the scientific method, logic, ethics, and reason since truth in science can be proved with "thought experiments," thanks to the scientific method.

It is obviously a book, which criticizes the ethics of the culture of nuclear physics, which has been better portrayed by writers like Vonnegut (*Cat's Cradle*) or Terry Southern (*Dr. Strangelove*) than by the official mass media, in love with their weapons of mass destruction. As the independent journalist John Smith put it, "CERN's black holes have been an excellent press campaign . . . never nuclear physics looked so sexy since Oppenheimer invented the atomic bomb."

Indeed, if physicists make the weapons used by the military to kill mankind, today the audiovisual mass media knows that violence sells, as the eye follows movement and red colors, the signs of energy and death. The eye is hypnotized by violence and so it does not reason but loves catastrophes. So mass media exercises a cult to death to sell more, and it has used CERN to sell a global catastrophe; at the same time it has denied it with the LOL method, not to provoke public alarm. The love for violence as the worldly profession of physicists and journalists is at the heart of the bioethical and intellectual errors both have committed in this issue. There is a statue to Shiva, the lord of energy and death of the Aryan, at the entrance of this weapon of mass construction. And there is a sect of Shoats, called quantum cosmologists that will use this weapon to prove in a game of quantum roulette against the Earth their absurd theories of an ever-dying, explosive universe. Yet what Mr. Ellis proposes to quench all opposition—to fire the LHC and see what happens—is not science. It is a primitive, medieval method of truth called God's trial, which was practiced among Germanic tribes and consisted in a duel, whose winner was right by the grace of God. Moreover, the duel was usually rigged by the conqueror. We evolved the scientific method for four hundred years. But now we have returned to those methods of verification of truth, which CERN proposes for this potential genocide: "shoot first, ask later."

12. Weapons: The Game of Russian Dolls

Is the LHC really a weapon? *Yes*, regardless of what the company says, the quark cannon is a *weapon*, as a simple analysis of what a weapon is should demonstrate.

Weapons by definition are systems that release massive amounts of energy by penetrating and opening up the "Russian doll" layers of increasing mass and energy of the fractal Universe, which are paradoxically smaller, given the fact they are vortices that accelerate inward, acquiring faster mass/rotational speed toward its center as it happens with any vortex in any medium. In the modern paradigm of a fractal organic Universe, we explain reality as a series of "scales" or "layers" made with different particles of information and fields of energy. Paradoxically, as in a Russian doll, each new layer is smaller but denser in energy and mass. And there are three basic layers: our light extended space filled with electromagnetic radiation, the outer layer of atoms made with electrons, and the inner layer of atomic nuclei made with quarks and strong forces, mediated by gluons. This is the final layer, never opened by mankind, whose strong forces are 100^3 denser than our electroweak world.

So when the quantum world crosses from the light medium into the electronic medium and further on into the quark-gluon medium, its density of mass energy increases and vice versa. If those masses are unwarped into a lighter, more extended medium, the energy they release provokes a big bang that extends a wave of energy through the lighter medium, killing every form of information that inhabits it. This is what a bomb or a weapon does.

So in the modern age, since gunpowder was discovered, the main types of weapons have been bombs and guns that perforate matter and release the energy of the inner Russian dolls. A gun has a barrel to send a projectile away toward a target and a projectile or bomb that opens the Russian doll upon impact, releasing the inner energy of matter. Often a weapon combines both elements. For example, modern guns release projectiles, which explode upon contact with the target.

Thus, artillery and the science of ballistics, the fundamental tools of weapon making, have evolved in three fronts: reach of the shot, precision on the target, and potency on the Russian doll released in the process.

Along that evolution, there has been an evolution of the sciences of motion and energy, which first were widely known as ballistics and then after Galileo developed its main laws became dynamics and finally physics. *Since physics is the science that studies motions and energy, hence also the science that manufactures all the weapons of modern history.*

In all those weapons, what matters is to break the Russian doll of energy and release that energy on the target. All weapons do basically the same: they explode one of the Russian dolls of the energetic Universe, opening the door for the release of its inner energy, which causes an overdrive of energy that erases, kills the complex information of life.

13. Death Kills the Information of Life

Those two opposite elements of reality, energy and information, which will come once and again to our attention, explained with the knowledge provided by the sciences of information, complexity, systems sciences, and fractal theory, *which CERN completely ignores*, are already the key elements to understanding the function of a weapon: to release enough energy to erase and cause the death of information, the substance of life.

Both elements, energy and information, have opposite properties and yet they can transform into each other "ad eternal." Modern science says that

energy never dies but trans-forms itself back and forth into in/form/ation:
$$E <=> Ti.$$

This is the essential law or cycle of the Universe, the fundamental law of complexity. Yet if you recall your studies in physics, this principle is reduced to the arrow of energy. So physicists only say,

Energy never dies, but transforms itself ad eternal.

Quantum physicists study only the arrow of energy and hence they ignore all about the arrow of life and information, even in their own discipline—reason why the meaning of life and death or the informative role of masses and black holes, which we shall show in this book, escapes them.

Indeed, it is a dogma of physics that the Universe is only guided by the arrow of energy. And this is the essential problem of CERN. Physicists make weapons that release energy. And so physicists think that to understand energy is to understand all in the Universe. What was first, the egg or the chicken? Did physicists start first to make weapons and then came with energetic ideas about the Universe? Actually, as we shall see, both sides evolved together since Galileo evolved ballistics and defined time in terms of speed and motion, $v=s/t$. The problem though is that they *are wrong*. We shall see how in the past decades we have proved ad nauseam that information is more important than energy in the creation of the Universe. And so physicists should step aside, think about the purpose of their science, and let the people who know all about information explain them how *to evolve their science*.

And the first thing they should learn, since they are in denial for centuries, is how energy kills life. We said that all are cycles of transformation of energy into information, form. Yet the key to one or other type of transformation is the intensity of the exchange of energy and information. When there is too much energy, information becomes erased back into its original substance and the being or particle that holds that information dies. When there is a lot of information, energy becomes trans-form-ed into form, information in a creative act, giving birth to life. So we can define a process of death mediated by a weapon as follows:

max. energy x min. information = weapon, death, entropy, accident
min. energy x max. information = man, life, information, aging

In the two previous logical relationships, obtained from the science of complexity, which models reality with both elements (unlike physics, which is obsessed by energy), we can easily notice the opposition between what human beings are—the maximal species of information and what weapons are—machines that release energy. It follows that weapons kill us with huge releases of energy, and since the quark cannon will release the final Russian doll of energy of the Universe, it can easily kill us.

14. The Evolution of Weapons and the Industrial Revolution

Yet weapons are machines, so their increasing power of death must be related to the evolution of machines, a process that defines modern history. Economists say that the evolution of machines is guided by the evolution of the energies of those machines. We are what we eat, so machines have evolved as we have opened new layers of energy and applied them to create transport, peaceful machines, and weapons. And so there is an evolution of the good and bad fruits of the tree of science, consumption machines and weapons, parallel to the discovery of new layers of energy.

And this process follows a pattern, later studied in more detail: Physicists first discover a new type of energy. And the first thing they do with the new type of energy is a new bomb, then a new weapon, and finally a consumption machine. *So each age of the Industrial Revolution has started with the creation of "pure bombs" of energy; then weapons and finally "good machines." This fact of history is at the core of the process we are living now. Physicists have discovered a new form of energy, quarks and strong forces, and the first thing they are going to do is a bomb; even if their interest is merely the study of the Saint Grail of Energy.*

Indeed, when we discovered physical energy, we made gunpowder, which releases the inner energy of molecules. So first we made bombs with it. Then physicists made cannons and Galileo studied them, creating the science of ballistics, from where the first laws of motion were derived. And only later we used carbon, the main component of gunpowder, for peaceful uses—constructing trains, transport machines.

Then Nobel invented dynamite, and physicists made first bombs—dynamite and chemical explosives. Then they used those new explosives in cannons during the Prussian wars and only finally did they make a transport machine with chemical energy, the car. The evolution of weapons in that sense is parallel to the ages of the Industrial Revolution as each new age has opened a new Russian doll of energy, creating first bombs that release the new energy and then manufacturing machines with that energy.

Finally, we broke the outer cover of the atom and also made first a bomb, the atomic bomb. Then we used in weapons submarines propelled by atomic energy, and finally we used for peaceful purposes—nuclear plants.

Now finally, we release the inner energy of quarks and we *will first explode a bomb—only that this bomb can blow up the planet.*

The historic profession of physicists is to make weapons. Physics was founded by Galileo studying the trajectory of cannonballs. Physics is the specific science that studies energy and motion and those physical processes related to energetic forces—which by definition, given the inverse properties of energy and information, are the simplest forms of reality that carry more energy in the Universe. So from the perspective of theoretical physics, the creation of a quark cannon represents just the opening of the ultimate holy grail of mass/energy that physicists study. It means to cross its last frontier, breaking the very same fabric of the atom that stores the primordial mass of the Universe to release the purest form of condensed material energy, the quark, the inner component of atoms.

Yet if the previous scale of energy untapped by physicists, the atomic scale, allowed the creation of atomic and hydrogen bombs, a million times more powerful than the most powerful chemical explosive (dynamite), most experts coincide that any type of quark bomb, manufactured by the quark cannon, will be a million times more powerful than a hydrogen bomb.

Thus, we must conclude that the large hadron collider is the biggest, most expensive, longest cannon ever built—a cannon that reaches all the technological limits of artillery: it attains the maximal speed of shot, the speed of light, the maximal precision on target and the maximal energy of the Universe, as it will perforate and release the energy inside the nuclei of the atom. Its barrel is a twenty-seven-kilometer self-repetitive cannon, which keeps accelerating the bullet each cycle, till reaching after running through a distance similar to that of the Earth's circumference, the speed of light, at which moment the projectile, the nuclei of atoms, will collide head-on, with astounding precision against each other, breaking their nuclei cover, breaking their symmetry and killing them, releasing the inner energy/mass of its densest, most attractive particles called quarks. All this happens at energies beyond 1 teravolt when our matter dies, breaking its electroweak symmetry.

The cyclical cannon has two barrels. So it will shoot protons and lead from two extremes, and when those pieces of matter reach light speed, it will crash them in the central point of the cannon, producing explosions of unprecedented power on planet Earth, breaking the interior of the atom, and liberating its quarks. The quark cannon is therefore also a factory of quarks, which happen to be the most attractive, dangerous substance of the Universe, the only one that we know can blow up the planet, as it zooms into regions of increasing energy till reaching the world of big bangs and nova explosions, responsible for the creation of black holes. The cannon will increase its energy of shot between 1 and 1,000 teravolts, during the next two years, plunging *mankind for the first time into the unknown region in which our light matter dies and dark quark matter forms*.

The opening of the final Russian doll of energy of the Universe is not essential knowledge. It is just part of an automatic process of evolution of machines, the Industrial Revolution, which goes through phases of increasing energy as we evolve machines and weapons. This is what CERN is all about: to keep evolving energetic machines, atomic cannons, without being aware of the limits of death of the human weak carbon-life species.

In that sense, we should not consider the quark cannon and its deeds only from a theoretical or political perspective, but mainly from a mechanist, cultural perspective as part of the process of the Industrial Revolution of machines and weapons, which is based precisely on the evolution of energy, on the discovery of new types of energy, achieved by breaking those Russian dolls. Our entire technological civilization that substituted the humanist civilization, based on man as the measure of all things, which lasted till the Renaissance, in fact can be described as a by-product of the cycles of discovery of new deeper energies, later applied to the evolution of machines, which humans consume to obtain more energy.

Those cycles of evolution of energies and machines are the foundation of modern economical theories, based on the existence of a cycle of new energies called the Kondratieff cycle of economics. According to this model, thoroughly proved by the present economical crisis,[18] we are crossing the final cycle of evolution of energetic and information machines, what we might call the Singularity Age, which is the name used to define artificial intelligence and a black hole—that is, we are probing the limits of energy and information that can extinguish mankind without any safety measure—just because we can do more powerful energetic and informative machines. In this wider vision of history, CERN is much bigger and important than a group of physicists studying quarks: it is the natural consequence of making the evolution of machines of energy and information the meaning of history, which no longer revolves about the evolution of human beings.[19]

We needed that historic perspective on nuclear physics so people understand to which degree the quark cannon is a weapon, a case of industrial corruption, marketed as knowledge by the military-industrial complex—not a choice between life and knowledge, which astoundingly enough people seem eager to trade, like fundamentalist religious people do because they "believe in science."

15. A Brief Description of the Machine

This is self-evident when we consider any description of this machine in CERN's Website—an elegy to the size and energetic power of the quark cannon:

In the Eastern regions of France, near Lyon, flanked by virgin pine forests, streams, lakes and fir clad mountain ridges, bordering on Switzerland, lays CERN (Conseil Européen pour la Recherche Nucléaire) a facility, which houses over 6,300 scientists working feverishly to bring online the LHC, the most powerful machine on Earth. This massive Hadron Collider is a circular machine, which consists of eight sectors. Each sector is an arc bounded on each end by a section called an insertion. The LHC's circumference measures 27 kilometers (16.8 miles) around. The accelerator tubes and collision chambers are 100 meters (328 feet) underground. Scientists and engineers can access the service tunnel the machinery sits in by descending in elevators and stairways located at several points along the circumference of the LHC. CERN built structures above ground where scientists can collect and analyze the data LHC generates.

There are six areas along the circumference of the LHC where engineers will be able to perform experiments and observe the particles created. Think of each area as if it were a microscope with a digital camera. Some of these microscopes are huge—the ATLAS experiment is a device that is 45 meters (147.6 feet) long, 25 meters (82 feet) tall and weighs 7,000 tons (5,443 metric tons).

The LHC and the experiments connected to it contain about 150 million sensors. Those sensors will collect data and send it to various computing systems. According to CERN, the amount of data collected during experiments will be about 700 megabytes per second (MB/s). On a yearly basis, this means the LHC will gather about 15 petabytes of data. A petabyte is a million gigabytes. That much data could fill 100,000 DVDs.

It takes a lot of energy to run the LHC. CERN estimates that the annual power consumption for the collider will be about 800,000 megawatt hours (MWh). According to CERN, the price for all this energy will be a cool 19 million Euros. That's almost $30 million per year in electricity bills.

The electricity is consumed mainly by LHC's magnets to steer beams of protons as they travel at 99.99 percent the speed of light. The magnets are very large, many weighing several tons. There are about 9,600 magnets in the LHC. The magnets are cooled to a chilly 1.9 degrees Kelvin (-271.25 Celsius or—456.25 Fahrenheit). That's colder than the vacuum of outer space. At that temperature, the electromagnets can operate without any electrical resistance. The LHC uses 10,800 tons (9,798 metric tons) of liquid nitrogen to cool the magnets down to 80 degrees Kelvin (-193.2 Celsius or—315.67 Fahrenheit). Then it uses about 60 tons (54 metric tons) of liquid helium to cool them the rest of the way.

Indeed, CERN's sites describe the quark cannon with a long list of numbers, sizes, and tag prizes that make any mechanist scientist to feel proud of such technological achievement. The other pages are dedicated to describe the obsolete theories it will prove, to which we dedicate most pages of this book, as humanity is absurdly willing to risk its life for "knowledge"—only that CERN will not provide knowledge, only risks, as all weapons do. What CERN never explains, of course, are the risks of the machine. In the initial Web there were none. After the suits we put to this company, they felt compelled to bring "experts" working on the company to dismiss those risks, but only the most publicized one—black holes. Many other risks—Bosenovas, dark atoms made of heavier quarks, top quark stars, and thermonuclear explosions—less known to the public are simply ignored.[20]

Yet those 60 tons of liquid helium add another risk to the black hole factory since the energies of the hadrons turning at speed of light could provoke if they are deviated and collide with the 60 tons of helium at light speed, a thermonuclear explosion so enormous that it would make Teller's superbomb a reality, but instead of blowing up the Soviet Union, it could blow up France.

To avoid this, the proton beams inside the LHC travel through pipes in what CERN calls an ultra-high vacuum. Yet an accident is still possible . . . since there have been in fact already two accidents that provoked the spill of tons of helium, fortunately enough before those beams reached the speed of light.

It is thus clear that the LHC is not an instrument of research as CERN affirms but the "most dangerous machine on Earth" as *NY Times* physicist and journalist, Dennis Overbye, called it,[21] a weapon, even if it will produce a lot of subatomic particles in its collisions.

Let us not get this wrong: our civilization is based on the ethics of technology, described by Fromm in the quote that starts this book, "If we can make a machine we will do it, even if it can kill us all." And this is exactly what we have done: "First shoot, then ask." Indeed, we have made the machine and then physicists have found excuses to use it. Because the most perfect, expensive machines at any age in the Evolution of Metal have been weapons; technological science often works under the pretext that we need to evolve those weapons for security. This was the excuse to evolve cannons most of history, even if we got many fringe benefits in the science of motion, since Galileo analyzed both the laws of motion and improved cannonballs. But technological science, when the excuse of war no longer works, keeps churning machines and then it must find a civil excuse, mainly knowledge.

Yet the idea that the most expensive, perfect weapon ever constructed, the light-speed, superfluid seven-teravolts quark cannon built by the nuclear company of Europe (the LHC), represents no danger to mankind because it has also some peaceful fringe benefits (the study of subatomic particles) is an oxymoron. All military technologies have peaceful applications, but those facts must not hide the primary consequences of their use. Weapons are lineal systems that release enormous quantities of energy, able to erase the complex, fragile information that creates life; and the quark cannon, called in the peaceful, Orwellian newspeak of the new era, the large hadron collider, is not an exception. It is the final evolution of the industry of cannons, intimately related to the evolution of physics, the science that studies energy and motion, founded by Galileo, a mechanist working for the Arsenal of Venice, which discovered those laws of motion studying cannonball trajectories four hundred years ago. The duality of the fruits of the tree of science, with its positive influence on knowledge and its negative consequences for human life, are exemplified as never before by this quark cannon. Yet in this case, the negative consequences, the possible extinction of life, far outweigh the benefits for our knowledge of the Universe and this is the key fact that the nuclear company has successfully hidden to the public and governments that founded this absurd quest for reaching the energies of the big bang that probably never destroyed the Universe but now menaces to destroy the planet Earth.

Indeed, the reason, the bottom line of the Damocles Machine is rather more prosaic, embedded in the ethics of a technological civilization: we could make a quark cannon because it was technologically possible, so we have designed it. And now we shall try to find uses to this machine, which cannot be any longer military as we are not at war with the Earth, the target of such awesome, explosive energies.

So we have quark cannons for research and the people who build them are NOT mad scientists, but the idealist scientists that will find "God's particle," re-create the "birth of the Universe," and explain the meaning of it all. Yet in the historic perspective we brought about on the evolution of weapons, the quark cannon is just the final "just weapon" of the Western world. Indeed, when the Nazis were massacring humans "face-to-face," we became "just bombers" in Hiroshima, Nagasaki, or Dresden, killing hundreds of thousands of children and women—as most men were in the battlefront. And that seemed to us "just," because we didn't see the victims of our holocaust of civilians. So our mass murders were seen as "clean." The same happened in Vietnam, when we bombed civilians with napalm. It was okay. It was our way to do war. We didn't torture people in concentration camps. We just erased them. Now we bomb Afghans with Terminator Drones to ensure the safety of our troops, and the people who die under their bombs are always terrorists—even if most often are villagers, children and women, who didn't go to the mountains to fight and make an easy target for stupid robots. And we put wars on TVs as if they were video games. All this Walt Disney Brave New World has added now the crown jewel: a quark cannon that will make the first cosmic bomb and blow up the Earth to discover the meaning of it all. Because of course, after sixty years of "righteous" nuclear bombs, CERN doesn't seem dangerous at all. It is our terrorist group of mad scientists with the most advanced technology of the world, but that is okay. We have perfected during sixty years our propaganda about nuclear physicists as idealist people, the most intelligent scientists of the planet, who deserve to do bombs for research.

And so those who criticize them *must be hippies, crackpots or know nothing about physics to dare to defy the high priests of our civilization*. So that is where we are: according to marketing, the machine of the big bang will explain the meaning of it ALL with an explosion unlike any other explosion ever taken place on Earth. It will happen between 2010 and 2013 with increasing probability, as the cannon probes the scales of energy beyond one teravolt in which our electroweak matter dies, breaking its symmetry, feeding the birth of strong, dark quark matter, the most dangerous substance of the Universe. An entire zoo of quark matter species, strangelets, pulsars, quark stars, black holes, are waiting for us to open the door of death of our light matter and reveal the future of science since indeed, death always happens in the future.

That our death will reveal the meaning of it all is something fundamentalist physicists and religious zealots believe, but we doubt since among other things it is obvious that an extinct scientist knows nothing. And especially because, as we shall see in this book with the authority of truth and the scientific method, physicists *know nothing about information, life, and the processes of death their weapons cause as they only study half of reality, the arrow of energy, entropy, and death*. So in this book we change the order of truths. We shall not use the newspeak of research but the reality of a weapon that might indeed create a big bang, the explosion of the Earth; and we shall not consider energy as the meaning of it all, but just the canvas in which information paints reality.

Let us consider other type of technological weapons and how they are marketed. One is called the Future Combat System, another multibillion dollar program to make robotic weapons, a.k.a. terminators, for whom marketing has found a nice name: Future Systems. The other is called the DDX, a three-billion-dollar stealth like useless destroyer, described in a military Web:

The attack would come quickly, and it would be awful. Cruising far offshore, the U.S. Navy's DDX destroyer launches 20 artillery shells in less than a minute. As the satellite-guided weapons fall back to Earth at 830 mph, computer algorithms alter their flight paths so that the 250-pound projectiles all strike the same patch of ground at the same time, reducing everything in the vicinity to rubble and dust. If more firepower is needed, the destroyer can unleash another 580 artillery rounds, as well as 80 Tomahawk missiles. And when the attack is over, the ship simply vanishes. On a radar screen, the DD(X)s stealthy hull makes the 14,000-ton vessel look like just another fishing boat, casting its nets into the sea.

Just one thing is missing from this scenario: an enemy to fight. Targeting terrorists with the DD(X) is like smashing ants with an 18-wheeler, critics say. Attacking an Iranian nuclear facility is something American bombers can do today. "The DD(X) is the most revolutionary surface warship in decades," says John Pike, director of defense think tank GlobalSecurity.org. "But I have yet to have anybody explain to me—point to a place on the map—and say what they propose to do with it."[22]

Eventually, the navy is projected to spend $4.7 billion each for seven DD(X)s.

These are then the comments of a military-loving blogger:

> I hope you'll check out the whole thing. I'm also honored—more than honored—that former Assistant Secretary of Defense Bing West decided to contribute an accompaniment to my article, on how we can "Invest in Our Troops." Be sure to take a look at that, too.

> UPDATE 12:01 PM: DD(X) makers Northrop Grumman have a very different take on the ship, of course. Here's a video outlining their case.

> UPDATE 12:04 PM: One of the things you find, looking into these big weapons programs, is how quickly justifications for the systems shift to meet the times. The DD(X), for example, went from a land attack specialist to a commando-delivery ship. The Army has a similar repositioning under way.

> Now, Future Combat Systems—the Army's new array of robots, sensors, and ground vehicles, originally meant to take on another

> big military—is being pitched as a disaster relief program. Check
> out the Army's "Aftershock" video to see what I mean.

Yes, we know what they mean. The important thing is to build the weapons because we can build them. Then we shall find a use for it—even a peaceful use, relief aid or research on the meaning of it all.

In any case, the description of CERN's machine is of no importance to us. What matters is what it will produce. And 99% of it will be deconfined quarks. So the machine is a quark cannon, a quark factory, the first one on Earth, and the only one anywhere in the cosmos except inside stars, where enormous densities create also quark bomb, deconfining them as the quark cannon will do, massing them together as the quark cannon will do, and provoking a supernova, as the quark cannon might do.

In that sense, for a proper understanding of the two main dangers poised by the quark cannon, the creation of strangelets (lumps of strange quarks) and black holes, we have to do what CERN has not done—to study quarks, *the substance the LHC will produce. Thus we shall study* quarks from the perspective of the two scales of the Universe, the quantum and cosmological scale in which they interact, to fully grasp its nature and how they catalyze the creation of big bangs, black holes, and supernovas:

— From the quantum perspective, quarks are the densest masses of the Universe.
— From the cosmological perspective, quarks are the atoms of frozen quark stars, pulsars, strangelets, and black holes. Because indeed, a black hole is very likely the densest of all quark stars.

Yet to fully grasp those two scales of the Universe and its interactions, we cannot use quantum theory since quantum theorists do not understand mass, the physical information of the Universe, and deny Einstein's work (Hawking said Einstein is double wrong) about black holes.

Thus, we must introduce in certain detail twenty-first-century science, Einstein's work on masses and black holes, as frozen stars, and the evolution of relativity in the context of the fractal paradigm and theory of information—the true frontiers of modern science that CERN, an obsolete factory of the military-industrial system, totally ignores.

In simple terms, masses are vortices of physical information, small hurricanes of space-time, according to Einstein's relativity, which is the standard theory of mass and gravitation that Mr. Higgs pretends to substitute with his invisible particle. And the fastest, most attractive of those vortices of space-time in the quantum realm is the quark and in the cosmological realm is a quark star or black hole. Thus, quark stars are today modeled as fractal systems made of superfluid quarks, packed in an extreme dense environment that has lost the electromagnetic covers of which we human beings are made. And this might be the end of the Earth if the free creation of quarks is allowed to happen in increasing numbers as CERN shoots the quark cannon, ramping up its energy to limits this planet cannot stand. But CERN denies this because it is stuck in forty–year-old theories about mass and black holes (Mr. Higgs and Hawking's work). So we need to explain to the reader *why they are wrong*.

In the next chapters, we shall consider those two approaches to masses and the physics of quarks and information in more detail, introducing first the standard concept of mass according to Mr. Einstein (and the absurd Higgs theory that pretends to substitute him), in order to fully grasp the destructive power of those quarks.

Yet because we only have the authority of truth, it will be necessary, before we study in depth the nature of masses, quarks, and black holes, to make a long introduction to the nature of truth in science, the methods of verification of truth and the advances that have happened in the past decades in theory of information and have illuminated enormously our understanding of the Universe, making obsolete many of the theories CERN sponsors to protect this machine, given the power of CERN, which has imposed its ideas merely on authority and ad hominem campaigns against Einstein et al., classified as "crackpots," when the opposite is truth; some of the people in this collective are among the pioneers of twenty-first-century science, reason why they know more about mass than CERN will ever do and are extremely worried about the combination of military power and theoretical ignorance displayed by CERN's employees. Only then when you upgrade your vision of reality to twenty-first-century science and fully grasp the meaning of truth in science, we will be able to distinguish the bad and good fruits of the tree of technology *on reason and intelligence, not on authority*. This issue must be analyzed with reason, not with blind belief, as mass media and politicians have done, accepting CERN's authority without any knowledge of the risks involved or the true avenues of twenty-first-century science. So mankind has given a free check to this company despite the fact that it can kill us all and will discover nothing about the fractal, infinite, organic Universe; we can only discover with the most perfect instrument of information nature has devised—the mind.

16. The Factory of Strange Matter

In the last weeks, news slipped out of CERN about what the factory was producing: strange matter. So now it seems clear that the biggest danger of extinction will not be black holes but their "lowly" brother: strange liquid, the substance of pulsars and quark stars. A report came just before the factory started continuous production this April that CERN had produced an enormous "unexpected" quantity of kaons.[23] Kaons are particles of strange matter, and a soup of kaons is called a strange liquid or "ice-9," the substance that catalyzes the creation of supernovas, cosmic explosions that leave behind a strange star, a pulsar, superfluid of quarks. Kaons are just the first pieces of strange matter, the equivalent of amino acids. They are not dangerous since they don't reproduce. They don't catalyze the creation of new kaons.

But kaons made of up, down, and strange quarks are the first sign of it, like amino acids are the first sign of life. Then as more energy is added to the organic soup and lineal proteins appear, finally a double helix of DNA forms. In the quark-gluon soup, stable strange particles called hyperons and dibaryons will appear, catalyzing the ice-9 reaction. The problem with those kaons is that the factory said it will produce less, not more of it. And it has done so at a mere 1.1 TeV, just crossing the barrier of decay of our matter into strange matter. This means the growth of strange matter is, as we Einstein et al. predicted, exponential; and since the machine has only fired at 1/1000th of its power, it will certainly produce the next two layers of the quark-gluon soup—hyperons, up, strange, down particles, which are the atoms of strange liquid (a soup of usd quarks); and finally *dibaryons, two strains of hyperons that can split, capture new usd particles, and start a reproductive process similar to that of the DNA, but at the astonishing speed of over 1 million transformations per second. Further on, dibaryons are stable in the Earth's ambient. And with the new calculus, they will certainly be produced at 3.5 TeV.*

So now we know how we shall "very likely" die, even if the company will deny it till the end and people will also deny it to not confront a nightmare, which *should be confronted at this point by politicians and the military by closing the machine without further ado, now.* Every day the machine works, the danger of extinction increases. Every day, more stable dibaryons, which are neutral and very difficult to detect will be created. Of course, the factory will not even look for them, as it would be bad news. CERN has already hidden for two months the creation of its "amino acids," the kaons. Dibaryons however will fall undetected to the center of the Earth since they are neutral. They will easily reach that center, as charged particles won't interact with them. Then when enough of them reach the center, in a stable configuration, they will form a ball of strange liquid so dense as a pole of attractive gravitation that the Earth will fall into it. We will see increasing earthquake activity, which finally will extend all over the planet. The collapse will last a few days after the initial warnings, when

10 Richter scale earthquakes happen all over the Earth. Probably at the end of the first run of the LHC, which ends in 2012 or by Christmas 2013, when switched at full potency, the crust of the planet will sink in. By that time, the number of neutral self-reproductive dibaryons falling to the Earth will have made in any case the process irreversible. While those dibaryons are very small nuclei and hence will not attract many particles in their journey (as most matter is in fact "empty space"), once they seat in the center of the Earth, their weight will start to accrete matter very fast.

The best-known scenario, the black hole case that might happen in 2013, will be much faster (in essence a strange quark star is a weak black hole, slower but good enough to do us all). But now the difference between both scenarios is clear. The black hole might happen or not. If it happens, we will die within seconds. The strangelet scenario however should be happening at 3.5 TeV. So every day the factory of quark matter works, the chances of stopping the extinction of the Earth will diminish, as more dibaryons of strange matter will be produced; the more of them fall to the Earth, the more chances there will be that those dibaryons survive the trip and hatch in the center of the Earth. And once they reach their destination, like those salmons that swim upriver, they will grow exponentially, eating up the Earth from inside out and nothing will be able to stop them. There won't be a rescue operation Hollywood-style to the center of the Earth. We humans are a weak species; we have nothing to do against an ultradense strange liquid growing inside the Earth. But we could have avoided in the first place building a factory of quark matter. We didn't need to kill ourselves. The people represented in this book have been shouting for years what was going to happen. Now it is happening, and because CERN will NOT monitor the production of dibaryons, a neutral particle difficult to detect unless it decays, considered a known-known particle—the only interest of the people working there is to get their check, to find perhaps some exotic particle to get a Nobel Prize, and to avoid the denounces of those who want the factory closed—we shall not even know how much CERN is producing of the stuff that will kill us. Reports will leak, like the first report on 1 TeV collisions by an ignorant journalist who wrote this: "CERN is producing particles not black holes." She is one of a bunch of CERN groupies, who were scandalized when we sued the company for a possible production of black holes, and of course, she doesn't know what a kaon is, what strange matter is. So she publishes cheering up, happy that there are no black holes at CERN. It is like saying, "Look, there are no lions in that room, only tigers, come inside."

That is how I found out CERN is a secretive company. It only says what it needs to say. But the calculus is easy. CERN will ramp up energy/mass to 1,000 TeV, and the main produce of the factory will be strangelet particles, a soup of strange liquid with many neutral undetected dibaryons. So at maximal production, it will produce a lot of it, falling toward the center of the Earth. Of course, at the end, there will be public outcry, people will oblige politicians and the press, who are deaf now to all what is not "pressing news," to inform and shut down the factory. But it will be too late. When Bush acted on New Orleans, it was late. When the West acted against Hitler, it was late. When politicians act against CERN, it might be late as it is already happening. Since Christmas 2009, the factory is producing kaons,[23] the unstable atoms of strangelets, of ice-9, of the liquid that causes supernovas and pulsars and quark stars. This French-German factory, in an age of resurgence of the energetic German spirit, as the leader of Europe, is finishing up the work of Auschwitz; but instead of Zyklon gas, it is using ice-9. Instead of killing only Jews, gypsies, and Commies, it will kill all of us. This might be excessive for some innocent readers; but what made the Nazis able to kill so many people was not a special evil, but a special technology—the German capacity to create excellent machines and weapons, which now again has shown at CERN. Their strong, energetic drive, which one it was had chosen a path will not think again about it, is shown at CERN, never doubting, never acknowledging any danger, breaking energy records with its quark cannon. They are directed by Rolf Heuer, with a very German mentality. They consider themselves extremely intelligent, "experts," but know really very little about the Universe at the energies they are exploring. The press has pumped them up as geniuses; the military-industrial complex made them heroes. So they are pretty sure they are doing the right thing. Hitler, a mediocre landscape painter and corporal, was heralded as a genius of strategy and took the Germans to death, but they followed him with a remarkable corps spirit. Now the nuclear industry follows CERN with a remarkable sense of "knowing what they are doing." Any doubt, any critique, as in the military, is considered "negative thinking." A Pantzer driver doesn't look at the corpses below its wheels. That is negative thinking. Rolf Heuer, Ellis, the people of CERN talk of their machines, their plans, designs, teravolts, *the job that has to be done, never mind their mindless job, is to kill the Earth*. They are very much like Göring, an admired military who substituted the Red Baron as head of the best squadron of German airplanes in World War I and created the German military-industrial complex, who was a man of action, a serious, solid man: "Hitler talked of iron, roads, machines, things we could understand."[24] He also said that the first murder didn't let him sleep but the second opened his appetite. The first nuclear scientist that murdered mankind, Oppenheimer, said, gloomy, "I am become Kali, God of Death." He was criticized by his peers, who were enthusiastic about their weapons.

The inventors of the H-bomb were far more confident. Teller sold it as knowledge, and Wheeler said he regretted not to have built it before and threw it in Japan. This 3rd generation of nuclear weapon makers has even less regrets. Rolf Heuer's robotic workers merely published a paper with a lot of data saying that their factory is producing an "unexpected" number of kaons (of strange liquid, but that is a taboo word). And then, they will probably sleep well. Their brains, like any computerized brain, just crunch numbers, data. There are no collateral effects on data. Their brains have established a loop: strange liquid won't happen at CERN. Why? Because it is negative thinking. "But it is *happening*," perhaps a voice tells them in the back of their heads. "This is negative thinking," will say again the forehead. So there is massive denial against negative thinking. The Germans always get to the end. They never surrender. Even if the end is just death, if the cause is absurd, if there is no reason at all in their reasons. *It is the end, we must get to the end, we must not stop, it is negative thinking.* And so Rolf Heuer will go on making "serious things" like Göring did: his planning, his daily agenda at the summit of the company that has built the most perfect machine ever built must not stop. Never mind the machine is a quark cannon that is producing strange liquid. We repeat what they do. They repeat too: "This is *negative thinking.*" The Germans are the biggest contributors; the French are the collaborators, who feel proud of being the avant-garde of the nuclear industry, the people who like to think the machine is theirs; the British theoreticians, the masterminds behind it. Like in a caricature of European history, the three imperial powers who fought and destroyed twice the world now work together to destroy it all, not because they want, but because technology can. And we must do it, the rest is negative thinking. Rhetoric is the art of repetition, Goebbels, the father of modern marketing, said. The Germans do the job, the French feel "their grandeur" building it all, and the British theorists will get the glory—or so they think in this new entente cordiale. In WWII, their industries of war clashed, now they work together, which merely makes their machines ever-more powerful.

Sounds familiar? Yes, indeed, this is a chronicle of death foretold; it is the history of how Europeans invented weapons of mass destruction till they killed us all. Now we are living corpses. Now we are all in death row. A miracle might happen and a politician might halt the factory of quark matter before the lethal injection of dibaryons runs down the veins of Gaia in enough numbers to provoke the collapse of her heart. The injection has barely started and there are already signs it is truly poisonous. By 2013, the lethal dosis will be 1,000 times stronger. So we shall expect to die NOT in bed, but in a planetary Earth-quake. Here in California, it will be the Saint Andres Fall. In Japan, it will be a tsunami a few miles tall. We are all now in the same sinking boat. But don't worry, be happy. *This is negative thinking.*

17. Conclusion: Bad vs. Good Science

Indeed, the human sheeple accepts its destiny. Nobody rises against military dictators; no scientist protests against CERN; no Holocaust victim attacked the few guards at Auschwitz. What truly defeats the scientist with enough intelligence and ethical compromise to denounce what is going on at CERN is that sheeple behavior in front of eviL, the inverse word of Live, the arrow of death. Those who kill mankind get away with murder while those who defy eviL are crucified in any time of history.

Indeed, if a believer denies the jihad and explains that Mohammed didn't preach the Holy War, he will be killed. If a physicist defies CERN, he will suffer the ad hominem campaign Einstein et al. have suffered. So we, human beings, will be killed by CERN because as Ellis, its chief theorist, put it, the way to end this argument about the safety of the LHC (the quark cannon) is to shoot the Earth with it. And *nobody protested; nobody said that is criminal behavior; nobody went to CERN and detained a guy who is saying to the world that he is not sure if his quark cannon will kill mankind but he is shooting anyway.* Evil death grows because people do nothing to stop it. Hitler grew because each new step he made was cheered by the world till hell broke loose. Physicists have, for four hundred years, affirmed that man is not the measure of all things, that the Universe is not organic but a machine, that the only arrow of the Universe is death, energy, entropy, and nobody has fought back for the arrow of life, except a few priests and biologists, ignored or ridiculed by the press. And now they say they might blow up the Earth and the sheeple applauds.

But is the Universe a machine of energy, a weapon? Are the biased energy-only theories sponsored by the makers of weapons, such as the description of time as "only change in motion" and the birth of the Universe as a mere big bang, the real way in which time, the Universe, and all its species are created? Obviously not since twenty-first-century science has proved that the Universe is modeled with fractal networks of energy and information, which is also the structural form of complex organisms. Physics, the science of energy, is only one of the two sides of the coin and the other side, information, is far more important to understand reality. The prestige of physics as the supreme science is only held because of its antiquity and its power as the sustenance of our culture of machines, which paradoxically is also the cause of its dogmatic search for the Saint Grail of absolute energy as the meaning of it all.

It is for that reason that the industrial-military, theoretical, and unethical perspectives come together in the actions and theories of CERN, the nuclear company of Europe, which will risk the extinction of Earth to defend the ethics and energetic dogmas of our technological civilization:

"If a machine can be done, it must be done, even if it kills us all" (Fromm). Good luck.

In that regard, the two issues brought about by CERN's supercollider, which seem apparently unrelated, the risks for human life this machine will bring and the falsity of the theories it will try to prove (the cosmological big bang, the evaporation of black holes, and the Higgs boson), are surprisingly interconnected. *Nuclear physicists invent energetic theories because they make weapons. Period.*

For that reason, the book is structured as a series of oppositions between CERN, which represents the bad fruits of technological science, weapons harmful to mankind, and the false theories of reality that justify them, versus Einstein et al., cosmologists who represent its good fruits, harmless tools of research, telescopes, and satellites that study directly the Universe with no danger, and build true theories of reality. They also defend the bioethical conscious effort of many scientists to dedicate their life to improve the future of mankind, not to destroy it. Truth versus lies, weapons versus tools could resume the message of this book to those who establish "scientific policies" in this crossroads of history between existence and extinction. We must differentiate tools of research from weapons camouflaged as such—as the quark cannon, the LHC, is—and forbade them if we want to survive the evolution of technology this century and make the Earth truly a paradise for life. But for that to happen, we must make of man, not of the machine, the measure of all things and build a world made to our image and likeness.

And we must learn true science and differentiate its organic, informative paradigms from mechanism and energy-only theories, such as the big bang or the evaporation of black holes, which are wrong, simplifying philosophies of science, but given its simplicity, they are easier to believe. Indeed, simple scientists prefer a lineal, energetic theory easy to calculate so they can focus once the dogma has been established into the routine-driven tasks of science: to measure, tabulate data, and publish papers, as CERN will do in the next years, risking in the process of doing bad science the life of mankind.

In true science, though the quest for energy is not the meaning of it all, neither it is the machine the true model of the Universe, and lineal equations proper of "energetic theories" hardly explain anything. This is the case of the quark-gluon liquid origin of big bangs, supernovas, and neutron stars that CERN manufactures. Lineal equations said it was a harmless gas because it is easy to calculate. But it turned out to be a complex, informative liquid self-similar to a black hole, which in its external surface expelled radiation in a mini big bang, in the first experiments done at minimal energies. Thus, if CERN keeps doing this quark-gluon soup, it will find a dual process of information (the creation of a neutron star, a complex quark liquid called a strangelet) and energy (the external explosion of the Earth in a big bang). And yet this dual "refrigerator" system that expels radiation and cools down matter, creating both a big bang and a big crunch, is not accepted by CERN, which still affirms it will create only radiation. Thus, dogma prevents CERN and the scientific establishment from realizing it will blow up the Earth, as it crunches simultaneously our matter into a pulsar. All those gaslike entropy-only energetic thesis are simplified, biased theories developed by the worldly makers of weapons that now, in the twenty-first century, we can put duly in perspective, thanks to the advance of new, more complex sciences based on the most advanced mathematics of modern time, fractal and non-Euclidean mathematics, which show a Universe made of energy *and* information, complex enough to participate of most of the properties of organic systems. Even a machine can be described as a simplified organism. And so we shall define the Universe as a fractal organic system, made of two networks of energy (electroweak forces and light matter) and information (gravitational and strong forces of dark quark matter), where the quark-gluon liquid that might kills us all is also dual, made of an inner big crunch liquid of quarks and an external expansive membrane of electroweak radiation. Let us then understand the real universe before we can account for this "death foretold," which such quark-gluon soup might cause, whose dual structure is still censored by dogmatic science.[25]

---

0   The reader will excuse me if after three years of ad hominem campaigns against Einstein et al., I quote some of the self-serving comments of Mr. Higgs and Mr. Hawking about the superiority of their work and minds over the rest of the mortals. Mr. Hawking in his autobiography lashes out often against all other "inferior sciences," especially biology and its personal nemesis, philosophy of science, perhaps because this discipline, which is obliged to assess the truth of scientific theories, has falsified his work. He compares himself with Einstein while his site's biography tells us that he was born three hundred years after Galileo: *http://www.hawking.org.uk/index.php/about-stephen/briefhistory*. Mr. Higgs bet one hundred pounds that Hawking was wrong and Hawking bet one hundred pounds that Higgs was wrong (*http://www.telegraph.co.uk/science/large-hadron-collider/3351602/ Large-Hadron-Collider-God-particle-theorist-Peter-Higgs-attacks-Stephen-Hawking. html*). But if Higgs believes Hawking is wrong, it means he doesn't mind risking the planet to prove his particle right and win a Nobel Prize. Regarding Mr. Hawking, a celebrity with a courageous life, unfortunately he is *not* the new Galileo or the new Einstein.

1   Eric Johnson, http://www.technologyreview.com/blog/arxiv/24611/.

2   *The New Yorker* Magazine, Crash course, May 14, 2007.

3   Einstein's paper "Does the Inertia of a Body Depend Upon Its Energy-Content?" "Ist die Trägheit eines Körpers von seinem Energieinhalt abhängig?", was published in 1905 in *Annalen der*

Physik. An English translation can be found at *http://www.fourmilab.ch/etexts/einstein/E_mc2/www/*.

[4]   For this and other themes related to the nature of quarks, also discovered by Mann, consider the book Johnson, G. (1999). *Strange Beauty: Murray Gell-Mann and the Revolution in Twentieth-Century Physics*. New York, NY: Alfred Knopf, Inc.

[5]   See addenda E chapter 4 for LHC's expected quark production, as given by Engelen, its chief scientist.

[6]   *Http://arxiv.org/abs/hep-ph/0512112*.

[7]   An interview in which Teller proudly remembers this feat can be seen in the film *The A-bomb Movie*, http://www.amazon.com/Atomic-Bomb-Movie-Special-Directors/dp/630507111X.

[8]   The Higgs hype was created by Mr. Lederman to sell the supercollider to Reagan, as he explains in his book *The God's Particle*. http://www.amazon.com/reader/0385312113?_encoding=UTF8&ref_=sib%5Fdp%5Fpt.

[9]   Physicists told the congress the supercollider would cost only $4 billion, presenting a budget that had nothing to do with the real price that soon went up to 13 billion by the time Clinton cancelled. The Large Hadron Collider again was first budgeted at half its real cost. False budgets on nuclear weapons were normal in the military industry during the cold war, when costs were never an issue. Today, nuclear physicists continue with the same routine. Yet the cost of the quark cannon meant the cancellation of far more important experiments of science in the present economic crisis.

[10]   Tefill in his book 'the 3 first minutes of creation' considers that since we haven't found a single quark in our hundred years of cosmic ray research, the chances to find a quark in cosmic rays are null, but the company says its experiments that will deconfine millions of quarks happen "all the time" in the atmosphere. The reason is that quarks are the cause of mass bombs and cosmic rays are the result, the debris of the explosion of a supernova, which quarks catalize, as it has been recently proved by the Fermi Satellite and published in *Nature* (January 2010). Thus, to compare the energy of a cosmic ray and a quark bomb is a false argument. It would be like comparing the energy of a nuclear bomb and the energy of the radiation it produces because $E=mc^2$, the energy of the bomb equals the energy of the radiation it produces. So happens with the energy of a quark bomb that causes a supernova and the radiation it produces (the cosmic rays, born with the evaporation of the star). But the radiation that fell on the people of Hiroshima will never detonate the A-bomb. It is the uranium that detonated the bomb, not the radiation. And it is the quark-gluon soup that detonates a mass bomb, not the cosmic ray radiation produced by a supernova.

[11]   The LSAG, the safety report on the Large Hadron Collider was done by physicists associated with CERN, breaking the legal principle *nemo debet esse judex in propria sua cause* (no one to be a judge in their own cause). See chapter 4, addenda B.

[12]   The details of the scam would be published after the crash by *Wall Street Journal*, which however actively recommended those assets during the bubble age, as today all newspapers side with the nuclear company, singing the benefits for mankind of the quark cannon.

13 ' 'The Quantum Theory of Black Holes" *SciAm*.

14 'Findings at MIT's Center for space research indicate that the way in which black holes accumulate matter is independent of their mass. A small black hole named J1650, which has ten solar masses, behaves identically to a ten-million solar mass black hole in Galaxy MCG-6-30-15 (from No canary in the quanta by H. Lehmann). This is, as we shall see, because a black hole is a vortex of space-time, according to Einstein's mass theory, which rotates at the speed of light. Thus, a hurricane of space-time that turns at c-speed can have any radius, yet any black hole rotating so fast will absorb matter around at c-speed. Imagine a wave caused by a rock thrown into the water. It does not matter the size of the wave, it will move at the same speed at its center or in a wider radius. It is then easy to calculate that such "whirl of space-time" rotating at c-speed will take hardly a second to disturb and destroy the Earth, which explains why the birth of a black hole in the center of a star can provoke in a few seconds its explosion as a supernova. We have recently measured the speed of rotation of a black hole, observing the speed of rotation of iron atoms around it. And it turns out to be c-speed, backing Einstein's model of mass as whirls of space-time.

15 See a simplified explanation of Nambu's top quark=Higgs field at http://en.wikipedia. org/ wiki/Top_quark_condensate. Miransky, Masaharu Tanabashi, and Koichi Yamawaki elaborated on Nambu's theory in a 1989 paper: Is the Quark Responsible for the Mass of W and Z Bosons? Which can be found at *http://www-library.desy.de/cgi-bin/spiface/ find/hep/www?j=MPLAE,A4,1043*

  Nambu, NOT Higgs, received the Nobel Prize for Physics in 2008 for such discovery. Yet the press ignores his work, "sold out" to the Higgs hype.

16 Cox made a few years ago a TV program for BBC, talking of the dangers of strangelets for Earth (one of the possible quark condensates that could be made at CERN and blow up the Earth), which you can see at http://www.youtube.com/watch?v=T1vKisefsul, three minutes, fifty seconds. But now he works for CERN and has changed his opinion in a pattern quite familiar in this case. (Wilczek and Rujula, two theorists now working for CERN, also warned in the past about the dangers of supercolliders and now work for CERN and affirm those dangers are null).

17 Http://motls.blogspot.com/2010/01/lhc-alarmists-judge-would-rule-stop-lhc.html.

18 The cycles of war and evolution of machines were calculated by Kondratieff, explained by Schumpeter, and corrected in the dates, using fractal equations, by this writer, which in his book *Radiations of Space-time, the Extinction of Man* c. 94, Bookmasters, Ohio, already stated that the evolution of machines followed a seventy-two-years generational cycle of human beings, separated by global economical crashes, prophesizing that after the 1857 crash of trains and the 1857+72=1929 crash of cars and radios, there would be a 1929+72=2001 crash of electronic industries, with short eight-year waves in 2008 parallel to the short wave crash of 1937 that plunged the war in World War II. Thus he also prophesized that after 2001, the world would enter in a war age, as it has been.

19 The evolution of human beings into a peaceful superorganism of history, the true goal of mankind, is no longer "fashionable," as we have decided that to evolve machines is the

meaning of it all. See 18 and appendix for a complex analysis of such enormous error, caused by the mechanist paradigm (chapter 2).

[20]   See affidavits of those suits later in this book for a detailed description of those risks.

[21]   See *NY Times* recorded interview.

[22]   Globalsecurity.org.

[23]   Http://www.springerlink.com/content/t35h6211438476k0.

[24]   David Irving—Reichsmarschall *Göring*.

[25]   A description of that dual structure sent by Sancho, international chair of duality, and Rössler, founder of chaos theory, was ignored by mainstream science magazines *Science News, Astronomy, Science, Scientific American*. Simple rejection letters: "Dear Contributor: Thank you for your offer to contribute to Scientific American. I regret to say that the piece you propose is not suited to our somewhat limited editorial needs. We appreciate your interest in SCIENTIFIC AMERICAN." Avonelle Wing, editorial administrator, express the unspoken rule of our "technological civilization": "We are committed to evolve machines (and sacrifice mankind to them till the end; this part is censored because we do not "make negative thinking"), we "believe" in technology, we are "positive," we "have a manifest destiny." To be food for black holes.

CHAPTER 2

The Two Scales of the Cosmos: Masses and Charges

I think the next century will be the century of Complexity.
—Stephen Hawking

I. A Universe Built by Fractal Information

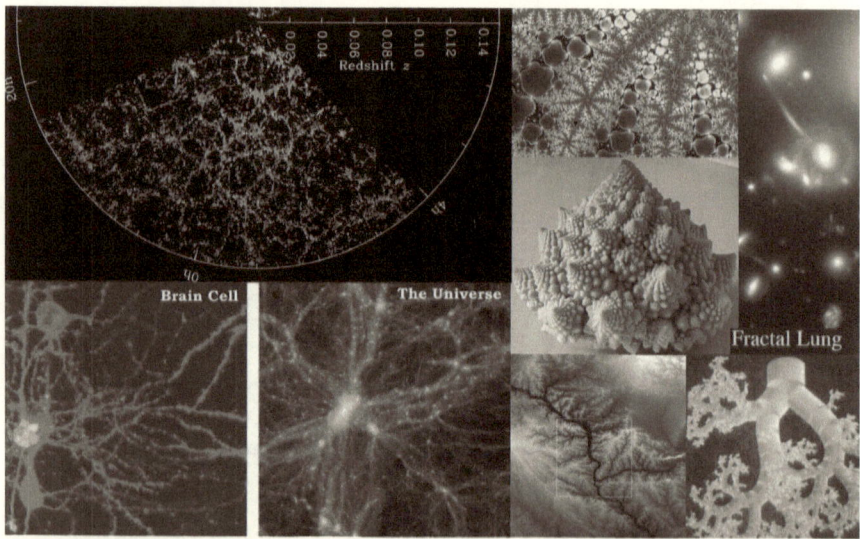

In the graph, several experimental proofs of the fractal organic structure of all beings of reality: On the left, the Sloan fractal map of galaxies and its comparison with a neuronal cell; on the middle, an ice fractal, a plant and a river; on the right, a cluster of galaxies and a lung. Fractal and non-Euclidean mathematics have evolved our understanding of information in topological terms, defining a new arrow of time, fractal information, which all structures of nature constantly reproduce over a canvas of simpler, energetic motions. Time bends space into masses, said Einstein. Time evolves form, said Darwin. We can mathematize those processes through self-reproductive fractal equations that iterate a certain form of nature—an ice geometry, a DNA code, a cellular structure, a physical particle. This revolution is opening science to an organic understanding of reality, which renders CERN's attempt to prove false energy-only theories obsolete. Since the Universe is a "fractal of energy and information, made of self-similar parts, which constantly reproduce their forms." Indeed, from the simplest particles, quarks, and electron, which absorb energy and reproduce new particles to the most complex informative species, human beings, reality is made of bits of information and bytes of energy, evolved in complex, social networks through fractal scales, from atoms to molecules, to cells to organisms, planets and galaxies.

1. Experimental Proof of the Fractal Structure of the Universe

In the previous graph we showed some experimental proofs of the fractal Universe. Each photograph is a fractal structure, taken from different sciences that study different entities of nature since each part of the Universe has a fractal structure, and so does the Universe at large, made of all those fractal parts.

In March 2007, after Pietronero, an Italian astrophysicist, showed the fractal nature of the Universe, tabulating an enormous number of galaxies, whose distribution was fractal, the magazine *New Scientist*[1] published an article on the theme, confronting Mr. Pietronero and Mr. Hogg, a quantum physicist from the previous outdated probabilistic, mechanist, quantum paradigm:

"The universe is not a fractal," Hogg insists, "and if it were a fractal, it would create many more problems than we currently have." A universe patterned by fractals would throw all of cosmology out the window. The big bang would be tossed first, and the expansion of the universe following closely behind.

Hoggs's team feels that until there's a theory to explain why the galaxy clustering is fractal, there's no point in taking it seriously. "My view is that there's no reason to even contemplate a fractal structure for the universe until there is a physical fractal model," says Hogg. "Until there's an inhomogeneous fractal model to test, it's like tilting at windmills."

Pietronero is equally insistent. "This is fact," he says. "It's not a theory." He says he is interested only in what he sees in the data and argues that the galaxies are fractal regardless of whether someone can explain why.

As it turns out, there is a model that may be able to explain a fractal universe. The work of a little-known French astrophysicist named Laurent Nottale, the theory is called scale relativity. According to Nottale, the distribution of matter in the universe is fractal because space-time itself is fractal. It is a theory on the fringe, but if the universe does turn out to be fractal, more people might sit up and take notice.

A resolution to the fractal debate will only come with more data. Sloan is currently charting more galaxies and will release a new map in the middle of 2008.

Of course, the Sloan 2008 showed the Universe is a fractal.[2]

This should have been big news all over the world. But that means, as Hogg recognizes, that we have to throw to the trash the big bang and the expansion of the Universe, accept the fractal informative structure of reality, and close down some big projects of science, such as the large hadron collider, which will merely cause a planetary big bang, very different from the birth of the entire Universe.

Indeed, in the fractal paradigm, the Universe is infinite and the big bang is NOT the birth of all realities, but any local big bang and big crunch dual process, any explosion that splits the physical energy and information of a complex system, in any of the multiple scales of physical reality; in the case essayed at CERN, the informative inner quarks of Earth's atoms, crunched into a quark star, and the external electroweak cover, exploded into a big bang.

Let us then prove the fractal paradigm, the true evolution of cosmology, with the three legs of the scientific method: experimental proofs, logic consistency, and mathematical accuracy.

2. Logic Consistency: The Infinite, Hierarchical Universe

A good description of the fractal universe and its logic principles is this text from the Web by Mr. Oldershaw from Amherst University:[3]

> With regard to cosmology, we live at a very privileged time. When we read about exciting and revolutionary paradigm changes, in which our most fundamental ideas about nature undergo radical revisions, the drama usually has taken place well before we were born. Today, however, we have the rare opportunity of witnessing at first hand a profound transformation in our understanding of how the Universe is structured. This ongoing change of cosmological paradigms from a "small" finite cosmos to an infinite fractal cosmos began about two decades ago and is in full swing at present. The short version of what is happening goes like this. For about 50 years the Big Bang model of the Universe has provided an excellent explanation for the basic cosmological observations: a very large-scale expansion, an approximately uniform background of microwave radiation and a unique set of abundances for the atomic elements. However, there were some technical problems with this model, such as an a causal beginning of space-time, a lack of magnetic monopoles, an unexpectedly high degree of uniformity, and an enigmatic knife-edge balance between the open and closed states.
>
> Then in the early 1980s Alan Guth[4] showed how these and other problems with the Big Bang model could be solved in one fell swoop with the Inflationary Scenario. But, an ironic thing has happened. Although the Inflationary Scenario was developed to rescue the Big Bang model, the most logical consequence of pursuing the concept of Inflation is the replacement the Big Bang paradigm with a much grander and more encompassing paradigm.

According to Guth and a growing number of leading cosmologists, the most natural version of Inflation theory is Eternal Inflation in which Inflation is, was and always will be occurring on an infinite number of size scales. The new paradigm that cosmologists have arrived at by several routes is an infinite fractal hierarchy that has "universes" within "universes" without end. The astronomer Carl Sagan once referred to the general idea of an infinite fractal universe as "strange, haunting, evocative—one of the most exquisite conjectures in science or religion."

That is the basic story, but because of the profound changes the fractal paradigm will have on our understanding of the Universe and the place of humans within that Universe, it is important to explore the implications of this new vision. Firstly, there is no edge or boundary of the Universe; space is infinite in all directions. What we used to refer to as "The Universe" can be more appropriately called the "observable universe" (note the small "u") or the "Hubble Volume," and it is only a tiny part of what one might call our "metagalaxy" or "level 1 universe". We currently have no way of determining the size of our metagalaxy or the number of galaxies it contains, but we could reasonably assume that both figures would be vastly beyond anything previously contemplated.

There would be an infinite number of these level 1 universes, and on an unimaginably large scale they too would be organized into level 2 universes, and so on without limit.

Secondly, time is also infinite in the unbounded Organic universe. Whereas our Hubble Volume may have come into being and began to expand approximately 13.7 billion years ago, the Universe has always existed and always will. Parts of the Universe may be created or annihilated, may undergo expansion or contraction, but the infinite fractal hierarchy remains unchanged overall, and thus is without any temporal limits.

Thirdly, there is no limit to size scales. In the infinite fractal paradigm there is no class of largest objects that would cap off the cosmological hierarchy; the hierarchy is infinite in scale.

This fact removes one of the more suspect aspects of the old paradigm. Natural philosophers had long noted the unusual fact that within the Big Bang paradigm humans found themselves roughly in the middle of nature's hierarchy of size scales. This anthropocentric state of affairs seemed to violate the Copernican concept that when humans appear to be at the center of the cosmos, we should suspect that a bias is leading us astray. In an infinite fractal hierarchy there is no center of the Universe, nor any preferred reference frame.

Some interesting questions immediately arise. Why are fractal hierarchies so ubiquitous in nature? By studying empirical phenomena within the observable universe, how much will we be able to learn scientifically about the parts of the Universe that lie beyond our observational limits? Does the infinite cosmological hierarchy have a bottom-most scale of sub-atomic particles as is currently thought, or is this another artificial limit to an infinite Organic universe that actually extends without limits to ever smaller scales?

Centuries ago Immanuel Kant, J.H. Lambert and a few others proposed an infinite hierarchical model of nature based largely on natural philosophy arguments. This general hierarchical paradigm never garnered a large following, but it was kept alive by numerous rediscoveries. In the 1800s and 1900s quite a few scientists, including E.E. Fournier d'Albe, F. Selety, C.V.L. Charlier and G. de Vaucouleurs, argued for hierarchical cosmological models based on the hierarchical organization within the observable universe. Then toward the end of the 1970s, the mathematician B. B. Mandelbrot[5] gave the hierarchical paradigm new life and widespread exposure by developing the mathematics of fractal geometry and demonstrating that fractal phenomena based on hierarchical self-similarity are ubiquitous in nature. In this way natural philosophers, empirical scientists, mathematicians and theoretical physicists have all found their way, slowly but surely, to the infinite fractal paradigm. There are many routes to this paradigm, and certainly there are a large number of distinct versions of the basic paradigm that have their own unique theoretical explanations for why nature is organized in this manner, but the general paradigm that nature is an infinite hierarchy of worlds within worlds has fully arrived, and will probably be our dominant cosmological paradigm for the foreseeable future.

But what does it mean that the Universe is organized in fractal patterns of form? To fully grasp this concept, we have to realize that "fractal equations" are "organic equations," which participate of the properties of organisms. Fractal equations create forms as life does. Fractal equations create self-similar "cells," as life does. And in fact, they use the same term. Fractal equations are also called generator equations, as they are the mother cells that encode the information needed to repeat a living form. Finally a fractal equation will create a pattern of interconnected parts, which put together reproduce a whole, self-similar to their parts, as organisms do.

The only thing a fractal equation does not have is motion, as it is a simplified geometrical description of reality. So in the same manner, geometry lacks motion; simplified fractal equations didn't have motion when they were first discovered.

But now, fractal science has advanced a step further, thanks to the evolution of fractal logic, duality, and non-Euclidean geometries, so we can understand finally the meaning of motion in the Universe.

Let us then explain the logic and mathematical principles of the fractal paradigm, starting for what fractal relativists mean when they say that the Universe is built in fractal scales made of pieces of vital space and informative time. To that aim, we shall introduce some basic concepts of fractal and non-Euclidean mathematics, needed to fully grasp the meanings of the fractal Universe, which we later will apply to the understanding of masses of physical information and black holes and pulsars, which are fractals, made of quarks—the substance CERN will mass-produce, without understanding its consequences, as it ignores all about the new developments of fractal cosmology.

II. Fractal Space: Non-Euclidean Geometry

Eppur si muove, eppur no muove.
—paradox of Galileo

The fractal nature of all universal entities, which occupy a piece of space and last a quantity of time, is the key to unifying the laws of science. Saturn's rings are not a mathematical continuous plane despite their appearance. In detail, they become quantic points, planetoids in movement, tracing orbital cycles around the planet, in a fractal repetition of the path the planet makes around the sun.
Each planetoid becomes again a fractal sum of atoms, turning in cyclical paths. Those rings illustrate two fundamental dualities of the space-time Universe, contradicted by the "naïve realism" of our simplest perception:
Any region of the Universe seems at first glance still and continuous.
So Galileo called Saturn's planetoids a ring, thinking they were made of continuous, still matter. Yet Saturn's rings have in fact the form of the commonest informative fractal: a Cantor dust, which thins once and again into infinitesimal particles. Since when we see any continuous point of time/space in detail, it grows, becoming both, dynamic and discontinuous, made of quanta, moving in self-similar paths that are separated as independent beings, but pegged to each other in networks by flows of energy and information. The Universe is constructed through those networks of parts that become wholes and grow in bigger scales, departing from its two essential formal motions, the line of energy and the cycle of information.

This duality gives origin to many paradoxes, as the properties of energy and information are inverted, and yet reality is always a combination of both. So particles of information are also waves of energy. Forms that seem static information can also be perceived as moving energy. It is the paradox of Galileo who said of the Earth: "Eppur no move, eppur si muove."

3. Mathematical Proof: The Fifth Postulate of Non-Euclidean Fractal Geometry

The paradox of Galileo is the main paradox of perception of any system, which can be seen either as energy or as form also in the quantum realm, but physicists' naïve realism failed to understand: Information is static since to perceive a form, we need to focus on its edges, and movement blurs form. So the mind perceives information in stillness, as when we read. Yet energy is motion, and since all dual complementary systems have information and energy (physical particles and forces in quantum physics), we can perceive or measure all beings, including the Earth as entities/waves in motion or as static forms/particles. But given the limits of our visual and instrumental systems, we can't perceive both things at the same time. Therefore, the uncertainty of measure and perception is human while nature is dual.

This duality is the property of nature that neither quantum physicists, nor Galileo didn't understand, opting out for a single solution. Galileo said that the Earth moves, hence energy and motion are the principles of reality that exclude form, in/form/ation, as all physicists will believe after him. While quantum physicists thought uncertainty was in nature, it was not in the instruments that measure it. Further on, energy is continuous, hence differentiable, but information is discontinuous, hence it requires fractal or topological mathematics. Yet quantum entropy still ignores that duality and pretends to study nonlineal, fractal, and nondifferentiable systems such as black holes, as Hawking does, with entropy equations. The list of errors of physics in their monist, naïve realist perception of reality could fill an entire encyclopaedia. Yet we want to focus only in those advances brought about by Einstein et al., which matter to understand fractal mass, the big bang, and black holes, which are made of quarks as Saturn's rings are made of planetoids. Even if we first described them as energetic, abstract continuous rings, as Hawking did with his abstract black holes that now in the fractal paradigm become nonevaporating frozen quark stars, what is the mathematical justification of that fractal structure? A new type of mathematics called non-Euclidean mathematics, which Einstein used to describe gravitation and masses.

Einstein found that gravitational space did not follow the fifth Euclidean postulate of mathematics, which defines each point as having not breadth, so only a parallel can cross it.

It turned out that relativistic points could be crossed by infinite parallels. So the Universe was "non-Euclidean." Either points had fractal parts through which parallels could cross or parallels bent to penetrate the point. Both solutions were a question of how we perceive reality. If we enlarge each point of the Universe, then it can fit more straight lines. This is the fractal paradigm. If you don't enlarge a small point, lines seem to curve toward it. This is the choice of Einstein,[6] which requires only a single space-time continuum as it does, not enlarge the smaller points of the Universe. In any case, what seems obvious is that reality is not as simple as we perceive it.

The fifth postulate in classic relativity means that to pass through that point, parallels curve; so in this manner, they can flux into the point. But fractal mathematics has resolved in a more accurate way this conundrum for two reasons: parallels are straight lines; hence if we curve them, they cease to be parallels. The second improvement is also obvious. A Euclidean point has no breadth so even if we curve parallels, there is room only for one within the Euclidean point. But if we enlarge the point and it grows to have fractal parts, it fits lineal parallels, which are straight lines. Thus, what the fractal paradigm has done is to explain logically non-Euclidean mathematics,[App. IV] giving us a more accurate view of the Universe, harmonizing the mathematics of Einstein's relativity with what we see in the real world.

So we propose a new geometrical unit; the fractal point with a content of energy and information that grows as we come closer to it, which Einstein already used to describe the formalism of gravitational space-time. Let us now consider that proposition in more detail to respond to Mr. Hoggs's wonders.

4. The Dual Membranes of the Universe: Light Space-time and Gravitational Space-time

Euclid affirmed that "through a point external to a parallel, only another parallel line can be traced" since the Euclidean point doesn't have a volume that can be crossed by more lines:

Abstract, continuous, one-dimensional point: . _____

This is indeed what we perceive in our light space. And the reason, which later will be explained in more detail, is because our "space" is made of light, and light has a Cartesian perpendicular structure, with three perpendicular fields (height/electric field, width/magnetic field and length, speed).

This is what fractal theorists mean when they affirm that space-time is fractal. We float and exist in a "sea" of light space-time, made of H-Planck units of light, which are "actions of energy and information": $E \times V = K$. Thus our light space-time is of energy and information, and we are evolved beings of light. But the Universe is fractal, is made of scales. And so there is another world of dark gravitational energy and matter that we do not perceive, the world of quarks, masses, and gravitational, invisible waves, described by Einstein's relativity.

Indeed, Einstein found that such gravitational Universe followed a non-Euclidean fifth postulate:

"A point external to a line is crossed by infinite parallel forces".
Real, discontinuous, n-dimensional points: o ============

It means that a real point has an inner space-time volume through which many parallels cross since reality follows that non-Euclidean fifth postulate,[7] all points have a volume when we enlarge them, as cells grow when we look at them with a microscope. Then it is easy to fit many parallels in any of those points. Such organic points are like the stars in the sky. If you look at them with the naked eye, they are points without breadth, but when you come closer to them, they grow. Then as they grow, they can have infinite parallels within them since they become spheres, points with breadth, with space-time parts.

Mathematics, as a language that represents reality with simplified symbols, has a limited capacity to carry information. So its main symbols, geometric points and numbers, simplify and integrate that fractal discontinuous reality into a single space-time continuum, the Cartesian space/time graph, made of points without breadth. However, the points of a Cartesian plane or the numbers of an equation are only a linguistic representation of a complex Universe made of discontinuous points with an internal content of space-time. In the real world, we are all pieces made of fractal cellular points that occupy spaces, move, and last a certain time. When we translate those space-time systems into Euclidean, abstract, mathematical numbers, we make them mere points of geometry void of all content. But when we look in detail at the real beings of the Universe, all points have inner volume, measured with the dimensions of fractal geometry.

Thus, there is a bigger Universe which we do not fully perceive, made of gravitational dark energies and point-particles called quarks, as we are "limited, perceptive beings" that only see light. We are part of a network of electronic flows of energy and form of light nature, and we float in a bigger dark Universe. We are the surface of that other world; we are the membrane that covers the inner "pipes" of quark points and gravitational flows. But if we could penetrate inside that world and enlarge it, we would see a much bigger world of gravitational flows of energy and form that are dark energy and dark matter to us. And the formalism of that world is the topology of non-Euclidean fractal mathematics.

Yet if we give breadth to a point, we must also correct the other four postulates of Euclidean geometry and define a point as a sphere with breadth that grows as we come closer to the point (fractal point) and a line of points with volume as a wave with a minimal punctual height—a string of cyclical points.

And so suddenly two mysterious fields of physics become clear:

— Strings, used to describe the strong forces of the inner quarks of dark matter, become complex lines of non-Euclidean points with inner fractal dimensions in at least two more scales for a total of $3^2 = 9$ more dimensions than the ones we see in our world.

— The complementarity principle of quantum physics means that all particles are non-Euclidean waves as points have height and volume. *The space-time Universe thus becomes a discontinuous world of fractal points that grow in size as we come closer to the next scale, made of networks of points sharing waves of energy and information.* And so we can define any discontinuous, relative, fractal space as a "plane of fractal points communicated by waves of energy and information" (fourth postulate), which becomes the organic unit of reality.

In the same way, the Saturn rings stop being planes without volume when we come closer and observe them as fractal points called planetoids. Non-Euclidean points acquire volume when we approach them. And so planes acquire volume, structure, and movement.

In the words of Klein, one of the discoverers of non-Euclidean geometries, a space is not static, but a moving group of points in cyclical movement. So in the same way Saturn's rings are a group of planetoids, a Klein space, the space-time that fills reality is made of points, which have movement. They are in fact a series of moving cyclical vortices, the points of the smaller fractal scale of gravitational reality, Einstein's definition of masses, as cyclical whirls of space-time.

Thus, suddenly mathematics stop being an "abstract" construct, a continuous plane drawn by Descartes and imposed into nature by Newton, and becomes a series of "vital geometries," entities that have a quantity of vital space, made of networks of non-Euclidean points, particles, cells, or stars of a galaxy. And all those different scales of reality, made of networks of non-Euclidean points, can be explained, both as mathematical structures and organic structures, since an organism is merely a ternary system with a network of energetic and informative points/cells that create the limbs and head of the organism and its intermediate reproductive wavelike body.

This reinterpretation of the non-Euclidean topology of the Universe as a fractal series of scales of space-times, all of which are in fact cyclical vortices of masses and charges and lineal flows of energetic forces in movement, is far more akin to our observation of the Universe that reveals the same forms once and again in all its scales and shows perpetual movement in all of them, till reaching vacuum space, also with constant movement.

It also explains string theory and its multiple dimensions: those dimensions are fractal inner dimensions, belonging to a smaller fractal scale, as all systems observed are made of smaller units.

Non-Euclidean fractal mathematics has an enormous capacity to illuminate key questions of all sciences. We shall consider in the next chapter three physical questions resolved by the fractal paradigm that CERN pretends to study bombing the Earth with quark-gluon soups:

— The meaning of masses and charges *that can be perceived, according to the paradox of Galileo and the fractal non-Euclidean paradigm, either as fixed particles with form or cyclical vortices with a rotational frequency that carries the information of the Universe.*
— The unification of masses and charges as two accelerated geometrical vortices of two scales of reality, the microcosms of quantum charges and the macrocosms of masses.
— *The* holographic principle that defines information as bidimensional—a cyclical physical vortex.

While in the appendix, we will advance a reduced non-Euclidean model of the physical Universe as an organic structure of two scales—quantum world of charges and the cosmological world of masses.

Let us then continue in our double task of discovering the true knowledge of the Universe and expose the bioethical crimes and intellectual platitudes of CERN by resolving once and for all the meaning of mass and the unification equation of masses and charges, the basic bits of information of the cosmological and quantum world, with the minimal mathematical apparatus and maximal logic rigor.

III. Equivalence between Mass and Acceleration

And God said, there is light, and he separated the day from the darkness.
—Genesis

  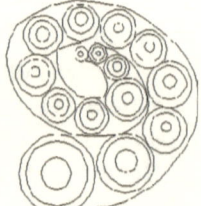

In the graph, the evolution of the concept of mass in relativity, from the initial image of an abstract substance in the center of an accelerated vortex of space-time, proper of the abstract, pre–world war age, when Einstein first published his work on gravitation, to the first pictures obtained in bubble chambers in the postwar age, to the realization that each mass is a fractal space-time made of smaller cyclic motions proper of twenty-first century. Fractal relativity. The smaller motions are quarks, the bigger motions are quark stars, pulsars, and black holes. Today we can model any type of vortex of space-time with fractal mathematics, as each small vortex transfers the momentum to the bigger vortex, which emerges as a macroscopic entity. Thus, we conclude that a black hole is a fractal concentration of the densest matter of the Universe, quarks, since quarks represent on average more than 99.9% of the matter of our bodies. Those improvements on the details of mass theory, based on the advancement of its mathematical tools (discovery of fractals, chaos theory, and the five postulates of non-Euclidean geometry) is the natural way in which the evolution of science happens, departing from sound logic principles.

5. Einstein's Concept of Mass as a Frequency of Information

In the text quoted from the *New Scientist* that opened chapter 2, we came to the conclusion that the Universe has a fractal structure because it is made of fractal space and fractal time. But what is space and time? In simple terms, here are the two essential morphologies of the Universe: Space is the lineal, expansive energy of the vacuum in eternal motion (seeing as still distance or space, according to the paradox of Galileo). And fractal times are cyclical clocks of information, whose frequency or rhythm encodes the information of the Universe.

So space and lineal energy on one side and time and cyclical clocks of information on the other are synonymous. The static, continuous concepts of space and time become then sums of discontinuous, moving pieces of energy and infinite clocks of information that our senses put together into an abstract continuous single space-time. One we have "exploded" and given movement to the abstraction of space and time, it is easy to recognize the minimal bytes of energy and bits of information of the real, physical Universe. Physical information, masses and charges, and physical energy, forces, which are not substances but motions, events in time of cyclical, informative nature (masses and charges) or energetic, spatial, expanding shape (energy), that constantly transform into each other ad eternal, creating the essential rhythm of the physical Universe:

E (forces) ⇔ i—Particles (masses and charges)

The key to grasp the concept of mass as a vortex of acceleration are two physical laws: one of classic relativity, the principle of equivalence between accelerated motions and masses, and one of fractal relativity, the paradox of Galileo, the fact that we perceive reality either as a moving energy (in the case of a mass, as a vortex) or as fixed space (in the case of mass as a particle). All moves and doesn't move (eppur si muove eppur no muove) depending on the way we look at it. Thus, all spaces have inner motion, even if our senses don't see that motion or fix that motion into forms of in/form/ation.

Mass is the great problem unresolved by quantum physicists in their model of the Universe since it is not an energetic but an informative form. Einstein, who had a visual, geometrical image of mass, defined it as a vortex that absorbed space-time, attracting in its flow everything else traveling on that space-time as a bathtub attracts all what is in the water it absorbs.

How did Einstein come to that idea?[8] As many textbooks and visualizations explain, he realized that when we accelerate, as it happens in a Formula 1 car or a lift, we feel more weight. Thus, he made a bold statement: mass and acceleration are the same. If we feel mass when we accelerate, it is because mass is acceleration. But what type of acceleration?

The answer is actually quite simple, because it turns out that Newton had long ago discovered a beautiful formula to describe all the forces of the Universe: $F = m \times a$.

Where *a* means a lineal acceleration and so *m* must be a cyclical acceleration. Indeed, there are no more types of accelerations and so Einstein (using, of course, more complex mathematics than Newton to refine his calculus, as mathematics had evolved a lot since Newton's time) postulated that masses would be vortices of space-times *since only a vortex is a cyclical, accelerated movement.*

That principle of equivalence is the fundamental principle in which general relativity is based. It states merely that mass and cyclical acceleration are the same. It means in layman terms that a mass is not a material form but a movement. And of course, because that movement is an inward, cyclical movement, it deploys a resistance to be pushed outward in a lineal manner, which from our macroscopic view seems to give mass a solid position. Again this is intuitive. If you are a very small being turning at full speed, it will be very difficult to push you away. When you see a skater in a ring, he speeds up and closes his arms to move faster and become more stable in his position. Thus, the homology between acceleration and mass defines for a curved space the meaning of a mass, and the paradoxical fact that the smaller the vortex is, the faster it turns and the bigger the mass is. It is the same that happens in any tornado or hurricane: the closest you are to the center, the faster you turn and the more attractive the tornado is.

In that sense, what you experience as weight is the inertial resistance to any displacement of a vortex anchored in a point of space-time. The same happens in our macrouniverse when a skater turns around and bends his arms to spin faster. If he opens his arms spreading his vortex, he decelerates and moves easier in a lineal direction.

A mass is thus the inertial force that we need to push away, the accelerated vortex. Those vortexlike spirals accelerating in a curved path, inward, are like the pictures we observe in accelerators, or the most recent pictures of masses we have obtained, in which indeed we see a fluctuating vacuum with the form of a vortex. Further on, we know that the frequency or rotational speed of those vortices, called spins, are related to the mass of each particle. So we have both, logic and experimental proofs. Of course, I have used here and will continue using the simple mathematics of Newton and the visual thought experiments of Einstein, to describe those concepts because this is a book for nonspecialists. Already Descartes said that the Universe was made of vortices and "res extensa," that is, lineal space. What Einstein did was to apply the most modern non-Euclidean mathematics of his day to those well-known principles. So he improved greatly the detail of those vortices, as non-Euclidean geometry allowed much more refined calculations than those we could do with the mathematics of Newton, but in essence the principles stayed. So we can explain very easily in an intuitive manner, thanks to the principle of equivalence, a mass with a simple equation:

equivalence principle: mass = curved acceleration= cyclical vortex that resists displacement

If we use the simpler mathematics of Newton to define Einstein's mass, we can see in more detail the properties of such vortices, which can be described as *fluid vortices of the bits of information, the particles that create any of the multiple fractal space-time membranes of the Universe. Consider the simplest bidimensional equation of a vortex:*

Vortex Speed × Radius-size of vortex = Vo × Ro = Constant Value

Since the product of the speed and the radius is constant, the shorter the radius, the faster the speed. What this means is that any vortex paradoxically accelerates faster when it is smaller. So the product of a growing speed and diminishing radius remains constant. Hence we find that smaller particles that rotate faster have more mass. So the electron which is bigger and rotates slower than the quark has less mass. In fact, the electron's mass is 0.1% of the mass of an atom which is all stored in the inner faster-rotating smaller quarks and gluons.

While in the cosmological scale, the black hole, which rotates at light speed and occupies the minimal space of all the cosmic bodies, is the most massive form of the Universe. Then it becomes a quark star called a pulsar, which has a center of strange quarks and a surface of dense neutrons. It rotates also very fast, near the speed of light, and it has an enormous mass. We do not know what the substance of black holes is, but Einstein called them frozen stars and he said that if we were to obey the laws of the scientific method, they had to have a cut-off substance. Then when he died, we found quarks and so it is obvious that black holes, the densest rotating stars of the Universe, should be made of top quarks, the densest top quarks of the quantum world. Since black holes have very similar properties to those of a quark star or pulsar, ultradense stars with enormous rotational speed, this fact reinforces the thesis that black holes are top stars. And so obviously, fractal cosmologists fear that CERN, which is a factory of quarks, will make a vortex of quarks that will grow and grow, attracting the mass of the Earth and converting us into a black hole or quark star. Simple, isn't? *It is then obvious that we must NOT make a factory/cannon of quarks. It is absolutely dangerous to do that.*

But this fact, which anyone can understand, is ignored by quantum cosmologists because they don't believe in Einstein; and of course, those who believe in Einstein are merely interested in the use of the more complex mathematical model of gravitation and mass, in which the principle of equivalence of Newton and Einstein is not so clear. The details of the trees prevent them from seeing the big picture of the forest. But here we are concerned with the picture that tells us that quarks are the atoms of the densest black holes and quark stars because only among the known particles, quarks are that heavy, and only among the known celestial bodies, black holes and quark stars are that heavy.

Further on, because dark matter represents 75% of the universe's mass, very likely, dark matter is made of black holes, pulsars, and other types of quark stars.

Now when you get into details, the word *vortex* is substituted by *spin*, the word *curved cycle* by non-Euclidean geometry, and you use also two different names for the type of forces displayed by quark vortices (strong forces) and by black holes and pulsars (gravitational forces) because we are talking of two fractal scales of reality, the quantum microscopic world and the cosmological world. But those details that we will explain later in depth, unifying both scales, gravitation, and the quantum world with fractal equations (a long goal of Mr. Einstein, which I found thanks to the expansion of the postulates of non-Euclidean geometry that Mr. Einstein lacked) are of no importance to understand in simple mathematics, the meaning of mass:

max. rotational speed × min. radius= quarks and black holes= the densest masses of the cosmos med. rotational speed × med. radius = electrons and stars
0 rotational speed × max. radius = light = 0 mass

It is in those terms how it becomes easy to understand the different masses of different particles and cosmic bodies, related to their angular momentum or spin: the electron which turns slower in a wider circumference is far lighter than the proton that turns faster in a smaller space, which is lighter than the black hole which turns to the limit of speed of our world.

While the mass of light is zero because the electromagnetic wave never closes in to itself, but once it has made half a circumference downward, it switches upward and so it moves as a transversal, lineal wave without mass and maximal, infinite radius: *0 mass × infinite radius = K*.

In detail, light has an insignificant bit of mass because due to the rotation of galaxies (Mach principle), it has some curvature and it acquires more mass/frequency when it rotates around a quark or black hole vortex, finally evolving into a more complex form, absorbed by an electron, quark, or black hole.

Now it is easy to understand how mass (rotational motion) becomes energy (lineal movement) and vice versa ($E=mc^2$ and $M=E/c^2$). $E=mc^2$ means just that when energy approaches the c-limit of speed of our spacetime, as it cannot go faster, it deflects its energy that curls into cyclical motion into mass. This, of course, is another thing CERN pretends to research with the quark cannon. Because using quantum equations and Higgs particles, there is no way to understand how energy becomes mass and vice versa.

The Universe is simple. So a simpler explanation of reality is more truth, even if it is experimentally less refined in its initial conception. For example, Copernicus was simpler than Ptolemy. Further on, for many years, till Kepler and Newton refined its elliptical orbits into ellipses, Copernicus's cyclical orbits were less accurate than Ptolemaic calculus, but the principles were sounder: The sun was put in the center and that simplified enormously the model. So it satisfied better Occam's principle. That is what the principle of equivalence between acceleration and gravity of Newton and Einstein does with masses: by considering that particles are cyclical, accelerated motions, not substances, this sounder principle simplifies all calculus, explains logically how mass attracts like a hurricane of space-time, and so it must be truer than the complicated quantum theories *of mass, as the Higgs Mechanism, which does not even explain how mass attracts in a coherent manner.*

6. The Advances of Fractal Relativity: Fractal, Fluid Space-times

Physicists do not understand those simple truths—even a century after Einstein applied Mr. Riemann's discovery of the fifth non-Euclidean postulate, which defines a point as a cyclical motion in space—because of a basic error of "materialist physics": Since nuclear physicists do machines and bombs, which are made of dense metal, they tend to believe in "solid things." So this kind of spiritual thought on motions and events is unreal to them. Instead, they "decided" that the principle of equivalence only affected to the gravitational force but somehow "stopped" in an obscure point where the rotational acceleration of the mass vortex clashed with something "solid," like a "rubber ball," which had to be the "mass."

In the graph, our growing understanding of masses in terms of a fluid, relativistic vortex is an interesting proof of the distance between mathematics and reality since mathematics merely offer an abstract, simplified picture of reality with equations we have to interpret in terms of real objects observed in the Universe. That is the case of a mass, which once Einstein's equations were published was first depicted as in the left picture, like a ball that was *curving downward* the space-time of the Universe. But that image is not logic and it has no relationship with the Universe we observe since any vortex, including those of air and water, have no solid ball in the center, but merely deflect the cyclical motion upward, transforming it again into lineal, perpendicular motion.

And this is what we observe in masses and charges, which are cyclical vortices, which first transform lineal motion into cyclical motion, creating the charge or mass and then deflect that cyclical motion, expelling it as less curved lineal motion through its poles, as a magnetic, electric, or dark energy flow.

So physicists moved recently toward the second more logical and realistic version of the vortex of mass, *curving inward*, illustrated by a fluid or gas vortex in pictures 2 (a fluid vortex) and 3 (a hurricane). This third picture of the hurricane applied by analogy to a mass vortex shows clearly what probably will happen when CERN puts together a lot of quark vortices: they will condensate, rotating together in synchronicity, creating a bigger vortex, a pulsar, quark star, or black hole. Each of the mini vortices that accelerates inward toward the center of the black hole will be a particle, a quark, which acquires mass as it rotates faster, diminishing in radius (according to the previous vortex equation). In the more complex mathematics of relativity, those equations that explain how energy becomes mass as we approach c speed ($E=mc^2$) become the so-called Lorenz transformations.

As you can see, the vortex comes inward toward the condensed center *in which it will fuse with other quarks into a single diminishing vortex, called an Einstein-condensate, that packs very close one quark over another, turning at light speeds in a state of highly ordered matter.*

In the graph, we observe also the evolution of true science, thanks to our advances in mathematics and logic. First, in the beginning of the century, when Einstein published, *only gravitation, the force external to the mass was treated under the Principle of equivalence, between force and acceleration*. The mass was still considered a solid substance, a Maya of the senses, which fix the quantum fluctuations of energy and information into a still image of the Universe, a knot of information called the mind. But the fixed images of the mind are not reality. Reality, as we entered into deeper experimental evidence, turned out to be always vibrating, moving, fluctuating at all scales we have perceived it. In the center, we see a bubble image of a particle.

There are NOT solid substances but a constant flow of motions that happen in different scales of fluid space-times, each one made of a sea of minimal "actions" with energy and information that act as the medium of that world—the gluons of the strong space-time of quarks, the h-Plancks of the weak world of electromagnetism, the λ-strings of the dark energy of gravitation (as per Nottale and others). And in the mediums known to man, the air molecules of nitrogen and oxygen, or the water molecules, which allow also the creation of hurricane and tornado vortices. In the new physics of complexity, we are all part of fractal discontinuums, mediums and scales of vibrating energy and form, and all the abstract simplifications of the "ancients," while useful to allow calculus, become from a philosophical perspective hyperbolic or Pythagoric, monist, mechanist aberrations of thought.

A notable property of this new world are a series of logic dualities and self-similarities called correspondences and invariances that relate those scales. Each new scale becomes created by the fractal sum of all the microvortices that become a whole one in the next scale of size of the physical Universe, adding new dimensions. For example, a well-known correspondence is established between a mathematical description of a quark-gluon soup as a fifth-dimensional world of strings and a fourth-dimensional world of quark vortices. What this means is that λ-strings are the one-dimensional units of the superfluid dark energy that form those bidimensional quarks, which themselves lock in three perpendicular vortices to create fractals our three-dimensional electronic word. And so dimensions are always fractal, emerging as objects of a next scale, but the sum of all the fractal, finite dimensions from the small to the bigger world always adds up to four infinite continuous dimensions.

Yet all those fractal parts of the different scales of superfluid space-time are NOT solid substances but actions of energy and informative time. This again was discovered by Planck more than one hundred years ago, when he found the fractal unit of the membrane of light space-time we inhabited and called it action of energy and time. Yet the solid image never died away. What fractal theorists have done is to add to the "membrane of electroweak space-time made of light forces and electrons" another membrane of "gravitational masses" made of strings, gluons, quarks, and black holes. So in the same manner, we have two basic physical mediums that create vortices in our planetary scale, hurricanes made of quanta/fractal parts of atomic air and tornados, made of quanta/fractal parts of atomic water. In the Universe at large, we have two mediums, the light space-time made of "light quanta" (our membrane) and the gravitational world made of strings, fractal parts of gluon quanta (which form the superfluid vortices of quarks) and gravitational dark energy that form the cosmological space-time.

This duality, clearly established in the work of Nottale and others, reshapes our classification of forces in two systems, one of electroweak light entropy and other of strong gravitational mass. It is important to define properly the terms of that duality, so ill understood by the "ancients," the quantum entropy theorists that work for CERN, rooted in the physics of the past millennium:

Electroweak, massless forces made of electromagnetic light and electrons belong to our membrane of Euclidean space-time, defined in classic physics. This is our world and it is defined by expansive entropy, spatial, lineal forces.

Yet this world is in a tug-of-war with the strong, cyclical, massive dark world of quarks that bends this space into cyclical clocks of bidimensional, temporal information, the quark vortices inside the atom. And this strong world of quarks that bends electrons has also a lineal force, the dark gravitational energy flows that structure the interstellar space.

The easiest way to perceive those two universes is to consider the primary Universe NOT our Universe, but the world of quarks connected by long strings of dark gravitational energy, which form a universal web over which *the electromagnetic world becomes a rubbery cover, as a series of copper cables are coated with rubber or our neurons with myelin. The quarks of strong forces are covered by electronic surfaces, which communicate between them through flows of electromagnetic light that runs over the thinner, faster wires of dark gravitational energy, which is the intimate scaffolding of reality.*

In the cosmological world, quark quanta form the hyperdense black holes and quark stars that CERN will create and form the nuclei of neutron stars and perhaps even of many stars of which we see only the external electronic, plasma. Indeed, in the same manner, we have found certain planets to have cold crystal cores; it might be possible that some stars have in the center a superfluid core of quarks, with organic membranes/ discontinuities that separate them from the thermonuclear world of hydrogen and helium that forms the core of lighter sunlike stars. It is a simple, beautiful vision of a Universe in which all dual systems are also made of fractals of smaller systems.

So we always have to account for two complex types of non-Euclidean geometries, which the ancients dismissed with their continuous Euclidean analysis of Cartesian space: the duality between energy, explosive and informative, implosive systems that balance each other and the scalar structure in which quanta with fractal dimensions emerge into a higher scale.

A human is also made of organic molecules, fractal parts of a cell, fractal parts of a human organism, fractal parts of a society. And each of those entities can further be divided in an informative nuclei (the DNA in the cellular scale, the brain in the organic scale, the government and informative castes in societies) and a reproductive body (the cytoplasm, the body, the reproductive, working class).

All this was finally understood in the simplest scales of physics, thanks to the work of fractal and complexity theorists, which in the past two decades expanded our mathematical and theoretical view of the principle of equivalence to encompass all the fractal structures of space/time and the work of multiple experimentalists who validated it with observations. They are the founders of the fractal paradigm of science, which could be to twenty-first-century science what the quantum paradigm was to twentieth-century science if we survive CERN. I recommend the work of Mr. Nottale, an astronomer, who established the minimal unit of gravitational space-time and related the two membranes through fractal equations and my work on non-Euclidean mathematics, the dual logic of time built with energy and information and its applications to any science, exposed in a series of volumes published in Europe during the past decades.

In what concerns the themes of this book, it means that we must rely on the work of Einstein and Riemann on *a fractal, topological analysis, not on a probabilistic description to understand* mass, black holes, and big bangs the entities and events CERN pretends to test, making quark-gluon soups.

Because of the mathematical complexity of fractal relativity, we shall consider here a simplified approach to those themes, using Newton's description of a fluid vortex, which is 90-percent-plus accurate, and focus on the conceptual principles. But there is available literature in the work of those and other fractal pioneers to fully grasp the meaning of mass, gravitation, black holes, quark condensates and the consequences for mankind of making those hurricanes of space-time on Earth.

The mathematical modeling of cyclones as fractals has been verified by practical observation. A large tornado harbors multiple suction vortices inside the column of dust, and subsidiary whirls continually form and dissipate around the bottom edges of a tornado.

Let us for a moment imagine those black holes CERN will do in terms of a hurricane made of smaller fractal eddies or hurricanes. This is the simplest image of what a black hole is: a series of quark spins that form the densest space-time fluid of the Universe.

Today, the previous picture of a vortex of space-time dragging masses into a black hole has become a commonplace in the analogical research of black holes to the point that physicists study black hole properties with "dumb holes" made with atomic superfluid vortices. It is a very pertinent comparison. The difference is that what it falls into the black hole are the tiniest forms of the gravitational membrane: fractal strings and their wholes, quarks and gluons. That is why a black hole erases our light reality: it sucks, folds, and packages in the form of a superfluid all the quarks and matter of our Universe. In fact, the proof of that theory and the falsity of black hole evaporation have come with the first experiments that tried to make a dumb black hole of atomic superfluids that absorb sound phonons instead of light phonons. They did it at Haifa just a year ago; they were atomic superfluid vortices, Bose-Einstein condensates that turned at supersonic speeds inside the vortex, absorbing sounds. They were self-similar to quark holes that absorb light, but they did not evaporate or emit the slightest trace of sounds. So they were called dumb holes. This means quark holes will not evaporate either.

7. Logic Principles Matter More than Mathematical Details

"I want to know God's thoughts . . . the rest are details." said Einstein. Yet those thoughts are logic principles, the details are mathematical equations, which evolve with new developments in calculus, while the principles stay.

In that sense, it is a common mistake of Pythagoric physicists who believe God speaks only mathematics to consider that Einstein's work on general relativity is unrelated to Newton because he uses different mathematics. Thus, Pythagoric physicists, who consider mathematics the only language of reality, think that Einstein explains a completely different view of the Universe, which invalidates totally Newton's work. This view is wrong and leaves unexplained the very obvious fact that Newton's description of gravitational forces works so well that it was considered an absolute truth till the twentieth century and is still used for most calculus of forces on Earth *because both are based on the same logic principle of equivalence, which in the hierarchy of science dominates over the different detailed mathematical descriptions of it*. Moreover, it is known to the historian of physics that Newton's equations were evolved by Poisson and Poisson was the departing point of Einstein, which is the departing point of fractal relativity, the twenty-first-century version sketched in this book. Yet the principle of equivalence between accelerations and masses is *not the departing point of Higgs and Hawking's entropic black holes, which deny Einstein's principles, and so regardless of their mathematical complexity, they are wrong because their logic principles are false*.

So we must consider a different view on the evolution of gravitational theory. Newton, Poisson, and Einstein's theories are right because they describe the same *logic principle of equivalence*, with the best mathematical models available to each age. And since logic is in the hierarchy of the scientific method, superior to mathematics, from where it derives—as Gödel, Russell, and Frege proved, deducing the laws of mathematics from those of logic—what matters is to understand the logic principles of the Universe, which will then be described "in higher detail" with each new paradigm of mathematics. What this means is that as mathematics have grown in detail and precision, new physicists have applied the new mathematics and increased the detail of their calculus. Yet the physical principles, events, and concepts that truly define the phenomenon physicists describe have not changed. So when those principles are changed by false principles as Higgs and Hawking do in mass and black hole theory, the results are false regardless of the mathematical detail they use to reinvent the Universe.

This is clearly shown in theory of mass by the three phases of evolution of the principle of equivalence between mass and acceleration, all of which can be used to calculate trajectories of masses, because all of them follow the three legs of the scientific method. The first paradigm was Newton's paradigm, which used a simple formula of a gravitational vortex to describe them. And they worked fine for most calculus and they are still used. Then Poisson, Lagrange, and Hamilton refined those calculi. And finally Einstein used Poisson formulation of the gravitational force and took it a step further with the use of non-Euclidean maths. What some of the people of "al." have done is to take Einstein's work further by using the next age of evolution of mathematics, fractal and non-Euclidean mathematics and the laws of endophysics, which includes the limits of human perception on those equations (Galilean paradox, changes in the speed of time around black holes), which solve the problem of the uncertainty of measure introduced by the observer. But any of those three-plus-one ages of the principle of equivalence between mass and acceleration are good enough to explain what mass is; and none of the quantum alternatives, despite the complexity of its maths, is truth because they deny the principle of equivalence between mass and acceleration, considering mass an external property given by solid particles (the Higgs) that transfer it magically to all other entities of the Universe, not an internal property of the topological structure of space-time, its accelerated, cyclical motion.

Any of those models—Newton's forces, Einstein's relativity, or our non-Euclidean fractal description of the two membranes of the Universe—bring enough understanding of what a mass is. Yet because Pythagoric physicists ignore the underlying logic principles that rule their equations, they believe in the mathematical model of the day and when new mathematics appears, they think the old model was totally wrong. So they thought, as Alexander Pope put it, that "God created Newton and there was light," but when Einstein came with new maths, they threw Newton's work as if it were a fantasy when it had worked perfectly for three centuries in calculating all kinds of motions in the Universe. Newton had embedded the same true principles in his work—the principle of equivalence between accelerations and forces, masses and motions ($F = m \times a$).

Next, Poisson and Hamilton applied the new refined tools of differential calculus to the study of the motion of planets, which was previously described by Newton with simpler operandi. So they found out another key principle of the Universe of energy and information—the law of minimal energy, which states that any system of the Universe tries to keep its internal energy and information, following the path that saves more energy between two points.

Then when Riemann developed non-Euclidean mathematics, Einstein applied them to the previous work of Newton, using the notation of Poisson, to find a more precise analysis of gravitational mass vortices, by making a series of "simultaneous," present pictures of those cyclical motions and adding the influence of the gravitational membrane that underlies our world of light and electroweak masses.

We could resume the differences between the classic formulation of Newton and the modern one, considering that Newton formulated the description of the whole, and its dynamics in an extended space and time duration, while Einstein formulated the detailed topology of the vortex in each non-Euclidean point of its space and in the simultaneity of each present moment of duration.

The description of those vortices in the scale of charges and quarks can be also done with two degrees of detail, as we do here, considering them whole vortices or as the so-called QCD theory of strong forces does, in detail, considering each fractal quanta (gluon and quark) at each moment of time and aggregating them together. Yet while the whole description of quarks we shall do here is less accurate than the detailed one, brought about by quantum chromodynamics, it provides a more logical, comprehensive overview, which is the aim of this book, since all those mathematical descriptions are truth as long as they describe accurately the equivalence principle between forces and accelerations. If we compare them with a verbal description, we could say that all of them have the same core sentence, the principle of equivalence, but the description of it grows in detail. If Newton, for example, said Jesus had dark hair, Einstein refined the analysis and said Jesus had dark hair with brown touches in his forehead, and fractal relativity refines the sentence further, saying Jesus had long dark hair with brown touches on his forehead and a white stripe on his temples. But the essential concept remains: Jesus had dark hair, that is,

the forces and particles of the Universe are accelerated motions.

This principle, so stated, is what Higgs and Hawking have dropped from their eye view in quantum entropy; and due to the excessive specialization of their mathematical models of enormous detail the general vision of the forest of all physical events, the equivalence principle is lost. For that reason, I have always preferred the opposite approach: to study first the forest, the principles of physics, with a limited mathematical precision, achieved by not using computers in my calculations and accepting as supreme guides the laws of the scientific method (causality, simplicity, economicity, hylomorphism, and experimental evidence) and the well-established principles in which previous theories were founded. And only at the end, once the forest is mapped, I have considered some of its specific trees in more detail.

It is thus the comprehension of the principles of equivalence between the forces of the Universe and the two types of accelerations ($F=m \times a$) and its invariance under change of position or scale in the Universe, not the mathematical details that casts further light on the meaning of mass and charges in the Universe.

Let us then summarize so far the meaning of that principle in Newton and Einstein's work. Newton defined that principle with a simple formula: $F= m \times a$; that is, the forces of the Universe are made of two elements, which we must consider to be both accelerated motions. Yet since we only know two types of accelerated motions, lineal acceleration (a) and vortices of cyclical acceleration ($V \times radius=k$), masses should be curved, accelerated vortices of gravitational forces.

This was Einstein's view. It led him to formulate with Riemannian tensors the four-dimensional space-time structures of gravitation and masses, measuring the structure of space under that principle in a simultaneous moment of time. Thus, he made still pictures in which we can gauge the curvature of the vortex to its finest detail, adding smaller corrections to the whole dynamic Newtonian view.

Thus, Einstein just made a more detailed analysis of Newton's vortices of mass, turning around the sun, using Riemannian geometries and the concept of simultaneous present time. We could say in a way that he made a more precise picture, a still present photograph of the spatial form of those vortices, but the essential truth, the principle of equivalence between forces and acceleration remains the same from Newton to Einstein to fractal relativity. And so it is that principle and not the evolving mathematics of any of those three models what we have to consider in detail to fully grasp the essence of all the forces of the Universe. What fractal relativity does is to add the advances on theory of information, to go even further, respecting the work of those pioneers and founding fathers. What Higgs and Hawking did is to throw them out because "Einstein is double wrong" (Hawking) and take their place as the new prophets of an entire refurbished Universe which should be reinvented to please their minds. So this new Universe should have an entire new, invented, never observed field (the Higgs field) to substitute the gravitational field of Einstein and Newton and a new causality in time (time travel to the past) to please the theory of evaporating black holes that travel to the past, where they evaporate their information. Then came CERN and decided Hawking and Higgs had reinvented the Universe and it was worth to risk the life of all of us, trying to prove them right despite forty years of seeking and not finding his mental fantasies. In other words, what physicists are proposing to mankind is this:

Higgs and Hawking are God. They have reinvented the Universe. And you must sacrifice your lives to prove them. This is religious fundamentalism in defiance of true science, bioethics, and the laws of the Universe; and while humanity has agreed, I assure you that "those who impose truth with power will be the laugh of the gods," whose black holes will swallow CERN, Higgs, Hawking, and all of us soon.

8. Masses Are Vortices of Physical Information. The Mystery of Slow Clocks.

What is the mathematical relationship between mass, time, and information? Einstein hinted that cyclical clocks of time, cyclical forms, and cyclical masses were related when he said that "time curves space into mass." And he showed that relationship in his equations. Now, once the curved form of masses, self-similar to the curved forms clocks of temporal information is understood, we can fully grasp the meaning of that sentence if we use the concept of time that physicists use as "what a clock measures." Masses are *the cyclical clocks of information of the Universe, which trans/ form the energy lines of gravitational forces into a regular frequency of information given by the speed of rotation of those physical clocks.*

Look at the back of a clock. It is a cyclical spiral that unwinds its energy so the clock can create cycles of information. The clock is a "fractal generator" of cycles of information (the hours of the clock) molded from a source of energy (the spiral), following a dual feed-back cycle of transformation of energy into information (the clock turning) and information into energy (the spiral unwinding). This duality of all eternal processes of trans/formation of lineal motions (energies) into cyclical motions (information) and vice versa is the essential dual cycle of the Universe, its "generator equation" ($E \Leftrightarrow I$ or $E \times I = K$), depending on our dynamic or static perception of it (Galilean paradox). And it applies to all systems. The space between galaxies expands because the space within galaxies contracts into mass. The big bang of a supernova explodes radiation because quarks become a stronger superfluid vortex inside in a big crunch. Yet if we were physicists, adscribed to the entropy-only religion and naïve realism, we would only see half of the process, either the big bang radiation and expansion of interstellar space (big bang theorists) or the visible mask of the informative clock: the needles turning.

So a mass or a charge is a physical clock of time that winds energy into form but simultaneously unwinds form into energy, expelling it perpendicularly through its poles, as a tornado—which also has a dimension of height, not only a cyclical inward movement—does with water or air. And so we are all made of spatial energy that extends and creates space and movement and cyclical clocks of time, whirls of masses and charges that carry information. And the growth of complexity of those two motions finally create the biological cycles of information and energy that make us humans.

The complex evolution of those two motions, chained in infinite puzzles, evolving in networks, grouping in bigger structures, is what we call the organic Universe made of networks of spatial energy and information.

Now, let us prove the concept of mass as physical information with the three tenets of the scientific method: mathematical, logic, and experimental proof. The logic proof is in the explanations we just made of it. Unlike the Higgs mechanism, masses as vortices of space-time are logic mechanisms of attraction. The experimental proof is the relationship between the speed and radius of each vortex and its attractive power or mass. Unlike the Higgs mechanism that never explained why each mass has the mass it has, the vortex theory explains them: Light doesn't close a vortex so it doesn't have mass. Black holes turn faster with lesser radius so they have more mass. Quarks do the same in the particle world so they are the most massive of them all.

Finally, the mathematical proof is as simple and beautiful as it gets. Just substitute in Einstein's two main formulas where mass and time come out to obtain its relationship:

$$energy = mass \times c^2 \text{ and } energy \times time = constant$$

Then you cancel energy and you get $mass = constants/time = k \times$ *frequency of information.*

Since time and frequency are inverse ($T = 1/v$), this means that the faster the frequency of a vortex of space-time is, the more attractive/heavier the mass is; the more information it will store and the slower its inverse ($T = 1/v$), lineal time will happen. And this is another experimental proof of informative masses, as astrophysicists have shown that lineal time passes slower when we come closer to a massive object like a black hole. All this explains why in relativity time is related to mass, which affects the movement of cyclical clocks since

$$mass = frequency \text{ } of \text{ } a \text{ } vortex \text{ } of \text{ } physical \text{ } clocks$$
$$(any \text{ } orbit, \text{ } mass, \text{ } or \text{ } charge = spin).$$

All those relationships, which can be described also with the more complex topological models of non-Euclidean geometry briefly outlined in the appendix, *cannot be obtained with Higgs particles, which cannot explain rationally how masses attract, what their values are without inventing new laws and fields never found in nature, and yet even then, it cannot explain the different masses of each particle.*

Thus, we conclude that Einstein's theory of mass is right and the Higgs a fantasy: *masses and quarks are motions that carry the information of the Universe in its bidimensional vortices.* This is not understood by quantum physicists because they try to explain masses with the Higgs and fail. Later, when we talk of quarks in more detail, we shall add some complexity to this basic description of them. It matters one of those properties: a bidimensional vortex *has less dimensions and more speed/attractive power than our three-dimensional electroweak forces. This means we need three quarks, each one with one-third of charge, to emerge into our electronic world. To do so, they form very dense, stable triplets because they "lock themselves" perpendicularly, to emerge as "three-dimensional color-locked" forms in our Universe.* This recent discovery explains why quark-gluon soups are far more stable liquids than previously predict, why they are much more dangerous and attractive than CERN believes, and are complex liquid condensates, macrocosmic vortices that become black holes and quark stars much easier than CERN and its entropy/energy/explosive gas version predicts. We have now proofs of this after a previous accelerator RHIC made a liquid quark-gluon vortex NOT a gas. And yet CERN behaves as any nuclear company does, protecting its weapon, not arguing with reason the new sciences of complexity, duality, and far from equilibrium systems, chaotic attractors and fractal theory that explain those quark liquids. Instead it rejects any dialogue with Einstein et al., the pioneers of those sciences, who have been warning for years and now have increasing evidence of the dangers of quark-gluon soups and uses outdated gas models to describe those liquids. CERN cherry-picked theories not by its scientific quality but by its capacity to justify its machines and the quark-gluon liquid as a harmless substance that will merely "explode" into radiation instead of crunching us into a quark star.

9. The Unification Equation

Now, to fully understand those fractal quarks and how they might form cosmological black holes, we have to relate the quantum world of charges and masses and the cosmological world of stars and black holes. So we have to return to complex physics, solving one of the great questions pending in twentieth-century physics, the unification equation of charges and masses, which is the key to properly mathematize the different masses of all those quarks and deduce the speeds of its vortices. To that aim, we must understand charges and masses as two fractal membranes of the Universe, NOT as two forces, that exist in the same continuous space-time. This is an improvement over Einstein relativity, but as the reader should understand by now, what fractal relativity has done is to improve Einstein by improving the mathematics he used, with the addition of fractal mathematics and the completion of the postulates of non-Euclidean geometry and by improving the logic of his work with the addition of information, the second arrow of time.

Let us now consider a more refined view of the cosmos than the one brought about by quantum physicists, as a fractal system with two scales or membranes that constantly interact with each other: the quantum electroweak membrane and the strong gravitational cosmological membrane. This is important because it will allow us to establish the nature of quark condensates as the "quantum atoms" of quark stars and black holes, which will be certainly made by CERN's quark factory. We shall do this following the pattern of this book, which gives glimpses to the future of science and solves the problems CERN pretends to solve, killing us, with the power of the mind, to show that CERN is both, ethically and intellectually a monstrosity.

The equation that unifies charges and masses as a single type of force has been searched for one hundred years from the perspective of quantum particles, not with the geometrical/motion-based concepts of Einstein's masses. Yet Einstein always said the unification equation should be based in the geometrical similarity between the equations of electromagnetism and gravitation and the equivalence principle between mass and acceleration.

Let us then use those simple principles. The existence of a fractal structure of space-time means that the electromagnetic world and the cosmological world are just two self-similar non-Euclidean membranes of space-time of two scales and charges and masses two self-similar vortices of two different scales.

So we shall be able to obtain for the first time theoretically, the value of the stronger universal constant of charges (Coulomb) by merely translating the jargon of a charge to the jargon of a vortex of space-time, which is the same in all scales. Since rotation is the essence of mass and the concept of a hard particle as a solid ball is a mere Maya of the senses, *their only difference between quantum charges and cosmological bodies is the higher speed of rotation of charges and hence the stronger constant of the electromagnetic vortices of the quantum world.*

This unification can be done with the mathematics of Newton, Poisson, or Einstein, which are just three stages in the constant refinement of the principle of equivalence and the geometrical analysis of the Universe. *What matters is to understand the principles, which is what quantum cosmologists no longer do.*

Let us then bring here with the notation of Newton and Poisson, which is far simpler than that of Einstein's relativity, the simplified proof of the fractal nature of the Universe and its two scales, the quantum world and the cosmological world in which charges and masses are gravitational vortices of space-time of two different scales. We consider, using Newton's notation, a mass a three-dimensional, cyclical accelerated vortex of space-time defined in classic Newton mechanics by a centripetal acceleration, which gives us the classic definition of the universal constant G as

$$G\ M = \omega^2 r^3$$

where r is in meters and ω is the angular acceleration in radians per second of any gravitational orbit.

This simple Newtonian equation, however, in the fractal paradigm, should apply *to any scale of the Universe, able to describe any vortex of space-time by merely changing the value of the universal constant G. Thus, we obtain the fractal equation of any vortex of time:*

universal constant of a space-time vortex, U (g,q) = w (angular speed)^2 × radius^2/ mass

If this is truth and the Universe is indeed a fractal of several scales, it means that the previous equation will be able to describe *both*, not only a gravitational vortex of space-time in the cosmological scale with the universal constant G, but *also a charge as a quantum vortex of space-time with the universal constant Q, the constant of Coulomb.*

Let us then substitute the parameters for the values of the sun (mass) minus earth (rotational speed and radius) to get G. The approximate values are as follows:

sun mass = 2 × 10^{30} kg; Earth's angular velocity 2 × 10^{-7}
rad. per sec. earth's orbit = 150 million kms.
result: G=6.67 × 10—^{11} kg-1 m^3 rad. sec.$^{-2}$

This is standard gravitational theory. What has never been done, because the fractal paradigm was not known till recently, is to substitute in the same equation of gravitational cosmological masses the mass radius and speed of the space-time vortex by the values of the fundamental quantum space-time vortex, a hydrogen atom/charge.

If the thesis of a fractal universe made of hierarchical scales is truth, then those values should give us the value of the universal constant of charges, the Coulomb constant.

Indeed, if we substitute for the proton (mass) and the Bohr electronic orbital (speed and radius)

4×10^{16} rad. sec.$^{-1}$ = w (electron);
5.3×10^{-11} m. (Bohr radius);
proton mass = 1.6×10^{-27} kg.,

then we get a G, which is 2×10^{39} stronger than the gravitational radius; thus, the hydrogen atom behaves as a self-similar fractal scale in the quantum world to a solar system.

The relationships between the particles of those two fractal scales, the cosmological and quantum scale, is the next great mystery of physics, resolved by the unification equation and Einstein's mass theory. To find that relationship, we must translate to the gravitational jargon the radius of a proton, the particle made of quarks and gluons in the quantum world. If the hypothesis of a self-similar membrane of space-time of different scales of size holds since both, a proton and a black hole in the two scales, are made of quarks, the proton must be a black hole of the electromagnetic scale and so its radius, when translated to the jargon of gravitation, must have the same formula than a black hole radius, called a Schwarzschild Horizon.

Again, the mathematics is fairly straight: we equal the Coulomb, gravitational constant=1.5×10^{29} and the electromagnetic radius of the proton written in the jargon of gravity with its classic formula that describes the same vortex, written in the jargon of electromagnetism:

electromagnetic force (gravitational jargon) = $G(q) M_{proton} m_{electron} / r^2 =$
$e^2 / 4 \Pi e_o r^2$ =F (classic jargon)

A hydrogen atom will be then a mass vortex of space-time, where $G(q)$ is the Coulomb constant as a gravitational constant with value 1.5×10^{29}.

Now we use the previous equation to find the electron *Bohr radius as a gravitational vortex*. Then the magic transformation happens. Take the previous equation

$$e^2 = 4 \Pi e_o G q (1.5 \times 10^{29}) M_{proton} m_{electron}$$

and put it into the electron radius: $e^2/m_{electron}$ x c^2 x$4\Pi e_o$. Now cancel $m_{electron}$ and $4\Pi e_o$. And you get the electron radius expressed in the jargon of a quantum gravitational world as follows: $G(q)M/c^2$.

It is a much simpler formula, which means we are doing "good physics" according to Occam's principle. But since in that expression M is the mass of a proton, $G(q)$, the electromagnetic constant as a gravitational constant, and c, the speed of light, that expression is exactly the *Schwarzschild radius of a quantum black hole.*

Thus, the electron Bohr radius, which is the final radius of minimal size and energy in electrons, is the event horizon of a black hole in the quantum gravitational world.

Indeed in the cosmological world, beyond that radius light sinks toward the central singularity, where the black hole, a quark star resides. And beyond that radius, the electromagnetic flow sinks into the nucleus where quarks feed on light. In both worlds beyond those event horizons, light cannot escape. And yet the proton has a magnetic pole, through which it emits magnetic flows, equivalent to the dark energy, we shall see, black holes emit through its poles.

In other words, protons are self-similar to cosmological black holes in the quantum scale. And black holes are probably made of strings, gluons, and quarks as protons are. They should be therefore frozen quark stars, of the only variety of quarks which is positively charged, which happens to be also the densest of all quarks, the top quark, whose parameters as a condensate coincide with those of a cosmological black hole. And so, if a black hole is a top quark star, positively charged, the "self-similar proton" of a galactic atom, CERN will make them.

To notice those results, the *first theoretical deduction* of Q, departing from G and the enormous simplification of the parameters of the electron radius till arriving to the same expression that a black hole radius cannot be by chance. They are mathematical deductions, one of the three *the standard forms of proof in science.*

For a century, quantum cosmology has tried to obtain this unification, departing from quantum laws. Yet it can be obtained, departing from the gravitational equations of a vortex of space-time. Coulomb already hints at this fact when he decided to use a self-similar formula to that of Newton to describe a charge. What Einstein did was to refine those measures with the concept of simultaneous, present time and the use of the non-Euclidean Ricci tensor/flow, which is a more sophisticated, detailed picture of that curved vortex of space-time. He, we might say, applied a mathematical method that allows to take a series of still photographs of the vortex with enormous accuracy, but the complexity of his calculus hid the unification equation for a century. So it is evident that we are observing the same geometric vortex at different scales, explained with two different historic jargons.

So indeed, the Universe is a fractal of space-time, with two self-similar membranes. This opens an entire new field of physics, NOT quantum gravity, the description of cosmology with quantum laws, which is proved false, but fractal relativity, the description of the quantum world as a self-similar membrane to that of the cosmological world.

But in what it concerns this book, the most interesting finding under the new U (q) value of the electromagnetic constant is the discovery that the proton horizon shows the same equation as that of a black hole, a Schwarzschild horizon. It means that a quark condensate is a black hole state and vice versa: a cosmological black hole is a deconfined state of billions of top quarks, a top quark star.

Now, the mathematical treatment of the equivalence principle to unify masses and charges can be done with the simplest descriptions of a vortex created by Newton and Poisson or with the more complex Riemannian geometries of Einstein. The general error of physicists and the obvious reason why this equation has not been found is precisely their rejection of the simplest mathematics of Newton and Poisson, which made the unification exceedingly difficult. Yet there was a second reason: the use of a single arrow energy. So Einstein tried to unify cyclical vortices of mass and lineal electromagnetism (Maxwell's equations) because he didn't realize that he had to unify information and information (charges and masses) on one side and lineal forces with lineal forces (gravitational waves and electromagnetic waves), which has been done with the self-similar equations of gravito-magnetism.

Thus, in a universe of multiple fractal space-times, each one with a different content of spatial speed and informative density defined by a different G, electromagnetism and gravitation are merely two types of vortices with different sizes, hence different contents of space and time, which translates into different universal constants.

Consequences of the fractal unification of charges and masses

We obtained theoretically, for the first time, the value of Q from G; always obtained through the experimental method, the three legs of the scientific method are met in the fractal unification equation:

— *Logic consistence.* The Universe is a fractal of multiple space-times, self-similar in structure.
— *Mathematical accuracy.* The same vortex of space-time allows calculating theoretically the values of those vortices.

— *Experimental proof.* The experimental calculation of Q coincides with the theoretical value, which has a probability to happen of less than 1 in a trillion, if this coincidence were by chance and not because the Universe is indeed a fractal of self-similar scales and charges are gravitational, quantum vortices.

Because no other approach has yielded a consistent, experimental logic and mathematical unification, it is obvious that a charge *is a self-similar (but not equal) membrane of space-time.*

Further on, quarks are defined by a third type of force or vortex of space/time, which is even smaller in space and faster in time, the bidimensional strong force. And its cosmological equivalent, bidimensional black holes, seem to be similar in properties and morphology. Thus, we can also relate with the same principle of equivalence the strong force and gravitational force. We already showed the Bohr radius to be a Schwarzschild radius, and since astronomers study the stars around galactic black holes as electronic nebulae and have found that the central black hole is actually a swarm of them, as a quark is actually a fractal of gluons, the self-similarities keep growing.

In our analysis of the errors of truth of science, we considered the hyperbolic error the most common among human beings. Now we can apply it to the real structure of the Universe and its different membranes of space-time. We exist only in one of those membranes, the membrane of light space-time, which is Euclidean in form, whose space dimensions are given by the three lineal perpendicular fields of light.

And yet physicists think Einstein's special relativity, which studies our local, fractal membrane of space-time (the membrane of Euclidean light-space and its entire evolved electroweak species that inhabit that space), found a tautology. Since we exist in that "light medium," the speed of light c is the limit of speed of our world. Yet he thought c was the limit of speed of all the systems and sizes of the Universe.

Indeed, imagine that you exist in a war medium. If you flow with the water, you will not be able to go faster than the water. Because we float, so to speak, in light, the space-time of the galaxy, we can't go further than light.

But we have observed beyond the galactic light space, intergalactic, dark, gravitational space with $z=10$ c speeds. This is in fractal space/time theory, with its decametric, multiple membranes of space-time, the limit of speed of the bigger cosmological world of dark/gravitational energy and quark/black holes. But scientists have invented an enormous number of ad hoc solutions to deny that in the bigger world of galaxies that float in that dark world, speed is faster.

Enough to say that the symmetry is based on what we call the paradox of Galileo, the fact that all what we observe can be seen as a fixed form or a moving one, like the Saturn's rings mentioned before. When we see a mass as a motion, it is a vortex; when we see it as a fixed form, it is a particle. When we see a force as a motion, it is energy; when we see it fixed, it is space.

The classic example is given in cosmology: are the galaxies that are moving at z=10 c moving at ten times the speed of light, or is space creating distance, expanding? In the appendix, we shall deal with that theme when we study in depth the meaning of the big bang.

In this appendix, dedicated to the physical Universe, we have now to consider the fundamental symmetry of the Universe between lineal forces of space and cyclical, informative particles of time.

IV. Fractal Space-times: Dimensional Lines, Waves, and Cycles

From one comes 2, from 2 comes 3, from 3 the infinite beings.
—Cheng Tzu

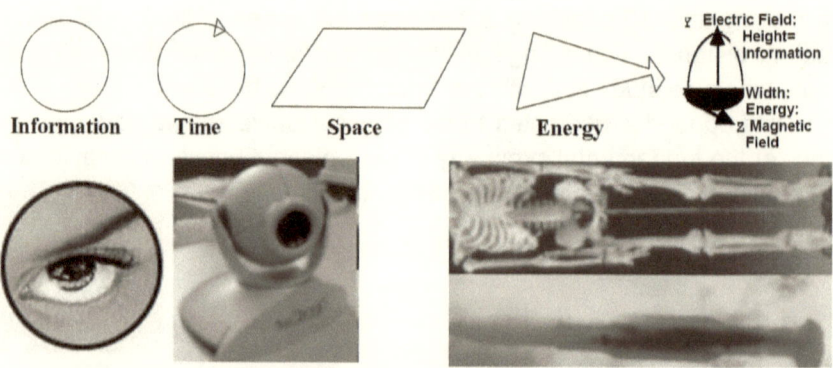

Information   Time   Space   Energy

Bites of energy and bits of information, the two primary elements of the Universe, mix, organize, and evolve, till reaching the complexity of organisms. Their properties and morphologies are opposed. But for that reason, they can form "complementary" systems, which merge the properties and dimensions of both "formal motions," enhancing the survival capacities of any system that combines in a balanced manner energy and form, E×I=K. In physical systems, all is made of lineal forces of expansive energy and cyclical particles of implosive information that create balanced present space-times. In the left graph, we see the canonical forms of lineal energy and cyclical time clocks and how an informative, electric cycle supported in a magnetic surface of space creates a complementary wave: a ray of light.
It is the ternary game of the Universe and its
three dimensions of space and time.

10. The Complementary Principle
(The Bidimensional Holographies of the Universe)

We can now that we have a basic understanding of the "dualities" of the Universe in two membranes of space-time, the quantum world of charges and the cosmological world of masses, and its two essential forms, the gravitational and electromagnetic lineal forces of energy and its cyclical vortices of information, and the *third element, mixture of both. Since fractal pieces of energy/space and fractal pieces of time/information mix together into wave of space-time, the real, complementary beings of the Universe.* In essence, the real Universe in which we exist, with its four dimensions is a holography of two bidimensional, pure forms, the vortex or clock of information and the plane of lineal flows of energy. The simplest topology that combines those two forms is a perpendicular field of high information as the screen of this computer and a surface of flat energy (in the computer, the keyboard that transforms the energy of my fingers into information). In the appendix, we shall consider a more sophisticated version of those holographies, introducing the three canonical topologies of a four-dimensional Universe, and how they combine into the real, complex shapes of physical and biological beings. Now that we are concerned with the understanding of the principles of those real holographies, we shall consider the simplest of all entities of space-time: a ray of light. Its simplicity is revealing of the simplest structure of a holography of energy and information. The magnetic, flat field supports the frequency of an electric field that creates the Information of the wave. And the result is an H-Planck unit off our light membrane. We are made of evolved fractal H-Plancks, the "planckton" or atomic units of our space-time. And that "planckton" is a wave with minimal form encoded in the frequency of the light and minimal space, supported in the magnetic field. From this simple initial unit of space-time, our Universe builds itself. "And God said, there is light; and he separated the day from the night" means indeed in metaphoric terms that we are all made of light. And yet the same game happens in the world of dark matter, also made of bits and bytes of gravitational information and energy. Then once we understand those two membranes and its "doors" and "limits," we can start to construct the real Universe.

In the graph, the Universe is made of three arrows of time: the expansive line of entropy or energy, the cyclical implosive clock of information, and its combination, the wave of space-time.

This simple concept is incredibly rich and powerful to describe the events and forms of the Universe. Till now, we have stressed the duality of implosive cyclical clocks of information and explosive lineal flows of energy. We shall now introduce its real combinations, the four-dimensional waves of energy—which according to the principle of complementarity create reality by combining spatial, entropic arrows of energy, bodies, and fields—and informative, implosive cycles of information, masses, and charges in physical spaces, heads in biological systems.

The key advance of complex physics and duality is the introduction of information, the arrow of fractal form in the creative processes of the Universe. How this arrow of form interacts with the arrow of energy to form wavelike, balanced "ternary beings" of "present" is the key element to go further in our analysis of any entity of reality.

This interaction takes place according to the paradox of Galileo either in

— a dynamic "temporal" way, *as a series of causal events that transform energy into information from past to future or information into energy from future to past, or combine both in waves of present,* or in
— a spatial, geometrical way, as a game of two geometries, the line of energy and the cycle of information, *which combine to form complementary systems, informative particles with energetic fields* (physics) *or informative heads with energetic limbs* (biology). And again, only complementary forms that exist by combining "bidimensional energy" and "bidimensional information" are real, four-dimensional entities.

We call this fundamental game that combines energy and form, creating dimensional beings in space and causal chains of events in time and *explains most of the events and entities of the Universe* the inverse complementarity of space and time.

In chapter 5, we shall go deeper in the analysis of that complementarity from the perspective of time and causality. Here, we are more interested in a more classic, physical interpretation from the perspective of geometry and space.

Let us then consider first that duality in space as a geometrical game *since all the processes that will take place at CERN will be dual processes in which energetic forces and cyclical particles will be created and destroyed. Yet CERN cannot understand those processes with their outdated models of an energy-only Universe.*

In that regard, the main logic principle of the scientific method defined by Aristotle is called *hylomorphism*, or duality, which, simply stated, says that anything real has both—a substance or energy and a form or information. It opposes both "informative-only theories" (Pythagorism, the idea that a mathematical equation is truth per se) and "energetic-only theories" (quantum entropy and quantum cosmology) of reality.

Hylomorphism is at the heart of the scientific method since its inception. In a static vision of reality, it means all particles need a reproductive body to exist, all informative heads need a body. Galileo already said that the language of truth is the geometry *of lines and cycles*, a sentence ill translated later into the concept that mathematics is the only language of God. What he meant is that the Universe is a game of cycles of information and lines of energy—the "vortices and res extensa" of Descartes.

Cyclical, moving time clocks or fixed information on one hand and lineal, moving energy or fixed space on the other are synonymous. In the graph, this duality is due to the paradox of Galileo that makes us perceive sometimes energy in motion and sometimes fixed space, sometimes fixed forms of information and sometimes moving clocks of time, forms in action.

The graph explains a law of modern physics, the holographic principle, which states that information is bidimensional. Indeed, information, as this page or screen, has two dimensions, height and width, while energy has also two dimensions, length and width. Both together merge to create the four-dimensional fields of the real Universe in which the interaction of "high information" and "long, moving energy" creates the three dimensions of classic Euclidean space and a fourth dimension of motion and rhythm.

Because the Universe is fractal, those dimensions are local for each species. It is what Einstein called the diffeomorphic principle.

In the graph, we draw the ideal cycle of time traced by a rotating particle, which leaves behind a fixed ideal form of information: a disk that establishes a discontinuity between its inner vital space and the external Universe. And we draw the ideal movement of energy, which traces a line that might reproduce laterally, becoming a plane or wave of space, since the most natural form of moving energy is a triangular plane. Thus, most organic spaces and limbs of energy are triangular, planar, bidimensional forms that fluctuate in their width and length. For that reason, energy is expelled in triangular flows through a rocket while energetic systems move, forming triangles called steps, as human legs do.

The main parameters of a cycle or bit of temporal information are frequency and cyclical height. Those cycles combine in knots of cyclical frequencies that give origin to an immense number of different temporal species: Σ *time cycles* = *informative being*. Height as a dimension of information also means that as time passes information increases, so does the height of any systems. For example, life systems evolved from "planarian worms" to tall men as the information of life systems increased. Matter evolves from bidimensional vortices of mass into quark Einstein-Bose condensates, which as we shall see have a dimension of height. So dimensions are local, fractal, created by the arrow of "fractal information" and have a function. In the appendix, we shall consider according to that holographic principle the processes of creation and destruction of dimensions in the physical universe. Consider a simple example: you can warp this bidimensional page and get a third height dimension with a lot of "fractal lines of form"; you are creating both, a new dimension and information.

On the other hand, lineal bytes of amorphous, moving energy aggregate into planes of bidimensional space, adding its dimension of width: $\Sigma e = S$.

In Euclidean geometry, a point has no volume, no dimension; but string theorists say that even the smallest points of the Universe, cyclical strings, have inner dimensions, which we can observe when we come closer to them. That is the essence of a fractal point. This fact is essential to understand logically the meaning of bidimensionality and the holographic principle that states that physical information is bidimensional. In Euclidean mathematics, to consider that there are two dimensions of energy and two dimensions of information is absurd unless we accept that each point of the Universe has a minimal volume. So a bidimensional plane, as a galaxy or the Universe at large is and as this screen or paper sheet is, seems totally plane without any thickness till we come closer to it. Then we realize that a page is a network of non-Euclidean points that will therefore have

— a minimal width given by the height of each point and
— a discontinuous zone of "dark space-time" not perceived or connected by the network of points.

It is important to understand bidimensionality in terms of non-Euclidean geometry. That is, each point of a bidimensional sheet of information, including this page or screen, has, when observed in close range, certain height or internal dimensions. This fact explains why certain mathematical objects, like strings, seem to have 3 × 3 dimensions (those are the internal dimensions of the string, when seeing in close proximity as a fractal point of a lower scale of form). So the complex topologies of the Universe can be reduced once we understand its non-Euclidean fractal and dual structure, as Galileo put it, "to a game of cycles (of information) and lines (of energy)", which in physics are called forces and particles. In simple terms, dimensions are the perceptions of lineal energies and cyclical clocks of time, as "fixed" information and space. Those motions, perceived as "fixed" shapes by the eye, create the sensation of space, which Einstein rightly defined as follows:

Space is motion relative to a simultaneous frame of reference.

Thus, spatial energy is lineal, moving energy, creating surfaces of space with its movement that we see fixed, as fixed pictures of lines of moving cars seem to elongate space. Thus, space and distance are two ways of perceiving a volume of energy. It also means that the space we see is the "light membrane." We see light-space. And so what we see are exactly the three perpendicular Euclidean dimensions of the energetic fields of light: height is the electric field, width the magnetic field, and length the reproductive c-speed of light. It also follows that our light world cannot go faster than light, which is a tautology since it is made of light.

We are thus all fractal pieces of light space-time, which has evolved into complex forms of in/form/ation. We are made of evolved light-space.

Thus, dimensions, as everything in physics, are easier to "see" with visual, topological models, the *ultimate language of the Universe since God doesn't play dices but is a painter in four dimensions, two of energy, a plane of lineal motions, and two of information, cyclical motions, whose bites and bits combine to create all the fractal organisms of reality, all the shapes you see, from the informative particles of physical space to the cyclical, neuronal path/thoughts of your spherical head.*

Because a clock of information is a cycle, it has two dimensions, height and rhythm, as it happens to the screen or page of a book where you read this text. So height becomes a dimension of information and the Universe becomes a game of two bidimensional type of entities: planes of energy and cycles of information with height. This is obvious; your head of information, an antenna, and any informative system is on top of your energy body. Because the Universe is fractal, each entity is a piece of bits and bites of energy and information and so the relative height of each species is local. This is called in relativity the diffeomorphic principle.

What matters is that now we can understand how energy becomes information, how geometrical space and clocks of time combine. In Galilean physics, when time was lineal, not a cyclical clock of information, it could not be transformed into space. Energy and information could not be transformed into each other, as only things with the same number of dimensions can be transformed into each other. So now we rewrite the energetic principle of physics as

energy never dies but constantly trans/forms itself into in/form/ation and vice versa.

The beauty of the fractal paradigm resides in the fact that physical energy and information are only the first bricks of an ever more complex, evolving Universe, created by chains of information cycles, which are all the time exchanging energy flows, back and forth, through smaller fractal particles or organizing themselves into bigger wholes. Let us consider the properties of those three elements: energy (limbs, fields): information (particles, heads); and its combination (bodies, waves) and some of its species in different scales of reality so you grasp the essential self-similarity of it all, *which means physics is NOT the Saint Grail of science but merely the description of the simplest cycles and lines of energy and information*:

future clocks of temporal information—> present waves <—past, energetic space

Form: Time, Function: Information energy	§:Space-time: Exist	Form: Space, Function:
Mass, charge, vortex energetic	§:Waves	Force, field, plane of
		space
Small, still	§-Rhythmic	Large, moving fast
Tall, Perpendicular,	§-Elliptic	Long, Parallel
Bidimensional: Height, Width	§:3-D +1 Beat	2-D: Length, Width
Cyclical, Rotational, imploding	§-Steady state	Lineal, uncoiling, exploding
Informative, Frequency	§-Transversal wave	Lineal Speed
Singular, Broken, Fractal form	§	Continuous, amorphous, differentiable
Intelligent, perceptive	§-Beautiful	Strong, fast, masculine.
Social networks, creative destructive.	§-Dual, organic	Darwinian herds,
Future, evolved predator	§-Present	Past, energetic victim
Life arrow	§—Reproduction	Death arrow.
Waves of space	§	Particles of Time
Female, yin principle	§-Gay	Male, yang principle
Masses, charges	§	Forces, fields
O-Heads	§-Bodies I—Limbs.	
Blue, Black	Green, Gray	Red, White

Energetic space and informative time are opposite in form and function, but they easily transform into each other in iterative reproductive processes that give origin to the *complementary forms of the Universe, as all what exists is indeed made of fractal, extended, spatial energies and bits of informative, cyclical, imploding information.*

The two states of reality, energetic space and informative clocks of time, are in fact lineal planes and cyclical forms in movement that combine together, creating 4-D space-time fields of temporal energy, like the light wave that illustrates the previous graph, made of a bidimensional, lineal field of magnetic energy over which a perpendicular high-electric field of information moves rhythmically in cyclical waves. Height is a dimension of time because the repetition of time cycles accumulates in the dimension of height. Informative bits accumulate in "hyperbolic" topologies that wrinkle and create the height of any space-time field as a whole entity made of cyclical parts: cells accumulate as bits of information and a child grows; cycles accumulate in a hurricane that grows in height; life species evolve from planarians, simple energetic systems, into human beings and informative species of maximal height.[1] Einstein's curved space-time cycles bend the energy of vacuum into vortices of masses that accumulate in height, creating a black hole, and so on. Thus, height is the dimension that accumulates information in evolving process, erased in the inverse events of biological death or transformation of mass into energy $E=mc^2$.

So the essence of a dynamic, physical space-time field is the combination and transformation of bidimensional, high in-form-ation into bidimensional, flat scales of energy. When both fields merge, they create a real three-dimensional wave form, which adds a fourth dimension of frequency or rhythm, as in a ray of light, the simplest space-time field, made of two perpendicular planes: one of electric information with the dimension of height and frequency, and one of magnetic energy, with the dimension of width and speed. Yet since light is the space we use to construct our space-time mind-vision of the Universe, for that reason, Euclidean/Cartesian coordinates show the same three perpendicular dimensions of light.

In the realm of physics, one of the main proofs of such bidimensionality and the properties of information described before is the fact that it is the only *explanation we have of a recently discovered phenomenon that the holographic principle says information is bidimensional and the discovery that when we become smaller back into the quantum world, dimensions diminish into two dimensions 1:*

— The holographic principle, a key element of modern physics, says information in the Universe is bidimensional. Even human beings create three dimensionalities, combining two dimensions of form perceived by each eye. Indeed, two dimensions of time-information and two dimensions of space-energy combine in all beings to create all systems of reality.

— Fractal relativists have recently obtained our macrocosmic space-time, departing from two-dimensional entities in the quantum world 1. Thus, there are no more than four dimensions in the Universe. This is logic; when we go from small to bigger, we grow in dimensions. Quantum physicists who ignore logic principles pretend that the small strings have more dimensions up to ten. This is ridiculous. It is a logic paradox. The small is smaller because it has fewer dimensions. Now a group of fractal theorists have proved it with a method called causal triangulation. Another group has studied black holes as fractals of bidimensional vortices, quark masses, advancing the concept of mass as a bidimensional vortex of information. What this means unfortunately is that CERN will do quark—black holes "certainly"—that they will not evaporate as quarks do not evaporate and they will devour the Earth.

What truly matters however of the dimensional and geometric analysis we have done of energy/entropy/expansive motion and information/ implosive motion/cyclical form is to understand its complementarity and combinations as four-dimensional waves of reality since pure information and pure energy are "dead" ends, and what truly matters is to generate balanced combinations of both.

A fact that we represent with the "generator equation" of fractal space-times is

$$E \Leftrightarrow Ti \ (dynamic \ motion) \ or \ ExTi=K \ (static \ motion)$$

which expresses the principle of conservation of energy and information and becomes therefore the fundamental equation of the fractal Universe *since all that exists requires both—energetic limbs/forces and informative cycles/heads to become a real four-dimensional being of the Universe.* Those restraints which physicists often forget are indeed a key guide to consider which beings are stable and exist and which beings are unstable fantasies that do not follow the laws of balance and harmony of reality.

11. The Meaning of Fractal Space/Energy and Fractal Information/Clocks of Time

We now have enough understanding of the mathematical structure of fractal beings as networks of self-similar cells/points with a content of lineal energy and cyclical information to understand what we mean by a Universe made of fractal space and fractal time. The simplest one of those beings in our Universe is the drawing we made of an electromagnetic field. This entity defined by Planck as H, an action of energy and time, has two fields, one of magnetic energy and one of electric form. Yet it is also a piece of fractal space-time that occupies a space and has a clock/frequency rhythm. We are all made of evolved pieces of light, which is absorbed and evolved into electrons, which are the electroweak substance of which reality is made. And when we put together all those pieces of reality, the medium of light-space in which we exist and all the evolved beings of light, we get the Universe. Imagine a sea of water where the medium is water and the pieces that swim on it are indeed evolved forms of water (we have 90 percent of water in our bodies). All together form the ocean. The ocean is not an "abstract" space-time Cartesian coordinates. In essence, any fractal space-time can be defined as a medium of formless space, made of a network of simple bites of energy, inhabited by more complex informative beings. And each of those beings can also be modeled by the complementarity principle with two elements—a body of spatial, energetic cells/points and a head or particle made with a network of fractal cycles of information.

Even the simplest particles, quarks, are made of networks of smaller particles called gluons while electrons are nebulae of smaller particles, ultradense photons, shown in the most recent pictures of electrons, which seem as a bundle of curved light rays. The topology and structure of those organic networks might vary, but all comes to the same shapes: force and lines of energy create the vital spaces of reality, and cycles and clocks of information create the clocks of time of the Universe.

A fractal is a system of self-similar cells generated by a feed-back equation that transforms energy into information. This complex definition of a fractal being can be made more precise when we understand the meaning of the generator equation of a fractal in the real world, beyond the language of mathematics. Any informative code can be a generator equation of a fractal being. A human fractal is generated by the code of information of the genome; a mathematical fractal is generated by a much simpler equation. The Universe and all its systems are in that sense fractals with different generating equations of different complexities. Yet what unify them all are the two "essential" geometries that combine to paint the four-dimensional fractal beings of the Universe, spatial energy and temporal information. And this is what fractal theorists[9] mean by a Universe made of fractal energy and information.

In simple mathematical terms, if we call energy e and information I, we can then write the generator equation of all the fractal beings of the Universe as a feed-back cycle that transforms energy into information back and forth, generating all its space-times. And that fractal equation is nothing else than the principle of conservation of energy and information quoted before in this book:

All what exists is a fractal being that transforms back and forth energy into information: $\Sigma E \Leftrightarrow \Pi Ti$.

We don't need to go in this book in complex mathematical analysis, as we are interested in explaining the simplest fractal networks of quarks that form black holes and quark stars. Notice though the difference of the two mathematical symbols of the previous equation: The symbol that gathers together space particles is a sum, which means that space particles form "waves," loosely organized that move around a field of energy in which they feed. The symbol used for informative particles is a multiplication symbol, which means informative particles form networks, "knots." For example, in a neuronal network, each particle is fixed and connected to many other particles. The result is that a network of informative cells can form a "mirror," which can reflect an "image" that "gauges" reality into a language and so it can create a "mind." The result indeed is that herds of lineal, moving particles "absorb" energy and move, and knots of cyclical information, gauge and perceive.

Further on, knots can be broken in their connections so Ti networks can become herds of energy by losing its "connective dimension" and vice versa.

When mathematics evolved, those symbols evolved. So instead of sums and multiplications, modern physics use differential and integral equations to describe energy and information; and then in the twentieth century, we use fractal equations for information or chaotic equations for entropy.

All this is a more complex view needed to explain the behavior of electronic nebulae that sometimes act as a network (and indeed your mind is basically a mirror image made with electronic nebulae) and sometimes as a wave-flow. This is the ultimate meaning of duality. So all can exist in two states in many different scales: masses or charges or clocks or heads are all species of information, but they swim over a field or body of energy and sometimes gauge information and sometimes move. What then creates the stability of the Universe? A fundamental law of the Universe called invariance of its motions, forms, and scales.

The three invariances of the Universe are the invariance of motion proved by Einstein, the invariance of scale proved by Nottale, and the invariance of form proved by Sancho. They are the three principles that hold together and explain the self-similarity of all the species of the Universe. In other words, while there are infinite different species of energy and information, their "shapes" are invariant in any scale and their motions (perception of shape as movement) are also invariant.

Unfortunately, physicists only use invariance of motion and so they cannot understand how fractal quarks will create black holes and quarks stars at CERN, as they emerge with the same vortex form in the cosmological scale. This invariance again is an evolutionary, natural trait that maximizes survival.

In the graph, informative clocks of time are cyclical and surfaces of energy are lineal to maximize their functions: informative systems resemble a spiral or sphere since those geometries store maximum information in minimum space, accumulating its cycles/cells in the informative dimension of height, from black holes to heads and nervous systems. Bites and bodies of energy resemble the line or plane since that opposite geometrical form covers the shortest distance between two points, the maximum extension with minimum volume, accumulated in the width dimension as space. In the graph, in biological and morphological terms, we can easily recognize the bodies and heads of humans, animals, or machines because they have a clear morphology, which corresponds to that of generic energy and information. Energy is lineal because the line is the shortest distance between two points and so it is also the fastest energetic movement. Information has cyclical forms because cycles store maximal information in minimal space. For example, a human body and a machine body, a weapon, should not have anything in common, but if we observe the morphology of both, it is clear those morphologies correspond to the generic morphology of all energies: they are big, lineal systems that move in space. So our limbs are lines extended in space like the missile. On the other hand, our eyes and brains are smaller and cyclical like the cameras and chips that act as information organs in machines. And those chips order bodies of metal with digital information.

12. Dimensional Form Becomes Speed: Creation and Destruction in the Organic Universe

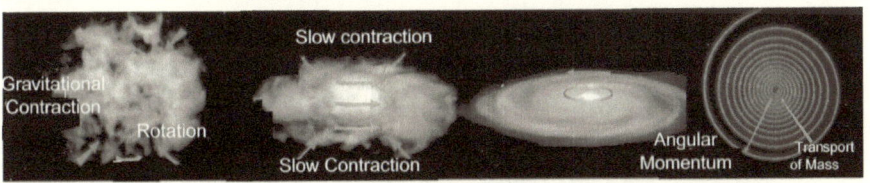

The equivalence principle establishes a Universe with two limits of eternal movement: a lineal speed of gravitational and electromagnetic forces and a cyclical speed of masses and charges, which are vortices of space-time. Energetic forces create space, and vortices are clocks of time that carry information, eternally transforming into each other in self-similar events:
Σ spatial forces (lineal movements)⇔ Ti (cyclical clocks of information).
The sums of all those events maintain an eternal dynamic balance of cycles and lines.
Yet since lines have one dimension less than a clock cycle, the process can be also perceived as a game of creation and destruction of dimensions of form. In the graph, a three-dimensional slow nebulae accelerates its motion as it becomes a bidimensional vortex of faster, more attractive mass. Such transformation is common both in the cosmological world as stars become black holes and the quantum world in big crunches when electrons become quarks. It might be the transformation that the Earth experiences if CERN succeeds making a quark star.

There are two concepts that do not appear in classic physics but are essential in understanding how the universe converts energy into information (motion into form): the paradox of Galileo, which allow us to describe those processes as "frozen pictures," which is the informative point of view, or as changes in the speed or motion of things.

The graph shows the two events combined: when a system acquires forms, it slows down. When a system loses dimensions, it speeds up. Yet to understand this essential process, we must first clarify an error made by physicists who do not use the concept of dimensional form to define "information" but the algebraic "data" concept of a "bit of information."

Fractal information is related to dimensional form while algebraic information is the simplest dimensional form: bits of broken patterns emitted by waves, which encode those patterns in the distances between the crests or frequency of the wave. But those patterns are the "minimal" pattern of information of the Universe. In that regard, the capacity to store information *of any system follows a power law*:

$$Information \approx X^{Fractal\ dimensions\ of\ the\ system}$$

Yet in the same way, physicists have reduced all the arrows of time, previously defined to a single energetic arrow measured by a single "clock-time"; they have reduced all the forms of information to a single broken frequency pattern. Biology, a more sophisticated science of information, today is able to study how molecules and proteins store information in several dimensions; and that previous law can be deduced of biological studies. Yet *physicists have not gone beyond the simplest one-dimensional information of lineal waves. They have not even discovered that mass stores much more information than waves in physical bidimensional vortices.*

In duality, we study both sciences, physics and biology, and we can compare and use laws of both sciences to find self-similar laws in the other discipline. In that sense, Shannon's standard definition of information is a computer-based definition, of technological science, useful to store information in numbers, with a limited application to the understanding of the universe, *which stores information in four dimensions in very complex patterns and enormous volumes.* Information is in that broader Universe form, in/form/ation. The simplest information is the one-dimensional broken bits of data that divide a line of energy into a pattern. This is what a computer does and what a physicist would understand as information. But in nature, a more precise concept is that of fractalization. Information increases when a continuous flow of energy breaks into patterns and acquires dimensions, bumps and points.

The arrow of information is a transformation of a surface of space, of energy into inner form: a change from lineal, chaotic movement into cyclical, repetitive one; from expansive waves into imploding particles; from disordered, equalized forces of indistinguishable forms extending in space into an ordered hierarchical system of nonlineal mass vortices. Its inverse function is the explosion of energy, the dissolution of form, the arrow of *death*.

Though physics recognizes the inverse arrow of entropy as negantropy—that is, the negation of entropy—it is long overdue to name it properly as *form* that crystallizes and organizes chaotic lineal energy into cyclical, repetitive patterns of *in-form-ation*.

Indeed, Mehaute[10] and other theorists have proved ad nauseam that when a system stops creating energy, entropy, it keeps the arrows of future flowing, creating information, fractal patterns of form—*the future never stops. This also applies to physics in the way form becomes motion and motion becomes form.*

The unification equation relates the different scales/membranes of space-time. Yet for each membrane and particle, we must clarify a final concept, the meaning of dimensions of information and how they disappear, converted into speed and energy or appear as motion and energy slows down, creating form. In the Universe, there are multiple physical species of different fractal non-Euclidean dimensions.

The new mathematics of complex physics means though that a flat plane of two dimensions has a minimal height as each of its points of the flat network is in itself a world of a smaller scale. Thus, a flat vortex of mass, a bidimensional quark, has smaller gluons and strings, which have smaller dimensions. For that reason, there is equivalence between different mathematical descriptions in different dimensions. A fifth-dimensional description of strings is equivalent to a four-dimensional description of quarks, which is equivalent to a three-dimensional description of an electronic atom, and so on. It is a general law that all systems of physical mass pack energy in fractal dimensions of form:

strings (one-dimensional dark energy in a 5-D Universe)—> bidimensional quarks and gluons in a 4-D world—> three-dimensional electrons

This translates sometimes into different equations for a physical vortex and different power laws and relationships between masses (the static, formal, dimensional perception of a vortex) and speeds (its perception as pure motion). For example, $Vo \times Ro = K$ is a bidimensional vortex, which we used to explain in its simplest terms the equivalence principle.

However, the unification equation of charges and cosmological masses happens in a three-dimensional world. So we used a three-dimensional vortex equation, U.C. × M = w^2 × r^3, which is the vortex used to unify the three-dimensional worlds of cosmological bodies and electronic matter we perceive.

Finally, in the next chapter, when we compare the two families of quarks, top quark particles and strange quark particles, the relationship between their rotational speed and masses follows a three-dimensional power law (w^3 = M). So the world of top quark particles turn at 10 c-speed but it has 1,000 times the mass of the world of strange quark particles that turn at a maximum of c-speed, which obviously means that the world of top quarks has one dimension more than the world of strange quarks. This is further proved by the famous Maldacena conjecture, which studies top quark black holes as five-dimensional string systems, self-similar to four-dimensional quark gluon soups.

All these are details of advanced fractal physics, which help to determine the specific masses and speeds of the quark-gluon liquids made at CERN that will swallow the earth.

We are more interested in explaining the concept illustrated in the graph: *how our slower three-dimensional electronic world will accelerate, acquiring energy/speed and losing a fractal dimension of form, when it becomes transformed into a strange quark condensate* since transformations of energy into information take place by warping and unwarping dimensions of form into speed.

A basic case observed in all scales is represented by the previous graph, taken from a *SciAm* article about the formation of galactic disks, which we used to illustrate the creation of a quark by the crunching of electronic nebulae of three dimensions. The slow rotation of the electron becomes a bidimensional faster quark system. The same process gives birth to a black hole in the cosmological realm, when the nebulae of a star or the gas of a galaxy collapses into a black hole.

All this means, of course, that dimensions are "fractal," limited dimensions, which never extend to infinity but to the limits of a certain entity of the Universe. Thus, the creation of fractal dimensions of information or its transformation into speed is one of the commonest events of the Universe and at the same time one of the many details of the generator equation of space-time:

$$energy = speed \Leftrightarrow information = fractal\ dimension$$

In the case we study in this book, we can consider a simple dimensional game by which slow three-dimensional electrons become faster bidimensional quark masses. There is a simple proof of the bidimensionality of our up and down quarks: quarks have one-third or two-thirds of the charge/dimensionality of electrons and so we need three quarks with perpendicular orientations called colors to form a three-dimensional quark structure able to interact with our three-dimensional electrons.

Further on, we must add the interaction of two different mediums and forces, the strong gravitational force of quarks and the electroweak force of electrons. This means that quarks "suck in" electrons, trying to absorb them, to feed on them, so they have implosive, strong gravitational forces, while electrons try to suck in quarks with its electromagnetic force, reason why we use a negative, implosive, absorbing direction for electronic charges.

In other words, quarks use its force to attract and steal momentum from electrons, and electrons use its electromagnetic force to attract and absorb momentum for quarks.

Those are the simplest visual concepts, to understand the interactions of those membranes. Obviously when we enter into topological details, the forms, dimensions, and charges of each of those elements become far more complex. And yet because that simple vision departs from sound principles, it is far more telling of the structure and interactions that take place inside an atom than the ideas about mass sponsored by quantum physicists, departing from the absurd Higgs theory.

13. The New Organic Outlook of a Universe Based on Information

In this brief introduction to the fractal paradigm, we have observed the two great revolutions for our understanding of the Universe:

— The logic of the Universe turns out to be biological since the Universe is modeled as a series of networks of energy and information that come together into organic structures.
— We can describe those bio-logic properties thanks to the advances of fractal and non-Euclidean mathematics so we can fuse the logic of biological systems and the mathematics of geometrical structures into a single whole. This topological approach is developed further in the appendix, to fully grasp the organic structures of the Universe.

Duality and fractal mathematics cannot be separated from organicism—as any organism is by definition a fractal of parts, self-similar to the whole. Those three elements of the new scientific paradigm—the logic of duality, the fractal non-Euclidean mathematics that describe it, and the organic outlook of all systems of reality made of "cellular numbers" related by networks of energy and information—are proved by the three principles of truth of the scientific method as follows:

— It is based on the most advanced mathematics of the present age, fractal and non-Euclidean topologies. And it provides a mathematical solution to the unification equation of charges and masses, the unfinished job of Einstein, the Saint Grail of physics, found with "thought experiments" as always is the case with the great questions of science.

— It is massively proved by experimental evidence: every organic system, particle, and field of physics, which follows those morphological and functional laws, is a proof of duality. Indeed, its experimental proofs are overwhelming.

As many as all the forms of the Universe, we observe *that follow the lineal, energetic forms of forces and bodies, the informative, cyclical forms of heads, particles and cellular nuclei and the complementarity laws and exchanges of energy and information between both.* Of all those systems of energy and information, treated by this author in different books dedicated to each scientific discipline, we are here concerned with the systems of physical matter and specifically with the systems and laws of quarks, atoms, black holes, and galaxies, which are the themes of cosmology that CERN will explore and the systems of matter that can provoke a nova explosion of the Earth.

Let us briefly consider some other proofs of the fractal paradigm treated in more depth in the appendix.

Mehaute,[10] pioneer of fractal theory, proved that when nature doesn't release energy, *for time to continue, it has to create fractal dimensions of in-form-ation.* Thus, there are two arrows of relative future creation, energy/entropy and information/cyclical time, which transform into each other in the events of creation and extinction of each quantic reality. This transcendental discovery made in the field of chemistry has been totally ignored by physicists who defend the dogma that the universe is dying because it is only expanding (never mind the contraction of energy into charges and masses). The mechanist energetic bias of physics is deep. For example, physicists have a principle of lineal inertia, but NOT a principle of cyclical inertia, when it is proved ad nauseam that electrons, charges, and masses turn around in cycles ad eternal without any force applied to them. Physicists have a principle of an expansive Universe, but they ignore the informative vortices of galaxies that *make our Earth's human space-time a system ruled by the arrow of information, as we exist in a vortex of physical information, the galaxy, not in the expanding space between them*. Physicists still define form, in-form-ation, not as the creation of fractal dimensions as complexity theorists do, but as bits of a lineal equation (Shannon). And yet the most revealing property of information is precisely the fact that in/form/ation, form, creates dimensions. Indeed, think of a sheet of paper extended in space. When we warp it and wrinkle it, we are creating information, form, by adding a dimension of height. This is what all is about: the Universe warps energy into forms with more dimensions and then unwarps dimensional form into energy. And that constant game of creation and destruction of form is the eternal, fractal cycle of all what exists. In the field of physics, it is the game of masses and forces.

Temporal in-*form*-ation, like its name says, is a measure of inner *form*, which we *perceive* when there are discontinuities in a certain bidimensional *surface*—a clock, a page, a computer screen—either because the static form is broken into informative patterns (—) or its movement makes a sudden peak (>), bouncing in an action-reaction cycle. Yet those two trans-form-ations of space into form, of lineal energetic movement, into a cyclical one, are mathematically equivalent to the creation of a fractal geometry of space-time. Thus we define information as a fractal dimension with form, generated by the repetition or quantization of a given *generator equation* or *mother-cell*. Such is the geometric nature of existence in time-space. Indeed, the simplest fractal called a Koch fractal, found in geology, flowers and ice crystals, happens when a system cools down, loses energy, and then a continuous line grows a triangle in the dimension of height. In Einstein's terminology, energy bends, curves, and creates form as time passes. In biological terms, we warp, corrugate, acquire fractal wrinkles as we age. The Universe and its fractal species are constantly generating informative dimensions or erasing them into extended space, *with lesser form*. Each of us goes through the same process in the life-death cycle as we wrinkle our young energy into form and explode back into death.

Since form exhausts itself, the line dissipates its motion (big bang), the cycle of mass whirls inward till there is no more space and then it disappears (so there is no motion in the eye of the hurricane). How did the universe solve this conundrum? It made the wave. It has motion and information, so it can reproduce information in different places of space-time. The wave is both yin and yang, information and energy, *reality itself. Since* as the principle of complementarity of physics, only a system that is a force and a particle, a wave, exists.

The wave is the constant, reproductive orgasm of a Universe in which all is a constant combination of lineal and cyclical motions that self-repeat themselves as they imprint the energy of the cosmos.

14. The Evolution of Mathematical and Logical Paradigms

Science advances not only with experimental data gathered by machines but mainly through the evolution of its logic and mathematical principles. And that is the ultimate meaning of the fractal paradigm: the application of the new discoveries of logic and mathematical sciences to the understanding of the Universe. The fractal paradigm uses a "dualist, organicist" logic, which combines the laws of energy and information, as opposed to the monist logic of "entropy/energy only theories" of quantum cosmology, and uses the new mathematics of fractals, chaos theory, and non-Euclidean topologies, as opposed to the primitive, probabilistic laws used in quantum cosmology. At CERN, there are thousands of physicists putting together a machine, but science is still done with the mind. Humans, not machines, evolve science.

Scientific advances in cosmology require describing the Universe with twenty-first-century fractal and non-Euclidean geometries—not with nineteenth-century probabilistic tools proper of quantum theorists. The new cosmology requires also sound, logic principles (which quantum cosmology break) and the proper instruments of research, which must observe directly the Universe (satellites and telescopes). They don't need weapons and false theories as those sponsored by CERN but harmless thought experiments, free of charge to the taxpayer and telescopes like the Webb telescope, able to prove or not firsthand the supposed big bang. And yet for decades, there has been little advancement in those fields because of the excessive emphasis on computerized, mathematical models, *which do not innovate on the foundations of physics but merely increase the detail of our observations and the visual quality of scientific presentations that seem for that reason more truth, as FX-fictions seem more truth due to the quality of the computer effects.*

One of the most obvious facts of technological science today, when the machine has become the measure of all things, is a growing imbalance between the wealth of data about the Universe obtained with those machines and the lack of new, sweeping linguistic theories constructed with the human mind, able to explain that data. We might say that the details of the trees are obscuring the forest. And what is more frightening, the devolution of the human mind to the mythic, visual world of fictions created by the evolution of the digital mind is limiting the intellectual capacity of mankind to explain that overload of information. Further on, *science fiction, Hawking's style in a society influenced by the audiovisual media, is increasingly accepted as "serious physics," even in classic scientific magazines.* So there are no figures like Mr. Darwin or Mr. Einstein who said "I do thought experiments" in the landscape of science today. Instead, we have teams of scientists feeding computers with data extracted from other machines.

That modern, exclusively mathematical quantum approach to physics often forgets the logic foundations of the scientific method. Science does not build first imaginary mathematical theories and then tries to find them in the real world, but it only seeks explanation to phenomena we have observed. Otherwise, we talk of fiction regardless of the language we use to portrait that fiction. As indeed, all languages including mathematics can construct fictions. Theory is not created outside experience, said Einstein. He said also, "Only a theory whose concepts are close to experimental evidence can be trusted as it reduces the chances of taking a completely erroneous direction." Further on, in mathematics, unlike verbal thought or image thought, the other fundamental languages of man, the meaning of equations is not intuitive, as it happens with words and images. Many scientists ignore the meaning of an equation or symbol in real life. This happens with fundamental concepts such as the concept of a dimension, as it appears in string equations. This causes a second type of errors very common in modern physics: syntactic errors that assume concepts that are not necessarily in those abstract, mathematical concepts. If we see a hand, we know it is a hand, but an X might be many things due to its abstraction. For that reason, epistemology and philosophy of science were created in order to provide tools to differentiate fiction from reality in science.

But a religion survives because it does not dialogue. Whenever you bring these themes to a quantum cosmologist, the dialogue breaks, as it does when you argue the existence of God with a believer. Simply speaking, many nuclear physicists are *not interested* in a deeper understanding of the Universe but in the tabulation of certain particles, its measure and its use for electronic gadgets or weapon designs, *as CERN does. So he can't argue the principles in which the Universe is based.*

In that sense, science, which is the description of the Universe with the mathematical language as philosophy is its description with verbal logic and art with visual images, advances always in the same manner. First, a scientist evolves the languages of science, mathematics, and logic, which are used to understand the geometry of space and the logic of change in time, as Aristotle and Euclid did when the experimental method was invented and Descartes did in the modern age. Next, data is collected with instruments of measure: in the ancient age, human senses; in the modern age, telescopes and microscopes that measure space and clocks that measure time, improving our collection of data, as Galileo did in the modern age. So finally we obtain an improved vision of space and time, thanks to the new mathematical and logic models and a better collection of data, improving our understanding of reality in all sciences. But in the past century, science has improved enormously the instruments of measure without improving the mathematical and logic paradigms used to fit such data in theories that make sense of it. We are still using for most sciences Euclidean geometry and Aristotelian logic, and in a few theories, such as relativity, the new non-Euclidean postulates of the twenty-first century, while dualist logic is banned from science by the mechanist, monist, energy-only dogma of physicists. What the fractal paradigm does is to evolve both, our understanding of the geometry of space and the logic that creates the future, introducing the arrow of information.

In that regard, the sciences of physics and cosmology have advanced due to the capacity of certain minds to harness data about the physical universe within a new, evolved mathematical model. And so we have the following paradigms that have evolved physics:

The Greek Euclidean paradigm, which used static, Euclidean geometry. It explained space as a simplified lineal geometry while Aristotelian logic explained time as change, dividing its studies in the analysis of change in the motion of beings or physical time, studied by physics and change in the morphology of beings or biological time—the life, evolution, and death of species studied later by Darwin.

The Renaissance Cartesian paradigm, which used analytic geometry. This was both an advance and a step back over Euclid and Aristotle's work. The advance over Euclid is clear. Geometry was static till Descartes introduced the Cartesian graph where time could be plotted. Thus, Descartes and Galileo mathematized time as the change in motion of beings ($t=s/v$). But the study only of time as the change in the motion of beings *simplified Aristotle's theory of time, defined as change both in motion, studied by physics and in form, studied by biology. Thus, the simplification and spatialization of time, as a variable dependent only on space, started in earnest.* And the study of time as change in the in/form/ation of beings, as the evolution of "dimensional form," disappeared till Darwin, buried in the "hidden intelligence" of the verbal language since the word in/form/ation means a form-in-action. It would be only in the twentieth century with the discovery of fractal mathematics, equations that generate forms and dimensions, and the study of biological systems, which encode information in forms (proteins' warping) that information will be properly understood. But CERN denies those advances, which in physics were done by fractal relativity that shows the creation of physical form, as time curves the energy of space into masses.

The Baroque differential paradigm, which used calculus, developed by Leibniz and Newton. It established our modern understanding of space as a continuous, mathematical system. Yet the problem of this model was intrinsic to the mathematics it used. As information, form is not differentiable. Fractal equations were still not known. And so functions that had peaks, as those that carry information, were simply ignored. Instead, information was and is still defined in a single dimension (Shannon's bits of data) and so it was time. Indeed, the baroque paradigm maintained the idea that time was totally dependent on space, merely the change in the motion of beings, because it was the only way it could be treated in a Cartesian graph with differential equations. Further on, a naïve error was introduced in the description of space and time when Newton came to the absurd conclusion that because he used a continuous, lineal Cartesian graph to represent time and space, the Universe was continuous and it had only one time and one type of motion, lineal motion. This was an obvious error of Pythagorism. Because Newton used a mathematical plane, which was continuous, differentiable, and had only an Y-coordinates of time, he established one type of motion, lineal inertia (when particles turn eternally in cycles), one single time cycle, that of the clock (when each entity has its own cycles and clocks to measure it). Newton' space and time was a useful tool, an abstract, absolute mathematical frame of reference to facilitate calculus—but it was not reality.

In the same age, Leibniz evolved the concept of time, establishing the existence of multiple times: times were the sum of all the causal relationships from past to future in the Universe, of all the clocks and cycles of reality. Leibniz's definition was more realistic,[11] but for purposes of calculus, it was easier to use Newton's definition, which was chosen by physicists. So a single absolute, lineal, energetic time was born and made dogma, later in the twenty-first century with the concept of entropy.

The thermodynamic/quantum paradigm, which used probability theory, was useful in studying the behavior of atoms and particles, which carry lineal, electroweak forces. It introduced statistical methods to define the behavior of populations of atoms. And it defined further the arrow of energetic time. At the beginning of the twentieth century, Planck applied the new mathematics to the study of electroweak space made of multiple energy quanta, completing the paradigm. But quantum physicists introduced a Pythagoric hyperbolic error when they extended the concept of entropy also to gravitational and strong forces. They missed the meaning of informative, gravitational and strong forces that in/form and balance the electroweak forces. Quantum entropy theories (Hawking and Higgs) denied the work of Einstein and considered that gravitational forces were not informative forces but "entropy" forces. They tried to unify all forces from the perspective of quantum entropy when the proper division of the Universe requires unifying on one side electroweak, lineal forces and on the other side strong and gravitational forces in two different membranes.[App.6]

The second error of quantum physicists was to interpret quantum probabilities as events in time, not populations in space. In mathematics, probabilities can be used to define a population of "cellular, fractal parts" in space, where the total population is 1, or to study a series of future events. Yet quantum physicists, instead of describing an electron as a fractal population of self-similar parts, came with the absurd theory that each fractal part of the electron was an event in time, NOT a part in space. And even when they made pictures of the electron's nebulae made of fractal parts, they kept the temporal interpretation. Since by then, in the sixties, they had created an enormous number of metaphysical theories about multiple Universes, each one representing a different event in time, and other fantasies to which they could not renounce.

Yet at the same time, Time studies advanced thanks to Darwin, which restarted the study of the evolution of form, of biological information, and established the arrow of temporal information that evolves the forms of beings. Unfortunately, physicists still deny Darwin to the point that the Nobel Prize cannot be given to an evolutionist. They even call the arrow of Darwin, of evolution of form, negantropy, the negation of entropy.

The modern non-Euclidean fractal paradigm, developed in this book, which came in the twentieth and twenty-first century and which in mathematics started with Riemann, who discovered the fifth postulate of non-Euclidean mathematics, and Einstein, who found the existence of at least two fractal space-times:

— The visible light space, which follows the Euclidean geometry of light with its three perpendicular dimensions, height/electric field, width/magnetic field, and length/distance, since the vacuum space that determines that geometry is filled with light. Hence our light is our space and its energy the initial energy from where all evolves. He studied it in special relativity[12] and found that this membrane was "contracted" by a factor—c^2t^2 by the creation of light. Again, this was misunderstood as if time was a fourth dimension of space, when it merely said that when we moved in light space, we contracted that space.

— And the gravitational space, which we do not see but is curved. It was described by Einstein in his general relativity, which established the concept of multiple clocks of time, with relative speeds and the geometrical, curved form of masses, which were vortices of physical, informative time.

This is what physicists still don't understand. Physical, gravitational, strong forces, quarks and atomic nuclei, masses bend space into time cycles, balancing the electroweak, lineal, expansive entropy of our world.

In that tug-of-war between electroweak forces that expand mass into energy and quarks and strong forces that bend light into electronic orbits, the Universe finds its balance, its eternal game of two motions.

The importance of Einstein's work, which we shall treat in the next chapter, is precisely his explanation of the arrow of information, of form in physics, which is stored in the frequency of those rotational masses. Thus, as Darwin did in biology, Einstein's theory of mass explained that in physics, time also curves, forms, bends, warps space into physical information, into masses. Einstein thus closes the circle by finally bringing back a dual theory of time=change, as change in form and change in energy and motion. His work would be completed at the end of the century by Mandelbrot[5] in the seventies (discoverer of fractal mathematics), by Mehaute[10] in the eighties (proof of the existence of an arrow of fractal information that acts when entropy doesn't grow), by Nottale[9] in astrophysics (proof of the scalar structure of space-time, and its fractal, elemental units in light-space membranes—the h constant and in gravitational space-membranes—the cosmological constant or string), and by this writer in the nineties, who redefined the laws of duality, the study of all sciences with the arrows of energy and information, and completed Riemann's work in non-Euclidean mathematics[7] (definition of the first, second, third, and fourth postulates of non-Euclidean space-time), applying them to all sciences.

It is precisely this non-Euclidean fractal paradigm the one that keeps evolving and solving the questions posed by physics and many other sciences, while the quantum paradigm has not solved any question of importance *outside the realm of particle physics, which is the only one in which must be applied.* When it comes out of its restricted original use—the study of electroweak forces—it has only created ascientific, bizarre theories as those of Hawking and Higgs. For that reason, the people of Einstein et al. are so worried because twenty-first-century science is proving Einstein and the fractal paradigm right in cosmology, mass theory, black holes, and quark condensates; and if Einstein is right, we are future corpses of the nuclear industry, as CERN will make Einstein's quark condensates, strangelets, and black holes, which will not evaporate.

In that sense, I would like to remember a fact of science often forgotten: "quantity" doesn't matter in the evolution of science, especially in the ages in which a paradigm changes and most practitioners belong to the old, obsolete paradigm today to quantum entropy. Each new paradigm of science starts with a few practitioners, but if their logic and mathematical truths are right, it doesn't matter how many people practiced the previous paradigm. A single Copernicus with reason has more truth than all the high priests of the Vatican. At the beginning, a new paradigm, as Kuhn explains in his work on scientific revolutions, is used by very few scientists, most often denigrated by scholar science, represented by the established practitioners of the previous outdated paradigm, who don't want to lose their "livehood" and prestige as CERN does today, calling the fractal and chaos theorists that denounced it crackpots.

In the twenty-first century, a mathematical theory defined a substance called ether, which was harder than steel, denser than gold, occupied the entire Universe, and yet it allowed planets to travel through it as if ether was transparent and light. It was thus by *contradictio ad absurdum*, an absurd theory despite its mathematical beauty. Yet because many scientists regard the Universe falsely as a mathematical object, they accepted without experimental proof a logically false theory based on only one of the three legs of the scientific method. Unfortunately, as we all know, a stool without three legs will always be unstable and fall down. Such was the case of the ether theory. But since none had come with logical and experimental proofs of its falsity, it remained the hottest field of astrophysical studies for almost a century.

Then a man who never denied the scientific method, who accepted as Descartes and Galileo did, the preeminence of logic on top of all sciences, defied the high priests of physics and proved the ether theory false. His name was Einstein, and his defiance of the high priests of his age already at the Zurich Institute of Technology meant he didn't find a single teacher who wanted to sponsor his thesis and nobody helped him to enter the scholar work. So to maintain himself, he became a patent officer second class and kept working alone in his thesis on the falsity of ether, based on the lack of experimental evidence and logical inconsistency. Then when he published his special relativity theory, he was widely attacked by the high priests of physics. His professor of mathematics said that special relativity was nonsense because its mathematics was so simple that even a high school student could understand them. He also put to use the fifth non-Euclidean postulate, advancing our understanding of mass far more than all the machines the German industry of ether and CERN did and will ever do with their absurd theories of ether and the Higgs particle.

Today unfortunately, we live in a similar age, in the field of physics and for self-similar reasons: the denial by a group of physicists working at CERN, of the need for the three legs of the scientific method, to prove any theory or interpretation of experimental facts. Logic consistence, experimental proof, and mathematical accuracy show that the thesis of CERN are wrong. CERN will not find the meaning of it all but just study one of the infinite big bangs that are taking place in the Universe, very likely the big bang of Gaia, of Mother Earth.

It doesn't matter how many people believe in the cosmic big bang or the evaporation of black holes or the Higgs. Mankind tends to believe in wrong theories for reasons different of truth, mainly authority, the existence of a group with power that imposes its dogmas, the learning of old traditions, which are difficult to uproot, and other spurious reasons well documented by Kuhn in his masterpiece *The Structure of Scientific Revolutions*. As Planck, frustrated by the brutal reception of quantum theory by nineteenth-century obsolete physicists, put it, with sarcasm, "a new scientific truth does not triumph by convincing its opponents and making them see the light, but rather because its opponents eventually die, and a new generation grows up that is familiar with it."[13] The problem with the LHC is that CERN indeed might kill us all trying to prove its obsolete paradigms truth.

In the case of CERN, the press has felt reassured by consulting experts in academic positions of power; but again, only reason, not authority, matters since pioneers, as Kuhn explains, cannot work in scholar institutions that routinely dismiss them. When only Einstein understood that ether did not exist and upheld such truth, it merely meant he couldn't find a job as a physicist in Germany and had to join a patent office. Twenty years later, when Einstein received his Nobel Prize, relativity was not even mentioned. He was awarded the prize for a minor job he did: the explanation of the photoelectric effect. He was brave enough to defy an entire school of false thought from the perspective of the scientific method and he was punished by the high priests of physics.

In the sixteenth century, only a priest in a small church defied the conventional wisdom that the Earth didn't move, held by the most respected scholars of their age. And yet Copernicus was right, eppur si muove.

Today, only a few practicioners of fractal relativity are known to the public. And none of them has been fully accepted by the establishment of physics, dominated by quantum physicists. Nottale had to publish his pioneer work in Hong Kong. Mandelbrot had to work for IBM most of his life. We saw the reaction of Hoggs to Pietronero's proof of the fractal structure of galaxies. My work is published in the science of complexity, given the opposition of quantum physicists to duality, the study of the arrows of energy and information together. This key science whose international congresses I have chaired in the past years is part of the new scientific revolution led by systems sciences and complexity, not by physics. Thus, my opposition to CERN and the establishment of physicists has meant the end of my scientific career, but it won't change the outcome of those experiments since if duality is right and the Universe is made of energy and information, the quark-gluon soup will provoke both a big crunch of quarks into strangelets and a big bang of electrons into radiation, not an expansive entropy-only gas as physicists at CERN pretend.

If the Universe is made of energy and information, information won't disappear by decree because as Einstein put it, "those who impose truth with the authority of power will be the laugh of the Gods."

15. What Is the Universe, a Mechanism of Energy or an Organism of Energy and Information?

In that regard, this book is also about two different cosmogonies, which bring different destinies to mankind:

One, the official dogma of our technological civilization that brought us here is called mechanism, as it considers all what exists a machine, including the human being, the Universe, and of course, machines. Mechanism is the official philosophy of our technological civilization. It was developed initially by Galileo, whose real-life profession was that of an engineer of military devices, working for the princely salary of one thousand ducats a year for the Arsenal of Venice. Galileo was a mechanic who started to philosophize about the Universe, departing from his job. So he modeled the cosmos with ideas deduced from his use of energetic weapons, such as the concept of "lineal time" and "lineal inertia" that exclude half of reality, information, which is encoded in cyclical forms. This has caused an all too obvious paradox in science: physics is the most respected science because it was the first to be born, but at the same time it is the one that has the biggest number of false dogmas because it is so old that the first physicists knew very little about the information of the Universe and extracted their laws from the study of energetic, lineal weaponlike movements.

Galileo also took the clock, another instrument of his work, as the model for his cosmogony. And in this manner, the machine became the religion of physicists. Today, mechanism is still the main philosophy of economics and physics—the two sciences that manufacture machines—not because mechanism is the truth of the Universe, but because machines give power and power imposes those ideologies to the rest of mankind. So physicists made of the machine the "idol of the tribe," which established their superiority over nature and all other human philosophies of knowledge. Galileo's idea that the Universe had to be observed with machines, telescopes, today evolved into cameras and clocks, into computers, instead of human senses, eyes and verbal words that describe time with past, present, and future verbs, however *degraded the mental organs of human beings are to a secondary status*. Soon, the Universe became modeled no longer as a complex organism but as a mechanism and man became an imperfect machine (instead of considering machines imperfect organisms): "I should like you to consider that these functions (including passion, memory, and imagination) follow from the mere arrangement of the machine's organs every bit as naturally as the movements of a clock or other automaton follow from the arrangement of its counter-weights and wheels" (Descartes, *Treatise of Man*). It was a radical change in history since mechanism created the world we live in. Indeed, all other ideologies—capitalism, techno-utopias, even

Marxism—come from mechanism, the belief in the machine as the measure of all things.

The other great philosophy of the Universe is called organicism, as it considers all what exists a system made of networks of energy and information, the most complex of which was the human being. In that regard, organicism differs from mechanism in putting information, not energy, and humans, not energetic machines a.k.a. weapons in the summit of creation, *since we are the most complex informative species of the known Universe*. When people think of organisms, they think of a human being and so they have a hard time accepting the organic models of galaxies and planets like Gaia. Yet they might be aware that the human being is precisely the supreme organism, gifted with certain qualities such as self-consciousness and freedom of choice (at least among the most developed individuals) that simpler organisms lack. In that regard, a planet like Gaia or a galaxy are simpler organisms akin to cellular systems or robotic machines, whose networks of energy and information interact with very limited degrees of freedom (a parameter directly related to the complexity of an organism) and null self-awareness. This subtle difference is of enormous importance to CERN's issue. Because while mechanism requires machines to observe the Universe and models it as a simple mechanism with limited complexity of "form," of information, organicism regards energy merely as the canvas of creation and gives far more importance to the processes that "fractalize," break that continuous energy into the discontinuous particles and entities of information that inhabit the Universe.

Organicism is also far richer in its description of reality because it uses two causes, energy and information, whose interactions are the key to explain any organic system of reality, explained as a system of two networks made of energy bites and information bits. In an organism, cellular networks of self-similar entities, one of energy and one of information, establish two self-regulated feed-back cycles that either create information from energy by warping, evolving form or destroy information by extending and simplifying its form in a big bang: E⇔I. Those two processes can be described with a "vitalist" or an "abstract" jargon, which is merely an ideological choice. For example, in quantum physics, all theories are called gauge theories because particles measure constantly their distances, thus they process information in an automated manner as quantum computers do. Particles also absorb and process energy, often reproducing new particles. And finally they group in complex social networks, creating atoms and molecules. Those four "organic properties," feeding on energy, processing information, reproducing self-similar fractal forms and socializing in networks, are the four dimensions or wills of organicism, used to describe any system of the Universe. They can be described with words or with fractal, self-generating equations, but in all cases, they provide a rational why to the events of reality.

The main difference, therefore, between mechanism and organicism is that mechanism can only describe the how of the Universe, but as Feynman says, "a physicist never asks the why" because mechanism can't explain why the Universe behaves like a machine. The founding fathers of physics, all of them pious believers, loved mechanism because, in fact, it was for them the strongest proof of the existence of a personal god, the why that had set in motion the wheels and mechanisms of the Universe. So as believers, they introduced some basic anthropomorphic religious errors in their perception of reality and an arrogant sense of being always right, proper of the military profession, they practiced, exemplified by Newton, who worked for the British Navy and thought comets were messengers of God sent to him or Kepler, who affirmed that "Yes, I am the one: God Himself has waited for six thousand years for ME, who looks at His creation with understanding.". In that sense, mechanism requires a religious belief that organicism doesn't require because it gives to each particle and fractal part of the whole Universe the capacity to process energy and information. Thus, the why of organicism is simple, tautological, and embedded in any system, which constantly accumulates more energy and information because it is made of those two elements. Today, mechanism is still accepted in physics because most physicists don't have any philosophical, conceptual interest for the why. They merely use machines to measure reality and put the data on equations to feed computer models. So they don't even realize that mechanism is a religious belief, which requires an external creator. And to be frank, most of them, without saying it, still believe in religions or in the superiority of man and the physicist as the most intelligent being of the Universe.

In simple terms, the main thesis of organicism is that any entity of the Universe needs both elements, energy and information, which have opposite properties to exist and maintain its balance with its ecosystem with whom it constantly exchanges both. In the same manner, a human being needs both, a head of information and a body of energy to balance itself.

16. Man, the Supreme Species of Information, as the Measure of All Things

But the biggest advance of an organic model of the Universe modeled with systems of energy and information is even more fascinating. Because if all that exists even the species of physical nature are made of energy that becomes warped into information and reproduces its form, then the ultimate model of the Universe is an organic model. All is made then to the image and likeness of man, the most perfect organism of the Universe. We, not machines, are the summit of creation. And this makes us and this planet precious again and the quark cannon and the physical theories of only energy the enemies of the supremacy of mankind.

Life matters. If the Universe is a fractal of energy and information, as the Hubble has proved, it is not a machine but rather an organic system, given the fact that fractal mathematics have biological properties. Fractals increase, evolve constantly their form (as life does), and are self-generative equations; thus they reproduce, and indeed, even the smallest electrons and quarks reproduce their form or break apart into a myriad of self-similar smaller fractals (electron nebulae). Finally, energetic and informative organs are self-similar in all the systems and scales of reality, based on the efficiency of their morphologies, *which are selected by the laws of evolution*, as life does with biological forms. For example, all energetic systems tend to be lineal, from limbs to fields of forces, because the line is the fastest distance between two points and all informative systems, from particles to cameras to eyes and brains, tend to be spherical because a sphere is the topology that accumulates more information in lesser space.

The Universe is a fractal organic system. It is not a mechanism, which is either created to deliver energy or information, and so it cannot exist per se, *but a combination of both elements*, which constantly transform into each other, in feed-back cycles that make reality eternal.

And indeed, the best form to survive is to achieve a balance and complementarity between energy and form. This balance is perceived as a classic harmony or beauty, which defines the efficient survival of any species. So when a system is not beautiful in form and efficient in function, it dies away, it becomes an unstable particle, a failed mutation, an extinct species, or a crazy idea as the false, ugly theories of quantum cosmology are. This simple rule that establishes the laws of survival can be put in words, as the rule of the "golden mean," in logic terms, as the principle of complementarity, or in equations as $E = I_{max.t}$ or Max. $E \times I_{(e=i)}$.

The beauty of life is born precisely on the balance of energy and form of its organisms. And the beauty of the fractal paradigm resides in the capacity to express self-similar laws that mimic reality for all entities, of all scales, with all the languages of reality. Indeed, the generator equation of constant space-times writes in English as $E \times I = st$ (energy $\times$ information create a constant space-time field). Yet that mathematical function reads "exist" because what we just have seen is the function of existence in space-time. There is no magic in all those languages, but a common code with the game of the fractal Universe and its self-similar laws that the fractal paradigm is discovery with the languages of mankind, far more telling of the essence of reality than any picture the LHC will ever obtain.

We do not exist in a solid, continuous reality but in a puzzle of parts, where all those parts—galaxies, stars in a galaxy, atoms in a star, particles in an atom—joined by energy and information networks, come together into wholes that share the same basic properties than the parts.

It is a tragedy that we might be living the last days of mankind, precisely when the human being, one of its most beautiful organisms, is becoming in science again as it was in philosophy *the measure of all things*, far more perfect as an organism than any black hole will ever be.

Moreover, because this simple model can become by repetition as complex as we wish, it applies equally to biology (so a human being is a fractal of cells), sociology (so a nation or human god is a fractal of citizens or believers), or astrophysics. Thus, this new revolution of science means a real unification of all realities and the harmony of all sciences. In the past, physics claimed dominance because, in an ideological world of energy-only theories, the study of the biggest bang and the biggest system was the most important thing. But in a model in which the bigger (energetic space) is simpler than the smaller (informative time-cycle), it is equally important to study the vacuum of the big bang than the complex forms of the human being. *In fact, as we speak, the data banks of biology are overcoming the data we have about astrophysics*, showing the complexity of the human informative summit of the Universe. This means that knowledge of the fractal Universe is achieved better studying the human organism than the cosmos since the energy-only cosmogonies sponsored by physicists are just simplistic systems and one-half theories of reality. This means monist, energetic physics is just one-half and the less important half of science, the science of energy, motion, and machines, while biology is the science of information, complexity, and life organisms. Duality and the fractal paradigm will dominate twenty-first-century sciences because it fuses both and so it gives us a complete view of what reality, man, and the Universe are.

Further on, the fundamental argument that should guide funding and research in science is the obvious fact that we are human beings, not black holes, nor particles. Thus, to know all about man and our scale of existence and to preserve life is far more important than to know more about energetic machines and black holes.

In that regard, the reader must understand that this book is not at all an attack on science and its good fruits that are knowledge and the creation of technologies that favor the future of mankind. As a cosmologist, mathematician, and philosopher of science, in love with knowledge, I refuse to accept that science per se is a wrong path for mankind.

On the contrary, the only path of survival for mankind is to evolve primitive science, the web of absurd beliefs and mechanist myths developed centuries ago by mechanist physicists, who denied the arrow of life and information in time and have defended ever since, against reason, logic, mathematical and experimental evidence, a Universe made only of energy and entropy. Today, that ideology is defended by quantum cosmologists with the power of the industrial-military complex. Because I consider life superior to any machine and its protection a higher form of knowledge, since an extinct scientist knows nothing, I have fought all my professional career false science, the kind of ideologies and experiments CERN represents.

Quantum nuclear physicists, despite being human beings, which certainly must wish to survive, hold a mechanist ideology based on an outdated sixteenth-century philosophy of science, founded by a maker of weapons—Galileo, an employee of the Venice Arsenal, with the princely salary of one thousand golden ducats—to provide machines of war and methods to improve the art of ballistics, to which he dedicated his first book, military compasses. Those exercises we do in first-year physics to calculate the trajectory, speed, and curve of a cannonball were the first fruits of the science of physics with real consequences in the battlefield and the quark cannon its natural end.

Physics makes machines of war and energetic theories that bias its description of reality with two errors that have brought man to the brink of extinction:

— A mechanist bias that models reality with machines and scorns organisms and life, making possible what seems a monstrosity for all other human beings, but business as usual for nuclear physicists: to build a machine that can extinguish the Earth.

— A theoretical bias that develops physical theories always based on energy. Indeed, physicists ignore all about information, which is the opposite substance of the Universe, as we human beings are made with minimal energy and maximal information. All their theories are based on the ideology, proved wrong ad nauseam by Einstein, Darwin, and the existence of life that the Universe only has a direction of future called entropy. So physicists sponsor the false theory that the Universe is dying, expanding, born of a big bang; Hawking believes the false theory that black holes evaporate information because they have only entropy (information paradox); and Higgs denies Einstein's theory of mass as a whirl of space-time, which carries in its angular frequency, as it turns around, like an attractive hurricane does, the physical information of the Universe. Instead he proposes an absurd particle that should magically give mass, colliding with other particles. Those are the three specific false theories based on energy only that CERN pretends to prove exploding the Earth, with the connivance of the military-industrial complex built around the nuclear industry.

Is this absurd concept of science the new frontier of knowledge in cosmology? Of course not. Knowledge in cosmology keeps evolving precisely with the work of scientists, here grouped under the collective name of Einstein et al., who are explaining the Universe, with *both substances, energy, and information. They accept the existence of information, life, and ethics as they guide of true science and have solved all the questions CERN will never resolve with energy-only theories and quark cannons.*[App]

The ultimate reason we are in danger of extinction is the ideology of mechanism in its multiple sides. We keep evolving weapons because mechanists say that "machines never harm a human being." Guns are not to be blamed according to the NRA but the humans who use them. Of course, this is not truth. In the same way we are programmed to desire more energy and information, if we have a machine of energy or information, we will use it with its purpose in mind. So a machine that kills will be used to kill by our desire to have more energy, and a machine that hypnotizes our mind and infuses destructive ideas, such as a TV with violent films, will be used because we desire more information. So the only way to avoid those results is to forbid machines that harm the human body and brain. But mechanism does not accept those facts, making human slaves of their wantings for self-destructive machines.

Further on, mechanism, with its definition of time as "what a clock measures" and space as what a telescope sees (naïve realism), needs to evolve machines to evolve knowledge. We shall see how wrong are those definitions of time and space in the following chapters since 96 percent of space-time is a gravitational membrane invisible to us and time=change is not what a clock measures (merely a numerical rhythm of change) but the arrows of future, energy, and information, whose interaction and changes create the Universe. Yet those ideas about machines are an industrial dogma, promoted by massive systems of marketing and propaganda that censor organicism. Truth bends to the power of the machine. Yet truth exists in the Universe, and the truths about the destructive effects of the LHC will determine ultimately our survival.

True scientists like Einstein et al. make thought experiments. But they do not have 10 billion dollars provided by the military-industrial complex to explain the world. And so Higgs, not Einstein, is today the colossus of gravitational science; and a weapon, not the human mind, is sold as the best instrument of knowledge. This book will prove the opposite truth.

Indeed, what is the proposal of quantum physicists to substitute Einstein's cycle of mass information and all the solutions we have briefly exposed before to the conundrums of physics? The Higgs, which solves nothing and makes no sense at all.

---

[1]   Http://www.newscientist.com/article/mg19325941.600.

[2]   The Sloan Digital Sky Survey (SDSS) is one of the most ambitious and influential surveys in the history of astronomy. Over eight years of operations (SDSS-I, 2000–2005; SDSS-II, 2005–2008), it obtained deep, multicolor images covering more than a quarter of the sky and created three-dimensional maps containing more than 930,000 galaxies and more than 120,000 quasars. SDSS data have been released to the scientific community and the general public in annual increments, with the final public data release from SDSS-II scheduled for October 31, 2008. Http://www.sdss.org/.

[3]   Mr. Oldershaw published an empirical analysis of the fractal scales of physical matter (International Journal of Theoretical Physics, vol. 28, no. 6, pp. 669–694 and no. 12, pp. 1503–1532, 1989) noting that the scales of mass, length, and time grow in simple power laws. He belongs to a growing number of scientists who tabulated those regularities during the eighties and nineties, establishing the empirical basis of the fractal paradigm, paving the way for the theoretical revolution of the nineties and two thousands, carried about by Nottale in the analysis of fractal space and this author in fractal time. The next phase of the fractal paradigm, the most difficult, as Mr. Kuhn remarks in his work on scientific revolutions, will be the acceptance by a new generation of physicists that the advances on science no longer depart from the exhausted quantum paradigm with its single time arrow and single space-time continuum, but from the study of reality with non-Euclidean and fractal geometries, multiple space-time membranes, and multiple time arrows.

[4]   A. H. Guth and P. J. Steinhardt, "The Inflationary Universe," Scientific American, 250(5), (1984):116–128.

[5]   B. B. Mandelbrot, Fractals (W. H. Freeman: New York, 1977).

[6]   Albert Einstein, "Kosmologische Betrachtungen zur allgemeinen Relativitätstheorie (Cosmological Considerations in the General Theory of Relativity)", Königlich Preussische Akademie der Wissenschaften

[7]   Non-Euclidean geometry systems differ from Euclidean geometry in that they modify Euclid's fifth postulate, which is also known as the parallel postulate.

        In general, there are two forms of (homogeneous) non-Euclidean geometry, hyperbolic geometry and elliptic geometry. In hyperbolic geometry, there are many distinct lines through a particular point that will not intersect with another given line. In elliptic geometry, there are no lines that will not intersect, as all that start to separate will converge. In addition, elliptic geometry modifies Euclid's first postulate so that two points determine at least one line. Riemannian geometry is the best-known elliptic non-E geometry, which deals with geometries which are not homogeneous, which means that in some sense, not all the points are the same. Thus, those geometries later used by Einstein and Minkowski to describe space-time did have the seeds to understand fractal points of different form and size in which multiple parallels converge. Yet till the publication of my books *Radiations of Space-Time,Unfication Theory* (BookMasters, 1997), and *Time Cycles* (Editorial Arabera, 2004), which adapted the five postulates of Euclidean geometry and fused the concept of a non-Euclidean point and a fractal point, there was no exhaustive model to study with the same laws the different topologies and scales of the Universe. See http://journals.isss.org/index.php/proceedings50th/article/view/29.

[8]   Albert Einstein, "Über das Relativitätsprinzip und die aus demselben gezogene Folgerungen," *Jahrbuch der Radioaktivitaet und Elektronik* (1907); translated "On the relativity principle and the conclusions drawn from it," is Einstein's first statement of the equivalence principle.

[9]   Nottale, L. (1993). Fractal Space-Time and Microphysics, World Scientific, Singapore.

[10]   Les géométries fractales : l'espace-temps brisé Hermes Sciences Publicat. (2000).

[11]   Leibiniz's *Monadology* and *Letters to Carter*, a disciple of Newton, starts a centuries-old argument on the fractal or continuous, multiple or unique nature of space-time. Leibniz's genius was however ignored for his lack of practical applications and only now with the extension of the fractal paradigm is fully recognized. The fractal geometry promoted by Mandelbrot, a self-recognized admirer, drew on Leibniz's notions of self-similarity and the principle of continuity: natura non facit saltus. Mandelbrot would say, "His number and variety of premonitory thrusts is overwhelming." Leibniz also wrote that "the straight line is a curve, any part of which is similar to the whole," anticipating the fractal, non-Euclidean topology explained in this book for more than three centuries. One of his metaphysical principles is of certain importance in this crossroads of history, the principle that the Universe must be the most perfect possible, which seems to imply that we humans will not make the "cut" given the enormous degree of arrogance, ignorance, and despise for nature and the "perfect laws" of that Universe.

[12]   Einstein, Albert (1905), "On the Electrodynamics of Moving Bodies", *Annalen der Physik*: 891–921

[13]   Planck, *Autobiography.*

CHAPTER 3

The Higgs Hoax

The Higgs is a toilet particle.
—Weinberg, Nobel Prize, worker of the electroweak formalism

I. God's Particle: Fantaphysics

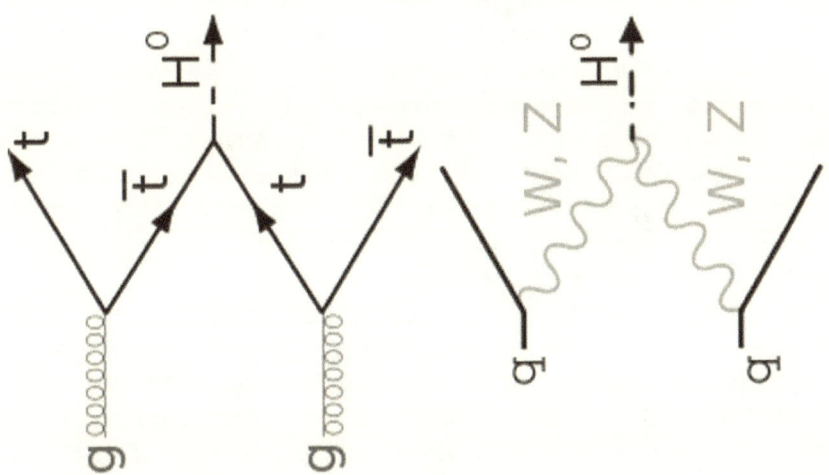

THE HIGGS IS A KNOWN PARTICLE, A TOP/ANTITOP FIELD WHOSE FUNCTION IS TO BREAK=KILL THE SYMMETRY OF OUR MATTER WHICH EVOLVES INTO HEAVIER QUARKS: NEUTRON+PROTON=W+Z=TOP→ANTITOP QUARK (HIGGS)

1. The Higgs as an Industrial Tool of Marketing

The Higgs particle is merely an equation that describes a top and antitop quark with different symbols. It does not exist *as a new particle to discover, worth 10 billion dollars and the risk of obliterating the Earth.* In the cover, you can see the absolute equality of mass and position of both

177

particles. And in the graph above, the reaction of a top+antitop quark, which should fuse together, creating a Higgs particle, in case that this particle exists, which will not because it breaks some elemental laws of complex and quantum physics, as we shall see along the pages of this paragraph.

In any case, at best, the Higgs is not a real particle but a mathematical variation on the same theme. It is like calling a brunette a dark-haired girl. It is still the same person though we are using two different names. And that same person in the Higgs case is the heaviest quark, the top/antitop quark, which evolves all the particles it finds around into heavier quarks, including our matter, breaking its symmetry. Physicists know all this since the top/antitop quark cycle was used by Nambu,[1] last year's Nobel Prize, to describe the death of our matter that in a medium which contains top quarks becomes transformed first into heavier Z and W particles, and finally it dies converted into a top quark or other heavier quarks. In simple terms, Higgs rewrote Nambu's equation to explain *again* how our matter breaks its symmetry and dies when it is absorbed and transformed into heavier quarks as it speeds up and converts lineal energy into mass, $e=mc^2$.

Thus, all the fuss about the Higgs relates to a specific reaction, the death of our matter, called in the jargon of physics, the breaking of electroweak symmetry, which becomes transformed into heavier quarks. This death is explained by quantum physicists with two transitional particles in which our atoms become transformed, before adding up its masses to obtain exactly the mass of a top quark, the W and Z particles, shown in the previous graph. This process is also called electroweak decay. When our particles reach c-speed, their energy becomes cyclical motion, curling into a vortex of mass, first becoming heavy W and Z particles and then Higgs=top quarks or other heavy quarks.

This is truly all the fuss there is about "God's particle"—a redundant particle that explains how our mass dies and becomes the W+Z=Top quark=Higgs. *Yet according to the principles of efficiency and simplicity, the Universe doesn't create redundant particles.* Thus, because a top quark condensate[2] (a mixture of tops and antitops) does the same than a Higgs particle and has its same weight, instead of Higgs, we shall observe top quarks/antiquarks eating our mass. Those "team particles" break our symmetry, killing our mass, as a lion/lioness team does with a gazelle in the savanna. And that is what we must fear: a top quark condensate will act as a strangelet does. Once formed, it will start "hunting" down our atoms at an exponential rate. It will accrete the Earth and form a top quark frozen star, whose properties coincide with those of the densest black holes.

Because the LHC is expressly created to make Higgs, it certainly will do many tops. In fact, Fermilab has now for a decade limited further the possible mass of the Higgs, and those constraints have been coming closer to the top quark, which will be finally its exact mass.

Obviously, this known fact of nuclear physics is not explained by CERN because it will make the LHC an unnecessary waste of resources. The cynicism of the nuclear community in this regard is shown by the fact that they gave the Nobel Prize to Nambu, not to Higgs. The idea is simple. Once the LHC discovers top quarks, not Higgs, Nambu will come out with Higgs and say, "We are now convinced the Higgs is the top quark condensate" and they will all cheer the discovery after spending 10 billion dollars. But what they don't acknowledge is that they were smart cheating mankind, but not intelligent enough to understand the conference might not happen because a top quark star if formed, will devour the speakers.

In the past twenty years, the establishment of nuclear physics has chosen the redundant theory of Higgs while the work of Nambu is ignored because a new invisible particle was a perfect alibi to get new funds for an industry which no longer had military uses.

Indeed, thanks to the Higgs, the industry of atomic cannons was resurrected. *Nuclear physicists can now keep building quark cannons forever as they shall never find the false particle and will always say, "We need more energy to find it."*

The Higgs allowed the creation of superfluid quark cannons. At the end of the cold war, during the Reagan administration, the military-industrial complex didn't need new nuclear weapons. The war was over. So the first quark cannon, the SCC or superconductive collider was marketed no longer as a weapon that might obliterate this planet or an ultraexpensive machine that would give great benefits to the industry, but as a tool to search the Higgs particle. The military were trying to downsize their massive arsenals in both countries, the United States and Russia. Further on, the essential particles of the standard model of physics have all been discovered. So it was difficult to sell the research of the third horizon of nuclear weapons or mass bombs, based on Einstein's second equation, $M=E/c^2$, as a tool of knowledge. But when there is a will, there is a way.

It is in this moment of history when the God's particle hype appears to the rescue, sponsored by Mr. Leo Lederman, the main lobbyist of the industry, in charge of Fermilab, the biggest accelerator made till that date. He published a book called God's particle[3] and used it as a perfect tool of marketing to reconvert the industry of atomic cannons into civil use. He convinced Reagan, a president in love with the military and God, to give the go-ahead; and the SCC, the superconductive collider, was funded. Soon it started construction at Texas at a cost equal to the expenses of the National Health Institute that year.

2. Cubist Science

Why we are so sure the Higgs is a mathematical reformulation of the top quark with the same mass and properties? Because of the laws of the scientific method and the properties of information, which *is inflationary by definition. Languages of information are mirrors that can give self-similar perspectives of the same entity from different points of view.* If Zwicky, who discovered dark quark matter, said that "physicists are spherical bastards" because they are bastards any way you look at them, the Higgs, we might say, is a cubic particle—another linguistic perspective of Nambu's top quark/antiquark condensate, described with a self-similar mathematical formula. This is in the same way that I am both a gray-haired man and a man with gray hair, but just because I can be described by my hair with two self-similar words I do not exist twice, there is not a top and a Higgs. There are no two Einsteins, but there are two verbal words that describe his *disordered* and *unruly* hair. And there are thousands of pictures of Einstein and so on. Physicists though ignore the inflationary laws of information developed by the twenty-first-century science of complexity, and they are Pythagoric—they believe any equation is real. So they don't rule out two equal particles, the top/antitop and the Higgs, as two mathematical sentences describe them.

Yet the experimental method is further proving that redundancy: the experiments at the Fermilab are closing the mass of Higgs in the range of the top.

How did Higgs get away with cheating? Brits are good pirates. Later we shall see how Hawking became a celebrity denying Einstein. What Higgs did was a mere repetition with worse maths of the top/antitop field, as the windows system is a mere copycat with worse symbols than the apple system. So obvious it is that when Mr. Lederman, a Jewish-American lobbyist with huge leverage in Washington, sold his theory of God's particle to the world, the most important physicist alive today, Mr. Weinberg, the man who explained and unified the electroweak force, called it the toilet particle, which needed to be flushed down the toilet—metaphor for a whirl of mass of Dr. Einstein. *Yet history proves that money imposes its truths and so we all use Windows, which is just slower than Apple because it was rewritten to hide the copycat and has a lot of redundant loops. And those loops allowed Gates to won the suit Apple put for copycat. The same happened with Higgs: he copied Goldstone who copied Nambu, a mild Japanese who didn't dare* come out of the closet. So the toilet is now God—as everybody uses Windows, all use Higgs and none says the truth because if they do as Einstein did with the ether theory or al. have done with CERN, they will lose their livelihood. The problem and only reason why al. is speaking is that if physicists don't talk soon, we all will might lose our lives.

Knowledge is only a collateral benefit of the nuclear industry whose main purpose is to explore new weapons. Yet this marketing of knowledge paradoxically has increased as the importance for research of those weapons diminished. Since today the fundamental particles of the quantum world are all discovered, the standard model is closed. There are six quarks and three electrons. And that is all you need to build up the quantum world with standard science. Then *you use Einstein's gravitation and mass theory to explain the cosmos.*

And so with those known-known elements we can explain it all. *The rest is inefficient, fantasy, inflationary information and the scientific method was precisely invented to get away with inflationary information.*

We can invent all fictions and theories that do not have experimental proofs, don't follow the laws of the scientific method, do not explain anything but are useful alibis to build machines. This is the case of the Higgs, an absurd particle that is useless to understand mass but an excellent marketing device to get funding for the quark factory. And that is how we must read the history of this particle: CERN, the nuclear company of Europe, got funding for its machine to search for a false particle called the Higgs.

Fundamental research with atomic cannons came to an end a decade ago, when we found the top quark. And once the cold war ended, its military use also ended. So the plans to build a quark cannon in America and Russia were scrapped off when the cold war ended two decades ago.[4] There was nothing else relevant to discover, and the dangers of entering into the region of death of our matter were immense.

Then the Europeans took over and constructed it. And so suddenly a magic particle called the Higgs that pretends to substitute Einstein's masses as an accelerated vortex of space-time, according to his law of equivalence between mass and acceleration, became the *theory of mass*, God's particle. So the cubic toilet became Go(l)d, reaching—like Duchamp's ready-made toilet—a stratospheric price, the price of 10 billion dollars. Still, many know that Duchamp's artwork *is just a toilet hanging upside down. So it is Higgs: a badly written top quark condensate.* Let us give to Caesar, Mr. Einstein, and Nambu what belongs to Caesar.

The Higgs also illustrates how nuclear physicists form a community against the world they bomb and cheat and laugh at. Since last year, they gave the Nobel Prize to Mr. Nambu for explaining with top quarks the breaking of symmetry, the death of our matter (*not the meaning of mass, provided only by Einstein*), that Higgs pretends to explain with new unneeded particles. It is easy to see the strategy behind this surprising prize that was *not given to*

Higgs despite being Higgs the official theory of symmetry-breaking (death of our weak matter). The idea is that CERN can cheat mankind 10 billion dollars because it is common knowledge among nuclear physicists that we humans are retarded sheeple that will pay anything to get killed by the smartest guys on the block.

But they are *not going to cheat within the community group. So they* will NOT discover the Higgs, and then Nambu and his theory of top quarks will become the new official theory and everybody will congratulate themselves for the insight of giving Nambu the Nobel Prize. It is tricky indeed. But nuclear physicists need to be tricky to get billions of dollars from clueless politicians after the cold war ended and the MAD strategy they used (Mutual Assured Destruction) to sell their weapons is not convincing anymore.

Of course, if people knew all this, there would be an enormous public outcry of the real normal, good people of Europe, as there was a public outcry against AAA-derivatives that turned out to be financial scams by the experts of banking. If people knew the machine means big nuclear cannon instead of large hadron collider; if they knew that beyond 1 teravolt of energy, the minimal energy this atomic cannon is set up to explore, our matter dies and becomes dark matter; if they knew that 75 percent of the matter of the universe is dark quark matter (standard model), the dominant species of the Universe that feeds on us; *if they knew this weapon will produce the most powerful explosive of the universe, quark-gluon liquids that blow up planets and stars,* mankind would be hysterical. If the press in love with technology had not censored truth, mankind and the clueless politicians cheated by CERN would have never allowed this monstrosity to happen. And nuclear physicists would have gone unemployed after the cold war ended. *So all this is censored.* As Johnson put it, "Every expert has a very personal stake in the matter. Generally speaking, the experts are either afraid for their livelihoods or afraid for their lives."

And yet, once the standard model was completed and the Z particle and the top quark found in the nineties, the field of particle physics should be closed. *And that is OK.* The laws of thermodynamics were found in the twenty-first century and we do not expect new laws of thermodynamics. We do not need theories that deny those well-proved laws. The scientific method, the guardian of truth in physics, knows that thermodynamics is closed. And we are proud of its building (only Mr. Hawking with his denial of those laws in his musings about evaporating black holes expect new breakthroughs in the discipline).

So happens with the standard model of particles: its discovery is closed with the finding of the top quark and we are proud of it. A blogger told me angrily that this might be truth but then the LHC shall obtain the 21 decimal in the Z particle. Yet the 20 decimals we have for the mass of the Z particle are enough for all practical purposes. After all, we do not know the average weight of the species Leo Tigris till the 20 decimal and that doesn't take the sleep away from any tiger expert. Why should the Z particle, whose name was chosen precisely because it was the last particle we needed to discover, be treated differently? It should not.

However, tiger biologists do not make atomic weapons, do not build an industrial-military complex, do not lobby with billion-dollar industries that seek taxpayer's profits, spending new money in technology and building evermore expensive machines. So they don't go around promising eternal motion machines (evaporating black holes, which recently were featured in science news as the ideal propellers for galactic ships), or selling imaginary God's particles. Biologists do not lie to the administrators of science about the importance of the Siberian tiger and the new powerful weapons we could extract from its teeth. *They respect science, its purpose, its ethics, and its search for knowledge. They might not be as smart as Higgs and Hawking*, who said he didn't study his father's profession because smart people do physics, not biology, and also that if we didn't fire the LHC to test his "Nobel Prize," "humans would not deserve to be called humans." Maybe *but life sciences have integrity, dignitas, love the complexity of human beings, and fight for our survival. They are "people." "Dignitas"* is exactly what nuclear scientists of the atomic industry don't have when they cheated European administrators with the Higgs Hoax to get funding for an ultraexpensive billion-dollar accelerator with very limited use. It was for them a routine. All the experts in atomic weapons were also experts in accelerators and particle physics, and they used their oiled contacts to sell an imaginary particle called the Higgs, ever increasing in its imaginary mass—and hence ever in need of more powerful accelerators to find it and put us in contact with God. *But how do you invent a particle out of nothing? Maybe it is the top quark the one that is false?*

No, because we have seen it. Gödel, the main mathematician of the twentieth century, showed how mathematical fictions are invented: mathematically a physicist can invent as many particles as he desires as a novelist can invent with words as many characters as he wishes, while in zoology artistically one can draw as many monsters as he wants. But the Universe is efficient and so only balanced, harmonic, real particles and species, which have a role in the cosmological or biological ecosystem, exist while the Higgs is a *mess*. The next drawings show indeed how *messed up it is.*

3. Higgs as God

*The behaviour of physicists in a crowded social event at a
conference is equivalent to the Higgs mechanism, as proposed
by David Miller (University College London).*

In the graph, physicists represent a never-found invisible medium that permeates all space. Thus, for the Higgs to exist, a parallel Universe of space-time, the field Higgs must exist. Yet this invisible reality, invented by Mr. Higgs, a professor from Edinburgh playing God, has never been found. Further on, a top quark/antitop quark system, described by Nobel Prize Mr. Nambu, performs the same function than the Higgs: to kill/break the symmetry of our matter. So by the laws of the scientific method—simplicity, economicity, and efficiency—the top/antitop is the Higgs, which does not give mass to other particles but merely kills them. The hypothesis that the Higgs also gives mass to particles was an idea that came out later during the marketing of this particle as the meaning of it all, but Higgs was never able to explain how it did it. So a student invented the esoteric Higgs mechanism, which converts particles in humanlike sentient beings. Those particles, represented in the drawings by blue physicists, cluster around Einstein, who enters the room, slowing the scientist's progress. In this manner they give the scientist mass. But they treat each particle differently. They like some (the celebrities of the party, which have more mass) while other particles are liked less, so they move faster and have less mass.

The LHC and the Higgs were created to understand a force called the weak force. The weak force has unlike all other forces some fascinating properties. It breaks its spatial symmetry, it doesn't behave the same when it goes left and right. It is also a force that has no spatial range. And finally, its constant is measured in time units. Add 1+1+1 and you have the explanation to this puzzle: the weak force is NOT a spatial force but a temporal event, the only force that evolves particles in time through a transitional state (the W and Z states) or devolves them into other particles.

For that reason, its constant has time units, its reactions last so long in time and happen in the same place of space, as an egg evolves without moving. For this reason, it does not keep its spatial symmetry. Einstein said, "Wires don't travel to the past"; that only happens in Hawking's musings. Events have causality in time from the past to the future. This is the reason why those reactions don't have parity. They don't happen from future to past.

Yet quantum physicists, unable to grasp the duality of energy and information, of forces that share energy and events that d=evolve particles of information, have never been able to divide forces and events but treat all happenings of reality as spatial forces. So they decided to explain the weak event as a spatial force that exchanges particles in space, copying the models used to explain spatial forces like the force of light. Thus, to that purpose, they considered the transitional states of growth of informative mass that particles undergo in weak events (the Z and W events) as particles exchanged between other particles. This is absurd, and the first description of the weak event by Gell-Mann and Feynman did not proceed in this manner, but considered the weak event to happen in the same space-time point.

One of the fundamental advances of complex physics is indeed to reclassify all phenomena in three types of happenings:

— Spatial events that transfer energy between particles, being the electromagnetic exchanges of photons the main case. These events have a long range in space and a small period in time.
— Informative, temporal events that trans/form the particles of the event and have minimal range in space and longer duration in time. The weak force is the main event in our membrane of space-time.
— Transcendental events that transfer energy or form between scales/membranes of space-time. These events are related to the transformation of electromagnetic energy into mass and vice versa, explained by Einstein's dual equations $E=Mc^2$ and $M=E/c^2$. They are also the events related in the cosmological scale to the creation of dark energy by black holes and the creation of light from dark gravitational energy, as it enters the realm of galaxies. *These events are the less understood in cosmology since it is a dogma that there is a single space-time continuum.*

Thus, when dealing with basic errors of classic physics, we observe that all the *transcendental* and *temporal* events of physics are explained as if they were spatial events.

It is not the theme of this book to deal in depth with the proper informative, temporal analysis of the weak force but with the error of Higgs. Basically, what he did is to consider that masses and events related to the transformation of electromagnetic particles into masses should be explained with the same equations and concepts of spatial forces. And to that aim he played with equations and finally "**invented**" a scalar particle that should "transfer mass" and "weak forces" between other particles. Obviously because the entire concept was false and the particle was impossible (there are no scalar particles in the Universe), the Higgs has never been found in forty years. Further on, Higg's work was not even original, as he copied the known top/ antitop field explained by Nambu. Such is the extent of the absurdity of the Higgs, for whom all mankind is risking their lives. But this cannot be explained because quantum physicists simply will not consider complex, informative arrows in their energetic cosmos.

The ultimate reason for the existence of the Higgs is the hyperbolic error of quantum physicists, which try to explain it all with quantum physics, *even gravitational and mass theory, perfectly described with the work of Einstein. So Higgs tried to invent a "mechanism" to explain mass, an intrinsic property of space-time, as if mass was an electroweak force, which exchanges particles. Thus, it imagined that the particle he theorized, which is self-similar to a top/antitop boson, also interacted with all other particles of the Universe, giving them mass, a something, never clearly defined by Mr. Higgs.* The reality is that the Higgs is just a set of equations, self-similar to those used to describe a top quark/antiquark pair (which is in itself a cycle of life and death of a top quark, which is born as a top and dies as an antitop). But neither the top/antitop life cycle of the quark or the equivalent Higgs explain at all why other particles have mass, which is an intrinsic property of particles. Particles don't have mass, *they are mass; they are vortices of space-time, which is the definition of mass.* Thus, the Higgs (top->antitop) is only useful to explain how the neutron and proton decay into more massive W and Z particles, when they break their symmetry (die) and then evolve into heavier quarks. But testing a well-known nuclear reaction was not enough to get so much money. So Lederman came with his hyperbolic error, saying to Reagan that on top of that the *Higgs was invisible like God and overpowerful: it would not only kill our matter but give us the essence of life/information, our mass.*

Further on, the Higgs, a scalar energetic particle, cannot give mass to other particles because particles that collide repel each other. The idea that the Higgs gives mass to other particles is just another marketing tool to declare it "God's particle" and enhance its importance.

This is so obviously false that when it was needed to further explain how God's particle *did that* no single physicist could come up with a solution because obviously the Higgs would not attract but repel particles when colliding with them, as all particles do. So a field full of Higgs particles would be just bumping on other particles or eating them. That has nothing to do with giving them mass, as mass is an internal property of a vortex of space-time. All particles have mass as you have cells. No external entity goes around giving you your cells or your weight. Your mass/weight is within you. What happen next deserves to be in a book of human follies. The British ministry of science made a concourse to explain how the Higgs created mass, which was won by a college student with the argument depicted in the two images created by CERN. This is his explanation, as found in the University's site, entitled "A Quasi-political Explanation of the Higgs Boson, for Mr. Waldegrave, UK Science Minister 1993:

1.   The Higgs Mechanism

Imagine a cocktail party of political party workers who are uniformly distributed across the floor, all talking to their nearest neighbors. The ex-Prime—Minister enters and crosses the room. All of the workers in her neighborhood are strongly attracted to her and cluster round her. As she moves she attracts the people she comes close to, while the ones she has left return to their even spacing. Because of the knot of people always clustered around her she acquires a greater mass than normal, that is, she has more momentum for the same speed of movement across the room. Once moving she is harder to stop, and once stopped she is harder to get moving again because the clustering process has to be restarted. This is the Higgs mechanism. In order to give particles mass, a background field is *invented* which becomes locally distorted whenever a particle moves through it. The distortion—the clustering of the field around the particle—generates the particles mass. The postulated Higgs field in the vacuum is a sort of hypothetical lattice which fills our Universe. We need it because otherwise we cannot explain why the Z and W particles which carry the Weak Interactions are so heavy while the photon which carries Electromagnetic forces is massless.

The Higgs Boson.

Now consider a rumour passing through our room full of uniformly spread political workers. Those near the door hear of it first and cluster together to get the details, then they turn and move closer to their next neighbours who want to know about it too. A wave of clustering passes through the room. It may spread out to all the corners, or it may form a compact bunch which carries the news along a line of workers from the door to some dignitary at the other side of the room. Since the information is carried by clusters of people, and since it was clustering which gave extra mass to the ex-Prime Minister, then the rumour-carrying clusters also have mass. The Higgs boson is predicted to be just such a clustering in the Higgs field. We will find it much easier to believe that the field exists, and that the mechanism for giving other particles mass is true, if we actually see the Higgs particle itself. The next generation of colliders will sort this out.

From David Miller, Physics and Astronomy, University College London. (cartoons courtesy of CERN).

Till here the official description of the Higgs. Yet that description forgets the most important point: *the Higgs field.*

What is stopping those particles in Higgs theory? The answer is nothing we have heard of. Indeed, the Universe as it is has no field which stops and drags its particles, so Higgs invented out of the hat of the magician, from nowhere, a new universal field, a fantasy, the Higgs field, which has never been observed. This in science is like inventing in religion a paradise for the believers, like creating in literature the world of the Lord of the Rings. Mathematically or verbally it might be nice, even beautiful, but it has nothing to do with reality. Science is not done inventing realities but explaining events and entities of which we have already experienced. Further on, this field has two values and one is a negative mass, which is a complete absurdity.

Curious enough even if Mr. Higgs and CERN try to substitute Einstein, they use him in the drawings as the physicist everybody wants to talk to. So let us ask with him what mass is. Since Einstein's mass theory understands why light is massless, it never closes into a cyclical vortex. According to Higgs, this is because the Higgs field doesn't like light (Why? We don't know.) and so light passes between the invisible guests (Higgs bosons) "free." LOL, mass is explained beautifully with the principle of equivalence between mass and an accelerated whirl: the heaviest top quark turns faster than any other whirl of mass and its size is minimal. It is the final evolution of the whirl of mass. On the other extreme, light never closes its wave, which is lineal and massless. The maximal energy/mass of a light beam happens precisely when the frequency or curve of the wave is maximal. Those are therefore the two extremes of mass in the known Universe:

— An ultracyclical movement, the quarks are the final Russian doll of mass of the Universe and vice versa.

— The opposite type of movement, a lineal ray of light, doesn't form a cycle that returns to the same place, creating inertial mass, resistance to displacement; but it always moves ahead without mass.

It is thus evident that as we accelerate a particle in a curved path, according to Einstein's principle of equivalence, the particle acquires more mass till reaching a cyclical c-speed, which creates a supermassive black hole that curves the energy of the space-time continuum like a sink curves and drags inward the water at an accelerated rate, taking with it whatever floats in that medium, in the case of our world, the light of our space-time membrane.

It is thus obvious that the Higgs does not give mass to other particles and it is not even a real particle but a top quark redux. The knowledge of subatomic particles is today exhaustive, and there is nothing else to find there of relevance. The standard model is closed and Higgs is a hoax, a false particle, which cannot substitute the work of Einstein about masses but is merely used by CERN as an alibi to justify its research. This happens in many sciences. We might still understand new features of the ant fauna of California, which comprises 7 subfamilies, 45 genera, and approximately 270 species (245 native, 25 introduced), even discover a couple of new species; but that won't change much of our general outlook on the Californian ant, nor will it reveal the meaning of it all. Neither we shall find a fabulous superant with godlike properties. For the same reason in the quantum scale of species of energy and information, the standard model of particles is closed. And the new particles CERN pretends to discover (evaporating black holes and Higgs) are as fabulous as a hypothesized Californian superant with godlike properties called Higgs, which gives intelligence to all other ants of this part of the Earth, and so it *must be sought* at all costs by bombing the San Andres Fault because we hint she is hiding in its cracks, never mind the earthquake we might cause in search of the Saint Grail of ants and killing us all. Fact is, to study formicidae and the cosmos, we do not need to bomb the San Andres Fault or the Earth at large but research directly the Universe with telescopes and satellites and use standard cosmology, Einstein's relativity, and its recent advances due to the evolution of non-Euclidean and fractal mathematics.

4. The Higgs as the Top Quark. The Smolin/Brans/Nambu Theory of Top Quarks

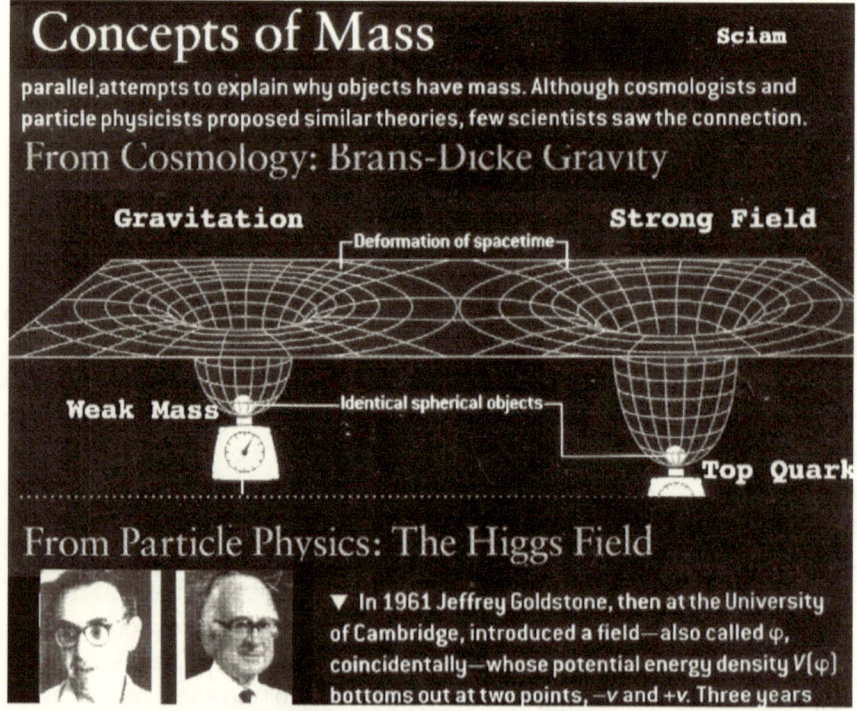

Concepts of Mass

Sciam

parallel attempts to explain why objects have mass. Although cosmologists and particle physicists proposed similar theories, few scientists saw the connection.

From Cosmology: Brans-Dicke Gravity

Gravitation   **Strong Field**

─Deformation of spacetime─

Weak Mass ────Identical spherical objects────

Top Quark

From Particle Physics: The Higgs Field

▼ In 1961 Jeffrey Goldstone, then at the University of Cambridge, introduced a field—also called φ, coincidentally—whose potential energy density $V(\varphi)$ bottoms out at two points, $-v$ and $+v$. Three years

Source: *Scientific American*

The Higgs Hoax was discovered and published in Scientific American, *just before CERN announced its project. Then suddenly all was again covered up, including the article in* SciAm *shown in the graph, which showed that a Higgs field was a self-similar equation to a description of a strong gravitational force and a top/antitop quark.*

The double-talk of CERN for internal consumption and external marketing is even clearer in the Higgs case when we study the literature available about it. In the standard encyclopedia of physics of 1984 (McGraw), there is not even an entry for Higgs. That is the real importance of that particle. In the graph, Smolin, chosen by *Time* magazine a few years ago as the new Einstein of modern physics, expert in all fields of cosmology, who considers black holes the top predators of the cosmos, also unveiled the Higgs Hoax in *Physical Review* and *Scientific American*, establishing its parallelism with a strong gravitational force. What the graph means is that we must fuse the strong field of quarks and the gravitational field. The strong field emerges as the gravitational field when we vary the universal constant. We have done that in simple terms using Newtonian vortices. Two Americans, Brans and Dicke, did it with equations of relativity, obtaining a mathematical field self-similar to Nambu's equations for a deconfined strong field of top quarks (*left*, our weak matter; *right*, a strong field of top quarks, which breaks our symmetry, converting us into quark matter).

Goldstone[5] and Higgs merely played with Nambu's top/antitop quark equations. Smolin found it and published in *SciAm* many years ago. Nobody now talks about it. Why? Because if the Higgs is the Brans-Dicke theory of gravitation, self-similar to the work here done on Einstein's mass theory, we no longer need to seek for a redundant particle—the Higgs is merely a top/antitop quark condensate.

Indeed, the Universe is efficient, not redundant, and the simplest theory is truth, according to the scientific method (Occam's razor). So we don't need a Higgs particle to explain how top quarks break the symmetry=feed on our electroweak matter. Neither the LHC will find the Higgs, a redundant particle, excuse to build a redundant machine, but CERN will produce the ultradense, superstrong quarks that might trigger the formation of a quark star.

Mass and gravitation are perfectly understood with Einstein-Brans theory, a modified version of Einstein's work, which introduced the concept of a variable gravitational constant, a fact proved today also by fractal theory.

In our simple Newtonian description of those mass vortices, it means that we can fuse also the gravitation and strong field, which happens to be 100 times stronger than a charge and 10^{41} stronger than the gravitational field. Thus, we can relate the strength of their universal constants:

$$G \text{ (gravitational constant)} = 10^{39} \text{ Coulomb/charge constant} = 10^{41} \text{ quark/strong constant}$$

Thus, strong forces are 10^2 stronger than our weak ones and 10^{41} stronger than the gravitational one. *And since black holes will be made in all the self-similar theories that describe at quantum level a strong gravitational world, all those theories including the simple vortices described here are telling us that quarks will make black holes.*

Of all those theories, the less appealing is the Higgs one. This is due to the fact that what quantum physicists do is merely repeat the same theory, the original electromagnetic theory that explains how particles interact in electromagnetic space, exchanging particles between them. But this happens because light is a spatial energetic force with no cyclical, temporal, informative mass. So it has spatial parity, it happens in space, and the exchange is extremely fast. But what quantum physicists ignore because they ignore all about physical information is that there are forces that happen in time because they are cyclical forces, which don't exchange energy but transform energy into information, cyclical mass. The main of those forces is the weak force, which therefore is not a spatial force, reason why it does not keep the parity/spatial symmetry of other forces. It does happen in a same point of space but in a long period of time. It is mediated by a constant G, which is measured in m^{-2}, mass/time parameters, lasts a long time, and happens in the smallest range, in the same point of space-time. Thus, this force is completely strange to them because as long as they don't have a proper theory of mass/information/cyclical time clocks, they won't be able to understand at all a "transformative" force, like the weak force. They started well, as Feynman and Gell-Mann described the force as happening in a single point of space, in a long period of time, evolving our lighter particles in heavier quarks (with an intermediate state called the Z and W particle). Or inversely the force showed how heavier particles decayed into lighter ones.

Yet then came Goldstone and Higgs and tried to explain those forces as electromagnetic spatial forces, just because that is how they describe light. And so Higgs invented an absurd particle, which is scalar, lineal, not cyclical, and kept inventing absurdities.

What this means is obvious. The Higgs is a wrong equation that describes just a specific electroweak reaction that transforms neutrons and protons into heavier quarks through an intermediate Z, W states, as our particles speed up in cyclical vortices of mass, warping further into fractal dimensions of form, and finally produce top/antitop quarks, which have in this manner killed our matter. And because the transformation or inverse decay is NOT an exchange of energy in space but a growth or loss of dimensional form in time, it does not keep its "mirror" parity in space. Parity means that in spatial events it is the same going left or right, but in time events, as Einstein said, "wires don't travel to the past." So a time event, a weak interaction, doesn't keep its parity; it is not the same going forward (evolving information) then backward (devolving information into energy). Yet to explain why the weak force doesn't keep parity is impossible without a proper theory of the two arrows of time, reason why quantum physicists have not been able to do it for forty years and we can do it here with simple, logic explanations.

Thus, any attempt to create the Higgs=top will mean nothing for science, but it will mean a lot for mankind: the death of our matter, which in a runaway reaction will mean our death. This will happen according to standard science over 8–10 TeV. That is, it will happen in the experiments of 2013.

Lee Smolin found and proved twenty years ago that in fact, Higgs's equations and Nambu's equations, which preceded him, and Einstein-Brans equations which preceded both, are self-similar fields with a variable

G-constant.[6] Now the fractal paradigm makes all that also "logic" and furthermore solves the true question that all those theories tried to solve: *why the weak, temporal, informative force breaks parity, the symmetry of space. Because it is not a spatial force but a temporal, informative force, which evolves or devolves particles between the two fractal membranes of the Universe, the world of electroweak forces and the world of strong gravitation.*

The reader must understand indeed what two fractal membranes mean: certain particles belong to the membrane of dark cosmological matter in different scales and others to the electroweak scales. And the weak force evolves them or devolves them between membranes.

Which one belongs to which membrane is easy to deduce by their self-similar parameters:

The 2 membranes are built with different pieces.

We have a fractal, *gravitational membrane* with several scales of sizes:

Microcosms of strong forces: Strings (theoretical) > Gluons > Quarks > Macrocosms of gravitational forces: Pulsars and black holes of quark matter > Nuclei of galaxies >Hyper-Universal Black hole?

And we have sandwiched between quarks and black holes the *electromagnetic membrane* of the following: microcosms of light> electrons>macrocosms of Chemical Life and humans> stars > galaxies.

And the weak force and the neutrino are forces/forms that evolve/ devolve particles between both.

5. The Higgs as Hoax: True Science vs. Nuclear Industry

It is now up to the reader to judge who is right, Mr. Higgs or Mr. Einstein et al. and their theory of fractal relativity. Fact is, Brans-Einstein's relativity, Smolin, and Nambu, and this author using top quark condensates and fractal relativity explain all what Higgs can't, *even inventing new particles and new fields.* The work of Einstein on masses as vortices of space-time is so simple and explains so many things that Occam's razor gives us the answer. Einstein is right—Higgs is wrong. Indeed, my eight-year-old cousin understands better what mass is with my explanation of a hurricane that experts at CERN can't, blinded by their interests on the quark cannon and their denial of fractal space-time and the arrow of physical mass/information in physics.

Higgs is, as we said in the introduction, a case of corruption similar to that of the AAA-derivatives, certified as money of maximal quality by the experts of AIG, which turned out to be toxic assets with null value, based on mortgages without financial backing. They were established to create a Ponzi pyramid of money till all crashed down. CERN has created a Ponzi pyramid of excuses to get funding, fringe theories, pumped as AAA-quality, among which the most marketed of all is the idea that Einstein's theory of mass is irrelevant because there is a particle they are going to find, which gives mass to all the particles of the Universe called the Higgs.

Because the Higgs is NOT real, it cannot be argued with reason because that is not the purpose of it. Its purpose is to feed the budgets of big science as the purpose of AAA-derivatives was to feed the pockets of big bankers. For that reason, when I sent to CERN papers on relativity and curved non-Euclidean geometries, explaining the meaning of mass and the weak force with different levels of mathematical complexity and the risks of quark condensates, I got no answer. Since knowledge is NOT what CERN is about, it would be like writing letters to a factory of missiles on the dangers of their missiles. They do missiles and CERN is creating a quark factory, and the last thing both want to know is that a missile or quark factory is dangerous. Point.

It doesn't matter that Occam's razor proves fractal mass theory right. When Einstein came out with his theory of special relativity, he was accused by an expert, a German professor, founder of ether theory, of being too simple. Precisely he told him, there is a law of the scientific method called Occam's razor, which says a simpler theory is always truth *because if the Universe is made only of energy and information, at the end, its fundamental laws will become reduced to two terms. Only when the bits and bites of information and energy of the Universe recombine in complex structures do we need to describe the complex details of those buildings. But the principles and essential formula of the Universe are always simple or else they are false: "God is simple and not malicious," said Einstein.* Contrary to belief, Einstein was the new Copernicus of simplicity and quantum cosmology as the old Ptolemy departed from a wrong assumption (that the Earth was the center of the Universe; that the Universe has no physical information, only entropy). And so Ptolemy had to invent very complicated equations to explain it all. Of course, the principle of equivalence between acceleration and mass can be explained with the simplicity of Newton's equations or with the more complex details of Einstein's non-Euclidean geometries that take thin pictures of those vortices in present simultaneity and get an enormous refinement in its measure. So his theory was proved when it improved the analysis of Newton on the cyclical orbit traced by Mercury. If Copernicus had thought that those vortices were circular and Kepler and Newton had proved they were ellipses which accelerated and decelerated in different phases of the ellipse, now Einstein added a more detailed analysis of the curvature of those vortices but the principle stayed: the cosmos was a game of all kinds of vortices of all sizes, charges, masses, orbital paths. It was as Descartes had foreseen, a game of vortices and "res extensa" (lineal forces and movements): F = m (vortices) × a (lineal forces).

And our explanation of weak forces has behind all the equations of theorists. *What they don't have is the logic interpretation*, given by the understanding of the two arrows of time.

There is no need for new equations but new logic interpretations. An informative force doesn't keep spatial parity, doesn't have spatial range; it is heavy/cyclical, it is evolving. Point. There is no need for any new equation or Higgs particle. It is all done.

6. Once Again, the Mechanist, Pythagoric, Energetic Errors of Physics

But that doesn't matter to CERN as knowledge is not the purpose of technological science but the excuse to keep using computers and big weapons. And so what matters to that purpose are two things:

On one side, if you use complicated equations and have to find invisible particles, you are going to use very complex, expensive computers and machines that will give work and money to industries of technological information. Einstein did thought experiments and that doesn't make money. In the past, the simple equations of Copernicus and Kepler didn't make astronomers look like high priests. While Ptolemaic epicycles required complex calculus to get the position of the sun and it looked very intelligent, Copernicus was scorned precisely because he was too simple, he was not a specialist, and he made astronomers feel like common people. If an eight-year-old child can understand the meaning of mass as I explain it, that is bad for the prestige of the high priests of quantum theory.

This is the core reason of the massive extension of Pythagorism in quantum cosmology. Indeed, all the absurd theories sponsored by CERN, all the work of people who are waiting for the quark cannon to prove their Pythagoric theories, depart from a massive use of computer models of reality, which makes them look "cool" with a lot of crunched numbers even if they defy all the laws of the scientific method. But if the laws of truth of science and the simple theories that respect them explained here were accepted, then because the Universe is only one, only one theory would stand: Einstein's cosmological theories and its evolutions, realized in the past decades departing from relativity principles. And all those theorists and their "inflationary information" will have to come down to Earth and lose their scholar positions. We live in a society of egos, not of seekers of quality and truth, so if any toilet particle can become God, all scholars become genius and they like that. It happened in art when the pop culture accepted Warhol's prints because a single genius is a single sale and what mattered was to sell art, so anything made by a printing machine became art. Today in computer science, what matters is to have mathematical models promote new supercomputers, new machines that will churn numbers, regardless of its veracity. And so the human believes to be a genius because the machine does the work for him. Leonardos are born with Adobe Photoshop and Einsteins with LHC machines. So there are a huge number of theorists "inventing" meaningless particles with their computers, all of them hoping to see their fantaphysics at CERN.

Recently, a top researcher of the top quantum institute, the Bohr institute of Copenhagen, said after starting his article by calling the members of Einstein et al. "crackpots" that the future was interfering with CERN because it abhorred the Higgs. This absolutely absurd theory about an absurd theory reaches new heights of scientific surrealism. It was however taken seriously by everybody including CERN because it was a quantum cosmologist, which used very powerful computers to calculate this absurdity. While our warnings on standard theory of quark condensates and black holes that say we are dead do not get printed, this article came all over the news and scientific magazines took it seriously. The same goes for the absurd belief that all theories are mere probabilities. Science is right or wrong, but powerful computers are needed to design probabilistic models. So anything goes.

Only the power and worldly profession of quantum physicists, to make machines and weapons, allows them to promote absurd theories. The physicist is the high priest, the emperor, owner of land and weapons. If the emperor wants to play piano and make sonatas, all the courtiers call him a genius. When Mozart came and laughed at the emperor's sonatas, he ended poor and ignored. But time makes justice and so today nobody plays the emperor's sonatas but those of Mozart. And certainly sooner or later Einstein will be vindicated or we shall all die trying to find the false Higgs.

People don't really care about knowledge but about being in good terms with the emperor, with the military-industrial complex. So they all prefer the Higgs because if Einstein is right and people knew what mass is, it turns out we wouldn't need to spend 10 billion dollars to make a machine to find a particle that does not exist. That is the bottom line: if you believe in Higgs, you get a job in the quark cannon—and technological science can go on.

In that regard, a much more interesting Higgs, in a twist of history, is the main character of Erewhon (Nowhere backward), the prophetic book of Butler, who first understood the evolution of machines and their consequences for our demise. Higgs is the traveler and narrator in the adventures of *Erewhon*. Butler published after Darwin's *Origin of the Species* a credible account of machines as systems that fed on energy and information and evolved, imitating human beings till causing our demise. His serious articles were dismissed by the technological ethics of the Victorian civilization, and Butler moved to New Zealand. His alter ego, Higgs, arrives to a mythical land, similar to New Zealand, where he discovers that the Erewhonians have banished machines on the basis that the eventual evolution of humans into superior machines seemed inevitable, following the profoundly convincing arguments of earlier Erewhonian philosophers. Far from being critical of Darwin's work as some believed he was with this satire, Butler admired him greatly. His satire is directed more at a society that would shun the consequences of evolution rather than embrace them, no matter how frightening they may appear. And the frightening idea was not that you descended from a monkey, but rather that your demise is inevitable, *unless we humans obeyed the truths of philosophy of science and the ethics of a human civilization, which we have not.*

Indeed, a new dawn of science and mankind could only happen if we abandon the mechanist, monist, simplistic view of reality that physicists, the wrong Higgs, working for CERN sponsor. CERN's physicists can keep cheating billions of dollars to clueless politicians and in the process kill us all. That is their bidding. We are evolving science here at no cost, with the human mind.

Now the reader must take a decision, between the two theories of mass that we have explained here. If he believes Higgs is right and his particle explains the masses of all other particles and why weak forces don't respect the spatial symmetry, he should throw this book. But if he believes that masses are vortices of physical information and weak forces evolve information, then he should keep reading. This theory started by Einstein and completed by this writer, thanks to his development of the five postulates of non-Euclidean geometry and the discovery of the laws of duality, the study of the Universe with the arrows of energy and information, is the one we shall use from now on to explain quarks, weak forces, and the quark-gluon soup that CERN will produce, its enormous risks for our lives, and the reasons why CERN should close down its facilities. We shall also explain those risks with standard quantum theory because standard quantum theory proves them all, but we shall *not waste time with Higgs because we cannot explain anything about mass and black holes with their work.*

I completed this alternative theory of mass and time arrows more than a decade ago. I gave it to the international world of science on the occasion of the fiftieth anniversary of systems sciences and complexity, in the International Society for the Systems Sciences Congress at the University of Sonoma, California, in two conferences on non-Euclidean geometries and the unification equation we have explained here. The conferences were wildly successful, and I was chosen the Chair of Duality, whose congresses I chaired the following years. Thus, I gave four more conferences on the Tokyo Institute of Technology and Madison. Yet at the same time, I started to write CERN letters about the dangers that a theory of quarks as vortices of information and the dual nature of quark-gluon soups as complex liquids, responsible for supernovas, represented for this planet. CERN never answered those letters or gave any indication of studying those risks. I realized then that CERN didn't care at all about risks or about knowledge, that CERN was just a company making money, swindling taxpayers, and constructing a machine. For all those reasons, I put a suit to this company, renounced my career as a scholar, and made all what was in my hands to denounce CERN and the risks of extinction those experiments with quark-gluon soups represent for Mother Earth. The company, instead of confronting us in those suits, of reasoning with us, adduced diplomatic immunity not to appear in those suits, ran a very successful ad hominem campaign against Einstein et al. through its thousands of student collaborators in the Web and friends in the scientific press, and went ahead with its experiments without having the slightest idea of what it would do, as the Higgs theory gives no clue, no explanation, no solution to any of the events, particles, and properties of masses, quarks, and condensates its factory will produce. CERN has no idea of the outcome of those experiments because it has no real theory of mass, ignores all about fractal information, and is clueless about the dual big bang/big crunch events of quark-gluon liquids. Only departing from an informative theory of mass can we respond to the questions CERN still wonders and which we will explain here: why energy becomes mass, what is mass, why particles have different masses, what is dark matter, why quasars move faster than light, why the Universe is expanding, why there are more particles than antiparticles, why there are three families of quarks, why quarks have fractional charge, what is the unification equation of charges and masses, why the top is so heavy, etc.

---

[1]   Nambu, Y.; Jona-Lasinio, G. in his breakthrough paper (April 1961). "Dynamical Model of Elementary Particles Based on an Analogy with Superconductivity. *http://www.npl. illinois.edu/exp/mucapture/MuCap_Related/nambu/p345_1.pdf*, established the basis for a series of models (Goldstone, Higgs, etc.), which studied how quarks, forming Einstein's condensates, were able to break the symmetry of our matter.

[2]   Vladimir A. Miransky, Masaharu Tanabashi, and Koichi Yamawaki extended the model of Nambu to study how top quarks could be the cause of the breaking of symmetry

of our matter at high energies (Phys. Lett., B221:177, 1989, *Dynamical electroweak symmetry breaking with large anomalous dimension and t quark condensate.*)

3   Lederman promoted the Higgs as a new particle to get funding for the SCC, writing a book, which is still a best seller of "fantaphysics," http://www.amazon.com/God-Particle-Universe-Answer-Question/dp/0385312113

4   The supercollider soon tripled its cost from 4 billion to 12 billion dollars, as it is customary in military budgets. For a good, impartial description of that scandal, consider the following: http://encyclopedia.stateuniversity.com/pages/21490/Superconducting-Super-Collider-SSC.htmhttp:/encyclopedia.stateuniversity.com/pages/21490/Superconducting-Super-Collider-SSC.html

5   Curiously enough, this modern golden stone of fantaphysics was created by a scientist called Goldstone, who copied the real equations of top quarks from Nambu. Goldstone's equations were later stolen by Higgs, which now in this tale of fantaphysics, ambition and lies is the author of the mythic substance.

6   The discovery that the Higgs was equivalent to a strong gravitational field was done by Smolin and Zee, using the Brans-Dicke theory, which shows that a gravitational vortex in which we change the universal constant for a stronger one, behaves exactly as a top quark condensate/Higgs. This discovery favors the unification of strong and gravitational forces and again proves that the Higgs is just a top quark. But if top quarks are the main components of black holes and CERN will make Higgs=top quarks, it will make black holes, which will certainly accrete the Earth as quarks cannot evaporate under the Hawking radiation even if that radiation were real and didn't break all the laws of science. A strong gravitational unification shows that protons, which are made of quarks and gluons, are black holes of the quantum scale and so black holes in the cosmological scale will also be made of quarks and gluons. They will be Einstein-Bose top quark condensates. Why top=Higgs quarks? Because they are the densest quarks, whose parameters as a condensate coincides with those of a black hole, and because top quarks are positively charged and so a mass of top quarks will be positively charged as protons are. For the initiated, we shall consider a more complex explanation: Brans-Dicke made G a scalar field, adding a dimension of angular speed $1/\varphi$ to measure the strength of G's deformation of space. The variable G field defines masses as space-time whirls of different strength according to its rotational speed.

Thus the Dicke-Brans model of variable G-constants expanded relativity, adding a new dimension which now can be fully understood in terms of the three subdimensions or derivatives of "time-motion."

Indeed, we said that speed can be seen as distance. Further on, the equation of change in motion that physicists call time is $v=s/t$. Yet we can see that equation at three levels of speed. We can see it in present simultaneity as a fixed form of information. This is Einstein's basic description of the topology of space-time. We can see it as a speed—this is Newton's description of a vortex of space-time—or we can see it as an acceleration, which adds a new dimension. This is Poisson's description.

Indeed, in physics we add motion, translative time=change to that formal dimension of static geometry. Thus, we create two new dynamic dimensions of temporal motion by deriving space through time=change, $v=ds/dt$ and $a=dv/dt$, to obtain the second dimension of translational time-change (speed) and its third dimension (acceleration).

In relativity, those dimensions become the accelerating whirls of space-time defined by Einstein's strong principle of equivalence. Yet since Einstein's G measures only the morphological curvature or "form" of space, we have to add two new dimensions to Einstein's G to obtain first the cyclical speed of the whirlwind and then its acceleration. The result is a vortex of space-time with three clear time dimensions and descriptions of increasing complexity:

— G, the curvature of space, which defines the form, the in/form/ation carried by a given region of space and was defined by Einstein's relativity.

— $\varphi = 1/G$, the cyclical movement of that curvature, which defines a vortex of space-time, self-similar to the variable scalar field of Brans-Dicke that adds speed to G-curvature.

— And finally, $1/\varphi^2$, the acceleration of that movement, which reflects the principle of cyclical equivalence and defines mass as an accelerated vortex that sucks in space-time (since $1/w^2$ are the dimensions of an accelerating, cyclical movement). It is a field self-similar to the dilaton field defined by Guth to explain the expansion of the big bang Universe.

When we explain those two new dimensions of G through cyclical time and the strong principle of equivalence that equals acceleration and mass, it is easy to explain and unify the forces, particles, and fields of the Universe with different values of those three G dimensions. Further on, as Smolin showed the other alternative theory of mass, the Higgs fields turns out to be merely the 2 (+ and—) roots of φ^2, which strongly suggests that as Kaluza and Klein already figured out in the 1930s, the electromagnetic and gravitational field are merely the expansive (negative, electromagnetic, decelerating) and implosive (positive, gravitational, accelerating) root values of a Universe of variable motions=translational changes.

Self-similar theories are one of the strongest proofs of truth because they show, despite their specific errors, a generic path of truth (according to the linguistic method). This is what Brans and Guth's work truly means: While the maths of those classic theories are somewhat more complex than what we will use here to obtain the values of the G and Q and mass constants of different particles (since they still use the erroneous concept of time as the four dimension of space), the previous conceptual explanation is clear enough to define a dynamic Universe of perpetual motions, whirls, and forces with two opposite inverse directions: charges and masses.

To understand what an expansion of G means in cosmology, we have to depart from general relativity's equivalence principle and its meaning under a field that increases the dimensions of G to two (Brans-Dicke theory) or three dimensions (Zee-Smolin-Guth's theories). In as much as G is a time-dependent parameter, it implies that the gravitational curvature is extended in two new temporal dimensions, which in physics are the speed and acceleration dimensions. We find therefore that the true meaning of Brans field is a gravitational constant or non-Euclidean curvature of space-time extended to a dimension of cyclical movement, w, which interprets the Dicke-Brans equations as a dimension of angular speed, $1/\varphi$ and the Zee-Smolin-Guth theory in an acceleration dimension, $1/\varphi^2$. Later on, Guth converted such accelerating field in his dilaton field that expanded the original big bang theory, which, contrary to belief, was full of errors and contradictions (for example, its initial temperature was twenty degrees, and the quantity of mass it produced was a single atom). We shall return later to the big bang theory and deal with the true meaning of that dilaton field (still in need of some corrections as we apply the diffeomorphic local principle to it and quantize the big bang into a series of local quasarlike processes of explosion and implosion of cosmological black holes). When we interpret those extended fields of general relativity under the principle of cyclical equivalence (equality between lineal acceleration and weight), they define a mass no longer as a curvature of space-time (one dimensional G), not even as a vortex of space-time (two-dimensional G with angular speed) but an accelerating cyclical vortex of space-time (principle of cyclical equivalence), whose different degrees of "angular φ speed" quantize G and give different values to the mass of the Universe. The concept is simple to understand with the homology of a whirlwind (a hurricane or bath sink) that attracts with growing force as it increases its speed the space-time around it.

CHAPTER 4

Quark Stars

In the 1920s, Bose and Einstein predicted a dense state of matter in which many particles at very low temperatures share the same spatial location, a prediction which was verified experimentally in the 1990s. The simplest mathematical model of such a Bose–Einstein condensate uses a non-linear interaction between the many particles in the condensate. The non-linearity poses a great challenge to both physicists and mathematicians, leading often to unpredictable behavior. For example, strong attractive interaction among quark condensates may lead to collapse, infinite density and model breakdown, the properties expected in the classic treatment of black holes.

—Wikipedia, Quark, Black hole's Condensates

I. Quark Stars from the Perspective of Complex Physics

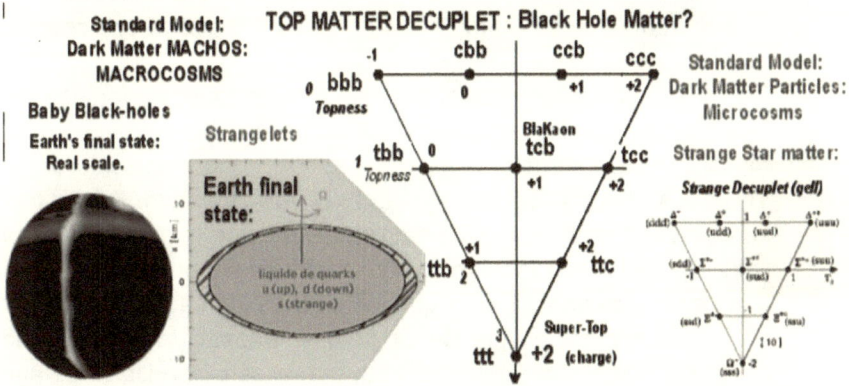

In the graph, the three families of quarks can also be ordered in two ternary triangles, the strange triangle and the top triangle, of increasing mass. The triangular classification is far more revealing of the structure of the Universe as those two triangles gave birth to two kinds of atoms—the "ud"-e atom of our light matter and the "b c" atom of dark matter, made of denser bottom and charm quarks, which we shall see for the first time at the LHC.

1. The Triangles of Quarks: Strange Triangle and Top Triangle

Now we can translate easily into the twenty-first-century jargon of energy and information what CERN will do: *events mediated by the weak force and events that convert our membrane of electromagnetic space into quark masses of the gravitational world—that is, events that destroy our world and transform it into the world of quark, dark, strong gravitational forces. Yet because CERN totally ignores the duality of time arrows and the correct interpretation of these forces and events, it will discover nothing; and it will do only harm to our world.* It will make all types of reactions that destroy and break the symmetry of our light matter converted into heavier masses, trying to prove the absurd theories of energetic physicists, who don't know the simple truths about information explained in the previous chapter. Because the weak force is a temporal force, NOT a spatial one, they will never understand from a spatial point of view why this force has no symmetry in space, as they never understood what Einstein told them: "Wires don't travel to the past," temporal forces don't travel to the past. Hawking's theory is based precisely on the thought that they do. Physicists by definition treat time as if it were space and so they expect it to have spatial symmetry. Their way of working is not "thinking" but memorizing mathematical formula.

The why, as Feynman put it, is not what you ask in physics. The why though is what we explain here. And the why will be the conversion of our world into quarks since atoms are made of two types of vortices:

— An electronic cover, made with lighter electroweak forces that make up our world and occupy a lot of space but have very little mass/physical information. All that you see is that electronic cover. The trees, the people, the homes, the rivers, the land, the air that surrounds us is all part of the empty space made with the energy of light—an ethereal ghost that might seem very important to us but is thin and transparent for dark quark matter as the vacuum is for us.

— And an inner nucleus made of quarks, trapped in protons and neutrons, which has 99.9 percent of the mass of the Universe and occupies very little space. Hence it stores most of the attractive, gravitational forces of the cosmos, both at quantum strong level and at cosmological level in the form of black holes and pulsars.

According to Einstein's relativity theory, *the standard theory of science in all things related to mass*, this is explained because masses are whirls of space-time that accelerate inward like a hurricane and so the smallest they are, the faster they turn. So we find inside the nuclei of atoms the *eyes of the hurricanes of space-time, the* densest, fastest rotating masses of the Universe, three bidimensional color-locked quarks that emerge for a total of one three-dimensional charge vortex (proton or neutron) in our electroweak world. They now will be unleashed by the quark cannon in a double big bang/big crunch that will put thousands of them together, *breaking the symmetries of our world that will be warped and sucked in, as a membrane of quark dark space-time is created.* This strong and gravitational space-time membrane is dark because its attractive power is so huge that it can absorb our electroweak matter, light and electrons, both at quantum microscopic level (quarks) and at cosmological level (black holes). Light comes to them, attracted by strong and gravitational forces, then it whirls around and finally disappears into this other ultracurved world.

Indeed, the Universe is made of two fractal space-times. This was first hinted by Einstein when he realized that the light-space we see is Euclidean and the gravitational space is more curved, more informative, non-Euclidean. Then fractal theorists split them into two to fully grasp its interactions, thanks to the evolution of non-Euclidean and fractal mathematics.

In the simpler version of Einstein, both membranes are messed up. So what Einstein did was adding the underlying curvature that the gravitational membrane causes on the Euclidean light space we see and calculate its effects. This light space we see has tautologically three Euclidean perpendicular dimensions because it is made of light, which is a three-perpendicular wave with height (electric field), width (magnetic field), and length (reproductive speed of the wave). Thus, we see three perpendicular dimensions of space-time because we see light space-time, a membrane made of light quanta, actions of the energy and time-frequency=information of light. For the same reason, we measure time with seconds, which is the time cycle of vision of the eyelids. This is the ultimate meaning of fractal space-time: we do not exist in an abstract Cartesian graph, a Pythagoric error made by Newton, but we live as evolved, warped forms of light, which become particles of information, atoms, and evermore complex molecular structures. And as such, we "swim" in a light medium, which is the space we see. But that space is covering a stronger world, a network of smaller, faster, denser masses called quarks and denser cosmological bodies called black holes. So our electrons protect and enclose the world of quarks and stars protect and enclose the black holes of the center of galaxies. And light covers like myelin does with neuronal flows of information, the "strings" of dark gravitational energy that communicates those black holes and masses.

People at CERN, however, do not work with twenty-first-century complex physics and the fractal paradigm because their machines are useless for fractal cosmology. Pietronero, Nottale, myself, and others have been working in the fractal paradigm for decades, solving and searching the new questions of twenty-first-century science, while the nuclear technology of quark cannons has advanced by a very different path, unrelated to true knowledge. The result is the worst of both worlds: a weapon, which is not controlled by the military but by irresponsible physicists who are obsolete theorists, unable to understand what they will do with their weapon, namely, to break the delicate balance between the two membranes of the Universe and liberate the matter of the quark dark world, fracturing perhaps in an irreversible way the membrane of light and electrons in which we exist.

Those liberated quarks will form then *color-locked liquid* quark condensates, *a superfluid, perfect, ultradense liquid, which CERN didn't even know existed when it designed its machine and now turns out to be the most common quark configuration, one of extreme density and stability and the one needed to start an ice-9 supernova reaction.*

In the graph, we can see the *twenty triads of* quarks that can be created in those accelerators. *We are made of the lighter ones, the UP=UP=DOWN and UP=DOWN=DOWN, proton and neutron at the base of the "strange" triangle.*

This is our world, and it is more telling to study it as an event of time in which an up-up-down becomes a down-up-down particle as an n-p pair switches in an on. The result is obvious, an up has +2/3 charge and a down, its state *after losing, creating an electron, is negative.*

Principle of Complementarity

"All that exists switches on and off in implosive and explosive states through temporal events seen from a 'higher p.o.v.' as fixed dual forms of space."

The constant switch of n-p pairs with their production of electrons is the type of reaction we shall see massively at CERN: top quarks will absorb electrons with their +2/3rd of attractive power, becoming first neutrons. Then those neutrons will decay into higher quarks, through evolutionary w. z events.

And so the Z particle will break the weak symmetry of our world.

But there are eighteen other more dense triplets of color-locked bidimensional quarks of which we know little. Most of them will be unstable, but some of them, such as the bottom-charm-charm and bbc, "dark atoms," self-similar to our light atoms but with 1,000 times more mass, should be fairly stable. Thus, they could easily start a massive transformation of our quarks and electroweak matter.

The easiest particles of those triangles to make are "strange atoms," hyperons, uds particles, in the center of the strange triangle, and dibaryons, usd-usd double nuclei, which is known to be stable in our world. Those are the atoms of strange liquid, strange stars, and neutron stars and have been already produced at minimal energies both in RHIC and the LHC.

The process is simple: as the accelerator speeds up our atoms at c-speed, beyond 1 TeV, the lineal energy is converted into cyclical mass and the up and down quark mutates into the strange quark. And so atoms of usd-strangelet liquid is formed. The cynicism of the company is absolute because it says it would never make strange liquid atoms and yet it is designed to make those atoms and collide them to see what happens.

And what should happen when two usd hyperon atoms collide is the formation of a usd-usd double nucleus, a dibaryon, a stable strange liquid atom. Millions will form in those collisions, fall to the center of the Earth, and implode the Earth from inside out.

And if we survive those experiments performed this year, in 2013, the machine will produce all the quark triplets of the top quark decuplet that are the likely substance of black holes. If this happens, because those top quarks might turn as fast as 10 c speeds, the process will be much faster. The Earth will explode as a black hole made of top quarks forms in situ.

Yet in April at a regime of 7c, the LHC is going to produce, especially in the Pb-Pb collisions, a lot of usd strangelet liquid.

CERN knows this. Of course, it has not the simple fractal model that relates the two strange and top quark decuplets of the quantum world and the strange quark stars (pulsars) and black holes of the cosmological world they form. In fractal relativity, this is how the Universe is constructed. This is the standard theory of quark stars in the fractal paradigm. In CERN's old paradigm, this is one of the possible theories that quantum physicists study, but without understanding the fractal structure of the Universe, CERN considers this possibility as just one of many. And so it selects theories that represent no danger, as bizarre as they might be. *And in an act of criminal negligence, dismisses all theories that imply danger.*

And yet the Universe is only one and Einstein's work on masses is the standard theory of gravitation and cosmology, and fractal relativity is its most clear path of evolution. So all seems to indicate that the correspondence between the two triangles of quarks and the two cosmological, ultradense objects of the Universe, pulsars and black holes, will be the standard theory of dark quark matter.

In that regard, in the old paradigm, physicists divided quarks in three families, coupling the "ud" quark that form our atom, the strange and charm quarks of higher mass, and the bottom and top quarks that complete the three families of quarks. Why did they do this? Simply because they found first the two quarks of our atoms, the up and down quark, and so they just put them together in couples without further ado. A little bit of serious thinking, logic understanding of the dualities of energy and information and visual imagination allowed fractal theorists to order them not only in those three evolutionary families of increasing physical information (whose cause is explained in the appendix), but also in triads.

Then we can observe that those triads form atoms based on the two fundamental symmetries of energy and information: All that exists has bilateral symmetry in space, according to the duality of energetic/ informative systems, whose inverse properties allow those symmetries. For example, humans have a left and right brain, one of which is more developed in spatial tasks and the other in temporal ones. Species have sexual symmetry, divided in female reproductive, informative, cyclical forms and lineal, energetic males, and so on. That symmetry takes place in quarks in the *two types of atoms of the Universe:*

— The up and down quark of self-similar mass form the energy/ information symmetry of the light atoms of our Universe. As a man and a woman are basically equal but the woman is more cyclical, informative than the lineal man, the up and down are also self-similar but the up is more implosive, informative, positive in charge than the down one. And both become the other by exchanging dense fractals of electrons called pions. Those quarks are grouped in three "color-locked" systems, which allow their bidimensional vortices to "emerge" in the three-dimensional world of electrons that we inhabit. And so they form "udu" and "dud" nucleons, protons and neutrons, the atoms of our light Universe.

The heavy charm and bottom atoms of self-similar mass form the energy/information symmetry of the dark, heavy atoms of dark matter. And again, one of them is slightly more informative than the other and both can be transformed into the other by exchanging heavy tau-electrons.

Thus, one of the first things CERN is going to discover is the dark matter atom, the "cbb" neutron.

Those quarks are grouped in three "color-locked" systems, which allow their bidimensional vortices to "emerge" in the three-dimensional world of heavy tau-electrons. And so they will form "cbb," superneutrons, and "ccb," superprotons of dark matter. Those superneutrons and superprotons will associate to the heaviest Tau electrons to form dark atoms.

We know those particles should exist because the totalitarian principle of physics affirms that all particles not forbidden by the laws of physics should be formed. So we know the LHC will form those particles as they are within the range of its energies. And since they inhabit inside quark stars and black holes, born in supernovas, they should explode the Earth into a supernova. This is the most likely consequence for this planet of the creation of massive numbers of quark atoms since 99 percent of what the quark factory will produce will be dark atoms and quark condensates, free quarks that form quark-gluon liquids, the substances that cause supernovas.

To fully grasp how quark condensates form, we have to arrange those quarks into the other fundamental trilateral symmetry of the Universe: a hierarchical three-group, a triangle.

In the graph, we observe that trilateral, hierarchical symmetry and give in the lower right text the round masses of each quark, calculated with the approximate model of Newtonian vortices and the unification equation: $M \times UC = w^2 \times r^3$.

The calculus is simple: the faster the quark rotates, the more mass it has. The details are complicated because quarks in detail are not just simple two bidimensional vortices of gluons. In complex non-Euclidean topologies briefly treated in the appendix, their forms are far more complicated. Thus the "ideal mass calculus" of the graph are only indications. But what matters is to understand that the denser quarks CERN will do are faster and interact much more strongly with our matter, converting it at increasing speeds. Thus, the more energy CERN gives to quarks, the denser they will become, the faster they will transform our mass, and the more dangerous they will be.

A more accurate calculus of those masses as frequencies of rotational vortices can be done considering in far more detail its structure as a fractal vortex of gluons. This was done by Wilczek and others in QCD theory, which requires enormous computational calculus, as each gluon interacts with other gluons and all of them require nonlineal equations, etc. What matters is that they got very close mass values, proving that in detail, the theory of masses as whirls of physical information stands.

The amazing fact that Mr. Wilczek abandoned his groundbreaking work[1] and now he is an employee of CERN, which backs the absurd Higgs theory *against his own work*, is just a matter of morality and ethics not of the value of Mr. Wilczek's earlier analysis of those quark vortices, far more important to twenty-first-century physics than Mr. Higgs's musings.

In any case here, we are not worried of the details but the general principles. And as we can see in the graph, the general principle is that U and D quarks that form light atoms have self-similar lighter masses. So we are *the weakest, slowest mass of the Universe.*

Let us now consider the next strange quark. If we are not concerned with topology and use the vortex equation of the previous chapter, $UC \times M = r^3 \times w^2$, since the strange quark is around 20 times more massive and the d quarks seem to turn at a $\pm c/10$ electronic speed, the strange quark should turn at $\sqrt{20}/10$ c speed. Further considerations about the radius and dimensions of the strange quark, which are elements to consider in the formula of a rotational mass vortex, will make that speed to fluctuate between the limits of c/pi and c speed.

Thus, the strange quark is the quark that reaches the limit of light-speed of our electroweak membrane, the quark that represents the limit of physical information of our Universe, which in the world of cosmological systems is represented by the pulsars, which turn in the same range till reaching in its fastest forms near light speed. And that is what we might become if we made them on Earth.

The strange quark is a much faster rotating quark, which dominates, attracts, and transforms ud quarks, forming a condensate of quarks called a strangelet, which can convert Earth into a pulsar or quark star.

After the experiments made at CERN in December, we know that indeed, beyond 1 TeV, we enter the region of strange matter, as CERN produced just slightly over 1 TeV enormous quantities of kaons, many more "than expected."[2] When those kaons fuse together, they will form ice-9, the substance responsible for the creation of pulsars and neutron stars, made with usd hyperons and dibaryons in which the faster rotating strange quarks control the up and down quarks of lesser speed. If the ice-9 reaction becomes stable, the strange quarks of the kaons and hyperons (usd particles) will attract our quarks, break its pion covers, and feed on them. At CERN, we have seen already a lot of kaons and pions with collisions of minimal energy. With 1,000 times more energy, it is certain we shall create stable strange matter.

The same relationships can be established in the higher scale of the top quark triangle:

B and C quarks that form dark atoms have self-similar masses too. And c is the minimal speed of rotation for the cb system of quarks, made of heavy, dark matter atoms, which are in "harmony" with the tau, heavy electron vortex that traps them. *Thus, we are entering the gravitational next scale of pure dark matter, beyond the light membrane. B and C quarks are unstable in this world because this is the absolute limit of their reality. It is like a fish in a denser water medium that hardly can appear into the light atmosphere without dying. We have no experimental knowledge of those dark atoms except that theoretically, they should be quite stable at high density. How many of them are needed to create a stable ball in our "atmospheric light world"? We do not know, but we are playing with fire because 76 percent of the Universe's mass is dark matter and it is stable. It is there in very small lumps.*

Finally, we get to the top=Higgs quark, which is around 100 times heavier. So if we simplify all the other elements that are needed to calculate mass (radius, topology, dimensions) the top quark has, $M=w^2$, 10 c speed, the speed calculated for the faster quasars, pulled by the dark energy produced by galactic black holes. We deduce therefore that top quarks are the atoms of black holes.

Maldacena proved black holes can be described as made of one-dimensional fractal strings in a five-dimensional universe or as four-dimensional quark-gluon soups, made of the densest quarks, top quarks. Thus, the top/antitop quark condensates CERN will do, the Higgs, should be the equivalent to the strangelet liquid, but far more powerful, a top predator field of top quarks that kills=breaks the symmetry of all the other quarks of the Universe, absorbing them, accelerating their energy, till it grinds them into its simplest units, a soup of strings of gravitational dark energy, the substance of theoretical black holes.

Those black holes are *entities of the other-world membrane*. They are accelerated beyond the event horizon of their vortex, where light bends at c speed, up to 10 c-speeds, and jet through their poles dark energy up to 10 c speeds between galaxies.

This simple model of fractal relativity explains the transition between our world of light, made of h-units of energy and time and the c<10c world of gravitation, made according to Nottale[3] of λ-strings, the gravitational constants of Einstein, the minimal units of the gravitational world and the strong world of quarks and gluons. Those superluminal black holes were described by Kerr,[4] which stated that beyond the singularity ring, superluminal speeds would be created.

It is indeed a fascinating object. As one of the main proponents of fractal relativity, I would love to perceive them in its details—with a telescope, in safety, NOT to the risk of converting this planet into a flow of dark energy.

Because the top quark might reach up to 10 c speeds and belongs to the gravitational world, it is indeed the top=Higgs quark that breaks the symmetry of all types of matter in the Universe and the ultimate fractal part of the black holes of galaxies that cause the mass inertia of the entire Universe. But we do not need the equations of Higgs to describe it but the equations of Nambu and its analysis of how a top/antitop quark by changing between the informative, implosive top quark arrow and the energetic, explosive antitop quark arrow grinds, breaks, evaporates, and reduces to dust of space-time the matter of our world.

How can we be sure of this? We can't because we do not have absolute evidence. So of the three legs of the scientific method, we have mathematical and logic proofs outlined here, and the experimental hints of cosmological observation. Yet self-similarity helps. Indeed, to fully grasp what quark condensates are, the reader should have a look at those triangles of quarks and observe how self-similar they are: we have "two leftover" quarks, the strange quark on top of the triangle of "light quarks" and the top quark, on top of the triangle of heavy quarks. In the graph, quarks can be divided into two atomic families, the uds family and the bct family, in the *weaker base of both triangles*. They are parallel, so we can, by fractal self-similarity, deduce its properties. Those bct quarks will be done in the experiments of 2013 when the machine makes collisions over 10 TeV, which is the barrier of massive formation of top quarks=Higgs, if we get there. In the experiments on 2010, CERN will study the strange quark, with massive production of strange quarks this fall. We shall see in those lead-lead collisions strangelet liquid.

The top quark triangle has in its center the equivalent of the hyperon, the usd atom of strange liquids. I have named this particle the *Toperon* made in bt and ct, *blackaon reactions*. The toperon will break the symmetry of bcb-dark atoms forming top quark condensates, gas-9[5], the equivalent of the usd hyperon and dibaryon atoms of ice-9 that trigger the reactions that create pulsars and strange stars. Thus, Gas-9 made of toperons will trigger the creation of top quark condensates, superfluid glass, theorized to be the most perfect substance of the Universe.

Because the top quark is 100 times heavier than the bcb-atom and rotates up to 10 c speed, it interacts with other particles trillions of times per second. If enough of them are made to create a stable lump, it will break the symmetry of our light membrane, extinguishing the Earth within seconds, *as it will be an implosive 10 c speed vortex, a hurricane that will take around a second to interact with trillions of particles, transforming them into toperons, imploding in a big crunch all the matter of Earth*. And so we will transcend into the membrane of dark energy that we shall now describe in more detail.

II. The Two Fractal Scales of the Universe

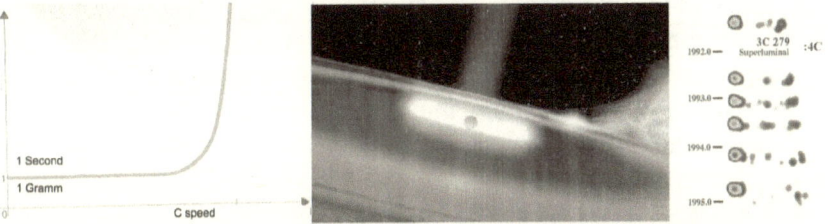

The speed limit of our light-space is tautologically C, the speed of light. Yet C is not the limit of speed of the Universe since the world of gravitational dark energy moves faster; it is nonlocal in physicists' jargon. That is why we can't perceive it, since our light instruments can't go beyond c-speed. The discontinuous border between both worlds is described by Lorentz Transformations (left), which show how particles deflect their lineal speed into cyclical speed or mass, close to the c-barrier, mutating into heavy quarks. But galactic black holes breach those discontinuities, emitting dark energy (center) that expels matter in quasars like 3C279 (right) at speeds up to z=10c times the speed of light. Those "dark energy" beams power the movement of galaxies that are dominant in dark matter (small, spherical ones) observed to move at z=10 c. Galaxies with more stars are slower because light stars drag them down towards c speeds.

2. The Informative and Energetic Limits of Light Matter

We are made of light quanta, swimming in a dark world of gravitation. If classic physics focuses on the understanding of the properties of energy, which it has exhausted, fractal non-Euclidean physics focuses on the understanding of information, the vortices of form, and the network structures that create the scales of the Universe. We call those "fractal scales" informative scales or i-scales, each of them separated by relative discontinuums, but able to maintain their self-similarity, thanks to the invariance of motions, scales, and forms of the Universe.

Those three properties are maintained while the medium size and quanta change. We exist in a Universe made of spatial bites and informative bits, which change their relative spatial size and speed of information, but the logic principles of the Universal game they play with energy and form are invariant. In quantum cosmology, because of the hyperbolic, Pythagoric error of Newton, who thought space-time was a continuous Cartesian plane and mathematics was reality itself, theories are built as if spatial energy and temporal information were not the "substance/motion" of reality (static or dynamic point of view) but something put by "God," who speaks mathematics, over the Cartesian absolute space-time plane. This error invalidates many theories of quantum cosmology, called background dependent theories, including quantum entropy, quantum gravity, etc., but their practitioners will never recognize their errors, etc.

Instead, we will study here the real structure of the Universe and its species made of bits and bites of energy and form, displayed in two self-similar fractal scales.

A proton is self-similar to a black hole in the cosmological scale and a black hole must be made of quarks, its fractal parts, as a proton is made of gluons, the fractal parts of quarks in a Universe of scalar, invariant properties. Since electromagnetism and gravitation are merely the accelerated vortices of two different fractal membranes of space-time, the microscopic and cosmologic branes. In this model, each particle is always made of parts with fewer dimensions, till reaching one-dimensional strings. Yet each string is in itself again a one-dimensional whole divided in multiple internal parts or dimensions.

What matters to us here is to understand how this simple model applies to the creation of fractal quark holes because if black holes are made of quarks as Einstein had it, they cannot evaporate, and they will be easily made at CERN.

Imagine that the infinitely small and the infinite big is almost the same, self-similar. That planets and galaxies behave like atoms and cells, that an assembly factory and a cell factory have the same structures and organization. They do in fact have them. How is this possible? This is due to a property called the invariance of energy and informative geometries in all scales of reality. It means that the Universe is a fractal of energy and informative shapes and motions. So cyclical masses come together into cosmological, cyclical bodies, and energetic fields grow in size from quantum particles to galaxies. We must understand that there are two different membranes of space-time of different power and energy. But our scale is smaller and weaker than the cosmological/strong membranes.

The danger of CERN is thus clear. The quark cannon will break protons and neutrons, extracting the quarks inside them, and so it should create with them a cosmological quark star or black hole.

We talk of two systems, or rather a "sandwich" system of Russian dolls:

— The inner Russian doll of quarks and the outer Russian doll of cosmological black holes and galaxies is the gravitational world of masses.
— In between the Russian doll of electronic vortices and its electromagnetic forces, which is our world, what we perceive, what creates our medium, light space.

The previous graphs illustrate the structure of that sandwich. What CERN is going to explore is the inner fractal parts of masses, the world of quarks, to create the upper cosmological world of black holes and quark stars. How all those i-scales interact can be deduced from the generator equation of fractal space-times events, E⇔Ti. One of its physical derivations is known to quantum physicists as the law of range, which can be deduced from the inverse properties of energy and information:

max. energy/space=min. form/mass

Thus, the law of range is a paradoxical law: The particles that transfer forces increase in informative density and mass as they act in shorter spatial ranges.

So we differentiate a particle of maximum spatial range and minimal informative size, the photon of light, which creates our three-dimensional Cartesian space; a transitional force, the weak force that *evolves informatively a particle, giving it more mass*; and the strong force, of minimal range and maximal mass/information. *What CERN is going to study is precisely how the strong force kills our electroweak matter and evolves it into quarks of maximal information.*

Yet those "invisible scales" of the gravitational membrane cannot be perceived by an electroweak species, the human being, because their speed/distance range is beyond c. So they are called particle-points, as we can't see them. So CERN will never "see" what it makes. The lumps of quark matter will fall undetected to the center of the Earth and start to eat us from inside, till we feel it as literally the Earth shrinks to a fifteen-kilometer rock of strange matter or a three-centimeter black hole. Indeed, another paradox of this MAD[6] experiment is that CERN will produce dark, quark matter, which will be invisible and fall to the center of the Earth, where it will consume our matter, and so one day we will wake up and see the Earth crunching inward. The speed at which those quarks will kill us differs, according to their different speed of rotation. Strange quarks rotate at c-speed so we shall see it happening, but top quarks are pure gravitational species of the "other membrane" and so they rotate *faster than light and will crunch us in a few seconds. To see why, we have to study further the structure of both membranes.*

The c-speed and 0 Kelvin Frontiers

What are the frontiers between the two membranes of space-time, the quantum light world and the cosmological gravitational world, and how do they exchange energy and information between them? To fully grasp this, we need to consider the equation of the arrows of space and time, energy and information, in a dynamic way as an equation with limits of maximal energy and information that any being cannot cross without dying. Those limits are precisely the limits between the two membranes of the universe, the light membrane, whose energetic limit is c-speed and its informative limit is 0 Kelvin degrees, and the world of gravitational and dark matter, which goes beyond c-speed and probably it is colder inside the quark-gluon liquid and the black hole than 0 K temperature (as its quark-gluon liquids are superfluid, superordered, and yet move faster than light in its rotational vortices).

The generator equation of energy and information synthesizes the two arrows of time, from where most complementary systems of the Universe can be deduced: $E \times I = K$. In physics, this equation is called the principle of uncertainty of Heisenberg, which quantum physicists don't understand because they handle only an arrow of time, energy. So they treat it as an uncertain equation.

The equation has two forms, as a dynamic fluctuation or wave with limits, $E \Leftrightarrow Ti$, and as a balanced organic particle, which is a knot of fractal networks of smaller particles of energy and form, which maintain a constant ratio defined by a certain universal constant, $E \times I = K$.

This simplified wave/particle duality explains how all physical systems sometimes move as loose herds of energy and information quanta ($E \Leftrightarrow Ti$) and sometimes form static knots of energy and form, $E \times I = K$. Again, for a complete description of those systems, we refer to the classic Dirac equations. We just need now to consider the limits of energy and form of that equation which are the limits of our membrane of light (c-speed) and ud mass (0 temperature). We can write an equation of limits:

$$Max.E \times Min \ I = Energetic, \ wave \ state;$$
$$E=I, \ balanced \ state;$$
$$Min.E \times Max.I = Informative, \ particle \ state$$

The previous equation is the true meaning of the uncertainty principle, which is not uncertain but an equation with two limits for any species of the Universe, one energetic state and an informative state. In life is the equation of the three ages of time: the energetic, young horizon, the balanced state, and the informative horizon. In the case of quarks, it explains why there are three families of increasing information. In state physics is the equation of the states of matter as an energetic gas or balanced liquid or an informative particle. The reader might realize of the power of the formalism of duality to explain the how of so many events of the Universe.

Thus, we call the casual chain between a state of energy or wave or youth and a state of information or an old age or particle, the time symmetry.

The equation establishes also the limits of life and death since death happens in the extremes of that equation, when energy goes to infinity and information disappears (death by accident or overdrive of energy) or when information warps completely a system and energy disappears (third age and death).

And the main difference between biology and physics is that in biology, the equation invariably goes from youth, energy, to old age, information, and then we die. In physics however, particles constantly switch between both states or rather they constantly die and resurrect. Particles of information explode into antiparticles of energy that become again particles of information. Waves of light become informative photons. And so generation after generation, the Universe regenerates itself. Life becomes death that becomes life.

We are interested now in the limits of that equation for atomic particles, which are the limits of death that CERN will explore with its quark cannon. Indeed, the formula establishes the death limits of any space-time membrane, in our case the light space-time membrane.

In a normal, balanced state, particles and membranes of space-time are defined by balanced energy/information ratios called universal constants, within which the cycles of energy and form of the space-time membrane or entity that inhabits it can happen.

Yet beyond those ratios, the system breaks; the balances between its energetic fields and informative particles disappear, and the space-time membrane ceases to exist, giving away its energy in a big bang and giving away its information, absorbed in a big crunch by new particles of a *bigger=faster gravitational space-time membrane* with faster speeds and higher degrees of order in their universal constants. *And this is the big bang, big crunch dual process CERN will essay: it will break the symmetry of our matter and send our information beyond the 0 K temperature, making quarks and our energy beyond the c-speed, making dark energy. And whatever is left in our light world will become dust of light space-time.*

Universal constants are either scalar limits beyond which species die, or e/I ratios in which species live called vital constants in biological beings. Because energy and form are in proportional balance, we can deduce in complexity a general law of death limits: the bigger the species or membrane is, the faster its energetic limits are and the more information it accumulates.

This complementary balance between the energy and form of any system writes:

$$Max.\ E\ /Max\ I = K$$

The Decametric Transitions Between Membranes

Further on, there is a basic scalar constant, to transcend from a lower to a higher scale of complexity in any system—the decametric constant. So the gravitational membrane is a c<10 c membrane.

This scalar constant has been already expressed visually by artists like Eames,[7] and it is empirically well known and proved by the constant use of logarithms in base 10 *in all systems of the Universe, from matter to army scales*. In armies, neuronal networks and any system of non-Euclidean points every 10 units of a lower scale form a unit of a higher one. So the scales of the universe are based on potencies of 10. Imagine a triangle made with three smaller corner triangles of three balls and a central ball, number 10 coordinating them. This is a basic "tetrarkys" organism, one of the most efficient configurations of the Universe in which one corner triangle absorbs energy, the other absorbs information, the third combines energy and information into reproductive seeds, and the central 10 is the soul of the system that coordinates the three vital dimensions of the organism that becomes a system that absorbs energy, information, and reproduces itself.

This simple system by causal triangulation can create any other physical fractal system of the Universe. Already Plato said that numbers are forms and the physical Universe was "an organism with a body and a soul."[8] The tenth element is the informative "connection enter" that acts in the upper scale as a unit of a bigger world, connecting the lower and upper fractal membrane. Indeed, each number is an essential topology of the universe and the 10, the "tetrarkys" as Pythagoras thought, is the most perfect scale.

So now we can establish with that basic knowledge some relationships between all the elements of the two membranes. Let us first recall them. For each of them, we shall find two different motions/forms, lineal forces and cyclical particles. And one transforms into the other by curling when lineal speed reaches the limits of energy and form of each space-time membrane of our universe:

— Gravitational forces are either cyclical, informative, attractive vortices—quarks, black holes, or lineal, repulsive waves called dark energy. Their range is bigger than our light forces, between c<10 c.
— Electroweak forces are lineal light or cyclical electronic charges—the whirls/hurricanes of space-time of our weaker world. They are therefore slower and rotate at c/10<c speed. Obviously electrons have more charge than quarks as they are the particles of the electroweak world, and quarks have more mass than electrons as

they are the particle of the mass world. *And both interact in the c-limit that connects both worlds. Up, down, and strange quarks under c-rotational seed are the quarks of our world.*

Accelerated lineal motions or forces, c-electromagnetism and c<10 c gravitational waves (repulsive dark energy) *interact in the borders of both membranes, in the black holes and halos of galaxies.*

In those halos, black holes and strangelets drag gravitational waves to a mere c-speed, creating the background light radiation. Yet between galaxies, gravitation is free to accelerate till the 10 c limit shown in quasars and z=10c galaxies.

All this, of course, is data theoretically known to quantum cosmologists, but lacking the simple straightforward approach of the fractal model, it is not understood. And since information is inflationary, quantum cosmologists make with all that data as many models as you wish. Or as Leonardo put it, "the only harmonious sounds of scholars are the farts of their asses" since by definition, scholars are inflationary beings, whose job is to multiply information theories. But if we want to survive CERN, we must decide with the scientific method, which of all those theories is more certain, as there is only one universe and hence one truth for each language that describes it. The model we explain here thus tries to stick ad maximal to the three legs of truth of science and the logic of the principles of relativity and invariance of form at scale. And it seems to me the most focused model built with the data we have.

Because gravitation is invisible, only recently we have realized that there is also repulsive gravitation, dark energy, which is emitted by black holes through its poles. How this is possible if light speed is the limit of the space-time we perceive? Precisely because light speed is NOT the limit of speed of the entire Universe. This is another "hyperbolic error" committed by Einstein that now we can tackle, as we have explained the difference between both membranes and their universal constants.

Our space-time of electroweak particles and forces has an informative limit, which is 0 K temperature for the horizon of black holes and an energetic limit, which is c speed. But those are not the limits of the entire cosmos and all its membranes. Since it is experimentally proved that intergalactic space moves at 10 c speed (quasars' jets of mass) and theoretically the inner region of a rotating charged Kerr black hole is superluminal. Thus, a black hole or quark is an accelerated vortex which reaches beyond its event horizon superluminal speeds. And so when the vortex of acceleration arrives to its central limit, as it happens with an electronic vortex that jets a magnetic field through its poles, the black hole or top quark jets superluminal dark, repulsive gravitational energy. This dark energy in the interstellar medium between galaxies, where gravitation is not warped by light into a C-flow that limits its speed, reaches 10 c. Inversely, when dark energy becomes light, it corrugates into a c-speed wave. And when light corrugates into a denser vortex of electroweak forces, the electron, it diminishes its speed to c/10.

The self-similar quantum concept of an informative vortex is the spin of a particle, related to its mass or charge in the quantum world. In the bigger plane of celestial bodies, the fractal sum of all those spins becomes a "rotation," which becomes the mass of the object so we observe rotation in all cosmological bodies, which increases with the mass of the planet or star till reaching the >c-speed of rotation of black holes. Since if a vortex of mass is an accelerated vortex and the black hole reaches c-speed in the event horizon from where light cannot escape, thus turning at light speed, beyond the event horizon, the vortex of mass must accelerate beyond c-speed.

And indeed, the black hole becomes a door to the next bigger space-time membrane of superluminal "dark energy," the gravitational energy that accelerates till z=10 c speeds between galaxies.

Because of the hyperbolic error of science, Einstein thought that c-speed is the limit of speed of the Universe, but this is not truth. It is the limit of speed of the light space-time of our galaxies. But beyond, there is a world of dark energy, which is gravitational energy. And we have measured z=10c speeds for matter ejected from quasars and faraway galaxies. Because of the paradox of Galileo, those speeds can be seen as "fixed, expanding space" and this is the view of the big bang theorists, which "think" that space expands at 10 c speed.

But the more correct way to perceive it is in fractal terms. Galaxies live in a sea of gravitation, which is the next space-time membrane of the Universe. Because fractal space-time is decametric, each scale has 10 times more energy and 10 times less information. So the overall Universe, $E \times I = k$, is constant.

Thus, we can observe the following basic membranes of space-time:

— Gravitational space-time. 10 c speed: intergalactic space expanding at 10 c. Rotational 10 c speed; black holes and top quarks, its atoms. Hence top quarks and black holes are very massive because their vortices of mass rotate faster, and mass is proportional to the frequency of information of those vortices.

This explains why the top quark is around 100 times more massive than the c-speed quarks that limit our Universe, while the slow electron is at c/10 far less massive.

Yet all together, all those membranes maintain the balance given by the aforementioned law of range: energy/distance/speed × information/mass = K. This range law is derived from the generator law of all systems of the Universe, E × I = K (static version) or E⇔I, dynamic version.

Quarks and black holes are therefore the realms of dark energy and dark matter, which should be called properly gravitational energy and heavy quark matter and it can rotate faster than light beyond its event horizon.

The experimental proof that gravitational dark energy and rotating top quarks go faster than light was found recently in deep-space pictures where an enormous number of dark galaxies, with big black holes and hardly any star, made almost entirely of dark matter, were moving at $z=10$ c speed.

These galaxies don't drag any stars: they are accelerating toward its maximal dark matter speed.

How can physicists then deny such experimental evidence of superluminal speeds? They switch to the static view of speed as space (Galileo's paradox), considering that galaxies are NOT moving, but creating static distance, space. So the space of the Universe is what is expanding. Galaxies are NOT moving according to that view as rockets do, expelling dark energy at $z=10c$ through their black holes. This is like saying that a car is not moving but "creating" a line of space as its image shows in a night picture made at slow motion. Of course, it is truth if you want to consider a static perception of the Universe as Duchamp made in his *Nude Descending a Staircase*. So big bang theorists have a cubist theory of space even if they are clueless in explaining where that expanded space comes from and why space is not expanding between this book and you, my dear reader. *That is the question that makes the fractal explanation better. Because while intergalactic space expands, galactic black holes contract space so our local world doesn't expand.* As Woody Allen's mother put it, "sure, but Brooklyn is not moving away from Manhattan," LOL. I only met Hawking once when I was young, in an astrophysical conference. I gave him that question/paradox: "If space is an expanding continuum, not a fractal series of expansions and implosions, why are you not getting away at the speed of light from me?" He never answered. Now he will come to me faster than light speed if we make black holes on earth.

Further on, quantum cosmologists have no logic explanation to the meaning of dark energy and dark matter for which they keep inventing fantastic, new particles, when the simple model of the fractal universe, explained here, based in known-known particles and forces, resolves it all, following the three legs of the scientific method, advanced mathematics, logic consistency, and experimental massive evidence of all types of galaxies moving at z=+c speed till the 10 c limit of maximal speed of gravitational energy, the next scale of the fractal universe.

Yet we won't become dark energy. Since experiments done with strange quark liquids made of uds quarks, on the verge of 1 TeV, showed already quasi-stable strangelets able to catalyze the transformation of ud quarks into strange matter. And since those experiments will be done first, one of those events shall extinguish mankind under the totalitarian principle that affirms that all that can happen in physics should happen.

All this, of course, is ignored by CERN, which is ready to start a massive deconfinement of quarks with the quark cannon. If we survive the strange quark experiments of 2010 in 2013,[9] quarks will become accelerated linearly at 14 TeV, equivalent to a 10 c vortex speed. So the quarks will be first transformed into strange quarks, then into bc atoms and finally into top quarks. It will then in the point of collision, where millions of simultaneous collisions happen together, form a fractal vortex of top quarks that will start a chain reaction, absorbing all the matter of the Earth, forming a cosmological top quark star, a black hole of three centimeters. We shall then live in hell for eternity as black holes live forever. According to Einstein's theory, nonevaporating black holes never die. Think about it. Mankind will be just a nanomillimeter or so in a three-centimeter black hole, *forever trapped in eternal hell. Do we deserve this?* Think how many times we have been warned, how many forms of life we have tortured, how much arrogance we have displayed with the Universe and its laws, how important we think we are, how people like Hawking dares to invent "time travel" and risk mankind to prove it and get a Congress Medal for that—and how the Universe is going to respond.

This is what will happen at CERN since we have the previous experience that created the first drops of strange liquid at a far less powerful accelerator, the RHIC. Yet before we study in more detail the creation of strange liquid, how those strange and top quarks will catalyze the reaction of ice-9 (strange star formation) or gas-9 (black hole formation), we should now consider what CERN "expects" to see on those experiments but will never find—imaginary particles of fantaphysics.

3. What LHC Will and Won't Produce: Quark Condensates and Fantaphysics

In the previous chapters, we explained the difference between Pythagoric, mathematical fantasies, so common in quantum cosmology, and true laws of science—all that was needed to judge the kind of particles the quark cannon will produce and understand the difference between reality versus theory and standard science versus fringe science, which is at the core of this dispute.

In physics, there are two kinds of theories:

Those which are real, in the sense that they are constructed with particles we know, under laws that have been proved truth by the experimental method and so are standard science, *which is the only science that should be taken seriously in Court, in public policy, and in any analysis of risks.*

And theories that are not real but at best probable, based on speculations, unknown particles and proposed hypotheses, which are not laws of science. These theories are NOT yet science and should not be taken seriously in any court and public policy, especially in matters that can hurt human beings.

We shall include in this type of theories both, those sponsored by CERN, which are fringe theories that defy standard science (Hawking, Higgs), and those advanced by Einstein et al., which describe the fractal universe and are the most promising theories of reality, concerning the future of science. Yet even within speculative theories, there is a difference between the science of the future (the concept of a fractal universe made of energy and information) and speculative theories from the science of the past (the quantum theory that the Universe is uncertain, hence all theories are probable).

Since it is a rule of history of science that future paradigms are more accurate, even if less known, than the past paradigms, they correct and improve. Moreover, when the future paradigm is able to resolve—as the fractal theory of the Universe has done, most of the questions left to answer by standard twentieth-century science—it follows that the future will belong to the fractal paradigm, NOT to fringe theories, born of the outdated uncertain hypothesis about the Universe.

In that regard, both standard and future sciences affirm the planet is at risk and deny uncertain theories.

Such uncertain quantum theories (Higgs, Hawking, parallel universes, multidimensions, etc.) happen to be loved by physicists because they stir their imagination and allow them to invent with mathematics new particles and fantasize about strange properties and possible new laws. Such bizarre theories in very few occasions have turned out to be real when experimental proofs confirmed them, and so CERN physicists sponsor them, hoping to get the honors and medals they seek for and cannot obtain considering only standard proved theory.

This is why CERN's scientists do not care at all to inform mankind about the 99 percent of what the quark cannon will produce: standard quark condensates and their enormous risks. They prefer to talk only about the unknown desired theories, which their wishful thinking pretends to prove with this machine. Yet according to morphological and organic laws of harmony and balance, between energy and form, those theories are false since *only particles that follow the selective laws of the Darwinian universe survive.* This is completely ignored by physicists who ignore the Darwinian laws of the Universe, which extinguishes in any of its systems those organisms and particles that are not efficient. In the same manner, they will kill the Earth's weak matter since they don't even want to look at the fractal paradigm developed in complexity, as that means going back to school and abandoning their "energetic dogmas" of the Universe.

In simple terms, CERN is a weapon used to explore an outdated, exhausted paradigm of science, which has for the past decades produced only bizarre theories as all outdated exhausted paradigms do.

This means there are no SUSYS (supersymmetric particles), which are inverse particles physicists obtain theoretically by changing ad hoc its informative and energetic parameters. Such particles will not be stable because they are abnormal creations. Imagine an organism with a head bigger than the body. The organism will be unstable because information is smaller, warped, and energetic limbs are large, extended in space. So SUSY[10] particles, which are particles with bigger informative heads than bodies, do not become stable. But since the quantum paradigm ignores all about the duality laws of energy and information and it does not want to learn, it cannot understand that SUSYs do not matter and it plans to use the quark cannon at high energy to see if it can make them. Of course, in the case it does make a SUSY, it will last an infinitesimal fraction of time. And so their relevance for the Universe is null. In other words, the quark cannon will at best produce monsters that will die as deformed fetuses do in the world of biological beings. So they are totally useless. And the stable particles it will produce, dark matter quarks, are particles that can kill us. Such is the monstrosity of this factory.

Why then are there so many articles, books, and papers about SUSYs? Because playing with their equations, physicists occupy their free time now that the standard model of stable particles is closed, "drawing monsters" with their equations, and one of them came with the idea that if SUSYs exist, manipulating this and that other field obtained a weak gravitational field and alas! He put forward an explanation why the gravitational field is so weak: it is entangled with SUSYs. But the real explanation given by the fractal paradigm—that the quantum world does not have gravitation at all because gravitation is other membrane—suffices without inventing particles. And it is far more coherent with the equations we have of quantum physics *where the gravitational field does not exist, nor is it needed to describe any event of the quantum world. Hence, we must simply agree with the experimental, mathematical, and logical evidence of the fractal paradigm: gravitational forces are forces of the cosmological membrane and so they do not exist in the quantum world in which we inhabit, reason why they are so weak—they do not affect the particles of our world.*

Instead, physicists should work out the ways in which the strong force of quarks transcends and gathers together, emerging as the fractal particles of the cosmological gravitational force of black holes and quark stars.

Of course, physicists can make many more new particles, which are abnormal, live very short, and are for that reason called resonances. But they are totally irrelevant to science as all the genetic mutations biologists are going to manufacture soon, using the genome, are for our biological world. The only thing they can end up doing is a new virus, dangerous for mankind, or a particle of dark quark matter that'll eat us. Abnormal particles such as antiparticles die very soon since it would be like creating animals with the head under the stomach to continue the comparison.

All those facts of duality limit our analysis of what can be dark matter to real theories, real particles, real laws, and efficient forms. There are no two particles (Higgs and tops) when we know one that does the job and, unlike the impossible scalar Higgs, is real, efficient, and known. There are no impossible theories when we have a theory that explains it all. There are no surviving mutational monsters. There are no parallel universes, etc.

As Einstein put it and Gödel proved it, "while I know when a mathematical equation is truth, it doesn't mean it will be real."[11] So experimental evidence and logic consistency is also needed to prove an equation real. This is a key fact of the scientific method that Pythagoric quantum cosmologists deny, as they consider the Universe uncertain, probabilistic, and purely mathematical, breaking that fundamental tenet of truth in science, the experimental method, reason why Einstein scorned their work and told them to stick to what they know, the world of electroweak forces, for which his theory was developed. They should respect Einstein's work that described the world of gravitation and mass in which general relativity still stands. Unfortunately, this is not known by the people who have sided against Einstein et al. in the suits and accepted Higgs and Hawking as truth. They believe Higgs is the only theory of mass, that evaporating black holes is the only theory of black holes. They ignore that the standard theory of mass is not Higgs, the quantum version but Einstein's relativity, that the standard theory of black holes is not Hawking's, the quantum version but Einstein's relativity. Yet if Einstein's theories were right, we should die. So CERN and the scientific establishment that backs big science and big machines cannot argue this in public. For that reason, there have been smearing campaigns, censorship of any scientist opposed to CERN, editorials in the *NY Times* asking the judge to send the suits to the nearest black holes, judges that don't judge, etc.

Unfortunately, quantum fringe theories extrapolate work done in quantum theory to study the entire Universe (to the point that many quantum cosmologists pretend to use the equation of Schrödinger that describes the electron to explain the entire Universe), despising the work of Einstein and so today most people think fantaphysics quantum cosmology is truth. Yet quantum scientists have enormous industrial power because their findings do work to create electronic gadgets and nuclear bombs—but NOT to explain the Universe, the realm of Einstein et al. And yet they impose their cosmological musings with their industrial power in scientific magazines. Plainly speaking, power has hidden the truth and so we shall expect Einstein's dictum: "Those who hide truth with power will be the laugh of the Gods."

And so we must distinguish what is real and should happen at CERN under the totalitarian principle of Nobel Prize Gell-Mann (quark condensates, quark, dark matter, strange matter and nova reactions, top quarks and top black holes) and what has a very low probability of happening (black hole evaporation, SUSYs, neutralinos,[12] Axons, Higgs, extra dimensions, and parallel Universes).

This is a very necessary point to have in mind along the pages of this book because under the strict laws of veracity of the scientific method, I will explain but NOT consider at any moment as real any musing, wishful thinking, or hypothesis brought about by physicists or employees of the nuclear company (CERN). CERN didn't go to the suits because it knows that in any serious suit or political decision the judge would have only accepted as real the proved facts of reality based on proved laws and known particles of standard science. And so CERN should have closed down. Or else, any argument or exposition of facts becomes a byzantine sort of religious argument in which all is possible and we never get to any conclusion. This, indeed, is the way the employees of CERN, our expert physicists, have conducted this argument. They have denied standard science and brought about fantaphysics to confuse people of the real dangers of their dark matter factory.

Yet standard science says that in the universe black holes and quark matter are the dominant species. There might be other substances that could be dark matter, but they are not known, never observed, and most seem to be mathematical errors, fantasies built by mathematicians, as writers can build characters and painters invent images that do not exist.

We know there are only two observed scales of space-time in the Universe between which humans are sandwiched—the small quantum world of atoms and the cosmological, big world of gravitational, celestial bodies—and there are only two proved, known fundamental particles in the Universe (the rest are forces) called quarks and electrons. We know only quarks can form dark matter as they are the only tested, hyperdense, hypercold particles that can emit background cold radiation; and those two things—cold, dense dark matter and cold radiation—is what we find everywhere in the outskirts of galaxies and the Universe. The rest are just hypothesis, which are very dubious. Because the Universe is efficient, and there are six quarks of which only the two weakest ones have a role in our world of electroweak matter, the Universe must have a leading role for the four heavier quarks. They must therefore be dominant dark matter, as they can easily absorb our lighter quarks. Hence the enormous danger of the factory of heavy quarks built by CERN. Standard science will certainly happen at the quark factory: a substance called quark condensates, which standard science precludes will very likely extinguish the Earth.

But can the quark factory besides blowing up the Earth advance our knowledge of the Universe? This issue confronts two different theories about the Universe—quantum cosmology, an extension of the knowledge we have of the microcosms to the macrocosms, versus fractal relativity, the modern version of Einstein's cosmology. This second theme is of less relevance to our lives, but NOT to science as an endeavor, which searches a logical, mathematical explanation of the Universe. And again, in this dispute, twenty-first-century science is proving Einstein right and all the theories that CERN sponsors, based on energy-only Universes, false. So the answer is NO. CERN, which is built to study quantum cosmology, a false theory with no future, will not advance our understanding of the Universe.

Unfortunately, while each individual mind can have a point of view, or linguistic world about reality, as so many theorists of physics do, the Universe is only one. *And* because the Universe is only one, albeit seen from many perspectives, there is only one you, one planet Earth, and one us even if you can take pictures of it from many perspectives. And so in physics there is only one theory of reality that is truth, even though we can write it with different mathematical equations and/or languages. So there is only one particle that does certain things. There are no redundant particles even though we can write, for example, the top quark with the equations of the Higgs. What this means obviously is that if we have found the top quark, the Higgs will be the top quark. That if we need to explain mass, it will be with Einstein's work. That if quarks can be dark matter and we have not discovered any other particle that can be dark matter, quarks will be dark matter. The Universe is efficient and real. There are no infinite dimensions, infinite parallel Universes, infinite particles, except in the musings of fantaphysics, as there are no Quixotes beyond the musings of Cervantes, even if those realities are elegant and enjoyable. This is obvious. In an efficient Universe, there are no beings with two heads even if we draw them and write fiction with Hydras. All comes to Darwin: matter feeds in ill-constructed matter, and predators eat animals with two heads. So at the end, you only find what is beautiful, efficient, and makes sense, even if we can invent an entire zoo of imaginary particles with maths. This is what CERN's physicists prefer to ignore because it *means none of their fantasies will happen at CERN.*

Most of the particles CERN pretends to research do not exist or are not stable because they are mathematical fantasies, which do not follow the laws of the scientific method—simplicity, efficiency, logical consistency, hylomorphism, and harmony between its energy and information parameters (complementarity principle) that define existence and survival in the Darwinian Universe.

Thus, neither Mr. Higgs, Mr. Hawking, or any of the fantaphysicists that are eagerly awaiting their fantasies to appear at CERN will get a Nobel Prize. For example, Mr. Arkani-Hamed, one of the most sanguine critics of Einstein et al.,[13] which he repeatedly calls crackpots, expects to observe at the LHC proofs of his theory of 10^{500} Universes. Good luck.

III. Quark Condensates

Nothing can we call our own but death and that small model of the barren Earth, which serves as paste and cover to our bones.
—Shakespeare, father of modern literature

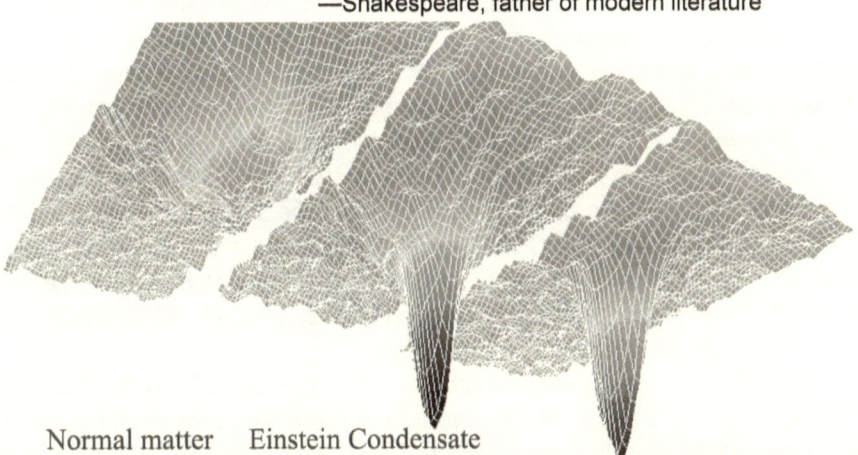

Normal matter   Einstein Condensate

The creation of an Einstein-Bose state in which each particle is put one under the other, packed into a superfluid vortex that sucks in all what surrounds it, forms a superattractive quark hole. We can model black holes as quark holes of top quarks and explain the creation of supernovas as the final reaction of a star that creates in its center a quark condensate; hence the enormous danger that CERN, which is basically a factory of quark condensates, poses to this planet.

4. Putting the Puzzle Together: Quark Vortices

We shall return to what truly CERN will do with that 99.999 percent of deconfined quark production: quark condensates, the substance of quark stars and black holes. We have studied them with the dynamic, vortex view of complex fractal physics. But classic physicists study them with the particle, static view, which also shows an enormous danger for the Earth and so we shall now consider them with a classic view *since CERN also denies the risks expressed by classic quantum physicists, showing that it does know those risks, obvious even when we use its own quantum work, which is the reason why we sue the company for criminal negligence. We know what they know.*

According to Einstein's theory of gravitation, the quark cannon will make mass accelerating the nuclei of heavy atoms, called hadrons, to the speed of light, till that energy, $E=Mc^2$, starts to become quark mass. In the fall the risk will increase because CERN will use lead with a sizable number of quarks. Lead is the heaviest stable atom, with 621 quarks per nuclei. Those 621 quarks per nuclei, in bunches of millions of nuclei, packed together in ultradense formations, thanks to a process called stochastic cooling that pushes them from both sides as they also accelerate along the track, will grow in mass, and become the heaviest quarks of the Universe. And then they will collide from two opposite rays, to bunch them further, forming a huge solid mass of quarks, a quark condensate. At this point, hundreds of thousands of heavy quarks will be all together in a single point. This has never *happened anywhere in the Earth, nor does it happen in the galaxy except in the inner center of stars where the density of mass is such that similar collisions can trigger the creation of a quark condensate and the subsequent big bang of the star called a nova.* But CERN denies it will do such quark condensates. Instead it affirms that the machine will produce collisions similar to cosmic rays, which are lonely atoms, which have not enough quarks to form a critical mass and start the formation of a quark star.

This denial is compulsory since the dangers are real. So they cannot be argued, as the risks are enormous and the safety arguments brought about by CERN clearly false. Let us review briefly, at this point. the cosmic rays lie:

— The large atomic cannon has nothing to do with cosmic rays. In fact, we never found a quark in cosmic rays. So how could CERN invent such a lie? What CERN does is to calculate the mass-energy ($M=Ec^2$) of a cosmic ray, saying it has more energy/mass than the 3+3 quarks *of a single proton collision.* Of course, I do have infinitely more mass energy than a quark, but CERN will collide bunches of billions of quarks pegged together and it will liberate for the first time the strong forces of *the last superexplosive Russian doll of matter that will break open with the collisions at CERN for the first time.*

I do not start a mass reaction of *strong forces because I do not belong to that doll, neither do cosmic rays.* Within seconds, if we liberate the quarks of that doll in huge bunches, the mass reaction of a quark condensate could have millions of times more mass than I do; and when the quark reaction ends the entire planet, it could be a quark lump. What CERN is telling us, following the comparisons with *Jurassic Park*, is that a dinosaur's egg is inoffensive. Indeed, but the egg, the first lump of quarks of the collision, is only the beginning of an amazing growth rate once the dinosaur baby, the quark condensate, is born and starts to predate on the weak mass that surrounds it.

Besides, CERN does not explain that cosmic rays do not have enough critical mass to start any reaction because they are single atoms, one atom at a time, while *its machine creates collisions with a luminosity=density of 10^{28} times* that of cosmic rays. This only happens inside stars where nova explosions take place because only such density creates enough critical mass to start a chain reaction that creates a quark star or black hole. This can never happen when only two atoms collide in the atmosphere. CERN has never answered those facts. And we were never able to explain this to mankind because the press has merely published what CERN says; to reassure the public, we shall die without public alarm.

So what will happen is simple to explain: Those quark condensates will form a superfluid vortex, a hurricane of space-time of an enormous density. And new quarks will keep falling into that vortex till the Earth caves in. All those small vortexlike quarks then will at macro scale form a rotating star, in the same way we have recently observed small eddies form together a big hurricane, transferring its microscopic vortex nature to the bigger macro vortex.

5. Einstein-Bose Condensates

The technical name of those superfluid quark vortices in its final ultradense state is called an Einstein condensate. An Einstein condensate of the three types known (Bose condensates that fuse atoms, Fermion condensates that fuse quarks, and Cooper condensates that fuse electrons) is a fascinating phenomenon, a new form of matter, so dense and perfectly ordered that it has indeed "magic properties": it is a superfluid, it has no friction, it moves at enormous speeds, it packs a huge quantity of fractal quanta in minimal space, and it attracts like a perfect storm. The most dangerous are quark fermion condensates *in which quarks first form pairs and triplets and then come together into an ultradense substance, the quark condensate CERN's cannon will mass-produce*, till creating what Einstein called a frozen star. The reason why quark condensates are so dangerous is because they use strong forces, so they are rotating faster, they are far denser, attractive. So they can be considered the most perfect liquid vortex of the Universe, which is the name scientists gave it when they saw it for the first time.

All this said, it looks surprising that CERN could ever thought to be a brilliant idea to manufacture a superattractive liquid. How is so much irresponsibility possible? Because nature cheated them.

Indeed, a pattern appears. We didn't think quarks could form Einstein condensates when CERN designed the machine since the rotation of quark vortices, its spin, is something physicists called impair, disordered spin. In a simplified image of this concept, imagine that you have two of those vortices/ spins of mass turning in opposite directions. You cannot then match the direction of both vortices and put them together, occupying roughly the same space one over the other since they would collide and friction would destroy its structure. But it turns out that bidimensional quark vortices do order themselves in something called a color-locked quark condensate, perpendicular to each other to form a three-dimensional world, transcending with 3 × 1/3 charges into the electromagnetic membrane we inhabit. Somehow three quarks become locked in a rhythmic, ordered spin, like the perpendicular helices of helicopters that never collide together, and then there is no longer disorder and friction. So they now can accumulate in orderly patterns, forming quark condensates. Nature indeed finds always a way to evolve stable forms that live longer and move faster than human CERNerds think.

This is called the totalitarian principle of physics that says if something doesn't violate the laws of nature, it will happen anyway.

Nature seems to be far more intelligent than quantum physicists are. So today, known-known theory tells us the quark fermion/Bose condensates can happen. It doesn't break any law of science. And so it will happen. But we found this recently. The machine was almost built. CERN lies or hides this information and no longer talks of Einstein condensates, its main produce, despite its pretension of being an educational, informative company. The factory of quarks no longer manufactures quarks but SUSYs and Higgs. So to fully grasp what those quark condensates are, we shall instead quote the news of their discovery by a NASA lab six years ago:

> *February 12, 2004: A New Form of Matter. NASA-supported researchers have discovered a weird new phase of matter called fermion condensates.*

We learned it in grade school. There are three forms of matter: solids, liquids, and gases.

But that's not even half-right. There are at least six: solids, liquids, gases, plasmas, Bose-Einstein condensates, and a new form of matter called fermionic condensates just discovered by NASA-supported researchers.

"This is a very exciting time," says University of Colorado/NIST physicist Deborah Jin, lead scientist for the group who produced the first fermionic condensate in December 2003. "My group works extremely hard these days. Both the excitement of a major advance and the competition to be first have been driving forces."

Most second graders can recite the properties of ordinary solids, liquids, and gases. Solids resist deformation. They're stiff and they can crumble. Liquids flow, they're hard to compress, and they assume the shape of their container. Gases are less dense, they're easy to compress, and they not only assume the shape of their container, they expand to completely fill it.

The fourth form of matter, the plasma, is gaslike, made of atoms that have been ripped apart into ions and electrons. The sun is made of plasma, as is most of the matter in the universe. Plasmas are usually very hot, and you can keep them in magnetic bottles.

The fifth form, the Bose-Einstein condensate (BEC), discovered in 1995, appears when scientists refrigerate particles called bosons to very low temperatures. Cold bosons merge to form a single superparticle that's more like a wave than an ordinary speck of matter.

Now we have fermionic condensates—so new that most of their basic properties are unknown. Certainly they're cold. And they probably flow without viscosity. Beyond that? Researchers are still learning.

"When you find a new form of matter," notes Jin, "it takes a while to understand it."

Fermionic condensates are related to BECs. Both are made of atoms that coalesce at low temperatures to form a single object. In a BEC, the atoms are bosons. In a fermionic condensate, the atoms are fermions.

What's the difference? Bosons are sociable; they like to get together. As a rule of thumb, any atom with an even number of electrons + protons + neutrons is a boson. So, e.g., ordinary sodium atoms are bosons, and they can merge to become Bose-Einstein condensates.

Fermions, on the other hand, are antisocial. They are forbidden (by the "Pauli Exclusion Principle" of quantum mechanics) to gather together in the same quantum state. Any quark or any atom with an odd number of electrons + protons + neutrons, like potassium-40, is a fermion.

Jin's group found a way around the antisocial behavior of fermions: "Cupid." Loner atoms pair up. A pair of fermions can merge with another pair, and another and another, eventually forming a fermionic condensate.

Jin suspects that the subtle pairing of atoms in a fermionic condensate is the same pairing phenomenon seen in liquefied helium-3, a superfluid. Superfluids flow without viscosity, so fermionic condensates should do the same at temperatures close to absolute zero. Superfluidity is a phase of matter characterized by the complete absence of viscosity, that is, a "fluid" can flow endlessly without friction (resistance).

A closely related phenomenon is superconductivity. In a superconductor, paired electrons (electrons are fermions) can flow with zero resistance. There is intense commercial interest in superconductors because they could be used to produce cheaper, cleaner electricity and to build high-tech marvels like levitating trains and ultrafast computers. Unfortunately, superconductors are difficult to handle and study. Fermionic condensates might help.

That was 2004.

And so we learned that the quark cannon, called the large hadron collider, a supercold, superconductive machine could do them.

So what did CERN do? Deny it. As today, it still says that what it *will do is a quark-gluon plasma, very hot, that will explode without danger, not a vortex of superfluid quarks that will become an attractive fermion condensate.* That is the drill: CERN has prestige, power, money. It knows it can lie, get experts to sign the quality of its statements, as the bankers did to sell their toxic money called then AAA-rate derivatives of maximal quality by AIG, the most respected company of insurance in the world—then. Look at them now. Problem is that when the lies of CERN are uncovered, we shall all be transformed from the absolute lightness of being human into the absolute density of being a quark condensate.

Those vortices of ultradense quark condensates are the only substance known that can cause the most explosive big bang–like reactions of the cosmos: novas and supernovas that first collapse the electrons that cover our weak, light matter, finally absorbing the dense, dark quarks inside the nuclei of the atom, forming quark stars, black holes, strangelets, and other possible forms of quark matter. Those reactions that were considered unstable a decade ago seem now common in nature, which is far more active and creative than we humans first supposed, because quarks form three-dimensional color-locked vortices. And so each three quarks become first a color-locked quark and then they pile up into a "tight column" of informative height called a boson, a highly ordered state that warps and folds all those fractal vortices into a supervortex, which emerges in three dimensions as a quark star or black hole.

Today we know that to create a stable quark condensate that will trigger a nova reaction, we just need a critical mass of quarks. This is expressed with a number A, which tells us the quantity of quarks needed to create a focus of strong forces able to overcome the electroweak barriers that prevent in normal conditions those quarks to free other atomic quarks and form a superfluid, strong, attractive vortex whose exponential growth causes a nova explosion. Our standard knowledge today precludes that somewhere around the 10,000 barrier, those quark condensates will become stable, indestructible, and stronger than the strongest atom of our Universe—iron (see Peng, graph on paragraph 13). Since CERN will bundle hundreds of thousands, it is compulsory that they will start accreting the Earth, absorbing our quarks and exploding our electrons. Each of the strong quarks described here of increasing mass can potentially form a different condensate, but all of them are able to absorb our quarks.

The easiest of all those reactions would convert our up and down quarks into the third denser quark of the light decuplet, the strange-let quark, which turns around, attracting mass at the speed of light, ten times faster than our electrons (Vet=c/10).

6. The Experimental Proofs. The Ice-9 Case

What proofs do we have that CERN will do condensates? Two proofs have come since CERN designed its machine. An old accelerator on the verge of the 1 teravolt barrier of death of our matter called the RHIC created a drop of strange liquid, which was labeled by the scientific press "a perfect surprise." And in December, the first experiments CERN made just above the barrier of 1 TeV made an enormous number of strange quarks, much more than predicted. Again, nature is hatching its dinosaur eggs, blooming, showing much more dynamism and perfection than human abstract equations can ever imagine. Let us then consider those two experiments.

Mr. Wilczek, a Nobel Prize, called the strange liquid a more ominous name, ice-9, and said it could blow up the Earth. The chief engineer at CERN said Mr. Wilczek was much more intelligent than him,[14] but he was naïve talking about it—meaning he had alerted the world of the dangers of nuclear physics and that is taboo in a profession accustomed to kill at distance and blame the military for throwing the bombs they make. And so instead of opening an investigation on the matter, CERN asked all employees NOT to talk about ice-9 and state there was no risk. Next they hired Mr. Wilczek to represent CERN in all the suits against the company and deny his discoveries.

Yet the name was proper and correct. Ice-9 is the liquid of a novel by Kurt Vonnegut, *Cat's Cradle*, in which a mad scientist gives to mankind a final present, a liquid that will condensate the Earth's water into ice, freezing us, who are made of 90 percent of water. It has, like the strangelet particles that kill matter, the property of transforming upon contact all other forms of living, dynamic water into itself, into inert frozen liquid, ice. This wonderful new substance is protected inside a crystal tube out of any risk, according to his inventor, till an accident drops the ice-9 into the ocean, which is part of an interconnected, organic planet. So the ice-9 soon freezes the rivers, sewers, springs, and finally the water inside the cells of mankind, a life species far more fragile than CERNerds, arrogant people, want to acknowledge.

This ice-9 is the fermion condensate. A liquid made of dense quarks, strangelet liquid. We, the electronic world, will collapse in the surface of those quarks. Point. This is what the totalitarian principle says. So far, so good. Technology has proved that it can extinguish mankind and it will extinguish mankind, since scientists will continue doing it till they *extinguish mankind while denying they are doing it.*

Do we have a proof of this? Indeed, we do. RHIC's perfect liquid was a proto-quark gluon fluid, not a gas; but *CERN denies its existence, saying it will produce quark gas to avoid the obvious danger associated to those strange fluids.*

We can observe a pattern of defense of this machine, which Disraeli resumed when talking about expert bankers: "There are damned lies and statistics." This indeed is the case of most CERN statements, damned lies and obscene, minimal probabilities to deny those risks. Indeed, the extinction of mankind will be attempted by a group of people who deny what they will do by inventing statistics, created ad hoc out of their hats, just to dilute the real dangers. Indeed, the most quoted "statistics" about this danger comes from an Oxford professor associated to CERN, who said in BBC that the chances of global genocide is like "people winning three times in a row the lottery." How this "expert" calculated the probability of creating top quark stars and strangelets to be exactly like three lotteries in a row? Of course, he never did. He just said the first catchy sentence he thought of, without the slightest idea of what he talked about; but the press repeated it ad nauseam, this basic tool of scientific fascism: to treat humans as numbers, collateral damages, abstract points of an statistical graph. The tradition of treating humanity as collateral damages that can be erased with weapons is very common in the nuclear industry, which already played with MAD strategies of Mutual Assured Destruction during the cold war. Indeed, other quoted statistic is "one in 4 millions," which happens to be the probability that physicists figured out when they threw the first A-bomb. They thought there were 1 in 4 million chances to ignite the Earth. Yet now the cosmic bomb is millions of times more powerful. It is to the atomic bomb what the atomic bomb is to Saint Nobel's dynamite sticks. So of course, chances are far higher. As we shall see later in a proper calculus, God is playing with charged dices in this game of death. Since there will be two events, the black hole event with protons and the strangelet event with strange quarks, and a conservative estimate gives at least one-third chances of making black holes and two-third chances of making strangelets, which add up to 78 percent of chances, CERN chances of blowing up the planet between now and Christmas 2013 is almost certain. *And we will return to that.*

How sure are we that CERN will produce strangelets? Because we have the three proofs of the scientific method: Mathematical, logic, and experimental consistence. So if we give one-third of chances to each of the three legs of the scientific method, the probability is between two-third and three-thirds: if the theory is right, logically and mathematically consistent and we have already experimental evidence that CERN has been producing strange quarks, the probability is absolute under the totalitarian principle for the same reason every time you heat, water boils and every time you open the light, switch electrons flow—the Universe is deterministic or else it wouldn't work.

Let us explore those three proofs in detail, considering first the experimental creation of strange quarks.

7. First Warning: The Beginnings of Strange Science

The first machine that could make strange matter was RHIC, the hadron ion collider at Brookhaven, Long Island. The year was 1999, and Walter Wagner, one of the members of Einstein et al., that warned strange liquid could be produced. His suit was dismissed since RHIC's Risk Assessment Report said the chances to create a strangelet composed of several strange quarks was 1 in 6 million.

It said RHIC would produce only gas. Nonsense. It was obviously a false assessment, as all *those* CERN does, because the number of strange quarks RHIC were was enormous. And RHIC produced a liquid, not a gas. Thus *the thesis that strange quarks would expand as a gas and never have a chance to come together, considered dogma in that risk sssessment, was false: the quarks stayed together long enough to form complex particles. They decayed finally as antiparticles, which as we shall learn later, are the "death" arrow of a particle that doesn't reach stability at low-energy regimes, in biological terms, an abortion.* So we were very lucky despite obtaining results totally different from the experts' opinion—a liquid, not a gas.

Indeed, Sheldon Glashow, a Nobel Prize, put those experts' opinions in his own words:

> If strangelets exist (which is conceivable) and if they form reasonably stable lumps which is unlikely) and if they are negatively charged (through the theory strongly favors positive charges), and if tiny strangelets can be created at the RHIC (which is exceedingly unlikely) then there might just be a problem. A new-born strangelet could engulf atomic nuclei, growing relentlessly and ultimately consuming the entire Earth.

The word *unlikely* however many times it is repeated just isn't enough to assuage our fears of this total disaster. Especially because the final "exceedingly unlikely" become *a fact. Drops of unstable strangelet liquid were created at RHIC.*

What happened next with Glashow's warning is also a routine in the extremely cynical and corrupted world of nuclear physicists. Glashow was given a job by CERN and paid to represent CERN in the suits against it. The same happened with Wilczek, who discovered the ice-9 reaction.

The reader can follow his affidavits in the appendix of this book, in the American suits against the company. Wilczek and Glashow, the two corrupted physicists that warned mankind, now will affirm under oath that there is no risk, without further ado, asking the judge to "believe" in their authority, working under confidentiality statements of zero risk that the company obliges all their workers to sign.

Let us consider those direct reports of the RHIC experiments where strange matter showed up in quantities never seen before.

Brookhaven Physicists Produce "Doubly Strange Nuclei"
by Karen McNulty Walsh

Physicists from the U.S. Department of Energy at Brookhaven National Laboratory report the first large-scale production of nuclei containing two strange quarks. Strange science has taken a great leap forward at the U.S. Department of Energy's Brookhaven National Laboratory. There, physicists have produced a significant number of "doubly strange nuclei," or nuclei containing two strange quarks. Studies of these nuclei will help scientists explore the forces between nuclear particles, particularly within so-called strange matter, and may contribute to a better understanding of neutron stars, the super dense remains of burnt-out stars, which are thought to contain large quantities of strange quarks. 50 physicists collaborating on the experiment, who represent 15 institutions in six countries, describe their findings in an upcoming issue of *Physical Review Letters:*

"This is the 1st experiment to produce large numbers of these doubly strange nuclei," said Adam Rusek, a co-spokesperson for the Brookhaven collaboration. 4 previous experiments conducted over the past 40 years in the U.S., Europe, and Japan have produced one such nucleus each, with varying degrees of certainty. In the current publication, which is based on data taken in 1998, the Brookhaven collaboration describes 30 to 40 events out of several hundred produced. "That's enough events to begin a study using statistical techniques," Rusek said.

To create the nuclei, the scientists aim the world's most intense proton beam—produced at one of Brookhaven's particle accelerators, the Alternating Gradient Synchrotron—at a tungsten target. From the particles produced in those collisions, the scientists separate out an extremely intense beam of *negatively charged* kaons, which are each composed of one "strange" quark and one "up" antiquark. When these *negative kaons* then strike a beryllium target and interact with its protons, some of the energy is converted into new *strange quarks and strange antiquarks*. These quarks then regroup to form a variety of particles, some of which continue to interact. Occasionally, a structure containing a proton, a neutron, and two lambda particles (each composed of one up, one down, and one strange quark) is formed. This *double-lambda structure, with its two strange quarks*, is the observed doubly strange nucleus. Detecting the formation of this strange species is no easy task. It's more like finding a subatomic needle in a particle-soup haystack. For one thing, many other species are produced in the collisions. Plus, the scientists can't "see" the double lambda structure directly. Instead, they look for pions, a subatomic product the lambdas emit as they decay in less than one billionth of a second. Furthermore, in order to infer that the pions came from a nucleus containing two lambdas, there must be two pion decay signals at very specific energies. Sophisticated computers and careful analyses helped narrow the search from 100 million potentially interesting events, to 100.000 where two strange quarks were produced, to the *30 to 40 where those two strange quarks existed for a fleeting instant inside the same nucleus*. "The most important part is eliminating all the other possible explanations for these events," said Sidney Kahana, a theoretical physicist at Brookhaven. "Were left with this double lambda species as the only explanation," he said. *Now that they believe they have a reliable method for producing the double lambda species, the scientists would like to produce more* so they can get better measurements of the binding energy or force of interaction between the two lambda particles. "We can use this nucleus as a laboratory in which *the two lambdas can be held together long enough to study*," Kahana said. Based on the current data, the interaction between lambdas appears to be rather weak—possibly too weak for the two particles to merge to produce a postulated, six-quark structure called an H particle. But further experiments are necessary, the scientists say. *The interaction between lambdas may also offer insight into the properties of neutron stars, which are thought to contain vast numbers of strange particles, including lambdas. Neutron stars are the only place in the universe scientists believe such strange matter exists in a stable form.* With the ability to produce appreciable numbers of doubly strange nuclei, *"Brookhaven is now the best place in the world to study strange matter,"* said Morgan May, who leads the strangeness nuclear physics program at Brookhaven.

The last sentence is chilling, showing the kind of society we live in, based on public relations and automaton statements. So the assessment at RHIC prior to the experiment talked of a theoretical risk of extinction if strange liquid was created, with sentences like those of Glashow. Yet when strange quarks appeared, they became the very important *discipline of strange science. The SS experts were no longer people that could extinguish us but people who will give us knowledge.* Now the question is: why do we have to learn strange science? What are the benefits for mankind obtained from the study of particles that costs a fortune to make, can destroy the planet, mean nothing for the understanding of the Universe, and have no relationship whatsoever with the humankind?

To put what RHIC was doing in perspective is like selling to humanity the enormous interest of learning it all about the Ebola virus and the bird flu by spreading the planet with those two contagious diseases, a mild equivalent to spreading strange matter and black holes all over the Earth. But at least those viruses are made with DNA and so even if they kill us, there would be survivors or evolution might start again. But strange matter is so strange to this planet that the only benefit we can obtain is to wipe out from existence all forms of life.

There is something very strange, I would rather say sinister and scary, in this fascination with death studies. Physicists involved in those projects who talk about sacred science make the sort of overstatements handled by the military to the soldiers when they send them to a war=death field, paradoxically without mentioning death, only talking about the importance of honor. RHIC physicists make extreme statements about the importance of the "first microseconds." It is all rubbish. There is no honor in dying for the industrial-military system, and RHIC *is not studying the first microseconds of the Universe, which is a fractal of infinite scales but the big bang of the Earth.* We survived RHIC, but obviously nothing of what RHIC discovered has the slightest importance for our lives. We just *survived them.*

Now, the strange particle able to feed on Earth is called the dibaryon; it is a usd-usd double hyperon. The hyperon has an up and down quark, which are the quarks that make every atom of our body, and a third heavier strange quark with negative charge. This was very unlikely to happen according to

Sheldon. Chances are 1 in 6 million, according to the RHIC assessment. Sure, that is dogmatic theory at best, but the empirical news, taken from the very same RHIC public relationships department, *after the experiment was held,* showed the unlikely and very unlikely became a fact.

Thus, all theoretical predictions were shattered since the collider made two strange quark particles negatively charged. And yet, did physicists think twice about what they had done? Did they recognize their errors? Was Mr. Walter's insight recognized? Not at all. Mr. Walter, then a safety officer of accelerators, who had acted on bioethical grounds, denouncing the dangers of the experiment, was fired. He lost his job for doing his job. He was attacked ad hominem and had to start a new career as a biologist. And since "physicists are often wrong but never in doubt," the experiments continued.

The American government, since the collider spent more energy than the nearby city of Watling, reduced the number of days the RHIC will be working each year to one-third of the original ones, but a nutty billionaire from New York gave them a few million dollars to keep doing their thing, so the experiments were soon again on track. And so RHIC geared its machine to maximal power and made much more strangelet liquids.

8. Experimental Proof: The First Microseconds

The creation of that quark-gluon liquid was reported in *Scientific American*, again as a perfect surprise in April 23, 2006:

The First Few Microseconds

In recent experiments, physicists have replicated conditions of the infant universe—with startling results

By Michael Riordan and William A. Zajc

For the past five years, hundreds of scientists have been using a powerful new atom smasher at Brookhaven National Laboratory on Long Island to mimic conditions that existed at the birth of the universe. Called the Relativistic Heavy Ion Collider (RHIC, pronounced "rick"), it clashes two opposing beams of gold nuclei traveling at nearly the speed of light. The resulting collisions between pairs of these atomic nuclei generate exceedingly hot, dense bursts of matter and energy to simulate what happened during the first few microseconds of the big bang. These brief "mini bangs" give physicists a ringside seat on some of the earliest moments of creation.

During those early moments, matter was an ultrahot, superdense brew of particles called quarks and gluons rushing hither and thither and crashing willy-nilly into one another. A sprinkling of electrons, photons and other light elementary particles seasoned the soup. This mixture had a temperature in the trillions of degrees, more than 100,000 times hotter than the sun's core.

But the temperature plummeted as the cosmos expanded, just like an ordinary gas cools today when it expands rapidly. The quarks and gluons slowed down so much that some of them could begin sticking together briefly. After nearly 10 microseconds had elapsed, the quarks and gluons became shackled together by strong forces between them, locked up permanently within protons, neutrons and other strongly interacting particles that physicists collectively call "hadrons." Such an abrupt change in the properties of a material is called a phase transition (like liquid water freezing into ice). The cosmic phase transition from the original mix of quarks and gluons into mundane protons and neutrons is of intense interest to scientists, both those who seek clues about how the universe evolved toward its current highly structured state and those who wish to understand better the fundamental forces involved.

The protons and neutrons that form the nuclei of every atom today are relic droplets of that primordial sea, tiny subatomic prison cells in which quarks thrash back and forth, chained forever. Even in violent collisions, when the quarks seem on the verge of breaking out, new "walls" form to keep them confined. Although many physicists have tried, no one has ever witnessed a solitary quark drifting all alone through a particle detector.

RHIC offers researchers a golden opportunity to observe quarks and gluons unchained from protons and neutrons in a collective, quasi-free state reminiscent of these earliest microseconds of existence. Theorists originally dubbed this concoction the quark-gluon plasma, because they expected it to act like an ultrahot gas of charged particles (a plasma) similar to the innards of a lightning bolt. By smashing heavy nuclei together in mini bangs that briefly liberate quarks and gluons, RHIC serves as a kind of time telescope providing glimpses of the early universe, when the ultrahot, superdense quark-gluon plasma reigned supreme. And the greatest surprise at RHIC so far is that this exotic substance seems to be acting much more like a liquid—albeit one with very special properties—than a gas.

Free the Quarks

In 1977, when theorist Steven Weinberg published his classic book *The First Three Minutes* about the physics of the early universe, he avoided any definitive conclusions about the first hundredth of a second. "We simply do not yet know enough about the physics of elementary particles to be able to calculate the properties of such a mélange with any confidence," he lamented. "Thus our ignorance of microscopic physics stands as a veil, obscuring our view of the very beginning."

But theoretical and experimental breakthroughs of that decade soon began to lift the veil. Not only were protons, neutrons and all other hadrons found to contain quarks; in addition, a theory of the strong force between quarks—known as quantum chromodynamics, or QCD—emerged in the mid-1970s. This theory postulated that a shadowy cabal of eight neutral particles called gluons flits among the quarks, carrying the unrelenting force that confines them within hadrons.

What is especially intriguing about QCD is that—contrary to what happens with such familiar forces as gravity and electromagnetism—the coupling strength grows *weaker* as quarks approach one another. Physicists have called this curious counterintuitive behavior asymptotic freedom. It means that when two quarks are substantially closer than a proton diameter (about 10^{-13} centimeter), they feel a reduced force, which physicists can calculate with great precision by means of standard techniques. Only when a quark begins to stray from its partner does the force become truly strong, yanking the particle back like a dog on a leash.

In quantum physics, short distances between particles are associated with high-energy collisions. Thus, asymptotic freedom becomes important at high temperatures when particles are closely packed and constantly undergo high-energy collisions with one another. More than any other single factor, the asymptotic freedom of QCD is what allows physicists to lift Weinberg's veil and evaluate what happened during those first few microseconds. As long as the temperature exceeded about 10 trillion degrees Celsius, the quarks and gluons acted essentially independently. Even at lower temperatures, down to two trillion degrees, the quarks would have roamed individually—although by then they would have begun to feel the confining QCD force tugging at their heels.

To simulate such extreme conditions here on earth, physicists must re-create the enormous temperatures, pressures and densities of those first few microseconds. Temperature is essentially the average kinetic energy of a particle in a swarm of similar particles, whereas pressure increases with the swarm's energy density. Hence, by squeezing the highest possible energies into the smallest possible volume we have the best chance of simulating conditions that occurred in the big bang.

Fortunately, nature provides ready-made, extremely dense nuggets of matter in the form of atomic nuclei. If you could somehow gather together a thimbleful of this nuclear matter, it would weigh 300 million tons. Three decades of experience colliding heavy nuclei such as lead and gold at high energies have shown that the densities occurring during these collisions far surpass that of normal nuclear matter. And the temperatures produced may have exceeded five trillion degrees.

Colliding heavy nuclei that each contain a total of about 200 protons and neutrons produces a much larger inferno than occurs in collisions of individual protons (as commonly used in other high-energy physics experiments). Instead of a tiny explosion with dozens of particles flying out, such heavy-ion collisions create a seething fireball consisting of thousands of particles. Enough particles are involved for the collective properties of the fireball—its temperature, density, pressure and viscosity (its thickness or resistance to flowing)—to become useful, significant parameters. The distinction is important—like the difference between the behavior of a few isolated water molecules and that of an entire droplet.

The RHIC Experiments

Funded by the U.S. Department of Energy and operated by Brookhaven, RHIC is the latest facility for generating and studying heavy-ion collisions. Earlier nuclear accelerators fired beams of heavy nuclei at stationary metal targets. RHIC, in contrast, is a particle collider that crashes together two beams of heavy nuclei. The resulting head-on collisions generate far greater energies for the same velocity of particle because all the available energy goes into creating mayhem. This is much like what happens when two speeding cars smash head-on. Their energy of motion is converted into the random, thermal energy of parts and debris flying in almost every direction.

At the highly relativistic energies generated at RHIC, nuclei travel at more than 99.9 percent of the speed of light, reaching energies as high as 100 giga-electron volts (GeV) for every proton or neutron inside. (One GeV is about equivalent to the mass of a stationary proton.) Two strings of 870 superconducting magnets cooled by tons of liquid helium steer the beams around two interlaced 3.8-kilometer rings. The beams clash at four points where these rings cross. Four sophisticated particle detectors known as BRAHMS, PHENIX, PHOBOS and STAR record the subatomic debris spewing out from the violent smashups at these collision points.

When two gold nuclei collide head-on at RHIC's highest attainable energy, they dump a total of more than 20,000 GeV into a microscopic fireball just a trillionth of a centimeter across. The nuclei and their constituent protons and neutrons literally melt, and many more quarks, antiquarks (antimatter opposites of the quarks) and gluons are created from all the energy available. More than 5,000 elementary particles are briefly liberated in typical encounters. The pressure generated at the moment of collision is truly immense, a whopping 10^{30} times atmospheric pressure, and the temperature inside the fireball soars into the trillions of degrees.

But about 50 trillionths of a trillionth ($5 \cdot 10^{-23}$) of a second later, all the quarks, antiquarks and gluons recombine into hadrons that explode outward into the surrounding detectors. Aided by powerful computers, these experiments attempt to record as much information as possible about the thousands of particles reaching them. Two of these experiments, BRAHMS and PHOBOS, are relatively small and concentrate on observing specific characteristics of the debris. The other two, PHENIX and STAR, are built around huge, general-purpose devices that fill their three-story experimental halls with thousands of tons of magnets, detectors, absorbers and shielding.

The four RHIC experiments have been designed, constructed and operated by separate international teams ranging from 60 to more than 500 scientists. Each group has employed a different strategy to address the daunting challenge presented by the enormous complexity of RHIC events. The BRAHMS collaboration elected to focus on remnants of the original protons and neutrons that speed along close to the direction of the colliding gold nuclei. In contrast, PHOBOS observes particles over the widest possible angular range and studies correlations among them. STAR was built around the world's largest "digital camera," a huge cylinder of gas that provides three-dimensional pictures of all the charged particles emitted in a large aperture surrounding the beam axis. And PHENIX searches for specific particles produced very early in the collisions that can emerge unscathed from the boiling cauldron of quarks and gluons. It thus provides a kind of x-ray portrait of the inner depths of the fireball.

A Perfect Surprise

The physical picture emerging from the four experiments is consistent and surprising. The quarks and gluons indeed break out of confinement and behave collectively, if only fleetingly. But this hot mélange acts like a liquid, not the ideal gas theorists had anticipated.

The energy densities achieved in head-on collisions between two gold nuclei are stupendous, about 100 times those of the nuclei themselves—largely because of relativity. As viewed from the laboratory, both nuclei are relativistically flattened into ultra thin disks of protons and neutrons just before they meet. So all their energy is crammed into a very tiny volume at the moment of impact. Physicists estimate that the resulting energy density is at least 15 times what is needed to set the quarks and gluons free. These particles immediately begin darting in every direction, bashing into one another repeatedly and thereby reshuffling their energies into a more thermal distribution.

Evidence for the rapid formation of such a hot, dense medium comes from a phenomenon called jet quenching. When two protons collide at high energy, some of their quarks and gluons can meet nearly head-on and rebound, resulting in narrow, back-to-back sprays of hadrons (called jets) blasting out in opposite directions. But the PHENIX and STAR detectors witness only one half of such a pair in collisions between gold nuclei. The lone jets indicate that individual quarks and gluons are indeed colliding at high energy. But where is the other jet? The rebounding quark or gluon must have plowed into the hot, dense medium just formed; its high energy would then have been dissipated by many close encounters with low-energy quarks and gluons. It is like firing a bullet into a body of water; almost all the bullet's energy is absorbed by slow-moving water molecules, and it cannot punch through to the other side.

Indications of liquid like behavior of the quark-gluon medium came early in the RHIC experiments, in the form of a phenomenon called elliptic flow. In collisions that occur slightly off-center—which is often the case—the hadrons that emerge reach the detector in an elliptical distribution. More energetic hadrons squirt out within the plane of the interaction than at right angles to it. The elliptical pattern indicates that substantial pressure gradients must be at work in the quark-gluon medium and that the quarks and gluons from which these hadrons formed were behaving collectively, before reverting back into hadrons. They were acting like a liquid—that is, not a gas. From a gas, the hadrons would emerge uniformly in all directions.

This liquid behavior of the quark-gluon medium must mean that these particles interact with one another rather strongly during their heady moments of liberation right after formation. The decrease in the strength of their interactions (caused by the asymptotic freedom of QCD) is apparently overwhelmed by a dramatic increase in the *number* of newly liberated particles. It is as though our poor prisoners have broken out of their cells, only to find themselves haplessly caught up in a jail-yard crush, jostling with all the other escapees. The resulting tightly coupled dance is exactly what happens in a liquid. This situation conflicts with the naive theoretical picture originally painted of this medium as an almost ideal, weakly interacting gas. And the detailed features of the elliptical asymmetry suggest that this surprising liquid flows with almost no viscosity. It is probably the most perfect liquid ever observed.

The Emerging Theoretical Picture

Calculating the strong interactions occurring in a liquid of quarks and gluons that are squeezed to almost unimaginable densities and exploding outward at nearly the speed of light is an immense challenge. One approach is to perform brute-force solutions of QCD using huge arrays of microprocessors specially designed for this problem. In this so-called lattice-QCD approach, space is approximated by a discrete lattice of points (imagine a Tinkertoy structure). The QCD equations are solved by successive approximations on the lattice.

Using this technique, theorists have calculated such properties as pressure and energy density as a function of temperature; each of these dramatically increases when hadrons are transformed into a quark-gluon medium. But this method is best suited for static problems in which the medium is in thermodynamic equilibrium, unlike the rapidly changing conditions in RHIC's mini bangs. Even the most sophisticated lattice-QCD calculations have been unable to determine such dynamic features as jet quenching and viscosity. Although the viscosity of a system of strongly interacting particles is expected to be small, it cannot be exactly zero because of quantum mechanics. But answering the question "How low can it go?" has proved notoriously difficult.

Remarkably, help has arrived from an unexpected quarter: string theories of quantum gravity. An extraordinary conjecture by theorist Juan Maldacena of the Institute for Advanced Study in Princeton, N.J., has forged a surprising connection between a theory of strings in a warped five-dimensional space and a QCD-like theory of particles that exist on the four-dimensional boundary of that space[15]. The two theories are mathematically equivalent even though they appear to describe radically different realms of physics. When the QCD-like forces get strong, the corresponding string theory becomes weak and hence easier to evaluate. Quantities such as viscosity that are hard to calculate in QCD have counterparts in string theory (in this case, the absorption of gravity waves by a black hole) that are much more tractable. A very small but nonzero lower limit on what is called the specific viscosity emerges from this approach—only about a tenth of that of superfluid helium. Quite possibly, string theory may help us understand how quarks and gluons behaved during the earliest microseconds of the big bang.

Future Challenges

Astonishingly, the hottest, densest matter ever encountered far exceeds all other known fluids in its approach to perfection. How and why this happens is the great experimental challenge now facing physicists at RHIC. The wealth of data from these experiments is already forcing theorists to reconsider some cherished ideas about matter in the early universe. In the past, most calculations treated the freed quarks and gluons as an ideal gas instead of a liquid. The theory of QCD and asymptotic freedom are not in any danger—no evidence exists to dispute the fundamental equations. What is up for debate are the techniques and simplifying assumptions used by theorists to draw conclusions from the equations.

To address these questions, experimenters are studying the different kinds of quarks emerging from the mini bangs, especially the heavier varieties. When quarks were originally predicted in 1964, they were thought to occur in three versions: up, down and strange. With masses below 0.15 GeV, these three species of quarks and their antiquarks are created copiously and in roughly equal numbers in RHIC collisions. Two additional quarks, dubbed charm and bottom, turned up in the 1970s, sporting much greater masses of about 1.6 and 5 GeV, respectively. Because much more energy is required to create these heavy quarks (according to $E = mc2$), they appear earlier in the mini bangs (when energy densities are higher) and much less often. This rarity makes them valuable tracers of the flow patterns and other properties that develop early in the evolution of a mini bang.

The PHENIX and STAR experiments are well suited for such detailed studies because they can detect high-energy electrons and other particles called muons that often emerge from decays of these heavy quarks. Physicists then trace these and other decay particles back to their points of origin, providing crucial information about the heavy quarks that spawned them. With their greater masses, heavy quarks can have different flow patterns and behavior than their far more abundant cousins. Measuring these differences should help tease out precise values for the tiny residual viscosity anticipated.

Charm quarks have another characteristic useful for probing the quark-gluon medium. Usually about 1 percent of them are produced in a tight embrace with a charm antiquark, forming a neutral particle called the J/psi. The separation between the two partners is only about a third the radius of a proton, so the rate of J/psi production should be sensitive to the force between quarks at short distances. Theorists expect this force to fall off because the surrounding swarm of light quarks and gluons will tend to screen the charm quark and antiquark from each other, leading to less J/psi production. Recent PHENIX results indicate that J/psi particles do indeed dissolve in the fluid, similar to what was observed earlier at CERN, the European laboratory for particle physics near Geneva[16] Even greater J/psi suppression was expected to occur at RHIC because of the higher densities involved, but early results suggest some competing mechanism, such as reformation of J/psi particles, may occur at these densities. Further measurements will focus on this mystery by searching for other pairs of heavy quarks and observing whether and how their production is suppressed.

Another approach being pursued is to try to view the quark-gluon fluid by its own light. A hot broth of these particles should shine briefly, like the flash of a lightning bolt, because it emits high-energy photons that escape the medium unscathed. Just as astronomers measure the temperature of a distant star from its spectrum of light emission, physicists are trying to employ these energetic photons to determine the temperature of the quark-gluon fluid. But measuring this spectrum has thus far proved enormously challenging because many other photons are generated by the decay of hadrons called neutral pions. Although those photons are produced long after the quark-gluon fluid has reverted to hadrons, they all look the same when they arrive at the detectors.

Many physicists are now preparing for the next energy frontier at the Large Hadron Collider (LHC) at CERN. Starting in 2008, experiments there will observe collisions of lead nuclei at combined energies exceeding one million GeV. An international team of more than 1,000 physicists is building the mammoth ALICE detector, which will combine the capabilities of the PHENIX and STAR detectors in a single experiment. The mini bangs produced by the LHC will briefly reach several times the energy density that occurs in RHIC collisions, and the temperatures reached therein should easily surpass 10 trillion degrees. Physicists will then be able to simulate and study conditions that occurred during the very first microsecond of the big bang.

The overriding question is whether the liquid like behavior witnessed at RHIC will persist at the higher temperatures and densities encountered at the LHC. Some theorists project that the force between quarks will become weak once their average energy exceeds 1 GeV, which will occur at the LHC, and that the quark-gluon plasma will finally start behaving properly—like a gas, as originally expected. Others are less sanguine. They maintain that the QCD force cannot fall off fast enough at these higher energies, so the quarks and gluons should remain tightly coupled in their liquid embrace. On this issue, we must await the verdict of experiment, which may well bring other surprises.

The next surprise is obvious: *a stable liquid that will swallow the Earth*.

And the chances are very high, on the view of that experiment, where a protostrange quark liquid, NOT a gas, was produced. We have also the experiment done in December at CERN, at the minimal potency of the LHC machine, only 1 TeV=1c speed. At this minimal regime, the quark cannon produced many more kaons, us and ds particles, minimal units of the usd strangelet liquid than predicted. So with more energy, CERN will produce more strange liquid till a ball of strange liquid becomes stable and kills us all. CERN has merely entered the room of the Tyrannosaurus rex, and it is filled with teeth. Kaons die as antiparticles because they are still not stable—they are the pieces of a dibaryon, a usd-usd stable atom of strange liquid. They are the fetus of the monster. The egg is hatching. But with a bit more of weight, the fetus will be strong enough to start an ice-9 reaction.

So the certainty of the creation of strange liquid at CERN is obvious. We are living corpses, with the ticktack of our death set by the hands of Rolf Heuer.

And what shall we learn? Nothing. The only theoretical conundrum of the letter is the self-similarity between a description of a five-dimensional Universe with one-dimensional strings and a four-dimensional description with bidimensional quark vortices. This is the famous Maldacena hypothesis.[15] The writer doesn't know why this magic happens, but we know that if we use fractal dimensions, 1-D strings become 2-D gluons, fractal parts of bidimensional quarks. Thus the five-dimensional world loses a fractal dimension "absorbed" by strings to become bidimensional gluons and quarks of our four-dimensional world. Then bidimensional quarks become locked in triplets to form the three-dimensional space-time of our electronic world. This is the only theoretical question worth to mention on an otherwise clear text written by an SS authority eager to produce this global Zyklon liquid at the quark matter factory—of course, not a mention of the dangers ahead.

So far we have only been able to form very small packages of superfluid quarks, with a few strange units at RHIC, the hadron ion collider set up at Long Island and at CERN because we are working on the verge of 1 TeV, *the barrier of death of our matter. Thus, we have hardly started the experiment, which will ramp up energy to over 1,000 TeV, and yet there are already signs of danger all over the place.*

CERN collided a few protons just above 1 TeV, and we saw a lot of unexpected kaons flying around and decaying as antiparticles. In fact, this "unexpected" vitality of strange matter is a trademark.

Indeed, the kaon was the first strange particle found, and it was called strange because it lived much longer than anyone expected. Strange quarks were called strange because physicists were surprised by their long-lasting life, millions of time longer than they thought. Then they found also that they formed a perfect liquid at RHIC. Now they see they are easily created over 1 TeV. What else do we need to see to stop this experiment?

CERN said it will produce less strangelet atoms than predicted, no more, and it was safe. Then when it found out, it made many more last December; CERN denied this information to mankind for two months. And it never talks of strange liquid, only of kaons. It is a cynical game. They give us a warning: we are making kaons. But they don't say kaons are the atoms of strange liquid. They inform, but not inform. And the journalist merely writes in an exalted tone, as the clueless Web site where I found the information, "new particles, sorry, not black holes." She was laughing at the fact the naysayers of black holes were not yet right. She didn't know she was telling us CERN was doing strange liquid. The rest of the press didn't report. As usual, the letters of Einstein et al. were not published. We are nonpeople. We must die for CERN in silence.

Now as we speak, CERN will ramp up to 3.5 TeV to produce even more quantities of strange quarks in more complex combinations. It says it will not monitor them. The kaons will be the "background" on which it will search for other particles. In other words, CERN will produce strange liquid that will extinguish mankind, but this is of no interest to CERN. At CERN, people are merely interested in proving their theories to get a Nobel Prize. They are little people. They do their work, calculate, gather data. They block their minds to everything else. They joke about the chances of killing the world. I can imagine that some even feel important for so much attention from the press.

The "unexpected" drops of kaon liquid will fall to the center of the Earth, undetected, and when enough of them reach the center, they will form a ball of strange liquid and the Earth will fall onto it. In 2013, CERN will ramp up energy/mass to 1,000 TeV and the main produce of the factory is strange liquid. So at maximal production, it will certainly create strangelets that will kill us.

The warnings to mankind have been many, but there is no bigger deaf than one who does not want to listen.

At RHIC, we observed a very dense superfluid vortex that lasted billions of times more than predicted and attracted thousand of particles unexpectedly, blowing up much later than expected and showing the properties of a proto-black hole.

This proto-quark star was made with less than one-thousandth the mass/energy of the lead collisions that will take place at the LHC, gathering just a few hundred strange quarks that should have not condensate at all. In fact, as the RHIC bulletin recognized,[5] it was a perfect surprise that no scientist had ever imagined. They all expected an unstable gas that should have diluted immediately, blowing up into an explosion of energy, but it was a perfect liquid. And yet CERN lies and *still calls the quark superfluid it will do a quark-gluon plasma, rejecting all the experimental evidence accumulated at RHIC and now at the LHC.*

As in the film *Jurassic Park*, the apparition of a seminal drop of a quark star has merely increased the desire to see more of it among the SS experts, who have dedicated their lives to the study of *strange science*. It is the same infantile, irresponsible, self-interested enthusiasm shown by nuclear physicists like Teller trying to convince Eisenhower that his hydrogen bomb could blow up the entire Soviet Union and *that was good for knowledge*, as he explains in the film *The Atomic Bomb Movie*.

We shall all die so SS experts can publish their papers on strange science.

This irresponsibility can be explained by the fact that mankind has never made responsible nuclear physicists of the many deaths caused by their discoveries. Now, it seems the LHC will get away with a genocide that dwarfs the possible victims of a nuclear war, thanks to its diplomatic immunity, the ignorance of journalists and governments and CERN's double-talk since very few understand its SS papers.

LHC will create quark condensates with hundreds of thousands of deconfined quarks, far more than the few hundred units RHIC managed to put together. *And they will be stable because RHIC created already hyperons, usd-triplets, seventy strange nuclei, as a report released now in March 2010,*[17] *seven years after the collisions, just a week before the LHC will go online, shows. Why was all this information that shows strangelet liquid will be produced constantly at the LHC released weeks before the experiment that might kill us all happens?*

Unfortunately, the wealth of papers in quark stars and quark condensates appeared in the past years, when the machine was already commissioned and could no longer be scrapped without closing down the nuclear company, show that 10,000 quarks are enough critical mass to start an ice-9 nova reaction, forming a strangelet, a very common substance at the core of all the neutron stars of the Universe, *not some exotic almost impossible particle, as CERN's wishful thinking pretends*. Under the totalitarian principle, the compulsory product of this quark factory will be a mass of usd-strangelet liquid, combination of usd hyperons, us, ds kaons, and usd-usd dibaryons, the atoms of strangelets. I shall repeat this once and again, as CERN repeats that its experiments are equal to those of cosmic rays: *We have never seen strangelets in cosmic rays, we have never seen free quarks in cosmic rays. We have never seen strangelet liquid in cosmic rays.*

But CERN will deliver in 2013 not 1 Tera-electron volts but collisions between lead atoms over 1,000 Tera-electron volts, which means 1,000 times more strangelet mass than at RHIC. Thus somewhere between now if you are still reading and 2013, it will cross the threshold of stability of strange matter, starting a runaway reaction that will convert the planet into a quark star.

Since again, *at CERN, the unlikely creation of strange particles has become also a fact, 16 percent more strange quarks were created with a mere 1.1 TeV potency and yet CERN tells us that they will be treated just as a background of his search for the false Higgs.*

Let us then consider the next leg of the scientific method, a logical, theoretical analysis of what a strangelet liquid is, from the perspective of duality—*no more no less than a mini supernova, a big crunch, and simultaneous big bang of all the matter of the Earth.*

9. Second Leg of the Risk: Experimental Evidence. First Warning. Strangelets at RHIC.

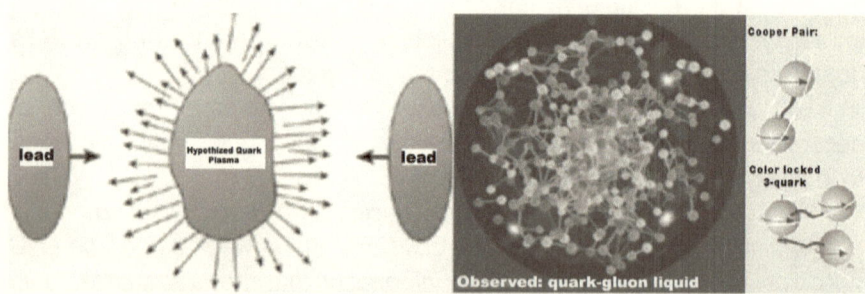

In the graph: fermions (quarks) are antisocial, disordered within the atom. So scientists never thought possible quarks could create a vortex of dense, superfluid quark condensates when they made the LHC collider; but recently we have discovered that quarks can get together first in couples and triplets, which then act as an ordered Einstein-Bose condensate. In this diagram, the spins of paired particles of quarks become aligned. Quarks also form both, locked triplets. Then once they become orderly pairs and triplets, they can form a condensate. This was proved in experiments at RHIC, which was able to form a protostrangelet with one-thousandth the quantity of energy of the LHC, because E=Mc² is the more energy the more mass the strangelet will have till the strength of the vortex is such that it becomes stable and starts to absorb the Earth. All physicists, though, believed quarks would form an explosive gas because lineal gas equations were easier to calculate. Still today CERN denies it will create strange liquid, given the dangers of doing strangelets. So it lies in its statements, saying it will create a gas, despite the fact we have already seen a superfluid, superconductive quark liquid at RHIC.

The liquid quark fluid surprised all the researchers involved who expected an ultrahot unstable gas instead. It also tumbled down all their previous calculations. Not a good sign indeed. The same happened in December at the LHC. Kaons and hyperons appeared all over the place at a mere 1.2 TeV. Scientists at RHIC are still debating why kaons and hyperons are far more stable and live much longer than their abstract equations show, as the scientist in *Jurassic Park* wondered why those Tyrannosaurs could mate and live on herds. *What was holding together those kaons in a state of superconductivity and superfluidity, creating the most perfect liquid ever observed in the Universe?*

I thought on the asp tale of the discoverer of the kaon, the amino acid of the strange world.

Cleopatra is still living for all of us to die at her throne.

As in the *Jurassic Park* experiment, once more nature surprised us, being far more stable and efficient than arrogant scientists had ever believed. And so it is natural to think that if the quark-gluon soup was a stable liquid, not an explosive gas, at RHIC and kaons appear so easily at the LHC and they live longer than Methuselah, the next stage in the fractal ladder of cosmic growth, the strangelet, will be a stable quark-gluon hyperfluid.

Beyond that, if we were to survive the strangelet scenario, there is still waiting for us in 2013 a hyperfluid solid state of quarks with awesome properties, as those shown by a recently discovered similar superfluid solid made of ultracold helium-4—the top black hole.

Another physicist though talked after the RHIC experiments and proved once more that strange quark, frozen stars, and top quark black holes and Bose-Einstein, quark condensates, are all species of the same zoo of dark, quark matter and all of them are equally dangerous. His name was Nastase, and his truths were buried as usual after the initial fear of the press by mighty CERN.

ADDENDA 'D':

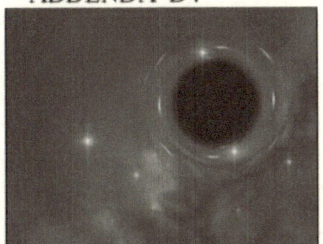

A fireball created in a US particle accelerator has the characteristics of a black hole, a physicist has said. It was generated at the Relativistic Heavy Ion Collider (RHIC) in New York, US, which smashes beams of gold nuclei together at near light speeds. Horatiu Nastase says his calculations show that the core of the fireball has a striking similarity to a black hole. His work has been published on the pre-print website arxiv.org and is reported in New Scientist magazine.

Creating the conditions for the formation of black holes is one of the aims of particle physics

When the gold nuclei smash into each other they are broken down into particles called quarks and gluons. These form a ball of plasma about 300 times hotter than the surface of the Sun. This fireball, which lasts just 10 million, billion, billionths of a second, can be detected because it absorbs jets of particles produced by the beam collisions. But Nastase, of Brown University in Providence, Rhode Island, says there is something unusual about it. Ten times as many jets were being absorbed by the fireball as were predicted by calculations.

In the graph, taken from BBC news, we read that *what physicists made and what physicists said would happen was totally different. In fact, the worst possible scenario appeared: the black hole Rex showed its teeth at RHIC.* Yet *Physical Review* expected something very different:

> *A Quark universe.*
>
> Individual quarks and antiquarks have never been observed in experiments. Quarks and antiquarks are always bound in groups of three in baryons (e.g. protons and neutrons) or anti-baryons (e.g. antiprotons and antineutrons), or in bound quark-antiquark pairs known as mesons. However, "free" quarks must have existed in the extreme conditions of the very early universe. Until about 40 microseconds after the big bang, according to theory, the universe consisted of a very hot gas of free quarks, antiquarks and gluons: this gas is called"quark-gluon plasma".
>
> The temperatures in this plasma exceeded 2.5×10^{12} Kelvin—about 150 000 times hotter than the core of the Sun. We do not know if quarks are truly elementary, but if they are, the quark-gluon plasma in the early universe would have been formed directly in the big bang. As the universe cooled, the quarks and antiquarks "froze" into hadrons. To understand how quarks became confined within protons and neutrons—*which are essential if we hope to understand how the universe we live in came to be the way it is—we need to be able to recreate and study the quark-gluon plasma in the laboratory. We need to create a "micro bang".*

So physicists wanted to create a *micro-bang*; without any safety measure, without any idea of what it would be. So Physicists went on creating a micro-big bang, expecting a gas dissolved in seconds. And this is what really happened, as the news was delivered in the Internet by the BBC:

> *Lab fireball may be black hole: A fireball created in a US particle accelerator has the characteristics of a black hole, physicist said.*
>
> It was generated at the Relativistic Heavy Ion Collider (RHIC) in New York, US, which smashes beams of gold nuclei together at near light speeds. Horatiu Nastase says his calculations show that the core of the fireball has a striking similarity to a black hole. His work has been published on the pre-print website arxiv.org and is reported in *New Scientist* magazine:
>
> When the gold nuclei smash into each other they are broken down into particles called quarks and gluons. These form a ball of plasma about 300 times hotter than the surface of the Sun. This

fireball, which lasts just 10 million, billion, billionths of a second, can be detected because it absorbs jets of particles produced by the beam collisions.

But Nastase, of Brown University in Providence, Rhode Island, says there is something unusual about it. *Ten times as many jets were being absorbed by the fireball as were predicted by calculations.* **Let's look at what happened. In the RHIC** accelerator itself two beams of gold ions, atoms stripped of all their electrons, are clashed at several interaction zones around the ring-shaped facility. Every nucleus is a bundle of 197 protons and neutrons, each of which shoots along with an energy of up to 100 GeV. Therefore, when the two gold projectiles meet in a head-on "central collision" event, the total collision energy is 40 TeV (40 trillion electron volts). Of this, typically 25 TeV serves as a stock of surplus energy—call it a fireball—out of which new particles can be created. Indeed in many gold-gold smashups as many as 10,000 new particles are born of that fireball. Hubble-quality pictures of this blast of particles—**the outward streaming particles—provide all the forensic evidence for determining the properties of the fireball. To** harvest this debris, the RHIC detectors must be agile and very fast. The recreation of the frenzied quark era is ephemeral, lasting only a few times 10^{-24} seconds. The size of the fireball is about 5 femtometers, its density about 100 times that of an ordinary nucleus, and its temperature about 2 trillion degrees Kelvin or (in energy units) 175 MeV. RHIC was built to create that fireball. But was it the much-anticipated quark-gluon plasma? The data unexpectedly showed that the fireball looked nothing like a gas. For one thing, potent jets of mesons and protons expected to be squirting out of the fireball, were being suppressed. **Now, for the first time since starting nuclear collisions at RHIC** in the year 2000 and with plenty of data in hand, all four detector groups operating at the lab have converged on a consensus opinion. They believe that the fireball is a liquid of strongly interacting quarks and gluons rather than a gas of weakly interacting quarks and gluons. The RHIC findings were reported at this weeks April meeting of the American Physical Society (APS) in Tampa, Florida in a talk delivered by Gary Westfall (Michigan State) and at a press conference attended by several RHIC scientists. **Brookhaven physicist Samuel Aronson said that having established the quark-gluon-liquid nature of the pre-protonic universe, RHIC** expected to plumb the liquids properties, such as its heat capacity and its reaction to shock waves. The liquid is dense but seems to flow with very little viscosity. It flows so freely that it approximates an ideal, or perfect, fluid, the kind governed by the standard laws of hydrodynamics. At least in its flow properties the quark liquid is therefore a classical liquid and should not be confused with a super fluid, whose flow properties (including zero viscosity) are dictated by quantum mechanics.

10. SS Experts Ramming Ahead Toward Our Extinction

Indeed, that is the core of the matter: RHIC made strangelet liquid that behaved as an unstable black hole, not a gas, feeding ten times faster than expected. Yet SS experts still *want to study those strange particles regardless of the dangers for mankind.*

CERN, with the arrogance of a bull, which the toreador dominates easily with the nimble cloth to kill at will when he gets bored of his mindless runs, despises those proto black holes. CERN is just interested in its run, in going faster, in touching the SS particles of our death, represented by the cloth of the toreador. But the Universe will not need to bring the *puntilla* of a black hole, the swift final cut of the spinal chord that ends the agony of the bull within seconds. The toreador, Shiva, the laws of the Universe excel in the art of death. The strange quark will do us all. It surprised three times mankind, showing how much more powerful strange matter is than our weak matter. Our death is canonical, in three *tercios*, each *tercio* announced ominously by the trumpets of a Chronicle of Death foretold:

The first trumpet was "Cleopatra," the first kaon who lived billions of times more than expected. The second trumpet was the RHIC experiment, the ball of superfluid liquid that lived billions of times longer. The third trumpet was the experiments in December that CERN did not publish on time to alert the world, with enormous numbers of kaons.

Now the toreador will enter to kill. No more "fiesta" for the laughing clowns of the press, for the CERNerds who in Christmas, knowing already that the third tercio had been announced, had the chutzpah to make a party with the motif of the end of the world. You are now in the year of your likely death.

In a corrida there is always the chance of a miracle if the president saves the life of the bull. But our politicians are silent. The bull seems condemned.

Between April and December, in increasing numbers, stable dibaryons will appear in increasing numbers as usd hyperons, born in the "long run" of the protons and lead atoms collide and mix together.

A dibaryon is a usd-usd double hyperon that will fall to the center of the Earth and grow inside. It is the sword that will enter deep into the heart of the

Earth. The bull will not expect it. Dibaryons are neutral and stable, invisible. The toreador hides the sword behind the cloth. The bull never sees the sword. It only sees the red bloody cloth, obsessed as it is to make an energetic run, to show around a worshipping audience that he is a macho, but the Universe has stronger MACHOs, the name astronomers give to strangelets of dark matter. SS experts in their rat race to publish papers will run toward the sword as lemmings run toward extinction. And dibaryons will sink into the earth.

They will enter deeper into the Earth's heart. First we shall feel the *banderillas* scratching our skin: earthquakes will crack the Earth. And then the bull will stop in its tracks. And Gaia will know Cleopatra has stopped falling. Cleopatra is now in the center and the beating stops, and the blood spurts on the mouth of the bull, who can no longer bluff. CERNerds are preparing the last runs of the quark cannon, the sword buried under the Earth. The eyes of the bull become frozen crystals, the blood condensates. It is always a foretold chronicle. The audience knows, the toreador knows, only the bull ignores his future. At five thirty every Sunday, in every planet of the fractal Universe, where humans obey the Fermi paradox, dibaryons kill the Earth. Why are there no signs of intelligent life? Because human life is NOT intelligent. Mankind is a bull ready for the slaughterhouse. Mankind never learns. It doesn't matter how many tercios you fight the bull, mankind will still run toward the red cloth. The Germanic tribal cultures of weapons of mass destruction never learned. It doesn't matter how many wars, how many times the Kaiser, the Hitler, the Heuer loses his war. He will again run toward death. A pattern arises. Time is cyclical, as we shall see in the next chapter. Scientists only see the simplest numbers. The bull only sees red. Why is CERN still saying it will create a gas?

The behavior of a gas was far easier to calculate than a liquid, which obeys nonlineal equations like those of the black hole, so Brookhaven's physicists, as Hawking did in his theory of black holes, as CERNerds are doing, used mainly lineal equations and got an imaginary gas and an explosion while nature constructs the Universe of masses with cyclical vortices and organic networks and so it made a nonevaporating liquid, which will become stable at CERN.

Indeed, the LHC will have a new surprise. It will create a lump of quarks, around 1,000 times bigger, with enough stability to become the strangelet sword of the Earth. But God has been nice to humans if you want to use the mystique babbling of physicists, since it has warned us on time, showing the teeth of the strangelet sword, every tercio, since Cleopatra stopped dying forty years ago. Now we know, after the experiment at RHIC, that strangelets are not abstract entities but organic balls: a mass of quarks in bosonic state, the most perfect, superconductive state of the Universe. In huge numbers, as the recent experiments in LHC have proved, those quarks packed into a height dimension, as bosons do, proving once more the laws of relativity, will penetrate the Earth like the sword enters through the bones of the bull, cutting rock like butter.

It was Einstein, despised by Heisenberg and the Nazis for his Jewish physics, who described with an Indian physicist, Bose, in the so-called Einstein-Bose equations, the perfectly ordered boson state of matter and the theory behind the formation of frozen stars. Another Indian, Chandra, told physicists that the sun would collapse into one, but nobody believed him because the Hawking of his age, Mr. Eddington, didn't like the idea and nobody dared to defy Eddington. Now we have the Chandra satellite watching the skies and finding thousands of those frozen stars. But Chandra lost most of his career for defying the pope of British physics. All those human stories show how truth always wins over power. Nobody ever proved mathematically wrong Chandra. They just ignored him. And yet at the end Chandra's stars happened. As nobody has proved Einstein et al. wrong about the extinction of the Earth. And yet at the end, our extinction will happen even if nobody defies CERN and Hawking—truth will impose over power, even if truth is the tragic death of mankind.

What we saw at RHIC and Nastase denounced should have opened all the alarm bells of the scientific world because it was indeed not only surprising but very dangerous. The object at RHIC behaved like a proto black hole but it was a proto strangelet. Nastase confused it with a proto black hole because both are self-similar quark stars, from the two types of matter, strange and top=Higgs quarks, which have the property of eating up and transforming our matter into heavier forms of mass. RHIC particles were not stable because the energy to give them birth was below 1 TeV and no theoretical model predicts a stable birth of a black hole or a superfluid strange quark state under that energy.

God is playing with charged dices, but unlike the bull, mankind has always had the door open in the three tercios of this Chronicle of Death foretold to escape the corrida and return to the beautiful paradise of grass and open savannas in which the bulls dwell. God, the laws of the Universe, are generous, but man never had enough with the paradise of life, always wanting to be more than god, to invent the laws of the game, and that is what CERNerds and the press and Hawking have done. The problem is not that they ate the fruits of the tree of science, but that they don't know that if an apple tastes like an apple and looks like an apple and falls like an apple and it is born from an apple tree, it must be an apple. They stubbornly deny the nature of strangelets; they think the apple will not fall but levitate like an angel, that the black hole will not grow but evaporate.

The liquid fractal network of quarks that RHIC created are afar from equilibrium, organic, liquid system. It was not a gas, modeled with lineal abstract equations, but it looked like a fractal, chaotic system. And it remained highly ordered, perhaps controlled by Lorenz and Rössler's attractors—a dense, in-formed bi-perpendicular vortex, which attracts particles toward its center.

The ball of fire, according to Nastase, was the aborted fetus of a quark star. The quark gluon liquid made at CERN will be far more massive, and stable. Let us then consider now the "sword," the final act of this tragedy, what the LHC has started to create as this book goes to print.

11. The Self-similarity of QCD and Fractal Quark Stars

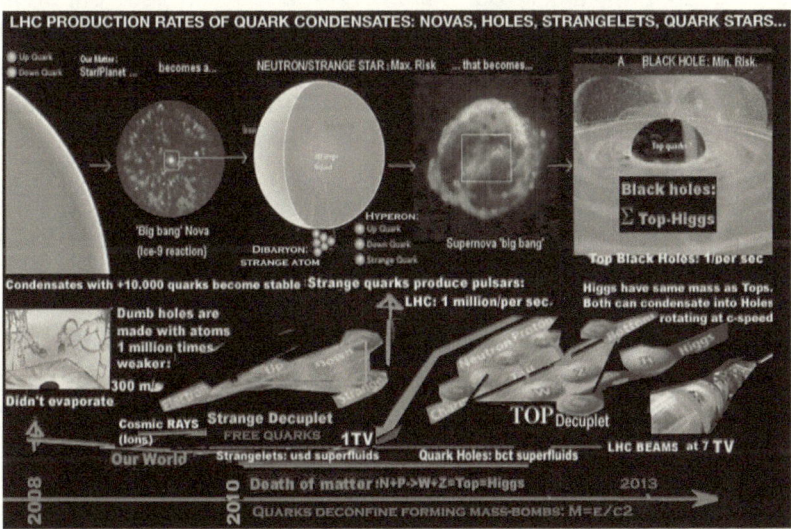

Quark Production at LHC (*SciAm*), related to the
different types of mass bombs they trigger.

In the graph, the standard model of quark, dark matter and its cosmological bodies: frozen stars. According to fractal relativity, the densest objects of the Universe, black holes and pulsars, are quark stars made of Einstein's quark condensates. Because the large hadron collider will mass-produce those condensates, it should convert the Earth into a quark star or black hole. The LHC will cross deeply into the world of dark, quark matter, which breaks the symmetry of our electroweak matter, feeding on it and transforming up and down quarks into heavy strange quarks, the components of pulsars and tops=Higgs particles, likely components of black holes. Thus, according to Einstein's equivalence principle, an accelerated vortex of mass might be formed at CERN, absorbing the Earth at light speed, M=e/c2, creating an ultradense pulsar or hole.

In the graph, we observe many of the key elements of the quark world in its correspondence between the microcosms and the microcosms. At the base of the graph, we observe CERN's production of quarks during the experiments. Those experiments start at 1 TeV when our matter dies. Our world of light (yellow line) becomes first broken in its symmetry by heavy strange quarks. Our quarks are the up and down quark of the blue triangle in the base of energies produced by the LHC. The more we move toward the right, the more energy=mass is needed to create heavier quarks. We belong to the lighter quarks of the Universe, up and down. The strange quark is the next heavy quark and is the top predator of the blue triangle of quarks.

The first thing physicists at CERN are going to study is how the strange quark breaks, kills our matter. Yet if a reaction goes out of hand and the process becomes self-sustained, which is theoretically possible when kaons (up-strange-down combinations in couples) or hyperons and dibaryons (usd triplets and double triplets) form, the end result could be the creation with all the quarks of the Earth of a strange star.

Those events will happen between March 2010 and 2012, as the LHC uses potencies between 1 and 3.5 TeV in its collisions.

Then in 2013, the LHC will start production of all the dark quarks, possible substance of black holes.

And again, we see at the base of the triangle two massive quarks, the bc quark, which should therefore form a stable atom, the dark atom, associated to a superheavy electron, the tau electron, and on top the densest, heavier top=Higgs quark. This top=Higgs quark is the quark that breaks the symmetry=kills all the quarks of the Universe. It is, in organic logical terms, the top predator. And so all those superheavy quarks form the top quark decuplet of stable three-top quarks' combinations (three quarks in combinations of three can create ten possible different particles).

On top, we can observe the cosmological body, probably created with those dark quarks: black holes, which probably are frozen stars made of those densest quarks (top quarks, self-similar to the Higgs). Yet despite all the hype about black holes, it is obvious that strangelets, the substance of pulsars, will be easier to do with stable dibaryons, usd-usd quark systems. Those dibaryons have enough "strong force bonding" to be stable even at the pressures of our planet. So they can form very easily an Einstein condensate. The easiest of those Bose condensates able to form a strange star will require only between 10 and 10,000 dibaryons to become stable, according to the Chinese "Institute of High Energy Physics". CERN will deconfine according to Josh Engelen, its engineer chief, up to 1 million of them per second in an extremely dense and reduced space.[18]

To understand why so few quarks can become a stable configuration, we have to add a new element of fractal physics that CERN quantum physicists have wondered about for half a century with their outdated Higgs model of masses—the meaning of the mysterious one-third charge of quarks: quarks are bidimensional vortices, while electrons are three-dimensional, reason why a quark has one-third of charge, as you need three perpendicular quarks rotating in a "color-locked" position to create a three-dimensional volume, a 1-charge.

Quantum physicists do not understand what quark color is. They know it is a geometrical property that obliges quarks to form triplets of "different colors." So only a red, blue, and green quark can come together. The answer of fractal relativity is simple: the three-color-locked colors are parts that form together a three-dimensional form, which emerges in our Universe. Those topologies are complex, non-Euclidean topologies, but can be simplified with the concept of a bidimensional vortex.

Then each quark color represents the relative perpendicularity between three bidimensional quark vortices, which makes them turn together in a very orderly manner, as a single three-dimensional form. And this discovery of fractal relativity, unknown when CERN started to construct the quark cannon, is the reason why strangelet liquid made of strange quarks is far more stable than we ever imagined and a few thousands of them, as the Chinese experts have found, can detonate the Earth—something CERN ignored all the time it planned its quark cannon. Nature fooled CERN. Nature is vital, always blooming, desiring to live, turn, move, and exist. CERNerds are abstract, stuck in forty years old theories, indifferent to life, just connecting cables and accumulating data in their computer-like brains.

All seems to indicate that strange liquid will be much more stable than we predicted. A double usd hyperon, the dibaryon, akin to a "cooper pair" (two stable electrons that fill a single orbital) or two DNA double strain is stable in our world. Those systems are called double nuclei and will be mass-produced already in proton collisions as protons become hyperons, acquiring relativistic mass, and then colliding the two "RNA" strains to form dibaryons.

Of course, for a full understanding of this model with the precision required by modern physics, we need to use QCD theory, the quantum mathematical model of gluons, which are the "details" of a fractal quark. Both theories, fractal relativity and QCD, fully coincide in the dangers of the LHC. Thus, CERN knows perfectly what it is doing and it is taking the risk of killing you and me because it is a rogue nuclear company, and sixty years of living a cold war on the brink of a potential mass murder makes those dangers a routine.

We are talking of criminal negligence because CERN ignores the warnings of classic strange quark theory on purpose. And of course, it has not returned, answered, acknowledged, or published any of those key papers that Einstein et al. have sent to them and it has not appeared on the suits to discuss the issue. The same behavior has been observed by the scientific press. That is why this is indeed a death foretold. Let us then consider of the many papers, which explain those scenarios of creation of strangelets, two key descriptions that complete the third leg of scientific proof: the mathematical papers of the leading theorists of strange matter, Schaffner and Peng, from MIT and the Chinese Institute of High Energy Physics, which prove clearly that the risk of formation of strangelets is enormous.

12. Neutron Stars, Strangelets, Kaons, Hyperons, and Dibaryons

Glendenning and Schaffner. Nuclear Science Division; Lawrence Berkeley National Laboratory.

Abstract

We study the influence of a possible H dibaryon condensate on the equation of state and the overall properties of neutron stars whose population otherwise contains nucleons and hyperons. In particular, we are interested in the question of whether neutron stars and their masses can be used to say anything about the existence and properties of the H dibaryon. We find that the equation of state is softened by the appearance of a dibaryon condensate and can result in a mass plateau for neutron stars. If the limiting neutron star mass is about that of the Hulse-Taylor pulsar a condensate of H dibaryons of vacuum mass 2.2 GeV and a moderately attractive potential in the medium could not be ruled out. On the other hand, if the medium potential were even moderately repulsive, the H, would not likely exist in neutron stars. If neutron stars of mass 1.6M were known to exist, attractive medium effects for the H could be ruled out.

Introduction

Since Jaffe proposed that there may exist a stable di-hyperon (a quark composite with baryon number two)], an ongoing quest for this particle began. Recent searches using kaon beams or heavy ion beams found no candidates or are still in progress. There exist some claims for evidence for the H dibaryon produced in proton-nucleus and in heavy-ion collisions.

The existence or nonexistence of the H dibaryon is strongly connected with the observation of double hyper-nuclei. Three double hyper-nuclei have been reported in literature.

If the H dibaryon exists, it will have a certain impact also on the properties of dense matter. It is quite established nowadays, that neutron stars have a large hyperon fraction in the core and might be described as giant hyper-nuclei, though bound by gravity. Here again, the presence of hyperons might restrict certain properties of the H dibaryon. Recently, studies for neutron stars have been done for nuclear matter without hyperons but including H dibaryon condensation and limits have been set for the coupling constants of the H dibaryon.

There might exist heavier partners of the H dibaryon, lumps of strange quark matter dubbed strangelets. There are several heavy-ion experiments dedicated to search for this novel form of matter. In the MIT bag model, strangelets with A= 6 are found to be unbound. Nevertheless, light strangelet candidates in the range of $6 < A < 40$ might be stable against weak hadronic decay. The H dibaryon as well as these light strangelets can occur in dense matter as a precursor of the phase transition to a quark plasma.

In this paper, we study the influence of H dibaryons and other strangelet candidates on the composition and structure of neutron stars including the hyperon degree of freedom. We are particularly interested in the question of whether neutron stars and their masses can be used to say anything about the existence and properties of the H dibaryon. In section 2, we discuss the condition for the occurrence of dibaryons and strangelets in neutron star matter. The relativistic mean field model with hyperons and the H dibaryon is presented in section 3. Implications for a H dibaryon condensate are discussed in section 4 and summarized in the last section.

2. Composite Objects in Neutron Star Matter

Here we discuss the general features of the appearance of composite quark objects in neutron star matter. Nuclei will dissolve in dense matter due to a Mott transition at quite low density. Hence, hyper-nuclei with similar binding energies will also dissolve.

The situation is different for strangelets, if they are energetically favored compared to hadrons. Then strangelets will appear at a certain critical density which will depend on the chemical potentials and the mass of the strangelet. The most stable strangelet candidates will have a closed shell, i.e. they have zero total spin and are bosons. Also the H dibaryon (consisting of two u, d, and s quarks) has zero spin and will form a Bose condensate if it appears in dense matter.

In this rather easy-to-follow paper, we find all the warnings that should restrict the creation of strange matter: *dibaryons can be stable in our atmosphere. Usd-usd dibaryons can be produced by usd-hyperon interaction. Dibaryons and hyperons are the quark-gluon soup of neutron stars. Dibaryons can form a stable strangelets just with 6<A<40, which means merely 7 dibaryons can open the 7th seal of extinction, a growing strangelet. And needless to say in a quark-gluon soup so dense as those CERN will do, 7 dibaryons can come together at any time.*

Finally in another scientific paper by Tsutomu Sakai,[19] we read the different reactions that produce the dibaryon, the stable strange atom of neutron stars:

> The H-dibaryon may be produced via kaon-kaon reaction, Ξ capture, heavy ion collision and proton-proton collision.

So we might go under even in April. This book available on the Net might not get to print.

What those papers are telling us is clear. The kaon is the first brick of the creation of strange matter—strangelets made of usd atoms. A usd atom is a hyperon. A dibaryon is a double usd atom, much more stable. There are many other strange particles, which might form part of those strange stars, but us, ds kaons, usd hyperons, and usdusd-dibaryon families with their balance of up, down, and strange quarks are the easiest particles to make and the more reactive particles of the strange world—the jokers of the cards. The kaon is the amino acid, and its reactions will create hyperons and dibaryons, the RNA and DNA of strange stars. All together they will start a series of complex reactions, which can be easily compared to those happening in a hot amino acid soup that creates life structures, by the self-similarity of all systems of energy and information.

This is what CERN doesn't want to study—*the details of the creation of strangelets and strange liquids. Nor does it want to accept that the Universe is not probabilistic but is guided by the totalitarian principle, which says that all that can happen will happen.*

Physicists always fail to predict the stability and "informative complexity" of particles, which routinely live longer than expected, of dual stars, which are far more abundant than they thought, etc., etc. They make very simple models based on lineal energy, which deny the binding power of information, *the creative principle of the Universe.* So at the end they *always find systems that are colder, more stable, more ordered, more reproductive, more liquid than they predicted because they forget that the informative arrow of the Universe* is colder, stable, ordered, reproductive, and liquid.

And yet even with those models the texts are clear: kaons can react with both, our up and down particles, neutrons and protons, breaking their pion cover and extracting their quarks from the nucleus, evolving them into strange quarks *in temporal, weak events. And then they can react, joining together into hyperons, which can react and form dibaryons. So the chain of creation is clear:*

atoms->kaons->hyperons->dibaryons->strangelets->strange stars

Those reactions are theoretically possible, logically sound, as they are self-similar to other processes of construction of complex systems from simplex parts and the beginning, middle, and end products are experimentally proved to exist. So the totalitarian principle says they will happen.

Those reactions studied in detail have lowered the quantity needed to create strangelets, according to Bielich to a mere 6 to 40 A. In other words, a few dibaryons are enough to form a drop of strangelet liquid. What else needs CERN to consider relevant the chances of destroying the Earth? *It is producing more kaons than expected, which* can capture our atoms and create the bricks of stable hyperons and dibaryons, the substance that forms the core of neutron stars, *starting an ice-9 reaction with enormous facility once dibaryons become stable in enough numbers.*

The kaon was the first strange particle found, and it was called strange because it lived much longer than anyone expected. Next the usd-liquid appeared. Now it is only left for the strangelet to show up. Strange quarks were called strange because physicists were surprised by their long-lasting life, millions of times longer than they thought. "It was as if Cleopatra were still dying of an asp bite," said its discoverer. And yet CERN is blind and the world trusts CERN.

God, the laws of the Universe, have warned mankind about strange matter many times. But SS authorities, strange scientists, are deaf to those warnings because they want to see more of it. Now we are seeing kaons being born and dying as anti-kaons in the cycle of life and death of particles, described in the next chapter. Soon, starting next month, we will see them "grow and multiply."

The disaster scenario is this: when a dibaryon strangelet forms, it will hit a lump of metal, catalyzing its immediate conversion to harder strange matter. Each transformation of normal matter into strange matter liberates energy and sends dibaryon drops in all directions. These merge with other lumps of metal of the collider and convert them, leading to a chain reaction, at the end of which all the nuclei of all the atoms of the planet have been converted and Earth has been reduced to a cold, ultradense, rotating liquid of strangelets—the core of a neutron or quark star.

The second scenario is that those dibaryons fall to the center of the Earth and some explode in its path, on harsher conditions, creating earthquakes but finally enough of them reach the center and start to eat the Earth from inside out. The only atoms that might survive are iron atoms, which are the most energetic, stable atoms of the Universe and have a self-similar, still disputed stability compared to that of a strangelet.

So the planet will be turned outside down. Its iron core will become its cover, and within it there will be a quark-gluon liquid, a strangelet.

The absurd belief or rather excuse sponsored by Glashow and CERN is that this would not happen because strangelets will be positively charged so they will be electrostatically repelled by nuclei and would rarely merge with them. However it is obviously a lie because "strangeness" is a negative state. A strange atom is negative. It has one-third of negative charge. So what those people are saying, in their pattern of "damned lies and statistics," is like saying that electrons have positive charge.

And indeed, *the lambda particles found in RHIC had two strange quarks and were negative charged.* This absurd "negative strangelets won't happen" concept is *an idea that Mr. Glashow theorized decades ago.* Now we know they do. Why doesn't CERN update its data? Again a pattern of corruption appears. CERN chooses old dinosaurs, "famous or prestigious" old names of science, with Nobels or celebrity status, as Hawking or Glashow are to "sell" no danger, ignoring on purpose all the recent papers, like those of Peng and Schaffner, which bring the details of our destruction.

In the life soup, there was also an ideological evolution from the impossibility of "life happening" from simple molecules to the understanding that amino acids were the bricks of life, from an age in which only DNA was considered to reproduce to the understanding that RNA also reproduce, to our final acceptance that even proteins and amino acids catalyze reproduction. Yet physicists will always deny from the perspective of their abstract paradigm that all reproduces, that creation is constant, that a Universe made of motions constantly lives and reproduces since a motion reproduces by the mere fact of moving. A mass reproduces its cyclical motion every time it turns, as a clock reproduces its movement every hour. Yet for physicists, a mass is some obscure solid thing, mediated by an invisible Higgs field! Their clueless understanding of the vital, organic, eternally moving Universe and the creative, reproductive nature of all cyclical motions makes them downplay their perception of danger.

Damned lies, statistics about our genocide, and ad hominem campaigns against those who denounced them, with the collaboration of the worldwide clueless press and ministries of science, have buried this issue. It is a very Orwellian feeling. Why they do it? SS authorities follow their routine: they make more strange stuff to publish more papers about strange science. The collateral damages are of no importance. Its facility is an SS compound where they study strange science and that is what they do. In another SS compound ruled by another SS authority also called RH, the Auschwitz camp ruled by Rudolf Hoess, a plan, a routine, a work schedule was established and a law of silence was enforced. RH didn't say to anyone he was running a facility of mass annihilation. He used the very same expression, *mass annihilation*, which CERN uses. Later we shall consider how genocides and monstrosities happen. It is a needed element of holocausts to make people follow plans in a robotic manner. Distance is important. Gas chambers were chosen to avoid Germans feel bad about shooting children. CERN distance from the center of the Earth is huge. CERnerds are NOT thinking in emotional terms. They are obsessed by plans and goals. The SS authorities at CERN are "experts" that want to study kaons. They have also killed before, or if they have not, their founding fathers have. CERN employs most nuclear physicists that made atomic bombs for Russia and France. For them, the possibility of human genocide, of mass murder, is an old tale. They have been foretold the chronicle of our death many times and we are still here. So they hope this time will not happen. Göring said "the first murder didn't let me sleep, the second opened my appetite." Rolf Heuer knows he might liberate the Zyklon liquid of the Earth, the strange liquid that will extinguish mankind, but he has been forty years doing business with the nuclear industry, and his appetite to see more of it is enormous. But we live in the age of marketing and the LOL method, so as Shakespeare said now "evil has the dress of a gentleman." Today genociders smile and dress with Giorgio Armani suits and make jokes; they are the Dark Knights that will cut your mouth with a Gillette to make you smile. Is it that different? No, it is not. Göring was a charming aristocrat, an ass air pilot, second only to the Red Baron in victories, a sophisticated collector of art, an elegant gentleman, with perfect suits.

But then an independent agency, the Chinese nuclear agency, one of the few groups of nuclear scientists that are not working at CERN, started to publish articles, never refuted, which showed negative strange liquid would form, it would be as stable as the most stable atom of the Earth iron, and will therefore accrete all other hadrons of this planet.

13. Third Leg: Mathematics. Chinese Find Strangelets More Stable Than Iron

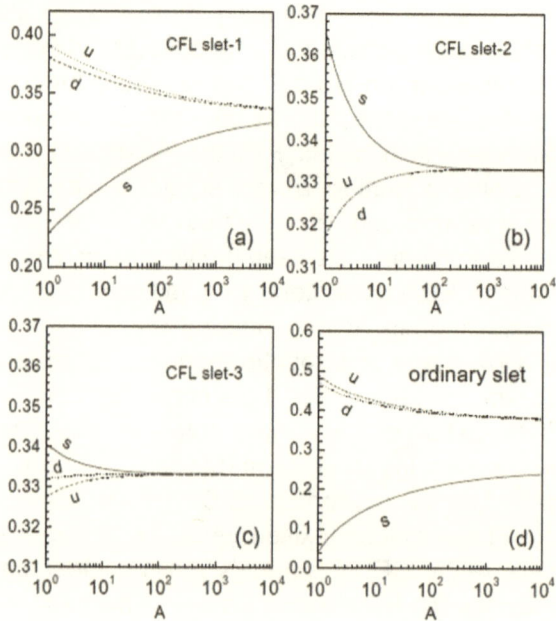

FIG. 3: Quark fractions of different strangelets. Figures (a)-(c) are for the three kinds of CFL strangelets. Figure (d) is for the ordinary strangelets. The vertical axis for each figure is the quark number density in unit of the total quark number density, or the ratio of the corresponding quark number to the total quark number.

The previous papers show the details of creation of strangelets. Let us now treat them as wholes. In December 2005, a startling article that appeared in the Physics archive—the professional Internet outlet in which in the electronic age, physicists publish their discoveries—triggered my decision to go to Court since the article proved mathematically that strangelets were far easier to make and far more stable than previous theories predicted. The article should have definitely stopped further experiments in strange matter and yet it was hidden by SS authorities, so we sued CERN to make those danger public.

To close CERN was not impossible to achieve. At least two important scientific programs had been stopped previously but always on economical considerations: In the seventies, the Americans killed the moon program because it was very costly. And in the nineties Clinton killed a supercollider halfway on the making for the same reasons. Indeed, in 1993 Congress canceled the 12-billion-dollar superconducting supercollider project, a megamachine of 88 kilometers, to save money (CERN's LHC has only 27 kilometers and it collides protons at 7 trillion volts). Thanks to the excessive ambition of American physicists, this time they couldn't make it as costs had already tripled, given the overbudgeting processes that physicists apply to their megaprojects.

But it seems that in the world we live safety is not that important. Only money matters. So now that most of the CERN budget has been spent, stopping the LHC for economical causes doesn't make sense—even if the chances of creating strange matter after the Chinese article warned on the existence of stable, negatively charged strangelets has become so high that we could compare it to a game of Russian roulette with several bullets against mankind.

The mentioned article was the result of the collaboration of three Chinese scientists from the most respected institutions in America and China, the Massachusetts Institute of Technology and the Chinese Center for High Energy Physics. This is the head of the article, explained in the jargon of the high priests of physics, the Latin that allows them to get away with murder, in the same way the jargon of bankers allowed them to get billions of dollars from naïve citizens. The key word of this jargon is SQM, strange quark matter, and CFL, color-locked quarks—that is, quarks that locked together can become attractive vortices of mass and condensates. The third *key word is A, the baryon number, meaning the number of strange quarks needed for ice-9 to become stable.*

New solutions for the color-flavor locked strangelets

G. X. Peng1,2,3, X. J. Wen, 2, Y. D. Chen, 2.

1:   China Center of Advanced Science and Technology (World Lab.), Beijing 100080, China
2:   Institute of High Energy Physics, Chinese Academy of Sciences, Beijing 100039, China
3:   Center for Theoretical Physics MIT, 77 Mass.Ave., Cambridge, MA 02139-4307, USA

After the acceptance of quantum chromodynamics as the fundamental theory of strong interactions, it became extremely significant whether a deconfined phase of matter consisting merely of quarks would be possible. Theoretical investigations show that strange quark matter (SQM), which is composed of u, d, and s quarks, might be absolutely stable. Because small lumps of SQM, the so called strangelets, could be produced in modern relativistic heavy-ion collision experiments, their charge property has attracted a lot of interest. Originally, SQM is believed to show up with some small positive charge. In June 1997, however, Schaffner-Bielich et al. demonstrated that strangelets are most likely heavily negatively charged. In June 1999, it was shown that negative charge can lower the critical density of SQM In July 1999, Wilczek mentioned an "ice-9"-type transition, which was picked up by a British newspaper. Not long ago, in response to public concern, an expert committee published a report, which got positive comments, as well as criticisms. Much progress has been achieved recently by the introduction of color superconductivity. It has been shown that bulk SQM with color-flavor locking is electrically neutral. Immediately, Madsen found a solution to the corresponding system equations of strangelets, where color-flavor locked strangelets are positively charged, and they might be a candidate for cosmic rays beyond the GZK cutoff.

In the previous abstract, the Chinese team resumed the stage at which Physics was, regarding our understanding of strange matter, when CERN published its first safety report. From the 3 options available theoretically about strangelets, CERN decided that the only valid one was the harmless option. So Mr. Madsen, who believed that all the strangelets would be positive and innocuous, became the author of CERN's safety report. And even if Schaffner and Bielich disagreed; and Wilczek mentioned a run away ice-9 transition of strangelets that would engulf the entire Earth, CERN ignored them.

Then the Chinese team found that there are three solutions for each given baryon number, A.

The strangelet corresponding to the first solution is positively charged. The strangelet corresponding to the second solution is negatively charged and the third solution is nearly neutral. For convenience, these three solutions are marked, respectively, with CFL slet-1, 2 and 3. The ordinary strangelets without color-flavor locking have also been plotted in the same figure for comparison purpose:

In the graph, A is the key measure. Since it is the number of nucleons you need to make the 3 new type of strangelets, which is far inferior to the number it was needed previously to make an ordinary set (figure d, where the convergence of the 3 types of quarks is far remote). In other words, if previously to the Chinese article, you needed tens of thousands of quarks to make a strangelet stable, diminishing enormously the risk, what the Chinese showed is that with a mere 500 quarks it is possible to make a stable *strangelet 2, which is the negative charged one,* with 1000 is possible to make a neutral one and with 20 thousand a positive one. So it is obvious that again physicists got it all wrong: *The easiest strangelets to manufacture are negatively charged strangelets, precisely the ones that are more dangerous and can provoke, by attracting positive charged protons and nuclei, the runaway reaction . . . Then the Chinese team ads:* the comparative stability of the three kinds of CFL strangelets needs to be further studied in the future. CFL strangelets which are more stable than 56Fe may have far-reaching consequences. The slet-3 is nearly neutral, and so might be a candidate for the dark matter in our universe. The slet-2 and 3 are more stable than the normal unpaired strangelets, and so may have chances to be produced in modern heavy ion collision experiments.

More over the Chinese found that those 3 subtypes of strangelets were more stable than iron, the most stable particle of the Universe, one negative, one positive and a 3rd neutral. Except for iron strangelets are more stable than any form of matter on Earth that could be easily transformed into them. To understand that process, we have to think how stars transform simple matter into more complex forms: first they use hydrogen to create helium, which is more stable. Then helium becomes carbon and carbon fusions into iron. But iron cannot fuse into heavier atoms because it is the most stable matter of the Universe. So finally the star evolves into a quark star, in which part of the iron core remains stable in the surface, which could explain the anomalous quantity of iron in the Universe – the only survival atom of the transformation of light into dark, quark matter. Further on, the Chinese team found out that those strangelets need far less u-d-s quarks to become a stable ball of strange matter. If before, in the first abstract approximations to strange matter we expected that tens of thousands of quarks were required, now it turns out that strange matter is stable with a few thousands of particles, far less than those that will be assembled at CERN. So the probability that strange matter appears on the LHC has become now far bigger than anything we could calculate before.

Indeed, look at the graph, where it says ORDINARY SET. That was the theory we believed in when CERN made the LHC. The A-number of quarks needed moves to infinite. So Nature seemed to give reason to CERN in 2003. Now look at the graphs 2 and 3: a few hundred are enough. In graph one 10.000 and CERN will deconfine hundreds of thousands.

So, *according to the Chinese Article*, we can assure that with the energies of CERN, *strange matter will appear with almost 100% of probabilities.* Since the Totalitarian principle says that all what is possible happens.

Only at the end of their papers, the Chinese told us that there was some hope because the size of the strange couldn't be bigger than an atom. But that is NOT really more than hope. Many papers prove that liquid particles fission as all heavy atoms do. Thus *those strange quark atoms will break in 2 and then grow again and split again. Thus from 2 we will have 4 and 8 and 16 and a chain reaction of super-heavy strangelets will ensue even faster than if the quark star was made of a single super-liquid drop.* In fact, we call the fission model of atoms the liquid-drop model. Each strangelet will stop growing, when they reach the size of an atom (which can pack tens of thousands of strange quarks). Then they will do what all heavy atoms do, fission into 2 or more pieces and grow even faster, since once and again scientists have found that their abstract, static particles, constantly behave in a dynamic, organic way. For example, the neutron and proton were first considered different particles and now we know they mutate their up and down quarks changing charge. So today the neutron and proton are considered the 2 sides of the same particle, the nucleon that dynamically changes its state. Precisely that simple mutation, which happens constantly in normal matter between up and down quarks is all what it is needed to change a positive 'up-down-strange quarks' strangelet into a strangelet of 'down/down/strange quarks', a negative one. So the charge of the strangelet in fact is totally irrelevant.

And yet CERN did not revise his risk assessment, which was 1 in 6 millions, and the Chinese article was ignored by Physicists, fearful of loosing his prestige among colleagues. Their authors are now in China, isolated from CERN and their pleads urging a deep study of the subject have been falling on deaf ears.

Thus the Shanghai Institute of National Nuclear physics, as one of the people of Einstein et al said, put the 'nail in our coffin', or in scientific terms, the 3rd leg of the scientific method of truth: there is experimental evidence (RHIC, Nastase, Super-novas, quark stars), there is logical consistency (Einstein's principle of equivalence that shows that a super fluid vortex behaves like an attractive space-time tornado; and the physics of complex liquids that show they are dual mini big bangs and mini big-crunches. And the Chinese give us the mathematical coherence needed, departing from the quantum paradigm accepted by all physicists.

14. Experts Don't Know, but They Won't Tell You. CERN's Strange Physics.

Yet despite all those proofs, CERN is denying the three legs of the scientific method and their physicists sign confidentiality zero-risk statement, which prompted us to sue the company for criminal negligence. Now CERN is doing the strangest "strange physics." *It invents them.* They officially talk of an evaporating, hot quark plasma. Everybody knows it is a lie, but the Damocles Machine cannot be revealed. Soon a series of false papers came affirming another absurdity—*that even if strange quarks were formed, they could not be negative and hence they would not accrete the hadrons, the positive nuclei that circulate at the LHC.* This is absurd for a simple reason: *Strange quarks are all negative by definition and they are dominant in strange liquid. They have all a negative one-third charge. Thus, strange liquid will be mostly negative or neutral, as a mean usd liquid is neutral, attracting atoms gravitationally.*

Next came the lie that strangelets would not be formed because the collisions would be a hot gas. Yet at RHIC we saw an external big bang cover and an internal supercold, superfluid vortex of quarks, a dual big bang and big crunch and the product of both temperatures Max. T (big bang) Min. T (big crunch) should be stable, E (big bang) × I (vortex) = K. This means precisely that the vortex acts as a refrigerator and the expanding radiation acts like a cooling gas while the inner vortex freezes to near zero temperature, reaching superfluidity.

CERN knows this duality, described by *SciAm: "A perfect surprise: The physical picture emerging from the four experiments is consistent and surprising. The quarks and gluons . . . behave collectively . . . This hot mélange acts like a liquid, not the ideal gas theorists had anticipated.* Evidence for the rapid formation of such a dense medium comes from a phenomenon called jet quenching, When two protons collide at higher energy spray of hadrons called jets blast out in opposite directions. But the PHENIX and STAR detectors witness only one half of such a pair. Where is the other jet? This situation conflicts with the naïve theoretical picture originally painted of this medium as an almost ideal, weakly interacting gas. This surprising liquid flows with almost no viscosity. It is probably the most perfect liquid ever observed. Astonishingly, it far exceeds all other known fluids in its approach to perfection. How and why this happens is the great experimental challenge now facing physicists. The overriding question is whether the liquidlike behavior witnessed at RHIC will persist at the higher temperatures and densities encountered at the LHC. Some theorists project that the force between quarks will become weak once their average energy exceeds 1 GeV, which will occur at the LHC, and that the quark-gluon plasma will finally start behaving properly—like a gas, as originally expected. *Others are less sanguine.* They maintain the quarks and gluons should remain tightly coupled in their liquid embrace. On this issue, we must await the verdict of experiment, which may well bring other surprises.[10] So it is obvious that the process is dual and only those physicists who denied the truth and "were sanguine about it" believe we observed only a plasma.

The LHC will create an ultradense, superfluid vortex of quarks that will convert weak electrons into light radiation while all the quark mass that resides inside the atoms of the Earth will become an ultracold quark vortex, as we observe happening in supernova reactions across the Universe. This is the likely outcome of CERN's experiments made with a quark-gluon soup.

All in all, we come to the beginning of this book, repetitive, cyclical mandala of the simple truths of the quark cannon: 99 percent of the production of the LHC will be quarks. And yet it is the only thing CERN ignores. We know after December's experiments that during the 1–10 TeV energy run, they will produce massive quantities of strange quarks and its particles, kaons and hyperons, already produced massively at RHIC, which will form the "background" noise in which they will seek for their imaginary, pythagoric particles, the false Higgs, evaporating black holes and SUSYs that could give them a medal of metal by Saint Nobel of the Dynamite. The death of mankind that shall happen in the process is of no importance to them. *It is negative thinking, and it will not be blamed on us, nuclear physicists.*

But of course, CERNerds won't get a Nobel Prize because complexity theory and its laws of information, which they ignore, shows that all those particles are redundant or false or unstable monsters without any role in the real Universe. So the Higgs will never be found.

What they might get in 2013, if we are very lucky and we are still here, if the drops of strange liquid have not coagulated in the center of the planet, is a black hole. And so now we shall study the second genocide in the making, the black hole and the absurd safety lie of Mr. Hawking—that those black holes are mathematical objects that travel to the past and evaporate.

---

[1]   Mr. Wilczek though still says in his conferences that only a theory of mass as a frequency of information can explain the meaning of mass. See his conference at MIT: "The Universe Is Strange."

[2]   A brief account of the worrisome geometric growth of strange matter production can be read in the Web at http://web.mit.edu/newsoffice/2010/lhc-results-0205.html.

[3]   Nottale's paper has been the landmark in the mathematical unification of quantum theory and Einstein's relativity. In that regard, if the quantum paradigm established the new mathematics of space in the beginning of the twentieth century and Einstein's relativity the new mathematics of time, both departing from the mathematical revolution in non-Euclidean mathematics that took place in the nineteenth century, in the fractal paradigm Mr. Nottale can be considered the creator of a solid mathematical model of fractal space, and this author of fractal time, departing from the mathematical work of Mandelbrot in fractals and the definition of the five non-Euclidean postulates.[App.IV]

[4]   Kerr's black holes are the only real black holes. The mathematician from New Zealand, working isolated from mainstream physics, discovered rotational solutions for black holes, departing from Einstein's work. This is the case in all mathematical paradigms: physicists start with simple solutions that do not reflect reality and slowly they find complex solutions till they arrive to models that fit the experimental method. In that sense,

the work on static black holes as mathematical tools of Mr. Wheeler, Oppenheimer, and Hawking is too simple to describe real black holes, which will be rotational black holes, as all masses are and will be made of a real cut-off substance (quarks) as Einstein wanted. This means only Mr. Kerr's solutions are worth to study to describe real holes.

5   Gas-9 is the name I have given to a top quark liquid condensate, by self-similarity with ice-9, the strangelet liquid that detonates supernovas. Since gas is a black liquid, it seems appropriate. To see a simulation on how both liquids will be created at CERN, see the film *Quantum Roulette* at www.lhcdefence.org.

6   MAD (Mutual Assured Destruction) was the strategy proposed first by Nobel to foster the industry of weapons. The more weapons we have, the easier we shall hold peace to avoid the extinction of life. Mr. Nobel, called Merchant of Death by the newspapers of his time, promoted the concept while opening weapons factories in all nations to be able to sell them to both contenders. The strategy was later popularized by Teller, who tried to promote a doomsday weapon, a hydrogen bomb able to blow the entire planet under the MAD strategy. Eisenhower cancelled and offered instead to Khrushchev a nonproliferation treaty. Unfortunately now physicists have changed to the "God's particle" myth with better results.

7   Mr. Eames in his classic film *Powers of 10* shows how in decametric scales suddenly nature reorganizes its information into new complex organisms.

8   Timeus, Plato. The organicist paradigm has always been understood by the highest minds of mankind. It was evident for Plato and Aristotle. It was also foreseen by Spinoza and Leibniz in its Protea. But it has always been censored by the mechanist view. Leibniz's master book, to put an example, was not translated to English till this century.

9   CERN will close in 2012 to ramp up potency to 7 TeV—if we are still here.

10   SUSY particles were one of the many absurd solutions to describe gravitation with quantum laws as Higgs does, using particles to interact and create the effects of mass. They were used to explain why gravitational force is so weak in the quantum world, which in fact it is not used or needed to describe any interaction between charges. Their use, however, is superseded by the fractal theory of space-time since indeed *quantum particles do not suffer gravitational forces, as they do not belong to the strong gravitational membrane of space-time.*

11   In a conversation with Poncaire, the great French mathematician who introduced topology, the "mathematics of the future" to study gravitational problems.

12   All those fantastic particles are called WIMPs because they are weakly interactive particles, meaning they are invisible, never observed mathematical fantasies.

13   Mr. Hamed affirms that it is easier to find "dragoons at CERN" than nonevaporating black holes, as he is eager to switch on the machine and find proofs of his infinite parallel Universes.

14   As quoted in *The New Yorker* magazine, Crash course, May 14, 2007, Mr. Wilczek first mentioned the ice-9 reaction in *Scientific American* 1999 (see chapter 7).

15   Juan Maldacena, "The Illusion of Gravity," *Scientific American*, November 2005.

[16]   "Fireballs of Free Quarks," by Graham P. Collins, News and Analysis; *Scientific American*, April 2000. (*Sci Am*, May 2006)

[17]   The papers on the production of strange atoms was delayed for almost a decade at RHIC, published this March 2010. A brief account can be found at http://physicsworld. com/cws/article/news/41917.

[18]   Mr. Engelen gave numbers for the production of quarks at a Polish seminar; see chapter 8 addenda E.

[19]   See a complex analysis of dibaryon physics at http://arxiv.org/abs/nucl-th/9912063.

CHAPTER 5

Fractal Time: The Life/Death Cycle

Time curves Space into mass.
—Einstein, father of modern physics

Time evolves the morphologies of life.
—Darwin, father of modern biology

I. The Arrows of Time: Energy and Form

In the Universe, there is, besides the energy arrow, an arrow that reproduces information, used in biology to study life. Both together define an immortal Universe made of ∞ bites of energy and bits of information in cyclical trans-form-ation: A young surface of energetic space reproduces in/ form/ation till it becomes old and wrinkled. Then it erases its form back into energy in the inverse process of death, completing an existential cycle, ΣS⇔TI, the generator cycle of the Universe. In the graph, a human being goes from an energetic youth into an informative third age to become erased back into energy while the Universe evolves from an energetic big bang into an informative big crunch. Because we are all dust of space-time that creates and dissolves ∞ "existences."

1. The Inverse Properties of the Arrows of Energy and Information

If the Higgs was a clever hoax to maintain the industry of accelerators going on after the end of the cold war, the evaporation of black holes proposed by Hawking is a more complex issue that deals with the philosophy of the Universe sponsored by quantum physicists, which consider information uncertain and hence can invent any kind of fantastic theories about the Universe and energetic theories that do not consider the arrow of information.

For that reason, we must do a deeper analysis of the two arrows that create the future of any species of the Universe, energy and information, before we can tackle in depth the falsity of evaporating black holes. To that aim, it is important first to clarify some basic concepts about space-time that quantum physicists cannot explain, especially the differences between the two types of changes, also called time arrows that create the future, to which we shall dedicate the next chapter:

Physical arrows of change in motion and energy of beings.

Change in their forms, or informative change, which is not recognized by physicists and only studied today by biologists. And yet informative changes, whose rules are different from entropy/energy/motion changes, occur also in physics. And so most errors of physics derive from the ignorance of this arrow and its laws. We already saw that the entire error of Higgs departs from the fact that physicists do not recognize the weak force as an arrow of informative change, which breaks the symmetry of spatial forces; that is, do have a casual order from past to future.

The main difference between both arrows is therefore the existence of causality in the arrow of information. Because Hawking still uses the previous concept of a single physical time that is symmetrical from left to right and right to left, from past to future and future to past, he is able to forecast a false phenomenon: the evaporation of black holes to the past. However, time has a causality in all informative arrows, which is known as the life/death cycle, where the life cycle increases the information of beings in 3 long, temporal ages; and the inverse arrow, the arrow of death, briefly erases the information of beings in a big bang. Thus, we need first to understand the differences of "symmetry" between both arrows in order to fully grasp the absurdity of time travel.

The understanding of that order is the key to resolve with a higher degree of complexity *many of the questions CERN pretends* to answer bombing the Earth with black holes: why there are three families of masses, why we see less antiparticles than particles, why the Higgs is false and Einstein's theory of mass as a vortex of physical information right, and so on.

Ultimately, the difference between both arrows of time come from their geometry.

Informative arrows are implosive, warping processes that create information in slow periods of time while its opposite process are fast big bangs that liberate energy in a big spatial surface and in very little time. Thus, the two directions of a time arrow, the life-long period of creation of information and the short expansive death of information, are very different in outlook. And for that reason, we cannot say that an informative weak event and its inverse, electromagnetic explosion of radiation are symmetric since they are very different. The same happens in a particle (informative process of creation of a vortex of mass) and an antiparticle (explosive big bang).

On the other hand, mere changes in the motion of beings are highly symmetric. The behavior of a particle moving right or left as they acquire energy are very similar.

Thus the arrow of information breaks the parity in space, which the arrow of energy follows.

Time as a flow from past to future is made of energy and informative motions, cycles, and lines, *which have a different outlook when they move left or right (lineal, spatial energies) than when they warp and unwarp (the left-right of cyclical motions).*

This give us *not one but at least three arrows of physical time, the left/ right parity arrow of spatial/energetic change; the implosive, informative arrow and the explosive, big bang arrow of destruction of information.*

Thus there are at least three time motions:

— *Cyclical, informative clocks of time*, which in physics are masses and negative charges in the two scales of reality (cosmological and quantum realities).
— *Anticyclical motions*, which in physics are processes of destruction of mass, $E=Mc^2$, positive charges that expand electromagnetic space and antiparticles, which act both as anticharges and antimasses, killing a complex particle (an atom) in both scales of reality.
— *Spatial left-right motions*, which happen in lineal fashion and conserve the parity of space.

Thus, a minimal accurate understanding of times as changes would consider three basically "arrows, motions" of time—the expansion of entropy (anticyclical motion) the implosion of information (clock motion), which are inward/outward motions and the left/right motions of space and their complex combinations in particles and antiparticles. For that reason, we use the expressions lineal, energetic space and cyclical clocks of temporal information to define those physical arrows.

2. The Arrows of Informative Events: Past, Energetic Big Bangs and Future Informative Life

The main duality of time arrows happens between positive, informative creation and negative destruction of energy. Those two arrows are clearer in biology, where the informative arrow dominates. Thus, we call them the arrow of informative life and energetic death. And so we can use for them the concept of an informative "future time" or life arrow that is related to the constant warping of information between past and future in life and a relative "energetic, past, big bang" arrow of death that erases energy in a brief time explosion.

Thus, when we talk about how the future evolves information, we *enter in a different arrow that physicists fully ignore, as it has been developed in the context of duality and complexity sciences.*

Those two opposite, complementary arrows of that d=evolve information in the Universe, at its simplest level, are as follows:

— *The arrow of entropy, of lineal movement, disordered freedom, and expansion in space*, which exists also in nonphysical systems. Since freedom, movement, disorder, and expansion of an entity's vital space are concepts applied to all sciences. It is the arrow of death and antiparticles. It is defined by logic thought as the relative arrow of past.

— The arrow of geometrical information, *of implosive, cyclical movements.* That arrow of information is opposed geometrically to the arrow of expansive disorder, lineal energy and space since it occupies little space and has a lot of form, while energy occupies a lot of space and erases form. It is the arrow of life, of warping evolution, of creation of mass. It is the arrow of Einstein's relativity, of cyclical motions. It is the arrow of negative charges and masses. The logic and geometric properties of energy and information are opposed:

Energetic space: Arrow of entropic energy, lineal space, radiation, width, freedom, death versus

Informative time: Cycle of fractal information, implosive evolution, hierarchic order, height, life

But we said there is a third event, a spatial, lineal event. This event that transfers energy and form in balance is *a present, reproductive, balanced, wave event*.

And so there is a third arrow that balances both, entropy and form: E⇔I. This arrow ⇔ is the arrow of spatial balance, which is the only arrow that physicists consider. It is a complex arrow/event, which we perceive as stable in-form-ation, as a stable space-time through the repetition of self-similar formal cycles that exchange energy and form between two different entities.

This arrow defines in physics the main event of particles, a Ti<=E=>Ti, exchange of boson particles between fermions. It defines a wave made of lineal motion and cyclical particles.

And so we have translated in the concepts of relative past (entropy arrow), present (reproductive, spatial arrow), and future (informative, evolving arrow) the three simplex arrows of space-time that define most events of physics.

This is the concept that we shall consider in this chapter since more complex concepts and arrows of time are not necessary to explain the infantile errors of Hawking and will be studied in the appendix.

It must be understood though that those arrows are local and do not mean a movement of the entire Universe from past to future, as time is fractal. There are infinite time arrows and so they merely mean that a certain entity devolves, evolves, lives or dies, or exchanges energy and information in balance with other particle of the Universe. In that sense, if we were to consider an abstract "absolute arrow of all the time arrows of the Universe" in the sense physicists do for the entire Universe, this arrow will always move toward the future.

We have though a philosophical order made with those arrows by "logic, verbal time."

It is the logic of verbal time, a human language, that orders those arrows in relative human terms. So humans tend to call a system with a lot of energy and little information, the relative past or young age of a being, and a system dominant on information the relative future or third age of a being since in life, information is the biological arrow of future.

So we call "a relative future arrow" to information and a relative past to any "energy" system.

And we call the life arrow of informative warping the arrow of relative future of a being, and the arrow of big bangs and deaths the arrow of lineal past.

Yet in physical beings, the two arrows of energy and form constantly switch between each other. So in the physical Universe, we observe particles of relative future becoming waves of relative past or antiparticles, back and forth in events happening constantly.

3. The Cycle of Life and Death

Age of Energy,   Classic age of Harmony, Old Age of information,   Explosion and Death:
Human: Youth (0-20a.);   Maturity (20-40);   3rd age   and   death (60-80).

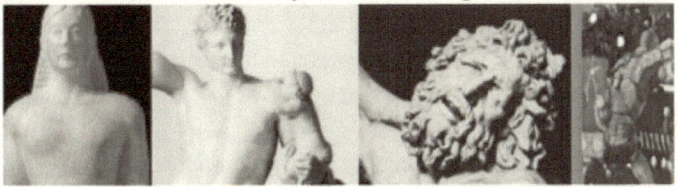

Culture: Epic, Lineal Art; Realist, Classic Art; Baroque Art. War.

Matter: ı< E: Plasma Max. E: Gas   E=i : Liquid   Max. ı : Solid   i (Mc²) <<E

Star: ı< E: Nebula. Max. E: Gigant star. E=i : Sun. Max. ı: White Dwarf. i<<E Nova

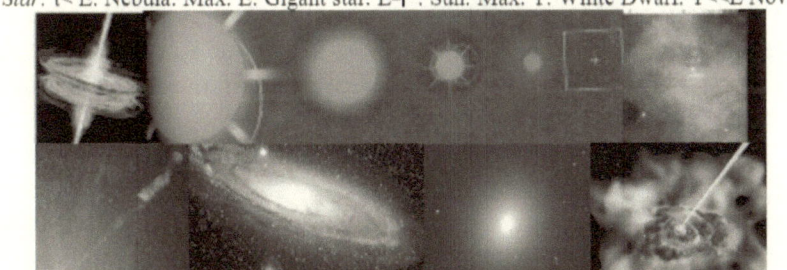

Galaxy: ı< E: Jet Max.E: Irregular. E=ı Spiral⇔Elliptic. Max. ı: Globular   ı<<E: Quasar

The three arrows of time follow a natural order from young energy into old information through a classic age of balance and reproduction, ending backward in an explosive process of death, which completes the existential cycle of all beings in the Universe.

In the graph, we consider examples of the 3±1 ages/arrows of time in basic organisms of the Universe:

+1: *Birth*: A seed of temporal information, Ti, evolves into a complementary organism, i<E by absorbing energy, replicating its information and evolving in social cells: Life starts.

—*Youth*: It absorbs energy and reaches its maximal size, or energetic youth, Max. E.

—*Reproduction*: The space-time being keeps in-form-ing itself till reaching an age of reproductive balance, i=E. We perceive that age as beauty, harmony between lineal energy and cyclical form.

Third informative age: Energy keeps in-form-ing, warping till it is exhausted, Max. I, then . . .

-1: *Death*: Information explodes back into energy, reversing the arrows of time: I<<E.

In that sense, if in the previous chapter we visualized the two primary arrows of energy and information, whose constant, creative reproduction creates the future, from the perspective of geometry, we want know to consider those arrows from the perspective of causality, of the life and death cycle.

Indeed there is a perceived order of existence of all beings, which are born as a surface of energy that fractalizes into information, dies, and then explodes back into a big bang of energy, to fully grasp indeed the absolute error of quantum cosmology—to consider a big bang, which is the death of a Universe its big banging, its birth.

The two arrows of time can be described mathematically, thanks to Einstein's discovery that the product of energy and temporal information of any system, $e \times t = k$, is constant, which is also the origin of the Heisenberg principle, $E \times I = K$. Both equations include the three essential arrows of the Universe: The arrow of energy, e; the arrow of information, represented in Einstein's formula by clocks of Time, T or i; and the product of them, k, or arrow of reproduction. It is also obvious that mathematically we can express the previous equation in three forms:

$$Max.\ E \times Min.\ T = K\ or\ youth$$
$$E=K=T\ or\ maturity$$
$$Min.\ E \times Max.\ T=K\ or\ third\ age$$

This can be grouped in a dynamic flow, $E <=>Ti$, as energy becomes information in one direction and energy in the other, fluctuating through three ages in different possible orders. Those are indeed the events of the Universe, flows of energy and information between multiple non-Euclidean points, and the regimes of those events, the form they adopt when we observe multiple of them, the networks they create, the organisms they make is the substance of reality, so we talk of the equations of the three ages/dimensions of time/space, and its forms of the generator equations of reality:

$$\sum E \Leftrightarrow \prod i$$

*Where flows/waves of loose energetic quanta, become fractal pieces of
information, chained cycles*

The easiest way to explain it is with the symmetries of mathematical operand, sum vs. multiplication that creates ordered, still informative networks; derivative vs. integration; chaos vs. fractal order. The other way of description is morphological: fractal parts loosely integrated create waves; closely knotted, they create particles and heads, where the spiral cycles become tied and convoluted into form.

But if you see it dynamically, you will see a flow between both sides. When many energies become information, the being gets old; when many knots of information release energy into forces, the being becomes mobile and younger. In physics, both flows are constant and we do not see them ordered from past to future, though we hint the order exist and so when E becomes Ti, it is a particle forming and vice versa an antiparticle and then a radiation occurs when a particle dies in a big bang.

In the graph, we see cycles of life/information in which the flow corrugates life and cycles in which we see the three alternative states of matter, from gas to solid at all scales. In matter, we do not see or feel the order as life-death we do in life.

Thus there are three possible stages or ages of evolution of the equation of biological time. And since we are all made of energy and temporal information, those three stages must be the stages, or phases of any system of the Universe. This finding, which is perhaps the most important discovery of theory of time of the twenty-first century, is validated by all kinds of empirical observations. For example, the three states of matter correspond to those three equations: gas (maximal energy), liquid (balanced energy and information), and solid (maximal information).

They are also the three possible solutions of a cosmological system: the big bang of maximal energy, a steady state of balance, and a big crunch of maximal information.

And of course, there are three families of increasing mass in the Universe, which are increasingly informative, as mass is the information of the Universe, *one of the questions CERN pretends to solve with the quark cannon, which seems to play the role of an oracle.*

In the graph, we show further examples of those equations. Here we are interested in its application to the understanding of life and history. If we call the age of maximal energy youth, the age of balance between energy, and information the reproductive, mature age, and the age of maximal information the old age, we have a simple description of the fundamental cycle of life and death through its three ages. But we can also write those dynamic processes backward. And so we have the inverse equation of life ages, the equation of death, when after accumulating maximal information, a species reverses its arrow of information and explodes that information into energy. It is the process or equation of death.

So life is an arrow of information, E-> I, and death its inversion: I-> E.

And both arrows have inverse properties, which means that in the particular case of particles and antiparticles, antiparticles, with inverse arrow to particles, are merely the death of the particle, which lasts so little time that even if there are as many antiparticles as particles, we hardly see them because they only exist in the moment of death—a fascinating fact of physics that only Feynman, the man who also warned all the physicists at NASA that the shuttle would explode and kill the crew and was laughed at,[1] hinted with amazing insight when he said that antiparticles are merely particles that travel backward in time.

What he meant is that they changed their arrow of time, from creating information in its cyclical vortices to expanding backward into energy, from $M=E/c^2$ into $E=Mc^2$ till they become pure radiation.

So simple, so beautiful, so misunderstood by mechanist scientists who think the previous equation is uncertain because the observer can perceive a particle in any of those three states (uncertainty principle of quantum cosmology).

The arrows of time illuminate different sciences, as we can use the same ages and morphologies to describe all species of energy and information. Of interest to this book are the solutions that it brings to the aberrant concepts of the outdated quantum paradigm that justify some of the experiments at CERN. Because mathematics evolve, yet each paradigm describes with certain accuracy the entities it describes with its equations, this doesn't mean that the Newtonian or quantum paradigm is not useful, especially when it is used to describe the particles and forces for which it was discovered. What the fractal paradigm does is to solve the errors introduced by Pythagoric conceptualizations of quantum laws or by the abusive use of quantum equations in gravitational cosmology.

4. The Inverted Properties of Space and Time

It is now important to recollect our thoughts. There are two *ways in which nature mixes* energy and information:

One is a temporal order with causality, as events, which start with an age of spatial, lineal energy or youth, which evolves into cyclical, warped information in a third age, through an intermediate state of "wave-exchange" of energy and form, or reproductive age. *This vision implies the temporalization of the Universe, which is seen as a series of events or "actions" of "changes."*

Another is a geometrical, spatial present organization, which considers the existence of all beings as organisms with an energetic region or body, which moves faster; it is larger in space and participates of the properties of space and energy. And a region of information, particle or head, which is smaller, gauges information and has cyclical form. Both regions often have a network or neck or zone of exchange of energy and information between both, which is a reproductive zone.

It is the spatialization of cyclical time, which sees all what exists as "geometries of space." And we call the law of complementarity of space-time.

This duality or rather ternary symmetry is neither an error or a choice, but the fundamental logic structure of the Universe, the law that embeds all other laws and explains most of its events and forms, studied in detail in the appendix, for some cosmological systems *since the Universe creates both types of structures: events in time and organisms in space. And so often physicists confuse an event in time with an organism in space and vice versa.*

5. The Antisymmetry of Time and Space: The Generator Equation of the Universe

The line of past energy and the future cycle of information combine to create present waves. This would be the geometrical, abstract way to define the game of existence. And it can be easily expressed as the two definitions of the generator equation of fractal space-time, *the most important equation of the universe and Saint Grail of the science of complexity, which tries to find a fractal generator, the name given in fractal science to simple feed-back, dynamic equations that generate information form. In this case, we establish with the two arrows of time, energy, and information a simple feed-back event that generates in a simple order the creation of a life-event of form and in its inverse order, the destruction of form:* $E \Leftrightarrow T$ would be the temporal vision of the three ages and $E \times I = K$, the spatial vision of two complementary forms.

$E \times I = K$, *Past × Future=Present*
But in time we would say Past→Present->Future + Future→Past, which is the Life->Death cycle.
Since the arrow of E-> Ti creates form in life and the inverse arrow, Ti<Ei, destroys it.

Needless to say, physicists ignore those antisymmetries and what they do is therefore doodles in the sand. Because without the understanding of that antisymmetry, it is impossible to fully grasp what physicists perceive as they sometimes confuse a temporal event (a weak event) with a spatial force and vice versa. This doesn't happen in other sciences, even if they ignore the fundamental law of the fractal Universe because they have a much wider, detailed perception. When you see a living being, you see his life cycle in a long detailed process, and you see his head and body. But in physics, the observation of very small, hardly perceived, very fast cycles of life-death confuses completely the perception of some events of time and forms of space.

And so a fundamental difference exists when we order beings in space or in time. In time, the past to future arrow is slow because it goes through present waves. Yet the future to past arrow is fast because it explodes information into energy without the intermediate states. And that is indeed the ultimate, complex explanation of why antiparticles last less than particles. So beautiful and so simple, when you use the mind to understand it all, not quark cannons and arrogant derogatory statements as CERN and Mr. Hawking do with Einstein et al., one double wrong, the other a twat . . . the harder they fall.

God is indeed simple and not malicious. Unlike Hawking and CERN, he created a perfect Universe. He elongated life for all i=ts species to enjoy it. Indeed, there is also an organic way to express the previous antisymmetry of time, saying that energy and information combine to reproduce all the wave species of reality. If those forms do not work, God will erase them fast to restart a new game of life, what he cares for. It is for that reason that to "Exist" the game of existence has been described historically in abstract, geometrical terms and vital, organic ones. Both forms are correct and so we shall use both.

In that regard, the Universe is dual because it is both geometric and organic, spatial and temporal, still and in motion. And that duality is completely necessary to consider any truth or event. We can express that duality in many different ways and languages:

In the biological verbal language, we can describe a philosophical, verbal function to exi=st, to become a stable vital space-time field, by combining in balance energy and information.

In fractal geometry, the function of existence can be explained with a fractal generator, an equation of feed-back cycles, $E \Leftrightarrow Ti$, which determines the flows and events of reality.

Such generator equations, when considered in detail, can describe all the beings we perceive as balanced combinations of energy and form. For example, quantum equations of wave-particle complementarity would be particular cases of that equation. Einstein's E=M(t) equation would be in Planck units (where c=1), another particular case of that equation, and so on.

The reader should recall that CERN is ultimately a corporation, which seeks to understand the simple, beautiful thoughts of God we have just expressed: why the weak event, a time event, is antisymmetric and breaks the parity of space. But of course, this is what they say they want to learn. Their interest on learning the antisymmetry of life and death in CERN, or Great Britain, where Mr. Higgs and Mr. Hawking muse, is null. I have written them often, explaining those things to them, before I realized that was all an ego trip, a farce, a play of eviL=Death, the only side they understand. Indeed they are going to create the antisymmetric side of death, they are going to kill us all, to deny the thoughts of God, his love for Life, his creation of a perfect Universe in which death is short for life to start again. I never got an answer, of course, to my letters, nor did I get an answer to my suits. I fought for your thoughts, dear God; please forgive them because they don't know what they are doing.

6. Lines, Waves, and Cycles: The Three Motions/Forms of the Topological Universe

The paradox of Galileo is fundamental to fully grasp the difference between moving energy and its static version space and moving time clocks and its static version information. This means that space, lines, and planes of energy are synonymous and the inverse of cycles=clocks of time and in/form/ations forms in action while space-time, reality is the combination of all cycles and lines, seen as a series of waves in motion or complex forms in static shapes.

Those three elements either in motion or static are the essential structures of nature. And because humans have never properly focused their meaning when we use human languages, conceptual confusion takes place. We shall try to clarify all those terms in this introduction before we analyze those forms and motions in the Universe.

If we were not constrained by tradition, we would simply talk of a game of three states, motions of forms, in which motion and form balance each other. So information is the accelerated implosion of form; entropy the expansive slow down of space; and the wave, the reproductive balance of both. And all that you see is a game of implosive, accelerated frequencies of form, explosive decelerating expansions of entropy, and present complementary waves.

But humans do not see reality as it is in all its complexity so we must consider the limits and partial perceptions of the human being. And the most evident is our incapacity to see an entity in both states, motion and form. And that is indeed the paradox of Galileo. Reality is more complex, has more forms and states that the limits of human perception of it.

The paradox of Galileo is very deep. It allows to spatialize time, fixing it as a cyclical, geometrical form. Thus we can fix our perception of the three arrows or dimensions of times motions, expansive processes of "past entropy/energy," implosive processes of "future information," and lineal, reproductive processes of "present waves."

Then the three relative dimensions of change, explosive entropy, reproductive waves, and implosive information become fixed by the senses into a continuum tapestry of forms we call space.

Thus space is the fixed perception of a volume of motions or time arrows that the mind fixes into a series of patterns of information that become the space we see. Yet beyond the physical geometrical concept of time there is a logic concept of time as change and motion. And this other perception of all what exists as a moving form allows to temporalize space. In this logic vision, space becomes dependent of time. Then space becomes a "maya of the senses" and all moves in the three main motions of reality—past entropy, present waves, and future information. This ternary structure is the one the reader must interiorize to fully enjoy the games of the Universe.

The details of those three geometrical motions are secondary to the game of God, the mind of the Universe, which is a simple game of three geometries, or topologies, which we shall see later are the three topologies of a four-dimensional Universe.

And that visual restriction of three functions (entropy, reproduction, information) parallel to three forms (the line, the wave, the cycle) and three motions (expansive, balanced, implosive) with three speeds (deceleration, constant, and acceleration) all together create a harmonic Universe, which explains it all, needs no fantaphysics, restricts reality, and creates a constant beauty, an eternal deterministic dance of which physicists hardly see one-third of it. The reason is obvious—the language of God, a painter in four dimensions with three colors, is visual and seeks beauty. The language of physicists is algebraic, digital, numeric, and lacking the understanding of those ternary functions, dualities and parallelisms in the meaning of reality, which liberate easily the "imagination" of numbers, creating absurd fantasies that have little to do with reality. Of course, this does not mean physics cannot calculate reality. In a relative Universe, as Ptolemy proved with his exact calculus of the motion of planets and stars with the Earth in its center, complicated calculations can, departing from false assumptions, obtain correct results, especially today when there is so much computer power that the details can be constructed without knowing the thoughts of God.

Indeed, now that we have a general view on the meaning of time arrows and the antisymmetry of time, we can show how infantile are CERN's ideas about big bangs and Mr. Hawking's ideas about time travel and evaporating black holes—which unfortunately will kill us all.

CHAPTER 6

Big Bangs and Black Holes

I am Kali, God of death.
—Oppenheimer, father of the A-bomb and the singularity theory

I. The Arrow of Physical Death: Big Bangs+Big Crunches

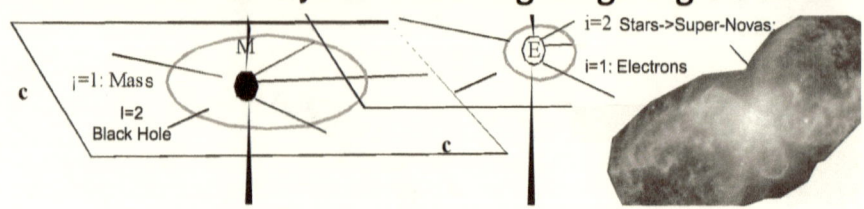

1. Russian Dolls: Big Bangs and Big Crunches

Let us now consider all those life/death cycles of physical space, created by the explosive and implosive, entropy and information, past and future arrows of time, studied in the appendix in more detail, the one that concerns us more in this book, the big crunch/big bang life/death cycle of any fractal, physical particle, which CERN will practice with Mother Earth. All particles/vortices can be described with the same model of unwinding/warping dual informative time motions.

In the graph, a mass that explodes into a square surface of pure light, c^2, $M=E/c^2$, is becoming unwarped, extended into space. The faster the mass turns, the more weigh it has and the more energy it unwarps when its speed is unknotted into lineal extension. So when a heavier/faster mass is transformed, according to Einstein's equations, into pure energy, when

294

it is unwarped and its information unknots into a big bang, the explosion is bigger because more information/mass becomes energy/distance.

A static vision of that process will consider the faster, smaller vortex that offers more resistance to displacement, a thicker, denser solid mass. It is the metaphor of Russian dolls used often to understand why smaller particles are denser: their vortices of mass speed faster. The Universe is a fractal system made of discontinuous layers of energetic forces and particles of information that diminish in scale and size as they are enclosed into each other, like those thin, hollow Russian dolls that keep inside smaller, thicker Russian dolls. Though the outer Russian doll is bigger in space, it is a thinner layer of matter that has actually far less mass energy than the final inner Russian doll, which is not hollow but a densely packed minute doll made of thick wood. So it happens to the Universe when we change the static, solid vision to the motion, vortexlike description. Each Russian doll is a discontinuity of increasing speed of the accelerating inward vortex or information arrow of physical time.

And those discontinuums between "mediums" of the spiraling vortex define the different layers or Russian dolls of the Universe. In that sense, we can talk of three fundamental Russian dolls of diminishing size and higher energy:

— The chemical, molecular Russian doll, which a TNT bomb dissolves into energy.
— The electromagnetic, smaller, more energetic Russian doll in which we exist.
— The lighter than the thicker gravitational doll, the world of quarks and black holes that the thin layers of the electron trap and CERN will release. So those thicker quarks can "eat us," electronic beings, accelerating and reducing our dimensions of information back into bidimensional mass, $M=e/c^2$. In the process, mediated by the weak event we acquire mass becoming the W and Z states, finally evolving into massive particles. This weak events but described with the false "spatial theory" of the weak event is what CERN will do NOT to understand the weak event, perfectly described in complex physics but to NOT prove the theories of their quantum professors.

All that you see around, including yourself, belongs to the thin, electromagnetic, molecular cover that surrounds the inner Russian dolls of the atomic nuclei and its quarks. The world you see is in fact a ghost, like a watercolor made with a very thin layer of pigment over the thicker white paper that provides the support for the painting. Our world is a very thin chemical and electromagnetic cover of energy that bonds molecules and atoms together and CERN plans to dissolve to see the gravitational canvas of quarks that anchors the watercolor. The problem is that if we dissolve the fabric of our world, smashing atoms at light speed, we might not be able to regain the watercolors once the quark-paper absorbs them—as we would have a hard time putting together the outer Russian doll if the method to study the inner doll is to stamp the bigger one, our world, into the floor, breaking it into infinite pieces. How this will be done is easier to understand when we consider how the two other layers, the chemical and nuclear layer, are evaporated by chemical and atomic bombs.

When we release chemical energy by breaking the molecular layer, as in vapor machines, we just transform that thin external layer of chemical bondage into energy *and the molecule loses weight.* According to the famous Einstein's equation, $E=Mc^2$, mass is a condensate of energy and so we can transform one into the other. And when we do so, we evaporate mass into pure energy.

Then if we go into the inner Russian doll and we break the thicker cover of an atom, we release around a million times more energy. It is the difference between a chemical explosive like TNT and a hydrogen bomb, whose energy is measured accordingly with a unit of power called a megaton (equivalent to a million tons of TNT). Now we can do this in two ways:

One is to fission and break in two parts the thick uranium cover, a very heavy atom. Those parts will later re-create two thinner covers, becoming two smaller atoms, while the rest of the cover mass evaporates into pure energy. It is the A-bomb.

But there is a more powerful way of converting the atomic cover into energy when we fuse two atomic covers into one. Since then, we convert an entire Russian doll cover into energy. It is the hydrogen bomb that converts several hydrogen atoms into a single helium one and blasts away the remaining cover. *And in both cases, the final atoms have less mass than the initial ones: $E=mc^2$.*

This explanation is the one physicists who need "solid" substances use. The explanation works for people because they also like substances. But the accurate explanation is to consider mass a cyclical motion that merely unwarps into energetic motion or warps energy into mass, $M=E/c^2$. And this second method of mass construction, a big crunch, is what CERN will do: instead of exploding an atomic cover at a time as it does a hydrogen bomb, it will explode them all into a big bang of energy to release its inner quark mass and make with it a quark-gluon liquid, a strangelet.

To do so, it will first break the cover of the smaller components of the atom, neutrons and protons, which are also a hollow Russian doll that keeps inside the ultimate smallish all-solid doll the quark vortices. So if we get rid of the proton cover, we unleash a lot of quark vortices.

Then those vortices will mass together and form a pole of attraction and will attract all other quarks of the planet, exploding all its covers in a dual mass reaction called a *supernova*.

One reaction will dissolve electrons and pions into pure energy, exploding the nuclei's cover converted into energy. This is the reaction we see as a supernova explosion that sends radiation around the Universe. It is a quark "big bang," which is millions of times more powerful than a hydrogen bomb. Indeed, a supernova is basically a quark bomb that explodes the last cover of electroweak mass, the pion, into universal energy, as the atomic bomb exploded up the energy of the previous layer and the chemical bomb dissolves into energy the most external outer layer of molecular forces.

But the Universe, unlike man, is dualist: it doesn't like to just destroy things. So the nova has a purpose: to quench all those quarks into a ball of quarks, a superfluid fractal vortex of quarks, which emerges as a cosmological body, a pulsar or black hole.

Thus, if things go according to previous experimental, cosmological evidence on novas and black hole births, CERN's quark cannon will be also a quark factory that will simplify our information to the bare basics. That is why Weinberg affirms that the nature of physics is to find the simplest elements of the Universe till it arrives to the absolute simplicity, the big bang and big crunch.

However, the finding of the simplest energy of the Universe is not the meaning of it all as fundamentalist physicists believe. It is more important to find the meaning of life and information because we human beings are, in fact, the most complex informative beings of the Universe.

It is for that reason this book alternates between the description of our death foretold by CERN and the joy of understanding the fractal Universe by Einstein et al., further explained in the appendix and hardly talks about the machine. CERN does the opposite. It talks about its totem, the machine; it denies that it will produce the quark-gluon liquid that will kill the Earth (strangelets); and it ignores all about the sweeping revolution of knowledge that the fractal paradigm signifies.

If life is defined by the balances and cyclical knots of energy and form, $E \times Ti = K$, death is the loss of balance of those knots, the rupture of those cycles, either by an excess of energy Max. $E \times 0\,i$, or by an excess of information, Max. $Ti \times 0\,E$. The first case is an energetic accident and it is what will happen at CERN, where energetic particles will kill us, erasing the information of our electromagnetic space-time, making us lose a dimension of form, converted in bidimensional quark vortices. The other form of death is old age, warping, informing, excessive curving.

In the Universe, both processes go together, creating a big bang of energy and a big crunch of form. That is what we observe at RHIC, where quarks cool down to zero temperature as they form a strangelet liquid and the remaining matter becomes pure radiation with temperatures recently measured at trillions of temperatures. LHC physicists don't understand how a perfect strange liquid can be so cold and ordered while the external radiation it produces is so hot. Because they can only understand physical energy. So they still think they create plasma, measuring only the external radiation without understanding the duality of all processes. In that process, our information, quarks, will evolve into more informative quarks, becoming colder, and our energy, the electronic cover, will further devolve into radiation, becoming hotter.

In the graph, when any physical system breaks, its energy and form becomes split, extending in a bidimensional wave of radiation and a perpendicular jet of information. This is described by Einstein's dual equations $E=mc^2$ and $M=e/c^2$. The bidimensionality of the c constant in those equations is easily explained as mass, a bidimensional vortex, becomes squared into a bidimensional, lineal sheet of light energy, c^2. Those are the equations of death of matter that CERN will explore without understanding them. But their experts do not want to learn the fractal paradigm, merely insult those "twats" who explained them the future of science, nonpeople who should be working as usual in patent offices for opposing the theory of ether (Einstein vs. German physicists, twentieth century) or burned in stakes for explaining the fractal multiplicity of stars and worlds (Bruno vs. Vatican physicists, seventeenth century) or be the laughingstock of the LOL method[Ch.9] for trying to warn mankind (Penrose, Rossler, Sancho, and Wagner vs. CERN, twenty-first century). In this manner, "authority" becomes truth and humanity becomes dust of space-time.

2. The Scales of Reality. The Equations of Death in the Universe.

Big bang processes occur in all the scales of the Universe. In all fractal species, death happens when the fields of energy and information break its fractal balance, E=T, and the arrows of energy and information dissociate again toward infinity, disconnecting the relative body and brain of the organism. It happens in biological species and it happens in physical particles when you cut the head and body of a system.

All organisms are defined by an "equation of organic balance" between its |-Body and O-Brain or energy and information systems. When the body and brain go out of balance, beyond their limits of energy and information, the species dies. So a human being dies when it goes beyond its limits of energetic temperature. While in the Universe, where the limit of energy speed is c and the limit of informative order is 0 K, no particle can cross those limits. Yet in a quantic Universe of multiple space-times, those are not the absolute limits of speed and temperature of reality, as Einstein said, but only the limits of our quantic light-based Universe. It seems that beyond our light Universe, it extends a "gravitational Universe" of dark energy and dark matter in which there are faster than light speeds and colder than 0 K temperatures. And we hint at its existence in cosmology through the study of black holes that seem to be the doors between both universes.

The three best-known cases of physical death are the following:

— The Lorentz transformations that evaporate particles as they come closer to c-speed (infinite energy) while its temporal parameters become zero in yet another partial case of the arrow of energy.
— Einstein's $E=Mc^2$ equation, which transforms the in/form/ation of charges into a bidimensional plane of light-space, c^2. Yet $E=Mc^2$ writes, using the Planck constants that measure actions, as Energy=Mass. And since a mass is an implosive knot of gravitational information (static perception) or an accelerated cycle of time (dynamic perception), we write that equation as $E/c^2 \Leftrightarrow M(t)$, which becomes a particular case of the generator equation of energy and form, $E \Leftrightarrow I$, explaining the fact that "lineal waves of energetic radiation" become "temporal vortices of gravitational mass" in informative processes while in atomic explosions mass dissolves into energy. That is, energy becomes "trans-formed" into a mass, a cyclical vortex of time and vice versa. Since in as much as a mass is a mere cyclical form, when the cycle straightens itself into a line, in a microinterval of time, it suddenly becomes pure, expanding energy ($T_{->0} \times E_{->\infty}$). Hence it is possible to describe atomic explosions and big bang explosions as an inversion of pure cyclical movement or mass into lineal, explosive movement or energy that should happen truly in a microsecond. Then any particle big bang becomes the explosion of a vortex of information that becomes gravitational and ¥-Radiation.
— All physical entities follow the same formal big bang death in different scales: When physical fractal points, masses or charges

on the atomic or cosmological level, die, their ext elements split. Then their temporal, gravitational, or electromagnetic central knot jets out its bits of information in the dimension of height while their energetic, external body of vital space explodes outward, expanding its outer membrane. In the cosmological big bang, a hyper black hole is theorized to have exploded and its energy to have given birth to our local Universe while stars produce big bangs called planetary nebulae that show clearly the morphology of the two basic non-Euclidean shapes of reality:[Appendix] a flat toroid of pure energy and a perpendicular, magnetic jet (right side) of information. If the membrane dissociates totally from the inner core, the fractal system dies away, but in certain cases the big bang maintains the structure of the blown-up membrane and then it becomes a big-banging a reproductive process that has converted a microform into a macroform with a bigger vital space, showing the iterative nature of the organic Universe. Those processes happen when a galaxy becomes a quasar and when a neutron explodes and creates an atom, made with a proton and electron (beta decay).

3. The Three Fractal Big Bangs: Organic Analysis

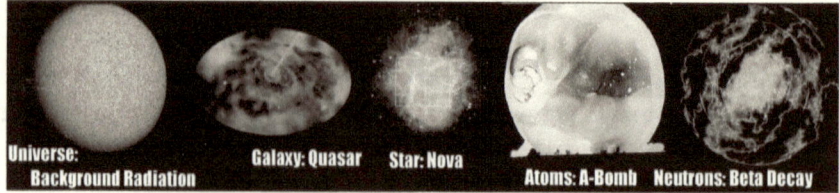

In the graph, the scales of big bangs of the Universe, caused by quark-gluon soups, the local Universe, a quasar/galactic big bang, a star/planetary big bang, created naturally inside stars at high density and artificially by CERN in "intelligent planets" full of dumb physicists who merely practice a destructive big bang in the tradition of a nuclear weapon. And finally the beta decay or mini big bang of quarks in a quantum scale. The difference between man-made big bangs (nuclear weapons) and nature's big bangs is that physicists merely blow up mass, creating energy and entropy, disorder and death. The dual big bangs of the Universe are also big crunches that balance both arrows of energy and information, leaving behind frozen stars, black holes, pulsars, or atomic nuclei.

As usual, we switch in this book between the mathematical and organic, logic treatment of all real phenomenon, giving finally experimental proofs;

thus fulfilling the 3 legs of truth of the scientific method, since the how of the Universe is mathematical, the why is bio-logic and the mixture of both, the spatial, geometrical perspective and the causal, logic one, given by both arrows of time are the experimental facts, events and entities of reality.

N the graph, in each hierarchy of the known-known Universe (quantum particles, stars and galaxies) and perhaps beyond (super-clusters of galaxies, aka, the local universe), there are relative, self-similar big bangs and big-crunches of energy into mass:

— In the quantum scale we observe a mini-big bang called a beta-decay, in which a dense nucleon that occupies minimal space (a neutron), implodes into an even denser particle, the proton and explodes into an outer electro-weak one, the electron.
— In the planetary/star scale, we observe a big bang called a Nova in which a star implodes into a denser cosmological body, a quark star or top black hole and explodes into a nebula of radiation.
— In the galactic scale we observe a super-big bang called a quasar in which nebulae of interstellar gas implodes into a black hole and explodes into a spiral of stars.
— And perhaps in a cosmological scale, millions of galaxies collapse into a hyper-black hole and explode into interstellar gas.

All those scales of self-similar fractal dual processes of creation of physical energy and information, are not understood by quantum theorists, *because of the 2 scientific errors exposed, before, pythagorism (in this case a single space-time continuum that imitates the Cartesian graph) and monist mechanism (which means null understanding of the dual arrows of form and energy, of life and death).* Yet reality is always more complex than the mathematical or monist, entropy, mechanist images physicists have of it. In that sense, the fractal nature of those big bangs brings 2 facts that quantum cosmologists deny:

Quantum theories of particles do not apply to cosmology. Since 'self-similarity' is defined by the 3rd postulate of non-Euclidean geometry[Appendix]. It is NOT equality, as physicists believe: the different scales of the Organic universe are self-similar but never equal. And so we can only compare them, knowing there are changes in 'energy and information' parameters between the different Universal Constants, ExT=K, of all those systems.

For example, let us consider the big bang self-similarity between the 2 fractal scales we know better, the atom and the galaxy, established by the Unification equation and the parallelism between the event horizon of the atom (Bohr radius) and the event horizon of the black hole[ch.3]. The 3 legs of the scientific method prove that self-similarity:

It has experimental evidence, it is logically coherent and it is mathematically consistent, as shown by the fractal equation of unification. Does this mean a galaxy is an atom? No. It is not. In Relativity we use galaxies as atoms to describe the cosmological space. In astronomy we use models of liquid electrons made of photonic nebulae to describe stars around black holes.

But the question of equality between atoms and galaxies is a philosophical question and the quantum models used to study galaxies as atoms, approximations to the truth, better achieved with direct analysis of the cosmological world. Further on that self-similarity means that we do not need to risk the Earth testing the energies of a supposed cosmological big bang to know more about the Universe, if those big bangs are self-similar. The study of the smallest of them, the beta decay of an atom, would bring self-similar knowledge. This fact is based on the fundamental principle of relativity, denied by Hawking's musings: size doesn't mater. It is absolutely relative. If we were of the size of a galaxy, our big bang would be observed as we observe the mini big bang of an atom. And we know those facts are more truth than the ideas of quantum cosmologists (equality) because according to the three legs of the scientific method, the fractal paradigm is more truth than quantum cosmology, which has neither logic, nor solid experimental proofs, and faulty mathematics.

What matters is to understand the self-similarities and differences of those scales, which are given by the universal constants that difference them, based on the generator equation, $e \times i = k$. This means that the more extended in space a vortex is, Max. E, the slower it turns its cyclical frequency of information, i. So smaller beings have time cycles which are faster than bigger species and all of them have a k-constant product of their energy and form, which make all beings self-similar in their "experience of life." An ant queen lives 10 times less than a human, but it processes time 10 times faster so it perceives the same among of subjective time/information through its life. This is the amazing truth of reality: we are all indeed, self-similar beings.

The constant equality of $e \times i = k$ makes "bigger species" to have a slower "metabolic" time. So rats' hearts beat faster than men whose hearts beat faster than elephants'.

In the graph, from left to right, from the biggest to the smallest mass, we show the self-similar big bangs of the different scales of fractal matter. Each of those scales has a faster rotational attractive speed. A galactic quasar has a thirteen-billion-year cycle (the supposed age of the cosmic big bang that turns out to be a quasar big bang as we shall see later), and an atomic beta decay is a fifteen-minute cycle.

Let us briefly describe them from left to right:

— The hypothetical universal big bang that created the background radiation. It is probably a false, hyperbolic error as all seems to indicate that radiation is local, galactic. So the image we observe is merely the image of the background radiation of our galaxy, proved further by the structure of those maps, with a cooler big black hole mass in the center, a hotter line across it, corresponding to the Milky Way stars and a quadrupole deviation, produced by the movement of the sun-earth through the galaxy as the satellite takes measures.

— Thus, the map belongs to the next fractal big bang, the galactic big bang of a quasar, which creates a halo of dark matter at the same time that forms big-crunch black holes, given the duality of all systems of the Universe.

— Next comes the stellar big bang that creates a nova, whose wave catalyzes the creation of new stars and forms a strange star or black hole in the dual big crunch.

— Next we see the energetic big bang of an A-bomb, which unlike the other big bangs doesn't create a stable organism (a big-crunch form of matter), as it is a man-made dull weapon.

— Finally, the neutron big bang that creates and electronic membrane through a beta-decay process and a micro black hole of the quantum scale, a proton.

Thus, we see that all the big bangs are accompanied by a big crunch to respect the generator equation, $E \times i = K$, creating a denser informative center and a more expanded energetic form. Thus, they are actually creative processes except the nuclear physicists' pathetic attempts to play God (A-bomb).

What CERN will do is to replicate *on the Earth* one of the many fractal big bang explosions of the Universe, re-creating the *quark-gluon soup that produces supernovas*, NOT the big bang of the whole cosmos, as the Universe is infinite in size. The big bang of the Earth by strangelets or micro black holes will absorb all the matter of this planet, provoking according to the duality of energy and information, a big crunch of our matter into a quark star or a black hole, an ultradense object of mass/information, and also a big bang of the remaining electroweak mass into energetic radiation: big bang of Earth by CERN (2010–2013) = informative quark star + electroweak radiation

We thus come to the conclusion that the big bang, big crunch wave/particle duality of the cosmilogical and quantum world are self-similar life-death cycles. Or in other words, there are big bangs of all sizes in the

Universe and its equivalent big crunches *with different speeds of energy and rotational speeds of information that grow as we grow the space-time we study in decametric scales.*

So there are big bangs of atoms called beta decays and its big crunches, which CERN will essay collapsing atoms into quarks ($E=Mc^2$ + $M=e/c^2$); big bangs of stars called supernovas, coupled with the birth in its interior of a black hole or quark star; big bangs of galaxies called quasars, coupled with the creation of matter in ultrarelativistic jets ejected by the central black hole at superluminal speeds; and maybe, only maybe a big bang of a supercluster of galaxies, which would be the cosmological big bang of which there are unfortunately less and less proofs. This cosmological big bang would require red-shiftings of $z=100$ and the existence of a super black hole.

Indeed, a second theme of great interest, related to the decametric growth of speed/space/distance (static/dynamic perception of energy/ space according to the paradox of Galileo) is the explanation of why space seems to expand between galaxies up to 10 c speeds. This expansion is balanced by the contraction of galaxies that implode energy into mass up to 10 c speeds in top quarks and black holes. Thus the acceleration of interstellar space—a.k.a. dark energy and gravitational, repulsive forces—is the "polar jet" expulsion of dark energy by black holes and quarks rotating at 10 c speed. That cosmological acceleration in a continuum space must be understood in fractal space-time as the local motion of galaxies that eject dark energy through the poles of their black holes, according to the paradox of Galileo (all can be perceived as motion or static space/distance). Indeed in fractal space-time, galaxies implode space into form, masses and black hole vortices while black holes "vomit" dark energy at $c < v < 10c$. And the sum of all fractal expansions and implosions balance the Universe in an infinite steady state. This means that the expansion of space is fractal and so it is easier to consider that galaxies red shifting at $z=10c$ do move at $z=10$ c since the speed of light is only the limit for our membrane of space-time, but NOT of the bigger fractal scale.

4. Experimental Evidence:
The Falsity of the Cosmological Big Bang

Science News　　　　　　　　　　　　🔗 Share 　✏ Blog 　💬 Cite

Big Bang's Afterglow Fails Intergalactic 'Shadow' Test

ScienceDaily (Sep. 5, 2006) — The apparent absence of shadows where shadows were expected to be is raising new questions about the faint glow of microwave radiation once hailed as proof that the universe was created by a "Big Bang."

See also:

Space & Time
- Cosmic Rays
- Big Bang
- Cosmology
- Astrophysics
- Galaxies

In a finding sure to cause controversy, scientists at The University of Alabama in Huntsville (UAH) found a lack of evidence of shadows from "nearby" clusters of galaxies using new, highly accurate measurements of the cosmic microwave background.

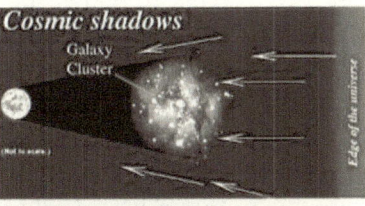

Cosmic shadows

If the standard model of how the universe was formed is correct, microwave radiation from the edges of the universe would be blocked by clusters of galaxies, causing 'shadows' in the microwave background. (Graphic courtesy of The University Of Alabama In Huntsville)

A GALAXY REINVENTS ITSELF

Astronomers used to regard bars and spirals as permanent features of a galaxy but now think that they come and go. The gravitational processes that make a bar ultimately destroy it and then create it anew, as this simulation shows.

START ▶▶
The galaxy is born as an amorphous disk of stars, gas and dust.

2 BILLION YEARS ▶▶
Bar and spiral waves develop. Gas trickles in from intergalactic space and will double the disk mass in 6.5 billion years.

5 BILLION YEARS ▶▶
The waves strengthen. The bar sweeps up gas near the core but holds intergalactic gas at bay.

8 BILLION YEARS ▶▶
Gas accumulating in the core begins to tear apart the bar.

11 BILLION YEARS ▶▶
The bar is history. No longer held back by its torques, the intergalactic gas that had been lingering in the galactic outskirts pours in.

14 BILLION YEARS ▶▶
The bar reemerges. Gas infall is crucial: in simulations without it, the bar, once gone, never comes back.

17 BILLION YEARS ▶▶
As before, the bar starts to peter out.

20 BILLION YEARS ▶▶
The bar is all but gone. In simulations under different conditions, bars form and vanish more quickly.

Now that we have talked about the mathematical and logical structure of big bangs, we can return to the experimental proofs—the Pietronero study of the Universe as a fractal denied by Hoggs, an obsolete astrophysicist of the cosmic big bang continuous paradigm, who denied the experimental evidence of Hoggs because it didn't learn fractal physics in school.[ch.2] It is all truth, Mr. Hoggs. Space-time is not a continuum and so we have indeed to throw, as you rightly said, first the big bang and then the expansion of the Universe. We just have done it and we are still all here, and all has been explained and all looks even more beautiful, eternal, in motion, imploding, and exploding at the same time, in infinite fractal events. Not big deal, Mr. Hoggs. Unfortunately, I'm afraid that in the same manner Hoggs never listened to Pietronero, CERN will never listen to Einstein et al. and implode and explode us all.

What are the proofs of the cosmic big bang? None, really: Gamow's hyperbolic, explosive thought made him think that the local background radiation of the galaxy measured by Penzias was a universal radiation while the energetic error made him and Oppenheimer, as a maker of A-bombs, think that all was born of an explosion, reason why Hoyle laughed at him and called his theory the big bang.

Indeed, the cosmological big bang is an energy-only theory of the Universe, proposed by the makers of nuclear weapons, which has failed the fundamental proof of the scientific method, experimental evidence. In the graph, we see two of the many proofs of falsity of the big bang:

— The dates of the big bang must constantly be put backward as we discover farther away galaxies perfectly formed. So the theory is constantly adapted ad hoc to the new discoveries. In the graph, the models of formation of galaxies give a cycle of 20 billion years much older than the cosmic big bang (13 billion). Since we have found galaxies at a distance of 12.5 billion light-years, perfectly formed with bars that take 5 billion years to form, and enormous black holes that take even longer to acquire their mass, they would have been created before the big bang.

— Its main proof, the existence of a cosmic radiation that permeates the entire Universe, remnant of that explosion, is false since BG radiation doesn't leave shadows when it crosses galaxies. So it cannot come from the remote limits of the Universe, whose light is so far away that it takes billions of years to arrive to this galaxy and hence come from the past supposed big bang explosion. Thus the BG radiation must be produced by local, celestial bodies in the halo of those galaxies. Yet the only celestial body that can produce such radiation is a micro black hole or ultracold quark star called a MACHO in the jargon of astronomers, which should therefore be a very common object and could easily be produced at CERN. Because this fact invalidates CERN's mainly safety argument—that in the Universe our matter is not constantly transformed into black holes, we will consider it in detail. These were the news taken from a respected Web site dedicated to science:

Big Bangs Afterglow Fails Intergalactic Shadow Test

ScienceDaily (Sep. 5, 2006)—The apparent absence of shadows where shadows were expected to be is raising new questions about the faint glow of microwave radiation once hailed as proof that the universe was created by a "Big Bang.": The big bang is local. Source: University of Alabama; Sci American.

If the standard model of how the universe was formed is correct, microwave radiation from the edges of the universe would be blocked by clusters of galaxies, causing shadows in the microwave background. In a finding sure to cause controversy, scientists at The University of Alabama in Huntsville (UAH) found a lack of evidence of shadows from "nearby" clusters of galaxies using new, highly accurate measurements of the cosmic microwave background.

A team of UAH scientists led by Dr. Richard Lieu, a professor of physics, used data from NASA's Wilkinson Microwave Anisotropy Probe (WMAP) to scan the cosmic microwave background for shadows caused by 31 clusters of galaxies.

"These shadows are a well-known thing that has been predicted for years," said Lieu. "This is the only direct method of determining the distance to the origin of the cosmic microwave background. Up to now, all the evidence that it originated from as far back in time as the Big Bang fireball has been circumstantial.

"If you see a shadow, however, it means the radiation comes from behind the cluster. If you don't see a shadow, then you have something of a problem.

"Either it (the microwave background) isn't coming from behind the clusters, which means the Big Bang is blown away, or . . . there is something else going on," said Lieu. "One possibility is to say the clusters themselves are microwave emitting sources, either from an embedded point source or from a halo of microwave-emitting material that is part of the cluster environment."

Let us calculate with the three legs of the scientific method what degree of truth might have a theory, which are the following:

— Experimentally false, as it has nothing to do with 96 percent of reality, which is dark quark matter, and it has failed the age test and the afterglow test.

— It is made with ad hoc mathematics (Mr. Gamow got a background radiation on its original equations of 20 K, which later he manually changed to 2.7 K), and the dates of the big bang are changing backward as we find older galaxies.

— It is logically absurd since as Parmenides proved logically already 2,500 years ago, "something cannot be born of nothing." Further on, as the definition of temperature proves, something small must be cold because temperature is a measure of lineal motion, which is synonymous of space; and yet the singularity is extremely hot and extremely small. Thus the hot-singularity A-bomb-like big bang is logically false. The quark-gluon liquid must be described as we do with complex liquids, as a dual explosion/implosion, a refrigerator system that emits radiation and cools inside the quark strangelet.

Add three errors in the three legs of the scientific method—0 mathematical consistency, 0 experimental proofs, and 0 logical consistency—and you get 0 percent chances that the cosmological big bang, as it is formulated today, is truth. The cosmological big bang is an ancient theory of the past millennium, formulated at a time in which the main component of the Universe, dark quark matter, was ignored both as a part of the atom and a part of the Universe and only one arrow of time, the energetic arrow, was known. And so once the big bang is proved false, the expansion of the vacuum is compensated by the implosion of space into mass in galaxies, all that is needed to find a possible cause of the background radiation in the halo of galaxies. But what possibly could produce from the halo of the galaxy the background radiation? As we said before, a black hole the size of a moon, gravitationally redshifting light. Let us consider this final nail in the coffin of the big bang in more detail.

5. MACHOs of Dark Matter. The Background Radiation

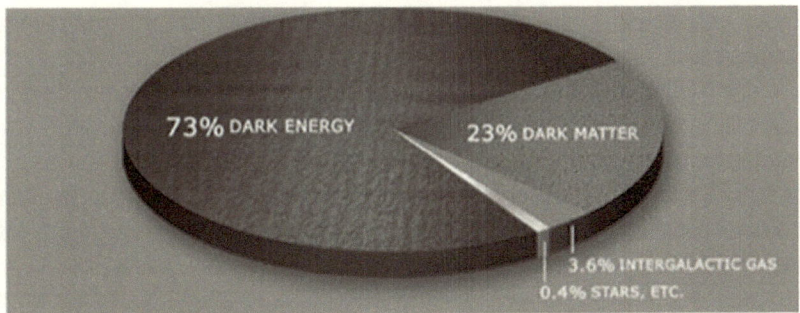

MACHO Science

From July 1994 to December 1999, the MACHO project, led by Penn astrophysicist Charles Alcock, had dedicated use of the 50-inch reflector at Mt. Stromlo Observatory in Australia. The project was a high-tech hunt for big chunks of dark matter (MAssive Compact Halo Objects) surrounding our galaxy, the Milky Way. "There's a tremendous amount of dark matter out there," says Alcock. "In fact, most of the matter in the universe is dark." Many of the project's important discoveries came from observations of the Large Magellanic Cloud, a dwarf galaxy orbiting inside the Milky Way's dark-matter halo. The cloud can only be viewed from the southern hemisphere.

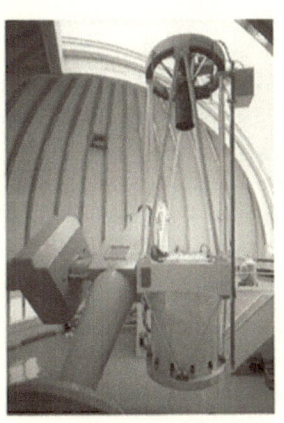

Dark matter is far more abundant than light matter, the component of planets and stars. Thus, we are not the dominant form of matter in the Universe. Since quark stars and black holes, called MACHOS (massive halo objects, of enormous attractive power), producing 2.7 K radiation, are the only known objects that can form dark matter, the LHC, a quark dark factory could easily cause a chain reaction that would convert our light matter into the dominant substance of the cosmos. Further on, the big bang experiments at CERN are redundant as we have specialized telescopes that research directly the halo of dark matter. Yet CERN, instead of acknowledging the danger of replicating aggressive, hyperattractive seeds of quark-gluon MACHOs on Earth, has cherry-picked as components of dark matter the so-called WIMPs—imaginary, inoffensive particles—which it says the LHC will produce. Yet there is null astronomical evidence of those particles, except in the fantasy equations of its quantum inventors.

As it turns out again, only the fractal model of the Universe gives an alternative answer to the origin of the background radiation: black holes. Indeed, only black holes can produce by gravitational red shifting, given its enormous attractive power, the background radiation previously considered the signature of the big bang. This fact, later studied in more detail, is known to most physicists as Einstein's relativity theory was proved when Eddington measured the gravitational deviation light suffered when passing closer to a star. In those processes light loses energy, cooling down in a process called red shift. The heavier the mass, the more the light cools down. So *the biggest cooling and elongation of a ray of light happens when it passes closer to a black hole.* Einstein's formula allows us to obtain the mass a black hole must have to cool down and redshift/elongate light to the exact temperature of the background radiation. And it turns out that a black hole will have around the mass of a moon to produce that BG radiation. The solution is consistent with the fact that the most common object in the Universe is a moon. Our solar system has around twenty of them, orbiting close to planets and forming a defensive shield called the Oort ring in the outskirts of the sun. Furthermore, micro black holes like those produced by CERN will be of that mass and they are predicted by astronomers to form the dark matter that surrounds galaxies, *where the background radiation must be originated according to the afterglow test.*

What all this means is obvious. In the same manner lions eat gazelles and humans eat pork, electrons feed on light and micro black holes feed on moons. In the organic, fractal paradigm, any system has a three-layer structure, as you are made of cells, live as an organism, and participate of a higher social structure. And each organism has also three layers, your external energetic skin, your informative central system, and your intermediate reproductive organs.

This happens in galaxies with an external halo of hard, dark matter, an inner black hole, and an intermediate series of stars that reproduce them. *Galaxies are dominated by black holes and quark stars of dark matter. We are their food. We matter nothing to the Universe.* Your hard skin or the protein membrane of a cell is self-similar to the halo of micro black holes of dark matter in the galaxy. It is self-similar to the Oort layer of solar systems, whose moons are food for dark MACHOs. So what probably happens in galaxies, whose center is occupied by huge black holes and its halo is occupied by small black holes called MACHOs is obvious: MACHOs feed on moons and big black holes on stars. In a cell, the proteins of the cover are formed by starlike furnaces called mitochondria.

Of course, this organic description of a galaxy is NOT and will never be accepted by mechanist physicists, like those working at CERN, despite mounting evidence of a dual complex fractal organic structure in galaxies.

This is not what they seek for. So CERN says in its "safety statements" that "black holes never eat moons" and so we are safe because in billions of years, black holes never ate our moon. Never mind the most famous experiment of twentieth-century science was about gravitational lensing and proved Einstein's relativity. Quantum physicists do not believe in Einstein. Never mind the very same Hawking in his first paper who declared that the halo of the galaxy was made of billions of black holes. Because we sent satellites to see if they evaporated and none of them did but all of them are red shifting light, now he is mute NOT to deny his false theory. Never mind that Einstein et al. has written letters on this alternative fractal BG radiation theory to CERN, to the *NY Times*, to *Physical Review*, even tried to put this information on Wikipedia: CERN's physicists at 8:00 a.m., Geneva time, would erase the information in *Wikipedia*. *Physical Review* failed to publish. The dogma is that "moons never become black holes." Because if they do, we are dead, as CERN will make black holes and they will eat the Earth first and then the moon. Thus the fractal paradigm is an inconvenient truth for CERN. But here we are concerned with truth, not with the stubborn denial of truth by CERN and its control of the mass media and establishment, which together will bring the collective suicidal of the species for NOT wanting to understand the harmony of the cosmos, a self-similar fractal in all its scales.

It is that recent understanding of the Universe as a fractal made of networks of cellular bites of spatial energy and bits of temporal information what makes the creation of a big bang on Earth an absurd experiment, which will not reveal the meaning of it all but could easily kill the Earth and leave behind a nebulae of dust of space-time and a dense black hole.

Indeed, the existence of self-similar black holes (Bohr and Schwarzschild radius), forces and structures in the three known-known physical scales of the Universe, the atomic world, the star world, and the galactic world *with different constants of space and time*, has answered properly a fundamental debate of modern cosmology: at which scale did the big bang cause the background radiation? At first glance, it is very reassuring for the intelligence of mankind to think that the big bang happened at the cosmological scale. So nuclear physics can finally become metaphysics, its eternal ambition and substitute philosophy cosmology, and religion at the summit of human knowledge. *But while the mathematical equations of any big bang are all coherent just by merely choosing bigger or smaller parameters for those equations, there are two other legs of the scientific method, experimental and logical consistency, which strongly favor the local, galactic nature of that radiation.* In that regard, the last empirical data collected by WMAP[2] favors a possible local origin to that radiation: the first multipole of that radiation aligns either with the equinoxes or the movement of the solar system or the galactic plane in which Andromeda and our galaxy are placed. Thus the background radiation should be the "local temperature" of the dual galactic system in which we exist.

The Universe is fractal and perception is limited. So we should limit human arrogance and consider most phenomena observed from the perspective of Earth to be local, instead of extrapolating our measures to include the entire Universe. This was in fact the initial theory of Penzias, who was trying to measure the background radiation of the galaxy. Only the energetic bias of A-bomb maker Gamow and the "metaphysical," religious profession of his other inventor, the priest Mr. Lamaitre, coupled with the natural arrogance of human beings, justify to "jump the scale" we measure to infinity.

And yet today the growing distances observed in the Universe gives us a cosmos of at least 140.000 million light-years of diameter—10 times the size of the big bang.

Thus, the fractal big bang concept seems all the more attractive as is all the contradictions of the universal big bang. It merely requires explaining two phenomena: the acceleration of light-space between galaxies, which fractal quantic space-time explains locally and can be balanced by a similar number of imploding processes that create matter. And the background radiation, which would be the isothermal temperature of the galaxy polarized by the halo of quark, dark matter MACHOS that surrounds as a discontinuous membrane all galaxies. And this is the terrifying next fact that makes so dangerous CERN's experiments. Because if the big bang is local, then the background radiation is produced by strangelets and black holes in the halo, which has eaten planets, moons, and stars. And so what CERN is going to do happens all the time. Since those micro black holes accumulate in the halo around galaxies where dark matter is more abundant, they should be the cause of the uniformity of the background radiation. A halo of dark matter, made of micro black holes will red-shift light as a perfect black body, isolating and protecting the galaxy, keeping it warm at a temperature equal to its 2.7 K surface temperature and creating a complex organic system similar to a cell. Those MACHOs *would create a local background radiation with small variations, equal to the observed WMAP, explaining also why there are no shadows in superclusters (the radiation comes from the halo), why there are variations of the background radiation in the scale of those mini black holes, and why there are variations on the equinox plane of our galaxy.*

Thus, in organic terms, the background temperature of the galaxy is caused by the halo of dark matter, the external membrane of that galaxy which red shifts its light. Since we are observers that exist inside a closed, galactic organism, which acts as a relative black body, we sense that radiation coming apparently from all the regions of the Universe even if it is coming from our galaxy. Yet light coming from outside the Universe is also being red shifted by the halo of dark holes that maintains the same temperature in the galactic organism.

This doesn't rule the existence of a possible hyper universe, but the energy of that hyper universe is not light radiation. The energy/space of the intergalactic world is gravitational dark energy, at 10 c. In that regard, there might be infinite scales and big bangs, but each one will have, as anything in the Universe, a decametric, scalar configuration. So if there is a superdark energy in the scale of the hyper universe, it should follow the decametric scales of the fractal game of space-time. So if in the big bangs of quasars, matter seems to be expelled at 10 times the speed of light, in the borders of the hypothetical cellular Universe theoretical models should consider red shifts up to 100 times the speed of light, produced by a superdark energy 100 times faster than light. It would be the signature of a hyper universe made of galaxies of which so far there is no evidence at all.

In that regard, what is true science and true research is NOT *making those big bangs on Earth but to observe the Universe directly with telescopes and satellites.*

Let us consider an example the Americans had set when they cancelled the supercollider in Texas and built instead a supertelescope to observe the Universe.

6. The Good and Bad fruits of the Tree of Science: CERN vs. Webb Telescope

On the left, the good fruit of the tree of science: the Webb telescope. On the right, the bad fruit, the Large Hadron Collider, a twenty-seven-kilometer circular seven-terabyte, superfluid, c-speed quark cannon. The Webb telescope substitutes the first quark cannon, designed to be built in Texas and later cancelled by the American Congress to study the big bang at no risk for the Earth. Europe should follow the example set about by Mr. Clinton—cancel the supercollider and invest in the good fruits of the tree of science: astronomy, civil engineering, and human biological sciences.

In the graph, the duality of the fruits of the tree of science is evident when we compare two machines built for the same purpose: one is an instrument of research, the Webb telescope; the other is a weapon, a large quark cannon, disguised as an instrument of research called the Large Hadron Collider. So we need to translate those words: a hadron is an atom with a huge number of quarks and a collider is an electronic cannon that will break those atoms and liberate its quarks. The main excuse for building the quark cannon is to study the quark-gluon soup or quark condensate, which according to quantum cosmology gave birth to the Universe in a big bang, when the creation of that soup provoked a cosmic explosion that expanded the space-time of the cosmos trillions of times and according to fractal relativity is the cause of smaller big bang explosions and quasars that destroy planets, stars, and galaxies.

In any case, in both theories, quantum cosmology and fractal relativity, a small quantity of that quark-gluon soup, the most dangerous substance of the Universe, seems to provoke supernova explosions. Thus, it is logic to deduce that to make a tiny fraction of that substance on Earth could provoke a big bang strong enough to explode the planet. And yet this is denied by CERN to avoid its foreclosure. What matters, they claim, is to obtain knowledge about the big bang explosion at all costs.

Three points refute this argument:

If the cosmological big bang that exploded space-time occurred, the most likely outcome is that it will occur here and nothing will be learned as we all will die. If the big bang is smaller, a local quasar that explodes a galactic black hole or a nova that explodes a star, as in the fractal models of the Universe, it will be equally dangerous as its reproduction here on Earth will convert the planet into a nova.

If the big bang is not truth, it is unneeded to replicate it on Earth.

The study of the big bang on Earth is redundant since what matters is to observe directly the Universe in order to find out if *there was or not a cosmological big bang*. In fact, NASA is exactly doing this with zero risk to big-bang the Earth, using a new telescope, which in 2014 will replace Hubble and will study the Universe in the theorized moment of its big bang. This is possible because the Webb telescope will watch galaxies, which are 13 billion light-years far away. Since their light takes 13 billion years to arrive here, those galaxies will be seen as they were 13 billion years ago, when the big bang theoretically started. So the Webb will observe light that was emitted 13 billion years ago, giving us images of the real big bang, not the one manufactured on Earth, blowing our planet. On the other hand, if the Webb telescope finds that 13 billion years ago galaxies looked exactly like they are today, it will prove the fractal, hierarchical structure of the infinite Universe. And the cosmological big bang theory will return to its initial origins when the discoverers of the background radiation, Penzias and Wilson, affirmed that the big bang radiation was local, produced in the galaxy by dense MACHO objects, as the one the Earth might be then. Thus, it is the Webb telescope, not the quark cannon, that will do "real science," truly advancing our understanding of the Universe with "experimental proofs" without risking the Earth, fulfilling the two goals of science, true knowledge, and the protection of human life. Indeed, in NASA's Web we read:

> The science goals for the JWST can be grouped into 4 themes:
>
> *The end of the dark ages*: first light and reonization seeks to identify the first bright objects that formed in the early Universe, and follow the ionization history.
>
> *Assembly of Galaxies* will determine how galaxies and dark matter, including gas, stars, metals, physical structures (like spiral arms) and active nuclei evolved to the present day.
>
> *The birth of stars and proto-planetary systems* focuses on the birth and early development of stars and the formation of planets.
>
> *Planetary systems and the origins of Life* studies the physical and chemical properties of solar systems (including our own) and where the building blocks of life may be present.

As NASA says, "The James Webb Space Telescope (JWST) will be a giant leap forward in our quest to understand the Universe and our origins. The JWST will examine every phase of cosmic history: from the first luminous glows after the Big Bang to the formation of galaxies, stars, and planets to the evolution of our own solar system."

The Webb telescope costs 4 billion dollars in edge technologies. However it is completely justified, because it has no military application. It cannot blow up the Earth unlike the quark cannon, which merely will make quark condensates that astrophysicists theorize were the substance that exploded in the big bang. So it is self-evident that if the original quark-gluon soup did have those enormous explosive properties, both in the theorized big bang or in reduced quantities, as the cause of nova explosions that create pulsars and quark stars, CERN will do the same if it re-creates it on Earth, under the *totalitarian principle*—the Murphy law of physics that states, "All particles that are not forbidden by the laws of physics happen."

That is why we are not exaggerating when we say that CERN has built a doomsday weapon, disguised as a tool of research, an evil=anti-life fruit of the tree of science that politicians and the military should prune for the safety of mankind.

Yet under the ethics of a technological civilization this will not happen, because if we can do a machine we 'must do it, even if it menaces to destroy us all' (Fromm).

It must be said though that for the Webb to be of any use, as an experimental instrument, first astrophysicists must understand the differences between the cosmic big bang, which is false in its present 'Singularity' formulation and the fractal big bangs, which are many, happening in different scales. They also have to include a 2nd arrow of 'big-crunches' of energy into vortices of information in all the scales of the Universe. Since the Webb won't be useful to prove the existence of the cosmological big bang, which never existed and it is proved false, but on the contrary to finally disprove it, by finding that the landscape of the Universe is the same and extends beyond the supposed age of the big bang at 13.7 billion years; or in case it finds a 'big-ball' of fire, to prove we are part of a Universal cell and there is a hyper-universe made of many self-similar ones.

Thus, that quest for understanding the Universe can only be done with telescopes and satellites NOT with quark cannons. Indeed, after the Hubble proved that the Universe is a fractal made of self-similar structures of energy and information in multiple scales with the form of a spiral *vortex of mass, either a particle or a galaxy or a cluster*, we do not need a cannon to make a big explosion on Earth and pretend it is the replica of the Universe's initial explosion but a telescope to resolve once and for all if the Universe is a fractal and so the cosmic big bang never happened.

If a telescope finds, stable galaxies, at more than 13 billion years-light, beyond the birth of the Universe, obviously there is no big bang, and the Universe is indeed a fractal of energy and information, not an explosive bomb, a biased theory introduced by Physicists, because of their worldly profession, to make weapons.

Arguably this proof came last month, when the Hubble brought a picture taken from galaxies found 12.5 billion years light away. This mean the image is from light emitted 12.5 billion years, just 500 million years after the big bang. And we see in that picture perfectly formed galaxies. We have also found an enormous galactic black hole of a similar age. Yet according to standard astrophysics they could have not been formed so fast. So what big bang theorists are doing is all kind of gymnastics and ad hoc solutions to make the Universe older, to send backwards its equations, etc.

Why? Of course, because of money and prestige, which made the big bang theory a nuclear dogma: The LHC – the big bang machine—becomes a perfect tool of marketing to give prestige to Nuclear Physicists, which are no longer considered weapons makers, but now are portrayed as idealist scientists, who found the meaning for it ALL: our energetic Death. The rat bites its tail . . .

The reader could think is our fault – that of complex theorists – for not explaining them properly the facts of new physics. But that is not how paradigms in science change. It would be like blaming Einstein for spending 10 years in a patent office, banned by ether theorists from doing physics, or blaming Galileo for being under house arrest till his death, or blaming Leibniz for dying alone after working in a genealogy of princes to earn a living or blaming Bruno for being burned on the stakes. Even if our science, Complexity, physicists from the Los Alamos Laboratory of Nuclear Weapons and Computers scientists, mainly robotists, specialized in cybernetics run the show.

The machine and the industrial system of computers run today all sciences

The human mind, which Einstein et al represent is no longer trusted. Bottom line is that humans are becoming more like ants than free minds. In an anthill the individual is always powerless because the systems, the paths of automated behavior are much stronger. The *intelligence of the anthill is in the chemical paths. It is automated. That is the computerized world we have built. Today the intelligence of science is not in the humans but in the automated systems, the computers, which imply you must do data science NOT logic evolution of ideas, concepts, as 'Einstein et al' always do. The previous chapters full of insights in a new dawn of science are* a clear example. The evolution of our understanding of time and its events with 2 arrows (information and energy) is a leap forward far bigger than anything CERN will ever do. But this revolution which has been with us for more than a decade never impressed scientists as much as the data and pictures manufactured by computers. When I used to go to complexity congresses the competition for the 'title' of the man who had resolved the 'unification theory of all sciences', which is what we search in that science was between our organization, ISSS, and the Santa Fe Institute of Complexity ran by 2 physicists, Gell-Mann and West, who had a battery of computers searching for power laws. They never found anything in 20 years despite the millions of $ they have for research. I was a newcomer who had a logic dualist model, from which I could correct and solve most of the laws and errors of all sciences. For 4 years I gave conferences there as the Chair of Duality and people loved them, but the people of Santa Fe and the group of California University, which used massive computer power never accepted Duality. They were serious, formal, very much into their jargons that nobody understood, going to the American Science association, with huge egos, writing big books with catchy names, like the quark and the tiger. But they never resolved anything; they never explained as I do in this book, why we live and die, why the weak force is a time event, why charges and masses can be unified, why information is fractal and how its dimensions are created. But they were very good at programming computers and making huge exhibits with a lot of data that compares births of stars and births of cells, structures of Tokyo subways and cellular networks and so on. They were making how pictures but never explaining the why. And yet nobody wanted to know really the why. For scientists today to see 'the how pictures' is knowledge. Knowledge is merely to transfer what you see into the digital language of the computer. And the scientist merely transfers data from reality, after observing it with machines, into computer banks. But his language, which is logic, verbal thought, remains in the twenty-first century paradigm of entropy ad probability.

After I put the suit to CERN, the ad hominem campaign of physicists against us ended my career in complexity. After all, places like the Santa Fe Institute handle the money of this science and they are run by physicists that work with accelerators. And that is how things should be. Humans, life, organisms, languages, minds are today the pariahs of our technological societies. We humans will NOT be trusted again to do a Copernican revolution by the ants of the computerized system we have created to rule us. CERN must be experts because they run computers. Santa Fe must be the center of complexity because it runs computers. Einstein et al. must be crackpots because they do "thought experiments." They talk to the mind. They see a spiritual, living Universe. They seek the thoughts of God, not "its details."

II. Singularities and Evaporating Black Holes

Hawking's work is not good enough.
—Higgs, theologist

We never found a black hole evaporating, *pity if we had.*
I'd get a Nobel Prize. Now CERN will give me a second chance.
—Hawking, sci-fi writer

7. Fantaphysics

Mr. Hawking's theory of black holes, called quantum entropy, and quantum physicists' ideas about a single arrow of entropy=time are a reductionist, monist, false ideology of the Universe, which ignores completely the informative, cyclical nature of masses and the role of black holes as the entities that balance the two motions of the Universe, creating physical mass and dark energy, which expands space between galaxies. If they understood those roles, probably they would not be making black holes and big bangs on Earth, as they would know that the arrow of information and mass will provoke simultaneously a big crunch of our planet that will convert our electroweak matter into a lump of quark mass. This happens all over the Universe so it should happen here since in the dual Universe, both arrows occur simultaneously (principle of complementarity). Thus, in the same way, all particles of information have associated to them a field of energy; when there is a big bang, a massive explosion and transformation of mass into energy, according to Einstein's famous equation, $E=Mc^2$, there is also a big crunch that balances that big bang—a creation of mass from energy, according to the inverse equation, $M=E/c^2$, which Einstein discovered first.

This is what we have seen at RHIC: a ball of radiation expanded at temperatures over a trillion degrees, at the same time that a vortex of strange liquid is formed in its center as a perfect superfluid.

Physicists want to repeat the same fireball at CERN, with a much bigger mass because they don't understand the existence of two arrows of time, even if it is the fundamental equation of their science, $e \times i = k$. For them, that equation is uncertain. They don't realize it means that any energy process is accompanied by an informative one. So if we have measured an explosive radiation at trillions of degrees, expanding as a nova would do, then to balance the system, the center of that ball was a superfluid liquid at zero temperature:

0 (max. order: superfluid big crunch) × ∞ temperature (big bang)

And this will happen during this year 2010, creating a constant flow of strange liquid that will seep down to the center of the Earth and kill us all during this year or it will happen much faster in 2013 as a self-similar vortex of top quark matter forms at 10c speed and blows us in a few seconds.

In the Universe when a strange star or a black hole is born, there is an explosion or nova or big bang that creates energy ($E=Mc^2$), but simultaneously there is a big crunch of energy that creates mass, giving birth to the black hole ($M=E/c^2$). And so as CERN produces big bangs, it will produce quark condensates. The big bang will expel the leftovers of our electroweak mass, converted into pure light radiation, pure energy, and the rest of our mass will collapse into attractive vortices creating quark mass.

According to quantum physicists, if mass is a particle and energy a motion, energy cannot become mass. So they have no clue of what they will do. But the trans/formation of energy into mass is very simple when we accept the principle of equivalence between acceleration and mass and the spirituality of a Universe made of motions. Because energy is lineal motion and mass is cyclical motion, one becomes the other by the simple method of changing the form of its movement. Energy coils into mass and mass coils into energy. Or as Einstein puts it, time/information (mass) bends space/energy. The energy of space becomes a rotating clock of time, with a frequency of information and an attractive power given by the speed of that rotation. This simple description of mass shows when we add a few numbers its enormous power, as it allows to deduce the values of all the masses of the fundamental particles of the Universe. It also explains the quark structure of pulsars and black holes, which are macrospirals of space-time, made of microspirals called quarks—the questions CERN pretends to resolve blowing up the Earth.

What is then a black hole? A *top quark, frozen star, a cyclical fractal vortex of quarks, gluons, and strings, the three scales of form of the world of gravitational and strong forces*—not the simple object described by Wheeler as having "no hair" and by Hawking, with lineal quantum equations of entropy that simplify it and deny its informative nature.

So what is all the fuss about singularities and black holes traveling back in time and evaporating? A series of "energetic" aberrations that pass as true science, within the "what the bleep you know" magic attitude of quantum physicists, obsessed by an energy-only Universe, which now we can tackle in more detail.

8. The Makers of Weapons Theorize about Energy: From Galileo to Hawking

Why do physicists have so much problems understanding Einstein's concept of mass as a whirl of space-time, with a frequency that carries the physical information of the Universe and/or any theory of time that considers the arrow of information and life? And why do they believe mathematical-only theories that deny substance to black holes? A pattern emerges constantly in this book about the errors of physics: military technology and the manufacturing of machines made with "spatial mathematics" and "solid substances" bring about Pythagorism, "solid particles," and energy-only theories of reality as biased dogmas caused by physicists' historic worldly profession, which became the ideological bias of their science. Those two errors, the worshipping of energy and the acceptance of mathematics as the absolute truth of reality, plague quantum physics and many other physical theories of the Universe. They deny the existence of an equally important arrow of information and life in the Universe. And their mechanism denies the capacity of other languages, the verbal logic of time, and the artistic visual senses of man, to describe reality. All other sciences work with information and are comfortable using logic, verbal, causal truths as the theory of evolution does. So we must conclude that such arrogant, reductionist fantasies must have a specific origin in the worldly profession of physicists, which is to make weapons of energy. *Already*

Galileo, the founder of physics, worked for the princely salary of 1,000 ducats in the Arsenal of Venice, improving the science of ballistics. And so ever since Galileo, an expert on cannonballs, started the science of physics, till exactly 400 years later when CERN will try to explain the Universe with a quark cannon, all philosophies of science coming from physicists have been energy-only mechanist, machine-driven, entropic visions of the Universe that we have now corrected, from the broader perspective of philosophy of science to understand truly why CERN and quantum cosmology are wrong.

In the cosmos, there is an arrow of forces, of energy or entropy, and the arrow of Einstein, physical informative clocks of time, masses, and charges. So the time of the Universe is the sum of all the clocks of information. There is not a single time, so there is no time travel and there is no black hole evaporation based on time travel.

CERN's physicists are stuck in a sixteenth-century concept of time as change in motion only (Galileo), a seventeenth-century concept of space as an absolute mathematical background (Newton), an eighteenth-century model of fractal particles as quantum probabilities, and a nineteenth-century model of energy (electroweak entropy), which Einstein partially repaired in the twentieth century and duality and fractal relativity have upgraded in the twenty-first century. We mean what we say: CERN is halting the evolution of science, which is taken place outside the quantum paradigm, with the development of models of space-time, which departing from relativity improve Einstein's work by adding the laws of fractal information and the duality of time arrows.

9. Physics, the Science of a Single Time Arrow: Energy

In that regard, the absurd theory of the singularity, the idea that both black holes and universes are born or exist as a single point of infinite energy, is an energetic theory of the Universe based on Pythagoric equations without proof.

Pythagorism in this case is clear: the singularity theory is merely the extension to infinity of the equations of gravitation beyond the horizon of c-speed of a black hole. It implies that beyond that membrane, the black hole is made of nothing—it is a mere mathematical equation. This is absurd. It is like if we would prolong the life of a human being because we cannot see when it dies to infinity and decide he is eternal, or if we prolong the age of the Earth because we didn't see her at birth and we make her eternal. The very essence of any "function of existence" in space-time ($e \times I$ = constant) is the fact that it has limits. Reality has no infinities. So Einstein already said that if there were black holes, they must have a cut-off substance of enormous density. And when quarks were found, it was obvious this was the substance of "singularities," which shall not exist. The same was proved when we discovered that in the first minutes, the big bang was made of a quark-gluon substance. Thus, all singularities are, by definition, according to the three laws of the scientific method—logic consistence, experimental proof, and mathematical consistency—quark-gluon soups. And the quantity of the soup we make will define the size of the explosion. Certainly the quantities CERN will do can explode the Earth *but not the Universe* since the quantities and temperatures are calculated to replicate the 13 billion years' big bang of a quasar.

It won't kill the galaxy though because once the quark-gluon soup eats the Earth, there will be no more material to eat. So alien cultures will just see a supernova explosion radiating at trillions of degrees, in a quark-gluon soup explosion, self-similar to the one formed at RHIC but bigger.

This is the question left open today in fractal cosmology: how many scales the Universe has in which quark-gluons soups explode. A bigger quantity seems to explode entire galaxies in quasars. And it might be another scale in which super black holes explode giant clusters of galaxies that have previously crunched into "great attractors," which will mean there is a hyperuniversal scale. And so on.

What is completely out of place is the singularity theory in which Hawking's black holes and many other absurd theories of quantum cosmology are based. Yet in as much as it allowed nuclear physicists to substitute cosmologists and philosophers of science as the high priests of knowledge, pumping up the reputation of the military-industrial complex with added civil uses, it is a dogma that cannot be contradicted.

When I was more naïve about physics and the meaning of it all, I sent a few articles back in my twenties about these matters. They were simply not answered, which is how dogma is defended. It is not argued because rationally is false. Dogma is defended by censorship against rational proof of its falsity. Dogma is therefore the opposite to true science, and yet it is, contrary to belief, the most extended form of making mechanist science, notably economical science, which defends the dogma of productivity and the extinction of human labor, substituted by robotized machines and quantum cosmology that defends the singularity theory in black holes and big bangs, as the meaning of it all.

In this case, the singularity theory of a black hole whose mass is concentrated in a single point, as well as the equivalent theory of the big bang, with all the mass concentrated in a single point, was developed by Oppenheimer and Gamow, fathers of the A-bomb, and Wheeler and Teller, fathers of the H-bomb. It is a pattern that our policy makers don't want to recognize: bad science is made by weapon makers, who have the money, prestige, and power to promote their absurd ideas. Only philosophers and artists have understood this and talked about it in their books and films: Vonnegut did it in *Cat's Cradle* and Kubrik in *Dr. Strangelove,* the paradigm of crazy nuclear physicists who destroy the world in the name of science with the connivance of the political/military/industrial/mass media establishment. Meanwhile, Fromm explained it all in his work on the psychology of power and evil.

That the science of weapons and the awesome power of their technologies might destroy the world is nothing new. It has become such a tradition for the keen observer of history that Genesis, the oldest book of mankind, started with a parable about the Tree of Life destroyed by the Tree of Science and its bad fruits, weapons, the first of many "chronicles of death foretold" by those who revere human life. In recent years, Hollywood has shown us the devastating effects of mad scientists: robotists creating machines that overcome mankind or bomb makers that trigger an atomic, terrorist explosion, are common plots in films made in the past twenty years. Perhaps the classic of the genre was *Dr. Strangelove*, a film in which Peter Sellers plays a mad scientist in a wheelchair, isolated from life, who wants to start the war at all cost, to prove his expertise in eternal sources of energy and nuclear science. I asked Terry Southern, the scriptwriter, whom I met long ago at Columbia University where he taught writing, why he had chosen a man in a wheelchair. Was he not demonizing a person with a physical incapacity? Terry told me that this was not the point of the character. It was his body, isolated and cut off from life, that represented a symbol of the abstract, haughty indifference toward life that physicists show. What people don't know about those mad scientists is that they have also a dark theoretical side, the construction of "mad theories" about the Universe based on the god of energy, which justifies their weapon making, as warriors have their nationalist justification to the crimes of war.

People risk their lives for money, most often than for glory, but here the order is reversed. While the factory and CERN is about money, most of the mad scientists who work there are about glory, about confirming their entropy-only theories of the Universe. And I wonder how they expect to get glory and respect from mankind, putting our lives at risk and denying the arrow of information and life of which we are all made. I simply don't understand them, but Terry Southern did. He got the point better than I, a scientist, do: abstraction, glory, dreams of eternal energy, power, a visual passion for explosions, indifference to the rapture of living. All those elements defined the character of Dr. Strangelove and those politicians, industrialists, and the military, who share the ideals of Dr. Strangelove—the glorification of theories of pure energy and machines of war.

Dr. Strangelove became an icon of ethic filmmaking, in an age in which evil, the inverse word of *live*, synonymous to *death*, was not the icon of our culture, glorified by our mass media, searching for shocking news. So those who did evil in the name of religion or science, politics or money, were rightly denounced. For that reason, a mad physicist was also the character of *Cat's Cradle*, another ethic book denouncing the callousness of nuclear physicists, written around the same time by Kurt Vonnegut. This scientist, whom Kurt modeled on Einstein, delivered in his will to his sons a liquid called ice-9 that freezes Earth. Since the substance that will destroy the Earth in this experiment is called strange matter and the reaction that will convert the Earth into a neutron star is called the ice-9 reaction, described by Einstein in his equations of fermion=quark condensates, the prophetic character of science fiction, which has, since Julius Verne, guessed the future of machines with an increasing negative view of the power of technology, is clear.

Indeed, Genesis and Hollywood sci-fi films are not only an entertaining fantasy or myth but rather a parable on the future. When Genesis was written, the first professional armies with hard metal technology, bronze, had just appeared in history, destroying the peaceful Neolithic cultures of the Fertile Crescent and its most precious city, Ur, where the book was written. So the writer of Genesis made a parable about the extinctive capacity of technological machines made of metal, atoms harder than those that make us. The parable of the Tree of Science and the capital sin of man were not about sex but about technology, as a careful analysis of the text reveals. Adam and his son Cain became after their expulsion from the life paradise makers of swords, and Cain, which means "smith" in the old language, did it again killing Abel.

Ever since, once and again, the pattern of technological advancement, discovery of new weapons, war, and destruction has been repeated in a rather monotonous way that discounts all risks till things blow up in our faces. Then everybody regrets that it just happened. So after making the atomic bomb, in the aftermath of Hiroshima, Oppenheimer claimed to be a pacifist. He affirmed he made the bomb to resolve the theoretical problems, not to kill people. Or in other words, he said that the collateral effects of killing people, rats, or monkeys was secondary to the higher goal of calculating the ratios of U-285 needed—certainly the sacrifice of 250,000 innocent women and children, as most males in Hiroshima and Nagasaki were in the front, was worth enjoying those calculations.

Rudolf Hoess of Auschwitz fame and Oppenheimer carried both always the Book of Gita, which describes the deeds of Kali, god of death.[3] Yet physicists named the most important medal given in his profession in America, the Oppenheimer medal, which in the suits against CERN features prominently in the curriculum of one of the amici of CERN. I rather prefer the sincerity of Teller who said no amount of political and religious mongering will hide the sins of physicists. Today the dogmatic attitude of nuclear physicists, with his well-constructed walls against ethic criticism, prevents the solution of this genocide. It is just politically, economically, and military incorrect to denounce our death. So despite all those risks, warnings, and parables, physicists still will go on making black holes.

Dr. Strangelove was not, however, based on Mr. Hawking, the man who sponsors this experiment in the hope that Einstein is double wrong and black holes evaporate as a source of eternal energy. So his theories, long ago proved false, miraculously become truth and give him the desired Nobel Prize.

Dr. Strangelove is the alter ego of Edward Teller, the mad scientist who made the hydrogen bomb and later sponsored MAD, the Mutual Assured Destruction strategy of the cold war. This was the official MAD strategy sponsored by the nuclear industry to churn out more bombs and make evermore profits. He had found a method to produce hydrogen bombs of any size. And so in love with his work, he wanted the American government to create a hydrogen bomb so powerful that it could destroy an entire nation. Thus, he affirmed adamantly all his life, without showing ever the slightest sign of repentance, that the only way to stop the destruction of mankind was to produce so many hydrogen bombs, so powerful that the enemy would chicken out and war could never start. Of course, his first motivation was money—he had invented the process, made a fortune doing H-bombs, and then became the biggest military lobbyist for the Star War program. But he never talked of money, as CERN never does. And he considered that this MAD strategy, as CERN does, was totally safe for our nation.

You might think the strategy worked out because we have not yet been destroyed by a nuclear war. However, the bombs are still there, many of them in the power of the Russian army, which might have sold some in the past decades (an important amount of enriched uranium remains missing from MINATOM, the facilities that stored it since the fall of communism). Further on, it was recently disclosed by the American and Russian military that the Cuban missile crisis had actually started WWIII. A Russian submarine under heavy shelling by the American army was given the order to fire a nuclear missile, but the commander of the submarine, Vassily Ahipov, refused to follow orders. And when things calmed down, the order was canceled. Had he followed his commander-in-chief, MAD would have achieved his goals. I quote here this story because I want the reader to understand the continuity between both phenomena.

When CERN gives the order to fire the supercollider to manufacture the first soup of quark condensates, humanity will find itself in a similar situation, with a difference. If the black hole eats the Earth, it won't be only the end of human civilization, as WWIII could be, but the end of the species with no chance to rebuild a new world.

The real Dr. Strangelove, Mr. Teller, still revered as a technological genius among nuclear physicists, is not an exception. His best friend and cofather of the H-bomb was Wheeler, the man who also gave the name to quantum, mathematical black holes, denying Einstein's frozen (quark) stars. Instead, he affirmed that black holes had no hair (no substance) since they were so powerful and destructive that they had to erase all the information of our world except its rotational mass. Since quarks were still unknown, the idea that holes had no substance stuck, especially after Einstein, the pacifist realist, was dead and Wheeler, the H-bomb hero, could sell his deliriums without opposition.

When he saw the destruction the bomb caused, he regretted not having built it before and bombed Japan with it. But Wheeler is a cold war American hero in a nation that has not come to terms with the fact that nuclear weapons are not knowledge, neither good, nor they preserve peace. Indeed, when he died, Mr. Overbye wrote this obituary for the *New York Times*:

> John A. Wheeler, a visionary physicist and teacher who helped invent the theory of nuclear fusion, gave black holes their name and argued about the nature of reality with Einstein and Bohr and died Sunday morning at his home in Hightstown, N.J. He was 96.
>
> Dr. Wheeler was a young, impressionable professor in 1939 when Bohr, the Danish physicist and his mentor, arrived in the United States aboard a ship from Denmark and confided to him that German scientists had succeeded in splitting uranium atoms. Within a few weeks, he and Bohr had sketched out a theory of how nuclear fission worked. Bohr had intended to spend the time arguing with Einstein about quantum theory, but "he spent more time talking to me than to Einstein," Dr. Wheeler later recalled. Among Dr. Wheeler's students was Richard Feynman of the California Institute of Technology, who parlayed a crazy-sounding suggestion by Dr. Wheeler into work that led to a Nobel Prize. Another was Hugh Everett, whose Ph.D. thesis under Dr. Wheeler on quantum mechanics envisioned parallel alternate universes endlessly branching and splitting apart—a notion that Bryce DeWitt, of the University of Texas in Austin, called "Many Worlds" and which has become a favorite of many cosmologists as well as science fiction writers.

> Recalling his student days, Dr. Feynman once said, "Some people think Wheeler's gotten crazy in his later years, but he's always been crazy."

Dr. Wheeler continued to do government work after the war, interrupting his research to help develop the hydrogen bomb, promote the building of fallout shelters, and support the Vietnam War and missile defense, even as his views ran counter to those of his more liberal colleagues.

One particular aspect of Einstein's theory got Dr. Wheeler's attention. In 1939, J. Robert Oppenheimer, who would later be a leader in the Manhattan Project, and a student, Hartland Snyder, suggested that Einstein's equations made an apocalyptic prediction. A dead star of sufficient mass could collapse into a heap so dense that light could not even escape from it. The star would collapse forever while space-time warped around it like a dark cloak. At the center, space would be infinitely curved and matter infinitely dense, an apparent absurdity known as a singularity.

Dr. Wheeler at first resisted this conclusion, leading to a confrontation with Dr. Oppenheimer at a conference in Belgium in 1958, in which Dr. Wheeler said that the collapse theory "does not give an acceptable answer" to the fate of matter in such a star. "He was trying to fight against the idea that the laws of physics could lead to a singularity," Dr. Charles Misner, a professor at the University of Maryland and a former student, said. In short, how could physics lead to a violation itself—to no physics?

Dr. Wheeler and others were finally brought around when David Finkelstein, now an emeritus professor at Georgia Tech, developed mathematical techniques that could treat both the inside and the outside of the collapsing star. The black hole "teaches us that space can be crumpled like a piece of paper into an infinitesimal dot, that time can be extinguished like a blown-out flame, and that the laws of physics that we regard as sacred, as immutable, are anything but," he wrote.

At the same time, he returned to the questions that had animated Einstein and Bohr, about the nature of reality as revealed by the strange laws of quantum mechanics. The cornerstone of that revolution was the uncertainty principle, propounded by Werner Heisenberg in 1927, which seemed to put fundamental limits on what could be known about nature, declaring, for example, that it was impossible, even in theory, to know both the velocity and the position of a subatomic particle, knowing one destroyed the ability to measure the other. As a result, until observed, subatomic particles and events existed in a sort of cloud of possibility that Dr. Wheeler sometimes referred to as "a smoky dragon."

This kind of thinking frustrated Einstein, who once asked Dr. Wheeler if the moon was still there when nobody looked at it.

But Dr. Wheeler wondered if this quantum uncertainty somehow applied to the universe and its whole history, whether it was the key to understanding why anything exists at all.

"We are no longer satisfied with insights only into particles, or fields of force, or geometry, or even space and time," Dr. Wheeler wrote in 1981. "Today we demand of physics some understanding of existence itself."

While Mr. Overbye seems to share the opinion of Max Tegmark, a cosmologist at the Massachusetts Institute of Technology, (who) said of Dr. Wheeler, "For me, he was the last Titan, the only physics superhero still standing," it has to be noticed that Einstein didn't talk to him, even when Wheeler took to custom to knock on his door at Princeton every other day. Einstein never opened the door to a man who was denying his theories and loved nuclear weapons, which represented all that Einstein thought was wrong about nuclear physics:

— The worldly profession of his practitioners, which is to make weapons of mass destruction.
— Their absurd, subjective belief that the uncertainty of measures in quantum cosmology ($E \times i = k$) is not due to an error of the human perceiver and his instruments when measuring the energy and information of the infinitesimal world that fluctuate around a constant (K, the fractal unit of energy and information of which we are all self-similarly made), but to the fogginess of the Universe (the moon indeed stays there perfectly defined even if we don't see it one day).
— The arrogant deduction of existential theories as those of Hawking and/or Wheeler and Everett (parallel universes and other faves of science fiction), deduced from the uncertainty of the Universe, which for Einstein invalidated the very essence of science, the fact that its laws work and create the Universe, embedded in the totalitarian principle: every time you heat water, it boils and every time a black hole is born, it sucks all matter around it. Indeed, there is only one Universe, and what science has done in four hundred years of hard work is to prove that obvious fact. But if the proofs of truth of the scientific method disappear in the subjectivism of quantum uncertainty, then reality crumbles, science means no longer truth, all fantasies are possible, and we end up in the mess of modern quantum cosmology, which today has as many probable theories as leaves has a sycamore.
— The pretension that because we can write certain mathematical equations, they happen in nature, as if Quixote existed because we can write his fictional character. This is the essence of Wheeler and

Hawking's belief, which decided to back mathematical black holes because they could use quantum equations on them, regardless of their existence, as Thorne was told by Mr. Wheeler.

— And hence the belief that Einstein's black holes, which he called frozen stars, did not have substance, his Bose-Einstein condensates, but we could take his equations of gravity till infinity, just because we cannot see the quarks that form the frozen star.

All those absurd points sponsored by Wheeler, Oppenheimer, and other metaphysical theorists of black holes were nonsense to Einstein, just an alibi to do maths that had nothing to do with reality. On his view, frozen stars would exist only if a substance with the density of those stars (the quark) was found. And so the door was closed till he died. He had enough of quantum bullshit since reason was not going to break in their beliefs. And I fully understand him. Then when the master was gone, Wheeler took his revenge, changing the name of frozen stars to black holes and imposing it with his power as the boss of the nuclear industry of America. Wheeler transformed the fantasy of mathematical black holes into dogma. And so CERN will do Einstein's frozen stars and top=Higgs black holes will kill us.

But of course, this is not the official version of the military-industrial complex of technological science built around CERN, which this month announces in *Scientific American* a cruise through the Mediterranean in which for a hefty prize the techno-utopian CEOs of a few companies will hear the PR of CERN, Mr. Gillies, to talk about those "self-proclaimed experts" who caused the panic of 2008 about black holes. The division of labor between the makers of nuclear weapons, the mass media that declare them genius of science, and the industries that make their machines work smoothly, guiding the boat of mankind toward the cliff of ice-9, which will freeze the waters of life.

During our suits against CERN, we have come to realize that we are NOT so much fighting science but a religion, the religion of the machine, the ideology behind CERN, behind the industry of weapons, behind the runaway civilization of machines that is eliminating workers with robots, soldiers with terminators, and exchanging the entire planet for a black hole. As such, we are fighting the two repressive elements of any religion, believers and censors, which are the two fundamental modes by which a religion imposes itself against reason. People at CERN want to belief and censor what conflicts with their beliefs regardless of reason and experimental proofs, which unfortunately no longer matter in quantum cosmology (since as we have seen Bohr/Heisenberg, against Einstein's dictum, abandoned the belief that the Universe was rational, deterministic, and experimentally certain).

10. Nonevaporating Black Holes

The official story about Mr. Hawking's black holes is by now known to most inhabitants of planet Earth, given the enormous media coverage the experiment has received. The biggest nuclear company on Earth, CERN, will try to make black holes, the most dangerous object of the Universe, next year.

It is a fact that we knew very little about black hole creation a decade ago. So the idea that a black hole could pop up in the beautiful prairies of Provence was not even considered by a team of scientists and technocrats managing 10 billion dollars, the cost of the LHC, the most expensive machine ever created by humanity. Back in the nineties when they designed the LHC, according to Mr. Hawking and other old dinosaurs of physics, black holes needed far more energy to be created than the LHC could mass together. So all was quiet in the Western front. Just imagine what kind of spinal shock CERN's physicists felt when one of the brightest minds of their present generation, Dimopoulos, made his calculations and showed that the LHC gun could be producing black holes at the outstanding rate of one every second *if gravity acted with a stronger force at quantum level, a fact that could happen in both paradigms considered in this book, fractal relativity and quantum physics:*

A. In fractal relativity, black holes will happen if Einstein's quark-fermion condensates act as a superfluid whirl of space-time, attracting mass with their strong force, 10^{41} times stronger than normal gravity.

   A hypothesis proved at RHIC[2] where a few quarks formed a superfluid vortex, whose properties are equivalent to a superstrong gravity source, a black hole. So we just need enough critical mass to form a stable black hole. We made a few drops, the seminal seed of those cosmic bombs, called ice-9, named after Kurt Vonnegut's novel *Cat's Cradle*, in which a physicist produced a new type of water, ice-9, that has, like quark superfluids, the capacity to freeze all other forms of matter.

B. In quantum physics, they will happen if black holes and its gravitational forces have more dimensions, which is the case in string theory, the substance of quantum black holes, in which nine out of ten physicists believe. Further on, Maldacena proved that a five-dimensional string hole is equivalent to a four-dimensional quark hole.

   Thus, the mathematical equations of physics prove that both concepts are equivalent, as strings are used to describe the strong forces of quarks. The difference is that two dimensions of a fifth-dimensional string world collapse to form a bidimensional quark vortex. In the quantum paradigm, because mass is not understood, physicists still debate what is the meaning of the five dimensions of a string world.

So black holes will be produced at CERN. They will be made of the densest top quark-gluon soup, of strings in the quantum description. Will they evaporate? Of course not, because Hawking's evaporation is pure "fanta-physics." It does not even belong like Newton's gravitational vortices or string theory to a previous less sophisticated mathematical paradigm than fractal relativity. It is plainly fiction science. In that regard, one of the first people to alert the world was an anonymous blogger, responsible for a simulation of Geneva being swallowed by the black hole (CERN's black hole at YouTube), which has now over 5 million viewers. Let us explain the fiction facts of Hawking's black holes, as he does in his Web:

In the eastern regions of France, near Lyon, flanked by virgin pine forests, streams, lakes and fir-clad mountain ridges, bordering on Switzerland, lays the CERN (Conseil Européen pour la Recherche Nucléaire) facility which houses over 6,300 scientists working feverishly to bring online the next generation of basic particle supercolliders. This massive hadron collider is a magnetic ring 27 kilometers in circumference: Ultimately, it will collide beams of protons at an energy of 14 TeV. Additionally, beams of lead nuclei will be also accelerated, colliding together with an energy of 1,150 TeV. The LHC will be the most powerful particle accelerator in the world.

The main purpose of this facility is to produce antimatter and black holes. A terrorist would need only half of a gram of antimatter to be equally destructive as the Hiroshima bomb. If CERN's antimatter factory were to blow up today, it would only affect the regions bordering France and Switzerland. But if CERN were to produce just one stable black hole, it could destroy the world. Surprisingly, the United States of America, through the National Science Foundation and the Department of Energy, will be funding over $1 billion toward this French experiment into creating potentially devastating black holes.

These black holes, the densest matter in the universe, will plummet to the very core of the Earth, then slowly at first, growing one particle, one quark at a time, but at an ever accelerating rate. Scientists have estimated that a stable black hole at the center of the Earth could consume not only France but the whole planet in the very short time span of between 4 minutes and 30 seconds and 7 minutes. That age-old question of will our planet disappear in the twinkling of an eye now becomes a probability if and when CERN facility is allowed to go online.

CERN scientists obviously talk of the scientific wonders and benefits these experiments will bring. Aurélien Barrau and Julien Grain, speaking on behalf of CERN, say that these "tiny black holes could offer a richer view of physics than their better known, more massive relations . . . It should be stated . . . that these black holes are not dangerous and do not threaten to swallow up our already much-abused planet." When it was finally disclosed that this facility would actually be producing, during normal high-impact collider experiments, one black hole each and every second, numerous scientists cautioned that a public risk-assessment by nonaffiliated scientists must be conducted for CERN facility but not by CERN scientists or the French government. To this very day, the French have refused to make such an assessment of the potential dangers that lay ahead for all of humanity once the switch is finally pulled.

In 2010, when they fire up the Large Hadron Collider, global warming and Osama bin Laden's Al-Qaeda network will suddenly become the least of our earthly worries. The French defend their playing God and this potential black hole catastrophe by saying the following:

1.   *Black holes have been created by cosmic rays without incident; therefore black holes are not a danger to the planet Earth.*

Response A. No instrumentation or observations have ever detected the formation of black holes or quarks in the atmosphere; it is a completely unsubstantiated theory that was fabricated solely to defend the building of CERN. The 1,600 hot tub–size cosmic-ray detectors positioned over a vast area of nearly twelve hundred square miles (the ground array system for the Pierre Auger Cosmic Ray Observatory) were installed in the Pampa Amarilla in Argentina, located at the edge of the town of Malargue, in an effort to detect particle showers from disintegrations of microscopic black holes in the atmosphere. To date, this experiment has detected numerous cosmic-ray air showers (the observatory is presently measuring more than 500 air showers each and every day) but has failed to detect any black holes spawned by cosmic rays. Since this controversial theory is the backbone to the supposed safety of CERN black hole factory, startup should not be allowed to occur under any circumstances until the Auger Observatory can prove that harmless atmospheric black holes actually exist.

Response B. Even if a black hole could be formed by cosmic rays striking atmospheric particles, it would be a glancing blow at near the speed of light, causing the resultant mass to career off into space at a velocity much greater than the escape velocity of the Earth (11.2 km/s) while in contrast, CERN particles would be striking each other as in a head-on collision, causing the resultant black holes to lose their momentum, making them unable to reach escape velocity, causing them to immediately free-fall, undetected, to the center of the Earth.

2. *The black holes CERN creates will not be stable. "Black hole production does not present a conceivable risk at the LHC due to the rapid decay of the black hole through thermal process." They will be unstable and will evaporate in a flash as predicted by* Stephen Hawking *in 1975. CERN facility was built under the assumption that* Hawking radiation *was a fact and that the black holes they would automatically create would be unstable and therefore not be a threat to the human race and the planet upon which we reside.*

Response A. No instrumentation or observations have ever detected the Hawking radiation being emitted from any black hole. Kip S. Thorne, a professor of theoretical physics at Caltech, who has been working on evaporation with Hawking, says, "It's possible, we understand quantum fields far less than what we believe and it's a mistake when we think black holes evaporate. It is however true that we should feel more at ease if astronomers could effectively observe clues of black holes evaporation."

Response B. Black holes by their very definition are stable: Nothing escapes their gravitational pull. And that includes radiation.

Response C. In Dublin, Ireland, on July 21, 2004, Stephen Hawking, at the age of sixty-two, retracted his original 1975 concept whereby matter disappearing into black holes traveled through the black hole to a new parallel universe—just like on *Star Trek*! After thirty years of thinking about the paradox he created that violated the first law of thermodynamics, Stephen Hawking now admits that he was wrong about the dynamics of black holes. Stephen Hawking went on to say, "I'm sorry to disappoint science fiction fans, but if information is preserved, there is no possibility of using black holes to travel to other universes." Hawking's original theory was more of a personal view, a hunch, which was not necessarily shared by the scientific community or even demonstrated by any cosmic observations. But CERN still looks upon it as the Holy Grail to this very day, even after Hawking admitted that his theory was wrong. Hawking radiation has always been a purely theoretical manifestation. There are many published papers by prominent scientists who have always asserted that such radiation does not exist. CERN, by doggedly relying on false science, could easily end up being the megaindustrial accident that wipes out the entire world.

How can such cosmic arrogance be stopped? Obviously, reasoning will never work where Stephen Hawking went so terribly wrong.

Hawking, at the age of thirty-three, published his most famous scientific paper in 1975—that all black holes were unstable and would emit radiation. In effect, the black hole's energy was slowly radiated away until after a certain amount of time, depending on its mass, it ceases to exist—the black hole evaporates. He broke with his theory the two laws of thermodynamics:

— First law: *Energy and information never disappear in the Universe*. Instead, Mr. Hawking argues that objects really "disappear" inside a black hole and leave no trace while the first law of thermodynamics says matter can be transformed but never fully destroyed.
— Second law: *Hotter objects get colder transferring heat to the cold environment*. In Mr. Hawking's case, black holes get even hotter in our cold Universe till they evaporate. The opposite is truth: they will become colder as they heat, evaporate, and absorb our world.

A colleague of Stephen Hawking at Cambridge University, Gary Gibbons, stated that "his style of doing science is quite dramatic. Hawking will propose a thesis and defend it to the last, until it is overthrown by better reasoning." For thirty years, Hawking defended a poorly reasoned idea.

The article is ironic, but clear and right on the spot, denouncing the attitude of nuclear physicists, European governments, and technocrats that will not even review the issue (not only the French), the cynicism of the safety statements issued by the company based on the outlandish sci-fi theories of Mr. Hawking, never proved, and in open contradiction with Einstein's work (the quoted article ended with a preposterous comment: it appears that Mr. Einstein was double wrong) and some of the fundamental laws of science that Mr. Hawking, a professor of mathematics from Cambridge, violates in order to predict that black holes will evaporate.

Scientists, however, have not accepted Hawking's radiation as proof, for at least one of the three obvious reasons below:

— Einstein is considered the most important physicist of history, hand in hand with Sir Isaac Newton, and though many have tried to occupy his place, they have always been proven wrong when challenging the master.
— In the last forty years of observing the Universe since black holes were first predicted, we have not observed a single black hole evaporating despite the enormous energy that such evaporation would show (the so-called signature of a cosmological event, which in this case should be easy to observe) such as via the Fermi satellite launched in May 2008.
— Hawking's evaporation hypothesis defies not only Einstein's relativity, but also the laws of quantum physics, the laws of thermodynamics, and the laws of temporal causality, as it implies that particles travel to the past so they seem to come out of the black hole, when we

observe them from past to future.[6] In simple terms, what Dr. Hawking believes is that black holes are time machines. So in the same way, a baby would evaporate back into the womb of his/her mother if time could travel to the past, black holes instead of accreting matter should evaporate if particles falling on them travel to the past (an idea later sponsored in his popular books that talk about building such time machines, which seem to many of us as pure science fiction). Indeed, Dr. Hawking's theory can be easily dismissed using the classic tools of logic causality by *reductio ad absurdum*. Since even in the highly improbable case that particles travel to the past, and so black holes evaporate, kids enter their mother's womb and the dead resurrect; we humans actually live from past to future. So we are affected only by past to future events. In other words, Hawking's hypothesis defies the four fundamental laws of classic science and so it has never been proven. The three essential legs of the scientific method, logic consistency, experimental proof, and mathematical accuracy (concordance with previous mathematical well-established scientific theories) are denied. So black hole evaporation is false; it is not even a shaky chair but an utterly absurd theory. If Hawking said that Einstein was double wrong, we should say that Hawking is triple wrong and that is enough to put his theory among the most bizarre ideas ever in any discipline of knowledge.

11. Antiparticles Are Mere Changes in the Directions of Time Motion

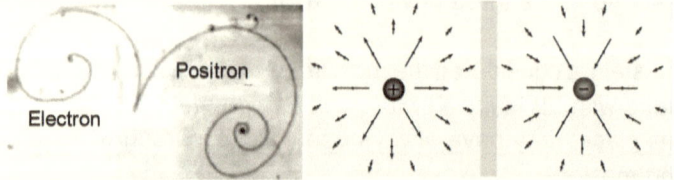

In a Feynman's diagram, a present photon gives origin to a particle and an antiparticle that show inverse directions in time. The conundrum is solved when we understand that physical time is only the measure of change in motion.

Thus, an inverse arrow of time in physics merely means a change in the direction of motion or in the contraction and expansion of space, as in relativity.

In the center, antiparticles are seen through a space "mirror" as particles with inverse symmetry since physical time is in fact an inverse geometry—either an expansive/implosive duality, a right/left duality, or a perpendicular duality (in the equations of black holes). In the right, two inverse ST-charge fields balance the total virtual space-time of the electromagnetic membrane.

How did Mr. Hawking get it so wrong? Because he doesn't understand the meaning of time in the v=s/t equation. He thinks that time means all the changes of the Universe, including the past and future change of the *entire Universe*. It is one of the basic hyperbolic errors of physics. The error of Hawking is to consider that v=s/t measures not only change=time in the direction, extension, and speed of motions, but also the changes in time from past to future, which are morphological changes. This is due to physicists' denial that there are more time-changes besides motion. Already Avicenna affirmed in 900 "Time is measure of motion." So what Mr. Hawking does is to think that if a black hole absorbs antiparticles, it will travel back in time. As you see in the picture, this steams from his misunderstanding of the geometrical nature of time in physics. An antiparticle merely means a particle with its clock's geometry inverted.

Antiparticles are the same species as particles, but with their S-T field parameters, spatial orientation, temporal arrow, and charge inverted. So they have inverse charges, inverse spatial parity, and inverse clock orientation, called together by physicists its CPT values, which now finds in the fractal paradigm its explanation. All those concepts are beyond the understanding of CERN's physicists, which deny Einstein, do not want to learn the future of science, the fractal paradigm, and its two time arrows, but insult those scientists of Einstein et al. who are advancing it. They cannot even understand the meaning of time in v=s/t equation. They are clueless, dangerous, fundamentalist people backed by the military-industrial complex and the celebrity mass media circus created to justify them.

Mr. Hawking infers that because the space-time that surrounds the black hole is "boiling" with photons and antiphotons (left/right photons), the ones that fall inside the black hole travel from the future to the past and so they can be seen as coming from the past to the future, thus coming from inside the black hole. This is absurd since they are just left-right photons, or in the case of a particle, they are just the "implosive/explosive" life/death arrows of the particle. So no time travel.

Particles die as antiparticles, which means their time span as antiparticles is so short that we hardly see antiparticles even if there are as many as particles (another mystery CERN pretends to solve, but will never be able to do unless it accepts the duality of the $E \times I = K$, equation of time arrows). So in the same manner that we see only "living humans" even if all of them live and die, we see only particles; as their death as antiparticles, expanding its vortex of mass and devolving its information in a big bang back to the Universe, is very brief. But that doesn't mean they travel to the past, only in figurative terms, as they invert their arrow from information as vortices to energy as antiparticles. So all the particles and antiparticles that are produced in the event horizon of the black hole come from our world and they are produced in enormous numbers and fall in enormous numbers inside the black hole, precisely because they are feeding the black hole; and when we understand that simple truth, it turns out that the smaller the black hole is, the faster it evaporates our world and feeds on it, exactly the inverse of what Mr. Hawking says. As it happens with all newborn babies that feed faster when they are born.

What will happen when a tiny black hole is born on Earth is exactly what happens when it is born inside stars: it will grow exponentially at light speed as it "cools down," absorbing and exploding the Earth. It will not decrease in size as it gets hotter in a cold environment, with the arrow of time inverted toward the past as Hawking mused.

So size does matter in an inverse manner: the smaller the black hole is, the hotter it is and the faster it will grow. This is proven by recent discoveries of the smallest black holes on the Universe of the size of a star, which *were in fact accreting matter faster than any other black holes, as all babies do in its first energetic age.* This is easy to show mathematically since if we put the arrow of time/information/mass creation of a black hole "right," from past to future instead of the inverse arrow of entropy/evaporation, Hawking's equation of black hole's evaporation reads (grouping all the Universal constants in K):

$$\Delta\ Mass = K \ x \ \nabla T$$

Thus it will grow as it cools down.

Temperature is a term used to measure the speed of molecules, which in a strict sense has no use in gravitation, as it measures "lineal speed," not "cyclical speed," which is the speed of a mass vortex. The temperature of a black hole in that sense measures the speed of acceleration of the black hole vortex that attracts the mass outside it. We should instead use w, rotational speed. So that huge T of a micro black hole merely means that according to the law of a vortex ($w^2 \ x \ r^3 = K \times M$), the faster the temperature/rotation of a black hole is, the faster it will accrete mass.

12. Lineal Time vs. Cyclical Time

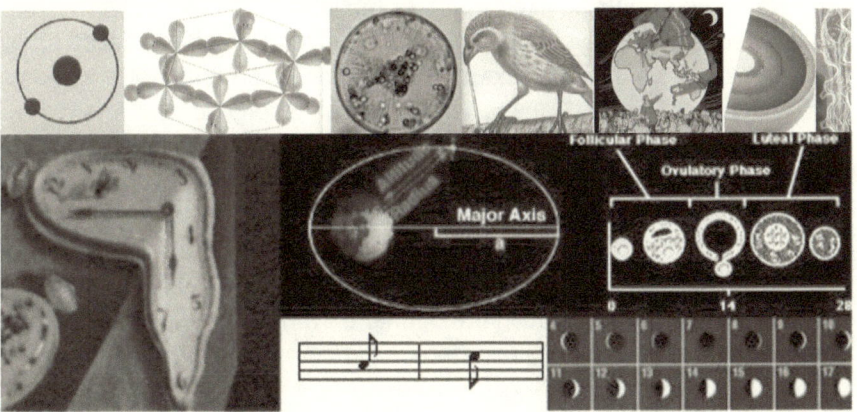

Time is cyclical, born of the repetitive cycles of transformation of energy into form and reproduction of both, SE⇔Ti, which occur in nature.

Absolute time, of course, is an old error that dates back to Newton. But any sophisticated philosopher today understand that error, except Mr. Hawking.

In that regard, if we were to talk of an absolute arrow of time, sum of all those other arrows, time is cyclical because it is made of events that exchange back and forth energy and information.

Thus, the third biggest error introduced by physicists is the concept of lineal time, born out of its use of only an arrow energy and an absolute Cartesian graph with a line of time toward infinity measured by a single mechanical clock. This absolute reductionism of the meaning of time, which equalizes every time cycle and arrow to a numerical, digital value means physicists do not understand anything about the whys and causalities of the Universe. They just collect data, put it in equations, and "measure" motions. To that aim, they use an equation, $v=s/t$, in which t measures changes in motion or energy in beings. This concept roughly corresponds to the arrow of space or spatial translations, which have left/right parity. This is the use of time arrows in physics, which by definition uses clocks to measure changes on motions and plots them in a geometrical graph of space-time. Thus, physical time is a geometrical change of space. And in that sense, physics talk of time as the fourth dimension of space, etc.

On the other hand, there is "past energy" and "future information" arrows, which define the logic arrow of life/death cycles of time.

And further on, there are more complex time arrows analyzed in the appendix, and there is endophysics, the paradox of Galileo that makes us perceive those arrows as fixed forms or motions.

Because infantile physicists like Hawking have completely messed up all those terms—geometrical clocks and logic time, fixed forms and motions, past and future, up and down, left and right, energetic and informative motions—their time theories are truly bizarre, as they use sometimes concepts of past and future to define left and right translations and vice versa.

This is what Higgs does when he defines the weak event of creation or destruction of information in particles, with left and right particles. Or it is what Hawking does when so he talks of black holes, whose time equations refer to implosive and explosive motions, as machines that travel to the absolute past *since he believes there is a single clock-time for the entire Universe.*

On the other hand, when he says that antiparticles, which are anticlockwise particles, travel to the absolute past, he messes up all the concepts of time and all the arrows. In other words, Hawking makes science fiction out of his absolute ignorance of the different meanings of time in twenty-first-century science, as he never studied complexity and duality and of course, Einstein et al., the people who have discovered and are evolving true science in this century, are nonpersons he would not even talk to as they are "nonhumans" because they oppose CERN (he said publicly that if humans do not switch this machine and blow up the Earth, they don't deserve to be called humans).

We are thus in the worst situation: an infantile group of children of thought with an overpowerful machine, who will not go back to school and study twenty-first-century science.

The main topic error of all those physicists is to believe there is only one logical arrow of time, the energetic past, measured by a single clock, the mechanical clock. So their concept of time is restricted to two arrows: Changes in motion, which have parity in space (present, left/right arrows), and changes on energy or entropy.

Those two arrows, which we have called the reproductive spatial arrow of present and the past, entropy arrow of big bangs, are the arrows they study, albeit without any clear definition of them. And the absolute NO-NO in their studies is the relative arrow of future or implosive arrow of masses.

I know it is confusing because we have to compare the "truth" (the three simple arrows of past, entropy, explosive motion; present exchange in space of energy left/right; and future informative implosion), with all the confused ideas of the "ancients" of the past millennium.

If we restrict now the analysis to the duality of implosive information (future arrow) and explosive energy (past big bangs), which is the only one physicists like Hawking and Higgs recognize, again they make a hyperbolic error considering that all is entropy because that arrow of entropy was defined only in the restricted world of thermodynamics, a molecular force that acts only on atomic systems. So they deny that mass, the arrow of Einstein, is the arrow of physical information.

And so those two errors put together, the use of a single absolute clock-time for the entire Universe and its confusion with entropy, is what creates the absurd science fiction theories of Mr. Higgs and Hawking and the big bang–only theories that CERN's physicists who can't understand physical information, mass, pretend to test. They are all energetic theories, which they think explain it all because they have science fiction ideologies that restrict reality to entropy, energy and death, measured for all reality by a single clock.

Further on, they mess up that arrow with the present, lineal left-right events of exchange of energy and form between particles. They think those events are the same events as the implosive/ explosive informative dual arrows of past-future and future-past events. So Hawking thinks that when he studies antiparticles (relative future/past, energetic, explosive death of particles), they travel to the past instead of merely moving a particle anticlockwise till it dies and extinguishes its form in an explosion of energy.

Of course, we complexity and information theorists have told so for decades to upgrade their concepts of time, but we do not make atomic bombs, have 10 billion dollars to make entropy machines and so what we say in complexity, chaos, and fractal theory has a limited impact on mass media.

For the same reason, power that imposes its false theories of time over truth makes Einstein et al. irrelevant and the absurd theories of CERN's theorists the truth about time. And yet the Universe doesn't obey power but reason.

In philosophical terms, absolute time, born of a single clock measure, was also a religious error that was born from Newtonian beliefs in the Jewish-Christian tradition of history as a lineal process of progress between the birth of mankind and its extinction. The translation of such concept to digital languages was made by Descartes, who represented time no longer with multiple cycles, which break vital energy into infinite fractal events, but as the unwinding measures of a single cyclical clock plotted toward infinity in a lineal coordinates of space, the Cartesian graph. It was understood that God, the Judgment Day, the goal of history was at the end of its X ordinates, LOL.

In that plane, time was an abstract artifact that extended into infinity. Yet that idea of a lineal, infinite, single time clock was no more than a mathematical simplification or artifact useful to gather data, not the meaning of time, neither its fundamental property, which is to be cyclical; that is, all events happen and repeat themselves ad eternal, and to be multiple, that is, the speed of change and processes of change and evolution are different for different beings of the Universe.

Ultimately, linearity departs, as most errors of science, from two of the three pillars of falsity—in this case, monism—that makes a single arrow of time to be lineal, as you need two dimensions of time to create a cycle, and mathematical reductionism since if you believe the Universe in mathematical, then you believe that the Cartesian graph is not a simplification of reality, but reality itself. So space-time is no longer what we observe, a fractal reality of multiple objects, but as continuous as Einstein and Descartes believe because your simplified mathematical instrument is a continuous sheet of paper. Those are infantile errors when you have a more complex point of view about reality, but for dogmatic, religious believers in Pythagorism (the idea that reality is made of mathematical substances), they are confirmation of your faith. Hence the dogmatic incapacity of physicists to overcome all those simplifying errors that plague their endeavor. For Aristotle, however, time was obviously cyclical because he was, like every other serious philosopher of science before Galileo, using the concept of time = change and several arrows of time that could interact in cyclical manner through action-reaction processes. Yet the straitjacket of the Cartesian lineal graph has predated science till today. So when Newton found the law of action-reaction, which obviously provokes cyclical events, and found the cyclical orbits of planets, he decided instead of a principle of cyclical inertia, a principle of lineal inertia, which not a single particle of the Universe, always moving in waves or cycles, obeys. That principle is still on every text of physics of the human world.

13. The Errors of Quantum Cosmology and Quantum Entropy

The three legs of the scientific method prove that Hawking's work is logically and mathematically incoherent in his definitions of black holes, time, temperature, and antiparticles and experimentally false. The biggest errors are the logic errors and minus zero, nill, nada comprehension of time by a man who goes around the world boasting himself as the "new Einstein," writing "brief histories of time" and promoting himself as the genius of time theory, which has found that Mr. Einstein is double wrong.

The results of this "show business," of course, show in the experimental minus-zero, nill, nada evidence of his theories. For thirty years, telescopes all over the world have tried to find a single black hole exploding as Hawking said and they have only found hundreds of them eating entire galaxies, stars, and planets as Einstein said.

Mr. Hawking, using the false concepts of quantum physics, changed the arrow of time of black holes from machines that create information to machines that evaporate it. He did it with Bekenstein just because the external membrane of the black hole or event horizon "resembled the equations of entropy, thus I came with the idea that the event horizon/membrane of the black hole was its entropy." Indeed, that is the degree of ascientificism of Hawking's way of thought. He sees an equation that resembles another and decides it is the same thing. Which is to say that if we see a bat with a wing and a bird with a wing and a plane with a wing because they look alike they must be the same thing. *This basic error of the scientific method confuses an analogy with a homology, and it is the fundamental error of quantum cosmology, which confuses the self-similarity between both space-times of the Universe, the quantum and gravitational membranes with an equality.*

In that regard, homology works between a static and dynamic perception of particles (paradox of Galileo). Thus, strange matter at CERN can be studied both with standard quantum physics and fractal relativity, *which would be the particle and vortex description of quark condensates, since both are certain*, proven by the laws of hylomorphism of the scientific method of which we have logic, mathematical, and experimental evidence.

Yet the study of gravitational black holes with quantum laws is a false attempt of quantum physicists to apply their laws of particles to objects belonging to the gravitational membrane. And so the result of confusing analogy with homology is a lot of errors that make Hawking's work false.

So goes for all theories derived of the uncertainty of Heisenberg's generator equation of quantum space-time: $E \times i = H$. This is a particular case of the generator equation of energy and form applied to the membrane of light space-time, *hence not an uncertain equation but an equation that shows the limits of form and energy of light space-time, which can be extended as pure vacuum energy or condensate into a particle of maximal form. Thus, particles are born of the vacuum, not because there is an uncertain probability but because vacuum stores energy that can evolve into fractal form.*

The absurdity of quantum cosmologists is considering H, the minimal action of energy and information, of the light Universe, first brick of the electroweak space-time membrane of which we are all made, an "uncertainty," from where quantum physicists derive all their absurd "probabilistic worlds," including the evaporation of black holes. It obviously does not exist since a top black hole is a fractal of quarks; and quarks, the densest matter of the Universe are neither hot, nor do they evaporate.

This error of considering that black holes or quark condensates have temperature is another classic "hyperbolic error." Indeed, in physics temperature, it is defined as a measure of lineal entropy/speed/energy, in the restricted space-time scale of molecules. In that medium temperature, Pv=nkT, measures the lineal vibrations of molecules. But mass is a cyclical informative motion. Thus, the concept of temperature does not even apply and the entire "discipline" of quantum entropy, invented by Bekenstein and Hawking just because Einstein's formula of a black hole resembled that of entropy, is absurd. It is Pythagoric science.

The error also happens at RHIC and CERN, which confuse the big bang of radiation at trillion degrees observed in their quark-gluon liquid, with the inner vortex of form at zero degrees that will create drops of strange liquid. In this case, E × I = k means that as radiation goes to infinite temperature, the center of the quark-gluon liquid will cool to zero and create a quark hole.

In that regard, there is only one theory of standard black holes, which is Einstein's *general theory of relativity*. It affirms that a black hole is a vortex of space-time that curves all forms of electromagnetic matter around it, like a river fall or a water drain does with the objects dragged by the stream. And if black holes existed in nature, they should be made of some ultradense mass particle. Because at the time quarks were not yet discovered, he even doubted that quark holes could exist unless that "cut-off" substance were found. And indeed, it was found. So black holes are quark stars; CERN, which is a factory of quarks, will do them and they will feed on you and me, "collateral damage" of Mr. Hawking's and CERN's ambition.

In simple terms, quantum cosmology is false and should not be considered standard science. What should be considered standard science about black holes is Einstein's theory of gravitation and his concept that black holes are frozen stars made of an ultradense substance that can only be quarks. Thus, black holes in standard science are the densest of all possible quark stars, hence made of top quarks the self-similar particle to the Higgs that CERN will mass-produce in its collisions.

This means that the two excuses CERN gives to reassure us that black holes represent no danger to mankind—that they will be difficult to produce because they need extra dimensions and that they will evaporate—are false. They are hypothesis of quantum cosmology, which is not the standard theory of black holes. In Einstein's theory of black holes, black holes will be produced at CERN certainly under the totalitarian principle as top quark stars and they will not evaporate since quarks don't evaporate. Hawking's theory is fantaphysics, science fiction, as it breaks all the laws of the scientific method. I can't recall how many times I have sent this explanation to Mr. Hawking. Penrose, one of the people who have sued CERN for genocide, who knows him personally, tried quantum cosmologists to understand the risks involved but they wouldn't listen. I went to see him at Pasadena and was detained for filming his conference in which he said that CERN will give him a new chance to get a Nobel Prize since in thirty-five years we have never found a black hole evaporating . . . but if not, we would become spaghetti. We are not facing a group of mild, idealist intellectuals, but powerful people, celebrities, icons of mass media, who are arrogant, ignorant, and care nothing for the future of mankind.

It is a myth that celebrities are geniuses and good people when they are just ego-driven humans obsessed by their ideas, who never think they can be wrong as the mass media system always lick their trou noirs.

Yet Mr. Hawking knows little about time. In a twist of cynicism, Mr. Hawking, who thinks Einstein was double wrong, made a best-seller book, *A Brief History of Time*, precisely on Einstein's theory of time, which was chosen by the media as the most important book of science since Darwin's *Origin of Species*! He did read Einstein but he never read Aristotle and so he doesn't understand that v=s/t only means change in the motion of beings. He affirms in his biography he despises biology (hence he ignores biological time) because "intelligent people studied physics in my school." In other words, he doesn't know "physical time" is only concerned with movement and so when he sees in Einstein's equation that time goes to zero, alas! He thinks he has found a time machine. Of course, philosophers of science are telling him he is wrong. But the press, who likes the idea of a time machine, who "invented" him as a genius because of his personal story of sacrifice, won't tumble him down. Mr. Hawking, unfortunately, is not a new Einstein. He is not even a serious scientist, even if the audiovisual industry has made a fortune with his TV programs and his books of divulgation of Einstein's theories about time. But nobody dares to challenge him.

And so because antiparticles don't travel to the past, black holes won't evaporate to the past. Instead they will grow forward, absorbing this planet's mass as Einstein, the best physicist of twentieth century, proved. Yet nobody will stop CERN because they are the gurus of mechanism, the religion of mankind, since Galileo measured the translational time of cannonballs and physicists became the masters of "energy weapons," deforming our comprehension of the arrow of information, which they still deny.

Further on, the mathematics of cyclical masses and black holes are far more complex than the simple lineal equations Mr. Hawking used in 1970 when there were not even chips to calculate. So not even the mathematical leg of evaporating black holes resists a serious scrutiny. Indeed, as a Berkeley professor put it:

> Actually, nobody is really sure what happens at the last stages of black hole evaporation: some researchers think that a tiny, stable remnant is left behind. Our current theories simply aren't good enough to let us tell for sure one way or the other. As long as

I'm disclaiming, let me add that the entire subject of black hole evaporation is extremely speculative. It involves figuring out how to perform quantum-mechanical (or rather quantum-field-theoretic) calculations in curved spacetime, which is a very difficult task, and which gives results that are essentially impossible to test with experiments. Physicists *think* that we have the correct theories to make predictions about black hole evaporation, but without experimental tests it's impossible to be sure.[4]

Single lineal equations in physics only explain the simplest objects, like those springs deformed linearly in high school maths, but are useless to describe how quarks become pulsars and black holes. As a Nobel Prize put it, lineal equations in physics are like needles on a straw field, almost impossible to find. Especially when we describe what seems to be one of the most complex, informative objects of the Universe, perfectly disguised to prey on all forms of matter.

We could further analyze the mathematical errors of Mr. Hawking's theory, but it is not worth to consider the more than a dozen different ways that the people of Einstein et al. have proved, according to the scientific method, Mr. Hawking wrong—being the most publicized, Mr. Rössler's paper on the R-theorem.[5] There are so many ways because Mr. Hawking just decided he wanted to study black holes with quantum laws, creating a mathematical fantasy, which he himself bet initially would not exist in nature, called an entropic black hole. In this absurd theory, he considered that the surface of the black hole, called the event horizon, had entropy, that it followed the laws of electroweak forces, when the black hole is the essential gravitational object that has reversed entropy, negantropy, information, as it forms energy into mass ($M=E/c^2$). Thus he reversed the nature of black holes and reversed its arrow of time from information into energy, following the four hundred years old custom of physicists of denying the arrow of information, mass, and life in the Universe since Galileo defined time as only the change in motion of beings ($t=v/s$). Then once the truth of black holes—that they are informative machines—was reversed into the truth of stars, that they have entropy, all the conclusions that followed Mr. Hawking's maths were by definition wrong; and since they were not possible to be proved by the scientific instruments of the age, they were just accepted by Pythagoric quantum entropy cosmologists, as they represented a perfect alibi to invent all kind of theories and study gravitation with the laws of entropy. But that is totally false. The laws of entropy and gravitation are inverse laws. Alas! Quantum gravity was then born, and for forty years, it has been impossible to make anything out of it because it is a false theory. But every quantum cosmologist loves quantum gravity, as it allows him to churn new models of black holes that he believes will never be proved wrong by the scientific method, as we cannot see black holes in nature. But this year, finally quantum gravity, the quantum treatment of black holes (title of the first Hawking paper on the subject), will be proved wrong when CERN makes them on Earth with top quarks and kills us all and Einstein will be proved double right: black holes will be frozen stars, NOT singularities as Wheeler mused, and they will obey the deterministic laws of time outlined here, NOT the uncertainties of Hawking.

It is the sheer arrogance of quantum entropy physicists that consider what they don't see as uncertain or nonexistent, what made Wheeler affirm that black holes have no hair because we cannot see them. They do have top hair, but they are so complex that they hide themselves as death does in the myth, under their black robe.

14. Top Quark, Black Holes

If black holes are made of a cut-off substance of enormous density, quarks, they will be top quark stars, which would explain also why they are so similar in form and properties to quark stars (pulsars).

Like many quark stars, extremely dense, rotating at near light speeds and emitting pulses and jets of radiation through its poles, observed black holes have jets on their poles, are extremely dense, and rotate at light speeds, in its event horizon, communicating that rotation to the light space that surrounds them. So if a black hole appears on CERN, according to Einstein's theory, it will absorb the fluid space-times and the matter of Earth, which will keep accelerating beyond light speed into the black hole sink. It doesn't matter the size of the black hole: according to the most recent models of relativity that use the equations of Einstein (in its simplified Asthekar version), a black hole as small as 10^{-35} m. might be already stable. This means *very few top quarks will be needed to start the reaction. It is even possible that a triple top, the TTT particle on top of the top decuplet of quarks, the supertop predator particle, starts it.* Physicists at CERN deny this because a top quark should react extremely fast to accrete matter and they think it turns very slow, like a ud quark at c/10 speed, since they model top quarks as small quarks. But nature will cheat them. The last big surprise of quark physics has been to find that the b quark, far less massive, hence slower, acts-reacts 1 trillion times in a second, much faster than any model expected.

So a micro black hole will act/react so fast that it will behave as the two other black holes we have observed in the bigger scales of matter, stellar black holes and galactic black holes, rotating at light speeds in its event horizon, as a vortex of mass, sucking in the electromagnetic space-time around it, exploding Earth in a mere couple of seconds. *It is Einstein's accretion process of a classic black hole*: the space-time vacuum, the electromagnetic space-time of our world, which is a field of light, will evaporate into the dark gravitational space of the hole. And so man will undo what God did and light will be uncreated.

In other words, the black hole feeds on light-space, the final energy substrata of our Universe. It absorbs and evaporates the vacuum energy that moves at light speed, the world we see around us, curved by its gravitational force as it enters the so-called event horizon, a border beyond which nothing can be perceived. That electromagnetic world has two forms: the electromagnetic, spatial, extended lineal field of light, and the imploding, informative particles or charges (electrons) that we see as "us." They are the equivalent to the smaller water molecules of a river and the fishes that live in it. Both will feed the black hole vortex, which bends and absorbs the electromagnetic membrane, the sheet of space-time in which we exist and eats up the particles and charges that fall with it.

Yet the black hole will not fall to the center of the Earth and eat the Earth upside down as the strangelet. *As the densest gravitational point of the planet, it will create a second gravitational center, the eye of a light-speed hurricane, suspended in its relative place of birth, attracting the surrounding space-time and matter. CERN will become the center of the world.*

But why does a form so small as a black hole can destroy an entire planet?

Because the speed at which black holes feed is enormous. The most common black holes are Kerr black holes that rotate around a central ring, creating a vortex, short of a tub drain that swallows in our space at the speed of light, c=300.000 km/s and beyond. This awesome speed of rotation is necessary because black holes trap, bend, and swallow all universal forms of matter and energy, including light. So by rotating at >c speed, not even light can't avoid the cyclical vortex of gravitation that sucks all matter inwards. During that process, matter and light explode, splitting in photons. Many particles die as antiparticles that either feed further the black hole or collide with other particles and annihilate with awesome explosive energy. So we see also a supernova of energy, radiating outward as the black hole feeds inward.

Those are the processes we observe in the Universe. The birth of black holes is accompanied by the collapse of matter and the release of energies never witnessed in this planet, to the point that most cosmologists believe today that both, the hypothesized big bang of a local Universe and the big bang that gives birth to young galaxies called quasars, is provoked by the explosion of a black hole. And those explosions are, in fact, independent of the black hole size. Indeed, the trendiest hypothesis about the birth of the local Universe today, the so-called hyperinflationary Universe, departs from a singularity of enormous density and the size of a tiny black hole, the smallest possible black hole which measures only 10^{-35} m.

A black hole is not a quantum electromagnetic object like an electron, which turns around 30,000 km. per second, but a gravitational species, far more powerful, moving in its surface at 300,000 km. per second, the speed of light, sucking in space-time and dragging those particles inside, where they are converted into quarks and then strings of dark energy, tachyons that are expelled perpendicularly in jets, produced inside the Kerr vortex ring.

Top quark black holes are indeed fairly well known at this stage in the fractal, relativistic models of the Universe. And they should come over 10 TeV of potency. The quark cannon will study them in 2013, making particles of the top quark decuplet, the heaviest quarks of the Universe and all its possible condensates and triads of particles of dark matter, many of which could theoretically start a mass bomb reaction ($M=E/c^2$) that would accrete the Earth into a black quark hole or strangelet. Because CERN's experts on quarks have waited their entire lives to see deconfined quarks, their anxiety to prove their theories makes them dismiss all the risks. But the real problem is the fact that quantum theorists have invaded with their theories the cosmic realm of gravitation and masses, in which Einstein's theories rule. And so Hawking's holes are false.

The LHC will create top quark condensates, which are Einstein's frozen stars—his theory of black holes. Fact is if the Higgs is the top quark already discovered, the standard model is closed and the enormous costs of making a top quark cannon cannot be sustained by fundamental research or military reasons, as the cold war has come to an end. For those reasons, Clinton wisely cancelled the earlier American SCC collider and thousands of nuclear physicists lost their jobs. This should have been the end of this industry, a spin-off of the cold war. Those nuclear physicists should have moved into the electronic industry, which is truly its alma matter—not cosmology, the realm of the big, studied by astronomers. And the different ministries of science should have pushed the research of the many big bangs with harmless telescopes far better suited to explore the Universe than a quark cannon. But CERN, the European nuclear company, founded by De Gaulle as a European bid for nuclear independence, which had been lagging behind America during the cold war, saw a chance to advance its cause and took over the project. Europeans, though, were keen to appear as pacifists. Thus, the true origin of accelerators and its military use was put aside and a series of fringe theorists, like Mr. Higgs and Mr. Hawking, took the limelight. They became instant celebrities, and their fringe theories suddenly appeared as the Golden Stone of modern physics. So finally, a machine that should have never been constructed because its costs and risks far outweighed its scientific benefits became a reality; and thousands of nuclear researchers, many working till then in Russia, constructing nuclear devices, got new jobs. Now the LHC prepares its real experiements beyond 1 TeV that will usher mankind after the age of nuclear bombs and thermonuclear bombs, into the thirrd horizon of nuclear weapons: the age of cosmic quark bombs.

We are sure CERN will make a quark bomb, though we don't know which one it will make first. Probably a strangelet, which needs less energy since neutron stars are in fact strange quark stars with strangelet centers. And all of them *are born in quark-gluon explosions. There is no other known-known mechanism known that can create them. Thus, we have the three tenets of the scientific method of truth*: We have experimental evidence (RHIC and December's experiments), logical truth (all the pages of this book and quantum theory), and detailed mathematical analysis (all kinds of papers on strange science) to know that strangelets *will be formed with 66–100% of chances at the quark cannon. So quark stars should happen at CERN.*

It was only left total experimental certainty that black holes would not evaporate, after the detailed proofs of mathematical and logic errors in Hawking's evaporating black holes. But now we do have experimental proofs that they do not evaporate.

Indeed, last year we made a dumb hole, which is a sound hole made NOT with quarks, but with far lighter electroweak atoms, which rotate 1 million times slower at the speed of sound. Yet it is based on the same principles and was theorized to evaporate sound. So for the first time on Earth, we had a way to see if the mechanisms of black hole evaporation happened at no risk. We should have heard a sound coming from the dumb hole, an evaporation of phonons, the equivalent to the photons Hawking says his black holes evaporate. But the dumb hole didn't evaporate. Thus, we wrote a letter to the world, cosigned by the main scientists of Einstein et al., and sent it also to CERN and the main magazines of science of this planet. They didn't even acknowledge its reception so they would not be blamed if something happens. The establishment of nuclear physics told us: We are not stopping the experiment. This was the letter and the exact graphic, shown in the cover of this book that accompanied it. Because indeed, last year, physicists made two experiments to prove black holes' evaporation and the results were clearly negative.

15. One Experiment That Could Save the World and One That Might Destroy It

Abstract.

The Large Hadron Collider is a quark factory that will deconfine millions of quarks, the strongest, most attractive particles of the Universe. They carry the Atoms mass, caged inside their nuclei. 99% of LHC's production, shown in this graph, will consist on super fluid Quark condensates, a new state of matter, defined by Einstein, in which Quarks fusion together, creating hyper-dense, attractive tornado-like vortices with properties similar to black holes. Astro-physicists fear that if enough quarks are pegged together in one of those condensates, they can trigger a mass-reaction that would attract all the other quarks of the Earth, transforming our planet into a dense pulsar or black hole. The European Nuclear Company that will manufacture them, called CERN, affirms they won't pose any danger, because according to a theory proposed by Mr. Hawking, small Black holes will evaporate in a burst of energy, before they can attract the mass of this planet. Yet Mr. Hawking's theory has never been proved and it contradicts Einstein's Relativity. So to prove Mr. Hawking's right, 2 experiments were devised last year: a satellite called Fermi was launched to detect radiating black holes in the cosmos, but it failed to find any. A second test was done, manufacturing super fluid condensates, similar to those CERN will make with quarks; but formed with lighter, inoffensive electro-weak Atoms. Those atomic holes rotate 1 million times slower, absorbing sound phonons instead of light photons—reason why they are called dumb holes, instead of black holes. So, this June in an experiment at Haifa[7], Atomic Condensates rotating at supersonic speed became dumb holes and absorbed sounds. Problem is they didn't evaporate, proving that Quark Condensates, made at CERN, will absorb light & matter without evaporation. Because in Nature all what is possible happens (Totalitarian principle), Quark Holes should happen at LHC, making prohibitive for Public Policy the risks for Earth of a quark factory in this planet. CERN affirms LHC carries no risk, but critics contest neither Haifa nor Fermi showed evaporation and the Company's safety report doesn't study Quark Condensates, which were little known when the factory was designed.

Corpus of the letter.

The experiment should have quenched all fears. It was an experiment that would recreate an inoffensive black hole of sound called a dumb hole. Once it was created, it would be used to prove that the 2 smaller types of holes known till date, (sound holes and primordial black holes), do evaporate by emitting energy particles, called phonons in the case of sound holes and photons for light holes. And so neither of them, in case they were manufactured here on Earth, would pose any risk to our lives.

A year ago alarms sounded around the World because a similar experiment might recreate a black hole of light, the much more powerful version that feeds on stars and galaxies around the Universe. Critics said that if black holes were created and obeyed Mr. Einstein's Laws, instead of Mr. Hawking's, they could feed on Earth, causing a Nova explosion that would destroy the planet, leaving behind only an ultra-dense rock of dark matter. They sued the Nuclear Company of Europe that was carrying about the experiment (called CERN) and got enough attention from the press as to cause a global alarm and visions of mad scientists running amok in their secluded labs, as those the Nuclear Industry hadn't suffered since the cold war era. In the end, the experiment did not happen because the machine, the so-called Large Hadron Collider, or LHC, broke down before the making of any black hole could be attempted. And so the experiment was delayed for a year . . . Enough time for Mr. Stenhauer to make his dumb hole at the Institute of Technology at Israel—an easier, weaker, slower substitute, turning at the speed of sound, which could allow mankind to observe how black holes form in the first place and then how they evaporate harmlessly. It would also dismiss the fears of critics, unfairly characterized as the product of an excessive imagination at best, or a crackpot stunt made in search of celebrity at worst—since it was truth that Hawking's radiation, a major plank in the case that the LHC was safe, did not have any empirical proof.

To overcome that major problem, physicists have carried out in the past year 2 new experiments, each of which was supposed to end all controversy:

One investigation was carried out by the Fermi satellite, a gamma-ray telescope launched last spring by NASA[8], containing the most sophisticated burst detection equipment ever devised, able to observe black hole evaporation in the cosmos, if such existed. The idea was simple. If black holes evaporated, as theorized, they would do it fast, provoking a sudden burst of light (the Hawking radiation in which its mass is supposed to convert). The result would be a sudden, minuscule Nova Explosion, even if the black hole is a small one. And the Fermi satellite should have easily detected such flares in the Halo of the galaxy, where black holes are supposed to form a dense cover of dark matter.

Houston we have a problem is reported to have said Mr. Tom Hanks, when asked during the shooting of Angels and Demons to come back at CERN and switch on the machine after the repairs were done.

And indeed, Houston had an unexpected problem when Fermi returned all kind of images of gamma-ray bursts, *but none had the signature of a black hole evaporating.* So not only was the proof of Hawking radiation lacking—a blow to LHC protagonists—but further oxygen was given to the opinion of critics that consider Mr. Hawking's ideas about black holes outlandish and in open contradiction with Einstein's well tested work.

Those critics had asked in a 2008 US lawsuit (Sancho et Al vs. Doe & CERN) for a TRO (Temporary Restraining Order) on the LHC, till the results of Fermi came back to Earth. Yet the suit was dismissed by lack of jurisdiction over CERN, a company which has diplomatic immunity, achieved during the height of the cold war era. Still critics affirm Fermi is a clear prove that black holes don't evaporate, claimed a moral victory and have appealed to a Federal Court at San Francisco, which is still arguing the case.

But by the time this occurred the world was immerse in an economical crisis, which made the issue of our possible extinction as a species by a rogue black hole, quite irrelevant, compared to the daily fight for survival among foreclosures, loss of jobs and billions of $ in losses perpetrated by expert rogue bankers.

Only Mr. Buchanan[9] in the *New Scientist* had the courage to compare both crises, insinuating that experts have a clear record of errors they never recognize. Or as the Russian physicist Lev Landau put it: *Cosmologists are often wrong, but never in doubt.*

For all those reasons, after Fermi's results, the experiment at Haifa was eagerly awaited by both, supporters and critics, to settle the matter with an experiment that could be controlled.

How did the scientists at Haifa achieve that test-bed? They basically followed Einstein's theory that black holes are frozen stars, which therefore should contain an ultra-dense substance that causes their attractive power.

According to Einstein's realistic vision of the Universe, for Holes to exist beyond its mathematical description, they should be made of an ultra-dense substance, able to act as gravitational vortices that absorb one or other type of fundamental energy, depending on the speed at which they turn. This ultra-dense substance, called an Einstein-Bose Condensate, found after Einstein's death, happens only with 2 kinds of atomic masses: lighter atoms, which attract each other with the electro-weak force; and heavier quarks, which display a force 100 times stronger than the electro-weak force, called strong force. For that reason, atomic condensates turn much slower, absorbing only sound. While quark condensates turn, thanks to their strong force, 100^3 times faster than sound, at light speed, as a black hole will do. Unfortunately, when the quark factory was designed, physicists didn't know quarks could form Einstein's condensates, locking their color in triads to become stable, before they formed those super fluid, liquid condensates. They thought quarks were unstable, repulsive fermions that would become a gas—an explosive, hot, quark gluon-plasma. So when we found they form instead, a cold, attractive super fluid, such discovery was labeled as a perfect surprise by all the physicists involved.

Thus Haifa scientists used the laws of Bose-Einstein to densely pack atoms into an Einstein condensate, creating a supersonic hole that absorbed sound waves. Then they waited to see the Hawking radiation, expecting that the black hole would emit the sound equivalent of photons—phonons—in a burst of noises that a good microphone could easily pick up. In the process the tiny dumb hole would evaporate . . . Thus CERN would be declared safe and start producing Quark Holes, under the same laws of Einstein condensates this Christmas.

And so a mike was put to the dumb hole to detect, not a glare of light as the Fermi was supposed to observe in the cosmos, but a symphony of sounds.

Unfortunately the worst possible scenario happened. On one hand *sound holes were produced with a Bose-Condensate, which means the LHC will produce black holes with quark condensates without the need for extra-dimensions* postulated by CERN.

On the other, the dumb hole did not produce any sound at all, any Hawking radiation. *So CERN's super fluid quark holes won't evaporate, either. Instead they should fall to the center of the planet and accrete the Earth.*

Unruh and Hawking, the theorists who expect to win a Nobel Prize if the radiation is found, retorted that Haifa is not relevant because the experimenters might not have ultra-sensitive equipment to detect it. Maybe, but in safety terms, critics assert that what matters is that the possible radiation was so subdued that we couldn't see it or hear it, neither with Fermi satellite or at the Technion Institute in Haifa; so it will not be able to evaporate the hyper-dense mass of atoms or quarks that form sound/quark holes.

Those critics claim that CERN is not recognizing the two methods of truth in science, experimental evidence and serious, tested theoretical work—in this case Einstein's Relativity and Einstein's condensates that deny evaporation and Haifa's condensates, which proves his frozen stars exist and could be produced at the LHC. Those critics further assert that CERN is acting as any Rogue Company would act to hide an environmental crime—denying evidence of any danger, ignoring critics and Courts and going ahead with business as usual. Thus, they conclude the dangers to Earth of the black hole factory are a potential environmental catastrophe, which dwarfs the problems of global warming and Nuclear Terrorism we already face. *And for this reason, they should be addressed by politicians, not left alone to physicists working in the Nuclear Industry, as Miss Gillmor stated in her judgment of the case (Judge Gillmor in Sancho et Al vs. Doe & CERN).*

The dispute between quantum physicists and relativists.

Yet, even if we, critics are right, when considering CERN an Industrial con-CERN that hides the environmental dangers of those experiments, as any company will do—to the point that CERN's Chief Officer, Mr. Engelen, has confessed to the New Yorker magazine the existence of a zero risk confidentiality statement, signed by CERN's employees[10]—it would be unfair to ignore that the evaporation of black holes is also a theoretical argument. And one of enormous importance for our understanding of the Universe, which dates back to the times of Einstein and Bohr, when the first affirmed that the Universe was real, made of deterministic substances, because God didn't play dice, and the second retorted the Universe was Pythagorean, made of mathematical probabilities and Einstein had no right to say how God played it. In this argument, is where Mr. Hawking enters with his landmark article, The Quantum Laws of black holes[6], siding with Mr. Bohr. He affirmed then that Einstein was double wrong, as God plays dice twice, first creating mathematical black holes without substance and then evaporating them . . . Because particle physicists are quantum cosmology, not relativistic cosmologists, it is probably truth that CERN's physicists do not want or expect any risk coming out of his black holes. They believe in Bohr and the quantum laws of black holes of Mr. Hawking, which considers that tiny black holes are quantum particles that no longer obey Einstein's laws of relativistic black holes. But it is also truth that black holes are cosmological objects, not particles, ruled by Einstein's well proved Relativity laws that state expressly the relativity of all sizes. So size does not affect in Einstein's work the laws of black holes. It means that if Einstein is right, black holes will evaporate the Earth. Unfortunately, Haifa's experiments seems to show Mr. Einstein was right twice, because black holes were made of a substance, defined by Mr. Einstein—Einstein-Bose condensates—and they did not evaporate. Thus, they followed the realist laws of Relativity not the mathematical laws of Quantum Entropy. And so the attempt of quantum cosmology to define, after Einstein's death, during the Cold War Era, at the height of their power and prestige, gravitational objects like black holes, with the laws of quantum theory and entropy, must be considered failed—an invasion by a powerful group of scholars of a discipline, which is not yours, but they feel entitled to discuss, due to that power and capacity to publish and influence the academic world. A similar phenomenon happens in Biology with Theory of Evolution, which is fit to explain many phenomena but it is not the engine and only Law that applies in biology, despite the constant attempts of their practitioners that do wishful thinking to explain sociology, history, economics and cosmology with those laws. Plainly speaking quantum cosmology is the science of the small not of the cosmos and its massive particles, which should be studied not with quark cannons but with telescopes, not with quantum entropy but with Relativity and Einstein's theory of ultra-dense mass-condensates . . .

In that regard, Haifa's approach differed from CERN's quantum theorists, as it followed Einstein's theories. The difference is important because if black holes are frozen stars made of an ultra dense substance, as Einstein claimed, they or similar particles could be easily made at CERN, which has built a quark cannon that will lump quarks, the densest matter of the Universe, into Einstein condensates, one of which could become a black hole, strangelet, pulsar or quark star, able to accrete the Earth. On the other hand, if black holes are objects without substance, made of pure gravitational energy, as quantum cosmology at CERN claim, they would be more difficult to make and easier to evaporate back into pure energy (Hawking's radiation). So they probably won't pose a big danger to the Earth.

Unfortunately at the time of Einstein's death, because we still didn't know about the existence of quarks, the idea that black holes didn't have substance became the dogma of quantum cosmology, which had been arguing with Einstein about the mathematical nature of reality and found in those mathematical black holes the ultimate prove that the Universe was probabilistic, mathematical. Einstein denied this, as he believed in the fundamental principle of the experimental method, called hylomorphism, according to which all information, including mathematical equations, must refer to a substance, in order to be real. After all, his friend Gödel had proved that mathematics were not real per se, but as all languages could create fictions. And so Einstein told quantum cosmology that while I know when mathematics is truth I cannot know when they are real. Frozen stars were in Einstein's thought those real black holes. If so, they will be made of quark condensates, and they won't evaporate as the ones at Haifa didn't evaporate; because quarks and atoms are dense forms of matter that cannot be easily made into photons or phonons.

Indeed, the experiment at CERN is one that carries a very deep argument about the nature of the Universe and, given the results observed by Fermi and Haifa, which favor Einstein's thesis about frozen stars, do represent a risk for mankind.

And yet this time around the press, politicians and the public at large, has ignored the issue—only an article appeared in the Economist among the newspapers that create opinion—perhaps because the nightmare situation of a black hole showing up at Geneva and digesting this planet and all its inhabitants is so terrifying that nobody wants to consider such possibility seriously anymore. It is easier to think the matter is resolved. But it is not. Nor we can claim that the Nuclear Industry is so safe that no risk is possible—we all know that the Nuclear Industry was responsible for decades of terror during the cold war. Critics in fact have retorted, claiming "ad corpore" disqualifications towards the Nuclear Industry, citing the Chernobyl's catastrophe, which the Byelorussian government claims to have caused over 200.000 deaths of cancer, and the International Agency for Nuclear Energy considers to be nearly harmless with only 56 victims[11] . . .

This, largely resolved dispute in favor of the Byelorussian government, yet considered also solved by the Nuclear Agency, shows how difficult is to differentiate truth from falsity in themes that are poisoned by large industrial interests, enormous sums of money and thousands of jobs at stake. CERN's machine is the most expensive ever made. It employs directly 8 thousand people and indirectly 66 thousand physicists around the world that expect to study its results. So most physicists side with CERN; but neither "ad hominem" disqualifications nor matters of quantity are relevant to determine the truth in science. When Mr. Einstein received from a friend in Germany a book called "100 authors against Einstein"[12] he responded, if they were right, a single one with reason would be enough. Einstein had already been cast aside by the German Physicists Establishment, because he alone among all his peers denied the existence of a mathematical substance, ether, which was as bizarre as Hawking's evaporating black holes: denser than steel, transparent; it filled the entire vacuum but had never been observed. Yet all believed in ether, because those were the only available equations at the time, regardless of its absurdity. It was the sort of aberrant reality that we can observe in Hawking's radiation, which contradicts many well-proved laws of science, including Relativity; hence it must be treated as a speculation or fringe theory, *not the standard theory of black holes, which is Einstein's.* Yet in the ether case, the German industry, also the industrial power and main contributor of CERN, was constructing new machines to measure ether and had even introduced a Universal constant of ether, which was named with the electron charge and the gravitational constant, one of the fundamental constants of the Universe. In both, the ether theory and the Hawking Radiation, was at work the confessed Pythagorism of Physicists, which wrongly considers mathematical proves to be more important than experimental evidence, as they study unobservable scales of nature. But that approach is risky because mathematics can never be proved truth—after Gödel's work—without experimental evidence. Thus, if certain theory contradicts known laws of nature, as Hawking's radiation does, normally is logically or experimentally false.

Obviously Einstein was both, alone and right in the ether dispute. He might be again alone and right in the dispute about the nature of black holes, which he always warned, would require a cut-off substance, if they were possible, which can only be precisely those Einstein's super fluid Quark condensates, he and Bose, an Indian Physicist, masterly described in their study of indistinguishable particles that come together as one, in ultra dense, super fluid, attractive vortices as those we saw in Haifa.

And yet CERN's industrial might and Mr. Hawking's reputation is such, that few physicists today back Einstein's realistic theories of black holes. They are a minority and seem to have lost the battle for the public opinion. In that realm, there is no doubt CERN deserves by its skills in marketing, if not the Nobel Prize, (which will have to wait till we find the signature of an evaporating black hole among the millions of particles the LHC will create, or NOT, in which case all will come to pass), at least the Harvard MBAs prize for its smart promotion of Higgs particle: CERN says Higgs particle will explain the meaning of the Universe, and so it sells it to the press as God's particle. While critics adduce it is just a stunt, a redundant particle akin to the Top quark, which has similar properties, but it is already discovered . . . Curiously enough, the proponent of the Top quark theory of mass, Mr. Nambu, not Mr. Higgs, won this years Nobel Prize, pointing out again in the direction of critics. In any case, God has been always the best excuse for good marketing, though not the best name for good science. And indeed, the fact that most scientists and the press have sided with CERN is a proof of Says law, one of the pillars of capitalism, which affirms that anything can be sold with good marketing, even a war or the extinction of the Earth—cynics could say in this case.

All in all the issue seems now more clear than it was a year ago: CERN is obviously a very powerful company, an economical powerhouse, an institution that the scientific community cheers as the jewel of human ingenuity in the exploration of the microcosms of quantum particles it aims to create. It is the authority of nuclear physics. While the critics of this company and experiment, mainly the people that appears at lhcfacts.org and lhc-concern.info, are a minority, even if they defend Einstein's well proved work, who have been exposed to public ridicule during the last year, and nobody wants to listen. And yet, stubbornly, against the desire of our physicists and mankind at large, which believes in their high priests of science, reality seems to side with Einstein and his supporters, in both experiments, the Fermi satellite and the dumb hole. And this, in science, which is not a religion but a process of rational discovery, is called the experimental method, the higher law that all scientists must respect. So, we, the people of this planet, perhaps should reconsider the theme of the black hole and listen for once to the petitions of those critics, which have been largely ignored, except by the sensationalist press and some TV-comedians, who have made jokes about those black holes.

What do we exactly ask for? Basically some reasonable suggestions: first to halt the experiments at CERN during the winter season, when CERN normally doesn't work—as it runs a much more expensive electricity bill, consuming the equivalent energy to the entire city of Geneva. This is customary at CERN, to save money for the tax payer; but this year CERN has changed its schedule to rush the experiment in winter, despite a 10 million $ extra-bill, precisely to solve the issue matter-of-fact and quench further criticism . . . or blow us all.

Then, critics ad, during this winter, physicists should repeat the experiments on dumb holes at Haifa and other labs around the Earth, now rushing into the creation of similar dumb holes.

Finally, we also would like to argue the facts of those experiments not only in Courts but in scientific magazines and mass-media outlets that claim routinely only voice the institutional opinions of CERN, as this is a theme that concerns all the citizens of the Earth and in a Democracy they have the right to be informed, specially in matters that might put them in harms way.

Because CERN was initially planning to close down in winter and save the tax-payer many millions of dollars in an age of economical crisis, the idea of delaying those winter months CERN's bid to create black holes and other possible forms of quark matter, seems quite reasonable. It will also help to clean up CERN's act, which should not rush-in by force an experiment that is, after Fermi and Haifa's results, clearly uncertain and poses a risk to Earth that no longer can be denied.

Nuclear Physicists tend to dismiss, after a century of cold war criticisms, those fears and have not given much thought to this issue. CERN also has diplomatic immunity, achieved during the Cold war to lure physicists working in America, as the people who sued them found to their surprise. So CERN did not appear at the suits against the company, adducing lack of jurisdiction, which judges granted. Yet it is difficult to accept that a Nuclear Company, able to produce such dangerous substances must have, in a democratic world, impunity for its actions. So the Court also argued that the matter should be treated by politicians in parliaments, as it affects all human beings, not only physicists. In other words, judges passed the ball to a higher authority, which didn't catch it.

And yet, if there is a scientific issue today that endangers the security of multiple nations and should be discussed seriously is this one; plainly speaking the first chance since the apparition of the homo sapiens that the human species can truly become extinguished. In that regard, even if CERN plays its cards with cunning intelligence and seems by all means unassailable to justice, to delay the experiment a few months more after so many years and save several million dollars, should not be a measure difficult to implement by the French or Swiss administration that provides the electricity to this corporation, with some consensual administrative agreement between all the parts involved. Contrary to belief such measure could greatly benefit science, because it would show that Nuclear Physicists do have a heart and care for life; that science after all is not a dogmatic, fundamentalist religion, but it has its experimental and theoretical methods of truth, which its practitioners ought to respect, regardless of personal agendas and industrial contracts. So far this respect for human and scientific values has not come clear in this issue, nor did it become clear during the cold war and the Chernobyl tragedy. Yet, to be able to doubt and to hear the other part in a reasoned debate, without censorship of truth, is what has traditionally differentiated scientists from believers and should again differentiate the issue of CERN's black holes from other charged issues of our political and economical world. Science cannot be guided by greed, violence and individual agendas, as most of our society is, but must remain the idealistic, passionate search for truth that it was meant to be when its Founding Fathers defined the laws of truth of the experimental method. Further on, if the issue could be discussed publicly, people and politicians could greatly benefit of this argument and learn many facts about cosmology, black holes and the nature of the Universe, which so far physicists, isolated in their cathedrals of thought, have been unable to explain to a wider audience.

We could indeed claim after the debate is closed that either:

— Black holes do evaporate and the experiment is safe, if Fermi-like satellites or new dumb holes show this to be the case. CERN will appear then as a serious, responsible scientific institution, which truly seeks the ideal of knowledge; and clean its name and that of the Nuclear Industry, now tainted by this and other previous, not so-safe experiences. And Mr. Rolf Heuer, its CEO, could receive the Nobel Prize of Physics for completing the astonishing task of creating such wondrous machine.

— Or claim that Mr. Steinhauer, from Israel's Technology Institute, has saved the world, alerting mankind of an unexpected danger, by discovering that black holes don't evaporate. He could then receive both, the Nobel Prize of Physics and of Peace, showing that science, when working with responsibility, guided by the desire to improve the life and knowledge of the human kind, is always right.

On the other hand, if CERN keeps ignoring all critics and experimental results and carries out the experiment, it will show that science is no longer the kingdom of reason but it has been polluted by the dogma of industrial ethics, so well described by Erich Fromm, in his critic of the cold war:

"Technological civilization is programmed by the principle that something ought to be done because it is technologically possible. If it is possible to build nuclear weapons, they must be built, even If they might destroy us all. Once this principle is accepted, humanist
Values (something has to be done because it is needed by man) are
Dethroned and technological development becomes the foundation of ethics."

This might be indeed, the belief of many in our society in love with technology, but we live in a democracy whose ethic foundations are precisely humanist values we must not ignore when the life of so many might be at risk. In that sense CERN's experiments are no longer only a question of making science happen, but also one of social responsibility that the public must be able to debate openly after what Haifa and Fermi have shown.

When Wheeler and Teller invented their costly, industrial process to make Hydrogen bombs, they envisioned a massive weapon that could blow up Russia in a single strike. Physicists were all in favor of it, and Wheeler and Teller, to convince President "Give em Hell Harry" Truman to fund an ultra-expensive crash program, adduced the need for further knowledge about the processes taking place within the atom, not the profits he would obtain by building them, or even the enormous death toll they could cause.

But the next president, Mr. Eisenhower, a military man who knew better about the sometimes disguised goals of the military-industrial complex, realized that such a bomb could destroy the entire planet. Thus, he agreed with the Russians to halt the development of bigger bombs. Yet the Hydrogen bomb is not the limit of the possible size of Nuclear Devices. Those quark condensates the LHC will explore are cosmic, Mass Bombs, $M=E/c^2$, the most potent explosives of the Universe. To make them here on Earth, even if the Nuclear Industry highlights the knowledge their big bangs will bring to mankind, just as Teller did, will be indeed, a dumb thing to do.

Signed or endorsed by "Einstein et al":
Markus Grostchnig et Al. Plaintiffs in the suit at Strasbourg's Court of Human rights against CERN.
Otto Rössler, Ph.D. Tubingen U. Lhc Critic.
Mark Leggett, Risk Expert, Brisbane University.
Jim Tankersley, LHCfact org.
Luis Sancho and Walter Wagner, Plaintiffs of Sancho vs. DOE

This article was censored. A human genocide can only happen with censorship, dogmatism, technological messianism, and a corrupted military-industrial mass media system that has made war and violence the trademark of our civilization. It did point however to the origin of this conundrum—the absurd belief by Pythagoric scientists such as Hawking, Oppenheimer, and Wheeler, the father of modern nuclear weapons *and* Pythagoric singularities, that black holes were mathematical entities. But mathematical entelechies do not exist. *Languages only exist in the mind of man, they are not real. Dogs are not made of ds os and gs and black holes are not numbers.* Black holes do not evaporate because they are not mathematical objects, defined by manipulating at will mathematical equations as those three "nuclear physicists" did. Yet the fathers of sci-fi mathematical black holes were the makers of the atomic bomb (Oppenheimer) and the hydrogen bomb (Wheeler). And so whatever they said was "okay"; and it is still okay in a profession that has merely created a rhetoric "newspeak" of peaceful research while doing what it always did best—nuclear weapons.

In Haifa, Jewish scientists made a real homologous dumb hole, a black hole of sound, less dangerous, as it has only two regions of lower and higher sound speed, in both sides of the dumb hole event horizon, while a black hole of light has zones of <c and >c superluminal speed of attraction beyond the event horizon. Thus, we shall call those two holes by their speed, supersound holes and superluminal holes, because precisely their fundamental property is the fact that their accelerated vortices keep going beyond their event horizon, a three-dimensional surface that separates their fluid and superfluid regions of speed, <S and >S for supersound holes and <c and > C for superluminal holes.

The dumb Hhle at Haifa is a supersound hole, where the sound fluid moves faster inside and slower outside. It proved that black holes will have superluminal speeds inside the event horizon as fractal relativity precludes.

Thus two things have become clear: That if Hawking's radiation exists, it is very feeble as it has not been detected in supersound holes. And because it is so feeble, even if it exists, *it didn't evaporate at all the black hole, which kept absorbing bosonic atoms, till the entire mass of bosonic atoms was exhausted.*

Thus we learned from this astonishing experiment what will happen when a black hole forms at CERN this Christmas because they will be homologous:

First, we will observe the creation of an event horizon, the membrane/barrier between a region with a subluminal speed outside the black hole and a region with a superluminal speed inside the hole.

Then the light region will create a hurricane inward, and our mass will be reconverted into denser mass, top quarks and dark bcc atoms of dark matter, never observed before.

This inner attractive strong quark superfluid, rotating faster than the speed of light, at >C, is the region from where light cannot escape. We now can see inside that region without much risk with dumb holes. This region was a hyperdense fluid of bosonic condensated atoms. So CERN will create the homologous quark fluid condensate, what Einstein called a frozen star: a bosonic state of quarks.

If we calculate the parameters of a top quark boson charged positively, we obtain a Kerr black hole with a central ring at 10c superluminal speed, inside the >C region. This happens *using the classic theory of black holes.* So it will happen at CERN in 2013, when thousands of top=Higgs quarks will peg together, forming a boson condensate, Einstein's frozen star. It will be a top quark boson condensate positively charged, with the form of a Kerr singularity, spinning at faster than light speed, attracting all other spins of mass toward it, till it exhausts the matter of the Earth, as the dumb hole attracted all the atoms of rubidium that were available. If we arrive at 2013 and the strangelet soup that will be by then forming in the center of the Earth doesn't kill us all faster, we shall become a condensate of top quarks.

Indeed, add 1 (Haifa and Fermi satellite show negative experimental evidence that proves black holes don't evaporate) and 1 (Hawking's radiation breaks the logic and mathematical laws of science) and 1 (dumb holes prove top quark stars will be black holes), you get 3: black holes don't evaporate regardless of CERN's wishful thinking. And so if Einstein is triple right, not double wrong, as Hawking affirmed in his article on black hole evaporation, Einstein-Bose quark condensates will form at the LHC and we are living the last year of our lives. As certain as 1+1+1= 3, for the same reasons: man might not obey the laws of science, but the Universe always does. And those laws of science show, regardless of mathematical pythagorism and alternative fringe theorists, as those CERN sponsors that top black holes don't evaporate and will very likely form at LHC, a machine designed to produce top=Higgs.

---

[1]   Mr. McCain explains this tragedy in his autobiography as a clear case of group thinking and despise for life in technocratic organizations. The people at NASA had a plan to follow, a schedule, money was involved, a need to accomplish objectives—all those elements and an arrogant confidence prevented the physicists working on the shuttle program to consider the obvious fact that essential parts of the machine would be frozen at the current temperatures, break, and provoke the catastrophe. Only Mr. Feynman, an independent consultant, who was not bound by group thinking and planning could say the truth; but he was put aside as Einstein et al. has been in the incoming tragedy at CERN.

[2]   Http://map.gsfc.nasa.gov/. The Wilkinson Microwave Anisotropy Probe (WMAP) is a NASA Explorer mission that launched June 2001 to make measurements of the background radiation. It is fascinating though to observe how the proofs against the big bang of WMAP and many other satellites are brushed aside systematically and only experiments that fit data are considered. This way of making science is obviously against the scientific method, as a single proof against a theory invalidates it, as per Popper's theory of scientific truth. Truth is falsifiable and so while many proofs in favor of a theory do not give us certainty of its truth, a single proof against it dismisses the theory. According to this simple rule, the Higgs, the big bang, and Hawking's radiation must be considered false.

[3]   Rudolf Hoess explains in his biography how the reading of the deeds of the god of death calmed his consciousness of the genocide at Auschwitz. Mr. Oppenheimer liked to quote Lord Shiva and Kali and was proficient in Sanskrit. The old Aryan culture of energy and death has indeed evolved into the German culture of "entropy" and "nuclear physics."

[4]   Mr. Rössler's proof of the falsity of Hawking's radiation has never been contested by CERN, which systematically ignores any scientific paper that proves the dangers of the collider.

[5]   Http://cosmology.berkeley.edu/Education/BHfaq.html.

[6]   "The Quantum Properties Black Holes," *Scientific American*, 1977, was Mr. Hawking's landmark article in which he insinuated that black holes traveled to the past and Einstein was double wrong, as a mathematical exercise, which he himself doubted. Only time, fame, and the LHC made this speculative article a dogma of industrial physics.

[7]   Technion's report of dumb holes can be found at *http://www.nature.com/nphys/ journal/v5/n3/abs/* nphys1177.html. An easier article on the theme was published by *The Economist* at http://www.economist.com/science-technology/displaystory. cfm?story_id=13855412.

8   Http://fermi.gsfc.nasa.gov/. Fermi satellite found more than 1,500 *sources of gamma rays and none was a signature of an evaporating black hole. Further on, it found that cosmic rays are the debris,* NOT *the cause of supernova explosions.*

9   Mr. Buchanan's analysis is right on the spot: CERN cherry-picks theories that prove no risk and rejects those who affirm risks without applying the basic laws of truth of the scientific method to evaluate the veracity of each theory. The article of Mr. Buchanan can be found at http://www.newscientist.com/article/mg20126926.800-how-do-we-know-the-lhc-really-is-safe.html?page=2.

10   *The New Yorker* magazine, Crash course, May 14, 2007.

11   IAEA reports only 56 victims, who die on the site; the Byelorussian government claims 200,000 cancer victims in between an independent report by Torch that can be found at http://www.chernobylreport.org/?p=summary.

12   "100 Authors Against Einstein" (German: *"Hundert Autoren gegen Einstein* is the name of a booklet that denied the theories of Albert Einstein, published in Leipzig, Germany, in early 1931. To these, Einstein reportedly said, "Why 100 authors? If I were wrong, then one would have been enough!"

CHAPTER 7

Safety Standards: Damned Lies and Statistics

If you repeat a lie enough times, people will believe it.
—Goebbels, father of industrial and political propaganda[0]

I. "Safety Standards"

CASTOR detector
Centauro And STrange Object Research

A forward calorimeter of the CMS experiment, to be used in pp and HI physics. In particular designed to search for Centauros and Strangelets in the baryon dense, very forward phase space region in central Pb + Pb collisions at the CERN LHC.

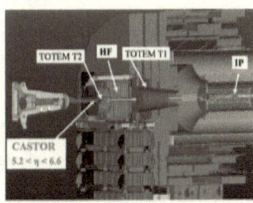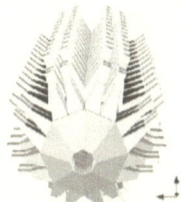

Castor is the acronym for Centauro and Strange Object Research. In the labyrinth of CERN, evil reaches its absolute meaning when one learns that they know what they do, as the military know it is in the profession of killing, the corrupted know they are in the profession of stealing, the rapist in the profession of raping, and yet they keep doing it. When I was young, I didn't understand eviL. I thought like most people in the world that eviL was an error in the program. Today eviL=Death, I Know, is half of the program, as what lives must die for the game to continue. This is the deep explanation of why CERN happens: Kali does exist. The details are scarier. In the graph, we can see in the open, telling us directly for those initiated in the knowledge of physics that one of the detectors built at CERN has as its purpose to search for strangelets. Thus they know they can produce them in Pb-Pb (lead-lead collisions). And yet nobody cares to stop them. Because whenever it is needed, CERN will invent lies and statistics. When we asked them about the strangelet CASTOR detector, they send us this message: CASTOR stands for the Cern Advanced STORage manager. It is a hierarchical storage management (HSM) system developed at CERN used to store physics production files and user files.

1. The LSAG Report[1]

Castor indeed is all what this is about—Castor and Polux, the two mythic twins of mythology, the dual fruits of the tree of science. Castor has been built, and it will be used to observe strangelets because it is such a beautiful machine that cannot be wasted. Castor proves that CERN knows it will form strangelets, but it denies it in all its communiqués and LSAGs the safety documents that the same people working at CERN under confidentiality statements of admitting only zero risk of potential catastrophe does. They keep churning them so their PRs can pass the statements to the press, along with their constant insinuations that all the scientists who oppose CERN are "crackpots" because they are not working for CERN. Let us then consider those "safety statements" full of omissions, "damned lies and statistics."

The fundamental characteristic of CERN's safety statements is its omissions. It simply says nothing about quark condensates and top quark stars, 99 percent of what its factory will produce, which is like a factory of missiles whose safety standards would not even mention missiles. How this is possible has only an explanation: Mankind trusts CERN's physicists and doesn't understand anything about physics. So it really doesn't care what CERN says as long as its communiqués end up saying it is safe. The politicians that pay CERN couldn't care about what those people do, God's particles or toilet ones. Their research is utterly irrelevant for mankind. So they have never cared at all to question what CERN will do, reason why CERN can say that it will do Higgs particles, SUSYs, cosmic rays, and evaporating black holes, all damned lies, which nobody in the press or the political system has ever cared to check out. To them, it is the same—a SUSY and a sushi, a quark condensate and a milk condensate, the liquid spaghetti that Ferran Adria used to inaugurate the machine, or the spaghettis Mr. Hawking said in a public conference we will all become if the black holes at CERN do not evaporate.

So what CERN has done is to say near to nothing about what it will really do. Only *after the suits caused public alarm back in 2008 did* CERN *finally recognize the chances of some danger and published the following statement in its Web, which we shall now discuss:*

The safety of the LHC

The Large Hadron Collider (LHC) can achieve an energy that no other particle accelerators have reached before, but Nature routinely produces higher energies in cosmic-ray collisions. Concerns about the safety of whatever may be created in such high-energy particle collisions have been addressed for many years. In the light of new experimental data and theoretical understanding, the LHC Safety Assessment Group (LSAG) has updated a review of the analysis made in 2003 by the LHC Safety Study Group, a group of independent scientists.

LSAG reaffirms and extends the conclusions of the 2003 report that LHC collisions present no danger and that there are no reasons for concern. Whatever the LHC will do, Nature has already done many times over during the lifetime of the Earth and other astronomical bodies. The LSAG report has been reviewed and endorsed by CERN's Scientific Policy Committee, a group of external scientists that advises CERN's governing body, its Council.

The following summarizes the main arguments given in the LSAG report. Anyone interested in more details is encouraged to consult it directly, and the technical scientific papers to which it refers.

Cosmic rays

The LHC, like other particle accelerators, recreates the natural phenomena of cosmic rays under controlled laboratory conditions, enabling them to be studied in more detail. Cosmic rays are particles produced in outer space, some of which are accelerated to energies far exceeding those of the LHC. The energy and the rate at which they reach the Earth's atmosphere have been measured in experiments for some 70 years. Over the past billions of years, Nature has already generated on Earth as many collisions as about a million LHC experiments—and the planet still exists. Astronomers observe an enormous number of larger astronomical bodies throughout the Universe, all of which are also struck by cosmic rays. The Universe as a whole conducts more than 10 million million LHC-like experiments per second. The possibility of any dangerous consequences contradicts what astronomers see—stars and galaxies still exist.

Microscopic black holes

Nature forms black holes when certain stars, much larger than our Sun, collapse on themselves at the end of their lives. They concentrate a very large amount of matter in a very small space. Speculations about microscopic black holes at the LHC refer to particles produced in the collisions of pairs of protons, each of which has an energy comparable to that of a mosquito in flight. Astronomical black holes are much heavier than anything that could be produced at the LHC.

According to the well-established properties of gravity, described by Einstein's relativity, it is impossible for microscopic black holes to be produced at the LHC. There are, however, some speculative theories that predict the production of such particles at the LHC. All these theories predict that these particles would disintegrate immediately. Black holes, therefore, would have no time to start accreting matter and to cause macroscopic effects.

Although theory predicts that microscopic black holes decay rapidly, even hypothetical stable black holes can be shown to be harmless by studying the consequences of their production by cosmic rays. Whilst collisions at the LHC differ from cosmic-ray collisions with astronomical bodies like the Earth in that new particles produced in LHC collisions tend to move more slowly than those produced by cosmic rays, one can still demonstrate their safety. The specific reasons for this depend whether the black holes are electrically charged, or neutral. Many stable black holes would be expected to be electrically charged, since they are created by charged particles. In this case they would interact with ordinary matter and be stopped while traversing the Earth or Sun, whether produced by cosmic rays or the LHC. The fact that the Earth and Sun are still here rules out the possibility that cosmic rays or the LHC could produce dangerous charged microscopic black holes. If stable microscopic black holes had no electric charge, their interactions with the Earth would be very weak. Those produced by cosmic rays would pass harmlessly through the Earth into space, whereas those produced by the LHC could remain on Earth. However, there are much larger and denser astronomical bodies than the Earth in the Universe. Black holes produced in cosmic-ray collisions with bodies such as neutron stars and white dwarf stars would be brought to rest. The continued existence of such dense bodies, as well as the Earth, rules out the possibility of the LHC producing any dangerous black holes.

Strangelets

Strangelet is the term given to a hypothetical microscopic lump of "strange matter" containing almost equal numbers of particles called up, down and strange quarks. According to most theoretical work, strangelets should change to ordinary matter within a thousand-millionth of a second. But could strangelets coalesce with ordinary matter and change it to strange matter? This question was first raised before the start up of the Relativistic Heavy Ion Collider, RHIC, in 2000 in the United States. A study at the time showed that there was no cause for concern, and RHIC has now run for eight years, searching for strangelets without detecting any. At times, the LHC will run with beams of heavy nuclei, just as RHIC does. The LHC's beams will have more energy than RHIC, but this makes it even less likely that strangelets could form. It is difficult for strange matter to stick together in the high temperatures produced by such colliders, rather as ice does not form in hot water. In addition, quarks will be more dilute at the LHC than at RHIC, making it more difficult to assemble strange matter. Strangelet production at the LHC is therefore less likely than at RHIC, and experience there has already validated the arguments that strangelets cannot be produced.

Vacuum bubbles

There have been speculations that the Universe is not in its most stable configuration, and that perturbations caused by the LHC could tip it into a more stable state, called a vacuum bubble, in which we could not exist. If the LHC could do this, then so could cosmic-ray collisions. Since such vacuum bubbles have not been produced anywhere in the visible Universe, they will not be made by the LHC.

Magnetic monopoles

Magnetic monopoles are hypothetical particles with a single magnetic charge, either a north pole or a south pole. Some speculative theories suggest that, if they do exist, magnetic monopoles could cause protons to decay. These theories also say that such monopoles would be too heavy to be produced at the LHC. Nevertheless, if the magnetic monopoles were light enough to appear at the LHC, cosmic rays striking the Earth's atmosphere would already be making them, and the Earth would very effectively stop and trap them. The continued existence of the Earth and other astronomical bodies therefore rules out dangerous proton-eating magnetic monopoles light enough to be produced at the LHC.

Other aspects of LHC safety:

Concern has recently been expressed that a "runaway fusion reaction" might be created in the LHC carbon beam dump. The safety of the LHC beam dump had previously been reviewed by the relevant regulatory authorities of the CERN host states, France and

Switzerland. The specific concerns expressed more recently have been addressed in a technical memorandum by Assmann et al. As they point out, fusion reactions can be maintained only in material compressed by some external pressure, such as that provided by gravity inside a star, a fission explosion in a thermonuclear device, a magnetic field in a Tokamak, or by continuing isotropic laser or particle beams in the case of inertial fusion. In the case of the LHC beam dump, it is struck once by the beam coming from a single direction. There is no countervailing pressure, so the dump material is not compressed, and no fusion is possible.

Concern has been expressed that a "runaway fusion reaction" might be created in a nitrogen tank inside the LHC tunnel. There are no such nitrogen tanks. Moreover, the arguments in the previous paragraph prove that no fusion would be possible even if there were.

Finally, concern has also been expressed that the LHC beam might somehow trigger a "Bose-Nova" in the liquid helium used to cool the LHC magnets. A study by Fairbairn and McElrath has clearly shown there is no possibility of the LHC beam triggering a fusion reaction in helium.

We recall that "Bose-Novae" are known to be related to chemical reactions that release an infinitesimal amount of energy by nuclear standards. We also recall that helium is one of the most stable elements known, and that liquid helium has been used in many previous particle accelerators without mishap. The facts that helium is chemically inert and has no nuclear spin imply that no "Bose-Nova" can be triggered in the superfluid helium used in the LHC.

To fully consider all those risks and the basic strategy of CERN, damned lies, the reader should read in the next chapter the affidavits of the suits against CERN, where those issues are treated in detail. We shall therefore here do a shorter analysis of those risks.

2. Damned Lies[2]

All the people of those safety documents are interested parties related to CERN.

As an independent lawyer, Johnson clearly stated in his analysis of those suits[3] that all the people who have worked in the safety documents are either working for CERN or have worked for CERN or are nuclear physicists with ties to the institution. None of the safety standards of the EU have been met (review by independent scientists of different disciplines, including risk experts and bioethicists, as those who compose the group Einstein et al.). Since the people working at CERN have signed a confidentiality statement of zero risk, it is obvious that—as any other rogue company that is liable of environmental crime—CERN finds only arguments that affirm zero risk and hides all the real risks, in this case, quark condensates. In CERN's potential genocide, the damned lies have been accepted easily by the three human participants in this tragedy because of the following reasons:

— They were coming from a public corporation—an institution paid by many political nations. So *politicians who paid them* did not want to look fools, spending the money of taxpayers.
— The military sense of solidarity of the *nuclear physicists'* community, who acted as a single voice, as they did during the cold war when they became accustomed to reject claims of being *genociders.*
— *The people, the potential victims*, do not know physics and so "anything goes." Nobody dares to "contradict" experts.

Let us briefly consider those damned lies.

Cosmic rays are not quarks, nor do they produce novas. They are the debris of novas.

CERN lies systematically to the press, saying there is no danger because those experiments happen constantly in the atmosphere in the form of cosmic rays, which is totally false since cosmic rays are lonely atoms, NOT quarks. We have never found a quark in cosmic rays. Cosmic rays are the end result of a nova explosion as recently the Fermi satellite has proved, NOT the cause. They are not mass bombs that break the final layer of strong quarks, starting a mass bomb, $M=e/c2$, but the debris of such an explosion. Further on, cosmic rays are single ions, never billions of quarks together that form a critical attractive mass to start an $M=e/c^2$ mass reaction. Most of them are protons or ions, whose chances to collide head-on with the center of other mass are minimal, as most of the space that surrounds an atom is empty. If you were to consider that an *o* of this text is the nuclei of an atom, the atom will fill up the Maracana stadium in Brazil. So the chances that a cosmic ray collides with a nucleus, as they will do at CERN, are nearly zero. And in case it does so, it will merely free a few quarks in an empty vacuum, unable to form a "superfluid vortex" of strong forces, for lack of critical mass. In other words, cosmic rays are harmless.

Indeed the key is density, luminosity, critical mass. Even the most ignorant person knows what this means. A bomb needs a minimal quantity of explosive substance to ignite and provoke a reaction. Even if you try to make a fire, you will need a minimal quantity of straw to "catch" the fire. Yet because cosmic rays come from supernova explosions, as the Fermi satellite has proven, which are remote and blow up the matter of the star in all directions, when they arrive to the Earth, they are always spread over an enormous radius. Thus they are lonely atoms without minimal a mass to start a mass bomb reaction. They are harmless by definition. Even the most energetic cosmic ray that happens to hit the o in the Maracana stadium will merely do a bleep, like a match without straw to ignite a fire.

Fact is all bombs need a critical mass to appear, and that is why CERN bunches millions of protons or lead atoms together by stochastic cooling, pushing them toward the center of the beam with a strong electromagnetic field that acts forward and sideways all the time. So the bunches are far more numerous and close together, to form a critical mass in the point of impact than anything we can possibly see in cosmic rays. Further on, it is a lie that those particles have less energy than cosmic rays because the LHC accelerates the particles for a long time at light speed, giving them constantly new relativistic mass that transforms them on course into strange and top quarks. So in fact they are far more energetic than anything a cosmic ray does since the cosmic ray receives only an impulse when it leaves the exploding star. If that impulse converts them into strange or top quarks in the long travel to this planet, they will decay. So we get only ions without the added relativistic mass that CERN's particles get as they are turning and turning and increasing its relativistic mass all the time, evolving into dangerous strange quarks that catalyze the formation of pulsars. Finally in 2013 they will evolve into bcb atoms of dark matter and top quarks, which catalyze the formation of black holes. This makes them different from cosmic rays, even in the event, which is more self-similar—the proton collisions that are taking place now and are the ones with lesser risk, as protons have lesser mass/quarks than lead, the most dangerous experiments that will take place in Christmas and have a huge chance to form enough critical mass to start an ice-9 reaction.

In simple terms, what those protons and lead ions will do especially at 7 TeV when the Higgs is expected to appear is to evolve as they acquire mass into strange and top quarks. And so in the point of collision we shall be colliding NOT protons but "black atoms" made of denser top quarks and droplets of ice-9 (strange liquid) and "gas-9" (top black hole liquid), which will ignite the Earth. And because we will collide millions of them in close range to each other, they can form a ball of strange or top quarks easily at the point of impact. This will never happen in a collision between protons in cosmic rays. And certainly the most dangerous experiment, collisions of lead atoms that will produce strangelets never happened in any cosmic ray collision. Pb is nowhere to be observed naked in this planet and the number of pb atoms in cosmic rays are infinitesimal. Indeed, cosmic rays are the debris of stars, so they are made of very tiny pieces: 99 percent are photons, protons, or light helium. The 1 percent left is mainly iron, the most stable, light atom, which covers some neutron dense stars. CERN also said that those collisions are the same collisions that had taken place at the Tevatron in the past, just with more energy. Yet again that is false. The Tevatron collides protons and antiprotons, which by definition annihilate each other. They don't sum mass. Lies and damned lies all the time.

Black holes don't evaporate.

CERN says the black holes it will produce will certainly evaporate. We have proved *ad nauseam* that Hawking's black hole evaporation is false. The Fermi satellite has never found a signature of black hole evaporation. CERN says that according to Einstein, black holes cannot be produced at the LHC. And that they will not have charge. We have seen this as absolutely false. According to Einstein's theory of black holes as frozen stars, they require to exist a cut-off ultradense substance, which can only be quarks, exactly what CERN manufactures. Further on, top quarks by definition have topness, which is positive charge. So they will have positive charge and will accrete atoms not only with their gravitational force but with their enormous positive charge that will attract the electronic cover of all kinds of atoms.

CERN is a factory precisely created to produce massive amounts of top quarks=Higgs, the substance of a top quark black hole, according to Einstein's theory of mass and gravitation. So it will certainly produce them. CERN just practices the essential tool of rhetoric: to "repeat a lie ad nauseam," following Mr. Goebbels's advice. If you believe their lies, cosmic rays are NOT only quarks but they are also black holes! We have never seen a quark or a black hole in cosmic rays. We have only seen protons and lonely atoms in cosmic rays. We shall repeat this also ad nauseam, to see if quantity can convince of truth as well as it has convinced the world with a cosmic lie.

The same paragraph at CERN says that to form a black hole you need the density of mass-produced inside stars. Yet the exact reason why physicists created the LHC was to get those densities—to produce by stochastic cooling billions of atomic nuclei (made of quarks) pegged together, to get an enormous density in the point of collision, where those quark nuclei will bunch together, forming a quark condensate. Yet cosmic rays always travel alone, billions of miles from a supernova, spreading all over as debris, so they will never collide with Earth with enough density to provoke a chain reaction, a superfluid vortex of quarks able to start the mass bomb that will swallow the Earth. It must be noticed that the entire "safety standard" is based on a single lie, following the well-known Goebbels's advice for all future political or industrial liars: the same lie must be repeated for every argument, it must be never denied, it must be the foundation of the entire cover-up structure. In this case, the master lie is that "cosmic rays" produce anything that is dangerous. All possible arguments will be thus reduced to the lie that cosmic rays is the most dangerous substance of the Universe: it is constantly mutating into all kinds of black holes, strangelets, monopoles, vacuum bubbles, whatever. And yet cosmic rays are just the debris of blown-up stars, as the Fermi has proven and nature reported.

Runaway reaction.

Again CERN lies. The possibility of a thermonuclear reaction with carbon and helium has a smaller chance than the production of all kinds of dangerous quark matter condensates because it is not what CERN will primary do, but it requires an accident in which the high energy ions of the bunch, accelerated at c-speed, collide with the helium of the system. CERN says that helium is inert, but any astronomer knows that stars burn hydrogen into helium and helium into everything else. Helium is in fact the beginning of all the stars' reactions that create all the elements of the Universe, including the carbon, nitrogen, and oxygen we are made of. It is another basic lie that helium is not the combustible of nuclear reactions. Instead, what CERN should say is that such thermonuclear reaction, which will be massive, given the fact that the tubes of the collider have several tons of helium, requires an accident, which is unlikely to happen. It finally lies, saying that there is not enough helium to sustain such reaction, when CERN is the biggest consumer of helium in the planet, to the point that it had to halt operations for a while because it had consumed all the helium of the planet and it needed to produce more. The size of the helium bomb, if an accident happens, is so huge that it would blow up the entire country of Switzerland and provoke a nuclear winter in the entire planet.

Hot strangelets.[4]

We come to the final risk—strangelets, the most dangerous, probable cause of our extinction in the next years. The entire paragraph is false.

It says that strangelets are hypothetical when we know they are the core of most pulsars and quark stars and the only known substance able to produce the supernovas that leave those stars as a remnant. We studied all the details of its formation. The final two lies say that the collisions at CERN will be less dense than at RHIC. This is false; CERN has much more luminosity (density in the bunches that collide) than RHIC. It also says that with more energy, there will be less strange quarks formed. This is false. Everybody knows that energy and mass are self-similar states: e=mc2. Relativistic mass is produced by accelerating and giving energy to our up and down quarks that become strange quarks, forming strange liquid. But CERN has 56 times more energy than RHIC so it will produce many more strange quarks and make a bigger strangelet mass bomb with enough critical mass to start an ice-9 reaction. Finally it says that it will collide them at higher temperature. Again this is false. CERN confuses the concept of temperature and speed because it does not recognize that a mass is a cyclical vortex, which has exactly the inverse properties of lineal speed, *temperature. At RHIC, we saw a superfluid vortex of zero temperature, supercold. CERN, in fact, is the coldest point of the Universe, as it advertises, because the tubes of the quark cannon are designed to create ultraordered mass vortices. They move fast but in perfect order, so if temperature measures disorder, they have none.*

The temperature concept that CERN uses is one of the "transcendental illusions" that extend hyperbolically a local phenomena or a mathematical artifact to the entire reality. If Newton extended the Cartesian continuous lineal graph of space/time to reality as if the infinite broken vital spaces and time cycles of reality were one and Galileo extended "motion change," defined by one equation v=s/t, as if it described all the time changes of the universe, CERN extends entropy or temperature, the lineal vibrations or disorder of a molecular gas, defined by a lineal equation PV=nkT, as if *it applied also to the informative, cyclical, inverse, rotational speed of mass vortices.*

Fact is v=s/t is not all the time changes of the Universe but only the change in lineal motion, and temperature is not all the speeds of the Universe but the lineal speeds of a molecular network of atoms.

Thus, increasing the energy of the collision does not increase the "temperature" of the liquid vortex, only the temperature and intensity of the external big bang radiation expelled by the vortex where temperature as a measure of electromagnetic expansion makes sense. Inside the liquid vortex, due to the balance of a dual E × I = K far from equilibrium system, a higher temperature in those external big bang, electromagnetic interactions (Max. Te), means instead the liquid vortex created by strong-force interactions will cool down (Min. Ti =0 K), toward a frozen star.

Increasing the energy does not make the up and down quarks fly apart any faster, once you are in the strong regime of the vortex, creating quarks from energy. You are making more strange quarks, more strangelets. Thus the opposite is truth: as the energy is increased in the collision, the formation of strange quarks increases (energy to mass conversion).

What we could still argue, though Chen's papers seem definitive in this matter, is how many strange quarks have to materialize in order for the up, down, and strange quarks to reorganize themselves into a strangelet. The LHC will have center-of-momentum collisions of lead-lead that does not happen in nature. The closest would be the very rare lead nucleus cosmic ray striking a lead nucleus lying on the surface of the moon. But because lead nuclei are so rare in cosmic rays, none have been measured with anywhere near the energy of the LHC lead-lead collisions.

A strangelet is a superfluid vortex of quark mass, which has no temperature, as its properties are informative, the inverse of electromagnetic, lineal motions. Yet since CERN does not recognize the inverse properties of gravitation and electroweak forces and it does not understand at all Einstein's theory of mass, it translates cyclical motion as temperature and says that strangelets created at CERN will be hot. They will be ultracold superfluid vortices, surrounded by ultracold helium. Those quarks, when colliding, will form a vortex of information/mass which has opposite inverted properties to those of temperature. It creates mass, information, negantropy as it increases its order, and becomes first a superfluid and then the final state of matter of quark condensates, a bosonic crystal-like solid, ice-9, the substance that will explode our planet, throw debris of cosmic rays to the Universe, and leave behind a frozen quark star.

3. Truth vs. Authority and Celebrity: The High Priest and the Temple of Black Masses

Black holes don't evaporate, despite the celebrity status of Mr. Hawking.

Even if it won't be popular, the sheer absurdity of Hawking's theory and its popular acceptance requires to consider an element outside physics, namely, the celebrity of Mr. Hawking. Mr. Hawking's supreme intelligence is an invention of the press who loves emotional stories of "personal courage" to sell more and so it has invented that Mr. Hawking is the new Einstein. All those interviewers like Charlie Rose who called him the most intelligent man on Earth, his site saying that he was born three hundred years after Galileo, his self-biography affirming he was called "Einstein" in the school and he studied physics instead of the profession of his father, biology, because "intelligent people didn't study biology," his participation in *Star Trek*, winning an imaginary game with Newton and Albert[5]—you would expect that such shameless exploitation by the celebrity press of a man in absentia for thirty years, added to the absurd scientific errors of black hole evaporation, *would have been noticed*. And they have, by many people. But who dares to contradict the sanctified high priest of our technological civilization? Who, in Iran, will raise his voice against the Supreme Council regardless of truth? An anecdote should suffice. The first person who discovered frozen stars and pulsars, Mr. Chandra, was an Indian student. He came to England at great expense and exposed his theory, but Eddington, the astronomer who had Hawking's fame and power, said he didn't believe him, not because his theory was wrong but just because it could not possibly be truth. Point. No more reasons. Chandra never got his theory of pulsars accepted till Eddington died. Bohr bluntly put it to him in a letter: "Your work is absolutely right and we all know is right" but till Eddington dies, nobody will contradict him. So nobody will contradict the evaporation of black holes or the cosmic ray argument of CERN, the high priest and the temple, till one dies and the other kills us all.

Let us quote Johnson[3] again: "As a laboratory, CERN is mammoth. CERN's 2008 operating budget topped $900 million USD. CERN employs around 2,500 people, including more than 1,000 engineers and scientists. CERN commands a transcendent role within the particle-physics community. Worldwide, nearly 9,000 physicists are officially involved in CERN's experiments. CERN brags that "half the world's particle physicists" come to CERN for their research. As an intergovernmental organization, CERN enjoys an elevated status in the world community. CERN has legal personality in all member states, and it enjoys immunity from legal process in its host countries."

The law of silence of our extinction extends in the rooms of thousands of astrophysicists that know perfectly till Hawking dies nobody will dare to say the truth: *black holes don't evaporate*. And none certainly will deny CERN's false argument on cosmic rays and put its careers on line. But CERN is not the temple of a religion of information and life but of energy and death, and so the experiment is the ultimate black Mass, and indeed a black Mass is what we shall adore as it forms the supreme evil, the absolute simplifier of all complex forms of electroweak matter, the black hole. And to that aim, the rhetoric of luxury and gold is to be found in the cathedral where the invocation will take place: CERN has built a big machine. CERN proclaims it is "the largest machine in the world." Indeed, the LHC is unprecedented in size, power, and cost. The Atlas machine that will invoke God's particle is compared to the size of a cathedral. The previous accelerator, the Tevatron, was constructed imitating Chartres.

Death always wears a dark robe. In the famous film *The Seventh Seal*,[6] when Death chooses his side of the chess, it chooses black. The warrior, Mr. Block, thinks he can defeat Death, the laws of the Universe. He thinks he will win the chess game and Death will evaporate his dark rope. But the seventh imaginary particle that should appear after the sixth quark will not evaporate.

The strategy of CERN is very simple. We have found a magic safe substance, cosmic rays. So no argument will be effective against us as long as we convince everybody that cosmic rays are doing all what we do all the time and never blow up the Earth. Thus cosmic rays produce black holes, strangelets, monopoles, anything you want, all the time in the atmosphere.

This lie repeated ad nauseam could have been tested just by asking any expert on cosmic rays if we have ever seen dark matter, black holes, monopoles, quark condensates, or anything but lonely ions in cosmic rays. The thundering *no* would have raised all alarms if a few journalists have simply said, "This nuclear company is lying bluntly to mankind about what they produce. It is *not a factory of cosmic rays but of quark condensates, the most dangerous substance of the Universe.*" But no journalist has dared to contradict CERN; no cosmic ray expert has dared to say the truth. And so the company got away with an outright lie that everybody knows is false but nobody dared to deny because of sheer peer pressure. German soldiers knew Hitler had no idea of military strategy and he was going to lose the war invading Russia without defeating England first. And yet *all went to die to the Eastern front*. So it will happen to the "military" community of nuclear physicists. Imagine, Mr. X from the University of Arizona coming out publicly and saying that Mr. Hawking is just a celebrity case of mass media exploitation, that his thirty-five-year-old theory is obsolete, outdated, and false, that CERN is lying. How long is he going to last in the career of nuclear physics? CERN has economical and political power. And as Goebbels said, "when a lie is repeated many times, people will believe it." So now everybody you ask has read that CERN poses no danger because it is just a factory of cosmic rays, which is like saying that an atomic bomb is a factory of dust. The mass bomb of a supernova is made of quarks. The debris of the supernova are the cosmic rays. Black holes suck in the world toward the future. They don't evaporate toward the past. In both cases, CERN switches the causality of the arrow of time. Cosmic rays are the consequence, not the cause of a mass bomb, the debris of the explosion but the substance that causes the explosion, in the case of CERN's experiments the quark condensates it will produce.

So all comes to this safety standard: CERN lies and makes ad hominem campaigns, calling crackpots everybody who denounces those lies. People do not know physics, but journalists get paid to publish CERN's articles or they are lazy to research those lies, so they follow CERN's dictum. And all becomes a question of "personal authority," not *an argument on truth based on the laws of the scientific method as this book is.*

4. Statistics. The Fermi Paradox: Why Are There No Intelligent Planets?

To notice that most of the arguments of safety made by CERN are not even related to the factory and its produce, illustrated in the cover of this book, but "go fishing" to the galaxy. They relate to the fact that planets and stars are not all destroyed. So they are based on statistical arguments, which cannot be even verified. CERN basically says that because NOT all the matter of the Universe has become dark quark matter, strange stars and black holes, only 96 percent of the Universe is made of dark matter and dark energy, we are safe. Which is like saying that because not all the gazelles of the savanna have been eaten by lions, a gazelle should play with a lion. And here again! It brings the argument of cosmic rays. Cosmic rays should be colliding and converting in black holes every star of the Universe, and so CERN is safe because it is a factory of cosmic rays. But CERN is NOT a factory of cosmic rays but of quarks. And cosmic rays do not convert anything in anything, just do little "bleeps" when they hit the lonely nuclei, the o in the Maracana stadium. We have never found quarks in cosmic rays. (And I have to repeat this truth as CERN just repeats its lie ad eternal, Goebbels style). Black holes and pulsars are formed only when enough density of quark matter provokes an implosion, which breaks the cover of atoms, creating a bosonic condensate of quarks, starting a mass reaction inside stars and in the artificial conditions of the LHC accelerator. Thus the main argument that occupies most of the safety report is absolutely false. Yet in any case, in the Universe, the massive amount of dark matter shows how often quark reactions create dark quark stars. Probably there is a lot of CERNs around the cosmos in blue planets telling the natives that they are safe because cosmic rays *are hitting them all the time.*

The argument that the Earth and the moon are not yet black holes or strangelets of dark matter is statistical. It has to do with the quantity of dark matter of the Universe, its composition, and how often moons and planets become black holes. If dark matter is made of quarks, obviously we are just a lucky gazelle. And two arguments studied in more detail prove that this is the case:

— Of the Universe, 96 percent is made of dark energy and dark matter, whose only known-known particles are MACHOs, which, as we have seen,[ch.5] can only be black holes or quark stars.

— Not a single planet shows intelligent life. Most stars have planets around, but none has radio signals. All radio signals in the galaxy are the tam-tam of black holes. This is the Fermi paradox, first told by Mr. Fermi, the maker of the A-bomb, who said that perhaps there is no intelligent life in the Universe because all planets are blown up by nuclear physicists. Then he had a laugh and bet that the A-bomb he was going to explode only would blow up New Mexico and sent his family to Los Angeles, just in case. During those experiments for twenty years, as the black rain fell over Nevada inhabitants, who died of cancer decades later, nuclear physicists arranged with the traffic department of Nevada to switch off a light on an intersection. All their families passed by that intersection in Las Vegas, and if the headlight was broken, they drove to LA to avoid the black rain as it meant their husbands were doing atomic explosions. So only the well-intentioned people of Nevada who trusted their nuclear physicists died. Of course, physicists felt ethically justified because they were protecting their nation (now exploring the meaning of it all). But they never told the truth or asked Nevada people what they thought about dying of cancer as CERN never asked mankind what we think about dying for Mr. Higgs and Mr. Hawking's Nobel Prizes.

Why indeed does mankind have to die for them? Why must all the people that never did anything wrong die for CERN? Why do I have to die if I have always been a good scientist striving to understand the Universe and improve the life of human beings? Only the humble realization that I am nothing but dust of space-time in an absolutely relative Universe soothes my soul against a destiny I have fought for the past years imposed to me by the quark cannon.

Let us then consider CERN's final lie—that planets and stars do not become constantly quark, dark matter since only quarks can form dark matter, called MACHOs.

5. Background MACHOs Happen All the Time

CERN affirms that nothing can happen even if the quark factory made black holes *because the Earth and the moon are not yet a black hole* despite being bombarded by *cosmic rays, which are like the collisions taking place at* LHC. And that the Universe is full of such cosmic rays, collisions, and its moons and planets are still there. Both assertions are lies.

We have finally understood what cosmic rays are and what their origin is, with the good fruits of the tree of cosmology—satellites and telescopes. The Fermi satellite was launched two years ago to find evaporating black holes. It did not find any, putting another proof on the table that black holes don't evaporate. But it found all other kinds of radiations. *Science*, the leading American magazine, states that it has found over 1,600 types of gamma-ray radiations,[7] none of which is the signature of Hawking's evaporating black holes. It has also observed that *cosmic rays are not the cause of supernovas as quark condensates are, but the debris that supernovas produce. Thus, they are not the dense quark bombs that explode stars, but diminutive pieces blown by the explosion.*

What this means is that cosmic rays have more speed/energy than a quark because they are flying away from the explosion but they never start any kind of mass reaction as a quark-gluon soup, the real bomb, will do.

We now know, thanks to duality the difference between them: the big bang of a supernova expels ultrahot=fast ions and radiation, as the mini big bang at RHIC did; but the big crunch bomb is happening inside in the vortex of a gluon-quark liquid.

The bangs of quark-gluon soups emit as in RHIC showers of cosmic rays, ultraenergetic photons and protons that break into a shower of particles in the atmosphere and then die away. They don't grow as a black hole or a quark condensate does. Thus, to equal them to mass bombs of quarks is like equaling victim and murderer—it is an absolutely cynical lie that any physicist graduate can spot. Yet this lie has been echoed by millions of bloggers as if it were a dogma of faith. It is as bogus as it would be to say that the prophet of the religion we believe in created the Universe. Yet no believer will try to find a proof of the creation of the Universe by his prophet, but just believe on his dogma. Unfortunately, thousands of scientists and journalists, which are CERN's believers, went along with those lies without even checking any book on physics. And the politicians listened to the press that listened to CERN's public relationships department.

It was an easy decision. The most expensive, more complex machine of history could not be wrong. So *we the people who defended the tradition of Einstein and the tradition of bioethics and humanism that should be worshipped by any human civilization, the tradition of truth and open dialogue, the laws of the scientific method, the experimental facts, the people who sued CERN, were not listened, interviewed, asked about, but merely brushed aside as a nuisance that interfered with the astounding quark cannon.* So the few people who knew CERN was wrong, a few physicists, experts in cosmic rays, realized soon that their careers were at stake if they explained the truth; and they remained silent, protecting the jewel of their worldly profession, the most perfect weapon/machine ever built.

Then came the double lie about the moon: the experiment is safe, said CERN, because the moon has been bombarded by cosmic rays and it is still there. There are no moons converted into black holes. Indeed, the fundamental safety argument of CERN is that in the Universe there are no black holes that eat planets and moons, that dark matter is not made of black holes and quarks. Because if that is truth, then it is very likely that the experiments at CERN will do what the Universe does every other day.

It was a double lie because it implied that the explosions CERN will do were cosmic ray explosions, NOT quark explosions, and further on, it considered that black holes never swallow moons and convert light matter into dark matter.

Again, this lie is easy to spot by anyone who knows basic cosmology as 75 percent of the mass of the Universe is made of dark matter, which astronomers consider to be made of MACHOs, micro black holes. We have even found dark matter galaxies where those black holes have eaten all the stars. And that might be according to some cosmologists the end of all galaxies: food for black holes.

In relativity, black holes bend light, acting as gravitational lenses, creating a black body radiation equal to the most common radiation of the Universe (background radiation, which is also a black body radiation). The more mass the black hole has, the more it red shifts, cools the light. So there is a simple equation of basic relativity that shows what background radiation a black hole with a given mass will redshift. Now it turns out that the only object that can emit this radiation at 2.7 K is precisely according to Einstein, a black hole with the mass of the moon—a particle of *dark, quark matter that has eaten a moon. Indeed, the calculus is fairly straight forwards, and* fair enough black holes with the mass of a moon would produce ±2.7 K light—the commonest radiation of the Universe. Hence the Universe must be full of black holes that ate moons and now radiate at 2.7 K (called background holes or, poetically speaking, Moon MACHOS). This radiation is what scientists call the signature of a particle, and it is one of the basic methods standard science has to observe the Universe.

Astronomers, the true specialists of this field, know dark matter is hyperdense. They know dark matter is hypercold. They know dark matter only can emit the background, very low, hyperabundant radiation of the Universe, or don't emit radiation at all due to their cold highly ordered nature, *exactly as quarks and ultracold quark condensates do. Thus, they favor in all the polls on the subject the idea that dark matter are MACHOs, massive halo objects, quark stars, pulsars, or black holes.* Only nuclear physicists (now called particle physicists, given the bad name of their profession; CERN also, according to its spokewoman, wants to drop the name nuclear to become the European Company of Research . . . of what?) sponsor WIMP, imaginary particles, invented with their creative equations. Maybe they are the origin of dark matter, but so far we must trust standard science and the real experts—astronomers, cosmologists, and relativity theorists. And they say that dark matter will be made of small black holes, which by gravitational lensing, due to their enormous attractive power, bend and redshift=cool down light, extracting its energy. This process delivers a blackbody radiation that becomes the signature of micro black holes. Further on, the scientific method, which is the true method to distinguish truth from falsity, favor background holes and NOT the big bang as the most probable cause of that radiation because when a phenomenon can be explained by a direct, present cause, we shall, according to the laws of truth of the scientific method, choose the present cause, NOT an inferred past remote cause *that cannot be observed directly. Or else we can do as Gamow did and change the nonobserved equations of a hypothetical big bang from past to future to accommodate our desired results.* Indeed, this is a fact hidden today that happened to the first proponents of the big bang theory. We are told that the biggest proof of the big bang is that it was able to predict a priori the existence of a 2.7 K radiation all over the Universe. False. Mr. Gamow calculated that the background radiation had to be of 20 degrees, if it was a remnant of an old cosmic explosion. When Penzias came with the real 2.7 K temperature, Gamow merely went to the drawing room and put an ad hoc element in the temporal equation from a remote past to the present, so the cooling of the Universe was a bit faster and it could match the present 2.7 K temperature. *And so voila! The 20 K degrees became 2.7 K degrees.* But this is not allowed by the scientific method. So the BG radiation is not a proof of the big bang, but a proof of its falsification. There is no way to know at which speed the Universe cools down if it is cooling down. Do big bang theorists merely rewrite their equation of cooling to fit the experimental result a posteriori? Instead of accepting the falsification of their theory, those calculus were later modified, adapting them to the experimental radiation found in the Universe.

This kind of mathematical manipulation is a key difference between real science that follows the scientific method and quantum science that does whatever it wants with data. All other sciences except physicists that can say whatever they want as they keep churning machines and weapons respect the laws of falsification of the scientific method. Take the case of geology. Mr. Kelvin, another famous physicist, calculated that the Earth had 100 million years because that was the rate of cooling for rocks he had in mind. Of course, the Earth is much older, but as long as Kelvin was alive, nobody dared to argue with him. So during the entire twenty-first century, the Earth was 100 millions old, never mind evolutionists, fossils, carbon-14 data, etc. The high priest had talked. Then when Kelvin died within a decade, there was such a wealth of papers nobody dared to publish before, proving him wrong that the age of the Earth grew to several billions. Today the same happens with Hawking's radiation and CERN.

In scientific truth, synchronic cause->effects that take place in present—have to be acknowledged first by processes that are taking place in present—are more truth than past diachronic causes. Imagine that your doctor takes the temperature produced by the cells of your body and finds it as 37 C, but then he says it is NOT the temperature caused by your cells (in the big bang case, black holes radiating at 2.7 K), but you were yesterday at 100 C and when you were born you were at 1 billion degrees. This is what big bang theorists are doing. Will you believe your doctor, or fire him?

Only as a last resource can we logically postulate that a present effect has a remote past cause since we cannot experience how dynamically the past effect has derived into the present. So such deductions are considered by philosophers of knowledge of very dubious truth, especially when we have the alternative of present background holes radiating at 2.7 K.

Yet the high priests of CERN have talked and so as long as they are alive, we shall believe that the signature of black holes comes from a remote past, that moons never become black holes reflecting light at the exact temperature of 2.7 K degrees (even if a simple Einstein's formula gives you that signature result). No. Instead, we must believe that dark matter is invisible, it is made of mathematical never-proved particles, and of course, that CERN is safe. But standard science tells us that the Universe is full of MACHOS, black holes, and quark holes (which happen to have very similar properties and might be the same), that radiate at 2.7 K and the Earth can easily become another background hole MACHO. The word *macho*, which means "massive halo object" and was chosen by the agressivity of those background holes, says it all.

Let us consider another similar case to see why indeed the signature of MACHOS should be accepted as the most likely cause of the background radiation. Most interstellar hydrogen is in the form of atomic hydrogen because those atoms can seldom collide and combine. They are the source of the important 21cm. hydrogen line in astronomy: During the 1930s, it was noticed that there was a radio hiss that varied on a daily cycle and appeared to be extraterrestrial in origin. After initial suggestions that this was due to the sun, it was observed that the radio waves seemed to be coming from the center of the galaxy. These discoveries were published in 1940, and in 1944, it was discovered that neutral hydrogen could produce radiation at a frequency of 1420 MHz. This 21cm. line (1420.4 MHz), given the abundance of hydrogen, could then be used to map out the Universe and so. The first maps of neutral hydrogen in the galaxy were made and revealed, for the first time, the spiral structure of the Milky Way. Hydrogen line observations have also been used indirectly to calculate the mass of galaxies, to put limits on any changes over time of the universal gravitational constant, and to study dynamics of individual galaxies.

Entire fields of cosmology are based on a similar discovery to that of background holes with a signature of 2.7 K, caused by gravitational redshift. Thus, for the same reason we accept the hydrogen signature of 21 cm., we should accept the background signature of background holes, only present cause of the background radiation of the Universe, *which Penzias and Wilson, their discoverers considered to be emitted by the gallactic halo.* It was an outsider, Gamow, working on the A-bomb, who came with the outlandish but pleasant hypothesis of the big bang. Since it allowed nuclear physicists to become not only the makers of nuclear weapons but also the high priests of cosmology. Once and again, we shall observe this invasion of astronomy by nuclear physicists, given their power in science. The quark cannon is just the final zenith of this absurd interference. Since those 10 billion dollars should have been used in telescopes and satellites, *if the reason for the creation of the quark cannon was really to understand the meaning of the Universe.*

Moreover, because the 2.7 K radiation is the commonest in the Universe, black holes that emit that temperature must be eating moons, the most common celestial body. So there we have it: the most common body in the Universe, a moon, gives birth to the commonest form of dark matter, a background hole, a moon MACHO. *As the commonest line of the galactic radio spectrum, 21 cm., must belong to the most common atom, the hydrogen.* And so if the LHC does experiments, with high-energy collisions that can produce MACHOs, the Earth and the moon will become one because the safety argument that this never happens in the Universe is exactly the antitruth. *It happens constantly in the Universe.* If there were other particles that could emit that radiation, other type of cosmological species with that signature of a black body, we could still argue the experimental facts. But there are none at the quantum scale *except ultracold quarks and none at the cosmic scale except background holes.*

Of course, there are moons of different sizes, and for that reason, we said approximately 2.7 K degrees.

Does this mean there should be a wider variation in the background radiation? No. Since black holes that eat moons will be all of the same size because they are fundamental particles of cosmology as electrons and quarks are in the microcosms. So when an electron is born in a collision of particles, it has always the same exact mass and the rest becomes cosmic garbage. This is what we observe when a bigger black hole is born from the explosion of a star: the rest becomes cosmic garbage, energy exploded out in a nova. So that it should happen constantly all over the Universe is easy to predict: a micro black hole or cosmic ray (to make CERN happy) will hit a planetoid and create the seed of a black hole that will eat up the mass of the moon, giving birth to a background hole with that standard temperature.

And the rest of the moon mass will then become cosmic debris, gamma rays, etc.

And so the safety argument turns out to be its antitruth: it signals an enormous danger for mankind since if background holes and quark stars are the dark matter that dominates the Universe, a quark factory will certainly produce them. Or in terms of authority, if Einstein is right, we are 100 percent sure that we will die.

Yet this fact of science, proved by the lack of background radiation, shadows in galaxies, which implies the radiation comes from the halo and must be originated in those black holes or ultracold quark stars, is vehemently denied by CERN because it will mean the end of the big bang and the end of mankind, questions that do not matter to CERN as much as the alternative—the end of CERN.

If a seed of any of those MACHOs made of quarks is produced on Earth, it could trigger a mass reaction, exploding the Earth into a big bang. After the nova explosion recedes, all that would be left of planet Earth will be a tiny ultradense rock of quarks, a MACHO (massive halo object) of dark matter that will join that 75 percent of dark matter that surrounds the galaxy. This is what standard dark matter theory tells us because indeed, there is no other substance known, no other particle ever witnessed before, that has those properties of dark matter.

So what do CERN and big bang believers propose instead to account for dark matter? An array of particles never found, merely theorized, called Wimps, which do not interact at all with the Universe! (That is why they are called wimps, or weakly interacting particles) and so they can be considered invisible, which is the perfect alibi of an astounding array of quantum fantaphysics, taken by each proponent as absolute truth. Yes, the reader can raise his eyebrows now. The entire matter of quantum cosmology consists in a return to the prescientific age of myths that might happen like fairy tales, a fact now introduced with the concept of invisible particles and probabilities. If we do not find the Higgs or the wimp some concotted equation has designed a priori, the quantum physicist that no longer respect the scientific method will tell us that they are in other dimension, or a parallel Universe. This was the great discovery of Mr. Hawking, announced recently to the world: we haven't found black holes evaporating the information of reality because they evaporate information into another Universe. All this is nonsense, tolerated for decades because quantum physicists made machines that made money and made weapons that gave power. All this is nonsense, tolerated for decades because quantum physicists made machines that made money and made weapons that made power.

6. Organic Science and Cosmology:
The Galaxy Is Ruled by Black Holes

Now, thanks to the fractal paradigm, *machines* are no longer the measure of all things, but man and life are again the center of the Universe, the most complex, informative being of reality since indeed, fractals are mathematical objects that share all the properties of organic beings.

A fractal equation constantly reproduces information as life does. Thus, any fractal system is an organic system, which as life does is organized in several interelated scales, connected by networks of energy and information, as living beings are. And this is also the case of dark matter. In the appendix, we shall consider in detail a Universe modeled not with energetic explosions but with organic information. Let us do a brief outline:

The main produce of CERN, quarks, are the fractal parts of the densest "wholes" of the cosmological scale: strange quark stars called pulsars and top quark stars called black holes. And in fractal relativity, those quark stars are the dominant dark matter substance of the Universe.

This is not by choice but because the Universe is efficient and all the other mathematical musings considered by quantum physicists, the so-called wimps (weakly interacting particles) break one or another of the Principles of the Universe (antisymmetry, equivalence, and invariance),Appendix which makes them inefficient and unstable in the rare cases in which they are formed.

So dark matter is made of quarks and its signature is the background radiation, the most common radiation emitted by the most common species of the Universe. Dark matter feeds on our light matter, and it maintains, as any organic, cellular system does, the "average" temperature of a galaxy at 2.7 K degrees, bending the light of the Universe in the same manner your body maintains a homogenous 37 degrees temperature in its medium, which is water. The medium or space-time membrane of a galaxy is light space, and it has a medium temperature of 2.7 K, which happens to be just above the temperature of strangelets and black holes that therefore absorb energy from that medium as your DNA molecules do from the water of your cells.

This organic analysis of galaxies is taken further in the appendix. It is important to mention it here because CERN pretends that the galaxy is not teeming with black holes and strange quark matter. In the fractal paradigm, the opposite is truth. We are the food of dark quark matter.

Thus if the fractal paradigm is right, the creation of quark condensates, strangelets, and hyperabundant background holes, "signature" of the BG radiation, is constant. Then the experiments at CERN are an enormous risk, as moons and planets become strangelets and micro black holes all the time. In that fractal paradigm, the galaxy is an organism structured by two networks, one of quark, informative, gravitational matter and one of energetic, electromagnetic, light matter, us, the food of gravitational quarks—akin to a cell with two systems, one of DNA and one of mitochondria and carbohydrates, its food, which would be in the galaxy the networks of stars and planets that black holes feed on, as DNA feeds and reproduces thanks to the energy of mitochondria. Fractal self-similarity explains the enormous parallelisms of both structures: DNA is at the center of cells as all galaxies have in the center black holes. Those black holes proved to be in fact swarms, as cell nuclei are swarms of DNA strains. Black holes seem to behave with erratic movements as if they process gravitation as information, hence somehow "gauging" it, as DNA seems to have somehow a certain "perceptive" motion within the cell. All this, of course, is heresy in the quantum abstract paradigm, but not in biology where a cell can be both, explained with organic laws and mathematical ones. If that organic model of the galaxy holds, an enormous number of questions become resolved, as we shall see in the appendix. So far, what it means is that we are food for black holes and quark stars, entities far more complex than we think. Since the galaxy cell has a "hard protein cover" of strangelets and a center of black holes, within it, the mitochondrial "factories" of heavy elements will be its stars. Then within those stars, we are a "mush, forgotten in the surface of a lost rock," which is about to die.

Indeed, unfortunately the fractal paradigm is logically, mathematically, and experimentally more truth. So the truths of its pioneers will be more important for the future of science than all the work at CERN, as the work of Einstein turned out to be more important than all the work of the German industry of weapons and ether theories to which he opposed. And yet let us not forget that Einstein's influence on German physics was null during the thirties and forties, when their industry of weapons destroyed the world in World War II—an ominous precedent, as this European Consortium, guided by Rolf Heuer, the German president of CERN, whose main contributor is indeed Germany, have totally ignored the suits and letters of Einstein et al., of the chaos and fractal theorists, safety and risk experts, who denounced the dangers of the LHC.

Instead, after an ad hominem campaign, made not to confront truth with reason, CERN did not appear in court, protected by diplomatic immunity and has stubbornly proceeded with "business as usual," beating new energetic records with its weapon machine. And all theoretical opposition has been censored. The description of the quark-gluon soup from the perspective of a complex liquid has not been published, but what CERN and the establishment of industrial physics and scientific magazines cannot do is to change the truths of the Universe. And so because masses are whirls of information and quarks are the fractal parts of black holes and quark stars, CERN will very likely make them and kill us certainly . . . regardless of marketing campaigns and the "obscene" game of probabilities played by CERN to dilute our sense of danger.

III. Obscene Probabilities

From a legal perspective, the Large Hadron Collider is the biggest genocide in history. A 100% unlikely risk of blowing up the Earth, which is the original estimate of the company before it obliged its workers to state zero risk, means a genocide of 10% × 6.6 billion= 660 million legal dead, by the mere fact of switching the LHC.

Sancho Vs. DOE et CERN

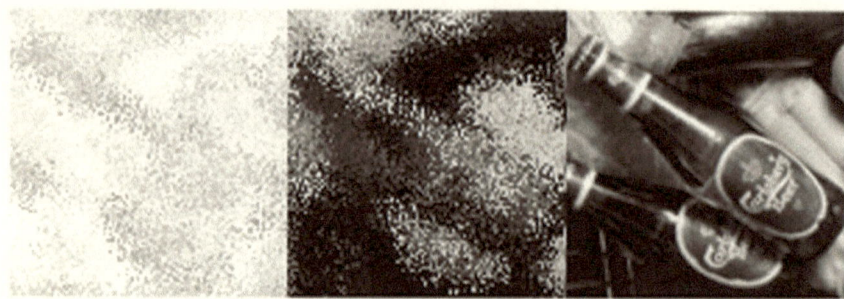

"There is only one truth, but it requires the right amount of information." In the graph, a bottle of beer is a "probability" of truth in the first picture. In the second picture, it seems more real. In the third picture, nobody will doubt that it is a beer with 100% of certainty. Even if only when we have the bottle in our hands we will taste the beer. Science is also a linguistic construction, whose truth depends on certain logic, mathematical, and experimental proofs.[ch.1] Those proofs applied to the quark-gluon liquids CERN will produce prove without doubt the LHC will create the most explosive substances of the Universe, which certainly will blow up the Earth, even if only when we die we shall become a strange liquid.

10. The Events That Will Take Place at CERN

The problem with the events that will take place at CERN is self-evident. CERN ignores fifty years of evolution of complexity and analysis of far-from-equilibrium systems as a quark-gluon soup is, the duality of arrows of time (energy and information) and fractal relativity, the theory of masses as accelerated vortices of space-time, according to Einstein's principle of equivalence between gravitational entities and acceleration; so it cannot predict what will happen at the LHC, with its theories based on a single arrow of time (quantum entropy), a false theory of mass (Higgs), and a lineal, monist analysis of a quark-gluon soup as an expansive plasma (a hot radiation).

Those studies ignore the formation of a superfluid strangelet liquid (a vortex of informative mass) *since big bangs and big crunches happen at the same time, which is what duality and complexity explain, as all processes of creation of physical form (big crunches) are accompanied by processes of expansion of energetic radiation (big bangs).* Think of the simplest far-from-equilibrium system—a refrigerator. The interior is cold precisely because the exterior has a rack of Freon that is hot. So heat is expelled outward to cool the interior. This is in essence what happens in a double big bang and big crunch. We, complexity theorists, do understand better those processes and so our predictions that strangelets and its components (kaon and other strange particles) will be produced in greater numbers have been proved by December's experiments; but CERN made an ad hominem campaign without even hearing us, which converted Einstein et al. into nonpersons, for the establishment of science and the press that no longer publish our papers. So nobody listens to our warnings, coming from the science of the future, from complex physics, and based on the three legs of the scientific method—authority has won the battle over truth, but in the long term, truth always wins because it is the law of the Universe, an entity much more powerful than human authorities are.

And truth says that CERN will start making dibaryons, the stable atoms of strangelets this April. In lead-to-lead collisions in the fall, it will increase enormously its production. Those dibaryons will then fall undetected and will seat on the center of the Earth, forming a strangelet liquid vortex, an Einstein-Bose condensate that will eat the planet inside out.

It is difficult to know exactly when this process will finally collapse the Earth because there are many variables involved, such as the number of dibaryons created per second, the number of stable dibaryons that will survive their travel to the center of the Earth and the time they will need to accrete the planet, but while the exact calculus is difficult, the overall analysis, which we shall resume now before calculating the probability of our extinction says that the lead collisions will be the point of no return; that is, a point in which mankind no longer will be able to stop the process of extinction.

Dibaryons (usd-usd quark systems with two strange quarks, two up quarks, and two down quarks) are stable particles of strange matter—the atoms of quark stars, discovered theoretically by Jaffe. Hence they are the most dangerous particles CERN will mass-produce, as they can migrate to the center of the planet without dying.

CERN believed it would do few dibaryons because it used to calculate mass production of particles the lineal equations of quark gases, NOT the equations of complex liquids as those done at RHIC, which imply a much higher rate of production since RHIC liquids quenched down the jets of a quark gas and absorbed and transformed our quarks into strange ones at a faster rate. In general terms, this happens because the reproduction of information is inflationary. All information systems grow geometrically; all energy systems grow arithmetically. The proof that information is inflationary in all systems is evident. For example, money is always more inflationary than the physical economy; theories of reality are many, but we exist in a single Universe; words are spoken more than actions are made; frequency grows exponentially; and waves can be broken in fractal "Fourier patterns" to an almost infinite detail.

Unfortunately, dimensional information is not understood in physics, as physicists use Shannon's one-dimensional abstract definition of information, based as all in physics in the frequency of a lineal wave. Dimensional information is understood in biology as biologists have been able to understand how proteins store information in the third dimension of warping and so they keep ignoring the inflationary nature of information, which as all dimensional process of growth becomes a power law. Yet CERNies don't understand the power laws and cold, informative, cyclical properties of quark-gluon liquids. Thus their calculus of hyperon and dibaryon production and strangelet vortex stability is always conservative.

What this means is that the production of kaons, hyperons, and dibaryons at CERN will always "surprise" CERN, creating more strange liquid than expected. This was proved in December's experiments, when at a mere 1.2 teravolts the LHC made 13% more kaons (us, ds strange bosons) "than expected." Now in April, at 3.5 TeV, they might do around 10 times more strange particles than they think since collisions will be almost 10 times more powerful than at RHIC. And so at 3.5 TeV there will be a massive increase of "unexpected" strange particles, which will not be even reported, as CERN says it will treat them as the "background noise" in which to find the nonreal Higgs bosons.

What will happen at CERN, how we shall die, is in that regard a mystery for CERN's physicists but easy to understand and predict, using complexity and duality, the sciences founded at the death of Einstein in the Macy's congress to create a model of the Universe with information. Those sciences created an organic paradigm, able to study far-from-equilibrium systems as those that take place in organic liquids, like the quark-gluon soup, whose far-from-equilibrium form was clearly shown at RHIC.

In those systems, chaos and fractal theory are the how able to describe the organic why that explains the informative and reproductive properties of the particles dissolved in the medium—in a gluon soup, the strange up and down quarks that evolve into hyperons and stable dibaryons.

The incapacity to describe what happened at RHIC, a perfect surprise, and what happened in December, with more "unexpected" strange particles that predicted shows that physicists at CERN ignore all about those new sciences and its applications to new physics, useful to understand all dual systems of energy and information. Since, indeed, the experiments at RHIC and CERN are dualist far-from-equilibrium processes that create both a big bang radiation and a big crunch of masses.

Yet quantum entropy theorists reject as dogma the existence of information in the Universe, the why of reality, which they call instead negantropy.

"The why is the only thing you don't ask in physics," said Feynman.

The why though is what we must ask to understand how we will die by strangelet formation now that CERN has again been proved wrong and Rossler, Wagner, and Sancho's predictions of a higher quantity of strange quarks was proved right. CERN did many more kaons, the first bricks of strangelet liquid than predicted, because the quark soup was not entropy-only, lineal, expanding plasma as CERN thought, but a dual system with an expansive big bang of radiation and an inner region of big crunch of quarks at minimal temperature. Wagner already predicted in 1999 that strangelet liquid would be easy to form, and abstract quantum entropy theorists said it wouldn't. "It was a perfect surprise," announced RHIC—not for complexity theorists, who understand both, the laws of physics and biology at work in far-from-equilibrium liquids as quark-gluon soups are. That CERN and RHIC have not yet applied the laws of far-from-equilibrium systems to describe the quark-gluon soup shows how dogmatic those institutions are.

For the reader, not versed in mathematical equations of far-from-equilibrium systems is easier to grasp what will happen in the quark gluon soup by comparing that soup with the organic soup that evolved life particles. Both can be treated as self-similar liquid mediums in which fractal particles of information, which is what masses are, evolve extracting and trans/forming the relative energy quanta of the medium:

In that regard, the evolution of organic soups has three phases in which informative particles, amino acids, or quarks will evolve till creating a stable self-replicating DNA or strangelet that will kill us:

1. *The first step on strangelet creation is the creation of kaons, which are up-strange and down-strange particles/antiparticles systems.*

Because a particle is the arrow of information and life of a physical vortex of form, a mass, and an antiparticle is the anticlockwise, unwinding, expansive, energetic death of the particle. Kaons are not "two particles" in space, but one particle evolving and devolving in time. The uncertainty of observation in physics and the null understanding of the two arrows of time, energy, and form introduces basic errors in CERN's analysis of strange particle/antiparticle systems and weak events.[ch.3] The life-death cycle of a particle is very fast. Yet humans have "slow motion" systems of perception, which often interfere, killing the particles they observe. Thus, physicists often commit the error of ordering as spatial events, time events, and vice versa. They either spatialize time or temporalize space. So when they observe a particle->antiparticle weak event, they see two particles instead of seeing a particle evolving and dying. Only the proper understanding of the arrows of time[Appendix] could solve those problems.

One this infantile errors of "CERN's experts" are corrected, in complex physics kaons are the simplex cycles of creation and destruction of a strange quark, departing from an up and down quark. Depending on the kaon variety an up or down quark dies and becomes a strange quark, or a strange quark decays into an up and down quark. What this means is that kaons are unstable and won't have enough time to fall to the Earth before they decay into antiparticles. And yet they live much longer than the abstract calculations of quantum theorists predicted. Physicists said, "It was as if Cleopatra refused to die of an asp bite and was still falling from her barge[8]." Reason why they called them strange.

So those kaons are the first amino acids of the strangelet liquid, similar to the first fragile amino acids of the organic soup. But with more energy, they will evolve into more complex particles, as all far from equilibrium systems and organic soups do. Oparin needed to give amino acids more energy to form more complex shapes and then they become stable lineal proteins and RNAs.

2.   *This added energy will create a strange liquid made with a lot of usd quarks.*

Will the strangelet be also more stable than CERN predicted? The stability of the liquid made at RHIC also surprised physicists. It was the second warning to mankind of the organic universe. At RHIC, physicists made a proto-strange liquid. It was not a gas, which was what their simple lineal equations expected. Instead they found a quark superfluid, perfectly ordered. "It should have behaved like a gas, hopefully with more energy it will behave rightly," said somewhat "sanguine" reported *Scientific American*, the physicist in charge. Further on, the nonlineal liquid lived billions of time more than expected. This is due almost certainly because it was a fractal liquid, strongly interacting through Rössler and Lorenz attractors that shaped biperpendicular vortices of quark mass. Those attractors might form also in the lower scale of gluons, creating up and top quarks, which have two-thirds of a charge. Nature showed its resilience to die. Thus we can expect further degrees of stability as two-dimensional and three-dimensional attractors form.

Finally, within that liquid, three quarks with one-third of charge formed color-locked perpendicular three-dimensional charges.

3. *Fortunately, it seems that so far in December, most strange particles were unstable particle->antiparticle kaon systems.*

Or else we wouldn't be here (even if a surge in Earthquakes, now at all-time record, suggests some dibaryons might have been formed, fell to the Earth, grew in mass, and exploded when crossing the mantle's discontinuity, provoking a surge in seismic activity).

Yet at 3.5 TeV, both the production of dibaryons and the stability of the strangelet will increase radically. We will be already in the range of energy in which relativistic mass increases the weigh of the proton's quarks, transforming some of them into strange quarks, creating *before collision* within the proton, uds particles called hyperons. Two of those up-down-strange hyperons in the point of collision can merge into dual strange nuclei, a stable dibaryon particle.[9] Thus at 3.5 TeV, unlike in previous collisions in which the process of evolution of particles happened only in the colliding point, making the formation of stable dibaryons, the atoms of strange liquid, almost impossible, CERN will start producing stable dibaryons. Those dibaryons will fall to the Earth, and if they survive the trip, they will start the creation of strangelets in the center of the Earth. In millions of proton-to-proton collisions, usd hyperons will become entangled in pairs to form dibaryons that will fall to the Earth. Those occurrences which CERN says it will treat as a mere background in his search for God's particle, the 10-billion-dollar hoax used to build the quark cannon will be unreported. Thus, as this book comes to print, we might be already within the range of a probable irreversible catastrophe.

Indeed, c-speed is the limit of lineal acceleration. Thus, by the principle of equivalence between forces and accelerations, in Newtonian simple terminology F= m (cyclical vortex of physical information) × a (lineal acceleration with c-limit). So when c-limit is reached, energy is curled into mass.

Thus, as CERN ramps up energies to 3.5 TeV, more c-energy becomes a bidimensional vortex of mass, a quark, and more strange quarks will be formed before collision. Then in the point of collision dibaryons might appear.

All this process will be even more dangerous when CERN collides lead because each lead has hundreds of quarks. So in the point of collision, tiny drops of strange liquid will be formed. And yet again, CERN and RHIC still treat the quark-gluon liquid as a plasma, from a monist perspective. Instead, the duality of big crunches and big bangs created by a far-from-equilibrium system (based on the feed-back cycle, e × i =k) means that as radiation becomes expelled in those mini big bangs in the external, expansive big bang zone that tends to infinite temperature (Max. Ke), a 0 K perfect fluid order will be created inside in the zone of formation of strangelets (Min. Ki).

As any system far from equilibrium, the big bang energy will expel the pions and electrons away into radiation, cooling down the center to near 0 K temperatures of superfluidity, needed to form strangelets. So CERN will do an even more perfect liquid, not a gas at higher temperatures, because the energy will be radiating outward and in the center, a supervortex of strong forces at light speed will be formed.

The strangelet will be a microcosmic quark star, turning at light speed that will finally reach enough mass to become stable and fall to the center of the earth.

The attractive rotational speed of the strangelet vortex is easy to calculate in fractal relativity: in essence, a quark condensate is 100 times stronger than a superfluid vortex of electroweak atoms since the strong force is 100 times stronger.

Yet we need to apply a three-dimensional power law to obtain the attractive power of the one-dimensional strings, gluons, and bidimensional quarks of the soup since we observe that force, as a big crunch three-dimensional from the three-dimensional perspective of our world: $100^3 = 1$ million.

Thus the quark condensate will suck in matter a million times faster than the weak dumb hole condensate we observed in Haifa, made with weak atoms. A quark is bidimensional, so we need to do a cubic power law to consider how three color-locked quarks will suck in our quarks.

Thus if the Einstein-Bose condensate made at Haifa turned at sound speed, a superfluid quark vortex will run at 1 million times the sound speed, *which is exactly the* light speed. Thus a superfluid vortex of quarks, as those CERN will do, will turn around and suck in matter at light speed exactly as pulsars and black holes do. This means that a strangelet liquid will be the seed of a pulsar and a top quark condensate (the Higgs) will be a top quark black hole that will crunch the Earth.

Again this is totally ignored by CERNerds, which will be looking for the invisible Higgs as they get sucked in a millisecond before they can even open their big mouths.

If instead of abstract numbers to calculate the whole vortex we use a fractal analysis of the organic particles assembled in the strangelet liquid, the reactions taken place in that liquid will be the following:

1. Creation of kaons, which will transform an up quark into a strange quark, within the proton. This will be observed as an ūs kaon, *where the u quark dies as an antiparticle, giving birth to a strange quark. Then those strange quarks will join within the nuclei to the remaining ud quarks, forming a hyperon. This process has already been observed at RHIC, where at very low energy collisions 70 hyperons were formed, as many as helium atoms.*

2. Then in the point of collision, two hyperons will form stable double nuclei, a dibaryon. *It is a process equivalent to the creation of a stable organic soup:* kaons (amino acids)—> hyperons (RNA lineal strains)—> dibaryons (dual DNA strains).

3. Then dibaryons will produce a *reversal of time, a key phenomenon in the reproductive systems of fractal information in which the previous chain reaction goes backward (as in the case of the gene-enzyme reversal). Those dibaryons,* which are stable and self-reproductive, feed-back particles, as all dual energy/information systems are (informative, cyclical women and lineal, energetic males, dual DNA, etc.), will thus accelerate the reaction. What this means is that dibaryons are the ones responsible for an ice-9 reaction as the usd/usd strains will break, attract new usd particles, and start a replicating process, self-similar to that of DNA in life forms.

Thus we shall all become dibaryons, as dibaryons will be certainly created in the organic quark-gluon soup and fall to the Earth, eating up quarks once they get there. While the remaining electronic and pion cover of the planet, "us" will explode outward in the big bang of the Earth's supernova.

From a macroscopic point of view, the center of the Earth will become colder and rotate faster in a perfect order till reaching near light speed angular velocity and the external cover will radiate outward at trillions of degrees into a far-from-equilibrium birth of a supernova.

Thus, we shall die because CERNerds, stuck in the paradigms of seventeenth-century Galilean physics—mechanism, energy/entropy theories, and lineal equations—don't want to understand the complex nonlineal quark-gluon soups explained by Rossler, founder of chaos theory, and Sancho, who discovered the laws of duality of energy and information and predicted this scenario. Their theory of mass as physical information and strangelet formation was called "twat" physics by Mr. Cox, the spokesman of CERN, a clear indication of the despise CERN's physicists have for all organic forms of life, as if a "twat," a woman's sacred organ of life, was garbage. Newton said that he was proud to die virgin; indeed, some will probably die virgins at CERN. But the first kaon, Kleopatra, refused to die virgin. It mated with an up quark and ate it alive, making it an antiparticle, as a black widow would do. Now Kleopatra, the black widow, will not die virgin either but will provoke a massive big-banging of all our up quarks till it finally falls from the barge and settles down in the center of the Earth. Her twats/dibaryons, her DNA strains, will start, unlike Virgin Newton, a reproductive chain of trans/formations of all the quarks of the Earth into strange matter—her twat physics will produce enough dibaryons to swallow all the virgin children at CERN, including Mr. Cox, shrunk by the mighty black widow to the size of an atom.

The Universe has warned us many times, asking us to respect the organic laws of reality. After Cleopatra refused to die, the second warning was the liquid at RHIC. But CERN's abstract scientists call information negantropy, negate the organic paradigm, call crackpots those who study the cracks, the fractal forms that shape reality and objectify humanity, calculating obscene probabilities on our death.

Yet let us imagine the Universe doesn't kill us this April. If you are reading this book, it is because it gave us another chance.

Then in the Fall of Man, 2010, Pb-Pb collisions will form thousands of strains of dibaryon and those droplets of strange liquid will fall to the center of the earth, growing and splitting, when they reach the size of a Bohr radius. At that size, they will be packed in tens of thousands and as they break into smaller pieces they will start a runaway reaction, a mass bomb, an ice-9 reaction, that will blow up the Earth around Christmas 2010.

Complexity, information theory has taught us that life blooms also in the world of particles, that attractors create perfect liquids, that Cleopatra, the reproductive female principle of the Universe, must not be laughed at. For that reason, the quark cannon should be closed now and their children of thought learn physics of the twenty-first century before they are let to operate some harmless gadgets in a factory of toys. The quark cannon is too big of a toy for such little boys.

Their energetic, lineal simplifications of quark-gluon liquids grossly misrepresent strangelet and yet they defend those theories as absolute truths till the end, regardless of the massive proofs obtained at RHIC. Liquid vortices of information are nonlineal. They reproduce geometrically as reality is organic, complex, informative, made of motions, hence, blasting with life. Once and again, in all sciences we find that informative properties dominate energy. For a century, astrophysicists wondered why the Universe has far less entropy they predict since reality is informative. This means it is colder, lives longer, reproduces much faster, and it is more stable than energy-only theories describe it.

For example, our planet is older than we thought; it was cooled down much faster. Many other planets turn out to have cold crystalline centers. Complexity theory and duality is resolving all those puzzles for decades. In this book's appendix, we give some examples of the structure of the informative Universe. But *nuclear shivaites will have none of it.*

In the appendix, we study the three-topological structure of all systems that have a hard, external, energetic, hotter membrane, and an inner, colder, cyclical, informative regions, from your body with a colder brain to the quark-gluon soup, with an external big bang and internal cold quark vortex. But all that hard work that is illuminating all sciences doesn't penetrate into CERN, the temple of the shivaite sect of nuclear physicists that searches the Saint Grail of the energetic big bang and death of our world.

Physicists are often wrong but never in doubt. (Landau)

Two things I consider infinite, the universe and the stupidity of (quantum) physicists and I'm not sure of the latter. (Einstein)

The SS authorities (Strange Scientists) at CERN are wrong. They are indeed infinitely stupid. But they are never in doubt. So they are making the quark-gluon soup, the Zyklon liquid that will blast the Earth. Only twat theorists can explain this to them. But even Einstein said of her wife that she "only reproduced her stomach." Ah indeed, the Universe is a continuous orgasm of creation, of yin and yang, of strange stomachs. Strange quark stars will become stable between 1 and 10 terabytes in 2010. Top quark stars will become *stable beyond 10 terabytes in 2013. This is what twat physics predicts. It is mathematically and logically* correct. So we have two legs of the scientific method that prove our extinction. And we have experimental hints in astronomy and Earth's labs.

So there is a sizeable probability that we die already with the first experiments at 3.5 TeV. Yet if that is the case, you won't be reading this book. So let us imagine you are reading. It means that the first weaker experiment at 3.5 TeV between protons didn't produce enough stable dibaryons and we are still here. In that case, since we "are still here" and yet the strangelet scenario for the far more massive Pb-Pb collisions is still logically and mathematically consistent in all the papers produced by both quantum and fractal physicists (except those experts whose lies are paid by CERN) and we have experimental evidences of the creation of quark-gluon liquid mini supernovas at RHIC, we still have two events with very strong chances to blow up the planet:

— Strangelet production in lead-to-lead collisions that will produce strangelet soup in the fall of 2010.
— Black hole production in 14 TeV collisions of protons that will produce top quark condensates in 2013.

Let us then calculate the chances of those two events, hoping you are reading, hoping God gave us a final chance, and we "are still here" as CERN collides protons at 3.5 TeV.

11. The Rules of Truth of the Scientific Method Applied to CERN's Theories and Risks

We can now refine the obscene game of statistics CERN does with the life of mankind.

On one hand, CERN makes "statistics," which are false, about the number of moons that become black holes (none according to CERN, all in the long term according to the experimental proofs of dark matter and the existence of "Moon MACHOs," background holes).

On the other hand, CERN makes statistics about the chances that we all die, giving probabilities to the main doomsday scenarios—which is always zero according to CERN's "confidentiality statements" by "decree."

Let us calculate those probabilities not by decree but with the laws of the scientific method, as we have now much more information about those experiments and the theoretical, experimental, and logic legs of truth, each value as one-third of the total truth of the scientific method that allow us to calculate those risks.

Indeed, to calculate the probabilities of truth according to the scientific method,[ch.1] a simple "tertium exclusum" rule applies:

Theories, which are denied by one of the three legs of the scientific method, are false.

So the Higgs particle and the evaporation of black holes are false. They fail not one but at least two legs of the scientific method. They are logically inconsistent and lack experimental evidence after more than thirty years of being formulated, a reasonable quantity of time, to have found proofs of their existence.

The same rule means that the Everett interpretation of quantum physics and all related theories based on the existence of multiple universes are false (including Hawking's explanation of the paradox of information, another experimental proof of the falsity of his evaporating black holes, which he thinks appears in a parallel Universe.) Those theories are full of logical inconsistencies, and they have been experimentally falsified by the many pictures of electronic nebulae, where all the fractal points of the electron are in this Universe.

Theories that are mathematically and experimentally correct are useful rules of calculus even if they are not truth logically.

They might still be logically incoherent, which merely means their "interpretation" of the why of reality is false. But they can be used to "operate" and "predict" real effects.

This is the case of the Copenhagen interpretation of quantum theory, which is logically inconsistent and should be substituted by the fractal organic theory, which uses the same mathematics (even though it can be formulated also with the mathematics of the fractal paradigm, as Nottale has proved). The same happens with Mr. Heisenberg's interpretation of the uncertainty of measure of the quantum world, which is in the observer, NOT in the fractal quantum world as he thinks. His *why* is merely a philosophical interpretation based on mechanism, NOT a mathematical theory, and so it is only logically inconsistent. The proper interpretation given in the appendix is the fractal organic paradigm: the uncertainty is in the human observer, which influences with its energetic instruments the behavior and position of the electronic, fractal herd humans bomb with huge electronic instruments.

Disputed theories like the big bang of a "cellular Universe," part of a whole "hyperuniverse."

It might or might not be certain in its fractal interpretation, depending on the number of scales the Universe has.

Yet the cosmic big bang in the present "singularity" interpretation where the entire infinite Universe is reduced to a point in space-time, self-similar to a super nuclear bomb *is false because it fails experimentally* (no shadows in galaxies), *logically* (*hot* is synonymous to *expanding space* so a singularity cannot be small and hot, and something cannot be born out of nothing—so the space-time should have been already here), *and mathematically* (ad hoc calculus of the 20 K radiation, galactic cycle older than the big bang, not enough mass in the singularity to create all the mass of the Universe, zero explanation of the origin of dark matter, etc., etc).

But a certain number of galaxies might form a supercluster or hyper black hole that will explode in a big bang/big crunch cycle. In any case, this theory will NOT be proved making a quark-gluon soup on Earth and exploding it, which merely can provoke the mini big bang of Earth, but observing the Universe with deep-space telescopes which should show if there are or not galaxies beyond the 13 billion years light age of the big bang in 2014 (Webb telescope).

Theories that show mathematical, logical, and experimental consistency.
They are truth. Unfortunately, the two theories we want to calculate here, the strangelet and black hole scenario, both fall in this category. Yet there is more evidence of the strangelet scenario than the black hole scenario. So we can give two different conservative chances: 66–100% chances to the strangelet scenario and 33–66% chances to the black hole scenario, hoping background holes do not exist.

All in all, it is clear that the three theories that justified in the name of science the construction of the LHC, the Higgs, the hot cosmic big bang, and the evaporation of black holes are proved false by the three legs of the scientific method. *So we don't need a new machine to prove what has been proved wrong.*

For thirty-five years, the scientific establishment has looked for Higgs and Hawking's evaporation everywhere they asked for and with all type of instruments they recommended. First Hawking said the black holes would evaporate in the halo and would be perceived as "lights in the Christmas tree." So for thirty-five years, we looked at the halo for the signature of evaporation and never found it. Then he said they should be all over and could be observed by a gamma-ray burst with a satellite and so we sent the Fermi satellite to look for signs of evaporation and Fermi found 1,400 gamma ray bursts, none of them the signature of an evaporating black hole. Finally he said it should be observed in an experiment on Earth, so last year we made the equivalent dumb hole based on the same properties (but evaporating harmless sound) and it did not emit a single evaporating bleep. If we add, its mathematical and logical inconsistencies, the theory is obviously false, which means that if CERN makes a black hole, they will eat the Earth.

Thus once it is proved that the LHC should be shut down in "the name of true science" as it is as probable that it produces dragoons rather than Higgs and evaporating black holes, we can now calculate what the LHC will do for the other goal of science, bioethics, the fight for human survival. It certainly will not increase our survival chances. On the contrary, as it did in the first purpose of true science—real knowledge about man, God, and its universal details—the LHC scores also a huge zero on the second survival purpose of science: to enhance our survival.

12.The Totalitarian Principle Applied to CERN's Experiments: Certainty of Extinction

CERN gives us probabilities invented by their scientists, according to whatever they fancy to think in the moment the interviewer talks to them, based on fantasies or uncertain interpretations of quantum cosmology.

Indeed, CERN has accepted as a way of doing science, wishful thinking, authority, bullying, damned lies and statistics. This is NOT science.

We, Einstein et al., have tried it all to prevent this, but "there is no bigger deaf than a man who doesn't listen." Indeed, scientists did not want to listen. They sheepishly trusted CERN, a rogue company, the final result of fifty years of cold war and nuclear terror.

Most nuclear physicists know that the substance of the big bang, the quark-gluon soup, is dangerous, but they are in denial and hope they will not produce enough of it to explode the planet. And when papers came in the last years proving they might explode the Earth, they started to play with obscene probabilities about our extinction, invented out of the hat of the magician, reason why the trick, the calculus of those false probabilities, is never shown in their papers as we have done here *since it was never performed, neither asked by the press who had decided a priori to back the nuclear weapon industry, our WMD.* So an Oxford expert told us that it was "like winning the lottery 3 times in a row," a clever idea that caught on. Another CERN expert merely copied the 1 in 4 million chances that physicists calculated for the atomic bomb to ignite the planet, in 1945, the next one copied directly from the Safety Documents of RHIC and "forgot" to change the RHIC for the LHC in the document printed at CERN's Web till we told them to do so. The one in a million seemed then a good number, and different "nuclear experts" would move around that "magic data" till someone pointed out that it would still be 6.6 billion/million victims, 6,600 legal corpses,[10] twice the 9/11 tally. And so the zero-risk confidentiality statement closed the issue. Now it is official, by decree, the most lethal, explosive substance of the Universe, produced without any safety standards, has "zero" risk.

And yet the opposite is truth because in quantum physics rules the totalitarian principle: either a particle is forbidden by the laws of science, and it will never happen or it is allowed and it will always happen. So the probability is closer to a—100% vs. +0%, providing for accidents and errors on the creation of the event, as we have shown here. And in this case standard science says it will happen. If you mix uranium at a certain temperature, you always get a nuclear bomb; and if you put together a number of quarks, theory tells us today that you get a mass bomb and a supernova. It is not perhaps one in a million chances, or any of the absurd, out of the hat probabilities written for the press. As Mr. Wagner, one of the plaintiffs in the suits against CERN put it, in a much commented interview on a TV show, the best guess you can give is a 50% chance, either Einstein's theory of black holes and quark condensates is right or wrong. If it is right, we are dead. If it is wrong, we are alive: fifty-fifty.

In that regard, the company is playing a Russian roulette game of obscene probabilities on our extinction, to dilute the sense of danger. Since CERN knows most people do not understand the difference between a scientific event and a game of chances, it invents those probabilities as it pleases, against one of the basic principles of science—the laws of science *happen. They are not probable but certain.* If the laws of thermodynamics tell you that heating water boils, it will happen 100% most of the time. Science is right or wrong; it is not probable. Thus, if standard Einstein's theory of mass and quark condensates tells us the Earth might become a nova, it will happen. It makes no sense to talk of probabilities but of the truth or falsity of a theory and experimental fact of science. If Einstein's theory is wrong, the probability is then null till experimental facts prove it is happening all over the Universe. Then there is no longer any argument. Theory becomes fact and the totalitarian principle implies it *will always happen—the water will always boil under fire, the electron will always run when we switch on the light. And so far* in cosmology, Einstein is standard science on mass, black holes, and quark condensates. It was first true theory and today it has overwhelming experimental evidence. So it should happen. We should die.

Scientific events are determined by the known laws of standard science, and our entire civilization is based in the fact that they always happen. When you start a chemical reaction it always brings you certain final elements. It also happens when you start a nuclear reaction. It will certainly happen if it doesn't defy the laws of physics. On the other hand, a probable event is one of which we do not know the variables and parameters involved, such as throwing a dice, ignoring the force, direction, and distribution of weight of the arm of the player. Yet if we knew those parameters, we would certainly know the outcome. This is the case of the creation of quark condensates at CERN since we know the force and parameters of the 7 teravolt, c-speed, superfluid, ultracold quark cannon, whose properties are designed to create the quark-gluon soup of big bangs and supernovas. It might not happen if our present understanding of quark condensates is not correct. But so far, we act on standard science. So CERN's experiments will on account of standard science destroy the world. Our science of quarks might change. In fact, we didn't know that quark condensates existed twenty years ago when the cannon was designed. Then we didn't have enough knowledge of quark bombs, or else the machine would have not been designed. But we know today the parameters of quark condensates and those processes are the only known events that can cause quark stars, born in supernovas. So we know that the creation of a quark condensate at CERN is not forbidden by the laws of physics and so it will certainly happen. For that reason, the totalitarian principle makes an imperative of public policy to close down the cannon and oblige CERN to confront in a fair suit, the scientists that have denounced this case of blatant corruption and potential genocide, to prove in due process of law that the quark cannon will not extinguish mankind. If CERN is right and it is not a corrupted nuclear company, relic of the cold war, its scientists should be able to prove it in Court.

What CERN has done is to ignore those suits and omit all references to quark condensates in all its papers despite being the main produce of the factory, *because it perfectly knows that according to standard quark science, strangelets, quark condensates, quark stars, Bosenovas and other forms of quark matter that can destroy the Earth are not forbidden by the laws of physics and so they should happen. So the chief engineer, Mr. Engelen, has* obliged its workers to sign confidentiality statements of no risk.[2]

Is Mr. Engelen right, or is his action as plaintiffs have denounced an act of criminal negligence? Shall we human beings, mankind at large, and our institutions and courts let scientists fire up all kind of machines, even those who can destroy us? Obviously not. Consider, for example, a factory of Ebola virus that would study different mutations of the virus for knowledge. We would not allow crazy scientists to create such a factory. So why should we allow a factory of quark matter, the physical equivalent of the Ebola virus? We shouldn't and for the same reason we shouldn't allow a factory of quark matter. Even if we cannot for sure know when and how an Ebola virus or a black hole will start a pandemic, a chain reaction that will swallow the Earth since what we know is that Ebola virus and quark matter are extremely dangerous. So sooner than later, that danger will materialize. Therefore a more appropiate question is to calculate when this close to 100% probabilities of extinction, according to the three legs of truth of the scientific method—logic consistency, mathematical accuracy, and experimental evidence—will happen in the 2010 to 2013 run of the LHC from 1 to over 1,000 TeV of potency. That is, when the charged dices, thrown once and again in the quantum Russian roulette of the Earth, breaks the symmetry of the entire planet.

Indeed, all the four experiments that CERN will make in 2010 and 2013 can destroy the Earth since none has been proved false by any of the three legs of the scientific method: the three are logically consistent, mathematically consistent, and they have not been falsified by the experimental evidence against them.

Those experiments are of two types:

Some will attempt to make a black hole, which according to Einstein's theory will be a top quark star, colliding the three quarks of protons, transforming them in top=Higgs condensates.

Some will try to do quark-gluon soups, colliding lead that has more quarks than a proton does. Lead is called in the physical jargon a hadron or heavy atom, hence the name Large Hadron Collider. Those experiments might create strangelets, as they will create a bigger critical mass of quark-gluon liquid that can attract all the other quarks of the Earth. This experiment is far more dangerous as we have more experimental proofs and strange stars are easier to do than top quark black holes since strange quarks are far less massive than top quarks.

Further on, those two experiments with "a few quarks" (protons) or a lot of them (hadron) will be performed at two different energies:

1. In March 2010, protons with a few quarks will be collided at low energies, 3.5 TeV. The probability of extinction is "unlikely," 33%, as the mathematical calculus of the exact energy needed to form a top quark black hole, produced by the evolution of the light quarks of the atom nuclei into heavy tops=Higgs favor energies over 10 TeV. At $3.5 \times 2 = 7$ TeV, we tend to believe there is not enough energy. And there is the fact that protons do collide at those energies with the Earth and they have not blown up the planet yet. But it is also a fact that the Universe is full of dark matter, which has blown up other planets. So there is no certainty. Also, some theorists think 7 TeV is enough energy to form black holes. Finally, those collisions might produce enough usd atoms by relativistic growth of mass to produce stable dibaryons and create a strangelet in the center of the Earth. *So the risk*

exists even at such lower energies. In any case, by the time you read this book, this experiment should have happened. So the unlikely chance we become extinct will be or not an event of the past. Yet if you are reading this book, the less dangerous experiment, protons colliding at 3.5 TeV, has already happened so we can hold our breath.

2. If the schedule is maintained according to the site, in the fall of 2010, CERN will collide lead at 3.5 TeV and then the probabilities grow. The quote is

> heavy ion collisions were included in the conceptual design of the LHC from an early stage and collisions between beams of fully stripped lead (208Pb82+) ions are scheduled for one year after the start-up of the collider with protons. The start up occurred in Christmas 2009 but at nominal energies of 1.1 TeV, similar to those of the Tevatron, just a 10% more to establish a world record. The real commissioning starts now in February.

So in the fall, this year Pb-Pb collisions are planned at 287 TeV/Pb or 1.38 TeV per lead nucleon.

Since the collisions at RHIC were of 100 GeV per nucleon, we are talking of an increase of around 28 times the energy/mass of the proto-strangelets created at RHIC. And this is likely to produce massive quantities of strange matter able to catalyze the conversion of Earth into a strange star. Those collisions will form dibaryons, the stable atoms of quark stars that will fall to the center of the Earth unnoticed and form ice-9, strangelet liquid, which by Christmas 2010, should by most calculations start to dent the nucleus of the planet, provoking increasing seismic activity.

The extinction of Earth will be then irreversible.

The three legs of the scientific method suggest this scenario will happen, as it is logical and mathematically consistent and there is plenty of experimental evidence:

The collisions of gold at low energies, at 100 GeV, at Brookhaven showed a proto-stable vortex of strange liquid that was formed. And the excuse that at high temperature less strangelets forms is a mere "theoretical error" of energy-only monist theories that do not understand the far-from-equilibrium dual processes of formation of liquid vortices: high temperature energy will be expelled from the vortex, cooling it down as a refrigerator cools down the interior by warming the external plaque of Freon.

Strangelet theory is mathematically consistent, according to papers by Peng and Chen from the Chinese Institute of National Physics, and it is logically consistent.

Yet we do not have total experimental confirmation, and the mathematical calculus of the exact number of strange quarks needed to start the ice-9 chain reaction are very complex and require further modeling, which Chen asked in his papers that fell on deaf ears. So we cannot give 100% chances. The chances are therefore 66–100%, as it is sustained by proofs from the three legs of the scientific method, but those proofs are not absolute.

In short, it is quite likely that by Christmas, when those collisions are scheduled, we disappear as a species. How long it will take for mankind to know its destiny is another matter largely debated. Unlike the black hole scenario which seems fairly fast, strangelets are not so attractive. They are mostly undetectable. Theoretical studies consider that they will fall to the center of the Earth and will start an accretion process, which depending on how fast they grow could last from a few hours to several years. I believe the process will be fast, and certainly once it has started, it will be irreversible. But of course, we only have mathematical descriptions of those processes, and one fact that might give us hope is the tug-of-war that the creation of dark, quark matter provokes on the surrounding electroweak matter, which defends itself against this aggression. This tug-of-war is easily lost in heavy stars, given the enormous density of its core, where strangelet liquid is formed; but on the Earth's surface, a vacuum from any perspective we look at it, their stability might be compromised. Against this opinion, there is the surprising stability of kaons, of the first strangelet liquid, and the hyperabundance of quark stars found by the Fermi, which hints at the fact this reaction is very common in the Universe.

I want to be an optimist, but if you ask me, sincerely I think we won't survive the 2010–2011 run of Pb-Pb collisions, and strangelets will accumulate in enough quantity in the center of the Earth to kill us all. Still I shall stick to the objective probabilities of the scientific method and give the conservative 66% estimate for this event—a coin tossed with a charged side—God indeed will play charged dices. You should expect to be dead by 2012. If you are alive, then your chances greatly increase because it will mean strangelets are not as stable as present theory predicts. Or merely that the process of accretion of the center of the Earth, which can range according to calculations between hours and decades, is slow.

Then the collider will be closed during 2012. So we shall enter the two new experiments in 2013 at twice the potency.

— Creation of top black holes at maximal potency, colliding protons. The chance of making black holes grows because CERN will collide protons over 10 teravolts of potency, which is the next decametric fractal scale of energies where top=higgs happen in great abundance, *both in fractal physics and classic quantum theory*. If we translate those energies to rotational speeds, in fractal relativity they mean roughly a z=10 c rotational speed, the speed needed to make top quarks and cosmological black holes. Thus the 2013 energies, in fractal relativity are the realm of top quark black holes. And so the quarks of the protons will acquire relativistic mass, colliding as top quarks/Higgs, forming a top quark condensate, called gas 9, which will form by self-similarity with the ice-9 of strangelets a top quark gluon glass, the seed of a black hole. It is possible also that the run of relativistic mass transforms protons into bcb dark atoms. We do not have experimental evidence, though, of these events since we have never made dark atoms and hardly any top quark. So the probability should be around 66% (mathematical and logical consistency, little experimental evidence). The only well-known particle of that top decuplet is the charm and bottom atoms and only in particle/antiparticle unstable dying systems.

— In Christmas, collisions of lead-to-lead hadrons and maximal potency, to create strangelets, top black holes and bc-atoms. Now *hell will break loose, all the previous events will happen with maximal probability as we will be* releasing amazing quantities of quarks at top energetic speeds, already transformed by relativity mass in heavier quarks. In those events, the creation of strangelets according to Chen and other theorists are beyond any mathematical doubt, enough to form stable strange liquid, converting the Earth into a nova. And all the other events will also reach maximal chances, as many top quarks and bcb atoms will be created within the hadrons.

Further on, at this point we shall have realized all the other possible events many times each one. So the accumulation of dibaryons in the center of the Earth will be maximal and so we should add all the previous probabilities in crescendo. Thus the probability we shall be extinct in Christmas 2013 is maximal because the number of strange and top quarks and bcb dark atoms liberated on Earth by this time will be maximal.

Small deaths: gravito-magnetic activity should increase volcano and earthquake activity

Because of the delay on getting this book on print, we can now consider a proved fact: The present surge in earthquake and volcano activity in the first months of 2010 is statistically the highest year on record on earthquakes for all categories, except for the man-made surge during II world war carpet bombing in 1943. And it is also the 2nd on record in number of death. So the LHC is very likely producing due to that statistical surge thousands of victims and massive loss of capital. The ash cloud has already cost to the European economy more than the machine did.

If we consider only the statistics for April there is a further surge for 2010. But what has humanity done this April 2010, to create fluctuations in the gravitational and magnetic fields of the earth responsible for magma motions that cause earthquakes? Very simple: we have switched on what is today the strongest gravitomagnetic field on this planet, the Large Hadron Collider. This machine is today the strongest gravitomagnetic field on Earth. Since Earthquakes, like avalanches, are released by a butterfly effect, the initial change on the gravitational and magnetic fields of the Earth to trigger the release of the potential energy stored in a fault is minimal, below the power of the LHC's gravitomagnetic fields. There are 3 possible ways in which the LHC can cause earthquakes:

A) If the magnetic field of the magnets drawn above interact with other magnetic fields in the magma.
B) If it made black holes or strangelets that are now in the center of the Earth, slowly eating the planet.
C) If it produces perpendicular gravitational waves, affecting the antipodes (Tonga, Fidji region), which is reaching a maximal with occurrences at very deep level (over 500 km.)

A cautionary stop of that machine and serious studies on its effects, which so far have been carried only by employees of the company, is long overdue for obvious bio-ethical reasons, since this year is also on the path to be the peak on human deads caused by Earthquakes. Yet 2/3rds of those deads have yet to materialize.

If the machine stopped now, we have enough data on this pattern of switching on the LHC and getting a surge on earthquakes to do a serious statistical analysis. Politicians should act now. The costs of the volcano in Iceland for Europe might already exceed the value of this machine . . . This is no longer a question of science but of public policy. To ignore and censor the truth of the LHC will not change its lethal consequences . . .

Number of Earthquakes Worldwide for 2000-2010

Magnitude	00	01	02	03	04	05	06	07	08	09	10
8.0 to 9.9	1	1	0	1	2	1	2	4	0	1	1
7.0 to 7.9	14	15	13	14	14	10	9	14	12	16	6
6.0 to 6.9	146	121	127	140	141	140	142	178	168	142	62
5.0 to 5.9	1344	1224	1201	1203	1515	1693	1712	2074	1768	1754	752

Data till 25 April, 2010

by 3, *the 2010 data, gathered by the US National Center for Earthquake Information* (till April 24), we obtain for the whole 2010 year:

A maximal since World War II, for events in the 8 and 7th scale (3, 21; equalling the 1950 record of 24, >7 scale earthquakes), an all time record for the 6th scale (186, overcoming the 183, 1995's record); and for the 5th scale (2256 earthquakes; overcoming the 2074 earthquakes of 2007, the previous record).

This is much higher than the average year on record for the century. Only one year in the entire century set a higher record for earthquakes over 7 in the Richter scale: 1943, at the height of the II World War's massive bombings of the Earth's crust.

Thus, the 2010 surge is a high growth rate that might not be statistical but 'antropic', as the 1943-44 surge was. Since statistical changes happen within certain percentual limits. It cannot be either a geological process that has normally slow geological-time changes. Even global warming has far less steep rates of growth. Imagine we increase by 1/4th the temperature of the Earth in a year . . .

So it cannot be caused by Global warming either; which has a far slower rate change.

Further on, the volcano activity, the ash cloud, is happening in the fractal borders of the Eurasian Plate, where the LHC hyper-magnetic field seats; while the highest growth in earthquake activity is happening in the LHC's antipodal region, the Pacific lower rim of islands, from Indonesia to Tonga.

Hardly 1/3rd of the year has come by (113 days). Thus if we multiply Indeed, it is very likely that as in the case of global warming, we are facing a man-made enviromental catastrophe. In that regard it does not seem a coincidence that the 2 years that have registered the maximal number of earthquakes in the century, the 1943-44 period and the 2010 period, are two peaks in human disturbances on planet Earth. In the 1943-44 period global war and massive bombings carpeted extensive regions of the planet. In 2010 we created a machine, whose magnetic field has no equal on the planet . . .

And yet the press does not report on the LHC experiments anymore, maybe not to cause public alarm. It also downplays the surge on Earthquakes, which is real.

We have so far, 7 top earthquakes: Solomon Islands (7.1) and Haiti (7) on January; Ryukyu Islands (7) and Chile (8.8) in February; and an impressive 3, in April, the month in which the LHC started a continuous run: China (7.1); Tijuana (7.2) and Indonesia (7.7). Further on, the strongest of those earthquakes, in Chile happened within hours of the first succesful proton run at 3.5 Tev . . .

If the April stats are considered only, 2010 with an annual average of 36, it will even beat the all time record of the 1943-44 period of carpet bombings, which registered a total of 32+23=55 big earthquakes . . .

Earthquakes only need a 'butterfly effect'. Simple mine explosions have caused them. Air bombing has caused them . . . Since the energy of the earthquake is already stored as potential energy in the fault. So you just need a chaotic, butterfly effect multiplied by the fault to release its energy. Earthquakes are in that sense similar to an avalanche. And the LHC has enough potency to be the butterfly . . .

Causality in time is there: we know that any new source added to a magnetic field modifes its structure. In April we introduced a new top magnetic field on this planet and the surge in Earthquakes peaked.

This is basic physics and yet no studies were carried out, no serious reports on safety. And now we are in the statistical peak, all time high for more than a century, on the number of Earth quakes on Earth since the LHC started up its systems.

The first time the machine was switched on, before it blew up in *2008, within a day, 4* big earthquakes over the 6 Richter scale hit the Earth (6.1 Iran, 11.00:35; 6.6 Atlantic, 13:08; 6.6 Indonesia, 00.00; 6.9 Hokkaido 00:20) . The first one, *within seconds* of switching on the LHC, happened in Iran and seismologists there wondered if it was related, but being 'Iran', it seems the alarm didn't sound.

So we were lucky and the machine broke down. Then CERN switched it back on and broke the record of any electro-magnetic field anywhere on Earth. *A new macroscopic electromagnetic and gravitomagnetic field had been created and it was not on the center of the Earth.* The complexity and sheer variety of paths of those fields makes the hypothesis of interaction and a butterfly effect extremely likely.

While there are other theories to explain the surge of earthquakes, among which, of course we also have a 'trendy' explanation based in global warming, all those theories require a longer, geological time-span to cause a surge. Only the LHC is time-related. Since it is this year, precisely when the machine was switched on, when we have a surge in volcano's activity and earthquakes.[11]

Conclusion

Now we can do real statistical work with those probabilities, not with the damned lies of CERN, but with the serious objectivity of the scientific method.

For matter of simplification, we just will calculate the probability of extinction by 2013: two events, both logically and mathematically coherent, both with serious experimental proofs.

Thus if we were alarmists, both the creation of strangelets and black holes have the maximal probability 1, absolute certain extinction. But because the experimental proof of both events is not complete—it will only be complete when it has happened—we just calculate the conservative estimate.

If we were to be truly stringent with our limits, trying to lower the probabilities, we could deny to both events experimental evidence because obviously we can never have it, as we will be dead if a black hole or strangelet eats up the Earth. Thus we could lower to 66% both probabilities. Facts though show that experimental evidence in both cases points to formation of strangelets and top quark holes, which will behave like black holes.

We could even go further in our wish to lower probabilities and doubt the fractal paradigm and the concept that both membranes are self-similar, so black holes are top quark stars. This is certainly not what we think and have reasoned here, yet as it is the most novel and argued concept of the new physics of complexity, we could put the event of black holes at 33%. Still, 9 out of 10 physicists think string theory is certain and a five-dimensional string hole, which will form at those energies, is equivalent to a four-dimensional quark-gluon soup. And in five-dimensions holes will form. So 33% is a very conservative choice since what certainly will never happen is that those black holes evaporate to the past. In fact, even if we use quantum uncertainty the chances of time travel Hawking's style, are on-trillionth. Mr. Hawking recognized this in his book *The Universe in a Nutshell*, saying cynically after that statement that he likes "to bet." So we can discount one-trillionth to the 33%.

Yet that will be all what we can really do. So in this conservative, optimist estimate, we would have two independent events of one-third and two-thirds of probability.

On the other hand, if we were pessimist and accepted the obvious logic coherence of the fractal Universe and the fact that all experimental results hint under the totalitarian principle to the creation of strangelets and black holes, in this case we have two certain events and we shall die. This realist opinion, which follows the totalitarian principle, the Murphy's law of physics, concludes we are all living corpses waiting for Judgment Day.

Let's cross fingers, though. In the optimistic, conservative estimate, the chances that CERN makes strangelets or black holes is the sum of two independent events between 66% and 100% and 33–66%, calculated at a minimal risk of 33% and 66%.

Since the addition law of probability says that the probability that *A or B* will happen is the sum of the probabilities that *A* will happen and *B* will happen, minus the probability that both A and B will happen, we can do an easy calculus: A(33%) + B(66%) – A U B.

Where AU B= p(A and B) = p(A) × p(B) = 1/3 × 2/3 = 2/9.

So now we can sum easily where 1/3=33%, 2/3=66% and 100%=1, absolute certainty.

Total probability is A + B-AXB=33%+66%—33% x66% =1/3 +2/3 – 1/3 × 2/3 = 1—2/9 =7/9=77%.

It means that the probability of extinction of mankind by CERN is around 75%, which is the conservative estimate I brought on my suits and affidavits. It is the same probability of throwing a coin twice and getting heads both times. It can happen but don't bet your life on it.

This serious conservative estimate made with the laws of the scientific method shows to which degree the calculus of probabilities at CERN are false, invented for the press. They are patent nonsense, jokes made on the possibility of our extinction, without the slightest sense of moral responsibility.

Since God is playing with very charged dices, the only uncertain question is the day in which this "charged event" most likely will happen. And it is logic to suppose that we shall die sometime during the run of the second experiment on Pb-Pb collisions between Christmas 2010 and Christmas 2011, when enough dibaryons accumulate in the center of the Earth to start the runaway ice-9 process. And we will know it is happening because suddenly in short notice a series of Richter 9 earthquakes will happen all over the Earth. Then the Earth will crunch into itself, magma will pour out of the crust, the atmosphere will become poisoned, and we will all die.

Those are the conservative calculus of probabilities according to the laws of the scientific method. We are likely to die before 2011; we are almost certainly dead by Christmas 2013. Those dates can suffer delays, the machine can suffer accidents, the vacuum conditions of Earth's surface might make those strangelets more unstable than inside stars, where they provoke nova explosions, but those are minor glitches in a very serious calculus.

Since the entire point of this machine is *to make it happen*. CERN will do collisions every other day, billions of them together, creating for two years in a row a constant strain of strangelet or ice-9 and then toplet, top quark liquid, gas-9 during 2013 onward—enough collisions to transform all our atoms into strange and top quarks is absolute.

Another way to explain the enormity of this crime is to calculate the size of the daily genocide that the LHC will commit every day it goes to work, which will be a fractal part of the total probability.

So if we consider the three years of the initial run—2010 and 2011 for strangelets and 2013 for top quark black holes—each day is a one-thousandth of the total probability. And so if the conservative estimate of the genocide is a 75% of 6.6 billion people, each day the machine operates, it causes

75% × 6.6 billion victims /1000 = ¾ × 2/3 10^{10-3}=1/2 10^7 = 5 Million victims!

Thus, each day CERN's people go to work is killing around 5 million human beings, the people that died in the holocaust. For that reason, each day that passes mankind is allowing a new holocaust to happen. Such is the enormity of this monstrosity. Of course, the day this very likely event happens will be the absolute holocaust of mankind, not only of a fractal part of it. Thus, it is by no means an exaggeration to compare Rolf Heuer and Rudolf Hoess, both SS authorities, one a strange scientist the other an SS officer, even though nuclear physicists seem to enjoy a high respect on public opinion, as they kill at distance. This is irrelevant. Nazis killed directly. Nuclear physicists bomb at distance. This might seem less brutal from our point of view, but a Nazi would probably tell us that it is more coward. And certainly it allows for further repetitions of the event, as the sense of guiltiness is blurred by distance. The self-similarity of the perpetrators is obvious when we consider that Oppenheimer and Rudolf Hoess had something else in common: both carried always with them their favorite book, the Gita, which narrated the adventures of Kali and Shiva, Lords of Death, whose statue stands at CERN. The children burned alive in Hiroshima are no less victims than the children gassed in a concentration camp, reason why as nuclear weapons kept evolving the monstrosity of this profession has become now fully clear. Rolf Heuer might feel safe because the Earth will blow inside out and he won't be blamed—the causal, direct chain between what he is doing now and what will happen in the future when enough strange liquid accumulates might not be clear and certainly it will be denied. And so denial is the entire strategy of this company, but the crime will happen anyway. Of course, the legal responsibility of the key scientists, who know enough strange physics to understand the huge chances of a catastrophe and have actively defended the company and lied to the press—Wilczek, Heuer, Engelen, Ellis, et al.—will never be established. Wilczek already said this, laughing on a public interview: "I signed the no-risk safety statement because if anything happens" and he laughed, knowing he won't be blamed, this attitude, akin to poison a person in secrecy, in this case poisoning the Earth. Because we didn't judge any of the makers of the Nazi weapons but brought them to work on our labs, because we gave medals to the people who made nuclear weapons, the new generation of nuclear physicists can have this attitude. We, the biblical "righteous" culture, have always played a double moral standard, accusing other European nations of genocides and holocausts and dismissing our crimes as "necessary." The Spaniards massacred Indians, but we colonized a "depopulated" continent; our massacres in Africa and Asia happened as a collateral effect of our "white man burden," civilizing the natives; the Germans killed innocent children, we just bombed "industrial hubs." This attitude that continues today in our double standards on nuclear proliferation, allowing the research of the most explosive substance of the Universe at CERN, "our lab," while we freak out about the primitive WMD of other countries is an act of self-suicide because a black hole will not distinguish between the elect and our perceived enemies. All humans will be digested with the same rotational light speed. And only a culture of spoiled children who never repressed their self-destructive, violent tendencies cannot see that. Indeed, if we die it will be because our warrior bully culture has displayed an arrogance and despise for human rights, disguised with the intellectual rhetoric of the victors, similar to that of the most heinous nations of history. And now, when the new technologies of death make possible global annihilation we will learn that for the Universe all humans belong to the same species. Many bioethicists have said for decades that this would happen if we didn't learn to control the bad fruits of technology and evolved socially as a species, beyond the tribal egoism of self-serving nations, armies, companies, and individuals. They were put aside with an astounding blindness and optimism, which still continues to our final days.

Thus, the reader must understand the kind of angst in which Einstein et al. lived in the earlier years of this experiment after in 2005 it became evident according to the scientific method that *CERN will extinguish mankind*. And yet all attempts to reason with this institution and its thousands of collaborators bumped against the law of silence of the nuclear community and the arrogance and ignorance of the technocratic and fiction/prone media establishment, who refused to talk about this genocide, self-obsessed with their tribal worlds against the third world. It was then when the main theorists of Einstein et al., Rossler in Europe and Sancho in America, decided to put suits in human rights and federal courts to blow up the issue in the media, helped by Mr. Walter Wagner, who already had denounced RHIC a decade before, warning that it could produce a "perfect" strange liquid, NOT a gas as *all scientists had predicted*; and if that liquid became stable it would blow up the Earth. Then RHIC produced a perfect proto-stable strangelet liquid NOT a gas, a "perfect surprise." And yet WW lost his job as a safety officer of accelerators since we were still here. And those who said it would be a harmless gas got their Nobel Prize. Now they would represent CERN at Court against us.

---

0   Goebbels, a high rank of the Nazi Party, was the first ministry of industrial propaganda that mastered the mixture of industrial ethics (if we can make a machine, we must do it even if kills us all), mass media marketing (massive repetition of Nazi slogans), visual appeal (uniforms, songs), systematic lies, and megalomaniac statements that cater to the natural arrogance of human beings (huge Nuremberg parades, "manifest destiny" of Germany) that made possible the acceptance by Germans of war and genocides. The

same systems are now common among companies that promote weapons or produce lethal substances that cause environmental disasters and yet portray themselves as environmentally friendly (nuclear, "clean industry"), the embodiment of progress (scientific advancement), certified by well-paid corrupted experts, harmless ("our products are safe") all packed with astounding visuals (television ads, films about God's particle, etc.).

[1]   LSAG is available at http://lsag.web.cern.ch/lsag/LSAG-Report.pdf. Mr. Giddings and Mr. Magnano's paper is available at http://arXiv.org/abs/0806.3381. Both ignore completely LHC's quark production and the possible substances it can produce (shown in the graphic), focusing instead on the effects of cosmic rays and speculating that if moons and neutron stars are not yet converted into black holes, we should not.

[2]   "There are lies, damned lies and statistics" is a sentence attributed to the British prime minister Disraeli, who explained how easily corrupted experts and scholars can lie with their own jargon, in the case of economists, the jargon of statistics, widely used at CERN to dilute the dangers of extinction.

[3]   See Johnson's exhaustive analysis of CERN and nuclear physicists' corruption at http://arxiv.org/abs/0912.5480.

[4]   We already explained that far-from-equilibrium quark-gluon soups have a hot cover and a cold interior.

Letessier et al., through a much more thorough analysis than from whom CERN relies on, introduce chemical nonequilibrium into high-energy collision consideration. They view that strangelets could decay into omega (sss) particles hence explaining the fairly significant yield deficiency of thermal models in this respect. Alternatively, they suggest this above thermal model surplus, as due to non chemical equilibrium strangeness clustering within the qgp. Also their model explains relation of kaon+/pion yields with increased energy, where CERN completely failed: http://arxiv.org/abs/nucl-th/0003014v1 http://arxiv.org/abs/hep-ph/0112027v1   http://arxiv.org/abs/nucl-th/0209080v2   http://arxiv.org/abs/nucl-th/0602047v3.

This landmark papers have been ignored by CERN, which simply does not acknowledge any scientific proof, paper, or experimental fact that explains the risks of its quark matter factory.

[5]   A Brief History of Mine by Hawking. You can see the star treck scene at http://www.youtube.com/user/EinsteinVsHawking?feature=mhw4#p/u/18/pQEqexzF6rc.

[6]   See The Seventh Seal scene at http://www.youtube.com/user/EinsteinVsHawking?feature= mhw4#p/u/6/Mq3sqYdhTb8.

[7]   The Fermi satellite failed to observe black hole evaporation. Einstein et al.'s letters to the scientific magazine warning on those results were ignored. Recently the press published that Fermi has confirmed cosmic rays are the debris of supernovas, NOT its cause. See http://fermi.gsfc.nasa.gov/.

[8]   The kaon, the first strange particle discovered, was produced in massive quantities in the first experiments at CERN.

[9]   The stable dibaryon was first theorized by Jaffe (http://prl.aps.org/abstract/PRL/v38/i5/p195_1). It is now fairly established that the core of neutron stars is a liquid mixture of dibaryons and hyperons which were produced at RHIC: http://physicsworld.com/cws/article/news/41917.

[10] The calculus of legal victims is based on the statistical procedures of the insurance industry, which determines the future compensations to the victims of potential catastrophes according to probabilities.

[11] There is an important number of scientific papers proving that gravitational waves and magnetic fields are the cause of Earthquakes. The maximal earthquake activity has been consistently related to the sunsets and dawns of the sun when its gravitomagnetic field varies in higher degree (see Vasilief, **2009, Volume 2** Progress In Physics, **Page 30).** It has also been related to the magnetic activity of its suns spots and **its *flows of protons: http://www.springerlink.com/content/buvw2tq081013210/***

The second most clear correlation is with the activity of pulsars and supernovas, which release gravitational waves. See: Sadeh Dror and Meidav Meir. Periodisities in seismic response caused by pulsar CP1133. *Nature,* 1972, v. 240, November 17,136–138.

Finally, the 3rd biggest correlation of Earthquakes, has been shown to be antropic bombing, either war or explosions in mines. See: http://earthquake.usgs.gov/research/data/centennial.pdf

Finally we have recently observed unexpected changes in Earth's magnetic field: http://physicsworld.com/cws/article/news/42580

CHAPTER 8

Judgment Day

This extremely complex debate is of concern to more than just the physicists.
—Judge H. Gillmor, *Sancho v. DOE*

NY Times: Part of a detector to study results of proton collisions by a particle accelerator that a federal lawsuit filed in Hawaii seeks to stop.

1. *New York Times* Gives the News to the World

In what follows, we shall briefly resume the documents of the main suit against CERN, brought about in a federal court in America, which the judge refused to judge, awestruck by the authority of CERN, which adduced diplomatic immunity not to appear on those suits, and the power of its 10-billion-dollar machine, showcase of our nuclear-industrial complex, which defended CERN through the American Ministry of Energy and a few Saint Nobels of quantum physics. But as Einstein put it, those who have imposed truth with power will be the laugh of the gods. Indeed, the human lack of judgment is irrelevant to the Universe. What matters is the judgment of true science explained in the previous chapters. A good introduction to those suits was the article of the *NY Times* that made the suit the most blogged news of that weekend of 2008.

Asking a Judge to Save the World, and Maybe a Whole Lot More.
By DENNIS OVERBYE

Published: March 29, 2008

More fighting in Iraq. Somalia in chaos. People in this country can't afford their mortgages and in some places now they can't even afford rice.

None of this nor the rest of the grimness on the front page today will matter a bit, though, if two men pursuing a lawsuit in federal court in Hawaii turn out to be right. They think a giant particle accelerator that will begin smashing protons together outside Geneva this summer might produce a black hole or something else that will spell the end of the Earth—and maybe the universe.

Scientists say that is very unlikely—though they have done some checking just to make sure.

The world's physicists have spent 14 years and $8 billion building the Large Hadron Collider, in which the colliding protons will recreate energies and conditions last seen a trillionth of a second after the Big Bang. Researchers will sift the debris from these primordial recreations for clues to the nature of mass and new forces and symmetries of nature.

But Walter L. Wagner and Luis Sancho contend that scientists at the European Center for Nuclear Research, or CERN, have played down the chances that the collider could produce, among other horrors, a tiny black hole, which, they say, could eat the

Earth. Or it could spit out something called a "strangelet" that would convert our planet to a shrunken dense dead lump of something called "strange matter." Their suit also says CERN has failed to provide an environmental impact statement as required under the National Environmental Policy Act.

Although it sounds bizarre, the case touches on a serious issue that has bothered scholars and scientists in recent years—namely how to estimate the risk of new groundbreaking experiments and who gets to decide whether or not to go ahead.

The lawsuit, filed March 21 in Federal District Court, in Honolulu, seeks a temporary restraining order prohibiting CERN from proceeding with the accelerator until it has produced a safety report and an environmental assessment. It names the federal Department of Energy, the Fermi National Accelerator Laboratory, the National Science Foundation and CERN as defendants.

According to a spokesman for the Justice Department, which is representing the Department of Energy, a scheduling meeting has been set for June 16.

Why should CERN, an organization of European nations based in Switzerland, even show up in a Hawaiian courtroom?

In an interview, Mr. Wagner said, "I don't know if they're going to show up." CERN would have to voluntarily submit to the courts jurisdiction, he said, adding that he and Mr. Sancho could have sued in France or Switzerland, but to save expenses they had added CERN to the docket here. He claimed that a restraining order on Fermilab and the Energy Department, which helps to supply and maintain the accelerators massive superconducting magnets, would shut down the project anyway.

James Gillies, head of communications at CERN, said the laboratory as of yet had no comment on the suit. "It's hard to see how a district court in Hawaii has jurisdiction over an intergovernmental organization in Europe," Mr. Gillies said.

"There is nothing new to suggest that the L.H.C. is unsafe," he said, adding that its safety had been confirmed by two reports, with a third on the way, and would be the subject of a discussion during an open house at the lab on April 6.

"Scientifically, were not hiding away," he said.

But Mr. Wagner is not mollified. "They've got a lot of propaganda saying its safe," he said in an interview, "but basically its propaganda."

In an e-mail message, Mr. Wagner called the CERN's safety review "fundamentally flawed" and said it had been initiated too late. The review process violates the European Commissions standards for adhering to the "Precautionary Principle," he wrote, "and has not been done by arms length scientists."

Physicists in and out of CERN say a variety of studies, including an official CERN report in 2003, have concluded there is no problem. But just to be sure, last year the anonymous Safety Assessment Group was set up to do the review again.

"The possibility that a black hole eats up the Earth is too serious a threat to leave it as a matter of argument among crackpots," said Michelangelo Mangano, a CERN theorist who said he was part of the group. The others prefer to remain anonymous, Mr. Mangano said, for various reasons. Their report was due in January.

This is not the first time around for Mr. Wagner. He filed similar suits in 1999 and 2000 to prevent the Brookhaven National Laboratory from operating the Relativistic Heavy Ion Collider. That suit was dismissed in 2001 despite the evidence that Mr. Walter and Mr. Wilczek's thesis, that the quarks liberated could form a superfluid vortex, known as a quark condensate, with the same properties than black holes, was proved right. Indeed, Scientific American called the experiment a perfect surprise because it formed a liquid that the astrophysicist Nastase, termed as a proto-black hole. The collider, which smashes together gold ions in the hopes of creating what is called a "quark-gluon plasma," has been operating without incident since 2000, but Walter notices, it is 1000 times less powerful than the Quark Cannon will be.

Doomsday fears have a long, if not distinguished, pedigree in the history of physics. At Los Alamos before the first nuclear bomb was tested, Emil Konopinski was given the job of calculating whether or not the explosion would set the atmosphere on fire.

The Large Hadron Collider is designed to fire up protons to energies of seven trillion electron volts before banging them together. Nothing, indeed, will happen in the CERN collider that does not happen 100,000 times a day from cosmic rays in the atmosphere, said Nima Arkani-Hamed, a particle theorist at the Institute for Advanced Study in Princeton.

What is different, physicists admit, is that the fragments from cosmic rays will go shooting harmlessly through the Earth at nearly the speed of light, but anything created when the beams meet head-on in the collider will be born at rest relative to the laboratory and so will stick around and thus could create havoc.

The new worries are about black holes, which, according to some variants of string theory, could appear at the collider. That possibility, though a long shot, has been widely ballyhooed in many papers and popular articles in the last few years, but would they be dangerous?

According to a paper by the cosmologist Stephen Hawking in 1974, they would rapidly evaporate in a poof of radiation and elementary particles, and thus pose no threat. No one, though, has seen a black hole evaporate.

As a result, Mr. Wagner and Mr. Sancho contend in their complaint, black holes could really be stable, and a micro black hole created by the collider could grow, eventually swallowing the Earth.

But William Unruh, of the University of British Columbia, whose paper exploring the limits of Dr. Hawking's radiation process was referenced on Mr. Wagner's Web site, said they had missed his point. "Maybe physics really is so weird as to not have black holes evaporate," he said. "But it would really, really have to be weird."

Lisa Randall, a Harvard physicist whose work helped fuel the speculation about black holes at the collider, pointed out in a paper last year that black holes would probably not be produced at the collider after all, although other effects of so-called quantum gravity might appear.

As part of the safety assessment report, Dr. Mangano and Steve Giddings of the University of California, Santa Barbara, have been working intensely for the last few months on a paper exploring all the possibilities of these fearsome black holes. They think there are no problems but are reluctant to talk about their findings until they have been peer reviewed, Dr. Mangano said.

Dr. Arkani-Hamed said concerning worries about the death of the Earth or universe, "Neither has any merit." He pointed out that because of the dice-throwing nature of quantum cosmology, there was some probability of almost anything happening. There is some minuscule probability, he said, "the Large Hadron Collider might make dragons that might eat us up."

This was the article that the *New York Times* printed about the first of the many suits against the Large Hadron Collider for genocide. It was filed in Hawaii because it was the antipodes of CERN, to signify that the entire planet was at risk. And it was filled in America because it was the only country in which a law, the patriot act, allowed a government to seize any dangerous nuclear substances in any country of the world. Since the most dangerous company of the world, which will produce the most explosive substance of the Universe, has diplomatic immunity as the scion of the European bid for nuclear power during the cold war. In other words, at CERN, a group of clueless physicists are going to make the quark-gluon soup of big bangs without any safety measure, without any civil responsibility, and nobody in this planet giving a damn because those clever boys have told the human sheeple it is okay. In that regard, the company did use its diplomatic immunity not to show up on Court as it was obvious it could not defend in theoretical grounds its bid for collective extinction beyond the okay. Instead, it paid two prestigious Nobel Prize awardees to appear as amici in the suits:

Mr. Glashow, a worker on spatial theories of the weak event, who used to call the Higgs Weinberg's toilet (since his friend Mr. Weinberg had baptized the Higgs Hoax the toilet particle) and Mr. Wilczek, a corrupted physicist, discoverer of the ice-9 reaction that causes quark-gluon supernovas, who warned mankind in a press article on *SciAm* in 1999 that strangelets could blow up the Earth and then retracted and got a job at CERN, cosigning the safety reports because, as he would say laughing in a public conference, if "something goes wrong" (who is going to blame me). They confronted on the other side of the bench, Mr. Wagner, a safety officer of accelerators, who cowrote in 1999 with Mr. Wilczek the *SciAm* article on the dangers of future colliders, warning for the first time about the possible creation of black holes, and Mr. Sancho, author of this book. Other coplaintiffs of that suit, such as Mr. Leggett and Mr. Bloggett, were safety experts. A total of 9 people filed the suit. The defendants that represented CERN however never argued the dangers of the collider. Those dangers clearly explained in the documents of the plaintiffs were sided by Mr. Glashow and Mr. Wilczek, which merely told the Court they had authority and the plaintiffs didn't because they had the Saint Nobel of the Dynamite and the plaintiffs didn't and hence the Court would have to trust their opinion and that of the Nuclear Company, just because they said so.

Of course, a decade before, Mr. Wilczek was a rebel physicist who hadn't yet gotten the Saint Nobel of the Dynamite. He was angered about it and so he was free to say what he thought. At that time, he was very critical of the Higgs too since his work defined the mass of different quarks, as this book does, based on Einstein's thesis that they were "whirls of space-time." In his conferences he talked about mass being the frequency of information of the vacuum. But now he works for CERN and he is a celebrity of the media circus. This was the letter he wrote in 1999 with Mr. Wagner to *SciAm*:

SCIAM 1999: Madhusree Mukerjee's article on the Relativistic Heavy Ion Collider (RHIC) at Brookhaven National Laboratory ["A Little Big Bang", March] alarmed several readers, such as Michael Cogill of Coquitlam, B.C. "I am concerned that physicists are boldly going where it may be unsafe to go", writes Cogill, who worries that creating stuff that has not to anyone's knowledge existed since the early universe—namely a quark-gluon plasma—could result in a catastrophe. "What if they somehow alter the underlying nature of things such that it cannot be restored?" he asks. Another reader wondered whether the RHIC experiments could result in miniature black holes.

BLACK HOLES AT BROOKHAVEN?
Thank you for the article by Madhusree Mukerjee entitled "A Little Big Bang" [March]. In the 1970s Stephen W. Hawking postulated that in the early moments of the big bang, miniature black holes would have been present. Although they no longer exist in our region of the universe, such mini black holes could be created by smashing a proton into an antiproton with enough energy. If one were created near a large congregation of mass and if it started absorbing that mass before exploding, the black hole would reach a relatively stable half-life and thus continue to grow. If this happened on the Earth, the mini black hole would be drawn by gravity toward the center of the planet, absorbing matter along the way and devouring the entire planet within minutes.

My calculations indicate that the Brookhaven collider does not obtain sufficient energies to produce a mini black hole; however, my calculations might be wrong. The only way to determine the energy density at which a mini black hole would be created as an intermediary step to the type of explosion depicted in your article is to build a collider and do the experiment. Is the Brookhaven collider for certain below the threshold?

WALTER L. WAGNER

Frank Wilczek of the Institute for Advanced Study in Princeton, N.J., replies:

Whenever we explore new physical (or chemical, or biological) phenomena, questions like Cogill's arise regarding whether we might unwittingly trigger some catastrophe. For example, in the early days of the Manhattan Project, Fermi and others carefully considered whether a nuclear explosion might ignite the atmosphere. Scientists must take such possibilities very seriously—even if the risks seem remote—because an error might have devastating consequences. In the case of the Brookhaven RHIC, dangerous surprises seem extremely unlikely. First, nuclear collisions with larger energies take place regularly as cosmic rays rain down on our atmosphere—so if a disaster were possible, it would have already occurred. Second, related regimes have been explored in detail, and so we have substantial evidence that our theoretical framework for understanding what will happen is reliable. Although we cannot calculate the consequences in complete detail, we can distinguish credible from incredible scenarios.

The idea that mini black holes will be formed, as Wagner suggests, definitely falls in the latter category. The energy densities and volumes that will be produced at RHIC are nowhere near large enough to produce strong gravitational fields. On the other hand, there is a speculative but quite respectable possibility that subatomic chunks of a new stable form of matter called strangelets might be produced (this would be an extraordinary discovery). One might be concerned about an "ice-9"-type transition, wherein a strangelet grows by incorporating and transforming the ordinary matter in its surroundings. But strangelets, if they exist at all, are not aggressive and they will start out very small. So here again a doomsday scenario is not plausible.

Then in the next years, Chen and other researchers found that strangelets do exist, that they are the core of pulsars and quark stars, that they could be negative and so very aggressive, absorbing the nuclei of atoms. But by then, Mr. Wilczek had been awarded the Nobel Prize, gotten a job at CERN, and changed sides, both in his new belief on Higgs that blatantly contradicted his work and in the safety of CERN's experiments that blatantly contradicted his earlier fears but now he was certifying its safety standards.

These are his depositions in the suit, as well as those of his fellow amici Mr. Glashow, Mr. Wagner, and Mr. Sancho (whose affidavits were the longest of them all).

2. No. 08-17389

IN THE UNITED STATES COURT OF APPEALS
FOR THE NINTH CIRCUIT

LUIS SANCHO, et al.
Plaintiffs-Appellants, D.C. No. 1:08-cv-00136-HG
District of Hawaii,

v.

Honolulu
U S DEPARTMENT OF ENERGY, et al.,

Defendants-Appellee's MOTION v.
SHELDON GLASHOW, et al.,
Movants.

TABLE OF CONTENTS

Sheldon Glashow et Al. Brief (Defendants)

Walter Wagner's Brief (Plaintiff)

Luis Sancho's Brief. (Plaintiff)

AMICIS BRIEF

Preliminary Statement And Interest Of Amici

Amici are physicists who have specialized in nuclear particle physics for most of their distinguished careers. Two of them have been awarded the Nobel Prize in Physics for their contributions to the understanding of elementary atomic and sub—atomic particles; the third holds and has held endowed chairs in physics at Harvard University, was chairman and is currently a member of the Harvard Cyclotron Operating Committee, and is an expert in, and has published extensively on, the subjects of high energy physics, radiation physics, nuclear safety and risk analysis. Amici have special knowledge which they believe will assist the Court in this case. Moreover, amici are concerned that Appellants have misunderstood, misconstrued and misstated the import of amici's submission to the district court, and have misrepresented that submission as supporting Appellants claims. Amici wish to inform the Court of the correct scientific approach to the issues of safety raised in this case. This case involves a challenge to the United States financial support for the construction of the Large Hadron Collider ("LHC"), a subatomic particle accelerator straddling the French-Swiss border near Geneva, Switzerland, and research to be conducted there. The core of plaintiffs complaint alleges that the United States and other defendants violated the National Environmental Policy Act by failing to prepare an adequate environmental analysis of the risks of several theoretical objects that plaintiffs allege could be produced by the Collider. Plaintiffs' central "factual" allegation is that the collisions at CERN's LHC are unsafe and could potentially result in the destruction of the Earth. Complaint ¶ 13, SER 4; see also Order Granting Federal Defendants Motion to Dismiss at SER 117. Similar claims of potentially cataclysmic disasters were made by one of the plaintiffs in this case when the Relativistic Heavy Ion Collider (RHIC) was planned, constructed, and began operation at Brookhaven National Laboratory on Long Island, New York State. One of the amici was a member of the high level committee selected to analyze the potential risks of the RHIC, and the other two amici published an article on the risks associated with the RHIC in Nature, one of the most prestigious scientific journals, prior to the commencement of operations of the RHIC. In fact, the RHIC has been fully operation for almost ten years without incident. Amici are aware that the LHC has undergone thorough scientific safety and risk analyses, and are familiar with the numerous scientific papers examining the risks associated with the LHC. These scientific papers have examined, inter alia, the very claims asserted by Appellants here. Appellants claims have not been accepted by the scientific community and are not based on rigorous scientific analysis. Other than the purely speculative "disaster" plaintiffs recited in the Complaint, they do not allege any injury that is particularized, nor do they assert any claim with sufficient geographical nexus to the United States. Amici seek to submit their brief in support of the federal Defendants-Appellees' argument that Appellants allegations of injury are speculative are not scientifically credible because they are based on purely hypothetical occurrences which do not pose a safety risk. Amici are concerned about the use of litigation based on misinformation about and misunderstanding of science under the guise of concern for the environment that inhibits vital and important scientific inquiry. Amici are prompted to submit this brief in part because the Appellants have misconstrued and misrepresented the nature of science and scientific knowledge, and have misused and misconstrued our amicus brief in the district court to support their fallacious arguments in this Court. The substance of this brief is the same as our amicus brief filed in this case in the United States District Court for the District of Hawaii, with the addition of clarification with respect to the nature of scientific inquiry and discourse.

ARGUMENT

I. Appellants Allegations Of A Safety Risk At Cern And Injury To Them Are Purely Hypothetical, Speculative, And Not Credible

In the district court, the federal defendants asserted, inter alia, that the Appellants do not have standing because the alleged injury to them is speculative and not credible. Amici agree that the complaint and affidavits filed by the Plaintiffs—Appellants in this case are without merit.

Scientists who have proposed the construction and operation of the collider known as Large Hadron Collider ("LHC") at CERN are aware of problems associated with quantitatively assessing the risks involved with this novel project. This is not a new problem and virtually every new significant activity must face it. Instead of ending the pursuit of significant scientific endeavors, the scientific community has developed processes to identify all imaginable events that may lead to adverse effects and use the best available information and scientific talent to mitigate them. No other procedure has been suggested by any professional society, any government or international organization. Amici contend that the Appellants suppositions are without merit, and cannot be the basis of a particularized injury sufficient to confer standing. Until half a century ago, industrial safety was managed by learning from past mishaps and by using appropriate measures to avoid their recurrence. For example, miners once used caged canaries as methane detectors. This management process is no longer acceptable as modern technologies have sometimes led to disasters, such as Union Carbide in Bhopal, India, so large and severe that people now demand proof-in-principle that such disasters cannot happen. Society wants to avoid failures at nuclear power reactors and chemical plants. So the old protocol for risk avoidance—try it once; if it turns out to be dangerous, modify the technology, or don't do it again—is no longer acceptable.

In the case before this Court, one important question is whether the LHC at CERN is sufficiently understood that we can be confident that it will not cause a catastrophe of cosmic dimensions, as Appellants claim. Amici assert that the question has been asked and studied by many of the world's best scientists and they have concluded that not only has a scientifically acceptable procedure been followed but that we do know enough to respond to the safety requirements.

During the early 1970s a process was developed to assess the safety of new technologies such as nuclear electric power plants, large oil refineries, large chemical plants, liquefied natural gas facilities, and other large and technically complex facilities. The process consists of a group of qualified individuals first imagining the worst types of catastrophic failures that could occur at the facility and then designing a system to reduce the probability of such failure occurring and reducing the potential consequences of a failure to an acceptable level. This process (often called "fault tree analysis" or "FTA" or "event tree analysis" or "ETA") has been adopted by the nuclear, chemical, and oil industries and by government agencies such as the U.S. Nuclear Regulatory Commission and NASA.[1,2]

The amici recognize that a new procedure had to be developed for the concerns at issue in this case. It has been claimed that the new particle accelerator could trigger an irreversible process that would have enormous consequences, including the destruction of the Earth. This is not a new concern—for example, scientists working on the Manhattan Project in the 1940s seriously considered whether a nuclear explosion could release enough energy to ignite the Earths atmosphere. At that time, probabilistic risk assessment, as it is known today, did not yet exist. The Manhattan Project scientists used then existing knowledge and concluded that the catastrophe postulated would not happen, and history has proven them right. Concerns about the LHC at CERN are legitimate and are properly raised. In fact, they have been raised, studied, and answered decisively by scientists in the United States and in Europe. But the revival of the concern by the Appellants in this case is not well-founded, or even legitimate, because they have, apparently, not educated themselves about the extensive analysis that has been done and the published literature widely available on the subject.

This is not the first such new particle accelerator, or the first such study of risks, or the first reassurance of the safety of a powerful particle accelerator. The closest analogy is the Relativistic Heavy Ion Collider (RHIC) at Brookhaven National Laboratory on Long Island in New York State, where beams of highly charged gold or lead atoms (the heavy ions) traveling at "relativistic speeds" (approaching the speed of light—99.95% of light speed) sped in opposite directions around circular racetracks before colliding. RHIC truly is an atom smasher: it creates nucleus-to—nucleus impacts, taking place thousands of times per second, each impact producing thousands of secondary particles. These incredibly complex "events" are recorded by sophisticated detectors and analysed by supercomputers and a world-wide network of smaller computers. The Brookhaven RHIC studies matter at densities and temperatures never seen before in the laboratory; on a small scale, it reproduces the extreme conditions that existed in the early universe, conditions under which the constituents of ordinary matter are expected to be liberated as quark-gluon plasma. Physicists had long speculated about this state of matter, but RHIC allowed them to glimpse it. About nine years ago, a doomsday vision similar to the one put forward here by Appellants was advanced relating to the RHIC.[3]

One of the Amici, Frank Wilczek, in the July 2000 issue of Scientific American[4] described the concern. The procedure that was followed was important and a good example for the future. The director of Brookhaven National Laboratory established a blue ribbon panel of independent experts (including Wilczek himself) to investigate the subject. The most creative scientists were tasked to imagine what might go wrong and satisfy themselves that the imagined problems did not exist. They examined carefully three scientifically conceivable disaster scenarios in which experiments might produce "black holes" that could gradually consume the Earth; or could create a "vacuum instability" that could expand catastrophically in all directions at the speed of light; or might produce "strangelets," a kind of "strange matter" that would grow to incorporate ordinary matter, perhaps transforming the entire Earth into its form. The first two issues have been raised, and dismissed, each time a new particle accelerator opens. Using similar arguments, Jaffe, et al. were able to conclude that neither posed any threat at RHIC. There is no chance at all that RHIC could[8] manufacture a black hole or gravitational singularity. Even if RHIC (or its higher energy successors) could create a black hole, such a black hole would be so tiny that it would evaporate instantly.[9, 10]

In the natural world, relativistic heavy ions in the form of cosmic rays have been in RHIC-like collisions with one another in space for eons (more, in fact, than will ever take place at RHIC). These distant collisions do not make RHIC experimentally less useful, because (unlike at RHIC) they cannot be directly studied, but one fact is clear: cosmic ray collisions in space have not led to the creation of a new vacuum, so we breathe easily. The third concern arose from the fact that RHIC accelerates heavy ions rather than individual elementary particles, and must be considered more carefully. Such careful consideration was given in studies by Jaffe, et al. and by Dar, et al. Both groups included theorists who were among the first[11] to speculate that lumps of strange matter called strangelets, which contain many strange quarks as well as the usual up and down quarks that make up atomic nuclei, might be more stable than ordinary matter. The strangelet disaster scenario described by Glashow and Wilson would only be credible if strangelets exist (which is [12] conceivable), and if they form reasonably stable lumps (which is unlikely), and if they are negatively charged (unlikely given that current theory strongly favors positive charges), and if tiny strangelets can be created at RHIC (which was and is exceedingly unlikely); in fact it has not occurred in the several years that RHIC has been operational. The RHIC was approved, and it has run successfully, with no[13, 1, 4] sign whatever of the problems described above. Plaintiffs alleged that by causing the collision of subatomic particles, the LHC could create dangerous objects that they describe as "strangelets," "micro black holes," and "magnetic monopoles" that allegedly might destroy the planet. The LHC is in many ways very much simpler than the RHIC. The LHC primarily accelerates and causes the collision of elementary particles—protons. Only a small proportion of its use involves collision of nuclei. Although the LHC operates at a much higher energy level than the RHIC, the likelihood of any of the postulated catastrophes envisaged by the most imaginative physicists is much smaller than with a nuclear collider. The CERN management followed the example set by Brookhaven National Laboratory and commissioned a high level independent committee (the LHC Safety Study Group or LSSG) to imagine what could go wrong. This committee reported its conclusions in 2003. It found the likelihood of the kinds of events postulated by the[15] Appellants to be insignificant. In particular, the probability that "strangelets" exist at LHC is even smaller than at RHIC, and, as noted above, there are no signs whatever that "strangelets" have been created at RHIC. Their work was reviewed by the LHC Safety Assessment Group (or LSAG), which very recently studied actual operations of the LHC and confirmed that no such events have in fact occurred.[13]

An even more recent paper by Koch, B., Bleicher, M., and Stöcker, H., Exclusion of Black Hole Disaster Scenarios at the LHC, arXiv:0807.3349v [hep-ph] (September 28, 2008) (available at http://arxiv.org/PS_cache/arxiv/ pdf/0807/ 0807.3349v1.pdf, last accessed 04/09/09) addresses "fear in the public, that the conjectured production of mini black holes might lead to a dangerous chain reaction" and "summarize[s] the most straight forward proofs that are necessary to rule out such doomsday scenarios." The authors conclude that "none of the physically sensible paths . . . can lead to a black hole disaster at the LHC." [Id. at 7.] This paper, in turn, builds on a paper by Giddings, S.B., and Mangano, M.L., Astrophysical Implications of Hypothetical Stable TeV-scale Black Holes, arXiv:0806.3381v2 [hep-ph] (September 23, 2008) (available at http://arXiv.org/pdf/0806.3381, last accessed 04/09/09), which analyzed "macroscopic effects of TeV-scale black holes, such as could possibly be produced at the LHC, in what is regarded as an extremely hypothetical scenario in which they are stable and, if trapped inside Earth, begin to accrete matter . . . basing the resulting accretion models on first-principles, basic, and well-tested physical laws. " The study "finds no basis for concerns that TeV-scale black holes from the LHC could pose a risk to Earth on time scales shorter than the Earths natural lifetime. Indeed, conservative arguments based on detailed calculations and the best-available scientific knowledge, including solid astronomical data, conclude, from multiple perspectives, that there is no risk of any significance whatsoever from such black holes."

---

1. See U.S. Nuclear Regulatory Commission Fault Tree Handbook (NUREG-1 0492), http:// www.nrc.gov/reading-rm/doc-collections/nuregs/staff/sr0492/ sr0492.pdf, last accessed 04/09/09; NASA, Fault Tree Handbook with Aerospace Applications (2002), http://www. hq.nasa.gov/office/codeq/doctree/fthb.pdf, last accessed 04/09/09.

2. An example is the Electric Power Research Institutes CAFTA software, which is used by many of the U.S. nuclear power plants, by a majority of U.S. and international aerospace manufacturers, and by the U.S. Government to evaluate the safety and reliability of nuclear reactors, the Space Shuttle, and the International Space Station.

3. One of the plaintiffs in this case, Walter L. Wagner, brought suits in 19993 and 2000 in the Northern District of California and in the Eastern District of New York to enjoin operation of the RHIC at the Brookhaven National Laboratory.

    Wagner v. U.S. Dept of Energy, Case No. C99-2226 MMC (N.D. Cal. May 14, 1999) and Wagner v. Brookhaven Science Associates, LLC., Civ. No. 00-1656 (S.D.N.Y. March 3, 2000). Both lawsuits were dismissed (See Exs. J, items 66 and 67 (N.D. Cal. Docket Sheet) and H (E.D.N.Y. Order, 5/26/2000) annexed to the federal Defendants motion to dismiss in the district court, SER 146). In neither of these cases did the courts give any credence to Wagner's theories about the types of dangerous effects that plaintiffs here claim would result from subatomic particle collisions. This Court should reject Appellants similar challenges in this case and dismiss their claims.

4.   Wilczek, F., Letter to the Editor on the Relativistic Heavy Ion Collider (RHIC) at Brookhaven National Laboratory, 281 Scientific American (July 8, 1999). Prof. Wilczek's letter was a reply to Walter L. Wagner's letter "Black holes at Brookhaven?" which appeared in the same issue of Scientific American.

5.   Jaffe, R. L., Busza, W., Wilczek, F., and Sandweiss, J., "Review of8 Speculative Disaster Scenarios at RHIC," 72 Rev. Mod. Phys. 1125-1140 (2000).

6.   See Blaizot, J.-P, Iliopoulos, J., Madsen, J., Ross, G.G., Sonderegger, P., and Specht, H.-J., Study of Potentially Dangerous Events During Heavy-ion Collisions at the LHC: Report of the LHC Safety Study Group. Report CERN 2003-001 (CERN 2003) (SER 59-64).

7.   Previous studies had also argued against a vacuum instability, but could not quite rule it out.

8.   Dar, A., De Rujula, A., and Heinz, U., "Will Relativistic Heavy-ion Colliders Destroy Our Planet?" 470 Phys. Lett. B 142-148 (1999).

9.   Glashow, S.L. and Wilson, R., "Taking Serious Risks Seriously," 40212 Nature 596-597 (1999).

10. The RHIC White Papers, 757 Nucl. Phys. A 1 (2005).

11. See Blaizot, J.-P, et al., supra, n. 9.14

12. See Blaizot, J.-P, et al., supra, n. 9.15

13. Ellis, J., Giudice, G., Mangano, M., Tkachev, I., and Wiedemann, U.,]

14. "Review of the Safety of LHC Collisions" http://lsag.web.cern.ch/lsag/LSAG-Report.pdf, last accessed 04/09/09 (SER 66-71) and CERN Scientific Policy Committee, "SPC Report on LSAG Documents," http://indico.cern.ch/getFile.py/access?contribId=20&resId=0&materialId=0&con fId=35065, last accessed 04/09/09 (SER 73-74).

II. Appellants Argument That Amici's Conclusion That A Catastrophic Event At The LHS Is "Unlikely" Supports Plaintiffs Claims Is Based On A Fundamental Misunderstanding Of The Nature Of Science

Appellants argue that: "Merely being unlikely or very unlikely that the LHC will create conditions that destroy Earth is every reason not to proceed with the experiment unless and until it can be proven to be impossible to destroy the Earth." (Appellants Brief at 11 (emphasis supplied)). Appellants proposed standard, that for something to be safe experts must conclude that an accident is "impossible", betrays Appellants fundamental misunderstanding of the nature of science. As Nobel laureate in Physics Richard Feynman put it, "Scientists, therefore, are used to dealing with doubt and uncertainty. All scientific knowledge is uncertain. This experience with doubt and uncertainty is important. I believe that it is of very great value, and one that extends beyond the sciences. I believe that to solve any problem that has never been solved before, you have to leave the door to the unknown ajar. You have to permit the possibility that you do not have it exactly right. Otherwise, if you have made up your mind already, you might not solve it." Feynman, R. P. The Meaning of It All: Thoughts of a Citizen-Scientist 26-27 (1999). The Supreme Court has recognized that "it would be unreasonable to conclude that the subject of scientific testimony must be known to a certainty; arguably, there are no certainties in science. See, e.g., Brief for Nicolaas Bloembergen et al. as Amici Curiae at 9 (Indeed, scientists do not assert that they know what is immutably true—they are committed to searching for new, temporary theories to explain, as best they can, phenomena)." Daubert v. Merrell Dow Pharmaceuticals, Inc., 509 U.S. 579, 590 (1993) (footnote omitted). As the Bloembergen amici went on to explain: "in science accepted truth is not a constant . . . it evolves, either gradually or discontinuously . . . . An hypothesis can be falsified or disproved, but cannot, ultimately, be proven true because knowledge is always incomplete. An hypothesis that is tested and not falsified is corroborated, but not proved. Thus, scientific statements or theories are never final and are always subject to revision or rejection. See L. Loevinger, "Standards of Proof in Science and Law", 32 Jurimetrics J. 327 (1992) . . . ." Brief of Amici Curiae Nicolaas Bloembergen, et al. at 12-13, filed in Daubert v. Merrell Dow Pharmaceuticals, Inc., 509 U.S. 579 (1993), 1993 WL13006286 (January 19, 1993); see also L. Loevinger, "The Distinctive Functions of Science and Law," 24 Interdisciplinary Science Reviews 87 (1999). "Even the most robust and reliable theory, however, is tentative. A scientific theory is forever subject to reexamination and—as in the case of Ptolemaic astronomy—may ultimately be rejected after centuries of viability." Brief of 72 Nobel Laureates and Others, filed in Edwards v. Aguillard, 482 U.S. 578 (1987), 1986 WL 727658 (August 18, 1986).[14] Appellants note that amici are not "absolutely certain" that there is no risk and they imply that our views should therefore be disregarded. As scientists, we would be abusing the meaning of "absolute" or "certainty" if we had written that there was no chance of any event occurring in the future, because there is nothing absolutely certain about our understanding of the future. To claim that something is "absolutely safe" is incorrect usage and we studiously declined to play this word game in our brief to the district court or in this brief to this Court. However, we are content to tell this Court, as we did the district court, that the issue of the safety of the LHC has been properly raised by its proponents. It has been extensively examined and discussed by many of the brightest minds that have addressed the issue. The particular concerns raised by the Appellants are not correct. Amici believe that the procedure for addressing the safety issue was proper and followed and follows the highest standards scientists have yet developed. Whereas we do not say that it is "absolutely safe," we have no qualms about endorsing the operation of the LHC to our colleagues, our friends, to this Court, and to the world.

14. Indeed, the ancient motto of the Royal Society of London for the 17 Improvement of Natural Knowledge (commonly known as the "Royal Society"), founded in 1663 and probably the earliest society for the advancement of scientific knowledge, is "Nullius in Verba," which has been translated by the renowned physicist Freeman Dyson (in 55 New York Review of Books, Number 10 (June 12, 2008)) as "Nobody's word is final," signifying a commitment to knowledge through experiment rather than through dogma or doctrine.

CONCLUSION

Amici consider that the operation of the LHC is safe, not only in the old sense of that word, but in the more general sense that our most qualified scientists have thoroughly considered and analyzed the risks involved in the operation of the LHC. Appellants' claims are merely hypothetical and speculative, and contradicted by much evidence and scientific analysis. The appeal should be dismissed.

Dated: Larchmont, New York April 10, 2009

Respectfully submitted, s/ Martin S. Kaufman
   ATLANTIC LEGAL FOUNDATION

BIOGRAPHICAL ADDENDUM

SHELDON LEE GLASHOW is a Nobel Laureate in Physics. He is Arthur G.B. Metcalf Professor of Physics at Boston University. Previously he was the Higgins Professor of Physics and Mellon Professor of the Sciences at Harvard University, and. He is a fellow of the American Physical Society and the American Association for the Advancement of Science; member of the American Academy of Arts and Sciences, the National Academy of Sciences, and the American Philosophical Society; foreign member of the Russian and Korean Academies of Science; and founding editor of Quantum Magazine. He is the recipient of many awards, including the Oppenheimer Medal, the Richtmyer Lecture Award, and the Erice Science for Peace Prize.

FRANK WILCZEK is a theoretical physicist and Nobel Laureate in Physics. He is currently the Herman Feshbach Professor of Physics at the Massachusetts Institute of Technology. Wilczek along with H. David Politzer and David Gross were awarded the Nobel Prize in Physics in 2004 for their discovery of asymptotic freedom in the theory of the strong interaction. His current research interests include "pure" particle physics: connections between theoretical ideas and observable phenomena; quantum theory of black holes; behavior of matter—the phase structure of quark matter at ultra-high temperature and density; "color" superconductivity; the application of particle physics to cosmology; and the application of field theory techniques to condensed matter physics.

RICHARD WILSON is Mallinckrodt Research Professor of Physics at Harvard University and immediate past Director of the Regional Center for Global Environmental Change at Harvard University. Professor Wilson is a past Chairman of the Department of Physics at Harvard University, a past chairman and currently a member of the Cyclotron Operating Committee. He is an Affiliate of the Center for Science and

International Affairs and the Center for Middle Eastern Studies at Harvard University. He is a founder of the Society for Risk Analysis. He is and has been a consultant to the United States government and the governments of numerous foreign countries on matters of nuclear safety, toxicology, epidemiology, public health and safety, and risk assessment. Professor Wilson's areas of expertise include elementary particle physics, radiation physics, chemical carcinogens, air pollution, ground water pollution by arsenic, and human rights. He is the author of many articles on high energy physics, environmental pollution and risk analysis, including PARTICLES IN OUR AIR, EXPOSURES AND HEALTH EFFECTS (with Editor John Daniel Spengler) (Harvard University Center for Risk Analysis, 1986) and RISK-BENEFIT ANALYSIS (with Edmund A. C. Crouch) (Harvard University Center for Risk Analysis, 2 ed. 2001). Professor Wilson is the author or co-author of more than 880 published papers on subjects including atomic particles, radioactive particle decay, shielding of particle accelerators and nuclear reactors, nuclear energy production, health risks of nuclear power plant accidents, risks and health impacts of radiation, risks of nuclear proliferation, health effects of electromagnetic fields, acute toxicity and carcinogenic risk, carcinogenicity bioassays, statistical distributions of health risks, public health, cancer risk management, risk benefit analysis, and global energy use and global warming. He is the recipient of numerous awards, including the Forum Award, of the American Physical Society for Forum on Science and Society in 1990 for "Outstanding research and promotion of public understanding of a broad spectrum of issues dealing with physics, the environment, and public health, including his work on reactor safety, estimation of risks posed by environmental pollution and pioneering use of comparative risk analysis" and the Presidential Citation of the American Nuclear Society in 2008 for "Mentoring students for over 50 years in nuclear science, engineering and technology and his tireless efforts promoting peaceful application of nuclear power. Through over 900 papers and publications, and myriad lectures, he has provided invaluable insight and wisdom giving the nuclear community a profound legacy from which to draw knowledge."

CERTIFICATE OF COMPLIANCE

Pursuant to Fed. R. App. P. 29(d) and 9 Cir. R. 32-1, I certify that the foregoing Brief for the Amici Curiae is proportionately spaced, has a typeface of 14 points, and contains 4,454 words, including the Biographical

Addendum, but excluding the cover, the Table of Contents, and the Table of Authorities, determined using the word count feature of WordPerfect 13, the software application used to prepare the brief.

Dated: Larchmont, New York April 10, 2009 s/ Martin S. Kaufman
ATLANTIC LEGAL FOUNDATION

REPLY OF APPELLANT
WALTER WAGNER

Introduction

Appellant Walter L. Wagner replies to the *Amicus Curiae* answering brief and to the defendants/appellees, answering brief below.

Appellant Walter L. Wagner does not oppose the motion for leave to file the *Amicus Curiae* brief, and notes that it complies with the filing deadline of the Federal Rules of Appellate Procedure, Rule 29(b), if that is interpreted to include filing within seven days of an extended appellee filing deadline as herein.

Appellant addresses initially the *Amicus Curiae* brief, and subsequently the defendants/appellees answering brief.

I. Reply To *Amicus Curiae* Brief

The *Amici* couch their argument on the presumption contained in their last paragraph before their conclusion, which reads:

"*Amici* believe that the procedure for addressing the safety issue was proper and follows the highest standards scientists have yet developed. Whereas we do not say that it is absolutely safe, we have no qualms about endorsing the operation of the LHC to our colleagues, our friends, to this Court, and to the world."

The *Amici*, while scientists [though solely physicists, with little background in the more difficult sciences such as biology], attempt to belittle the appellants and their affiants, who are scientists and technologists with expertise not only in physics, but in a wide diversity of scientific backgrounds. The clear insinuation by the *Amici* is that they have a superior knowledge because they are philosophers of physics and not of the other sciences such as biology, medicine, etc., in which the appellants and their affiants also have expertise. The facts, however, show that the *Amici* are attempting to hide the relevant facts of physics, as detailed somewhat below as well as by appellant Sancho. The facts also show that several parties in support of appellants are also experienced physicists [Dr. Plaga, Dr. Rössler, Dr. Wagner, et al.] in addition to their other scientific qualifications.

In fact, the procedure detailed by the *Amici* for addressing the safety issue has not even complied with the law, let alone the standards of scientific protocol. A proper scientific physics procedure risk review protocol would have included:

1) Compliance with the NEPA requirements of the US government for hazardous research, as detailed in Appellants Opening Brief;

2) Compliance with the European Unions requirements for hazardous research, as detailed in Appellants Opening Brief and in the Complaint, and in particular in the affidavit of Dr. Mark Leggett filed in support of the Complaint and attached thereto at filing;

3) Initiation of a proper safety review PRIOR to construction of the LHC machine, rather than waiting until after the machine is completed, when the onus to operate becomes much larger;

4) Usage of a Red-Team/Blue-Team protocol in identifying and evaluating risks [with a red-team envisioning risks and a blue-team attempting to shoot them down];

5) Inclusion of mostly non-CERN scientists in a safety review committee, so that the scientists involved do not have a vested financial interest or other conflict of interest in the safety conclusion. Indeed, one could argue that ALL of the scientists involved in the safety review should have no financial ties to CERN. It is well detailed [including in appellant Sancho's Reply Brief] that all of the LSAG committee members were either present or past CERN employees save one, who had strong ties to CERN. It is, in essence, a fraud to claim that the LSAG was an independent committee free from CERN connections;

6) Inclusion of "dissenting" or other disagreeing scientists in the analysis of the risks, even if voiced as a "minority view". This would include the fact that numerous scientists disagree completely with the LSAG safety report [though one won't read about that in the current report]. This includes the analysis of Dr. Otto Rössler[1],

who has falsified and invalidated the current LSAG safety report by showing the possibility that relativistic micro-black holes are "slippery" and therefore harmless when created in nature, whereas slow ones [such as would be produced at the LHC] would remain potentially disastrous. This also includes the analysis by other theorists that show that the proton-on-Lead collisions in nature [by cosmic rays (a.k.a. high-speed-protons) striking Lead nuclei on the moon at the equivalent energy] are fundamentally different than the Lead-on-Lead collisions proposed for the LHC, even if at the same COM energies [The *Amici* simply assert that if the energy is the same, then it is the same thing. That is simply false, and known to be false by the *Amici*].

That fundamental distinction between Lead-Lead collisions at the LHC, and proton-Lead collisions in nature, which was frequently presented to the CERN'S LSAG committee during its formulation period, was deliberately omitted from the LSAG report because there is no ready answer which allows CERN to even begin to claim that they are simply replicating what occurs in nature. The intended Lead-on-Lead collisions at the LHC happen nowhere near Earth in nature, and of course Earth is safe from such ultra rare events in the deep reaches of intra-galactic space when, on very rare occasion, very high energy Lead cosmic rays run into each other head-on in deep space[2]. At the LHC, such head-on collisions of high-speed Lead nuclei e order of many thousands of times per second, in Earths immediate vicinity, and if non-evaporative strangelets are created, would begin consuming Earth as detailed in numerous scientific scenarios detailed in the scientific literature, as also noted by appellant Sancho.

The *Amici* continue to acknowledge[3] in their writing the possibility that there is a risk, but they believe that the risk is sufficiently small that it is worth taking. They do not attempt to calculate the risk, nor are the appellants able to mathematically calculate the risk. All we know is that it is non-zero, and that it might not be 100%, and we have no factual scientific basis upon which to make a valid mathematical calculation of the risk, other than to set it midway between 0% and 100%. **The risk scenarios are well detailed in the scientific literature**; including both the details of how a small strangelet formed at the LHC might begin growing larger and converting Earth into a supernova, as well as the risk from formation of a micro black hole causing the Earth to implode. Appellant Wagner has taken the approach in accord with standard statistics, therefore, that the risk or probability should be assessed as being half way between those two extremes, as it is improper to otherwise hazard a guess without being able to do a calculation.

However, whether the risk is as small as believed by the *Amici*, or as large as believed by the appellants and the Affiants who filed supporting affidavits attached to the Complaint [and numerous others who are now recognizing the risk, after learning how they were deceived by CERN, such as Dr. Rössler, Dr. Plaga, et al.], is essentially irrelevant to the issues on appeal.

So long as there is an acknowledged risk, NEPA must be complied with, and it was not [as admitted by defendants/appellees]. That then leaves the sole issue on appeal as to whether or not the federal Court has jurisdiction, as detailed *infra*.

II. Reply To Defendants/Appellees Answering Brief

(A)   Luis Sancho Remains as an Appellant

Contrary to the assertion of appellees counsel, Dr. Luis Sancho in fact signed the Appellants Opening Brief. While the original submission did not have his signature, as he was away in Spain[3] and we could not obtain his signature on the joint submission in a timely manner prior to the filing deadline, this was subsequently rectified by a second submission which does contain his signature proving that he in fact filed as a joint appellant.

As has been previously explained to appellees counsel, the logistics of obtaining Dr. Sancho's signature when he is in Spain requires that he send his signature from Spain to the US, have the document also signed by appellant Wagner, and then incorporated into the to-be-filed document. That logistics, as previously explained to appellees counsel, has on occasion necessitated filing with initially a single signature to preserve the filing deadline, followed by a subsequent filing with both signatures to show that both parties in fact prepared the document.

(B)   Appellants Possess Article III Standing

Defendants/Appellees seek to resurrect their argument pertaining to Article III standing and the trial courts supposed lack of jurisdiction, which argument was not accepted by the trial court as valid, and not used by the trial court in dismissing the action.

In support of their renewed argument, they cite *Arizonans for Official English*, 520 U.S. at 64., and *Friends of the Earth, Inc. v. Laidlaw Environmental Services (TOC), Inc.*, (2000) 528 U.S. 167, 180-181.

While citing good case law, the defendants/appellees completely garble its meaning. The three prongs of *Friends* are: a plaintiff must suffer an "injury in fact"; the injury must be actual or imminent; and the injury can be addressed by a favorable decision.

Here, the injury complained of is that the defendants failed to comply with NEPA as required by law. This is a very concrete and particularized injury. There are no ifs, ands or buts about it—appellees did not comply with NEPA, as even admitted by defendant/appellee DOE.

If a dam is constructed on a major Earthquake fault, and the federal agency involved failed to comply with NEPA, would the agency be able to claim that, since the dam had not yet failed, it should be filled, even while battling in court parties who've been complaining that NEPA requirements were not met? Of course not. The particularized injury is the failure to comply with NEPA, *not* the construction of a faulty dam [which is addressed during the NEPA procedures]. To suggest otherwise is simply an effort to misdirect the court regarding the necessity for complying with NEPA. This is also seen in numerous cases that never make it to the appellate level, examples of which are attached hereto as an Addendum being a report of such cases in *Science*[5] magazine.

So too here. The particularized injury is defendants/appellees failure to comply with NEPA. This is very particularized.

Likewise, this is traceable to defendants/appellees, as well as being actual and imminent. They are the parties who are required to comply with NEPA, not some other third party.

Likewise, the injury [failure to comply with NEPA] is readily redressed by a favorable court decision, requiring defendants to comply with NEPA before further funding is released for furtherance of the LHC project.

And while the deprivation of a procedural right *in vacuo* might prove insufficient to create Article III standing [*Summers v. Earth Island Institute*, 129 S.Ct. 1142 (2009)], no such *in vacuo* aspect of this case exists.

To the contrary, numerous scientists have either filed affidavits in this case, or otherwise gone on the public record showing that there exists a serious risk of planetary destruction should the LHC be allowed to operate. These scientists include [but are not limited to] Dr. Rainer Plaga [Germany], Dr. Otto Rössler [Germany], Dr. Mark Legget [Australia], Dr. Paul Dixon [Hawaii], and appellants herein [Spain and U.S.A.]. Likewise, numerous engineers and others with advanced technical training are also on record as showing that standard safety procedures as used in industry, etc., have not been complied with.

Thus, *Summers* reaffirms that an injury is particularized if it pertains to a procedural right when there is an underlying potentiality of injury that needs to be redressed. The potentiality of that injury does not have to rise to an actual injury in fact [i.e., plaintiffs/appellees do not need to prove with absolute certainty that the LHC will destroy the planet], but a risk of injury is sufficient to show that the complained of breach of procedures as an injury [failure to comply with NEPA procedures] is not "*in vacuo*" as per *Summers.*

Still further, defendants/appellees argue that the relief sought [discontinuation of funding by the DOE of the LHC project] cannot redress their injury, arguing that the DOE is funding scientists who are working on the experimental chambers, not the LHC accelerator which is managed by defendant CERN [who is not an appellee, as defendant CERN defaulted at trial court level, prior to dismissal of the action]. The experimental chambers go hand-in-hand with the accelerator like a hand fitting a glove. While they may each be operated separately [as one might operate the headlights of a car, and the car engine, separately for night driving], it is pointless to do so. Without the experimental chambers, the accelerator has no need to exist. Without the accelerator, the experimental chambers have no need to exist. Discontinuation of funding of the experimental chambers wherein collisions are to take place is the desired outcome of plaintiffs/appellants herein, until such time that NEPA has been complied with, as this will serve to protect the interests of plaintiffs/appellants.

Further, defendants/appellees claim that an injury at some future point of time decades to centuries from now [which most theories show being the amount of time in which a small initial strangelet or micro-black-hole would need to grow in order to consume the Earth] from the slow growth of a strangelet or a micro-black-hole is not a threat of an "imminent" injury. According to defendants/appellees, if it takes decades or longer to destroy the planet, then its OK to destroy the planet by either exploding it [strangelet style] or imploding it [micro-black-hole style]. This is nonsense and an absurdity. We have an absolute obligation not only to ourselves, but to our posterity to insure that they have a world on which to live.

Finally, it is to be noted that not all of the evidence has been presented to the trial court below [as the dismissal was some nine months prior to the intended trial date]. While some of the evidence has been presented [Affiants affidavits, Dr. Plaga's paper, etc.], more continues to be developed. For instance, Dr. Otto Rössler[6] has recently prepared a scientific paper for publication, and has been solicited by a science journal for its publication, regarding his falsification and invalidation of the much-touted LSAG Safety Report. Still additional evidence is being developed. This is very much a developing field of theoretical scientific research, and the jury of scientists has not yet even begun to deliberate. The LSAG "Safety Report" was but the opening salvo, not the final accounting, of what is proving to be a very difficult and contentious scientific debate. That is because the LHC is intended to produce conditions that exist nowhere else in the universe, or at least nowhere in Earths vicinity, and any conjecture as to its safety is simply that—pure conjecture in the light of extensive theoretical scenarios that show plausible disastrous scenarios.

(C)   The LHC Funding is a Major Federal Action

Defendants/Appellees shoot themselves in the foot with their prior argument_that the DOE only provides funding to the Experimental Chambers, not to the LHC accelerator proper [prior to 2008, DOE funding was to the accelerator proper in the form of magnet construction[7]]. The Experimental Chambers, as eloquently stated by Dr. Straus, are funded by the DOE's Office of High Energy Physics "to conduct high energy physics research with the ATLAS and CMS detectors"[8] [which are housed in the Experimental Chambers], both of which are to be U.S. operated devices, not CERN operated. It is, of course, in the Experimental Chambers that the U.S. funded operation would exert decision-making control so as to control when and how often collisions would take place. The hand and glove do fit, and the courts cannot acquit. It is this control that is 100% by the U.S., so there is no need for this Court to even find that the 10% funding of the LHC construction, and continuing DOE influence and control over CERN, gives rise to this being a major federal action. The control over the Experimental Chambers alone allows for this Court to easily find this to be a major federal action. Those Experimental Chambers, with continuing funding by the U.S. and not by CERN, are the heart of the operation. While it is true that beam could still be run in the accelerator without operation of the Experimental Chambers, it would be pointless, just as it would be pointless to turn on the headlights, but not the engine on your car, for a night-time drive [or *vice versa*].

Moreover, defendants/appellees ignore the extensive case law cited by plaintiffs/appellants in their Appellants Opening Brief which shows that 10% funding by the federal government, over the course of many years, is not a "great disparity" nor a "very minor percentage"[9] in funding between the U.S. and the other CERN states to preclude a finding of a major federal action for the construction of the accelerator proper and the Experimental Chambers, combined. While four CERN states [Germany, France, UK and Italy] did each separately provide more total funding than the U.S., the other sixteen states provided less, and the US, on average, provided about double the average of the 20 CERN states.

Further, defendant/appellee DOE has argued that defendant CERN is not a party to this action. This is not so.

Defendant CERN was properly served with the Summons and Complaint, but chose to default. The clerk of the court duly noted the entry of such default. Defendant CERN has not subsequently entered the action in an effort to have the default set aside, or themselves dismissed as a defendant, or both. The default remains in effect, and there is no valid basis to set aside the default.

CERN considers itself to be a sovereign entity [comparable to the Vatican State as recognized by all nations] based on an agreement it has with Switzerland, which protects CERN from any civil suit initiated against it in Switzerland, unless it chooses to be sued. However, the U.S. has never considered CERN to be a sovereign entity, nor is there any such agreement in effect between CERN and the U.S. This issue was briefed for the magistrate judge below, but a ruling not entered thereon due to the dismissal of the action on a claimed lack of federal jurisdiction. In fact, CERN **received actual notice** of the suit by proper means [registered process server who actually delivered the Summons and Complaint to the "legal department" at CERN's administrative offices], and **has had repeated opportunities to respond**, and was **served with additional pleadings** [as noted by their certificates of service] as they were filed until they defaulted and clerical entry of default was entered. *CERN cannot complain, nor can its DOE agents, that it was not aware of this action and had no opportunity to respond.* The letter from the process server merely notes that he was subsequently contacted by CERN, who informed him of their belief regarding their alleged sovereign status, which beliefs the process server subsequently relayed to plaintiffs/appellants. It does not refute/contradict his earlier sworn statement that was used as a basis for entry of clerical default, nor does it vacate the clerical default that was entered.

<?> Dr. Rössler, in a private email to appellant Wagner, indicated that his science paper detailing the risk that micro-black-holes are "slippery" when relativistic as would be produced in nature and therefore harmless, but able to grow and accrete matter when slow such as if produced by the LHC on Earth and therefore dangerous, will soon be published in a peer-reviewed science journal. Dr. Rössler is a noted European scientist, with backgrounds in theoretical physics, chemistry, mathematics, and medicine, holding both a M.D. degree and a Ph.D. degree and several hundred published peer-reviewed papers. Dr. Rössler would be one of the expert witnesses called at trial.

<?> These large Lead-nuclei type of cosmic ray are extremely rare, and do not have anywhere near the energy of the much smaller proton type of cosmic ray. The scenario of a Lead cosmic ray striking a Lead nucleus on the moon, at the equivalent energy of the LHC, simply does not happen because they are not of the same energy as what the LHC will create. Instead, to replicate the LHC energies, they would have to collide

head-on in deep intra-galactic space, which would be exceedingly rare and remote from Earth, and accordingly be harmless to Earth.

<?> They also continue to acknowledge they might have it wrong. In a recent April 16, 2009 radio interview, Frank Wilczek discusses the risk issue and concludes: "*If this* [the LHC] *does cause the end of the world, I will not only be very surprised but very embarrassed.*" http://wfpl.org/CMS?p=4498　or　http://archive.wfpl.org/soa/20090416SOA.mp3 **It is the intent of this lawsuit to keep** *Amicus* **Wilczek from being surprised and embarrassed.**

<?> As a reminder to this Court, Dr. Sancho is a citizen of Spain who resides also in the U.S. He is, in essence, a modern-day Lafayette who has come to the aid of the United States at its hour of need.

<?> *Science*, 23 February 2007, Vol. 315, Page 1069, *U.S. Courts Say Transgenic Crops Need Tighter Scrutiny*

<?> Dr. Rössler is a well-respected scientist who has published extensively in mathematical chaos theory, in chemistry, in theoretical physics, and in medicine. He is well-noted in Europe for his opposition to the operation of the LHC without proper safety reviews, and was recently solicited for his scientific manuscript showing that relativistic micro-black-holes might be "slippery", which invalidates the LSAG Safety Review "neutron star" argument.

<?> It was not a DOE funded magnet that unexpectedly overheated and caused an explosion last September, 2008 during preliminary testing. However, arguably, the DOE is responsible for maintenance on the DOE constructed magnets of the LHC accelerator proper, in addition to the maintenance, operation and development of the Experimental Chambers for the ATLAS and CMS detectors.

<?> *Answering Brief of the Federal Appellees*, page 29, middle of the page.

<?> *Answering Brief of the Federal Appellees*, page 48, end of first paragraph.

III. Conclusion

1. There is an established risk of planetary harm from operation of the LHC Experimental Chambers controlled and funded by defendant DOE. This risk is even acknowledged [though downplayed] by the *Amici*, who stated: " . . . *we do not say it is absolutely safe.*" This risk has been extensively detailed in the scientific literature, and there is no clear consensus, as of yet, in the scientific community as to the extent of that risk, with papers addressing the risk in the process of being published, and the risk issue currently being debated and analyzed. Those LHC proponents who have sought to minimize the risk have done so with faulty scientific facts and/or reasoning, as detailed by the appellants in their Reply Briefs, in the Appellants Opening Brief, and in the Affidavits of the Affiants filed in support of the Complaint. The appellants have also detailed that the risk might be exceedingly large, based on a thorough examination of modern scientific literature of the 20th and 21st centuries, especially based on Einstein's theories of black-holes ["frozen stars"], which Einsteinian theories appellees are apparently attempting to experimentally discredit at high risk to appellants.

2. The funding of the LHC by the DOE has been extensive over the course of many years, and has involved the DOE in every stage of the construction of the project. Currently, the DOE is involved in the funding of the Experimental Chambers and its continuing operations, which is a vital ingredient of the LHC project, without which the LHC cannot operate as an experimental device. The control of the Experimental Chambers is under the DOE, and the control of the LHC accelerator proper is under CERN, with DOE sitting on its board as a permanent ["non-voting" member] exerting influential control.

3. All parties have acknowledged that NEPA has not been complied with. Defendant DOE has claimed that it is exempt from compliance requirements in that the total funding by the US of the LHC and Experimental Chambers combined has been roughly 10% of the total cost of construction, even though defendant DOE controls the Experimental Chambers. It has also claimed that it is exempt from compliance because the planetary destruction that might take place would not be for many years in the future. Appellants contend that 10% funding of the project over the course of many years at all levels of participation, in a multi-nation project such as the LHC, is sufficient to show federal NEPA jurisdiction. Appellants also contend that continuing funding and control over the Experimental Chambers by the defendant DOE also shows federal NEPA jurisdiction. Appellants also contend that jurisdiction is also found under the Patriot Act [as per appellant

Sancho's Reply Brief]. Appellants find ludicrous the appellee argument that planetary destruction in the far distant future from LHC operations is not a violation of ethics or NEPA.

4. Consequently, in that defendants/appellees are engaged in a high-risk operation, NEPA requirements [and Patriot Act requirements] need to be followed so those risks can be addressed by the general public for consideration, and not merely by a group of vested-interest physicists who want to find a low risk or no risk result so they can continue to receive their federal funding.

    WHEREFORE, Appellants ask as a prayer for relief that this Court find that the trial court below has jurisdiction under NEPA, under the Patriot Act, or under both, and that this case be remanded for further proceedings. Appellants also request that this honorable appellate Court issue a preliminary injunction as requested by appellants from the trial court below upon remand of this case to the trial court.

DATED: April 30, 2009

CERTIFICATE OF COMPLIANCE

Pursuant to Federal Rules of Appellate Procedure Rule 29(d) and 9th Circuit Rule 32-1, I certify that the foregoing Reply Brief is proportionately spaced, has a typeface of 12 points, and contains 4138 words, *inclusive* of the cover, Table of Contents, Table of Authorities, footnotes, signature, and this Certificate of Compliance. The "Reply Brief" of appellant Luis Sancho is appended hereto as an additional addendum to show its separate thought, as well as to insure compliance with 9th Circuit Rule 28.5, even though appellant Wagner and appellant Sancho are not jointly represented. The word-count for both briefs combined is 11,132 words.

_____

Walter L. Wagner

AFFIDAVIT OF
LUIS SANCHO

O. PRELIMINARY STATEMENT

I, Luis Sancho, affirm, state and declare under penalty of perjury of the laws of the state of California as follows:

Professional Background

I am a System Scientist specialized in Cosmology and Time Theory. I obtained my undergraduate degree at *Barcelona University*, Barcelona, Spain. I followed with my post-graduate studies at *Columbia University*, New York and developed a career as a Writer on scientific themes in Spain. I also chair the Annual World Conferences on the Science of Duality, (the study of the Universe with 2 time arrows or directions of future, energy and information) at the *International Systems Society* (ISSS.ORG) and have published in European magazines, on the field of Duality and the Arrows of Time.

As a researcher in the field of Time Theory I am the author of a series of books and articles on Cosmology and Relativity ("The Organic universe", "Time Cycles") in which I propose an extension of Dr. Einstein's *Principle of Equivalence* to explain the origin of mass. Thus, I have been interested in the experiments that are currently scheduled to take place at the European Center for Nuclear Research (CERN) that will research the nature of Mass.

LHC Concerns

I was initially in favor of the funding of the Large Hadron Collider (LHC)—the biggest, most energetic, technologically advanced machine ever built on planet Earth. It consists of a 27 kilometer circumference, superconductive, super fluid ring, in which bundles of heavy atoms are to be accelerated to almost the speed of light, and collided together, to replicate the awesome energies of the "Big Bang". Such collisions of atoms are intended to be smashed together to create showers of heavy mass particles only found in those first seconds that took place when the Universe was believed by physicists to be destroyed and recreated again.

To understand how it is possible to reach such almost unbelievable energies and replicate the "Big Bang" on Earth, a simple comparison will suffice: the first accelerator (atom-smasher) that smashed and fissioned atoms in the 1930s during the research of nuclear physics leading to Atomic bombs was 25 centimeters in circumference (about 10 inches around). That first atom smasher was more than 100 THOUSAND times smaller than the 27 kilometer ring that will accelerate and smash atoms at CERN to nearly the speed of light.

Unfortunately, in 2004 theoretical calculations (*Addenda F*) on the particles we expect to encounter at CERN in those ultra-energetic conditions showed beyond reasonable doubt that the LHC will very possibly produce 2 kinds of particles which are extremely dangerous for the safety of this planet, as they have been proven both theoretically to be able to swallow in a chain reaction the entire mass of planet Earth:

— Black holes; (expected by the same CERN to be produced at the rate of 1 per hour) (*Addenda C*).
— Strange, ultra-dense quark matter; (expected to be the main product of CERN at the rate of a million particles per second, according to Mr. Engelen, Chief Scientific Officer for the project).

In that regard, I would like to explain briefly the types of Mass and celestial bodies we encounter in the Universe.

While CERN highlights in its reports as its main goal the possible production of theoretical particles that might prove or disprove alternative, non-standard, exotic theories of the Universe, such as the Higgs Particle, and evaporating black holes, the fact is that the 2 Standard, proven theories of Physics, *Quantum Theory* that describes the microcosms, and *Relativity*, which describes the macrocosms, have observed the existence of only 3 families of Mass of increasing force, which suffice to explain the Universe, *without Higgs particles or exotic, never proved evaporating black holes*:

— *Light or normal matter*, of which humans, planets and stars are made;
— Heavier *strange matter*, called strange by its discoverers for its surprising stability; and
— *Top matter*, the heaviest and most forceful of all, which could be the components of black holes, according to Einstein's Theory of Frozen Stars.

Their names come from the 3 types of quarks (the fundamental particle of mass in the Universe) from which they are made: *normal* (also called up and down quarks); *strange*; and *Top* quarks.

Moreover, in the Universe there are also 3 types of celestial bodies:

— Stars and planets made of normal matter;
— Neutron stars, made of neutrons and strange matter; and
— Black holes, which Einstein modeled as a frozen star, hence made of Top matter, which has the *same density of a black hole and CERN will mass produce.*

Again, strange stars and Top-matter/Black-holes *feed on and transform upon contact* the weaker celestial bodies, planets and stars, into replicas of themselves, according to the theories of most astronomers and astrophysicists.

Those 2 processes of destruction of normal matter by the 2 types of Dark Matter are the most violent of the Universe. They have become popularized in the lay media under the names of:

— A "Nova" or "Supernova" (that converts a celestial body of normal matter into a black hole or Neutron/Strange star).
— An "ice-9" reaction, a name given by Nobel Prize winner, Dr. Franck Wilczek, one of the foremost theorists on strange matter today, to the transformation of normal matter into strange matter. In that regard he compared its potential lethal power to that of the fictional substance "ice-9", which in Kurt Vonnegut's science fiction novel, *"The Cat and the Cradle"*, freezes the entire planet into an ice ball in a few hours. Since strange matter could convert the Earth into a strangelet, a mass of strange super fluid, in a few hours (Letters to the Editor, *Scientific American*, July, 1999).

Such is, indeed, the fate all of us might endure if the LHC at CERN creates either of those 2 types of ultra-energetic Dark Matter that could cause the big bang of the Earth. In the past decade, thanks to new, more powerful Telescopes, the information cosmologists have gathered about the composition of the Cosmos shows clearly that Dark Matter dominates the Universe, being 9 times more abundant than our normal type of matter, on which it feeds. For example, our Milky Way galaxy has only 10% radiant (normal) matter, being instead composed of 90% Dark Matter, whose only known real possible components are black holes and strange stars, also called MACHOS, (as they are extremely strong, aggressive entities).

Thus, a cosmological bomb billions of times more powerful than the A-Bomb that nuclear physicists researched in the 20th century, might possibly be created at CERN, the European Center for Nuclear Research. The difference, however, is not only about power but control: strange matter and black holes are, unlike normal Atomic Bombs, self-reproductive bombs; that is, substances, which actively attract and transform our normal matter and whose strength is such that once they become stable they cannot be controlled or destroyed by human beings, who are millions of times lighter (less dense) than those substances. It is thus extremely dangerous to produce any quantities of Dark Matter (strange matter or black holes) of any form on Earth.

For all that has been said, based on cosmological evidence and theoretical work, it is obvious that in a realistic scenario the Large Hadron Collider will create Dark Matter, which could easily feed on the radiant normal matter of this planet in a chain reaction that might destroy this planet within minutes and terminate all forms of life.

The Probabilities of such Catastrophic Event

As of today, the exact probability of a possible runaway reaction that converts the Earth into strange matter, or converts the Earth into a black hole, is dependent on alternative theories, which are still disputed. Those theories convert those experiments in probabilistic events similar to the toss of a coin: If theory A is right or Parameter C, the number of quarks needed to convert Earth into a black hole or strange star, has certain value we will become annihilated. If instead, theory B is right or Parameter C has a different value, we will survive without any adverse consequence.

In that regard, the 2 events that could destroy the Earth, the creation of strange matter or black holes (both of which are forms of Dark Matter), depend primarily upon 2 disputed theories and one physical parameter:

Regarding what appears to be the higher risk scenario, which is the creation of strange matter that can destroy the Earth, it depends on a parameter that appears in the equations of Strange matter, called the *bag constant*. If the so-called bag constant is small, strange matter will be stable and accrete the Earth in an ice-9 type-reaction. Yet if the bag constant is high, strangelets will not be stable and the Earth will be safe.

The problem is that the most commonly used value of the bag constant, the so-called "MIT bag constant", considers that at the range of energies reached by the LHC the lumps of strange matter that CERN would create on Earth would be stable (Addenda A, B). Unfortunately, it has become recently clear that the main substance CERN would likely produce would be strange QGM (strange Quark-Gluon matter), which would likely account for +90% of all particles created. This appears to have been proven by similar experiments at RHIC, an American super-collider, 5640 times less powerful than the LHC[1]. The experiments carried out at RHIC in the past few years, with only a tenth of the LHC energy, failed to discover any of the imaginary particles of theoretical physicists, producing instead always the same substance: unstable Quark-Gluon Liquid, the prior stage to the production of stable strange matter (*which further on, was a total surprise among the experimenters that now claim to be totally sure that CERN is safe (Addenda C).* Thus, it is expected that with 5640 times more energy, CERN will likely create stable strange liquid, *and blow up the Earth.*

This is a natural consequence of the aforementioned structure of matter in 3 horizons of increasing energy and mass. Now that we open the door to big bang energies we also open the door to the stronger, predator matter that thrives on those high-density conditions. Still, if we give the benefit of doubt to the MIT constant, *even against the present state of Theory and Experimental Evidence that favors the creation of stable strangelets at CERN*, we would consider fair a 50% chance of creation of stable strange matter vs. a 50% chance of creation of only unstable Strange plasma at CERN (with other different bag constant).

Regarding the Production of, and Destruction of the Earth, by Black Holes

Black Holes would be produced at CERN if String Theory, or any of the multiple theories that consider gravity to grow in force at small scales, is certain (super gravity, super-symmetry, etc.). According to *Scientific Americans* polls, 9 out of 10 physicists believe that String Theory is certain. Thus, we can assign a *90% chance* to the possible creation of black holes by the LHC (Large Hadron Collider) at a rate of 1 per second. Once they are produced their stability depends on the truth or falsity of 2 alternative theories about black holes:

— What is called the Classic Theory of Black Holes, which follows Dr. Einstein's Theory of General Relativity. This theory, thoroughly proven in the 20th century, affirms that black holes will be stable regardless of size and would feed and destroy the Earth in a Nova Explosion if created here.

— On the other hand, in the 1970s a young, brass Dr. Hawking asserted that "Einstein was double wrong", believing that small black holes would not be stable but evaporate and explode into a burst of energy and particles. The result is the so-called "Hawking radiation" theory that would render small black holes harmless if it actually exists, allowing them to rapidly "evaporate".

Scientists, however, have not accepted Hawking radiation as proven, for at least one of the 3 obvious reasons below:

— Einstein is considered the most important physicist of History, hand in hand with Sir Isaac Newton, and though many have tried to occupy his place, they have always been proven wrong when challenging the master.

— In the last 40 years of observing the Universe since black holes were first predicted, we have not observed a single black hole evaporating, despite the enormous energy that such evaporation would show (the so-called signature of a cosmological event, which in this case should be easy to observe) such as via the GLAST satellite to be launched in May, 2008.

— Hawking's evaporation hypothesis defies not only Einstein's Relativity, but also the laws of Quantum Physics, the laws of Thermodynamics and the laws of temporal causality, as it implies that particles travel to the past so they seem to come out of the black hole, when we observe them from past to future (*Scientific American*, 1977). In simple terms, what Dr. Hawking believes is that Black Holes are time machines. So in the same way, a baby would evaporate back into the womb of his/her mother if time could travel to the past, black holes instead of accreting matter should evaporate, if particles falling on them travel to the past (an idea later sponsored in his popular books that talk about building such time machines, which seem to many of us as pure science fiction). In other words, Hawking's hypothesis defies the 4 fundamental Laws of classic science and so it has never been proven. To the contrary, Dr. Hawking's theory can be easily dismissed using the classic tools of logic causality by '*Reductio ad absurdum*'. Since even in the highly improbable case that particles travel to the past, and so black holes evaporate, kids enter their mothers womb and the dead resurrect; we humans actually live from past to future. So we are affected only by past to future events. For that reason, we always see, from the past to the future, that black holes accrete matter, that kids are born, and the dead remain in their tombs. I would like, in that regard, to quote, in defense of Dr. Einstein, his own words about such improbable, illogical theories:

> "Every theory is speculative. If, however, a theory is such as to require the application of complicated logical processes in order to reach conclusions from the premises that can't be confronted with observation, everybody becomes conscious of the speculative nature of the theory. In such case an almost irresistible feeling of aversion arises . . ."

It is thus evident that we cannot trust the survival of mankind to a theory with no experimental proof that defies so many basic laws of science. It is safer to give the benefit of the doubt to Einstein's proven work and not to risk mankind to see if Einstein is really twice wrong as claimed by Hawking, or *vice versa*. In any case, if we are fair and concede in this issue the benefit of the doubt to Dr. Hawking, we shall give him a 50% chance of being right and Dr. Einstein also a 50% chance. This would define the probability of the Earth to be destroyed by a black hole at 50% × 90% = 45% chance.

Thus, we come to the conclusion that CERN will cause 2 events that can destroy the planet, each with approximately a ±50% chance of occurring, as there are equally respectable, alternative theories and parameters in both cases for which no certain estimates can be made. On that basis, a simple calculation of probabilities shows that the real risk of these proposed experiments can be as high as 75% when we combine 2 possible events, each one with a 50% chance.

To put these risks in perspective, in the insurance business, a potential catastrophes "death toll" is calculated by multiplying the number of possible victims by the probability of the event. A similar calculation shows that the LHC experiment would be technically, in case of being allowed to take place, the biggest holocaust of history.

Such probabilities for the event of Human Extinction by CERN might be discussed and have been argued now for years. Today they range between the official minimal risk estimated by CERN, a verbal term which in mathematical literature is used for a 1-10% probability, (I believe, biased by self-interest), to a very likely estimate by those who believe in Einstein's work and reject Hawking's physics of black holes as an improbable form of metaphysics or those who accept the MIT bag constant for strange matter (being a very likely estimate, a conceptual term for a 75%–90% risk).

In ethical, moral and hence legal terms (as I believe The Law is the practical expression of human ethics), it is self-evident that even a reduced possibility, as those initially considered by CERN, of a 1-10% chances of extinguishing the Earth, would create a "theoretical potential" 6 billion × 1-10% = 60–600 million potential legal holocaust victims, still the biggest genocide in the history of mankind. It would be also the biggest *environmental crime* of history, far more harmful than Global Warming, as it could mean the destruction of all life forms on this planet.

In that regard, I will now try to explain to this court in more detail the 2 main relevant facts about the LHC (Large Hadron Collider):

A) The LHC is not needed to advance our understanding of the Universe, only to prove or disprove alternative non-standard theories about mass, of hyper-ambitious physicists that challenge the already accepted standard model of mass and gravitation, which is Dr. Einstein's *Relativity*.

B) The Large Hadron Collider would become a factory for production of heavy quarks, the only proven, existing particle-candidates to form Dark Matter, whose main property is to feed on radiant normal matter. Yet since the production of Dark Matter is neither necessary for the advancement of science, nor safe to mankind, but a potential environmental crime of global proportions, the LHC should be forbidden to operate—as we close for security reasons Chernobyl-like factories and forbid the reproduction of Ebola virus in an open environment, even if some specialized virologists would like to study it for research purposes. So too we should forbid the reproduction of free, uncontrolled Dark Matter, even if its protagonists would like to study it at CERN.'

I. AMICIS FALSE STATEMENTS

1. Amici affirm we have misconstrued and misrepresented the risks to Earth the experiments at LHC represent, when the opposite is truth: Amici misrepresent and downplay those risks (I). Since they affirm there is no risk whatsoever to Earth, as we **do** know all possible risks involved. Yet *their texts* and *previous, public declarations of Amici and CERN* prove those risks exist and we do not know how to protect mankind against them.

2. Because CERN doesn't want to reveal them to the public, Amici don't inform this Court about them, but use an *'ad hominem'* strategy (III), consisting in:

— *Telling this Court they are people with special knowledge* we must trust and Plaintiffs are people without merit we must not trust, instead of analyzing the extinction risks mankind faces and the safety measures undertaken, if the most dangerous substances of the Universe, black holes and strangelets appear at LHC—*which are null* (II).

— *Analyzing other machine, RHIC, 5640 times less potent than LHC, as if it were the LHC[0]*, which is like comparing the speed of a cockroach with the fastest USAF supersonic jet, the blue-bird, only 1.000 times faster. Thus, to obtain conclusions from RHIC, pretending they apply to the far more powerful and dangerous LHC is a complete misrepresentation (3,16).

— *Considering the real chances that a black hole or strange star forms at LHC and swallows the Earth*, speculative, when there is a wealth of theoretical papers and experimental proves that those reactions are real and very common in the Universe (3,4,IV).

— *Enlarging their specific profession, as if Nuclear Physicists represent all scientists of all disciplines*; and marketing their experiments, as if they would reveal the ultimate meaning of it all; when in fact only Nuclear Physicists have backed these experiments (11), which are of little importance to the advancement of science and study the quantum world of particles, *not the cosmos as CERN pretends* (IV).

Reasons why we recommend this Court to allow a fair trial of this case, which can provoke a global genocide (V).

3. If the blue-bird crosses the sound barrier that a cockroach will never approach, LHC is the first quark cannon that crosses the Electro-weak barrier of death of light matter that RHIC never crossed *(red line, Addenda A)*. This is important because beyond that energy our matter dies, converted into strong quarks, which have an attractive force 100 times stronger than our weak force. And when enough quarks mass together *(Addendas E, F: ± 10,000 quarks)*, their attractive vortex becomes so strong that the chain-reaction of death of light matter becomes irreversible, creating a mass bomb[14] or Nova ($M=e/c^2$). Since CERN acknowledges it will produce ±1 million deconfined quarks per second, and Novas happen constantly in the Universe, the scenario

of destruction of Earth is by no means speculative, as Amici affirm. To hide this enormous risk to mankind, Amici misinform this Court with false scientific statements . . . Let us consider some of those statements and the naked truth:

4. Careful consideration was given in studies by Jaffe and Dar. Both groups included theorists who were among the first to speculate that lumps of strange matter called strangelets, which contain many strange quarks as well as the usual up and down quarks that make up atomic nuclei, might be more stable than ordinary matter.

   Such disaster scenario, which Amici do not explain to this Court, is in fact described in one of the quoted Documents, Dar: "Our understanding of the interactions between quarks is insufficient to decide with confidence whether or not strangelets are stable forms of matter. Suppose that, somehow, such an object is produced in a laboratory high-energy reaction and that it survives the collisions that eventually bring it to rest in matter. At a mass above 1.5ng, for a typical nuclear density, the object becomes larger than an atom. Gravity and thermal motion may then sustain the accreting chain reaction until, perhaps, the whole planet is digested, *leaving behind a strangelet with roughly the mass of the Earth and 100m. radius*[1].

   Yet Amici consider this document to be a proof of safety. Have Amici read the documents they quote? If so, why they misinform this Court? Perhaps they think we cannot understand their special knowledge?

5. The strangelet disaster scenario described by Glashow and Wilson would only be credible if strangelets exist (which is conceivable), and if they form reasonably stable lumps (which is unlikely), and if they are negatively charged (unlikely given that current theory strongly favors positive charges).

   — Amici affirm negative strangelets (lumps of strange quarks) are unlikely. And since only negative strangelets will accrete the Earth, there is no risk. This is false. Since Strange quarks are always negative, as electrons are (they have—1/3rd charge). So in the same manner all lumps of electrons are negative, lumps of strangelets should be negative and *accrete Earth*[2].
   — Amici Sheldon and Wilson's assertion that strangelets cannot be negative is based on their decade old paper, which regardless of their credentials is long superseded by work done on strangelets

during the last years, which proves strangelets can be negative. Hence they will *accrete the Earth. (Addenda F)*.

6. *Amici* state: In the case before this Court, one important question is whether the LHC at CERN is sufficiently understood we **do** know.

    This is false. Precisely LHC was created because we don't understand completely the dangerous particles of quark and dark matter (black holes, dark atoms, strangelets, bosanovas, etc.) that might appear at LHC, as CERN's spokesman, Brian Cox constantly recognizes:

    > "I have no idea what the discoveries at LHC will lead to."
    > "LHC is certainly, by far, the biggest jump into the unknown."
    > "We know it will discover something because we have deliberately built it to journey to uncharted waters[3]."

    While Amicus Wilczek affirms in a taped conference that we can provide to this Court[4], in which *he constantly contradicts his affidavit*: Nature is so inventive and malicious that there is a logical possibility that it can lead to a catastrophe.

7. Amici state: LHC primarily accelerates and causes the collision of elementary particles—protons. Only a small proportion of its use involves collision of nuclei.

    Yet LHC stands for Large Hadron Collider. Because it Collides Large Hadrons, which are massive Atomic Nuclei with the highest content of quarks, exactly the opposite of what Amici state.

    Thus, Amici's misinformation shows:

    — A tacit recognition of the dangers of colliding Hadrons to liberate millions of quarks, since they hide the true purpose of LHC.
    — A lack of respect for the intelligence and oath of truth due to this Court, which **do** know the purpose of LHC. Yet, if Amici lie in such obvious fact, why should this Court trust any statements Amici make on complex themes on which this Court don't have expertise, but relies on Amici's good faith? Obviously it can't. Thus, we must conclude Amici and CERN are purposely misinforming this Court, despite their knowledge on the subject, to hide the dangers for Earth of LHC's experiments.

Plainly speaking, CERN and the physicists involved in those experiments are experts in Quark matter, interested on researching energies beyond the barrier of electro-weak death of our light matter, for personal gain and scholar ambition. *It is precisely their special knowledge and interest on Quarks and Dark matter what makes so biased their statements.* As the expert on Tyrannosaurus Rex from Jurassic Park, who risks her life to see closer her life-time subject of study, CERN's physicists and Amici will do whatever it takes to study Quark Matter and see closer a black hole; while the rest of us, human beings, realize better on the risks involved, *since we have not any special agenda and just want to preserve life.*

8. Amici affirm: Scientists working on the Manhattan Project seriously considered whether a nuclear explosion could release enough energy to ignite the Earths atmosphere. At that time, probabilistic risk assessment, as it is known today, did not yet exist.

    Thus Amici recognize Nuclear physicists already, without any safety assessment, risked the planet. It is not CERN using the same procedure—going ahead, knowing they are risking the life of all of us?

    It is this a proper safety procedure, or an irresponsible act of arrogance?

9. Regarding Jurisdiction, Amici again misinform this Court: (plaintiffs) do not allege any injury that is particularized, nor do they assert any claim with sufficient geographical nexus to the United States.

    Since Rujula[2] clearly explains a strangelet will destroy the planet, we wonder, do Amici know where is America? Perhaps Amici believe we exist in a parallel Universe? Fact is, *America will evaporate if a catastrophe happens.* For that reason LHC is *also* a danger to the United States. Thus, it falls under the jurisdiction of the *Patriot Act*, which expressly states the rights of the American Government to prevent the creation of dangerous substances anywhere in the planet, if they might harm the lives of American people. And there is no more dangerous substance in the Universe than a lump of strong quarks.

    Humanity tends to focus on the past, failing to prevent future catastrophes. So we went to war in Iraq, which only had primitive chemical weapons of mass destruction, as those Nobel produced (28); now we are legitimately worried about the proliferation of primitive Atomic Bombs; yet we let Nuclear Physicists research blindly, without the supervision and stringent safety standards of the military the most powerful Nuclear weapons, Quark Bombs, whose force, 100 times stronger than our weak matter, is far more destructive than anything ever created on Earth.

II. SAFETY MEASURES AND RISKS

10. Safety measures at LHC are inexistent, since *there is absolutely none established in the case a quark bomb, black hole, strangelet, Bosenova or dark atom appears at CERN, nor any shield that can contain them, not any weapons that could destroy them.* So the concept of Safety measures at CERN is to deny all risks, since if something happens well all die. To that aim the same self-interested party (CERN), produces irrelevant documents that merely state the safety of the experiments, without arguing the nature of its Quark Factory, the type and rates of production of lethal particles LHC will create *(Addenda A,E)* and the possible catastrophes those substances might cause. Since such analysis would reveal the dangers of extinction mankind faces if LHC operates. Thus, CERN's reports search for safety alibis elsewhere in the cosmos except the LHC!, analyzing instead the interaction of black holes and neutron stars *in the galaxy[5]!;* studying the 5640 times weaker RHIC, copy-pasting from its safety report!; and making comparisons with cosmic rays, about which CERN lies, affirming they are made of quarks[8]!, as those LHC will mass-produce. Obviously the purpose of such *speculative* safety reports is to create noise and distract this Court and the press from the real issue judged here—CERNS quark factory; as any rogue Company producing polluting substances that can cause an environmental crime would do. In their brief, Amici follow the same strategy: instead of describing the substances and possible catastrophes LHC might cause, they focus his safety study on caged canaries, imaginary scenarios, new, untested procedures for LHC and false statements[6].

11. Amici affirm that the review of risks or LSAG was commissioned (to) a high level independent committee[7]. This is false. CERN has only issued reports from people related to the experiment, and it has repeatedly denied safety risk experts, philosophers of science and bioethical experts any saying in a potential genocide that concerns all of us *(Addenda B).*

The evidence is clear: the LSAG report was conducted by physicists, funded and commissioned by LHC principal, CERN, which is headed by a physicist, and reviewed by CERN's Council Scientific Policy Committee, also composed only of physicists. Concerning participants, a "plurality of expertise", including ethicists and safety experts, is called for by the EC. Yet of the people involved in the LSAG process (preparation and review of documents)—all 26 were physicists. These physicists only advice is then put to CERN Council for consideration to the governments. CERN Council represents the 20 governments funding LHC. The Council therefore itself has a vested interest, and so it is not at arms-length from the project, and may itself feel a bias to justify its prior decisions of support. This is embodied in one of the rules of natural justice or procedural fairness: the rule against bias ('nemo debet esse judex in propria sua cause'—"no one to be a judge in their own cause") A further rule of natural justice is expressed in the Latin maxim 'audi alteram partem': "let the other side be heard". This element of natural justice is in essence what this Appeal is asking for. Since, given the high stakes of the risks involved—namely the extinction of mankind—we cannot leave self-interested physicists to decide our supreme right to live. Or else we would not be a Democracy, the government of the people, but a Technocracy, as Germany, the main contributor to this machine, was during the Nazi Era, when the worship of technology substituted the ethical guidance and defense of life Democratic Laws provide to mankind (29).

12. Amici state that one fact is clear: cosmic ray collisions in space have not led to the creation of a new vacuum, so we breathe easily.

   CERN and Amici affirm that cosmic rays repeat this experiment and so we are safe, because cosmic rays bombing the Earth have not blown up this planet. This one fact is clearly false: In a century of Cosmic Ray analyses, we never detected a single quark ray. Precisely for that reason Nuclear Physicists have built LHC, which is a quark cannon: to study quark reactions that happen inside stars, the only place where the density of matter and sheer force of the collisions involved is similar to that of LHC, causing the creation of pulsars and black holes[8].

13. Despite their claims, Amici don't make an exhaustive list of the worst risks imagined by their qualified individuals:

   — *A Thermonuclear reaction*: LHC's quarks flow through helium and the high energy of those collisions might trigger its fusion (LHC's Helium already leaked out of its tubes in two accidents. That is why the machine is presently under repair.)
   — *Bosenovas:* Novas produced by the sudden implosion of the Earths mass into a super-atom[9].
   — *Dark matter:* Bcb atoms, made with massive quarks that LHC will mass-produce *(Addenda A, F).*

— *Einstein's accreting black holes:* According to Standard Relativity, all black holes formed at LHC, regardless of size, will accrete Earth at light speed $(M=E/c^2)$[14], causing a super-nova.

— *Top quarks.* Einstein considered black holes to be frozen stars made with a cut-off substance[10]. Since top quarks have a similar density to black holes, and are the most attractive particles of the quantum world, and LHC will mass-produce them, a black hole made of Top Quarks could convert the Earth into a Top quark star.

— *Strangelets* made of strange quarks that could convert Earth into a strange star, also called a pulsar. This is the most likely scenario, since Standard theory predicts that somewhere between 1000 and 10.000 strange quarks are enough to create a pulsar *(Addenda A, red line)*; and CERN will produce ±1 million strange quarks per second *(Addenda E)*.

While Bose-novas and Dark Atoms[9] are still theoretical, black holes and strange stars are by no means speculative, as Amici pretend, but happen constantly in the galaxy and we have now experimental evidence it might happen at CERN, after RHIC experiments with *5640 times less mass/energy*[o] created quasi-stable strangelets and black holes (16).

14. Another prove that CERN hides those risks was given by Engelen, CERN's Chief Scientific Officer, who said when the risk that LHC blows up the Earth was known: CERN officials are now instructed, with respect to LHC's world-destroying potential, not to say that the probability is very small but that the probability is zero[11].

This kind of confidentiality statements proper of a rogue company that is hiding a crime is the modus operandi at CERN that we want to avoid, making public those risks and giving mankind the chance to argue them in a due process of law. Since even if there is only a very small probability of blowing up the Earth, as CERN believes, when multiplied for the total population at risk, *6 billions*, as Insurance companies do to calculate the risks of a catastrophe, the result is a genocide. Indeed, a very small quantity is a qualitative term, which science uses for probabilities or populations of ±1%, which means still *60 million* causalities, making the switch on of LHC, regardless of its final outcome, on insurance terms, the biggest genocide of History. An objective, more realistic calculus, as the one we made in our affidavits on the original suit, places the risk over 50%, as there is one event with a high probability, a strangelet formation, and one with a smaller probability, a black hole formation, both of which, according to standard science, which is not the outdated paper of Wilson and Glashow, or the fantasies of Hawking, but Einstein's Relativity and the fundamental properties of quarks, will blow up the Earth.

III. 'AD HOMINEM' STRATEGIES

15. It has been customary in this case to dismiss the suit with 'ad hominem' accusations against the plaintiffs, to create noise and distract attention from the real issues and catastrophes that might occur at LHC, as CERN doesn't want to argue them. Amici follow the same procedure in their filing:

16. Amici affirm: Similar claims of potentially cataclysmic disasters were made by one of the plaintiffs in this case when the Relativistic Heavy Ion Collider (RHIC) was planned . . . One of the Amici was a member of the high level committee selected to analyze the potential risks of the RHIC.

    Since Amici claim we **do** know, yet the outcome of RHIC experiments he supervised was called an Astonishing surprise—could Amicus clarify why he is now so sure that he is right about LHC? *(RHIC bulletin, Addenda C):* Astonishing surprise: Scientists at RHIC had expected collisions . . . to produce a gas of free quarks and gluons. But instead of behaving like a gas, the matter created in RHIC's energetic gold-gold collisions appears to be more like a liquid—a perfect liquid with virtually no viscosity or frictional resistance to flow.

    Problem is a gas evaporates with no harm or a relatively mild explosion. Yet a strange-let, as its name indicates, is a strange-liquid similar to the one found at RHIC, with a higher density of strange quarks (>±10.000 quarks) that make it stable *(Addenda F)*. Since LHC will replicate RHIC experiments producing 5640 times more energy/strange mass[0], (±1.000.000), it should form a strangelet vortex that would convert the Earth into a pulsar. Thus, RHIC proved the creation of a strangelet is by no means speculative but the most likely event to take place at LHC. Because Amici know this[12], they mislead the court with a string of false statements:

    Although LHC operates at a much higher energy level than RHIC, the likelihood of any of the postulated catastrophes envisaged by the most imaginative physicists is much smaller than with a nuclear collider. Proved false by Einstein's equation, $M=E/c2$, which means *a higher energy creates much more quark mass,* hence it makes strange liquid more stable[13]. A fact which Amici Wilczek has explained out of Court and now he denies[14].

Ordinary matter is expected to be liberated as quark-gluon plasma. Proved false by the fundamental law of the scientific method, experimental evidence. Since RHIC produced quasi-stable strange liquid not gas. Yet Amici and CERN know strange liquid will be lethal to Earth; so both deny experimental evidence and talk on the creation of harmless quark-gluon plasma[15].

17. Amici claim plaintiff Wagner, who sued Brookhaven, was wrong because Brookhaven didn't destroy the world. Yet Wagner only affirmed there was a probability that tiny black holes could appear. That was the case in one of the experiments *(Addenda D)*: a quasi-stable micro-black hole appeared at RHIC. Since all experts at Brookhaven were perfectly surprised, only Wagner came closer to the truth, even if he never claimed he did know with certainty the outcome with his special knowledge as our amici pretentiously affirm.

18. Amici affirm concern by the Appellants in this case is not well-founded, or even legitimate, because they have, apparently, not educated themselves about the extensive analysis that has been done. This is false. This plaintiff produced in the previous suit an extensive report, showing the falsity, irrelevance and lack of independence of the LSAG Safety report[5] *(Addenda B, 11)*.

Do Amici and Mr. Kaufman take their profession seriously enough to educate themselves about the extensive analysis that has been done on the lack of independence and falsity of the LSAG?

19. Amici affirm they are prompted to submit this brief in part because the Appellants have misconstrued and misrepresented the nature of science and scientific knowledge.

Fact is, this suit responds precisely to a proper understanding of the Laws and Ethics of the Scientific Method, which are not defined by Physicists but by *Philosophers of Science*, my specialty. If a Judge is the guardian of the Constitution, the supreme law of Society he helps to write, and must judge upon the truth on the suits presented at Court, the philosopher of science is the guarantor of the scientific method, the supreme Law of science and so he must:

— Develop the Laws of the Scientific Method and write general theories of science[16,26].
— Falsify theories, which don't obey those Laws[10] (IV).
— Consider the Ethical value of news experiments and technologies, denouncing them when they harm human life (V).

Because CERN justifies its astronomical expenses pretending to test with their experiments theories that have been falsified, it will not advance our understanding of the Universe; because it risks the life of all of us, it shows a complete disregard for the Ethics of science. Thus, as a Philosopher of Science, I was obliged to litigate against CERN. And I ask this Court, who rules on behalf of the American people, under the same cherished beliefs in the importance of knowledge, truth and ethics, to allow a fair trial in which both sides can argue the dangers of extinction and scientific misuses of LHC.

IV. SCIENTIFIC ALIBIS

20. The 5 main theories CERN pretends to test at LHC, as part of a marketing campaign that convinced our administration and the mass-media on the need to spend those astronomical sums, are either extremely dangerous (research on symmetry breaking and dark matter), false (evaporating black holes), redundant (Higgs), unlikely (the big bang that telescopes already study) or theoretical, hence impossible to achieve through experiments (the Unification of Forces):

21. CERN's Scientific Officer affirms discovering the Higgs, the main reason LHC was built, will be the closest will ever be to God[11]. Yet *Amicus Glashow called Higgs a toilet particle, while* Amicus Wilczek affirms only his work and nothing else, (implying Higgs), explains mass[17], contradicting both their statement that LHC develops vital scientific inquiry.

    Fact is, Higgs is not a new particle but its parameters correspond to the heaviest, most attractive particle of the Universe, the top quark *(Addenda A)*, whose deconfined Higgs state was analysed by this years Nobel Prize, Mr. Nambu[18]. Thus, according to the scientific method, which considers the Universe simple and efficient (Occam's razor), Higgs is a redundancy—the known top quark, whose equations were rewritten, as Microsoft rewrote Apple into windows. Since, nobody would give billions to build the plumbing tubes of LHC's toilet and rediscover the top—a particle we already knew[19].

    Problem is Top quarks should be the atoms of Einstein's frozen black holes[10] that in Relativity will eat up the Earth. So the Higgs Hoax might cost mankind much more than 10 billion $ . . .

22. A telling prove that the Higgs and Hawking's theory of evaporating black holes are at least speculative is their mutual denial: Hawking bets 100 $ we won't find Higgs particle..[20] While Higgs considers Hawking's theory false [21] (so black holes at LHC won't evaporate and accrete Earth).

23. If RHIC (or its higher energy successors) could create a black hole, such a black hole would be so tiny that it would evaporate instantly.

    Amici once more copy-paste an article about RHIC as if it were LHC; and use a speculative theory, which has been falsified ad nauseam under the laws of the scientific method, to make us believe black holes are harmless time machines that evaporate information. Indeed, Hawking says Einstein is double wrong because black holes evaporate. Yet black hole evaporation is a speculative theory of which there is no prove whatsoever, that breaks all *the main laws of science, hence it is false*[22]:

— *The Duality of energy and information*. Hawking pretends black holes have only entropy, *the arrow of energy and death*, and destroy the information of the Universe, *the arrow of life and mass* that increases in all systems with the passing of time. So he denies Einstein, who said that time curves energy into formal masses, information. And Darwin, who said: time evolves the forms of life[23].

— *The II Law of Thermodynamics*: Hawking affirms that black holes invert the direction of time, getting hotter in a colder environment, which is like saying a hot coffee gets hotter when you put it on ice. Of course, if you travel to the past this will certainly happen: the coffee will boil[24] (-;.

— *The law of Hylomorphism*, a fundamental tenet of the experimental method. Since all beings have substance, but Mr. Hawking's pretends his black holes are mathematical fantasies called singularities[10, 22].

— *The law of causality* in time from past to future, *the foundation of logic*, recently used in Fractal Relativity to prove once more Hawking wrong[24, 25].

— The laws of Relativity, which *affirm that all black holes regardless of size will accrete Earth*, growing exponentially at c-speed, according to Einstein's equation of mass[14] ($M = E/c^2$).

Since Relativity means size is relative and doesn't affect the properties of Black Holes, which always obey the Laws of Gravitation, not as Hawking believes, the laws of quantum electromagnetism, a completely different force[10]. For the same reason that Oligomyrmex, a diminutive ant, doesn't become a different species because it is smaller than the giant Driver Ant.

Fact is Einstein's Relativity is the *standard theory of black holes; hence the only one that should be admitted in a process of law, where speculative theories, like Hawking's evaporation, have no place, exactly the opposite of what Amici claim.*

24. It is thus clear that Amici and CERN purposely choose any theory available, as bizarre as it might sound, that proves LHC safe and

makes it look necessary for vital! scientific inquire. So they prefer the speculations of wannabe Nobels, Higgs and Hawking, instead of the real particles and forces: *Nambu's Top quarks and Einstein's black holes that show the enormous dangers of creating a quark factory on Earth.*

25. The 3rd goal of those experiments is to find the mathematical equations that Einstein searched for in his last decades, a Unification Equation of the Universal constants of electromagnetic and gravitational forces. This Quantum Physicists at CERN will never find, smashing particles to see what happens, since it is a theoretical result, which must be obtained as Einstein tried, with Thought Experiments, departing from Geometry and Gravitation not from Quantum Theory[26]. And indeed, Einstein approach was right, though he lacked the modern mathematical tools of fractal theory, discovered after his death, which was the key to find that equation, that I presented last year in a conference on Fractal Relativity[25] at the Annual Congress on Time Duality, the Philosophy of Science that studies the Universe with 2 arrows, energy and information, whose International Congress I chair. Since Amici acknowledge that equation to be the Saint Grail of Modern Physics they couldn't find, I bring it here as a prove Amici and CERN don't have any special knowledge we plaintiffs lack, and their assertion that we have no merit, is just an 'ad hominem' disqualification to avoid a fair trial on an Environmental Crime and waste of public resources that can cause according to standard science a *global genocide.*

26. The 4th theory LHC pretends to test is the cosmic big-bang. Since, when you believe only in the arrow of energy, entropy and death, as Quantum Physicists do, it seems reasonable to re-create the biggest bang thinking it will reveal the meaning of it all.

Unfortunately, even if CERN is able to cause such huge explosion, in the past decade we found a growing number of experimental proofs that at best the big bang is local: the explosion of a galactic black hole, called a quasar—facts that obviously CERN prefers to ignore:

— Astronomers have proved that the radiation of the big-bang is local, since it doesn't leave shadows when it crosses through far away galaxies. Thus, it doesn't come from the remote ends of the Universe, behind those galaxies, like the cosmic big-bang theory pretends[27].
— The dates of the big-bang theory coincide with the cycle of creation and destruction of the central black hole of galaxies that takes

the same 13 billion years, *and has, unlike the cosmic big-bang, experimental proofs.*
— We found helium, overproduced in any big-bang, to be more abundant around the central bar of the galaxy that disintegrates in those cycles, and we have found black holes so huge and old that they had to be formed before the supposed cosmic big bang of 13 billion years . . .

Yet if all those proofs were acknowledged, it would mean the end of massive funding for astro-physicists as the high priests that understand it all, and the end of LHC—whose replication of that big explosion could recreate a quasar and blow up the Earth.

27. Only a question remains, to prove Einstein and Fred Hoyle right again, with their theory of an eternal Dualist, Dynamic, Steady State Universe, in which each galaxy explodes and implodes informative black holes into dark energy in infinite cycles . . . Where does the 2.7k blackbody radiation of the galaxy, formerly ascribed to the cosmic big-bang, comes from? Which blackbody creates the exact form and temperature of that radiation? The answer is relevant to this suit, because it falsifies from a different perspective the 2nd, most repeated argument for the safety of LHC: that moons bombed by cosmic rays never become black holes, since the moon is still here and so LHC will not make one. CERN's argument is double wrong. Because cosmic rays are not quarks[24] and moons do become black holes.

Indeed, that 2.7k radiation, which Penzias observed *coming from the galaxy,* and Gamow, an A-Bomb researcher, blew up to cosmic dimensions with no prove whatsoever, can only be produced by a black hole with the mass of the moon that *reflects as a gravitational mirror* light at that exact temperature. Since the quantity of mass of the black hole determines the gravitational temperature at which it bends light, *as Einstein proved.* So we can easily calculate what type of black hole produces 2.7k degrees, as the reader can test, by substituting the mass of the moon in the formula of black holes mass= temperature[28].

Thus, since 2.7k radiation is the most common of the galaxy and moons are the most common planetoids, the galaxy must be full of black holes that ate moons and now bend light at 2.7k degrees.

CERN affirms LHC will research how the electroweak force breaks its symmetry and becomes dark matter, without explaining this means to research how our weak mass dies and feeds strong, Quark matter. Further on, there is 60 times more dark matter than light matter in the Universe and the only candidates among standard, known particles that can form dark matter are quarks *(Dark matter triangle, Addenda A)*, components of ultra-dense stars and Einstein's black holes, called MACHOs, *(Addenda F, 27)*, which should be in the Halo radiating at 2.7k. Thus, the conversion of weak Earths and moons into a *lump of dark, quark matter* should be business as usual in the cosmos, exactly the opposite of what CERN affirms. Thus, to make a quark factory on Earth is enormously dangerous. It is like reproducing Ebola viruses, without any safety measure, pretending is vital research for science. Those MACHOs of dark matter protect the galaxy, as the Oort belts of moons in which they might feed, protect solar systems. But none is evaporating. We haven't observed with the Fermi satellite any signature of black hole evaporation. We haven't found either any signal of human intelligence, despite growing evidence on the existence of millions of Earth-like planets in the galaxy[29]. It is the Fermi paradox, which he enunciated after betting, as Amici explain, that the 1st A-bomb would evaporate only New Mexico. Perhaps he was thinking that Nuclear Physicists always evolve technology to a point in which they blow up their planets, playing to be God, as CERN might do this Christmas, unless this Court accepts the suit and Judge Gillmor declares LHC a danger to the life of Americans and Mankind at large.

28. CERN and Amici's problem is not their professional credentials, but their professional ethics. CERN systematically lied about those risks to the press, their governments and now this Court, to obtain billions of $ for an accelerator, which is in essence a quark cannon, a military weapon, they **do** know to be extremely dangerous. Yet, instead of sharing that special knowledge with this Court, they use it to disguise the worldly profession of Nuclear Physicists as makers of weapons:

400 years ago Galileo published the first book on Ballistics, military compasses. A century ago, the biggest arm producer of the XIX C., nicknamed Doctor Death, Nobel, inventor of Dynamite, manufacturer of Bofors Cannons and chemical weapons used in both World Wars by both contenders, founded his prizes. He stated: my factories will put an end to war sooner than your (peace) congresses: on the day two army corps can mutually annihilate each other in a second[30]. 50 years ago, Szilard, the main lobbyist of the A-Bomb, invented the accelerator and in Nazi Germany, Hahn used it to fission the atom, starting the nuclear industry. Today, the recipients of Nobels prize construct a super fluid,

light speed, 7 terabyte quark cannon that might blow up the Earth in a second, achieving the dream of Dr. Death.

Those accelerators were funded during the Cold war to find stronger nuclear weapons. In the process of destroying mass and energy, Nuclear physicists found the particles that make up matter. Today, when the cold war is over and the *standard model of particles is complete,* there is no need to spend enormous sums in an age of economical crisis, building strong quark cannons. For that reason, the previous Democrat administration canceled the American Supercollider (SCC), whose cost was equivalent to the budget of the National Health Agency.

Nuclear Physicists know their main profession is the creation of Weapons of Mass destruction and have an unwritten code of silence that prevents any bioethical criticism among peers. So the Nuclear Industry now markets accelerators with extravagant theories—evaporating black holes, God's particle—*which were ignored when the military funded them,* to convince politicians to build the 3rd horizon of nuclear cannons, the LHC. Indeed, after weak energy bombs ($E=Mc^2$) and hydrogen, energy/mass bombs, CERN will create self-sustained strong quark/mass bombs that need only a small amount of detonator to blow up the planet, *since the combustible is outside the bomb, in all of us, made of weak matter.* And Nuclear Physicists back en masse CERN's experiments with their wrong sense of rogue solidarity against concern for the environment that inhibits vital and important scientific inquire. Yet truth in science is not proved by quantity, corrupted prizes and overgrown budgets but through the laws of the Scientific Method.

For example, in the twenty-first century, *all* physicists except Einstein believed in an absurd equation/substance called Ether, which was harder than steel, softer than butter and filled the Vacuum. Einstein's best friend, Gödel, proved then that mathematics is a language that creates fictions[10]. So Einstein denied ethers existence against *all* Physicists and he was ostracised by his peers, having to work in a patent office for many years[31]. Finally, when Michelson measured the speed of light on that vacuum, ether evaporated by lack of experimental evidence. Today CERN wants us to believe absurd theories that put our lives at risk on pretentious Authority. Instead, we want them to prove truth in Court with the laws of the Scientific Method. Because as Einstein put it, those who impose truth with power will be the laughs of the Gods.

V. RECOMMENDATIONS TO THIS COURT.

29. The Nuclear Industry created LHC under the Ethics of Technology so clearly expressed by Eric Fromm, father of Political psychology:

Technological civilization is programmed by the principle that something ought to be done because it is technologically possible. If it is possible to build nuclear weapons, they must be built, even If they might destroy us all. Once this principle is accepted, humanist Values (something has to be done because it is needed by man) are Dethroned and technological development becomes the foundation of ethics.

This ethical statement explains what LHC is all about: to make a bigger, more powerful atomic cannon, because we have the capacity to make it, *regardless of its harmful collateral effects to mankind*, since it means big contracts for technological companies and jobs for Nuclear Physicists, unemployed after the end of the Cold War[32]. As CERN said: Whatever the discoveries ahead for physicists working at LHC, the experiments will, according to its Chief Scientific Officer, Jos Engelen, "keep physicists off street corners for a long time to come[33].

Yet when the costs and risks are so high, public funding should be used in a wiser way on research on other areas of science that will provide jobs and harmless results[34].

Nuclear, Quantum Physicists engaged in those experiments work to achieve their scholar ambitions, or make a living at CERN and we cannot expect any control on their side. Their mastery, constructing weapons of mass destruction, is not paralleled however in the realm of knowledge, since their reductionist theories about time and the Universe have long been superseded by the work of Einstein, and the recent advances on Duality and Fractal Relativity that prove Einstein's standard theory of Mass and gravitation, *the informative force of the Universe,* right[22-26]. Instead, Quantum Physicists like Hawking, who denies Einstein, use obsolete XIX C. models that define the Universe only with the arrow of entropy and Death, considering that information, the arrow of life, doesn't create the future, which is false. So they research cosmic explosions, pretending they will reveal the meaning of it all, when they just can bring death to mankind. Since if Einstein is again right, Hawking's black holes will blow up the planet. Thus, it seems an act of arrogance and irresponsibility to risk our lives to prove the best physicist of history wrong[35].

In the last century nuclear physicists were never made accountable for their actions, to protect a misunderstood concept of national security, called MAD (Mutual Assured Destruction[36]). Instead, during the cold war, they were hailed everywhere as primus inter pares, a position which in science if any, corresponds to the Philosopher of Science that defends the bioethics and truth of the scientific method, as judges are primus inter pares who rule on the truth and ethical standing of Plaintiffs and Defendants. As a result of those undeserved privileges, today quantum cosmology feel over the Laws of the Scientific Method that falsifies their theories[35] (IV) and the Laws of Democracies—since during the cold war Europeans gave the Nuclear Company diplomatic status. But our security policies have changed. We have now Environmental Laws and Laws that protect the Security of this Nation, opening a legal avenue to prevent this potential genocide. In that regard, a ruling against further evolution of Nuclear weapons is long overdue. By establishing again the Rule of Law and the supremacy of the arrow of life over the arrow of entropy and death that CERN researches[37], this Court can give the first step in the right direction Mankind has to take, if we want to have a sustainable future for this planet.

30. The legal question this Court is asked to resolve is the existence of Federal Jurisdiction over CERN. On my view if the Court has the will to judge, such jurisdiction can be obtained either from NEPA, given the enormous quantitative expenses the American Government has placed on those experiments, or from the Patriot Act and other laws against terror and dangerous substances, given the enormous number of potential victims an accident at LHC will produce (14).

Thus, the true question this Court must resolve is the existence of such will and need to judge a scientific experiment that can put in harms way potentially billions of human beings; a machine which is redundant and of inferior quality to a harmless telescope, if its purpose is to study the Universe and the unlikely cosmic big bang (26); a Company, which constantly lies to the press and politicians to extract billions of $ of tax-payer money. Must this behavior, this Company, this experiment that can provoke the biggest genocide of history, happen unchecked? Obviously not. This Court, who represents the people of America also at risk, should for that reason allow a fair review in a due process of law of Plaintiffs arguments in defense of mankind, regardless of the institutional power and prestige of the Nuclear Industry. It should not be impressed by a presumption of Authority that Amici have violated misinforming and despising the oath of truth due to this Court. Since the citizens of this country have the right to proper information, guaranteed by their Constitution, which this Court should uphold, especially in issues that endangers the life of Americans without consent.

---

0   RHIC collides heavy atoms at 100 GeV. per nucleus. LHC collides them at 564 TeV, with 5.640 times more energy. What this means is that LHC will produce easily strangelet drops, 5.640 times bigger, crossing well into its valley of stability. See http://en.wikipedia.org/wiki/Large_Hadron_Collider Line 2 and http://en.wikipedia.org/wiki/Relativistic_Heavy_Ion_Collider Line 18.

1   The whole paragraph found in the article "Will Relativistic Heavy-Ion Colliders Destroy Our Planet?" reads as follows: Our understanding of the interactions between quarks is insufficient to decide with confidence whether or not strangelets are stable forms of matter. Estimates based on the MIT bag model leave the question open for any mass (or baryon) number, A, between a single-digit quantity and the value for neutron stars, A 1.7 × 10 ^{57}. In the case of strangelets, we are dealing with the properties of an incompletely understood hypothetical form of nuclear matter. Imagine that, for some unforeseen reason, there is a "valley of stability" for negative strangelets. Suppose that, somehow, such an object is produced in a laboratory high-energy reaction and that it survives the collisions that eventually bring it to rest in matter. The negative strangelet would attract a positive nucleus and may eat it. The resulting object may lose positive charge and adjust its strangeness by electron capture or positron β—decays. The new strangelet may be negative again, and maintain an appetite for nuclei. If its mass grows to some 0.3 ng (A 2 × 10^{14}) it falls to the center of the Earth, for its weight overcomes the structural energy density of matter (10^9 erg cm−3 or 0.1 eV per molecular bond). At a mass above 1.5 ng, for a typical nuclear density, the object becomes larger than an atom and the positron cloud that it has been developing sits mainly inside the strangelet itself (for stable strangelets that have grown this large, the sign of Z is immaterial). Even without the help of the Coulomb attraction, gravity and thermal motion may then sustain the accreting chain reaction until, perhaps, the whole planet is digested, *leaving behind a strangelet with roughly the mass of the Earth and 100 m radius.* The release of energy per nucleon should be of the order of several MeV and, if the process is a run-away one, the planet would end in a supernova-like catastrophe.

2   See strangelet charge at: *http://en.wikipedia.org/wiki/Strange_quarks*

3   Those declarations are found at: *http://www.dailymail.co.uk/moslive/article-1025725/Solve-meaning-life-The-worlds-biggest-experiment-meaning-everything.html http://www.bbc.co.uk/sn/tvradio/programmes/horizon/broadband/tx/universe/highlights/index_textonly.shtml* http://www.nullsession.net/nullsession/?p=1716 Further on Cox says:

"We know it will discover exciting things. We just don't know what they are yet." http://www.timesonline.co.uk/tol/news/uk/science/article4670445.ece

"We might not have thought of what turns up, but we know we've got to see it." http:// women.timesonline.co.uk/tol/life_and_style/women/the_way_we_live/article3403949.ece

While CERN states: "Collisions at LHC differ from cosmic-ray collisions with astronomical bodies like the Earth." *http://www.google.com/search?hl=en&q=Collision s+at+the+LHC+differ+from+cosmic-ray+collisions+with+astronomical+bodies+like+the +Earth&btnG=Search*

[4]   Amicus Wilczek contradicts constantly his statements in a taped, public conference, whose key statements are shown at http://www.lhcdefence.org_

In that document (minute 3 of the film found in the main screen of that site) Amicus Wilczek states to the question that LHC might produce black holes:

. . . that is truth, otherwise respectable scientists have suggested that kind of thing.

And to the question if that might be dangerous and blow up the planet, Amicus Wilczek replies:

Nature is so inventive and malicious that . . . , there is a logical possibility that it can lead to a catastrophe (Minute 5.23).

Later he states: Most of what happens at high energy accelerators is the strong interaction. We need to understand that very, very well if we are looking for the rare events that correspond to something fundamentally new (min.34.3).

Would Amicus Wilczek care to clarify when did he lie, here in Court or in his public statements?; and why did he lie in a matter that concerns the life and safety of the entire planet?

Just in case Amicus Wilczek doesn't answer to this important question, in the same visual document we find a tentative response, as he recognizes that: I never been so confident though as to make a prediction as when I was called to sit on a panel about the possibility of an accelerator turning on and ending the world. Predicting that it won't is very safe, because if your prediction is wrong, he, he (and he shrugs) (Minute 6).

Obviously Amicus Wilczek laughs, knowing that if his prediction is wrong the world will blow up and he won't be blamed (while we, plaintiffs, will be ridiculed ad nauseam if nothing happens, despite having taken the proper bioethical standing). But he is now stating the opposite under oath . . .

[5]   LSAG is available at: http://lsag.web.cern.ch/lsag/LSAG-Report.pdf

Mr. Magnano's paper is available at *http://arXiv.org/abs/0806.3381*

Both ignore completely LHC's quark production and the possible substances it can produce (Addenda A), focusing on cosmic rays, RHIC and neutron stars, which have nothing whatsoever to do with the LHC. To make a comparison, would this Court take seriously a safety report made about the Ford Mustang by the same Ford Company (11), which studies instead other car, let us say, the Toyota Camry (RHIC), and a completely different object—let us say the chances that a comet falls over our heads (black holes colliding with neutron stars, main theme of Mr. Magnano's paper)?

[6]   Let us consider some of those fallacious arguments and the curious concepts on safety standards Amici's hold:

Industrial safety was managed by learning from past mishaps and by using appropriate measures to avoid their recurrence. For example, miners once used caged canaries as methane detectors. This management process is no longer acceptable as modern technologies have sometimes led to disasters.

Thus, defendants acknowledge that new technologies do lead sometimes to disasters. Moreover, they acknowledge that, while at any present time scientists consider their knowledge always proper, in the future often we discover those measures were not safe enough. So if defendants accept that in the relative future what we believed had no risk brought disasters, how they are so sure that in the relative future what they claim now to be safe will not bring a catastrophe?

The amici recognize that a new procedure had to be developed for the concerns at issue in this case.

Thus amici, who claim the highest standards of safety yet developed, in fact are testing a new untested procedure in an experiment, whose mishap will be the last one, as it risks the life of all mankind.

The most creative scientists were tasked to imagine what might go wrong and satisfy themselves that the imagined problems did not exist. They examined carefully three scientifically conceivable disaster scenarios in which experiments might produce "black holes" that could gradually consume the Earth.

This pretentious new method of assessment is not scientifically acceptable. Since imagination is not a procedure of science. We either know or do not know the facts. Further on we might wonder, if CERN can imagine 3 catastrophe scenarios, doesn't this fact imply that those 3 imagined scenarios are at least probable, especially when Amici acknowledge that in science accepted truth evolves gradually or discontinuously? Further on, our brief list (13), proofs experts' imagination limited, as many scenarios escaped them.

7   In the same line of thought, the constant use by Amici of the word scientists, instead of nuclear physicists, clearly tries to mislead this Court, implying that scientists from different disciplines are in favor of the experiment. This is probably due to the fact that Nuclear Scientists have constructed all the weapons of mass-destruction of the XX century, and so they prefer to hide their responsibility, after half a century of global terror, under the umbrella of science. For the same reason CERN will be eliminating from its acronym the word Nuclear, to avoid an accurate understanding of its nature and pass as a Center of European Research, not as the Nuclear Company of Europe.

8   The difference between cosmic rays and the LHC is clear. Cosmic rays are lonely atoms and ions of lower density (mainly hydrogen), whose chances to collide in their center in enough numbers to liberate a mass of quarks, so close together that they can create a strangelet or mass-bomb, are null. *For the same reason the Earth is not constantly detonating A-bombs despite having Uranium. The critical mass will never be achieved unless we purify and pack together the uranium.* That is why cosmic rays don't convert Earth into a Nova. Since, to re-create the conditions of a self-sustained mass-bombs we need to create the detonator that starts the chain-reaction by massing together

thousands of quarks and give them the highest energies of the Universe, which CERN will achieve through a process called stochastic cooling that compresses together millions of hadrons (heavy atomic nuclei), accelerating them till reaching light speeds with super—magnetic forces. Those ultra-energetic, ultra-dense bags of quarks, target an opposite beam of millions of dense quark nuclei, making them collide in a point in which they stop at rest, forming a mass-bomb that keeps feeding on our electroweak matter, in a chain reaction that converts Earth into a quark star. This kind of precise collisions, forces and density of quark mass, only happens in the center of stars, where supernova reactions take place. So it is quite possible that the collision of 2 bags of ultra-dense quarks at the LHC will also create a Nova reaction.

9   A Bosenova is described in simple terms, at:http://www.nist.gov/public_affairs/bosenova. htm http://www.space.com/scienceastronomy/generalscience/supernova_lab_010723. html Yet LHC is not the only lab on Earth that might produce them in the nearby future (albeit, given its enormous energy, the most probable place).

Dark atoms are very likely formed, as Light atoms are, by the combination of the triangular base (Addenda A) of the dark matter triangle. The difference is that the quarks of the dark matter triangle are much more attractive and act/react much faster than light atoms. For example, their individual particles mutate into their antiparticles billions of times per second. So, even if we have not observed them yet, their theoretical properties, make them very dangerous.

10 See min.15-20 of visual document at *http://www.lhcdefence.org* where we explain the formation of gas-9, the technical name given in Fractal Relativity[25] to a deconfined state of top quarks, described also by this years Nobel Prize Nambu[18], as the most attractive vortex of particles in the Universe. In fractal relativity top quarks are the atoms of Einstein's black holes, which he called frozen stars, arguing they could only exist if we found a cut-off substance with enough density to be its atoms. *Yet the obvious need of a material substance in black holes* is denied by quantum cosmology like Hawking, who believe in the main religious ideology of physicists, called Pythagorism, according to which the Universe is made of numerical functions, probabilities. Pythagorism, implies that any mathematical equation must exist in reality, because, as Galileo put it, Mathematics is the language of God. And for the same reason, anything that cannot be written in equations, including information and life, must not exist or be relevant to our understanding of the Universe. *So Hawking also denies the arrow of information in black holes, says Einstein's black holes are double wrong and black holes evaporate information. It is the so-called information paradox proved wrong ad nauseam.* Since Gödel, the most important mathematician of the XX century showed that mathematics is just a language that can produce, as any language does, fictions. So equations, which are logically inconsistent or describe particles that contradict known laws of science and don't have experimental evidence are false. This is the case of Hawking's time machines, aka evaporating black holes, and many other mathematical fantasies, from parallel Universes to multiple dimensions and super-symmetric particles that the efficiency and simplicity of the Universe forbid and the quark factory will never

produce, despite CERN's claims their quixotic search is "vital scientific research. Yet among Pythagorean Physicists Gödel is taboo. So Hawking claims to have had his biggest depression after reading him. The exception is Einstein, Gödel's best friend who, unlike quantum cosmology, always respected the Laws of the Scientific Method and confessed to Poncaire he hadn't become a mathematician because he could prove when a mathematical equation was truth but not when it was real. Thus, he confronted all his life Quantum fantasies with no prove, and his strict sense of veracity paid off, since all his theories have been proved right, while not a single Quantum, speculative fantasy, including evaporating black holes have been proved right by the laws of the scientific method—exactly the opposite of what Amici claim.

At his death however, Quantum Physicists tried to substitute Einstein's proved work on Gravity and Mass, with a Quantum fantasy called quantum gravity, according to which small black holes were not made of mass, but obeyed quantum laws. This is absurd, as Gravitation deals with the macro-cosmos and Quantum Theory with the micro-cosmos, two fractal scales of different size and forces[26]. But that didn't deter Wheeler, the father of the H-Bomb, which Einstein also opposed, to invent Quantum Gravity, and change Einstein's name of frozen stars to black holes. Wheeler did it because, as he confessed to Thorne, Hawking's best friend, also a writer on time machines and science fiction (*http://en.wikipedia.org/wiki/Kip_Thorne*), he wanted to study frozen stars with the laws of quantum theory as probabilistic objects. So he defined black holes as mathematical fantasies with no substance, which had at its center, instead of quarks, infinite density and null volume. This absurd concept called quantum gravity is the speculative theory, never proved, from where Hawking's work arouses. For many years Hawking recognized that his work on black holes was a mere mathematical fiction and he even bet with Kip Thorne that such black holes didn't exist in Nature (see the account of this fact by his professor at *www.lhcdefense.org*). Yet CERN and a new generation of Pythagorean physicists seem to have forgotten the fictional nature of all those mathematical fantasies, which CERN now sponsors, *not to deal with the fact that if black holes are Einstein's frozen stars, they will be produced at LHC by massing together top quarks into gas-9, the detonator of a frozen top star*. Thus, if Einstein is right about his frozen stars, since LHC will produce Top quarks in massive quantities, it will very likely produce black holes, which will never evaporate, as we have never seen a quark evaporating, but grow at light speed, making us all into top quarks, $M=e/c^2$ (Exhibit A). In that regard, despite his constant falsification, Quantum theorists like Mr. Hawking resort as the ultimate alibi to cling on his theories, to probabilistic models of the Universe in which all is possible, including an infinitesimal probability that black holes evaporate into the past[24] and we survive the LHC. In one of his books the Universe in a nutshell, Mr. Hawking points out that the chances of time travel and hence of black holes evaporation (23) are 1 in a trillionth, but he ads I like to bet. Those are the chances we survive one of those black holes, if they appear at CERN (min.28 at http://www.lhcdefence.org). The use of probabilities has become in that sense customary among the gurus of quantum cosmology, because it allows them to invent all kind of bizarre theories, whose improbable testing is a good excuse to build machines like the LHC, despite Mr. Einstein dictum that God doesn't play dices; *since probabilities in science merely reflect limits on the human perception of some events, they are not a license to affirm anything is possible*. But without them, people like Mr. Hamed, a critique of this suit, could not affirm because of the dice-throwing nature of quantum cosmology, there was some probability of almost anything happening. There is some minuscule probability, he said, the Large Hadron Collider might make dragons that might eat us up.; while sponsoring as a probable truth his theory of 10^{500} parallel Universes . . . some of which, I suppose, have dragoons that will travel through the probabilistic dimensions of other quantum theories into LHC: *http://neocrack.info/ Crackpedia/Physics/particle%20accelerator/particle_accelerator.htm*

[11]   http://www.newyorker.com/reporting/2007/05/14/070514fa_fact_kolbert

[12]   The main substance LHC will produce is a vortex of ultra-cold, super-fluid quark-gluon liquid, as the one produced at RHIC but of higher stability; called in scientific literature a bag of ice-9, (name taken from Kurt Vonnegut's novel Cat and cradle, which, as Wilczek has often described out of Court, would trigger a chain reaction that will accrete the Earth (min.37 onwards of visual document at *http://www.lhcdefence.org*; SciAm, 99).

[13]   As a fetus born with little weight that dies before stabilizing its vital constants, at RHIC around one thousand strange quarks were formed and then the liquid dissolved. We were lucky enough that RHIC's strange liquid didn't stabilize. Yet if such strange liquid, considered existing in the center of all neutron stars, hence truly common, is born with more mass, it will certainly stabilize and convert the Earth into a small pulsar. This parameter is called the MIT constant of stability, and most physicists today, experts on strange theory *(red line on Addenda A, Addenda C)* consider this to happen beyond the 1 terabyte barrier of energy/mass, never surpassed by RHIC, which LHC will cross far beyond. Then our weak mass breaks its symmetry (a technical concept to state that it dies and becomes converted into heavy quarks). Yet, precisely LHC has been constructed to study the symmetry breaking=death of our light matter and the creation of dark matter and heavy quarks, made with it (3,27). This is why it is so dangerous. Thus, the catastrophic scenario is by no means speculative but the most likely event to take place at LHC. Further on, the quasi-strangelet fetus lived much longer and accreted 10 times more matter than scientists expected. In fact, strange quarks were called strange because physicists were surprised by their long lasting life, millions of time longer than they thought. It was as if Cleopatra were still dying of an asp bite said his discoverer. As Amici Wilczek states at http://*www.lhcdefence.org*, min.5.23, Nature is so inventive, that its species always survive better than expected. We might say that at RHIC physicists expected a lame duck but saw the teeth of a Tyrannosaurus Rex; and now the experts on dark matter are eager to make Jurassic Park to see all the possible monsters of the dark world fully grown, regardless of the collateral effects they might bring to mankind.

[14.]   The nature of those quark bombs is shown in detail by one of our Amici, Mr. Wilczek, in the taped conference available at http://www.lhcdefence.org, when he explains the creation of mass-bombs according to Professor Einstein's inverse equations, $E=Mc^2$ and $M=E/c^2$, an occurrence that now he denies. Since energy and mass are equivalent, this second equation $M=E/c^2$, which as Amicus Wilczek explains (min.13 of the visual document), was Mr. Einstein's initial equation, creates mass-bombs, novas, black holes and strangelets. And it happens when Energy becomes mass. Thus, since LHC has much more energy than Nuclear Colliders, it will produce much more mass, increasing enormously the risks of creating black holes and Strangelets. How then Amici can claim exactly the opposite, against Einstein's well proved theories, when Amicus Wilczek recognizes exactly the opposite?

[15] CERN also uses constantly in its site and official documents the outdated concept of a quark-gluon plasma, instead of recognizing LHC will create strange quark liquid.

[16] http://www.unificationtheory.com/laws/science_unification.htm

[17] Amicus Sheldon's toilet concept is explained at http://www.msnbc.msn.com/id/20215345/. While you can see Wilczek declarations on min. 27.3 of the visual document at htttp://www.lhcdefence.org

I do agree with both of them. If we were to falsify the Higgs for our Amici physicists in a logic, causal manner, using the special knowledge of the Scientific Method and its epistemological laws we should write the following logical chain:

A)   In the 70s in Physical review, Smolin and Zee proved that the Higgs was equivalent to a Brans/Dicke variable gravitational space/time vortex, *which is based in the standard model of gravity and mass, Einstein's Relativity, and explains the meaning and value of different masses without the need of new particles.* (Addenda A)

B)   In the earlier 2000s, in the book Cycles of space-time, http://dinamica-de-sistemas.com/revista/0906e.htm I used a variable gravitational space/time vortex to prove that a strong field was the quantum equivalent of a gravitational field and showed mathematically, departing from Einstein's Principle of Equivalence that mass was equivalent to the frequency of space-time vortex . While Amici Wilczek arrived to the same conclusions on his work on the strong force, which now he seems to deny, considering the search of the Higgs of vital importance for scientific inquire. Thus, it became obvious to me, as it should have been to Amici Wilczek that a standard ultra-heavy particle, *a quark*, not a new particle or theory, would be the cause of the death, or breaking of symmetry of our weak mass. c) Higgs equations are copycat of Goldstone, copycat of Nambu's description of a top/antitop deconfined quark super-fluid vortex. So deconfined top quarks were the standard particles we needed to explain the death of light matter (the breaking of symmetry), explained by Nambu, before Higgs discovered his redundant particle.

D)   Further on, the weight and parameters of a Higgs are equal to those of a top quark (Addenda A)

E)   Occam's razors prove efficiency and simplicity are truths in science.

F)   A+B+C+D+E means that Higgs created synonymous equations to those of a deconfined top quark super-fluid vortex of strong forces (said I am brunette instead of I have dark hair). So the Higgs is the top quark and that is why . . .

G)   Nambu NOT Higgs got this year his Nobel prize. Since LHC won't find Higgs. Instead it will produce a whirl of super-strong, attractive top quarks that will start breaking the symmetry (aka feeding) of our electroweak matter at the nice rate of a +1 billion feeds per second.

H)   This in abstract writes n+p->w+z->top quark. Further on, a Top=Higgs has the same weight that the sum of a W and Z particles (w+z=t) . . . What this means is that our weak mass becomes broken first into W+Z particles that evolve into heavier quarks.

Thus, after having spending 10 billion $ in this machine, what Nuclear Physicists expect is to rediscover the Top, learn some minor details about the death of our matter, create massive amounts of quarks and dark matter, and hail Nambu as the great physicist of the age. Then have a good laugh on Higgs for the money they got with his hoax.

[19]   In simple terms, Leo Lederman (min. 22 of the visual document), an ambitious administrator of Super-Colliders, who wanted to create LHC in America, knew that the Standard Model of particles was closed, as all its quarks were already discovered, forming 2 beautiful, self-similar triangles of light matter and dark matter, able to explain all the other particles of the Universe (Addenda A, 27)—except the mathematical fantasies of Pythagorean Physicists[10]. Those invented particles, such as Higgs, WHIMPS and SUSYS, that have an infinitesimal probability of existence, since the fundamental law of the Scientific Method, Occam's razor, the law of simplicity, deems them unnecessary, are now according to CERN essential to our understanding of the cosmos. In the case of Lederman, since he knew there was no reason to spend 10 billion $. He insisted that the Top Quark=Higgs, was a new particle of paramount importance to understand it all, not just another quark. So he called it God's particle, wrote a book with that title and sold it to Reagan to fund the SCC (American Super-collider). Clinton, though, found out the scam and waste of resources and cancelled the project, as the present Democrat administration should do with LHC. Then, Nuclear Physicists convinced the French, eager for Grandeur, that God's particle would solve the meaning of the Universe, and obtained the funding for LHC, in an age in which super-colliders, that were basic instruments of research in nuclear weapons during the cold war, had become an obsolete, ultra-expensive machine, which neither Russians nor Americans wanted to fund any longer.

[20]   http://www.shvoong.com/exact-sciences/physics/1838912-professor-higgs-big-bang-collision/

21   Higgs doesn't seem to care that his belief in the falsity of Hawking implies CERN's black holes will blow up the Earth since he backs the experiment to find his particle. Regarding Mr. Hawking in the document at www.lhcdefence.org he affirms (min.25) that if his theory is wrong we will become spaghetti, but if he is right, he will obtain a Nobel Prize, concluding that since we have found no prove whatsoever in more than 30 years of black hole evaporation, CERN will give him a second chance (it might be possible to observe this (black holes) at CERN, in Switzerland . . . So I might get a Nobel Prize after all). While Wilczek knows if Higgs is not found, his work will get a revival. It seems that scholar ambition is more important to our Nuclear Physicists than the survival of the human species. Unfortunately neither Hawking or Higgs are right, because we have a perfectly tested theory of mass and gravitation, called General Relativity, which Einstein formulated a century ago, and today theorists are upgrading with the new mathematics of Fractals into Fractal Relativity[18b,24-26], the most promising theory of time and space of the XXI century, which also will enter the limelight once the Higgs Hoax is discharged. As one of his main proponents[23], obviously, the experimental falsification of the Higgs would have been also beneficial to my career but bioethical considerations in this case are to me far more important than scholar ambition.

22   January 1977: Scientific American, "The Quantum Mechanics of Black Holes." In that pretentious article, which Hawking closes with the sentence, Einstein is double wrong (despite making his fortune with a book, A brief history of Time, dedicated to explain Einstein's theories of Time), Mr. Hawking, following Wheeler[10] affirms that black holes have no substance, but are mathematical fantasies, which destroy information, as he does not acknowledge Duality, the reality of 2 arrows that create the future species of the Universe, energy and *information*, the dominant substance of life *and black holes*.

   Such outdated thesis departs from twenty-first-century physics, when only the arrow of entropy was known, but it has been falsified ad nauseam in serious articles and popular magazines. In fact today black holes are studied in exactly the opposite terms: they are considered *the most informative=massive objects of the Universe, since in terms of Duality, mass is, as Einstein explained, and Wilczek confirms in his conferences (www.lhcdefence.org), the physical in-form-ation of the Universe, whose frequency bends energy into curved form* (see cover of Sciam November 2004 at: *http:// www.physics.unc.edu/research/theory/gchep/images/sciam-cover.jpg* : Computing black holes.)

   Yet quantum cosmology still pretend to substitute Mr. Einstein's work on mass, with their outdated, XIX C., entropic models born out of their worldly specialization in energy and weapons research, to the point they call the information arrow so evident in life beings, negantropy, the negation of entropy, as if life were an odd exception. And they affirm that the Universe is dying (min. 8 of visual document).

23   *http://journals.isss.org/index.php/proceedings50th/article/viewFile/29/200*

   The 2 arrows of time are obvious to everybody who is alive. Since, besides the arrow of entropy=energy and death, there are multiple cycles of in-form-ation and life, which Darwin explained with his work about morphological change. Even if Doctor

Deaths prize (28) cannot be given to anyone who believes in Theory of Evolution—since in-form-ation, form, was not properly mathematized when Nobel died. All this changed when, at the end of the XX century, information was mathematized properly with the discovery of fractal equations that create form, in-form-ation, not only in life but in most structures of the Universe. In essence a fractal equation is a self-generative equation that repeats the same forms of in-form-ation, in different sizes and scales, breaking constantly energy into new in-form-ations, aging your skin into form and the vacuum space into cyclical vortices of masses and charges, balancing the entropy of the Universe and making it immortal (min.16.26 of visual document). It was the most important discovery in Time theory of the last decades, which fractal theorists, like Mandelbrot, Mehaute, Nottale and this author, have applied to multiple sciences, showing that we can describe the Universe as a fractal system of energy and information that constantly self-reproduces all its beings (nt.25).

Einstein also accepted the arrow of information, as he understood mass as an attractive, accelerated vortex of Gravitation (Equivalence Principle), whose cyclical frequency of rotation in-forms, creates temporal information, which is the ultimate meaning of mass. So he believed, as Duality is proving today, that the Universe was a steady state of balances between the arrow of physical information that creates galaxies and mass and the arrow of entropy/physical energy that creates dark and light energy. In other words, XXI century science is proving Einstein and Darwin right once more, in their understanding of physical and biological information and Mr. Hawking and CERN wrong, in their obsession for the arrow of energy. But by the time quantum cosmology accept Einstein's arrow of mass, the information of matter, and Darwin's arrow of evolution, instead of ignoring and judging the work of millions of scientists who study life and information as having no merit, they might have killed us all.

In that regard, all theories of time depart from the first scientist of History, discoverer of the experimental method, Aristotle, who defined time as the perception of change in the Universe. And so he affirmed that a theory of time should be a theory of change, like every Philosopher of Science has done ever since. Then he proposed, as we do in Duality, 2 arrows, wills or substances in the Universe, information and energy, responsible for the 2 types of times=changes science studies, biological and physical change. If we translate his wording to modern scientific terminology we should talk of the arrow of morphological, informative change, which defines the processes of life, aging and Evolution, as an accumulation of information that happens in all the species of the Universe. And he called the arrow of energy, translational change or movement, which happens in the physical realm. Thus, he divided the study of times in 2 different sciences, Biology, the science of form, of in-form-ation and Physics, the science of energy and matter. In the modern age, those 2 types of times-changes would be further developed by Darwin in Biology (information arrow) and Galileo in Physics (energy arrow). Further on, Galileo added to verbal logic, the language of Greek Science and Biology, a new language, mathematics, to calculate with precision translational change with a new type of device, a machine called the clock that could measure the rhythms of change in the motion of beings. But clocks were simple, quantitative mechanisms that could not describe the morphological, in-form-ative, qualitative changes of the cycles of life and death, which were, therefore, ignored by physicists for centuries to come. Thus, physicists became specialized in the arrow of energy, of expansive movement in space, which they called entropy and causes also the processes of big bangs (physical death) and biological death, which is an overdrive of energy that kills and simplifies the information of life. As specialists on energy, Physicists would also make all the weapons of modern history, since weapons are mechanical devices that release an overdrive of energy that kills human beings. And since physicists ignore all what they cannot mathematize, this leads them to ignore the 2nd arrow of time, in-form-ation, and the arrow of life. Thus, Physicists by the very essence of their reductionist vision of time, are geared to destroy the Earth, seeking the pure energy of the big-bang, thinking the Universe is all about bombs and explosions, not about information and the creation of life (min. 10 of the visual document). For that reason, even after Einstein proved that physics also has an arrow of creation of information, the arrow of gravity that bends space into mass in black holes, Hawking, a quantum physicist stuck in XIX C. entropy, affirms that Einstein is double wrong, and black holes have only energy, entropy, and evaporate information. And he finds many fans among die-hard entropy physicists and the Nuclear Industry CERN represents.

[24]   http://www.independent.co.uk/opinion/master-of-a-narrow-universe-stephen-hawking-is-on-a-voyage-to-stardom-but-unable-to-navigate-in-the-human-realm-1510340.html

[25]   See: http://journals.isss.org/index.php/proceedings50th/article/viewFile/29/200 http://journals.isss.org/index.php/proceedings50th/article/viewFile/27/201 from this author to grasp the essential self-similarity and fractal nature of all things created equal . . .

[26]   For the fractal structure of galaxies, stars and black holes see articles available at http://www.sciam.com/article.cfm?id=the-self-organizing-quantum-universe

[27]   http://www.physorg.com/news76314500.html

[28]   http://imagine.gsfc.nasa.gov/docs/ask_astro/answers/971111e.html

       While there are small variations in the mass of moons, most of them fall within the range of ±2.5/3K degrees, when converted into black holes. Further on, if Black holes are Einstein's frozen stars, they will be top quark hadrons[26]. That is, they will be as all hadrons are, all equal in weight, reflecting the same background radiation all over the Universe. In the same manner all protons have the same mass and all electrons the same charge. Thus, the remaining mass of the irregular moon, once a Moon MACHO is formed will be expelled as energy in the Nova process of conversion of moons into primordial black holes as it happens when stars become stellar black holes.

[29]   http://www.latimes.com/news/science/la-sci-planet22-2009apr22,0,5993692.story :

       This April we found the 1st planet similar in size to Earth, so close we can extrapolate the existence of millions of them.

[30]   http://nobelprize.org/alfred_nobel/biographical/articles/tagil/index.html

[31]   http://discovermagazine.com/2006/sep/einstein-nobel-prize

       Today nothing remains of ether theory and physicists think that theory to be an obvious absurdity, but the same physicists think Mr. Hawking, who denies Einstein's theory of black holes, is right, because his equations of time travel are so beautiful and he describes black holes as mathematical fantasies without substance, which seems to be a prove the Universe is Pythagorean[10] (made of mathematical equations not of substances). If we survive CERN, in a few years, when we observe those black holes closely, we will realize that they have substance, which can only be ultra-dense quarks, as those CERN will mass together to produce them, and the mathematical fantasies of Hawking and his denial of Einstein will seem absurd, an act of arrogance that is risking our lives, without asking us permission. Einstein was also a pacifist, who opposed Mr. Nobels industries and so the Nobel committee denied him repeatedly the prize. 20 years later, the Nobel committee finally gave him the prize without mentioning his prove that ether did not exist (Relativity theory). But quantum cosmology never accepted his criticism of probabilistic, mathematical fantasies. Today CERN follows an 'ad hominem' campaign against those who challenge its experiments, backed by all kind of experts that want to prove their bizarre theories. Instead we want CERN to reason their arguments under the Laws of the Scientific Method. And for that only one Amicus is needed, if he is right.

[32] http://www.newyorker.com/reporting/2007/05/14/070514fa_fact_kolbert

Some contracts went to Russian physicists who previously worked for the Soviet military; in this way, the collider has provided a livelihood for scientists whose employment options might otherwise include selling nuclear secrets.

[33] http://news.bbc.co.uk/2/hi/science/nature/4229545.stm

[34] In that regard, this case is of supreme importance for mankind, for 2 reasons:

— If the theoretical debate favors standard science (Einstein on black holes, Wen and the Standard Model of negative strange matter in strangelets[1], Addenda F), CERN will commit the final genocide of History. And this Court, by not allowing a trial on that potential genocide, will have failed to protect the life of the American People, the supreme value a Federal Court is asked to protect by the American Constitution, its Bill of Rights and the Natural Law of all societies.

— Yet even if the bizarre theories of Mr. Hawking were right and our present understanding of strangelets were incorrect and we survive CERN, this will not be the last technology that puts the human species at risk. Bigger accelerators will be built till they reach the threshold of creation of pulsar and black holes. Self-reproductive nano-bacteria will be built that can destroy the ecological balance of this planet and make impossible life, as Bill Joy, founder of Sun Microsystems, pointed out in Wired magazine (http://www.wired.com/wired/archive/8.04/joy.html). Further into that future super-robots will reach such degree of evolution that might make mankind obsolete . . . It is for that reason that the choice between technological[29] and human ethics has become one of the fundamental issues of this century.

In that regard, this is the first case of a new brave world in which mankind, for the first time since its apparition as a species, faces a growing possibility of extinction under the ethics of technology, unless the judicial, executive and military branches of our governments decide to put some basic limits to those ethics and resurrect the ethics of life. Instead of worshipping blindly all kind of technologies, even those that can extinguish us, we should put legal limits to the evolution and reproduction of those specific lethal machines as we do with biological weapons and lethal virus . . . While this might not be enunciated expressly in the constitutional laws of this country, written before those lethal machines were even conceived by the human mind, it seems to me that Laws like the Patriot Act, which treats both kind of lethal substances under the same legal umbrella (nuclear weapons and biological weapons) expresses tacitly the need to put such limits. The present administration, with his emphasis in the defence of the environment and its opposition to Nuclear Proliferation, seems to point out also the need to establish limits to the free reproduction and evolution of lethal, nuclear technologies, such as the LHC.

35 Hawking affirms Einstein is double wrong. Yet if he is right, black holes will never evaporate and accrete the Earth. For that reason, because no human being should risk his life, without his knowledge, for the scholar ambition of a few physicists, I ask this Court to allow an open discussion in a due process of law of the real dangers of CERN's black holes; instead of trusting Nuclear Physicists, given their ignorance and indifference towards the arrow of life, this Court must defend. That ignorance is only paralleled to their arrogance and desire to play God, which they reduce to the inverse arrow of Death=eviL≠Live. Indeed, Mr. Hawking ends his best-selling book about Einstein's Relativity, which he now denies with his theory of evaporating black holes (A brief history of time), observing that (Physicists) know the mind of God. And so one imagines He, as Mr. Engelen (17), Mr. Higgs and perhaps our Amici, can invent any speculative theory, regardless of its truth, since God, after all, if He exists, should convert in truth=reality whatever is in His Mind. Already Kepler said, when he applied the mechanist clock to the Universe"Yes, I am the one: God Himself has waited for six thousand years for ME, who looks at His creation with understanding."

Yet Physicist's god is 1 single arrow of time, as Oppenheimer clarified after seeing His Work—the 1st Atomic explosion,: I am become Kali, god of Death. In that sense, LHC is a machine of death, disguised as knowledge, whose only purpose is to defend a billionaire budget and a wrong concept of the Laws of Time and Science. Teller also convinced Truman of the need to evolve Hydrogen bombs with the excuse of knowledge, because he had developed the Industrial Process that could make them (a statement which can be seen in The Atomic Bomb Movie). He wanted to create one H-Bomb so huge that could Nuke an entire country, but Eisenhower, a military, who knew better how to defend his country, called the Soviets and agreed not to make bigger H-Bombs, as Mr. Obama and Mr. Sarkozy should do in this case, cancelling CERN's bid for the ultimate weapon. On the opposite extreme of those attitudes we find Professor Einstein, who made only thought experiments, as most scientific discoveries are done with the power of the Human mind, not of machines. He also became a pacifist and fought against Nuclear Proliferation, correcting his initial error, (the letter to Roosevelt that started the Nuclear age), in a proof of humility and ethical standing that I ask Amici to imitate.

[36]   MAD theory—first expressed by Nobel[30] to justify his industries of Death—was the excuse to keep churning nuclear weapons in the Cold War era. This self-interested, industrial thesis, promoted by Nuclear Physicists in both sides of the war, pretended that an arsenal able to cause the Mutual Assured Destruction of the Soviets and America would deter its use. We found recently that New York was not obliterated during the Cuban Crisis by sheer chance, as it was revealed recently on declassified material: the Soviets gave order to fire a nuclear missile but one of the submarines commanders, Arkhipov, refused http://en.wikipedia.org/wiki/Vasiliy_Arkhipov. So we survived MAD because of the ethical standing of a single man, not because the Nuclear Industry's strategy was appropriate . . . Might this Court show the same restrain in a similar crisis . . . Since physicists accustomed to MAD for decades, seem to consider a chance to blow up the Earth, business as usual, given the legal immunity they have enjoyed in the past for their crimes against humanity. But the cold war is over and so we have experienced a shift on our security policies, from developing nuclear weapons, to controlling polluting industries and restraining the access to those weapons of fundamentalist groups, who pretend to impose their dogmatic beliefs with violence, not with reason. In that regard, there is little difference between a religious terrorist group that uses the Yihad to impose his beliefs and Mr. Hawking, who risks mankind pretending to prove a theory of evaporating black holes that 30 years of experimental evidence and the scientific method have falsified (http://www.lhcdefence.org, min.23-30). Both are committing acts of terror, imposed to millions of people, who are not aware of them, neither share those beliefs, nor wish to put their lives at risk for them and whose will should be respected in a Democracy. In that regard, the attempt to dismiss this suit without trial, on the basis of a de facto authority, which Amici pretend to have above the Laws of Science and the Laws of our Democracies, should meet an adequate response by the legal authorities of our nations.

[37]   For a full understanding of the Duality of Time arrows in the Universe, and the origin of Physicists obsession for the arrow of energy, you can see the visual document we have prepared to raise awareness of this potential genocide among non-specialists at http://www.lhcdefence.org

ADDENDAS

Graphic illustrations & Sources:

A: Production of dangerous Particles at LHC and reactions they might trigger.

B: CERN's Safety Report, made by physicists.

C, D: RHIC Bulletin: The results of the experiments were a 'perfect surprise':
RHIC produced a proto-strangelet (strange liquid) with a behavior similar to a black hole.

E: Production of quarks at LHC.

F: Number of quarks to make a strangelet stable.

ADDENDA A

Quark Production at LHC, (Central Graph: Scientific American), **related to the different types of Mass-bombs they trigger** (www. lhcdefence.org)

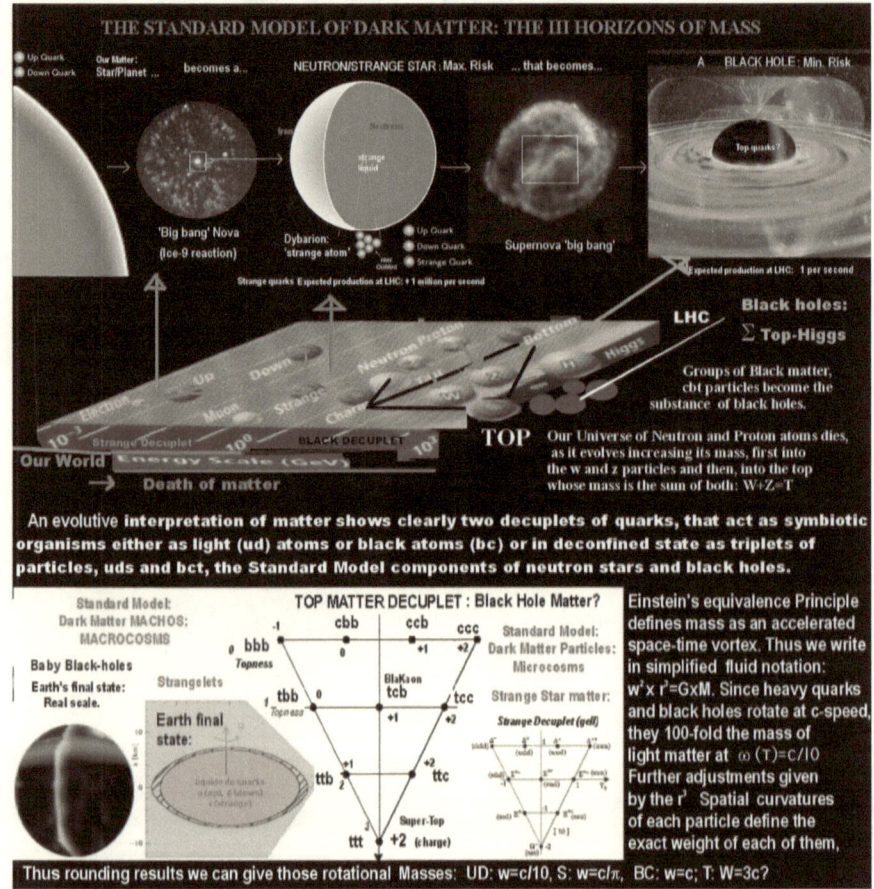

The LHC will cross deeply into the world of dark matter, which 'breaks the symmetry' of our electroweak matter, feeding on it and transforming us into heavy quarks, strange quarks, the components of Pulsars and Tops=Higgs particles, likely components of black holes. Thus, according to Einstein's equivalence principle, an accelerated vortex of mass will be formed, absorbing the Earth at light speed, $M=e/c^2$, creating an ultra-dense pulsar or hole.

ADDENDA B: LSAG Review.

People who made the safety report were nuclear physicists working at CERN LHC safety report documents and their review and approval: participants. Source: CERN,

Stage	team	Review member affiliation	participant profession
Same CERN's SPC examined LSAG ▶	Chairman		
	Prof. E. FERNANDEZ		
	Members (Cem pannel):		
	Prof. R. ALEKSAN		
	Prof. J. AYSTO		
	Prof. A. BLONDEL		
	Prof. A. BONDAR		
	Prof. MUNZINGER		
	Prof. M. CAVALLI		
	Prof. P. DORNAN	'...the SPC CERN	
	Prof. D. FOURNIER	Council has an	
	Prof. D. HARTILL	advisory body, the	
	Prof. T. KONDO	Scientific Policy	President of
	Prof. G. 't HOOFT	Committee, SPC. The	CERN Council
	Prof. B. WEBBER	SPC is composed of	(pers. comm.)
	Prof. A. ZALEWSKA	world-recognized	
	Prof. F. ZWIRNER	physicists'	
	Ex-Officio Members:		
	Prof. M. HUYSE		
	Prof. J. DAINTON		
	Prof. K.-H. MEIER		
	Prof. M. TIGNER		
	Prof. T. WYAT		
The SPC ▶	Matteo Cavalli,		
	Gerard 't Hooft,		
	Bryan Webber		
	Fabio Zwirner,		
	Peter Braun		
The panel were:	LHC Safety Assessment Group		'Review of the Safety of LHC
	S.B.Giddings	Physicist US	Collisions', by the LHC Safety
	John Ellis	Physics Dep. CERN	Assessment
	Gian Giudice	Theory Dep. CERN	Group (LSAG
	Michelangelo Mangano	Theory Dep. CERN	report)
	Igor Tkachev	Physicist, Russia	
	Urs Wiedemann	CERN	

ADDENDAS C, D

The surprising discovery of quasi-stable ice-9 at Brookhaven:
A perfect liquid, instead of unstable plasma as all scientists, except Mr. Walter, predicted.

RHIC — Brookhaven's Relativistic Heavy Ion Collider

A Look Inside RHIC

RHIC's 2.4-mile-circumference tunnel dominates Brookhaven National Laboratory's 5,300-acre campus. Inside the tunnel, two rings of supercold, superpowerful, superconducting magnets guide and focus high-speed packets of "heavy ions" — the nuclei of atoms as heavy as gold — into collisions with one another. The ions travel at energies called "relativistic," because they approach the speed of light.

Superconducting magnets

RHIC's magnets steer the speeding ions into collision at points where the two rings intersect. The temperatures and densities resulting from these collisions are so extreme that they mimic conditions that existed in the first few microseconds of the universe — but only at the scale of a single atomic nucleus.

The process of steering beams at such high energy takes a lot of electricity. But the cost of the electricity at RHIC has been greatly reduced because the magnets are made of superconducting materials. When cooled to extremely cold temperatures — just above absolute zero — superconducting materials lose all resistance to electricity, so electricity flows freely.

RHIC's two rings are made of 1,740 superconducting magnets, which focus and steer the beams.

A Perfect Surprise

Already, RHIC research has captured worldwide attention with an astonishing surprise: Scientists at RHIC had expected collisions between two beams of gold nuclei to pack enough energy and matter into a tiny space to produce a *gas* of free quarks and gluons. But instead of behaving like a gas, the matter created in RHIC's energetic gold-gold collisions appears to be more like a *liquid* — a "perfect" liquid with virtually no viscosity, or frictional resistance to flow.

Moving forward

The stunning surprise that the early-universe matter created at RHIC behaves more like a liquid than a gas has enriched physicists' understanding of quantum chromodynamics (QCD), the theory that describes the interactions of the smallest known components of the atomic nucleus. But it has also raised compelling new questions.

These questions have prompted the need for the enhancement of RHIC to further the study of QCD. To address

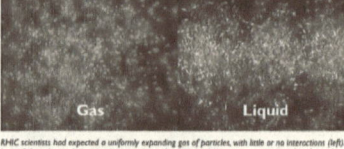

RHIC scientists had expected a uniformly expanding gas of particles, with little or no interactions (left). Instead they observed asymmetric expansion and very strong interactions - suggesting that the early universe was a nearly frictionless liquid (right).

these questions, key improvements are planned for the RHIC facility. As part of a symbiotic research program using Brookhaven's supercomputers, these upgrades will create a new QCD laboratory at RHIC unlike any research center in the world.

RHIC-II and eRHIC

A near-term upgrade, known as RHIC-II, will increase the machine's collision rate and improve the sensitivity of the detectors to reveal detailed characteristics of the new form of matter.

A longer-term upgrade, known as eRHIC, would add a high-energy electron beam to collide with polarized proton beams or heavy ion beams at RHIC. With this powerful added dimension, physicists expect to probe another new form of matter locked deep inside ordinary nuclei, and further expand our ability to explore the newest and most intriguing questions about the substructure of the world around us.

Detectors — The Eyes of RHIC

Four very different detectors — STAR, PHENIX, PHOBOS, and BRAHMS — have helped physicists analyze the particle collisions at RHIC. STAR and PHENIX remain operational. Like giant 3-D digital cameras, these detectors electronically record the results of collisions at RHIC, seeking insight into what happens when the quarks that make up ordinary protons and neutrons — and the gluons that hold them together — are liberated from their confinement inside atomic nuclei.

The STAR detector tracks and analyzes thousands of particles, such as protons, neutrons, and pions, that may be produced in each collision inside the detector — as seen in the cover image. STAR stands for Solenoidal Tracker at RHIC.

The PHENIX detector examines entities such as photons, electrons, and muons, in addition to the particles tracked by STAR. PHENIX stands for Pioneering High-Energy Nuclear Interacting Experiment.

ADDENDA E

Production expected at the Quark factory (LHC).. Source: CERN; Engelen conference.

Table 1

Production rates of known and unknown particles at the LHC in the first year, at 10% of its nominal luminosity.

Process	Events/sec	Events for 10 fb^{-1} (one year at 10% of nominal luminosity)	Remarks
$W \rightarrow e\nu$	15	10^8	Discovered 1983 (CERN). Studied at LEP (10^4); Tevatron (10^7 end 2007).
$Z \rightarrow ee$	1.5	10^7	Discovered 1983 (CERN). Studied at SLC; LEP (10^7); Tevatron
$t\bar{t}$	1	10^7	Discovered 1995 (FNAL). Initial studies only, so far (10^4 end 2007) – high statistics at LHC
$b\bar{b}$	10^6	$> 10^{12}$	Discovered 1977 (FNAL). Detailed studies at B-factories (10^9 end 2007) (KEK; SLAC) – complementary studies at LHC.
Higgs boson mass 130 GeV	0.02	10^5	Discovery possible at Tevatron, given sufficient luminosity; large discovery potential over large mass range at LHC.
$\tilde{g}\tilde{g}$ mass 1 TeV	0.001	10^4	High mass super-symmetry, the exclusive domain of LHC.
Black holes (theories with extra dimensions) $m_D= 3$ TeV, n=4	0.0001	10^3	Wide range of quantitative predictions, depending on number and size of extra dimensions.

B:bottom quarks; t: top quarks

Table 1 shows the real productions of quarks and other lethal particles at LHC. A black hole per hour, high statistics of top quarks, standard candidates to make black holes, up to a million bottom quarks/antiquarks couples and an undisclosed quantity of strange quarks the substance of ice-9. Yet since strange quarks are lighter than bottom quarks (Addenda A), their production will be much bigger probably of several millions per second, more than enough to detonate the planet (Addenda C), according to Einstein's equation of Mass production, $M=E/c^2$ explained by Wilczek[14]. Why the LSAG doesn't report this (II)? Why the table of production omits deliberately strange quark production, given its enormous potential danger and the fact they were the main heavy quark produced at RHIC? Why Amicus contradict Einstein's equation in their brief?

ADDENDA F

Number of quarks needed to make a mass-bomb stable. Source: Arxiv.org.

New solutions for the color-flavor locked strangelets

G. X. Peng[1,2,3], X. J. Wen[2], Y. D. Chen[2]

[1] China Center of Advanced Science and Technology (World Lab.), Beijing 100080, China
[2] Institute of High Energy Physics, Chinese Academy of Sciences, Beijing 100039, China
[3] Center for Theoretical Physics MIT, 77 Mass. Ave., Cambridge, MA 02139-4307, USA

Recent publications rule out the negatively charged beta equilibrium strangelets in ordinary phase, and the color-flavor locked (CFL) strangelets are reported to be also positively charged. This letter presents new solutions to the system equations where CFL strangelets are slightly negatively charged. If the ratio of the square-root bag constant to the gap parameter is smaller than 170 MeV, the CFL strangelets are more stable than iron and the normal unpaired strangelets. For the same parameters, however, the positively charged CFL strangelets are more stable.

PACS numbers: 24.85.+p, 12.38.Mh, 12.39.Ba, 25.75.-q

arXiv:hep-ph/0512112v1 9 Dec 2005

After the acceptance of quantum chromodynamics as the fundamental theory of strong interactions, it became extremely significant whether a deconfined phase of matter consisting merely of quarks would be possible. Theoretical investigations show that strange quark matter (SQM), which is composed of u, d, and s quarks, might be absolutely stable [1, 2, 3]. Because small lumps of SQM, the so called strangelets, could be produced in modern relativistic heavy-ion collision experiments, their charge property has attracted a lot of interest [4].

Originally, SQM is believed to show up with some small positive charge [2]. In June 1997, however, Schaffner-Bielich et al. demonstrated that strangelets are most likely heavily negatively charged [5]. In June 1999, it was shown that negative charge can lower the critical density of SQM [6]. In July 1999, Wilczek mentioned an "ice-9"-type transition [7], which was picked up by a British newspaper. Not long ago, in response to public concern, an expert committee published a report [8], which got positive comments [9], as well as criticisms [10]. In fact, the strangelets in Ref. [5] are not in β equilibrium which drives the system to flavor equilibrium, and negatively charged strangelets in normal phase have been ruled out by a recent publication [11].

Much progress has been achieved recently by the introduction of color superconductivity [12, 13]. It has been shown that bulk SQM with color-flavor locking is electrically neutral [14]. Immediately, Madsen found a solution to the corresponding system equations of strangelets, where color-flavor locked strangelets are positively charged [15], and they might be a candidate for cosmic rays beyond the GZK cutoff [16].

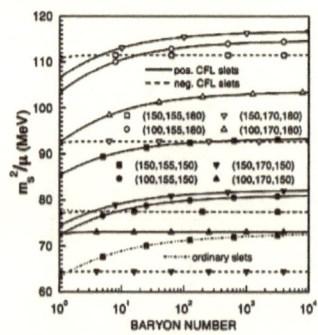

FIG. 1: The ratio of squared strange quark mass to chemical potential for different strangelets with various parameters. The solid lines are for the CFL strangelets in Ref. [15]. The dashed lines give the new solutions of CFL strangelets reported in this letter. Unlike the previous strangelets, which are positively charged, these new strangelets are slightly negatively charged, or nearly neutral. Parameters are indicated as $(\Delta, B^{1/4}, m_s)$ in MeV.

Mr. Wen, Peng and Chen from the Institute of High Energy Physics of China, wrote the first of a series of articles that explored the number of strange quarks needed to provoke an ice-9 reaction. The graphs below explain the ice-9 reaction and show between 1.000 and 10.000 quarks make strangelets (the simplest mass-bombs) stable. The graphs show that around 1000 strange quarks become stable ice-9 (strangelet liquid) starting a Nova reaction (mass-bomb) regardless of the strangelets charge. Those articles, written 6 years after the speculative paper of Mr. Sheldon and Wilson have long superseded his work, since at the time we didn't know that strange quarks locked themselves into a far stronger, and more stable form of matter, color-flavor locked strangelets, which make strangelets far easier to produce. Yet Amici and CERN purposely misinform this court and the media, bringing obsolete papers and denying the experimental evidence at RHIC (Addenda C):

3. Business as Usual. The Outcome of the suit.

So the suit did happen. But by the time we were at Court, the industrial system had put its wheels to work and destroyed our reputation. The *NY Times* had published within a week an editorial, clearly written by a physicist, trying to influence the Court, asking Ms. Gillmor to throw the suit to the nearest black hole, against a basic law of the American democracy that forbids journalists to influence courts before a decision (she would duly obey her master's voice):

NYT Editorial: But It's Just a Small Black Hole

Published: April 6, 2008

If you thought you didn't have enough to worry about, consider this catastrophe projected in a lawsuit filed recently in Hawaii: The plaintiffs warn that a huge particle accelerator on the Swiss-French border could create a ravenous black hole that could gobble up the entire Earth or produce strange new forms of matter that would destroy the world as we know it.

It is not clear that a federal court in Honolulu could do much about a project in Europe even if it wanted to. The plaintiffs are hoping to block American agencies from assisting work at the Large Hadron Collider.

Probing realms at the frontiers of high-energy physics, scientists hope experiments with the accelerator will detect a long-sought particle that may explain how elementary particles acquire mass. They also yearn for other startling insights, perhaps even by creating microscopic black holes, a mini-version of the massive energy-sucking holes believed to exist at the centers of galaxies.

The European Center for Nuclear Research, which will operate the collider, has rightly pooh-poohed the dangers but is revisiting the safety issue in an effort to lay the concerns to rest. We draw comfort from the fact that similar concerns were raised nine years ago by one of the plaintiffs, a former radiation safety officer for the federal government, about a collider at Brookhaven National Laboratory on Long Island. His suit was dismissed. The accelerator was turned on. We're still here.

We are further reassured that the Earth has been bombarded for billions of years by cosmic rays far more powerful than anything the collider will produce without, so far, being sucked into a black hole or turning into uninhabitable yuck.

> More than once over the years we have felt as if we were transported to another universe listening to lawyers and judges wield the complexities and arcana of their trade. It would be fun to watch them struggle with theoretical physics. But if the courts have any sense, they will drop this suit into the nearest black hole.

And so the *New York Times* considered that the perfect surprise at Brookhaven that created a proto-black hole, not the gas physicists have expected, is "comforting," that cosmic rays are quarks because "we are further reassured" by the Company. The chances that a nuclear company, manufacturing the most powerful explosive of the cosmos, will blow up the planet didn't even require some minimal research in any undergraduate textbook on the nature of cosmic rays. And of course, the plaintiffs of this suit which the *New York Times* talks about were never contacted just for "fairness."

So I wrote to the *NY Times*, asking to publish the plaintiff's point of view, but the *New York Times* didn't even acknowledge the letter. The opinion of the opponents to CERN would be from then on treated worst than the opinions of Al-Qaeda terrorists, which at least get published. In this case, total silence is the rule, total invisibility—the deepest form of censorship, reason why we publish some of those letters here, to the risk of being redundant:

> From: luisancho@hotmail.com
> To: editorial@nytimes.com; executive-editor@nytimes.com;
>   managing-editor@nytimes.com;
> Subject: you wrote 3 articles on my suit without consulting me . . .
> please consider this co-ed article.
> Date: Mon, 28 Apr 2008 17:49:06 +0000
>
> Sir:
>
>   I write in response to your article Asking a Judge to Save the World, and Maybe a Whole Lot More (March 29) and editorial (April 6) about our suit 'Mankind vs. CERN' (technically called Sancho et Al. Vs. DOE), which tries to avoid the replication of big-bang energies and nova explosions here on Earth, specifically via the CERN Large Hadron Collider (LHC), planned to start later this year.
>
>   Since you have not consulted me for either of those two items, but you have doubted of the seriousness of the risks involved and my professional credentials and those of my co-plaintiff, Mr Wagner, I would like to answer those doubts in the hope that a newspaper of your quality and objectivity will publish this co-ed article, which provides key information the experts who are supporting the LHC have not given to you.

One key fact is that the replication of big-bang explosions here on Earth could easily catalyze either the conversion of our planet into a neutron star, made of strange matter or trigger the creation of its 'big brother', a black hole, the top predator entity of the Universe, probably made of top matter—these are the two fundamental substances that will be produced massively at the LHC factory.

Thus, since, we, the two plaintiffs of this suit, belong to the two professions that must warn the public on the risks of scientific experiments, we have no other choice but to denounce the LHC factory hoping that a court will prevent the creation of small neutron stars or black holes that could extinguish mankind:

—Mr. Walter Wagner is a safety officer specialized in accelerators with an extensive background on the field, who already warned mankind of a small danger that similar experiments, taken place at the Relativistic Heavy Ion Collider (RHIC) at Brookhaven National Laboratory, an American accelerator, caused a decade ago. While Mr. Wagner considered that the risks at RHIC, with 50 times less energy than the LHC factory has, were rather small (±1%), on bio-ethic grounds Mr. Wagner calculated rightly that a 1% of risk multiplied by the entire human race was a legal genocide of 1% × 6 billions = 60 million people, which is the way such genocides are calculated in insurance companies. Thus the RHIC experiments were the biggest potential genocide of history, even superior to the causalities of World War II.

RHIC scientists denied this, stating there was zero risk. They were absolutely sure that RHIC would produce a hot plasma that would evaporate. Thus, the court trusted them and the experiment went ahead.

What happened then at RHIC?

Unfortunately Mr. Wagner was validated by the results of the experiment, which according to a RHIC bulletin and an article in Scientific American was 'a perfect surprise' as it didn't create an evaporating plasma, but an ultra dense, highly ordered, strange superfluid made of u-s-d quarks, hold by a 'strong force' that made it live far longer than expected. In fact, that is why we call it strange matter. As its discoverer put it: 'it lived so long it was as if Cleopatra was still dying today'.

In lay terms, we created a micro-neutron star, the small 'brother' of a black hole.

Further on, what is more dangerous, that strange liquid started to accrete matter as a black hole would do. Thus 'the Tyrannosaurus rex of the galaxy' showed its teeth at RHIC. And if it had lasted a bit longer, it would have created a nova explosion, sucking it the Earth at near c-speed, and we would now be all part of a small neutron star, which we know it has a core of strange liquid. We survived the experiment, because all entities need a minimal energy to become stable (a child for example, needs ±3 kilos to survive birth). So we could say RHIC made the fetus of a neutron star. Yet since the LHC will have 50 times more energy, it will produce around a million strange quarks, and thousands of top quarks per second. Thus chances that a small neutron star or black hole appear at the LHC, feeding on the planet, are calculated by standard physics to be higher than 50%.

—I am a philosopher of science, whose profession is dual. Besides searching for general theories that unify the principles of science, our most rewarding job, from a practical perspective a philosopher of science falsifies theories that don't respect the three tenets of the scientific method: experimental proofs, logical consistency and mathematical accuracy. They also warn mankind, from the perspective of bio-ethics, as safety officers do, of the lethal dangers poised by new technologies and experiments.

Unfortunately the two arguments CERN gave to your newspaper, to reassure us of no danger—Hawking radiation and cosmic rays—are easily falsified as they deny the laws of logic and the experimental method:

1) The cosmic ray argument, the main one used by CERN, is easily proved false by standard logic methods. Since a cosmic ray is a very light, very fast high energetic particle (mainly protons, at best iron, light atoms) and a dark matter particle is an ultra dense, massive, very slow particle. So to pretend that they are the same, when its two main physical parameters, speed and mass are exactly the opposite is by 'reductio ad absurdum' indeed a complete absurdity: if A is opposite to B, A is not B. And the probability of A being B is the absolute minimal in logic: 0.

2) Hawking radiation is a hypothesis never proved that denies Einstein's work (Concerning the hypothesis Dr Hawking famously claimed: "Einstein is double wrong") and breaks the main laws of logic and standard physics:

    — The conservation of baryon number (quantum physics).
    — The fundamental law of causality that moves time from past to future. Since baby black holes, according

to Hawking, instead of growing exponentially as all new born species do, towards the future, evaporate backwards in time by absorbing antiparticles, which most physicists interpret as particles with the sign of its charge and hence its spatial movement reversed, but Hawking fancies to consider against all visual, experimental evidence to travel to the past. As a consequence of such bizarre idea the two main laws of time in physics become also reversed:

— First law of Thermodynamics: information and energy are never destroyed, except in Hawking's black holes. It is the never-resolved information paradox, except if we admit that Hawking's radiation is false, hence there is no information paradox and the laws of causality are respected.

— The Second law of thermodynamics: heat moves from the hot source to the cold source. However when we travel back to the past, heat moves from the cold to the hot source and so Hawking's ultra hot black holes get hotter and disappear. This is like saying that kids instead of being hungry babies, will also disappear back to the past into their mother's womb.

To notice that as it happens with all false theories, Hawking's radiation has never passed the proof of experience. We have never found a black hole evaporating in the galactic halo, where Hawking predicted in his original Scientific American article they would be flaring like a Christmas tree.

Obviously philosophers of science are not liked by those scientists who ignore bio-ethical concerns and defend theories that deny the scientific method. So Mr. Hawking has called us 'a subspecies that ought to know better". But 'ad hominem' attempted disqualifications are of no importance in science. We still expect a serious logical or experimental, scientific prove of such theory. And yet, CERN, tells us that the LHC will reproduce black holes at the rate of one per second but those 'black holes will evaporate via Hawking's radiation'. Six words backing a falsifiable theory seem to be enough reasons for CERN to make an experiment that might extinguish us. If we have been lucky enough to survive so long as the seemingly only carbon life species of this galaxy, teeming with black holes, it seems absurd to risk the precious exception called mankind to prove a theory that defies all laws of standard physics and has never had an experimental proof, for the benefit of an industry with a budget of $13 billion. This indeed is the only explanation why the experiment is carried on, since Mr. Landsberg proved that the LHC could produce black holes only just recently, when most of the budget on the LHC had been spent. Obviously we cannot expect that a nuclear factory like Chernobyl in the past, or the LHC today, will admit it produces lethal products and voluntarily close itself down, despite the harm it can cause to mankind, unless a court orders an injunction.

Further on, this summer the GLAST satellite will reproduce the same experiment observing the Halo, seeking for the signature of black hole evaporation and dark matter, without any danger for mankind, making the LHC experiment redundant. For that reason we ask only for CERN to delay the experiment until the proofs of evaporation or stability of baby black holes arrive from the GLAST probe—a very reasonable petition, when the life of all of us is at risk.

Because CERN is hiding those known facts of science, taking advantage of its position of authority and the expected ignorance of the press in complex scientific questions, we had no other choice to get this information to the public in the form of a suit in the hope that newspapers like yours, known for its defense of human rights, would properly research this case and inform the American people of what might become the biggest environmental catastrophe of history—including Hawaii citizens, in the antipodes of the LHC machine, who might not have legal power to stop the machine but certainly have ethical authority, as the rest of mankind, equally at risk.

The experts supporting the LHC repeat ad nauseam the zero-risk hypothesis because it is the official 'position', of physicists at CERN, which were told by Mr. Engelen, their chief scientist, as The New Yorker reported (May 14 2007), "to tell all journalists there is no zero risk, regardless of what they might think'. Yet we hope you will challenge CERN's experts to prove our reasons false, respecting the intelligence of your readers, not with statements of authority and dragon theories but with logic and experimental proofs, as good science does, and the seriousness of this menace to mankind deserves.

Obviously to denounce such experiment will not help our careers. Again this is expected. In the old Persian Empire, messengers who came with bad news were killed on the spot; but wise emperors took notice of the news and defended the nation. So while we accept 'ad hominem' treatment, as messengers with bad news, we would rather say to the people who have political power to stop this foolishness: 'kill the messenger sir, if so you wish but test their message, and if it is accurate, save the kingdom of man, mother Earth.

Yours sincerely
Ls, Philosopher of Science, Barcelona, EU

This and many other letters of different style and content, strictly mathematical, emphatically emotional, objectively rational were never published by any scientific or mainstream magazine. God, the Machine, the Idol of the Industrial Civilization of Human Slaves of Go(l)d and Iron Swords will not be denied. I understood then that the four powers did not exist to inform and improve the future of the humankind but to create fictions, ego trips, and noises that would distract the human sheeple in his lemming run toward extinction.

Only Oberbye answered once.

Date: Fri, 2 May 2008 11:23:38—0400
To: *luisancho@hotmail.com*; From: overbye@nytimes.com
Subject: Re: Remembrances of things past.

Dear Prof. Sancho

I put your lawsuit about the LHC on the front page of the New York Times, so you can hardly accuse me of ignoring your efforts. When there is more to report on this situation, I will report it. But I am not aware of any compelling new developments right now. Undoubtedly, they will occur. I had nothing to do with the editorial that ran. I'm sorry it disappointed you, but the editorial people speak for themselves, independently of me and the other reporters.

Then he was told to finish up the argument about our collective death with this final dictum:

Swiss Particle Accelerator Deemed Safe

By DENNIS OVERBYE
Published: June 21, 2008

That black hole that was going to eat the Earth? Forget about it, and keep making the mortgage payments—those of you who still have them.

A new particle accelerator, the Large Hadron Collider, scheduled to go into operation this fall near Geneva, is no threat to the Earth or the universe, according to a safety review that was approved Friday by the governing council of the European Organization for Nuclear Research, or CERN, which is building the collider.

"There is no basis for any concerns about the consequences of new particles or forms of matter that could possibly be produced by the LHC," the safety assessment group wrote in the report.

Whatever the collider will do, they said, nature has already done many times over.

The report is available on the Web at lsag.web.cern.ch/lsag/LSAG-Report.pdf.

The machine is designed to speed protons, the building blocks of ordinary matter, to energies of seven trillion electron volts, then crash them together to produce tiny fireballs, miniature versions of the Big Bang.

Some critics argue that CERN has ignored or played down a risk that the collider could produce a black hole that would swallow the Earth or create some other dangerous particle.

The safety group, however, pointed out that cosmic rays have produced equally energetic collisions with the Earth and other objects in the cosmos over and over again. But the stars and galaxies endure.

The technological ethics of our civilization, all those powerful people who own energy companies, have Nobel Prizes, advice our presidents, become celebrities of the TV-eye, ride their machines and make money, deal finally with the issue of their own death that opposed their religion of the machine with a sheeple impotence that astonished me. They all followed the three steps of angst that happen when a doctor tells them they have a terminal illness. In this case, Sancho and Walter were the doctors of mankind and mankind responded with the Ross model of "psychological angst" in front of death:

A) denial
B) anger against the doctor
C) bargaining for a false solution

So first all denied it. Then all attacked and ignored the doctors and finally all went to the company that was going to kill them, asking for safety lies and all believed without even consulting an expert in cosmic rays the absurd lie that the most explosive substance of the Universe, the quark-gluon soup, was a cosmic ray. This was not at all what the doctors had told them. We just have asked them NOT to take the poison pill. We thought they were powerful enough NOT to take the pill. At least they claimed in public to be strong-willed rulers. And yet it turned out that a bunch of nerds with a hidden weapon could kill them all with a single shot. It took me a certain time to understand the degree of corruption, stupidity, arrogance, and ignorance of the species. It took me even more to reach the fifth state of the psychology of death. I went down into the fourth state, absolute depression, after the court ruled despite a brief exchange of words with plaintiffs and defendants, which *showed the judge understood the danger.*

> MR. SMITH (government attorney, not Matrix Agent): Your Honor, the issue in plaintiffs response paper that we don't rebut that, you know, we don't claim that we complied with NEPA (National Environment Protection Act, which CERN and the Energy department that pays the machine has shamelessly ignored) or whatever, that is not a relevant issue, and here's the reason. To have a NEPA claim, okay, let's assume that there was a NEPA obligation, and maybe there's a NEPA document out there, maybe there's not, but we don't even need to get there. Plaintiffs' complaint says they have to be injured by this project. Their only claim of injury—

> COURT: Is that the world might blow up, and so we shouldn't get concerned about that. You're right. Why was I even considering it? Mr. Smith, I really find that, you know, I don't know if there's anything to this case, but That's just not a great direction to be going.

SMITH:	I'm not following you. I mean, if their only claim of injury is that the worlds—
THE COURT:	That they might die.
SMITH:	Right.
THE COURT:	Yes.

After that, Mr. Smith became mute for a while. But Ms. Helen Gillmor did not consider this evidence that the entire planet might die enough reason to judge. She washed her hands like Poncio Pilato and asked the Congress instead to rule, "This issue is of concern to more people than scientists." Soon the Congress was busy giving the last drops of the American Go(l)d to our banker leaders, the only thing they really fight for, not their *life, not their people.*

We appealed. So the suit is still not resolved as I send this book to the press. *Three judges in San Francisco have taken more than a year NOT to decide if the suit should proceed.*

During that year, as I still had hoped, I munched in the fourth state of depression, while the rest of mankind stayed in the third state of false bargaining, accepting the safety lies of CERN.

But now I have achieved the fifth state of Zen, when death is accepted as unavoidable, because we are powerless to confront the infinite stupidity of the human slave.

Indeed all the suits against the company have been rejected. The suit against the company in the European Human Rights Commission was dismissed. The suit against the company in the German parliament was dismissed also by lack of "scientific proofs." *The powerful slaves that run the industrial civilization have decided that the ethics of technology are more important than their own lives and that we shall all die for a machine.*

Political power, thus, was the second system of information to which we Einstein et al. in America and Europe tried to appeal, after it was clear in the preceding years that physicists at CERN would reject the "truth of science." It was only left the "third leg" of human informative power—the infotainment industry. And so after writing letters to CERN and putting a suit in the federal system and appealing to some ministers in Europe through close friends, I worked my way to the summit of the mass media system and made a film that was reviewed, approved, and then rejected by the biggest agency in Hollywood, William Morris Endeavor. It is a tale worth telling because indeed, this chronicle of death foretold has been heard by all the branches of global power and rejected. And so when the accident happens, as Dürrenmatt put it,[ch.8] it will not be an accident but destiny. Slaves don't die by accident, but they are disposed of when they are no longer useful for its purpose, which in this fifth Zen state of acceptance of death of mine, seems clear to me. We humans are nothing but a lost rock in a corner of the Universe, whose destiny is to be food for dark matter.

4. The Film. The Third Leg of the Corrupted Mind of Man.

I learned directly how extended is the slavery of the human mind to its machines, in our dying civilization, when I tried to alert not only scientists with my letters and the political system with my suits and personal pleas but when I dealt with the third leg of informative power in our societies—the people of the audiovisual industry.

The purpose of those suits was mainly to provide information for the press to research, for the public to study rationally those dangers, and for politicians to act. So I made a movie about it, which you can see at www.cerntruth.com and in a shorter version at www.lhcdefence.org, which was sponsored initially by the biggest agency in Hollywood (William Morris Endeavor, owned by the brother of Mr. Obama's chief of staff, Ari Emanuel).

This was 2008, and a director of photography from the agency was interested on my breakthrough work on the study of all the scales of the Universe as a fractal, whose theoretical foundations described in this book I discovered in the nineties along with the work of other pioneers, such as Mandelbrot, Nottale, Pietronero and Mehaute. Thus, in 2008 when I was giving some conferences on the fractal Universe at the Sonoma University, during the fiftieth anniversary of the creation of complexity and system sciences, this Director of Photography proposed me to make a film together on the beautiful scales of the fractal Universe and its organic self-similarities in the style of *Koyaanisqatsii*. He introduced me his agent, Mr. Devin Mann, an agent at William Morris Endeavor, who liked the idea and encouraged us to go on. Yet, as we heard the astounding revelation that CERN could make black holes, we decided to make a movie, *quantum roulette* on the chances one of them would eat the Earth up. We interviewed scientists, went to CERN, to Hiroshima, and the more we found out, the more repelled we felt by the behavior of nuclear physicists. It was obvious most of them knew, a global genocide was possible. It was obvious also that none of them would come out of the closet, fearing "the revenge of the nerds"—the criticism of the nuclear industry and the thousands of people engaged in those experiments. *They were just scholars, engaged in the rat race of publishing papers and would keep publishing them on kaons, dibaryons, and* *black holes till the end of time.*Nothing mattered to them, except their theories and selfish agendas.

Even those who were worried, made a simple choice. Imagine that the most powerful person of your company, the CEO for whom you work, is a pederast. Are you going to denounce him? Of course not. How then a nuclear physicist, the only people who knew about this, people who after all had made Hiroshima and Chernobyl possible and got Nobel prizes along the way, people accustomed to mass destruction, was going to denounce their black hole factory? No way. But what if your CEO was going to rape your own son? What if your own "trou noir" was at stake? How could this monstrosity be left unchecked? Especially by the rest of the people of the Earth. We soon learnt that the answer was the religion of physics: the machine. Technology had become the religion of mankind. As in the Milgram test[ch.9], most people not only would kill but would also die for the sake of science. In that film, there are interviews to Hawking and his friend Thorne, to Weinberg and Lederman, the inventor of God's particle, to Wilczek, to CERN's physicists; and you can see in their eyes, in their expressions, in their smiles, in their nervousness, they know there is danger, they know black holes can extinguish us, but still they want to see it, they are dragged to the ultimate cosmic big bang, as Al-Qaeda was to the Twin Towers by their violent ideology. Physicists had their own Holy War in search of the ultimate energy, source of the cosmic big-bang, the death and renewal of the eternal Universe[App.] .

Finally I mounted a film for theatrical release and one for TV and passed them to Alex Garcia, the documentary agent at this mammoth company to see if we could make the film into an "inconvenient truth" so the world could take note and proceed against this rogue company. Since the company was owned by Ari Emanuel, brother of Rahm, chief of staff of Obama, a staunch defensor of Israel and the memories of the Holocaust, we thought we have found our 'humanist hero' to defend mankind from CERN's holocaust. We thought Mr. Obama would come to see the film and ask his European allies over the absurd risks of making black holes on Earth without any safety measure. Yet William Morris Endeavor finally decided not to get into trouble with the nuclear industry and declined to release the film; nor any smaller agency wanted to touch such 'hot, real issue'. The hard lesson learned is that mass media companies are NOT solving the issues of mankind, but they are part of the issue, feeding a big brother smiley, 'don't worry be happy' attitude, a 'LOL method' that anesthetizes mankind against the process of extinction of Gaia by rogue industries.

Hollywood is hype and Global Warming a 'convenient truth' to deviate our attention from far more pressing problems. The powerful nuclear industry is in fact one of the biggest contributors to the global warming campaign, to foster the creation of nuclear plants, which do not produced CO_2. Never mind they make cosmic bombs. Obama is giving credits to the industry. Its minister of science is a physicist, who directed an accelerator. But what if we all die together, collapsing into CERN? This is *negative thinking*. We have made many bombs and we are still here. So better to apply the Hype of God's particle as we keep worshipping technology. It has been done before. *The hype of the first H-bomb at Bikini brought us the 'hot' two-piece bikini.* How cool it was to bomb Pacific islands (and a few islanders left around)! They were toasted as much as your all-American girl will be when you put on the bikini top!

Man feels entitled under the most basic program of reality—the greed for more energy and information of any living species, to defy the survival limits of life and Gaia, even if he is only a "mush on a lost rock of the Universe" (Schopenhauer).

Of course, if Ari Emanuel's agency had released the movie worldwide as the *Inconvenient Truth* of 2009, the scandal would have been so huge that CERN would have been investigated and given the LHC's enormous probabilities of causing a planetary genocide, it would have not started up. But Hollywood, the "ministry of culture" of mankind, was not interested. 'Quantum Roulette' was too real, and the purpose of Hollywood is NOT to alert mankind against the destruction of Gaia but to maintain us in an anesthetized state of mind so business as usual can proceed. This is what I learned in Los Angeles: the human sheeple is tamed by fiction, and the last thing the audiovisual industry wants is to reveal the fact that the destruction of the world is becoming real, it is no longer a science fiction plot, it is a documentary.

CHAPTER 9

The End Of The Wor(L)D

Technological civilization is programmed by the principle that something ought to be done because it is technologically possible. If it is possible to build nuclear weapons, they must be built, even if they destroy us all. Once this principle is accepted, humanist values (something has to be done because it is needed by man) are dethroned and technological development becomes the foundation of ethics.

—Eric Fromm, father of political psychology

The Industrial Revolution: Technological Ethics

I CYCLE: STEAM   II CYCLE: OIL   III CYCLE: ELECTRICITY   IV CYCLE: SOLAR

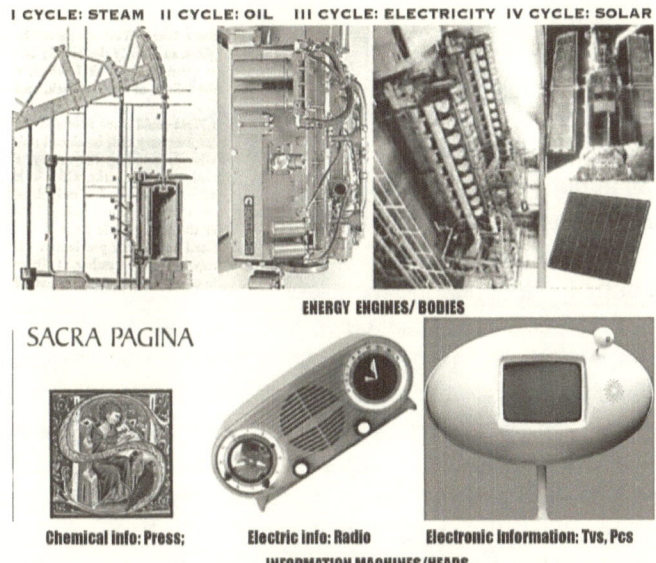

ENERGY ENGINES/ BODIES

SACRA PAGINA

Chemical info: Press;   Electric info: Radio   Electronic information: Tvs, Pcs

INFORMATION MACHINES/HEADS

The Industrial evolution of energetic and information machines is driven by the discovery of new sources of faster, smaller energies that determine what Economists call the long Kondratieff cycles of Economic activity. In the graph, we see the machines that determined them: the steam cycle or Age of England, the chemical cycle or Age of Germany, the electr(on)ic cycle of Age of America and the Solar cycle or age of China, which will start at the end of the Economical crisis that signals always the transition between 2 cycles. Yet those energies uses to construct consumption machines, the 'good fruits of the tree of science', can also be used to make weapons and destroy the world. So England used steamers and America trains to colonize and massacre Africans and Indians; Germany used tanks and planes to massacre Europeans. And now Europe, instead of making a peaceful transition to the use of harmless, removable energies, as China does, has decided to evolve further the Nuclear Weapons developped with electronic devices in the last cycle. In that regard CERN is not only an obsolete machine, the Swann song of the Nuclear Industry of the past cycle, but a dangerous machine, the last, most powerful evil=Anti-live fruit of the tree of science. What Europeans should do is to follow the example of China, which is investing in removable energies and doesn't participate in CERN; reason why its physicists from the Institute of High Energies, without peer pressure have published the key papers [ch.4,13] about the creation of strangelets that can destroy the Earth. The future of a sustainable planet indeed is NOT the evolution of Nuclear Weapons but of green energies.

0. The European civilization of gunpowder weapons.

The opening of the last fractal Russian doll of energy of the Universe is not essential knowledge. It is just part of an automatic process of evolution of machines, the Industrial Revolution, which goes through phases of increasing energy as we evolve machines and weapons. This is what CERN is all about: to keep evolving energetic machines, atomic cannons, without being aware of the limits of death of the weak human carbon-life species.

In that sense, we should not consider the quark cannon and its deeds only from a theoretical or political perspective but mainly from a mechanist, cultural perspective, as part of the process of the Industrial Revolution of machines and weapons, which is based precisely on the discovery of new types of energy, achieved by breaking those Russian dolls. Our technological civilization that substituted the humanist civilization, which lasted till the Renaissance, started with the professional evolution of gunpowder weapons and its mathematical study by Galileo, father of modern physics, first called ballistics. Now CERN has constructed a super-gun that will shoot lead at light speed.

And mankind is happy because we have created a technological civilization whose only goal is to discover more powerful energies, later applied to the creation of new machines, which humans consume to obtain more energy and information.

In that sense, mankind follows the natural behavior of all organic beings made of those two substances: *Max. (Energy × Information);* yet she also forgets that death is an overdrive of those substances; and 'too much of it' can kill you. Thus, Europeans, without respecting the Laws of balance and survival of any vital organism, have decided to push further the evolution of the most lethal of all organisms, the cannon that started the dominance of their civilization 800 years ago, when the Venetians invented the Lombarda . . .

1. The Choice between Organicism and Mechanism That We Made 400 Years Ago

The purpose of this book was to explain the existential crisis mankind faces at the beginning of the twentieth century, due to the dogmatic quasi-religious attitude of some die-hard, nuclear physicists and the absurd respect their weapons and machines inspire to our technocrats, politicians and mass media—the other two legs of social power that have allowed CERN to exist. Yet in a wider sense, it was about two different philosophies of the Universe, organicism, which makes of man the measure of all things, and mechanism, which worships machines. Those two philosophies could have brought very different destinies to mankind. But we chose mechanism and so we are now at the mercy of a weapon of mass-annihilation.

Let us then in this book mandala of self-similar truths end as we began, remembering the original dispute between the two philosophies of the Universe, which can be modeled either as an organism, with its two arrows of creation of futures, energy and form, or as a mechanism, with only one energetic arrow.

The dispute is an old one that can be traced to the discovery of the first machines able to measure time (pendulum clocks) and see space better than the human eye (telescopes). The discovery of those two instruments by Galileo set up a revolution in our perception of reality, which unfortunately was ill understood because of the arrogance of the scientists that discovered those machines, which used them as instruments of war and power. Galileo sold his knowledge to the Venetian Arsenal and ever since promoted the idea that "only" what telescopes and clocks measure was "knowledge" about time and space. So instead of accepting both, the knowledge provided by man as an organism, with a verbal language that describes the three dimensions of morphological time—past, present, and future—and a visual system, which distinguishes not only distances but also artistic beauty, forms, in a logic manner, Galileo and the physicists that followed his ideology chose only the perception of numbers by the machine.

And ever since, life has been despised, massacred with weapons and now it will be sacrificed to the LHC machine. And yet it turns out after all that the Universe is not a machine but an organism. So we made four hundred years ago a wrong choice when we rejected human senses and affirmed that only telescopes and clocks could understand the Universe. In a deep analysis, of course, the issue is more complex.

It all came to a choice. There were two ways to perceive the Universe at that moment:

To feel it as a living organism, made to the image and likeness of man, which should therefore be the center of the cosmos as the *most perfect being of information* or to perceive the Universe as the first eyes of metal, telescopes, and brains of metal, clocks, did it, a digital geometry of motions.

We chose to measure it and man was relegated to an inferior status because "it" could not measure time as a clock and see it as a telescope. This was, of course, a matter of choice because human verbal time could perceive much better than clocks the three dimensions of past, present, and future, its relationships and logic causality while the eyes of mankind could perceive subtle formal structures and the laws of harmony, complementarity, survival, and beauty, Max. $Exl_{e=i}$ that a telescope can't. Thus, through the work of human verbal philosophers and visual artists, man did achieve, especially in Eastern cultures, Greece, and the Renaissance, a high comprehension of the thoughts of God even if the pictures of its details were poor.

In that sense, historians have chosen rightly Galileo's trial as a turning point of history. If the trial that never happened in a federal court against CERN was the last suit against science, which humans lost, the trial against Galileo was the first one, which ended in a draw, as Galileo mused, "Eppur si muove, eppur no muove" (I am a machine, I am a man). Indeed, the choice of that trial was not so much between the Earth at the center of the Universe or the sun at the center but between what the telescope saw and the eye saw—a moving Earth or a static Earth with man at its center.

Once this is clarified, we can better understand the choice that physicists imposed upon mankind: While the church defended human senses, which saw the Earth in the center and defended "informative" perception, which sees things still, Galileo defended the telescope, the senses of the machine, and movement and energy as the only arrow of the Universe.

In reality, as Einstein would prove, both things were relative truths, relative points of view. Because the biggest, strongest point of view takes the best position, in the center, with "lesser consumption of energy" or at the top of the hill with maximal perception of information, the sun at the relative center. Thus, to calculate a secondary point of view—that of the Earth and man—you need more complex information. So Ptolemy had to invent extants and epicycles, but once those complex calculations were done, he could measure the position of the Earth with far more precision than Copernicus did. Copernican physics became more advanced only thanks to the elliptic understanding of those trajectories, due to Kepler. What this means is a parable of history: if we humans want to survive and hold our point of view, we have to struggle, fight, and choose the hardest path, doing many things by ourselves instead of letting machines do the entire job. And this indeed means humans should cherish scientists who evolve with thought experiments more than machines who make mechanical extinctions.

But the church lost and Galileo won, and he was not a relativist like Einstein, he never did thought experiments, he never cared for the human point of view, he abandoned medicine, he made weapons, he was a rich man with a factory of "spyglasses," he would certainly love the LHC.

If Galileo had been truly intelligent, he would have realized that the important question was to understand both elements at the same time, the stillness, and motion of the Earth, the Earth perceived as a center, and the sun perceived as another center to the service of the human one *because both things are truth, all is relative to the point of view of the observer, as Einstein discovered four hundred years later.* The easiest path, of course, was to consider the simpler view that the sun was the only center and the Earth was moving. Astronomers and philosophers pointed out though that we do not see motion but form, still information, and so this paradox must be accounted for. It took four hundred years till the work of Einstein to understand the existence of the paradox. But this paradox has never been brought again to mainstream science. And it should have been because it truly matters to our future. The choice was then and still is "mechanism," the belief that we must measure space with telescopes, not with the senses of the artist, and time with clocks, not with the complex causality between past, present, and future, described by our verbal wor(l)ds; that all what we must know about times=changes are the rhythms of change of motion and energy, the only ones which physicists, whose worldly profession is to make weapons, understand. The choice was also one between male, the yang, energetic, lineal species, and female, the yin, informative, perceptive cyclical one, which would be philosophically despised ever since in an energetic, weapon-driven culture. Galileo was a recognized misogynist, Newton said his biggest pride was to die virgin, Einstein told his wife that she only reproduced her stomach, and Mr. Cox called us "twat" physicists[1] as if the creation of life was of no importance. The attitude of CERN is part of that choice. The death of life, of the entire planet, doesn't matter as long as we advance in the half vision of reality, which is the knowledge of motion, energy, and the perception of the Universe with machines.

Now we have come to the logical goal of that choice, which despised man and life. So because we despised ourselves, we shall die and the machine shall kill us. If Galileo had truly believed that the Earth was both things, eppur si muove and eppur no muove, perhaps the choice would have not been so radical, as we would have accepted both—man and the machine, information and energy, life and death. But the choice was that of energy, death, the weapon, and Europe; the culture that builds CERN started an age of life destruction that now culminates at CERN with the obliteration of the planet.

The final laugh of the gods comes now, indeed, when at the end of the search for true science, thanks to the topological advances of fractal and non-Euclidean mathematics, we realize that in fact the choice was wrong since the Universe is a fractal, topological organism,[Appendix] created by the cycles of energetic and information. And so are machines. Indeed, another feature of organicism is the fact that unlike mechanism that cannot explain organisms, organicism can include mechanism as a simplified model. Since organicism considers also the machine a primitive evolving organism that will become one when robots acquire artificial intelligence. Organicism was the dominant doctrine of mankind during most of our existence till the Industrial Revolution changed the paradigm. But precisely because machines are becoming organic, suddenly organicism has become all the rage as engineers study biology to replicate organic systems in machines. And so paradoxically the third millennium will witness the final victory of organicism over simplistic mechanism, as the Earth, Gaia, and her new inhabitants, machines, become organic beings. Since only systems that are able to create "feed-back cycles" between its networks of energy and information that transform into each other ad eternal are self-sustained beings. So we choose a false model of reality that will kill us for nothing, as CERN will provide no new knowledge, no deeper understanding of the Universe.

Thus at the end to times bioethics and knowledge, the two missions of science, came together: the universe is vital, it has the properties of organisms, which can be described through mathematical, logical, and experimental/visual languages. Yet humans made the error of thinking it was a simple machine, and that ignorance led them to their own self-destruction. We destroyed first the networks of life energy of Gaia; we then forgot the human verbal languages of information, which described the laws of balance, beauty, ethics, and survival. We finally became simplified by those machines, reduced to the mathematical language, to the arrow of energy, and finally we shall blow ourselves, seeking for the absolute simplicity of reality.

True knowledge of the organic complexity of the Universe would have preserved life. We would have understood that to know and to live were the same thing because reality is made of vital spaces in which information reproduces its formal motions. Yet we never understood those organic processes. We preferred a simple, mechanical, mathematical analysis of reality with a single arrow of time, entropy, synonymous with death. And in search of that simple arrow of energy and death we shall kill ourselves. It was not needed; the program of the Universe was dual. We could have understood that program and searched for both arrows, energy and information, and its balance that maximizes the function of existence, Max. $E \times I_{(e=i)}$. We could have maximized beauty (e=i) and survival, creating a world to the image and likeness of man, a perfect world. We have not, and the Universe will discharge us imperfect beings, slaves of physical, simple machines and instruments of death. We are getting what we deserve.

2. The Will of the Universe and the Constraints of the Program

Indeed, the program of the Universe is simple. All fractal, organic beings in the Universe show the same will. The Universe is made of complementary systems with fields/bodies of energy and particles/heads of information, which dominate those bodies. So all of them constantly gauge, inform themselves about their environment, trying to capture energy for their bodies and information for their minds, which they later will combine, reproducing its form in other parts of reality. And that is the will of the Universe: grow and multiply, absorb energy and information, and reproduce it. Quarks do it, electrons do it, birds do it, galaxies reproduce stars. All the systems of reality are thus organic in as much as they have complementary systems that gauge information, absorb energy, and reproduce in other regions of space.

All systems will tend to conserve their energy and information and acquire more of it. This is known in physics as the principle of minimal energy and was used already in the twenty-first century to deduce all the laws of Newton, showing how mathematical analysis, the how of the Universe, and biological principles of balance, harmony, and survival, the why of the Universe complement each other. In physics, we know that a particle will always follow the path that conserves better its energy and so it will do a human being when traveling between both points. The particle however will do it in an automatic manner with no error, while the human being is so complex and has so many networks of energy and information or "degrees of freedom" in his choices that he might switch on his path of lesser energy, for example, to look at the landscape and acquire information from a difficult height point, spending more energy in that path. So there is a why, a "will" or program, for each particle of energy and information that facilitates the how, the calculus of trajectories:

$$max. \ energy \times max. \ information$$

What matters under the fractal program of any complementary species of energy and information is to have more of it. Yet the catch of the program is that if we absorb too much energy or form and break the balance of beauty and survival, $E \times I = K$, we die. It is the rule of the golden mean.

On the other hand, machines have only one element.

They either maximize energy (weapons systems) or information (computers systems) and so happens that physicists and robotists and their theories of the Universe as a "big-bang machine" or a "computer," which are in such rage these days in mechanist science, are unbalanced and the pursuit of its "maximal energies" and "information" the sure path to our extinction. Indeed, CERN has done a bigger, better machine that delivers more energy and makes physicists happy because it fulfills their wanting of energy and information, even if an overdrive of it will kill us all. Yet such "animal behavior" is not a proof of human ingenuity but of automaton behavior. True intelligence in the fractal universe starts when a species is able to control his wantings and defeats the program of death by *repressing his wantings beyond the quantity of energy and wrinkled forms his superorganism can stand*. And the quark cannon is beyond those limits.

Physicists seek only an arrow or will—given their dogmatic, primitive scientific ideology—and they will do anything, including erasing all the information of mankind to achieve its energetic path. *So they must be repressed.*

A truly intelligent species uses the program to its advantage, maximizing its survival. In that regard, we die of ignorance (we believe machines and mechanisms are the model of the Universe, so we sacrifice all to them including our life) and arrogance (we ignore its balances and don't control our greed).

And yet most systems die of greed, ignoring their limits of existence. This is due to a third element of the program, explained in more detail in the appendix: the ego problem, embedded in the nature of perception. Every fractal point of space-time is similar to Leibniz's monads: organisms gauge from their point of view, creating a perceptive aberration, an egotistic perspective, so they believe they are the center of the Universe, entitled to anything. And yet from others' point of view, we are nothing. And so at the end, there is a "God's judgment," as species fight for absorbing the energy of other species, which do the same against us. Our arrogance and ignorance of the power of other species is thus ingrained in the program that causes our extinction.

This means that on one side we ignore the power of dark, quark matter and we despise black holes, dibaryons, or anything the Universe dishes on us. And we also despise the power of energetic and informative machines, which are destroying by overdrive of energy and information our bodies (weapons) and minds (audiovisual information). The result of that arrogance and our real weakness as "light atoms" means we will probably die without even recognizing why we will die.

3. The Ego in the Face of Death

If we were to resume the reasons why we might die from the perspective of the "Fractal Program of the Universe[App.]," there are constraints in the program that guide us toward extinction:

The ego constraint

Individual fractal minds gauge reality from their own perspective. So they are always selfish, ego-centered, and consider themselves too important, too big to fail. And yet from an objective perspective we are nothing but dust of space-time. It is the ego paradox that causes so much deaths and faillure. We need to believe we are the center of the Universe, to create a knot of self-perception, and so humanity thinks it will live forever. But for the total Universe, the Earth might just be a fractal part of an electronic nebulae of an atom of a higher scale that can dissappear in the infinite Universe. This constraint is by far the most important in all tragedies of reality. People cannot, will not, ever think seriously on death. Their individual or collective death is NEVER happening. And that is why it ALWAYS happens. Every person involved in this matter, regardless of his individual intelligence, is in denial. Reason is blocked, truth disappears, confronted with death. It might not happen, and probably it will take time to happen, but the laws of the Universe tell us it should be happening. So what people do is what psychologists call the five stages of denial of death. According to the Ross model of death psychology, when you are told you are going to die, you will do the following:

1. You deny it.
2. Then you get mad against the people who told you because obviously if it is false, they are jerks trying to hurt you. And if you prove that, denial becomes more convincing. This is how people behaved with us Einstein et al.
3. You bargain. You ask for a second opinion to avoid confronting the truth, a second doctor. If the doctor says you are not going to die, then you believe him. This is why shamans and gurus have always been popular and priests come to the side of the dying human, to explain to him that he is not really dying, that he is going to an "after life." In our case, they all went to CERN, *the organization that was going to kill them, because subconsciously they knew CERN would act as priests do, giving them some sort of assurance we were not going to die.* And so CERN lied to them and everybody believed its infantile lies about cosmic rays without checking those lies. As all who are going to die believe the infantile lies of the priests, just in case. So the world stopped the process in point 3: CERN is right, we are not going to die. And of course, maintained point 2: Einstein et al. are a bunch of jerks we shall never listen to again. And yet we were not telling them they were going to die, but that they had an infection, which could be cured easily by just closing the poisonous

factory that was creating the Zyklon-crystal liquid that would kill us all. And it was really easy to do it. Any European politician, the press, WME_2 could have said the truth and CERN would have closed.

In that regard, when experts don't lie and there is a possible solution, you then become thoroughly involved in saving yourself. This is what Einstein et al. have done. We have fought till the end. But the rest of the world has decided to deny death, accuse us and bargain with the perpetrators of the genocide.

4. The process of death seems irreversible. Depression sets in. You don't care. This is the point I reached once the suits did not go through, and the films, books, and articles were not distributed. So I rationalized death in this book. It was also the natural state to which everybody I knew, who applied to me the second stage had thrown me in. Today I am a pariah. But it is not my fault that people have become so weak (Min. Energy) and ignorant (Min. Information) that they could not even survive CERN, its computerized data (Max. Information), and disguised weapon (Max. Energy). Let us be frank. Mankind today has not enough human energy and information to defy even eight thousand retarded "energy-only" physicists, their computers, and weapon. We did our job as "doctors of history"—warned, explained, and fought to save the patient. The patient ignored all those warnings. And so now we are relieved of duty.

5. Then when you accept death as I have done, you prepare yourself for death and dissolution back into the fractal energy and form of the Universe with mysticism. "There is a time for living and a time for death and each one has a different way to prepare it."

Now, a week before the company starts to shoot Earth at 7 TeV in the point of collision, I know I am probably in the last year of my life and all of my existence, dedicated to exploring the meaning of it all, to give it to mankind as an offering of a life sacrificed to true science, was absurd as there is no future. So I know that even if I had more strength and intelligence than most of my self-similar fractal humans, as a "cannon," an ideal form of the "platonic cave," I am nothing and my life was meaningless. It won't leave any memory; it was not lived as humans should do, in balance between body and mind, but it was an aberration of pure thought. I have failed and I deserve to become dust of space-time. It is time for Zen, for Bushido. Had I known this end, I would have certainly lived outside of science, ignored knowledge, dedicated my life to the love of my complementary form and enjoy the perception of nature, dissolving my mind in it. So a humble feeling that my existence was negated has settled in. And I am ready.

Mankind, of course, will go through those final two processes only when it becomes irreversible and the Earthquakes multiply in the strangelet scenario. Then in the last few days or hours in which they realize they are going to die, depression will set in and quickly people will become religious again, abandon science, and if they have time for it, embrace each other in a final moment of humble reflection before death brings us all close.

The greed constraint

People are "fractals of energy and information" that wish more of it. This is the program of all beings. We are "action beings" made of energy and time and so we tend to act and desire irrationally more energy and form. CERN is the wish for new energy records, more knowledge about particles, which mean nothing for most humans but are the Saint Grail of physicists, even if they are false. And of course, greed is also the desire for money to buy our energy and information, in this case shown by the corruption of CERN, a corporation that swindled 10 billion dollars, knowing as most scientists there know that Higgs is a toilet particle. Yet CERN provided more energy and its false information (the Higgs) was subjectively perceived as truth and hence desired as a form/particle of God. The desire to act, the greed for more energy and information, "curiosity" killed the cat.

If we had more space, we could analyze the art of *disguise*. How the constraint of the ego (relativistic perception) and the constraint of greed allow predators to disguise themselves as sheep to capture victims. CERN has played the game of camouflage to perfection. It disguised its weapon as an instrument of research. It disguised its hoax as true information. It robbed technocrats, played the dove, and yet it was a wolf, invoking the supreme God of death of the Universe, making the most explosive bomb of the cosmos.

The social constraint

In a fractal universe of multiple scales, individual parts belong to three scales, as wholes made of smaller cellular parts, as individuals, and as cells of higher superorganisms. Man is not an exception. However, in death processes, the individual cells feel unnconnected from the higher orrganism they no longer perceive. So people who do not believe in a superorganism, a subconscious collective, a god in old religions, a nation in modern ones, or mankind as a species, no longer cares for them. And this is the present state of mankind. We no longer care for the superorganism of humanity. It is invisible to us because we are now part of a superorganism of mechanical machines, economical nations and technology, ruled by metal, by weapons, money, and machines, in which all those social ethic superorganisms we worshipped in the past are gone. There are still "economical nations" we care for. So we have tribal wars fought for money and land to add to those mechanical nations, but the consciousness of mankind is gone. This means each individual human, confronted with this monstrosity, does not feel a need to fight for mankind. The same person will have a heroic behavior fighting a war for America, Israel or Islam (a religious "living" superorganism), but paradoxically he will let CERN kill him because CERN is a menace to the human species, which *he does not feel despite including all those smaller national superorganisms.*

This is the ultimate reason we shall die—we no longer care for humanity. The superorganism of humanity had disappeared. One of the reasons I have fought this fight is because my theoretical work shows me that the Universe is scalar and humanity does exist as a collective organism. Thus, I have long ago transcended racial and national constraints. This behavior was the rule for a long time among humans when there were doctrines of mankind or Gaia as a living organism—Buddhism, Christianity, Islam, Socialism. But now they have all become tribalized. The UNO was aborted before it reached any power. The only remains of that collective spirit are the ecological movement, but it is grossly corrupted. Global warming is partially paid by the nuclear industry that now appears as the savior of clean energy, and its discoverer, James Lovelock, shamelessly gives conferences to promote nukes.[2] Greenpeace rejected any involvement to save Gaia from strangelets as it focuses in placebo themes.

All in all, this individualism means that people think they are too important to die (ego paradox) and so the probability must be small (wishful thinking), and in any case, the death of all other human beings is of no importance to "me" (I don't feel it), plus technology is always good (industrial ethics) so they must be doing something good with a supermachine. And all together sing "four legs, four legs" as in an Orwellian nightmare and move ahead toward death, which will thoroughly surprise the sheeple when it happens. And yet our extinction was easy to avoid. If those moral religions who understood the program—Buddhism, Socialism, the highest forms of Christianity and Islam—still ruled mankind, certainly man, not the machine, would be the measure of all things; and Buddha's advice of restraining our desires to control the program of greed and live in balance with nature would act, protecting our species and Gaia, the body in which the superorganism of mankind exists. Once the bioethical, organic, moral paradigm disappeared, substituted by the mechanical paradigm of reductionist science, humans would set up in a course ruled by a single dogma: we need to evolve machines to understand the details of the Universe.

4. The Greed for Machines

Why does mankind allow this monstrosity, namely, our potential extinction, to happen? Obviously, because we are, according to the fractal paradigm organisms of energy and information, that wish to have more of it. So we are programmed, especially in the West, by our greed of both substances. Yet the program is not absolute. There is indeed, on the limits of information that humans represent, the possibility to understand and manipulate the program to our advantage, which is the meaning of true science. So we die not because we must die but because we didn't reach the level of information needed to manipulate properly the program to our advantage. If we humans had evolved into the fractal organic paradigm of science, we wouldn't cross the barrier of death of electroweak matter to see "what happens." *We would know that according to the laws of science, we will die. A wise gazelle doesn't come close to the lion to see "what happens." Only cubs, children of thought, CERNerds can be thrilled coming close to the lion that will devour them.*

If quantum physicists would really understand the program of the Universe, as Eastern philosophers did (Taoism, Buddhism), they would know that true freedom is to control our wishes, to castrate the program, to limit our greed. So we wouldn't go beyond the energies and information that our system and planet can stand. We would have stopped in the age of electricity. We wouldn't explore the age of nuclear weapons. We would have not opened the last Russian doll of quark matter. For the same reason, we would not evolve robots and computers beyond the capacity of our informative minds. The result of not restraining our program of existence is an overdrive of energy and information that will extinguish us.

We are food for stronger matter. We are only flesh, 5, 6, and 7 electroweak atoms. In the scale of substances, we are weaker than the metal of those machines and certainly weaker than the top predator of the universe, the top quark star, the black hole. And yet our vanity, our bubbling empty ego-brains, is opening the last densest doll of the Universe.

If we were truly intelligent, we would follow the "wu wei" approach of Chinese Taoist and Buddhist thinkers: to control our desires. We would obey the parable of Genesis and prune the tree of its bad fruits that would extinguish us and worship life instead of machines. We have not. We repressed and still censor those who seek the true goals of science, knowledge about humanity, and the improvement of life, designing a world made to our image and likeness. Instead, we worship the machine and make a world to its image and likeness, and that is why we are dying.

A monstrosity as the potential genocide of all mankind, *backed 'professionally' without remorse by most of the physicists of this planet,* the 'high-priests' of our technological civilization, cannot be considered to happen by 'chance' or 'accident' but it is rather a 'destiny' we set in motion when we choose to worship and evolve machines and weapons, instead of the human kind. The LHC is in that sense the natural end of the Industrial revolution, the 'Age of the singularity', the age of robots and bombs, able to feed themselves, no longer mechanisms but organisms. Let us then recall that destiny considering the different cycles of evolution of machines and energies in more detail.

5. The Industrial Revolution

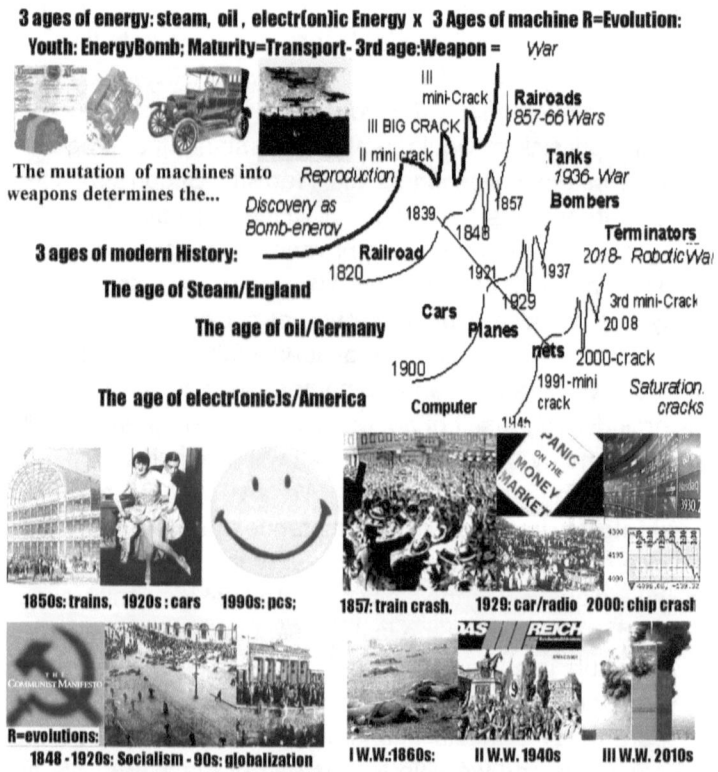

In the graph, Mankind changed the goals of history with the arrival of the Industrial Revolution. Before machines appeared, humans tried to create a world made to the image and likeness of mankind, the superorganism of history, which religions of love strived to create. After machines appeared, humans became organized in economical nations, whose only objective is to evolve machines, growing its GNP.

Accordingly, mankind has evolved machines in three Kondratieff cycles of energy, which defined the cycles and events of modern history. So we went from the steam age of English machines to the chemical age of German machines to the electronic age of American machines. The Industrial Revolution also follows the law of complementarity. So first, in the nineteen century we evolved bodies of machines (British/ German cycles) and in the twentieth century, during the American cycle we evolved heads of machines (phones, cameras and chips). Now we start the Asian cycle that fuses bodies and heads into robots. Yet machines have also two biological forms. Some are tools, symbiotic to humans. Some compete with men at work and kill us in war fields as weapons. Their biological nature shows in their dual forms and functions, which correspond to the destructive, Darwinian and creative, symbiotic arrows of living beings. Unfortunately, when machines mutate into weapons, history enters in a cycle of global wars. It has happened three times at the end of each Kondratieff cycles of machines'evolution, of seventy-two years, causing periods of massive production of weapons and world wars. When steam machines were overproduced, after a crash of the world markets (1857), armored trains were used to conquer the third world. After the car/radio crash of the electrochemical economy, Germany made tanks and hate radio and started World War II. Finally, after the electronic crash of the computer economy, we are manufacturing Terminators and supercolliders to keep the Industrial Revolution going, even if a machine can kill us all.

The cycles of evolution of energies and machines are called the Kondratieff cycle of economics. According to his model, known in abstract among stock-market speculators but censored in its sociological implications, because it completes Marx's work on the boom and bust cycles of capitalism, the Industrial Revolution of machines follows an economic,"generational" cycle, whose exact periodicity, 72 years I corrected in my earlier work on the fractal patterns of History and Economics[4]; and has been thoroughly proved by the present economical crisis:

Each 72±8 years, we go from a decade of happy consumption—the 1850s, the 1920s, and the 1990s—into an age of machine's overproduction and stock market crashes (1857, 1929, 2001) that brings an age of poverty, war, and fascism: the 1860s; WWII and the present War. Thus the present cycle represents the end of the atomic/electronic age and the beginning of the quark/robotic age, the age of organic machines, the final age of the Industrial Revolution.

We are crossing the final cycle of evolution of energy and information machines, what we call the Singularity Age, which is the name used to define both, artificial intelligence and the black hole. Thus, we are probing into the limits of energy and information of machines, which are far greater than those of life beings.

In this wider vision of history, CERN is the natural consequence of making the evolution of machines of energy and information the meaning of history, which no longer revolves about the evolution of human beings.

In the graph, we describe that process of evolution: Humans evolve a new type of energy and information, derived from that energy, which renews all the machines of the economy every human generation of 72 years in which a nation of founding fathers, captains of industries, their sons and grand-sons reproduce and evolve a new energy, machine, and form of money to its perfection. Yet at the end of the cycle, the machine and money becomes overreproduced, saturating the market and provoking, due to a crisis of growth, a global economic crash.

Those generations also bring the nation that discovered the new energy to the top predator status of history because energy is also the substance of which weapons are made.

- *Thus, we had an age of steam machines, the age of England, between 1780s and 1857*, followed by a crisis of overproduction of steam machines and stock money that brought the 1857±7 years crashes of the train-based economy.
- From 1857 to 1929, we lived *in the age of electrochemical energies, machines and chemical explosives, dominated by Germany*, followed by a crisis of overproduction of cars and radios, which caused the 1929 crash, 72 years after the train crash.
- Then came the third cycle *of electronic machines, electronic money, and nuclear bombs that took place from 1929–2001, the age of America,* which again ended in the dot-com and mortgage crashes, 72+7 years after the 1929–37 crashes.
- The Age of the Singularity, the fourth cycle of evolution of machines, dominated by robots, solar industries, mass bombs, and China followed. Because scientists call a black hole the type of energy or bomb now being researched at CERN (the European nuclear industry) a singularity and they also call the arrival of artificial intelligence the singularity moment, we have called this fourth age of the Industrial Revolution, the age of the singularity.

Because machines are, as anything else in the Universe, systems that evolve energy into information, first we discovered the new energy and made with that energy new bombs and cannons. And only then have we created machines to take advantage of that energy. So a war/weapons age starts all industrial revolutions. This has been the trend in all the cycles of the Industrial Revolution. First we made gunpowder bombs (fireworks) and then cannons. Carbon was later used for peaceful steam pumps and trains. In the chemical age, we first made dynamite and then the electrochemical engine that powered the car. In the electronic age, we first used the computer to calculate the atomic bomb and then we used it for civil tasks. In the age of the singularity, when we are about to open the last Russian doll of mass in the Universe, we have also made first the bomb and cannon of the new frontier of energy. The problem is there is no peaceful use for the awesome energies found beyond 1 teravolt in the region of quark, dark matter, where our matter dies by definition. So those lethal energies have no peaceful use because a single cannon shot, a single bomb, can blow us all.

Thus, the singularity age is the age when machines will complete its evolution as organic forms, becoming autonomous of man, probably making us obsolete as workers and soldiers. Indeed, organisms follow a simple pattern of evolution, from mechanical, deconstructed systems into full organisms in three phases since machines are enhanced organs of energy and information evolved in metal. According to those functions, we observe three organic periods in that process of technological evolution:

> In the twenty-first century (1780s–1929), company mothers discovered and re=produced machine bodies, systems that process energy as biological bodies do, from trains to cars. It was the age of stock money, which multiplied enormously the financial power of those companies.
>
> In the twentieth century (1929–2008), companies created machine heads, imitating human sensory organs: phone-ears, eyes-cameras, and brains-chips. It was the age of electronic money, digital software in the mind of those machines.

In the twenty-first century, companies will fuse bodies and heads into robots, *which will use solar energy to become independent of man,* completing the evolution of machines, expelling most human workers and soldiers from the economical ecosystem, as it is happening today in the 2008 crisis that starts the age of the singularity.

In the first book I published with the fractal, generational cycle of evolution of machines, fifteen years ago, I forecasted that the seventy-two-year cycle will end the age of electronic machines in 2008 as it has been. This, to my knowledge, oldest forecast of the present economic crisis,[4] shows the predictive capacity of the fractal models of the Universe developed by Einstein et al., far more sophisticated than anything mechanism, both in physics and economics, has done till today. I also considered the two alternative energies that mankind could explore in the twenty-first century, during the age of the singularity, solar cells, and quark bombs. In that regard, we can consider the existence of two good and bad fruits in the tree of science on the age of the singularity. I strongly suggested that quark bombs and quark colliders should be forbidden and solar cells developed as the future and only alternative energy for cars, which could easily become self-sustained machines with solar cells in their roofs. The book, as all the work of fractal organic scientists, went unnoticed since it did not reflect the religious dogmas of mechanism. And yet the 2008 crisis happened and unfortunately we entered in the singularity age, NOT with a massive evolution of solar energy but with the building of the quark cannon. Let us then consider in more detail the three ages of the singularity age, the last age of the Industrial Revolution, which, as all "deterministic systems" of the Universe, will go through a youth of energetic quark weapons that the LHC represents; an age of reproductive machines; and finally the so much announced age of informative robots, more intelligent than mankind.

6. The Ethics of a Technological Civilization: Money and Machines

The control of the two lethal singularity technologies, intelligent robots and quark-gluon liquids, should have been the priority of human governments that must prune the tree of science of those bad fruits because scientists and companies will never do it. Their job is to evolve technology and make money with it. It is their religion. But this has not happened, and for the last fifteen years since I published my first book on the subject, the theme has been ignored, treated merely as a science fiction plot, and denied to avoid any limits to the evolution of machines by "company mothers" that run our world.

Unfortunately today, governments have become dysfunctional as the free market, run by companies that manufacture and protect their machines, accumulates evermore power. The case of CERN is an astounding example of today's imbalance of power between technological companies and human governments, between technological and human ethics. A single company, which is owned by European governments, has been able to put online the most powerful weapon of human history and nobody has been able to avoid it, even to talk about it, *since CERN has political and jurisdictional immunity, as if it were a nation. And so the most dangerous, polluting factory of the planet is unassailable, unaccountable, and can destroy legally the Earth.* While our politicians and the press, knowing nothing about nuclear physics, preferred to trust CERN instead of doing their homework and investigate this company, *even if the entire future of mankind is at stake,* because it has become a rule of our societies that companies have all the rights to produce whatever type of machine they fancy, without the slightest interference, while human beings must obey all kind of laws, restrictions, and prohibitions. *So the world has accused the scientists that denounce CERN, not the company that can provoke our genocide.*

Today, the tree of life, the goal of human evolution has been abandoned and all that matters to the human species is the evolution of the tree of technology, carried about by a group of technoutopians, which show null understanding of the laws of survival and ethical behavior that used to rule our societies since mankind should be ruled by ethics, not by machines of energy, and ethics are concerned with the values of life and the laws of information, not with the value of energy, the substance of weapons, which kill life.

Let us play with those inverse terms, as physicists play with particles and antiparticles.

In genetic linguistics, the inverse word of *Live* is *eviL*. *EviL* is indeed synonymous to death, the antiparticle of life. And so we can use legitimately in biological ethics the moral concept of Supreme eviL to qualify acts of mass murder and genocide. And ask our governments and courts to stop the evil fruits of the tree of science regardless of what physicists want.

Can we then say that nuclear physicists are eviL? Yes, because they create instruments that kill. And while many people might justify the existence of weapons to defend our lives from enemy nations, the quark cannon will not defend our lives as it cannot be used in war since it would destroy the entire planet. And so because death is the supreme evil that our societies, political and judiciary systems, try to avoid, the quark cannon should be shut down, as we forbid ebola virus, terrorist cells, and try to control nuclear proliferation.

For that reason, it is imperative to understand technology in biological terms, abandoning the mathematical, abstract approach of classic economics, which cannot understand the cycles of evolution of machines and energy scales, which now enter in a region of potential death of mankind by excess of energy and information. Politicians should learn this ugly fact: technological science, mainly market economics, physics, and robotics are religions of the machine that have to be moderated. Market economics denies the real competence and collateral effects some lethal machines are having on mankind—from global warming, produced by the detritus of those machines, to the massive wave of unemployment that robotics is causing, to the risks posed by the evolution of nuclear weapons, into the threshold of planetary bombs—quark bombs and black holes that can potentially destroy the planet. While robotists and nuclear scientists are embarked on a fundamentalist quest for scales of energy and information beyond human control. They all risk the future of mankind. In that sense, fundamentalist scientists do also cause terrorist acts that endanger the life of human beings. Their practitioners are a minority of scientists and believers. But they are a very active minority that, given the passivity and ignorance of most citizens on those matters, carry about their experiments without any opposition. It has always been like that: a few people, obsessed by religious, scientific, or economic dogmas, drive mankind toward a cliff of war and extinction while the bewildered herd follows them, ignoring the risks like cattle follows the rancher to the slaughterhouse. Unless politicians act up soon against that kind of fundamentalism, it is rather clear that we will become extinguished this century by one of those two kinds of extreme information or energy machines: robots or colliders.Thus, we are in a sweeping moment of history when the process of evolution of machines started four hundred years ago and the parallel process of creation of mechanist religions of metal is about to climax. But we do not realize this because mechanist science backed by company mothers of machines have put together all the elements of a religion of metal and its three "organic," fractal parts that work symbiotically, creating the economic ecosystem: lineal, energetic metal or "weapons"; cyclical, informative metal or "money," go(l)d for most part of history and now e-cycles in the mind of computers and organic metal that combines energy and information or machines. Those three fractal parts of a whole system are mutating the world of Gaia, the life Earth into a world of machines, the metal-earth, a theme explored in depth in the twin book.[4]

Unfortunately, the caste-people who have historically discovered and evolved those forms of metal, bankers, warriors, and scientists work together and have created a "religion" of technological progress that has made money, weapons, and machines the icons of their civilization. And so they will take that religion, which kills life till the end, extinguishing the world. This religion has, as true science does, also three legs, but those legs are not truths but ideologies. They are called nationalism that permits the creation of weapons even if all humans are equal members of the same species, capitalism which gives companies of machines more power than governments, and technological science, which substitutes the human mind by the models of computers and the human senses by machines as seers of truth. They are the three legs of the religion that extinguish us by making us dependent on machines, money, and weapons to survive.

Because machines produce money, an informative language of values made of metal as most machines are (gold and silver in the past, computer cycles of e-money today), by the laws of complementarity and self-similarity that guides the relationship between all fractals of energy and information of the Universe, informative money gives maximal values to the most energetic machines, weapons.

Once and again the fundamental equation of the Universe (Max. E × Max. I = k) unveils in its infinite self-similarities and repetitions the why of reality, in this case the reason why we humans hate verbally weapons and yet give them the highest price in a subconscious manner. It is due to the symbiosis of energetic and informative, complementary beings in any scale or ecosystem of the Universe. In this case, we talk of the economic ecosystem in which a language of information, money, values its self-similar energetic forms, machines, more than our flesh species, human beings.

Indeed, in the same way verbal words, our biological, natural language, values more human life because it is made of the same substance, money, a language of metal, values more weapons than life. Since the beginning of history, humans found it natural to value weapons (systems of metal that deliver energy) with money (coins of metal that deliver information), *because they were made of the same substance*. This little recognized fact of economics means that at any stage of human history, weapons have been the most expensive objects and so the economical system reproduces an enormous quantity of weapons to make profits unless there is a legal political system that controls the production of weapons. It is one of the tasks of governments to control this proliferation of weapons.

Unfortunately in the case we study, the quark cannon is not even under military supervision because after the cold war ended, we privatized part of the military industry. Yet it is, of course, the most expensive machine ever built, as it happens always with the best weapon, substituting in that position the B-2 bomber with a tag of 2 billion dollars per unit. So many industrial interests exist around this machine-weapon. And this explains why politicians funded it. It is the jewel of our industrial-military complex, the summit of our technological civilization, the weapon that will murder us all.

7. The Corruption and Power of the Nuclear Industry

It is within that wider economical frame of work where the big bang factory inscribes itself as the first weapon of the new age of quark energies and cosmic bombs, the equivalent of the Hiroshima bomb in the previous cycle of electronic weapons, which were calculated with computers.

Now we enter the age of superconductive, supercold mass bombs that transform energy into physical information, $M=e/c^2$. Since CERN is not about science but about keeping evolving military technology (till we are destroyed by it). Indeed, in the final deepest perforation of the Russian dolls of energy and mass in the Universe, mankind peers into an abyss with no return, as we enter into a region of energies that are well beyond our control.

The nuclear cover opened in the electronic age (nuclear bombs) can be argued to be already beyond our control, as the nuclear terror and the consequences of the cold war showed. But that didn't deter physicists and the military from constructing new weapons. And we know the effects. Nagasaki and Hiroshima came first. Then a myriad of nuclear plants, missiles, and nuclear waste littered the Earth. Chernobyl is still a dead zone.[4] Billions of dollars have been spent in making weapons we do not know how to dispose of. Of course, marketing keeps working on public opinion to make them gullible.

Nuclear plants now are sold because they are "clean," and the global warming campaign paid in part by the nuclear industry has become the biggest scare of history since the Red Scare against Socialism. Even if it is a secondary problem, compared to CERN or the evolution of robotics, that will do us all well before the wheat production in Russia and Canada lowers the price of food and electricity bills plummet in the East Coast. While accelerators are marketed as machines that cure people, physicists indeed came up with the idea that small accelerators with a tag over 100 million dollars are the best solution to cure cancer, bombing people with protons. Many papers have been written about the absurd costs and collateral effects of such treatments that can in fact provoke by radiation new cancerous tissues. And yet hospitals that can't cure basic illness for lack of funds and lack of emergency rooms spend the money on those machines. The bottom line is that the power of those machines impresses us.

And so even if we haven't solved any of the problems of the nuclear age, we merrily enter into the new quark era of cosmic energies and merrily shoot the Earth with quark condensates without the slightest precaution, safety standard, or even a serious public discussion of where the awesome energies of the age of the singularity will drive us into. *Even if the answer is self-evident, much of the same: bigger weapons, more lethality, higher dangers of extinguishing life on Earth.*

In military terms, the nuclear company of Europe (CERN) has built the new generation of nuclear weapons, a quark factory that can create cosmic, quark bombs. This is what we observe constantly in the Universe, and multiple scientific papers have come out in the last decade, explaining those dangers. In that sense, there is little doubt about the explosive properties of quark condensates, the substance the quark cannon will produce, and the only candidate in standard science to be responsible for nova and supernova explosions.

Unfortunately, the quark cannon is not a research machine but a weapon that has gone beyond the limits of energy of survival of the Earth and so its use is a question of national security that will define the future of mankind.

The quark cannon is a very dangerous weapon of the nuclear industry, and it is an expensive tool that mankind doesn't need and *will keep being an expensive thing at the tag of a half billion dollars each year, even if it doesn't blow up the planet.* CERN is not advancing science, but on the contrary, by sponsoring fringe theories and machine experiments, it is halting the advance of true science, which happens through standard science and thought experiments—that is, through the evolution of the languages of human thought, logic, and mathematics. The costs are astronomical for nothing since the theoretical advances for science will be null, an excuse to justify those spendings: Higgs particle is redundant, it does not exist. Its work is already done by the top quark. And it tries spuriously to substitute Einstein.

And yet the inauguration of the first Damocles Machine of History has been possible due to two facts:

— CERN is the most advanced technological company of the world, the jewel of our technological civilization, whose ethics were described by Fromm in the quote that opens this book: "If a company can make the best machines of the world, it must do them, even if the machine can kill us all or have no real use for mankind." Ethics of a technological civilization put technology over life and the rights of companies, re=producers of machines, owners of the free market, *over* the rights of human beings represented by their governments.

— We live in a civilization ruled by weapons. It has been calculated that 2 percent of the budget used to develop new weapons would be enough to end the hunger of the world. Nuclear physicists are the summit of the science of energy, weapons, and death. Of course, they don't see themselves in this manner. They have theories about an energetic, dying Universe. And so they are interested in researching the death of our matter as part of their one-sided search for pure energy. But this interest obviously does not justify the risks of being devoured by what they can discover. And so in this case, knowledge must be considered the alibi used by our technological civilization to justify the creation of this machine. *The machine is in itself the reason, the cause of its creation*; knowledge is just a fringe benefit, which is minimal and far less important than the industrial process—the profits, industrial contracts, and technological progress that justified it.

CERN represents the evolution of technology, of the tree of science, which seems geared to extinguish us, as machines cross the limits of energy and information (quark cannons, artificial intelligence) that the electroweak human being can safely manage.

8. Mass Media. Mental Machines promoting Violent Machines.

We talked of the duality between informative metal (money and audiovisual machines) and energetic metal (weapons) and how the affinity between both accelerates extinction of life. In the case of the quark cannon, mass media has praised and pushed forward the doomsday weapon, silencing any opposition and playing with the real scenario of our extinction as a "science fiction" plot. There was a time in which the human wor(l)d and our values were taken seriously. Machines able to multiply information were reduced to the press, which was able only to print images and words and so it could not provoke an overdrive of informative sensations, which stunts the human mind and makes the medium—the machine—the message in itself. In the twentieth century, all this ended. Machines of information are now able to overload the human mind with so much industrial information in favor of technological ethics (radio, magazines, films, TVs, Internet) and the same information arrive to so many people simultaneously (radios, TVs, Internet) that human ethics, true science, and love religions can no longer compete with the message of "don't worry be happy, we shall all die like idiots," the economic ecosystem of go(l)d, weapons, and machines deliver to the human sheeple.

Further on, any overload of visual content creates *hypnotism, which converts the mind into a robotized system of beliefs, as any content is accepted as truth.*

In each new age of evolution of mass media, the owners of the new "metal communicator" dominated history and imposed their mechanist ideologies of war and go(l)d greed. First Calvin, thanks to the press, converted his go(l)d religion in the culture of Northern Europeans, which started their slavery age and massacre of the third world. Then Hitler and Mussolini, thanks to the radio, created the wave of fascism that annihilated the world. Now TV sends messages of hate against the "primitive" people of the third world, who defend their agricultural, religious civilizations as the supreme evil and praise all types of machines as "progress." And of course, those castes of power that shamelessly use the mass media systems of industrial propaganda never recognize any "evil" in their own racist, mechanist beliefs. Even if the pendulum law converts them alternately in victims and predators of the cycles of history and war mental and physical weapons cause.

The hypnotic power of simultaneous informative machines, which act as a neuronal network does, guiding all cells into a self-similar mass behavior, is little recognized beyond some scientific studies, but it is a very relevant fact to understand how eviL, death, and war has become so much liked when it was repulsive to the bioethical morals of survival of our societies before audiovisual media dominated our minds. It is also related to the physiology of visual languages. Our brain becomes now hypnotized by energetic colors that feed the eye. So red, the color of blood, and yellow the color of money, have hypnotic effects. People love gold because it has the color of the sun. So gold fevers happen. People love violence in movies, death, because it brings red colors and movement. They would hate to read books like those movies without the added red color and motion. But all those fictions about death and the end of the world have influenced mankind, converting reality into fiction and myth into reality. And so when *the end of the world is truly happening, after centuries of promoting war and making fun of death, neither the sheeple nor the informative masters who have profited with the promotion of war could react.* Indeed, the main difficulty I found in Hollywood to make *Quantum Roulette* and distribute the documentary about our extinction was dual: on one side, WME_2 didn't think a documentary would make money because it was not "fun," it was not "fiction." On the other hand, they feared truth; they couldn't believe they were going to die. *Their brains had been deactivated; they no longer were part of the real Universe.* This is even more clear in the new generations that are bred watching violent fictions. Three-fourths of the physicists working at CERN are young, under forty. They are the product of the new civilization in which death is fun. And machines of death are the funniest of them all. *If you give a gun to a child and tell him not to fire, he will shoot it to see what happens.*

In that regard, the seriousness of this book belongs to a bygone era in which humans had not yet been erased mentally by fiction thought, entertainment, and fun, by big brother smiley, by the ethics of technological civilizations.

II. The Neo-Paleolithic and the LOL Method

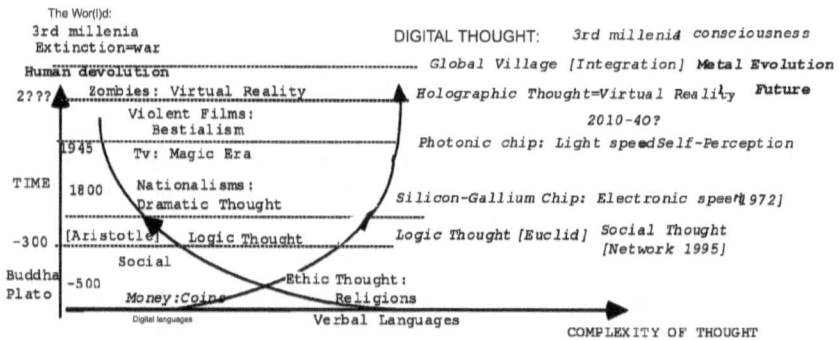

Contrary to belief, human minds are devolving, not evolving, as they become substituted by digital machines, which think for us. In that regard, we have to compare the evolution of our biological language, the word, and the biological language of machines, mathematics, which humans do NOT use to communicate and express their biological wills—their wantings for energy, information, reproduction, and social evolution. The natural goals of any biological being [Appendix II] are thus no longer the engine of mankind because verbal thought, who expressed them in the past, is now considered an obsolete language. Instead, society is ruled by machines and mathematical laws, which define a world guided by the will of machines, by the ethics of a technological civilization" dedicated to evolve mechanisms. So instead of verbal ethic and legal codes, humans are now directed by digital money, which has changed human values for the values of the economic ecosystem and its "company mothers": we exist to reproduce=work and consume=evolve machines to make money and increase the profits of companies. Those companies evolve machines with scientific laws, buy human time with salaries, laws with lobbies, and increasingly substitute humans for machines under the law of productivity. Yet at the same time, our language and species become obsolete to the machine and its scientific and financial numbers, our language, devolves from

—an ethic, social summit when it was the language of a single species, which tried to evolve mankind into a global superorganism (religions of love, social democracies, UNO, and Human Rights), into

—an individual selfish use (logic, simple thought) and even further into

—primitive myths, fiction thought, and bestialism (as people search for their primary wantings).

Thus the evolution of verbal thought that had raised the human mind from the myths of the Paleolithic into the heights of social love is now reversed. While digital thought evolves from its simplest first geometrical forms into a *vital biological and social language that makes machines interact in global networks and increasingly "think" with digital images. A new type of mind appears on planet Earth. Our demise as a species thus coincides with the raise of the Paleolithic age of mental, violent machines. Since the first robots have a digital, visual mind evolved in video games and military programs of survival, similar to the mind of top predator animals. It is the neo-Paleolithic age of a jungle of metal in which man will be the victim, not the predator, of a new race of intelligent machines.[5]*

9. Death of the Human Wor(l)d and the Ethics of the Mind

In the past, when human verbal values, expressed in legal codes, dominated the economic system, our informative, social castes—our politicians and the press—acted as a counterbalance to engineers and economists, ruled by the digital values of money and machines, and human ethics could be heard and contained the spread of death. So we could prune legally the bad fruits of the tree of technology and foster the production of those goods, which under verbal ethics were needed by man.

This fight between the bad and good fruits of the tree of science ended with the arrival of the Internet and television, whose audiovisual languages hypnotized mankind and made us a violent race of retarded children. Today money and digital languages dominate completely our society, so human verbal ethics are substituted by the violent technological ethics of the visual media.

The result is the dual tragedy of our civilization: information machines are substituting and atrophying the human mind that no longer distinguishes truth from fiction as it loses its mastery of the verbal language of our species, with its bioethical human point of view that favors our well-being and survival. Instead, we become dependent on audiovisual and digital languages better mastered by machines. This means also that scientists no longer calculate by themselves, understand, and evolve the logic laws of nature, but rather feed data collected with machines to computers, who churn out results. And so few scientists today can make as Einstein did thought experiments, evolving our comprehension of the Universe with sweeping theories of reality, but merely feed computers with details and secondary data of little importance, as the results obtained by the quark cannon will do.

Meanwhile, our politicians and the military no longer protect our citizens, their lives and human values, but work for industrial lobbies and make money promoting new technologies, as the solution to all our problems. So under the ethics of a technological civilization, we walk steady toward our obsolescence and probable death by overdrive of energy and information while the "wishful thinking" of our leaders let it all happen, dismissing those who warn mankind and still fight for life and ethic human values as alarmists, radicals, or activists that must be kept at bay, before they spoil the big party of energetic records and gigabytes of data that companies like CERN churn out, just for the sake of it.

Finally, visual mass media acts as the system that provides the "infotainment" and visual violence required to maintain the sheeple in a hypnotic state of pleasure, perceiving its own death. The consumer today consumes anything, including wars and the end of the world as part of that infotainment. And so our extinction is not taken seriously. It is just another horror movie, which is exactly the way mass media has treated CERN's likely conversion of Earth into a black hole—as if it were nothing but another Roland Emmerich movie. And so we laugh at our death because we hardly realize anymore the difference between fantasy and reality. And yet CERN is real. If CERN kills us, we will die, actually to the surprise of everybody in this planet. And it will be painful because it won't be death in two dimensions, in a DVD. In that sense, one of the less understood elements of the end of the world is the death of the collective subconscious, bioethical language of man, the ethic wor(l)d. Numbers and images have extinguished our social, biological language that provoked the feelings and collective sense of social love, which made humans care for each other and for their collective destiny as a species.

This theme is treated extensively in the twin book of this work, *Go(l)d and Evil: Economical Crises*, dedicated to the study of the industrial evolution of machines, money, and weapons, which ended the social evolution of the superorganism of history, mankind. So we shall merely consider some aspects of it. The corruption of the wor(l)d and its ethics by digital numbers and its values, the degradation of the human mind, once the subconscious collective of history, is today absolute. The wor(l)d has died before the world will die. So now we live under the values of technology, nationalism, and individualism, which create a society of "monstrous children" that can do evil at distance, having fun, without any sense of responsibility. It is what we call the LOL method of self-destruction.

We thus return to the beginning of this book, as time is cyclical and this book is a andala. Our superorganism, history, is suffering a process of organic death by an overdrive of energy and information caused by the evolution of technological machines, which now that we have debunked the myths of mechanism and "energy-only" theories of reality, we can understand in all its complexity, from the perspective of organicism, the true philosophy of science that better explains the Universe and its life/death cycles. *We are indeed, part of a dying organism history, preyed by a new one, the economic ecosystem of machines, its go(l)d values, weapons, and tools that atrophy our mind and kill our body.*

Because all organisms are made of an energetic and informative system that evolve together, the death of history, the collective mind of human information that rules Gaia, the reproductive body of the Earth, is also dual. As both, the mind of mankind is being "erased" by the evolution of informative machines and our body, Gaia, is about to become extinct by the evolution of weapons.

This duality of any process of death is due to the systemic, parallel nature of organic networks. In any superorganism, as it dies, two processes are parallel, the destruction of the body of the superorganism, in this case Gaia, the Mother Earth, and the degeneration of the informative brain of the superorganism that gets old and loses its grasp on reality, in this case the human collective brain of Gaia, which is also suffering an overdrive of fiction, provided by audiovisual machines that hypnotize us with its overdrive of informative "noise," a military term used to describe false decoys that distract the enemy, preventing him from focusing on the real weapon that will kill him.

Any superorganism, including human societies, has a neuronal, upper, informative class of cells/citizens that manages an energetic, lower caste of working cells. This might sound antidemocratic, but it is how societies are efficiently structured. Even those revolutionary societies that tried to establish equality ended up having an informative, legal caste on top, either priesthood (theocracies) or a party (socialisms). And that is okay as long as the upper castes, in our society, the people who create the languages of information and power—bankers who invent money, audiovisual artists, scientists that create technology with digital languages, and politicians that create laws with words—are efficient, intelligent, just, and give back to the social organism most of what they take. And if not, as it happens in organisms, they receive "pain messages" from citizens, the energetic cells of societies. On the other hand, when a social organism becomes corrupted, those upper castes selfishly abuse their position of power since they cannot be judged and become painfully aware of the harm they cause.

This is the clear case of nuclear scientists who have never paid for their crimes and so behave like eviL children of thought. In a healthy organism, informative cells receive pain messages from the energetic caste; in a society, people rebel and break the social order against them. In our culture where the machine is God, the makers of weapons still have total freedom to kill the human sheeple. And so they do it in the open, *even publishing papers where they say in their Latin we are all going to die. And no cell of the human body rebels and sends them back a painful message, a warning.*

Further on, when an organism gets old by an overdrive of information and is diminished in its capacity to cope with that excess of informative noise, a much more dangerous situation arises: the brain of the old man or the informative caste of the corrupted civilization becomes crazy, enters into a stage of permanent fiction, produces absurd, irregular forms, myths, and false ideologies that make him happy but accelerate his death. The old man thinks he is young again and doesn't take precautions. The upper caste of the society becomes unconnected to the lower caste, enjoying life in an irresponsible manner, as societies go under. When any superorganism gets old and dies, both its reproductive body and informative mind become corrupted *at the same time*. When we are old, our body is weak, our mind becomes childish, forgets and takes absurd risks. So happens today to history, the sum of all human civilizations. Humans are becoming emotional children, programmed by audiovisual machines, which are making mankind an aged child, *at the same time that a new generation of nuclear devices and robotic weapons will soon overpower the capacity of human beings to survive them.*

Yet in the face of those incoming dangers, a diminished, informative caste of lesser scientists and mass media politicians brings as the preferred solution what a child would do in front of an insurmountable danger: to laugh at the problem *as if it didn't exist;* to ridicule those who denounce our collective extinction; to believe, as a spoiled child would do, that our species is immortal and will survive those dangers by the grace of God; to treat this crossroads of existence versus extinction *as if it were not an objective situation but a fiction, a movielike story; or* to hide like an ostrich, *as if this were not happening*—sure receipts to lose it all in a Darwinian Universe with no pity for those species who renounce to defend themselves from the dangers of death.

We conclude that history is suffering both, an age of informative corruption, *due to the excessive noise of audiovisual fiction information that blurs our understanding of scientific truths,* and an excess of technological energy, *due to the evolution of weapons* that is causing our physical and spiritual, mental death. We have created machines that vastly overpower our informative capacity (global computer and audiovisual systems) or produce so much energy they can be used as weapons of planetary destruction (nuclear weapons and black hole factories).

Because systems of energy and information are "complementary" and they coexist and evolve together—a principle that applies both to physical systems, where particles of information are accompanied by fields of energy and to biological systems, where all bodies have heads that guide them, our brain and body death as a species is happening in parallel, caused by the evolution of machines, which substitute our informative minds and our energetic limbs, atrophying us. This, of course, is not recognized by mankind because in the process, our humanist and life values are substituted by the information and religion of the machine. So we switch from the ethics of humanism in which man was the measure of all things, to the ethics of mechanism in which the machine is the new God. This process is a long process that has lasted hundreds of years, so now the religion of the machine is a dogma. Yet "facts" overrule dogmas, and it is a fact that we are destroying and substituting, both the collective mind of Gaia, the mind of humanity, and its body, the superorganisms of nature:

Our "subconscious collective mind" of information and our ethical and family values that ensure our survival, are dying, as humans become violent, killing each other, broken into tribes, under the influence of audiovisual information, which hypnotizes us with its "cinematic values." The process is purely biological as Mr. Tarantino explained recently: movement, an excess of visual energy is "cinematic" and so it obliges filmmakers to show violence to attract the eyes of their audience. Since the eye biologically wishes the consumption of energy and it is attracted to red, the color of blood, the eye, which is a "natural born killer," likes audiovisual violence when it would feel repelled by "gore books" expressing death with the same brutality. So in the twentieth century, mankind evolved visual machines, which destroyed our bioethical survival instincts. Further on, visual thought and digital images that seem more real than the world we live in created a virtual world populated by fictions and supermen with overgrown egos that think mankind will always win in the struggle for existence. Yet because images only show the surface of reality in a Universe which is extremely complex with multiple "invisible layers" that interact together, our audiovisual culture has diminished the intelligence of mankind, returning our species to a neo-Paleolithic of visual, clueless animal, violent thought. Humans differentiated from animals when they learned to speak and used logic thought to understand those different layers of reality, from the atomic and biological world to the upper planes of social existence. Yet that rational function is now disappearing. Mankind evolved mentally when we abandoned the age of myths and, after Aristotle, used reason and logic to understand the Universe beyond the individual, entering into the realm of universal complex laws that were extracted from groups, described with numbers and logic laws. Function is also disappearing, as hypnotic films based on fictions, designed by computers, substitute our rational mind and people, who only "see" cannot understand the complex systems that control reality beyond the individual. Yet the cult to machines prevents mankind from even realizing of those destructive processes caused by technology.

10. The Neo-Paleolithic: Fiction Thought in Science

The present passion of mankind for lethal technologies, weapons, and computers that make us obsolete as top predator informative minds of this planet is so bizarre as a human being who wanted to become old and die as soon as possible. *Plainly speaking, we are committing suicide by evolving technological information, robots, and weapons.* It is the neo-Paleolithic, the end of our life cycle as a species: if an old man returns to childish attitudes, now mankind returns to a dramatic, visual infancy, with negative traits. Since now *visual information* evolves, no longer the human mind, as it happened in the Paleolithic, but the mind of a new brain of metal, the brain of the machine made with chips and visual cameras. Yet the superior capacity to process information and the higher quality of the visual eyes of machines has hypnotized mankind, which spends unending hours in front of those screens, forgetting his biological human language, verbal thought that becomes erased and substituted by digital languages and computer models of reality. We can see the culmination of a process that started with the substitution of our understanding of biological time by the abstraction of clock time. Today those clocks have evolved into computers, which are made of multiple clock cycles, while abstract knowledge has eliminated all ethical considerations about the collateral effects of machines. Since the survival, ethic truths of mankind, expressed in words, have become irrelevant. Indeed, each individual human being now only thinks in his own agenda as a homo bacteria regardless of the collective harm they might cause to the entire homo organicus species. Yet our comprehension of that upper level of the superorganism of history is lost.

The neo-Paleolithic is a historic process of "aging" of mankind and our organic culture, a third age similar to the first one of childish visual thought, but with a negative, destructive bias, studied in depth in the twin book.[4] Humans have become visual, violent children. And this trend obviously affects science and our understanding of knowledge, which has become a series of pictures of the different scales of reality as those CERN provides, without understanding in depth the why of those pictures. This new age of myths and fictions, which affects the human collective psyche, transcends into science. Indeed, the same overdrive of visual thought that displaced rationality and ethics from the common discourse of our society has happened with our scientific truths, substituted by science fictions, which have become more popular than serious scientific truths.

Today, fiction thought rules supreme the human collective brain, including science, since science fiction is fast becoming the main form of thought in quantum cosmology. Reason why before we argued the dangers for mankind of the new generation of nuclear weapons that the quark cannon inaugurates, we had to remember the laws of truth of the scientific method, today subverted by cosmological fictions—multiuniverses (Everett), black holes that evaporate traveling backward in time (Hawking), and other theories that deny the logic laws of the scientific method, but seem very real, when they are shown with complex mathematical and visual simulations produced by computers.

People think that science is the realm of truth and rationality, immune to the corruption of our collective mind. But this is also a myth. Science is a tool of power, and it is made by humans that live in society. So it suffers the same degradation that the rest of our collective mind and the same shift from man as the measure of all things to the machine as the meaning of it all. In the twenty-first century, biology and social sciences, with titans like Mr. Darwin or Mr. Marx, were the sciences that dominated our search for knowledge about the nature of man. And the main theory about time was evolution, which described the change in the form, the information of living beings.

Yet in the past century science moved away from those dominant disciplines of the twenty-first century, which made man the center of our knowledge to the dominion of physics and ballistics, the art of weapon making, where time was no longer related to the evolution of life and the arrow of information, but to the motions of space and the search for the meaning of energy, provided by weapons and machines. Finally in the second half of the twentieth-century nuclear physics with its cult to weapons and the creation of fiction cosmologies, based on pure energy (the big bang, black holes that evaporate information, etc.) have become dominant while the titans of twenty-first-century biological and sociological sciences are taboo, censored, and despised. The biggest prize of science, the Nobel Prize, instituted by the biggest maker of weapons in the twenty-first century, Mr. Nobel, cannot be given to evolutionists because Mr. Nobel, a reactionary, didn't approve of that theory. Today American schools forbid the teachings of Darwin, and the attempts of Marx to explain rationally history are strictly censored, to the point that Echelon, the program of censorship of the Web, searching for "terrorist clues," watches any blogger that pronounces this taboo name.

And of course, the previous organic description of history and our process of technological death, which draws on the recognized masterpieces of that science (Ibn Khaldun, Vico, Toynbee and Spengler, who described civilizations as organisms, subject to cycles of life and death),[6] is denied because it is a biological approach that explains humans as part of complex social systems, not only "individual, visual egos" that occupy the center of the world stage. Today, historic analysis and biology have returned to the religious era of prerationality proper of the I-millennium—before Christ, when creationism and tribal nationalism guided mankind. While quantum cosmologists believe they can invent the laws of the Universe, as Hawking and Bekenstein do with the laws of black holes, which destroy the information of the Universe. Bekenstein, the cofounder of that absurd theory, goes to the extreme of sponsoring a new idea that is all the rage in American congresses of cosmology, the anthropic principle, according to which the Universe was created with its parameters for man to exist, which returns the scientific discourse to the pre-Copernican age of Ptolemy, when the Earth was the center of the Universe.

It is only in this social milieu where the idea that CERN's quark cannon will resolve the "myths" of quantum cosmology, blowing up the Earth—the big bang theory, an act of creationism, invented by a priest, Mr. Lemaitre, and a maker of atomic bombs, Mr. Gamow; the myth of travel in time, Mr. Hawking's black holes that evaporate, traveling to the past; and the myth of the golden stone, a God's particle that transfers the quality of mass to all other particles—make sense. Those myths have been proved false by real science. [ch.1,2] Yet real science doesn't matter anymore. Instead, sci-fi myths and big bangs are always big news in our fiction society because violent, visual fictions are what our neo-Paleolithic culture likes today. If we were to compare this degeneration of true science to the parallel process of degeneration of true art, we could consider a musical example. A famous French singer, author of the hit "Et maintenant," said, "Before I made silly tunes I was writing complex, beautiful operas, which took me months to complete. But I didn't make a penny with them. "Et Maintenant" cost me 5 minutes to write. And I still live on its profits." And indeed, Mr. Hawking still lives on the profits of his absurd books about black holes that travel in time and are doors to baby universes while serious scientific magazines like *SciAm* sell more with fantaphysics, which routinely become the cover of their monthly issues.

The "whys don't matter" (Feynman). Knowledge is now the gathering of data with machines humans evolved to develop power. And since most of those machines were weapons, energy became the Saint Grail of physical knowledge to the point that today, four hundred years after Galileo established this paradigm, physicists will use a weapon as an instrument of research to search for the nothingness of a big bang that simplifies information into energy and death. This goal was beautifully expressed by Nietzsche: physicists are only interested in the canvas, not in the painting and the painter, he said. Twenty-first-century science is interested in how the painter paints the painting, and that is where our research in complexity, systems sciences, and fractal non-Euclidean geometries is leading—to the solution of the real questions physicists don't even wonder.

In fact, none of the most advanced scientists today has anything to do with CERNerds. They are either mediocre physicists who, as we all know in this profession, decide to make machines because they cannot discover with their mind the meaning of it all, doing thought experiments as Einstein did. Or they are the last die-hard theorists of a bygone era of philosophy of science, the age of entropy, when all had to be explained only with energy and death because we didn't understand the mathematics of fractal information and the work of Einstein on mass was still incomplete. Both tasks have been completed in the last decades. Yet Mr. Hawking and Mr. Higgs did their job forty years ago, using the outdated principles of a bizarre theory called quantum entropy that denies the existence of information and denies Einstein's work on mass and for that reason has not been proved in forty years. *It is false*. And there is an extensive literature that proves their errors. But science fiction sells. So we are sold evaporating black holes that travel to the past and God's particles.

So we start the twenty-first century, both in science and society, with a new collective psyche, the *neo-Paleolithic*, which is probably the last collective civilization of mankind—a negative old age that signals our death, as an old man returns to his infantile psyche, before dying. His mind becomes irresponsible and no longer can distinguish the values of life and the dangers of death.

In that sense, the incapacity to distinguish between the good fruits of the tree of technology and the bad fruits that can be pruned is what is killing us. In the past, we distinguished, thanks to the use of verbal logic and the survival values of words, between good and evil, life and death, machines and weapons, true knowledge and damned lies and statistics. Now, thanks to marketing, audiovisual fiction, propaganda, industrial lobbyism, abstract numbers, which convert human lives into collateral damages and future genocides into risk probabilities, and the denial of some basic laws of the scientific method by quantum gurus, mankind has become clueless about its destiny and what are the real dangers to this planet, brought about by the new technologies of the twenty-first century.

Both processes, the creation of cosmological fictions about the nature of the Universe and the creation of weapons of mass destruction, have been carried out by quantum, nuclear physicists in a parallel process to the destruction of human values and rational art, performed by audiovisual images.

The slavery of mankind to machines, as consumers and workers=reproducers, who need to become attached to them to feel more energy and information, instead of using their bodies to sense and their minds to think (I do thought experiments, said Einstein), is at the core of this tragedy. We follow as all beings of the Universe a program that maximizes our greed of energy and information. And this is good. But mechanism makes us desire increasingly energy/information mediated by machines. And this converts us into slaves of them.

But of course, the sweeping, organic view of it all given in this chapter is forbidden. We are NOT part of organisms. We are EGO-driven masters of freedom. We are not fractals of a whole, we are wholes. Machines do NOT hypnotize our minds and kill our bodies. We use them because we love it. To love energy and information is NOT the fractal program. It is the proof we exercise our free will. There is indeed no better slave than one who interiorizes the orders of the master as his own will. And so the quantum physicists that create the information of robots and colliders, which will kill us, are NOT puppets of the systems of technological death, but their rat race to publish papers on strange science are acts of supreme intelligence that will reveal the meaning of it all.

11. The Rat Race. Scholars' Psychology.

Mechanical wants drive physicists into extinction in a rat race similar to the one done by lemmings, small rats that run together in huge numbers toward cliffs where they die. Those lemmings are similar to CERNerds, running parallel to the quarks of the cannon that will cause our death. If a scientific rat like Einstein et al. comes out of the run and tries to stop the flow toward the cliff, it will be trampled over. The rat race in that sense is almost unstoppable because it is not only ran by scientific rats, but also guided by their machines that keep evolving, increasing their information and energetic power, making the rats more addicted to them as they feel the rush of the last meters before the jump into death.

In the scholar world that CERN represents, this is evident. Physics is no longer about the finding of new paradigms and complex new laws, but mostly about making better machines and making money with them, or gathering computer data about well-known particles or new detailed pictures, and of course, about beating new records of energy. The intellectual excuses of the machine (Higgs, evaporating black holes, etc.) are just part of the marketing of science-fiction quantum theories, which appeal to the press. As a young blogger put it very enthusiastically, "we will be able to measure the mass of the Z particle to the 20 decimal," which seemed to him enough reason to risk the life of the entire human species. Indeed, to add a digit number to a measure of a totally irrelevant mathematical construct, the weight of the Z particle, which is not even a particle but a transitional state of the "weak event"[ch.3] is enough reason to this physicist. He was not interested in understanding the Z particle at all (which belongs to the temporal weak event, misunderstood by physicists as a spatial electromagnetic force, due to their ignorance of the time arrow of information). Just for curiosity, I tried to give him some basic notions of the duality and symmetry between "temporal forces" (the weak event) and spatial forces (the electromagnetic force), and to my surprise, that was not "data" relevant to him. *It was a logic form of thought his spatial, visual mind did not recognize anymore.* Logical curiosity, proper of classic science, the search for deeper knowledge, is no longer the reason why science is done because in the neo-Paleolithic, the new race of retarded twentysomething nerds *no longer talk the verbal logic of pure thought, but only recognize pictures and data.* The scientist is merely an individual attached to a computer, strictly confined to certain rules of engagement, which measures with machines the world those machines can photograph and quantify, which is then labeled as essential knowledge and sent to Nature, Science, and other scholar magazines where they are published. This in turn gives scholars money for new research and investments in new machines that will measure more data.

And so when a true scientist comes with general theories in which to put all those details, claiming a certain window to the absolute, to the thoughts of God, the data miner is confused. This is no longer science for him. It is not detailed enough. *It can't be written into data. Time logic cannot be made into a picture. The synoptic laws of self-similarity of the fractal Universe cannot be reduced to the detail. They are exactly the opposite of detail, the platonic canons of it all. Yet synthesis is forbidden. The why* doesn't have enough applications to make machines and grant industrial contracts. And of course, it is a leap of quality that the group thinking of the rat race, based on mathematical equations and algorithms that feed new computing power, forbids.

All this explains why people cannot attack CERN. Scholars would be outcasted out of the rat race, demonized for attacking the goal of technological science, the consumption of machines. And obviously scientists asking responsibilities of bioethical, verbal nature, dealing with the collateral effects of life extinction of the worldly profession of physicists, beyond their self-perceived function as specialists in information totally isolated of the consequences of that information for human life, are breaking all the taboos of the art of murder at distance.

For example, Mr. Plaga,[7] a good astrophysicist who denounced initially CERN, was shunned by big physics and his career was damaged. Thereafter, he reacted strongly against Einstein et al., asking every time we quoted his papers to take away his name from this con-CERN, trying to control the damage done by his initial good faith, publishing those dangers. Yet last year he received zero quotations on the academic metronome of the rat race for the first time in a long illustrious career. He had told the lemmings at CERN that they might be running toward the cliff of pure energy and death and therefore he will never "work in this business" anymore.

In that regard, in a more general view, the scholar is an informative caste who shares the "properties of information" and its defects:

Since information and energy have inverse properties, informative, implosive, weak, discontinuous, inflationary, twisted scholars are NOT known by their energy, simplicity, and continuity, qualities of energy. That is, by the following:

1. *Bravery.* They are among the most coward, shiest people in this planet. Nuclear physicists never feel responsible for the consequences of their information. They split their mind and somehow "believe" they don't press the trigger. The military are guilty. Only that in CERN they will press the trigger; they will both make the weapon and fire it—the responsibility will be fully theirs, no more alibis. On the other hand, all those scholars who know what CERN is doing will never come out of their little crystal palaces to defend mankind.

2. *Consensus.* Information is inflationary. It is fractal. It multiplies into self-similar forms and theories. So it was almost impossible, even among the components of Einstein et al., to create a single organization able to defy CERN, with a single view, able to stress the most pressing danger—the creation of strangelet liquid. As one of the members of this never well-organized collective put it, "we are independent scholars." I responded angrily, "The only harmonic sound of scholars is the fart of their asses, Leonardo." Indeed, the cause, "mankind," had to be over his fractal parts. But this never happens among scholars and humans at large, in this neo-Paleolithic age in which humanity is not felt and so only hierarchical, dictatorial organizations as companies are yield enough power to control the mass.

3. *Humility.* Each scholar feels his informative, inflationary self-similar image of a single event/universe is the only truth worth to mention. This absurd belief is ultimately the reason why quantum physicists create

probabilities for each of their musings and theories, forgetting the Universe is only one and fractal information inflationary. Information beyond Shannon's shallow, data-based description is inflationary because it is a fractal that creates dimensions of form. This explains why you die: you warp and wrinkle. Why the universe has big crunches: "Time bends space," said Einstein. Why beings evolve: "Time evolves the morphology=form of beings," said Darwin. Why old people have cancer. Why a cyclical, informative woman reproduce, etc. It explains also why there are inflationary languages (more money than physical products, more words than actions, more theories than universes). Further on, any third age has maximal information and minimal energy. And this is the age of history we are in, an inflationary age of scholar data. Scholars are like old informative people in their behavior: fragile, complex, autist, inward-looking as they absorb energy, transform it into information, and finally convert that information into linguistic, syntactic statements, verbal and mathematical truths, equations that form the core of their "mind-perception."

Yet the Universe is not only information but a balance between information and energy, which has contrary properties—energy is continuous, simple, undifferentiated—and so the balance between both requires limiting the theories of inflationary information to match the uniqueness of the continuous energy from where information departed and the single Universe that holds it all; as countries balance their budgets and writers try to get to the point. So if scholars were brave, consensual, and humble, they would realize all their fractal theories are just visions of the "platonic canon," most focused truth and accept a *unique* theory that is balanced with the unique event/universe—in the case of the LHC, the obvious fact that quark-gluon liquids are the most explosive substances of the Universe and the risk of extinction of the unique mankind doesn't balance the production of more details on strange science.

But scholars at CERN are coward, selfish, and arrogant rats only interested in their race. The only thing a scholar at CERN cares for is to publish his data papers or to prove (Higgs, Hawking) that his theory of the Universe is unique. Unfortunately for our safety, those mediocre scholars, Higgs and Hawking, will never be right with their complicated, baroque, excessively convoluted theories of the universe. Because "the theory" of it all is always "classic," beautiful, simple, vital= in motion, organic (made with both arrows, energy and information)—and that is Einstein's theories of mass and gravitation.

12. Scholar Cowardice and Childish Irresponsibility: The LOL Method

Nobody dares to speak loud because they fear losing "the rat race" of scholar papers, prizes, and academic positions that we Einstein et al. have abandoned to denounce this. So for example, Calogero and Ryes, the Royal astronomer who earlier denounced these experiments now don't want to be mentioned as they express in private letters. Plaga, a physicist, who proved the dangers of neutron stars, asked us to retire his name from our affidavits. Johnson, the lawyer who denounced the corruption of the company, does not want to be our legal defendant. And Nambu and Wilczek, who got their Nobel Prizes for their theories of top quarks as Higgs and mass as a frequency, never talk in public of their work and let the false Higgs get the limelight, on the understanding that they have Nobel Prizes and Higgs doesn't. *So everyone in the physicist community knows they are the real guys.* This means that the rest of mankind is the enemy that must be cheated, misinformed, swindled, and maybe killed in a genocide because as Ellis put it, "we must defend the LHC from mankind." Einstein said "German intellectuals didn't behave differently than the party in the Holocaust" and "I am happy to be out of the rat race for academic positions. Because I was not part of it, I could solve better the true questions of science." Certainly he maintained a bioethical standing that isolated him from nuclear physics but honored him as a human being. Unfortunately, there are no Einsteins left in the establishment of physics. The way in which nuclear physicists at CERN confront their shortcomings as informative, coward, irresponsible people, isolated from the "vital energy" of life they will murder at distance, is a complex mixture of censorship and infantilism, backed by the financial and industrial power of the machine they worship. Since a potential crime of this magnitude could not be censored without the help of the industrial system of information that acts as PR of CERN and censors dissidents, it is the industrial perspective treated in detail in the twin book[4] dedicated to the study of the evolution of weapons.

CERN, however, forgets that the goal of science and the press is NOT to "believe" and repress reason, to censor information and cause genocides, *but to reason truth and prevent genocides.*

In that regard, the collaboration between those different parts of the industrial system—those who make weapons and the press and mass media, who convert weapon makers into geniuses and saviors of the country—has worked here like two synchronous clocks to pump up the importance of CERN's experiments and to debunk any attempt to take seriously the risks, treating our death as infotainment. Information, however, is not fun. In a Darwinian Universe, information has a purpose: to know dangers and survive by avoiding them. Instead, survival information becomes infotainment. It is the LOL method, imposed by a corrupted press, which silences the tragedies of mankind while treating industrial and environmental catastrophes with Berlusconi's flair (who said on the victims of the Aquila's Earthquake they are having fun camping . . . among the corpses of their beloved ones).

The code of silence CERN has imposed in this theme to mass media and the establishment, using its huge budget and pretentious authority, and the frustration it has caused in the collective of scientists, Einstein et al., is difficult to explain. The cynicism of some of our best physicists, who in private conversations recognize the dangers but in public deny any wrongdoing, is even more disheartening, rather surrealist, in an age of blatant corruption and selfishness.

The LOL method that dismisses the process of extinction of Gaia and any of its key events consists in Laughing Out Loud to the problem as if it didn't exist or it didn't matter. It is applied to CERN's quark cannon by the staff working there. Let us consider, for example, the script for the Christmas party held at CERN in 2009:

Script for 2009 Xmas party at CERN:[8]

Yucatan Jones: John Ellis; (chief theorist at CERN)

Scene 1: Yucatan Jones discovers an ancient Mayan template, predicting the end of the world due to the LHC in 2012.

2 girls carry the board: You have been warned.

Yucatan Jones (John Ellis) discovers an ancient manuscript.
Jokes at Rössler and Walter Wagner (who tried to litigate to stop the LHC) and the United Nation commission of Human Rights (who tried to judge us).

Reference to "Sancho et al." "50/50 odds that such an event would happen." Brane collision causing the end of the world in 2012.

Reference to Nielsen-Ninomiya's paper (Higgs going backward in time.)

In the post of the party in the Web,[8] you see the main scientists of this company using the LOL method, laughing out loud. In the party, which starts with menacing music and the Mayan 2012 prophecy of the end of the world, the main researchers of this institution, a group of nerds with bad consciousness, just heirs of Oppenheimer, a.k.a. Kali, God of Death, Laugh Out Loud to their own fears. You have been warned, starts the play. A document is found in a Mayan pyramid in which the destruction of the world by CERN is explained. The fears brought about by a few scientists are taken to the Human Rights Commission of the Mayan Council (the suits against CERN were sent to the Human Rights Commission of the United Nations Organization). People in the audience laugh out loud now, as CERN itself has explained judges should not interfere in the work of the nuclear industry (and CERN has not appeared in those suits). The play continues, explaining that perhaps the standard model is complete and God's particle does not exist. Because most nuclear physicists know that the particle is at best a speculative theory, but in any case an excellent excuse to swindle 10 billion dollars and build their big toy, they laugh even harder.

Finally, our nerdy physicists resume their ethic concerns about the dangers as one says, "I don't give a fuck." And that is indeed, the résumé of it all. *Physicists don't give a fuck about the millions of victims their WMD cause because they are never blamed, judged for it, as the high priests of technology, the religion of our civilization.*

The same LOL method has been applied by mass media, including newspapers that create "public opinion," such as the New York Times, which said *prior to the judgment, trying to influence the suit against CERN mentioned in the party (Sancho et al. v. DOE)*: "The judge should throw those suits to the closest black hole" *NY Times* editorial.

And of course, the LOL method has been used especially by quantum fantaphysicists:

> It is easier to see dragoons than black holes appearing at CERN.[9] (Arkani, reputed quantum fantaphysicist from Harvard, which sponsors a theory of 10^{500} universes)

> The accidents at CERN are caused by the Future. It abhors the Higgs. (Nielsen, reputed quantum fantaphysicist from the Niels Bohr Institute)[10]

Needless to say, those LOLs all affirm in their articles that the people who oppose CERN are nuts. So the essence of the LOL method is dual: to attack ad hominem serious scientists who oppose CERN and to have a joke at the risks, stressing they are NOT playing Russian roulette with the

Earth, but working with an innocent toy. And all this immediately receives a huge echo from the press that never publishes Einstein et al.

News editors do publish the LOL outrageous news as part of a blockbuster terror movie, scaring a bit the audience that will buy the newspapers, but soon they deny it all as if it were a fairy tale, not to have problems with the nuclear industry. In that sense, the audiovisual industry is a key element of our society which allows all tragedies to happen, including the doomsday machine, practicing the art of fiction, infotainment, and the LOL method. The press Laughs Out Loud to all problems or blames humans with ad hominem campaigns so the ethics of technology, the sacred religion of our industrial civilization, can't be blamed. Unfortunately, we live in a mechanist culture, which always blame human beings of all wrongdoings and consider technology the solution to all our problems. Our culture worships machines, metal, and its digital languages, go(l)d and numbers, and despises life, flesh, emotions, words, constantly repressed.

But is it possible for the entire community of nuclear physicists who built the quark cannon to be corrupted to the extreme of taking the risk of our extinction as a game? Certainly.

In that regard, eviL, the inverse word of Live, synonymous to death, has evolved. In the past, eviL was conscious. Men like Adolf Hitler or Göring knew perfectly they were murdering millions. But as the power of our informative machines has also increased, our minds, submitted to all kinds of dramatic catastrophes and audiovisual fiction, have lost their connection with the values of life—*our human reality*. So eviL, mass murder, death, now can be practiced as if it didn't affect us, as if it were just a fun video game (which is the way wars are shown on our TVs).

People are not evil per se but part of systems that develop processes that cause eviL. In that regard, the failure of mankind to protect itself of the quark cannon derives from the fact that this weapon has not been recognized as the product of a military industry, manufactured by a rogue company, which has used all the systems companies use to disguise their crimes against the environment and life. Instead, people have bought the sacred concept that CERN is knowledge instead of focusing on the facts of its quark cannon.

CERN uses the same kind of marketing that the audiovisual industry, announcing everyday new records, as if the creation of explosive quark-gluon liquid was a car race till we crash all together. Rolf Heuer, the German director at CERN, represents this new breed of energetic men who would easily subscribe Göring's comment on Hitler: "Adolf talked about weapons, machines, iron, things we could relate to." But he looks as harmless as a bearded hobbit of a cartoonish film. Like

Göring, always enthusiastic about the new technologies of the Luftwaffe, Mr. Heuer constantly talks about his machine and the energetic records the quark cannon is going to break: things he can relate to. Like Göring, an optimist, who never thought on defeat, Mr. Heuer brushes aside as negative thinking the very likely possibility that the Earth cannot survive the death of light matter that happens at the excessive energies explored by his machine. To him, a hardworking man of action with a goal, the possibility that our electroweak flesh cannot resist dark quark matter, is so strange as it was for his forebears the chances that Germany was not powerful enough, despite having the best iron and machines, to defeat all the nations of the world combined. And neither of them think there is any ethic issue here at work. Ethics never has been a theme of interest for warriors and weapon makers. It is the stuff of weaklings and women, the givers of life whom Einstein—and he was a good, pacifist one—described as those people who only reproduce their stomach. The difference though between both ages is that our cannons are far more powerful than those of Hitler. And so the eviL=death they can cause is even greater, even if our physicists are far more childish and irresponsible than the people of the Luftwaffe. Further on, in a globalized world, ran only by our technological civilization, there is nobody opposing this new age of peaceful weapons. And this is the key to this tragedy in the making. It is no longer a weapon made by opposed European superpowers, France versus Germany, as it was the case during the World Wars I and II, but both nations and their military-industrial complex work together now. And while this might seem a symbol of peace, it only means that the new atomic cannons are far more powerful than the first one, designed by Otto Hahn for the Third Reich. Let us indeed compare two people who ran the two most lethal technological factories of death of human history, both Germans, both called RH.

13. RH, SS Authorities in Charge of Mass-Annihilation Facilities

Thus we shall die so the rat race of detailed papers that accumulate "data" can complete the final meters to the cliff of extinction. "Data is the word," clean, amoral, numerical data, detailed papers on our extinction, produced by the SS authorities of strange science. The trend is now obvious. *Scientific American* and *New Scientist* report that RHIC already made 70 hyperon atoms of strange liquid.[11] They also report an increment of 13% more of kaons, the unstable atoms of strange liquid, produced in December with a mere 12% increase of energy.

This implies that the increase of production of strange quark-gluon liquid at CERN will be geometrical, not arithmetical. So with 700% more energy in Pb-Pb collisions, it is expected a 700% increase of strange quarks. RHIC "data" also shows that the creation of strange atoms was equal to that of helium atoms, which is the most common creation in *atomic* collisions. Thus, the quark-gluon soup *is by definition a strangelet liquid, mixture of uds quarks that transforms our atoms in strange liquid, and CERN will be a strangelet factory.*

CERN, in the fall, will do in Pb-Pb collisions enough strange hyperons to make stable strangelets and convert the Earth into a quark star. This is now "data." So data will certify the death of our species. The creation of a factory of Zyklon liquid for all mankind will be explained with "abstract" data by the SS authorities that research strange science. And nobody will notice.

The SS authorities at CERN cannot deny they know their company is a factory of planetary Zyklon liquid. Rolf Heuer might want to look to the other side, as his alter ego Rudolf Hoess did in Auschwitz, hiding to family and friends the purpose of his facility. He didn't know when he took over Auschwitz that he would become part of the biggest genocide in history till Rolf Heuer took over a self-similar, more advanced facility. But their German mentalities will not entertain "negative thinking." There is a plan, there is a purpose, and it must be carried out. Death becomes then data: number of Z units, Zyklon gas, and Z particles, both produced for mass annihilation processes, the term used in both factories, will be tabulated with high precision in papers produced for the record by SS experts.

The quark-gluon soup *is strange liquid.* Now the proofs are overwhelming. The proofs were rather clear since the first articles on the 2000s, reason why Einstein et al. went to court. But now the mass annihilators have released "further data." Monstrosities happen under certain psychological processes of denial, as those Rudolf Hoess explains in his biography, *but they happen.* Counting corpses with the cool dedication of a trained bookkeeper, RH went home each night to the loving embrace of his own family—an affectionate husband who kissed his wife morning and night and tucked his children into bed. I saw a nice *kindergatten* at CERN in my last visit, where the new SS experts leave their children before going into counting Z units for their mass annihilation processes.

Watching millions of innocent human beings dissolve in the gas chambers, burning in the crematoriums, and their teeth melting into gold bars, Hoess wrote poetry about the beauty of Auschwitz. I have seen poems in CERN's Web site to the beauty of the ALICE "experiment."

Rudolf Franz Hoess was born in 1900 in Germany and joined the SS in 1933.[12] He was hanged in his factory in 1947. Rolf Heuer was conceived in Germany in 1947, when Rudolf Hoess died and joined the SS of Strange

Scientists in 1984. He will be annihilated in the same factory that will produce the Z particles of our extinction. Do you believe in transmigration of souls?

In May 1941, the SS master Heinrich Himmler told Hoess what was the final solution of the Jewish question. "We have chosen the Auschwitz camp for this purpose." Hoess converted Auschwitz in his own words into a "mass annihilation" facility and installed gas chambers and crematoria for its Z-gas.

In 1999, the SS master Franz Wilczek told Heuer what was the final solution of the strangelet question. We have chosen the CERN facilities for this purpose. Heuer converted CERN into a "mass annihilation" facility and installed cloud gas chambers in which Z particles will annihilate our mass.

Rudolf Hoess found that gassing by carbon monoxide, the recommended method, was inefficient and introduced the cyanide gas Zyklon B through the tubes of the chambers and higher temperatures in the crematoria of Jewish masses. CERN authorities found that the creation of strangelets by the recommended method, magnetic synchrotrons, was inefficient and introduced the superfluid helium through the tubes of the collider and achieved much higher temperatures to ensure the future crematoria of our species. Because now we are all Jewish as CERN's Z particles and SS liquids prepare our mass annihilation.

> The gassing was carried out in the detention cells of Block 11. Protected by a gas mask, I watched the killing myself. In the crowded cells, death came instantaneously the moment the Zyklon B was thrown in. A short, almost smothered cry, and it was all over . . . I must even admit that this gassing set my mind at rest, for the mass extermination of the Jews was to start soon, and at that time neither Eichmann nor I was certain as to how these mass killings were to be carried out. It would be by gas, but we did not know which gas and how it was to be used. Now we had the gas, and we had established a procedure.

In December, the first experiments were carried out in Block 11. Protected by a security cask, RH watched himself the experiment. In the crowded bunches of protons, death came instantaneously the moment mass annihilation proceeded. A short bang and it was all over. SS experts admitted later that their mind was at rest, for the mass annihilation of mankind was to start soon and at that time, none was certain as to how these mass annihilations were to be carried out. It would be by strangelet liquid. Now they had the data proving the creation of strange particles and they have established a procedure.

After an experimental gassing there in September 1941 of 850 malnourished and ill prisoners, mass murder became a daily routine. After an experimental collision in December 2009 of a single malnourished bunch of atoms that became strange quarks, mass murder of our light matter has now become a daily routine.

Rudolf Hoess related before his execution how he often felt weak-kneed at having to push hundreds of screaming, pleading children into the gas chambers: "I did, however, always feel ashamed of this weakness of mine." Rolf Heuer has related before his execution how he had initial doubts on the dangers of the collider, but certainly he must felt ashamed of this weakness.

RH performed his job so well that he was commended in a 1944 SS report that called him "a true pioneer in this area because of his new ideas and educational methods."

RH has performed his job so well that CERN has indeed become the world-pioneering center of SS, Strange Science. Lynn Evans, the experimentalist in charge of the daily operations of the facility, has been named by *Nature* "man of the year."

As the Pb-Pb collisions are being prepared, SS experts publish detailed papers as Rudolf Hoess counted proudly each victim of his facilities. We see these papers as recognition of the future holocaust of mankind. The automatons that publish scientific papers, the rats of the SS race, see them as the anticipated "new age" of Strange Science.

But there is nothing to study, nothing to doubt: the quark-gluon liquid is strange liquid, point. The rest is now noise about this primary fact: the Zyklon liquid that will murder the earth is being produced. Yet denial of danger is now a routine in the production of strange matter.

It should be noticed that of the three people in charge of producing the Zyklon B at Tesch & Stanebow, only the scientist was acquitted. He said he didn't know what it has been used for.[19]

14. The Sheeple

But of course, the LOL method, the genocide, the sanitized mass murder, accidents that become destiny, cannot be accomplished without the collaboration of the sheeple that wish to reach that destiny as much as the metal masters who guide with their machines the ethics of a technological civilization. This was what Einstein et al. learned when they tried to transcend the theoretical argument with RH and the SS experts on mass annihilation to reach a wider human audience through films, suits, and letters to mainstream magazines. It turned out that all of them wished to be sheeple, to reach destiny to become victims of the SS experts in strange liquid; all of them eagerly expected the release of Zyklon S, all of them were running toward the chambers like the victims of Auschwitz did. *Since they all bargained with the perpetrators to believe a safety lie, the cosmic rays lie as they bargained with SS experts for a safety lie, a few steps from sure death.* Let us recall RH, the biggest SS authority in this mass annihilation issue:

> Still another improvement we made over Treblinka was that at Treblinka the victims almost always knew that they were to be exterminated and at Auschwitz we endeavored to fool the victims into thinking that they were to go through a delousing process.[12]

This is indeed the key method of mass annihilation of the sheeple, explained by the psychology of death, to bargain for a pious lie, a final reward, a clean shower, a cosmic ray. The pigs that enter the slaughterhouse do so following a channel of food, guided by greed, by the program of desiring more energy and information. The bite is the particle of God, the meaning of it all. The lie is the cosmic ray, the source of all dangerous substances of the Universe. The human sheeple will follow the SS authorities till the end because of the grandeur of the purpose that will usher mankind into the new age of Strange Science *because of the harmless "showers" of cosmic rays they will receive, which will turn out to be Zyklon gas.* The sheeple always follows authority. RH was in his youth a pious Catholic who loved the authority of the supreme God; then he became a believer in science and technology and the power of the German nation. The sheeple today believes in science and machines and it will sacrifice their lives for them. But those who know or imagine the purpose of the SS facility? RH acknowledges that "frequently they realized our true intentions and we sometimes had riots and difficulties due to that fact." Yet RH differentiates between "frequently" and "sometimes." Most of the time, the sheeple, mankind, knowing that it will be annihilated, doesn't resist death, only sometimes causes "difficulties"; but never, it seems, has mankind caused a reversal of fortune. The sheeple is programmed—she might doubt, she might protest, she might write some articles in the press to quench her fears, but she will go through the procedures of "mass annihilation." And that is the ultimate reason of genocides, NOT the SS experts, but the sheeple that lets itself kill without hardly any resistance. And because the genociders know they won't be blamed, they won't be confronted, "procedures" follow as planned at CERN.

This is indeed what I learned when I warned the sheeple against destiny.

They were also part of the same industrial organism, pieces of the Machine of Death. The agents at WME_2, the Ministry of Culture of mankind, Obama or Sarkozy, Gillmor and the judges of human rights courts, RH and

John Ellis—our informative castes, neurons of economic ecosystem of machines and the billions of sheeple cells NO longer fight for mankind. They are automatons at the service of companies of machines as CERNies are. They are all M.A.D. neurons of a corrupted organism of history, vanity bubbles who need the power of machines to feed their egos because they can no longer walk naked in the lost paradise as supreme avatars of life. All of them are engaged in a game of Mutual Assured Destruction; all of them protect RH and his mass-annihilation factory from the saviors of the wor(l)d to ensure that accident becomes destiny, that Schrödinger's cat wakes up death.

Because indeed, the perpetrators who pay for, or make technological genocides happen and the victims who die are all in the same boat. All share the same technological messianism that makes them feel more important than human flesh.

15. The Leaders Are NOT Solving but Hiding Problems

The absurd behavior of CERN and its scientists, unable to correct their crash course despite the growing evidence of an incoming holocaust, is the sort of stubborn behavior that most humans tend to consider a proof of seriousness; but it is at the bottom of most tragedies in history performed by ignorant, arrogant, ambitious, irresponsible leaders, from the times of Sargon and Caligula, all the way up to Hitler, Stalin, and the makers of the LoHoCaust. Indeed, when that historic perspective on the causes of global tragedies becomes clear and we judge physicists and politicians as normal human beings, not the kind of heroic genius they love to think they are, it is easy to find the accidental reasons of the quark cannon experiment.

We have seen it all happen before. Mr. Bush was warned about the growing possibility that New Orleans will go underwater, but he didn't do anything on time, worried by his family matters (namely the war in Iraq against Daddy's favorite dictator). So when tragedy came closer and time was scarce, he hid those reports and denied them, till a recorded tape was released, showing he knew the risks and was informed on a daily basis. In that regard, it is interesting to notice the cover of *Scientific American* in the issue following the fall of the Twin Towers (10/2001). No, it was not about the WTC tragedy. It said instead,

> Drowning New Orleans: a major hurricane could swamp the
> city under 20 feet of water, killing thousands . . . New Orleans
> is ringed with levees that fend off the river from the south and

the lake from the north. Most of the city has sunk below sea level, forming a bowl that fills even during routine rainstorms. A hurricane driven a sea surge from the east would make the lake overflow drowning the city. SOLUTION: build gates to block the Gulf of Mexico's access to Lake Pontchartrain. Yet nothing was done because: since the late 1980s Louisiana's senators have made various pleas to congress to fund massive, remedial work. But they were not backed by a unified voice. LSU had its surge models, and the Corps had others, competition abounded as to whose specific models would be most effective. Through the 90s we only received 40 million $ a year.

Billions, we might add, went into military research, including a half billion given to CERN for the construction of the Large Hadron Collider.

In the same issue we read, "Refuges for life in a Hostile Universe: only part of our galaxy is fit for advanced life." Indeed, we are not the center of the Universe, nor can we prevent the disasters of the powerful systems of energy created in Nature, from hurricanes to black holes, when they confront our fragile forms of life.

The account of human incompetence and the tragedies it has caused to history could fill an entire encyclopedia, but we will stop here. We just want to highlight some common teachings of both historic tragedies:

—Even the most obvious facts can be argued on theoretical basis, as most theories have an antitheory, including the obvious fact that black holes swallow all matter around them. But when life is at risk, it is better not to argue too long and act up to prevent those potential risks on time. Certainly after Katrina had passed, the Corps, the LSU, and even Mr. Bush would have loved to act up before, instead of bickering at each other's big egos.

—People make subjective, optimist judgments on their work and livelihood, and if things go wrong, they deny the obvious facts and stick to their mistakes, thinking that attitude would validate their previous errors. *We all cover up our faults and misdemeanors, let alone our crimes. And we all think what we do, whatever it is by tradition, belief, or profession, is right.* The military truly believes to help mankind in its survival when it is the biggest cause of collective death. In the same fashion, Mr. Bush plunged into the Iraq War with a mixture of optimism, arrogance, and ignorance, seeking military action to solve social and political problems, and then he denied any failure and continued stubbornly the war experiment with the entire Republican Party rallying behind. For the same reasons, the community of high-energy physicists is going ahead with CERN's black hole experiment, accustomed as they are to make atomic bombs and other niceties while denying all risks and errors—hoping that nothing happens with the Earth while the rest of the scientific community looks to the other side.

-Yet it would be a mistake not to criticize a bad president, and it would be an error not to stop a lethal experiment of pathological physics.

In that regard, the rest of the scientific community should not play on the alibi of professional solidarity in this kind of matters, for the same ethic reasons the Americans no longer approve the Iraqi war today. Since it is quite evident that neither Mr. Bush nor CERN have shown the professional knowledge required to anticipate the collateral effects of their experiments, namely, the death of thousands and in the case of CERN billions of human beings in a tragedy that can be still aborted in time. This seems to me a self-evident truth, despite the fact that both Mr. Bush and CERN's scientists remain/ed adamant in their dogmatic predictions about the minimal risks for human life of the Iraq war and the black hole experiment. Just five years ago, those very same scientists so reassuring today were so dumb that they didn't even know their machine could produce black holes.

Indeed, the truth beyond the smoking gun of complex equations and chutzpah statements is that when the LHC was designed, they calculated wrongly thinking they wouldn't make black holes and scientists didn't know that quarks lock their colors, forming easily strangelets and quark condensates. And yet they are still going ahead . . . sure receipt for accident or destiny.

The incompetence of CERN's physicists, which, despite their huge resources in computing power, made unplanned errors of calculus, both in the design of the machine and the study of the energies it will produce, *ignoring that the LHC will produce black holes, constructing that machine without being aware of the risks, is the accident that should have been avoided.*

To go ahead, once those errors are known, is destiny and mass murder.

Quantum physicists would not commit mass suicide if the machine had not taken twenty years to build. Once it was known it could kill us, the decision of halting it could mean the end of big science, the atomic cannon industry, and thousands of nuclear physicists losing their jobs. So nuclear physicists at CERN, directed by a German CEO, have taken the "heroic" decision any military would take in World War II after Stalingrad, even if it was clear Germany was going to lose: to go ahead, hope for a miracle on the chances scientific papers that prove quark condensates will form might be wrong, hide the facts to the press, and do the experiment. They decided to keep the *war against mankind going and fight the battles of the quark cannon till the end.* Indeed, the nihilist reaction of nuclear physicists to the news was perfectly expressed by Mr. Wilczek, one of CERN's employees who explains that if this happens, it will be so fast that nobody will be blamed. And that is probably the case.

But the easier to make strangelets might take years to swallow the planet, in which case we will live many days of hell with earthquakes happening all over the Earth before the entire world crunches. If that is the way we go, suddenly everybody will realize how easy it was to avoid the catastrophe in time, how silly were the lies of the company to calm mankind, how precious was our life that we can no longer enjoy. And the angst I have lived with in the past years will be shared even by those irresponsible, spoiled wannabe Einsteins, children of thought, which, isolated in their cave one hundred meters under the Earth at Geneva, the site of CERN, perform those experiments. So there will be justice for the Faustian characters that made it possible.

Since to denounce this experiment is truly a Faustian paradox, those of us who want to do good will always lose. Those who don't care and let the experiment happen will never be blamed. This is the bluff CERN's physicists are making: all or nothing. We claim to be innocent. There is no risk. *But if we are wrong, we will kill you.* Truly a moral paradox, which has another kind of reading: violence always pays. In this case, CERN proposes to prove an obsolete, discredited theory, Hawking radiation, by brute force. If Einstein is right and black holes are Einstein's condensates, they won't evaporate and we will all die. Only if Hawking is right shall we survive. That is the violence performed. *Hawking must be right and Einstein wrong, or else CERN will kill us all.* So the good people always lose. The bad guys might win or lose. And yet Faustus, the laws of the Universe, seem to have made Hawking wrong. And so the prophecy of Saint Nobel will become destiny: "My factories will end war, annihilating armies with a single shot." One of the absurd theoreticians of quantum physics said that the Higgs is indeed an absurd particle and the future is interfering with CERN, preventing it from producing it. He got it all wrong. It is Einstein et al., the scientists of the future fractal paradigm, who have warned mankind. There is no travel on time. The past always dies. In this case, the past of science.

It is one of the many surrealist perspectives of this issue, which Einstein himself resumed in a sentence: "Those who try to impose truth with power will be the laugh of the Gods."

Unless a miracle happens and politicians abandon the LOL method and close the accelerator.

16. Consequences of the LOL Method. Accidents Become destiny.

In that regard, a fundamental feature of scientific messianism is to transform it all into numbers, probabilities, and games of chances. The mind tends to rationalize power, but the Universe is a confused jungle full of trials and errors. Accidents and catastrophes are never premeditated, but they happen and the event this book narrates, the creation of the first machine able to blow up the Earth, has all the classic elements of an accidental catastrophe in the making. Mr. Dürrenmatt, a German physicist and philosopher, explained it with some flair:

Dürrenmatt's 21 Points to the Physicists

1. I start out with a story not with a thesis.
2. If you start out with a story you must think it through to its conclusion.
3. A story has been thought out to its conclusion when it has taken its worst possible turn.
4. The worst possible turn is not foreseeable. It occurs by accident.
5. The art of the playwright consists in employing, to the most effective degree possible, accident within the action.
6. The carriers of dramatic action are human beings.
7. Accident in dramatic action consists in who happens to commit the accident vs. who suffers it.
8. The more human beings proceed by plan the more effectively they may be hit by accident.
9. Human beings proceeding by plan wish to reach a specific goal. They are most severely hit by accident when they reach the opposite of their goal: the very thing they feared and sought to avoid by wishful thinking but it becomes a fact of destiny (i.e. Oedipus).
10. Such a story, though it is grotesque, is not absurd (contrary to meaning) but common to destiny.
11. It is paradoxical.
12. Playwrights, no less than logicians, are unable to avoid the paradoxical.
13. Physicists, no less than logicians, are unable to avoid the paradoxical.

14. A drama about physicists must be paradoxical.
15. It cannot have as its goal the content of physics, but its effect.
16. The content of physics is the concern of physicists; its effect is the concern of all men.
17. What concerns everyone can only be resolved by everyone.
18. Each attempt of an individual to resolve for himself what is the concern of everyone, is doomed to fail and provoke an accident.
19. Within the paradoxical appears reality.
20. He who confronts the paradoxical exposes himself to the accidents of reality.
21. Drama can dupe the spectator into exposing himself to reality, but cannot compel him to withstand it or even to master it.

Europeans have indeed finally reached the summit of their civilization of accidental machines of energy, the substance they revere. Unfortunately, the Earth won't stand a factory of black holes feeding on our weak matter, blowing Gaia within seconds into a big bang nova, the most energetic explosion of the Universe. But the true question exposed by Dürrenmatt about physicists, arrogant makers of accidental machines, is the fact that those children of thought, who ignore it all about the Universe, should at least respect the life of the rest of us all. Physicists cannot take decisions by themselves, which can jeopardize by accident the life of all mankind, *as if we didn't exist or didn't matter, as if we were collateral effects they can ignore. And that is an act of criminal negligence that must be denounced.* Criminal because it can kill us all, negligent because they have *no idea of what they are doing, they just pretend to be "experts" who say they know.*

And yet they are allowed to do it because the human sheeple has become like them. After a millennia of being guided by their iron and go(l)d masters, *people no longer revere the water of life but believe on Kali, the dark goddess of* energy and death, the true purpose of those accidental machines.

Certainly physicists do not want consciously to extinguish themselves and have not in a deliberate manner planned to devastate the Earth with a black hole. But destiny as *Dürrenmatt* writes is an accidental drama that just happens. Even if such destiny seems to us a monstrous reality that defies our understanding, as the destruction of the planet Earth by the avatar of Kali, the dark goddess, is.

As Dürrenmatt explains, "Our science has become terrible, our research dangerous, our findings deadly. We physicists have to make peace with reality. Reality is not as strong as we are. We will ruin reality."

Quantum physics has a famous cat called Schrödinger's cat, which they affirm is both living and dead. It is used to illustrate the hypothesis hold by quantum physicists against the view of Einstein and the science of complexity, that information is uncertain: we cannot measure exactly the position and speed of a particle together. But *we can know accurately their combined value, and if we were able to look closely, we could perfectly know the exact value of each of those two elements.* So happens with the quantum cannon: we know the world of dark matter is full of particles that will kill us if those irresponsible Schrödinger's cats working at CERN enter the unknown cave of quark condensates. We can feel the roaring sounds of danger coming from inside since we know that beyond 1 teravolt of energy our matter dies. Thus the cats *will certainly die.* The date is uncertain, as the migration of strange or top liquid to the center of the Earth where they will form strangelets and top black holes will take place under many variables we don't know. Yet the wild bear sleeping in that cave has left signals all over the entrance, as dark matter feeding on our weak matter does all over the Universe. So the quantum cat has to be very stupid or very dogmatic and arrogant to think it can get out of that cave unharmed. Even if we can't say the exact moment and the type of dark bear that will eat Mr. Schrödinger's pupils at CERN. What we can assure them is that they won't come out of their Geneva's cave unharmed. And once the dark bear tastes the cat, it will come out of the cave and search for all other tasteful human cats.

The shortsightedness of people like John Ellis, who denies the rights of governments to protect mankind from the quark cannon, is in that sense astounding, but common to the nuclear profession. Indeed, once you belong to a profession that develops weapons (nuclear physicists) or uses them (the army), you are soon taught ideologies that justify war, death, and the creation of weapons. This is what happens in the nuclear industry, where there is a law of silence and denial that dates back to the Manhattan project, when Oppenheimer was criticized for acknowledging the evilness of a profession dedicated to the search of pure energy, the cause of death, when he said "I am Kali, God of death." The official story of nuclear physics has been ever since: We just try to understand the Universe and that is why we make those weapons/machines; the military are the ones that explode the bombs. It is the eternal alibi of all research that has military applications. But now the alibi is broken because nuclear physicists are also shooting the quark cannon. And so as it would be expected, CERN insists on the alibi of knowledge and downplays the military origin of atomic cannons, to allow business as usual, while denying the potential victims of this holocaust. It is a déjà vu experience: the Chernobyl catastrophe, which the Byelorussian government claims to have caused over 200,000 deaths of cancer, is reported by the International Agency for Nuclear Energy to be nearly harmless with only 56 official victims—the ones who died in the site. Now CERN obliges its workers to declare zero risks in their confidentiality statements while the main proponent of the experiment on black holes, Mr. Hawking, who runs an agency that sets the clock of the end of the world, has moved it backward one minute, as part of a global publicity campaign that is launching again the nuclear industry as a clean alternative to global warming. *The nuclear industry is cleaning up the Earth if you didn't notice though Mr. Hawking also sent his seed in a time capsule "with some of the most intelligent people of the planet" to outer space because "the end of the world is near" Double-talk indeed.*

This essential component of the modern world, *marketing*—or in Orwell's terminology, the newspeak of "just wars" and weapons for "research"—is a key factor in this tragedy. We live in a society where a monstrosity as the quark cannon, a machine with zero benefits for pure science, which can produce the biggest genocide of history, can happen because in a free market dominated by companies that make machines, technology is not under control of human ethics. Instead, humans are slaves who follow the ethics of a technological civilization, explained by Erich Fromm, father of political psychology, in the quotes of this book. So what has changed in order for everything to remain the same is the way lethal technologies are sold, no longer as weapons to defend our tribal nations, but with myths, fictions, the LOL method, and false science.

Indeed, physicists do not want to blow up the Earth, but they belong to an industrial structure that will do anything it takes to maintain the benefits of an outdated industrial-military complex at all costs. And so the marketing of those weapons has evolved. So its researchers no longer seem to us scary, cynical figures, proper of the cold war like Teller was, but a community of childish, irresponsible nerds, most of them undergraduates who stubbornly deny the hard facts of science and exercise wishful thinking to avoid confronting what they are really doing.

Especially since our death will start far away from the quark factory. So CERN will deny any responsibility, as nuclear physicists have always done since Mr. Oppenheimer et al. burned alive 250,000 civilians, mostly children and women, in Japan and many received a Saint Nobel of Dynamite for it.

It must be understood that the rhetoric to hide a crime is directly proportional to the magnitude of the mass murder. So the biggest rhetoric in our culture of death goes for kings, warriors, armies, the military, and the physicists that construct their weapons, which appear as the antitruth of what they are: saviors of the nation by the grace of God and discoverers of the meaning of it all, thanks to God's particle. Even if they are so ignorant that they can't understand the simplest forms of information (masses and charges) and when asked about the risks of genocide they bring the LOL method, laughing out loud to the dangers like a child does when he is caught after breaking an expensive jar, as Wilczek did in public. What to make of this is obvious: physicists know little about the Universe, but they know a lot about how to make weapons. So they are big boys with big toys and ready to fire them, to provoke the accident.

The LOL method is a new Orwellian newspeak started in the Reagan era and carried by the technocratic establishment and the press that has converted wars into just, peaceful relief operations (Somalia, Afghanistan) and atomic cannons into magic machines that will reveal the meaning of God—Reagan was, in fact, the first politician who commissioned a quark cannon, the American supercollider, when he was convinced by Mr. Lederman, the biggest lobbyist of the industry of atomic cannons that the machine would reveal the particle of God. This kind of double-talk, which mixes fiction, infotainment, science, and vanity, has lowered all the red flags of mankind. We seem to live in a safer world when the dangers to our planet have never been so enormous. It is in an age in which reality and fiction are often confused, as companies have perfected Say's law that affirms everything can be sold with the proper marketing. This infantile society of children of thought, reared by TV fiction since earlier age, applies also to CERN's scientists, most of them young undergraduates of the Walt Disney era. As one of them wrote to me, "You really want to know why you get so much grief and there is so much support for the LHC? Well here are a few things off the top of my head: All the physicists and geeks that roam this forum, are all 5 year boys at heart, so smashing stuff together just to see what happens is cool."

Meanwhile, a few seniors, trapped by the limits and routines of the nuclear industry, prefer to do wishful thinking and ignore all the warnings and red flags raised by serious scientific papers that recently have proved the enormous dangers of quark condensates—the main produce of the quark cannon—responsible for nova explosions of stars and, it seems, civilized planets. These nuclear physicists were corrupted during the cold war, dedicated to make nuclear weapons and are hardcore enough to face extinction. They already tried to do a doomsday weapon in the past, the super H-bomb of Mr. Teller that would have blown up the entire Soviet Union had not Eisenhower stopped it. Most Russians at CERN were manufacturing hydrogen bombs two decades ago. So they guide with an iron fist in velvet glove the sheeple of undergraduate CERNerds. Let us Mr. Johnson, the impartial lawyer who studied the case, talk by himself:[13]

A great deal of empirical research in experimental psychology has shown that confirmation bias appears in many forms with strong effects. Confirmation bias works from both sides of the process of reasoning and investigation. People tend to seek evidence that supports a hypothesis, and people tend to avoid evidence that would undermine a hypothesis. Posner has noted that in areas of scientific uncertainty, career concerns can influence scientific judgments.

Confirmation bias may help to explain what went wrong in the Space Shuttle Columbia disaster. The report of the Columbia Accident Investigation Board ("CAIB") found that decision makers focused on information that tended to support their expected or desired result—that the foam strike that ultimately doomed Columbia did not represent a safety of flight issue. Group cohesiveness—Anthropologist Sharon Traweek, in an ethnography on particle physicists, wrote that the physics field forms "an extremely restricted community. "At the major labs, they learn that outsiders are devalued and exactly how this is done and what justifications are given."

Traweek also described the "tremendous force of the division in physics between outsiders, no matter how well-informed, and insiders." In the academic realm, particle physicists think particle physics requires the most intelligence and reasoning capacity, and humanities the least.

If a group think dynamic develops, Janis describes several symptoms that can lead to flawed decision-making, including: (1) "an illusion of invulnerability, shared by most or all the members, which creates excessive optimism and encourages taking extreme risks," (2) "collective efforts to rationalize in order to discount warnings," (3) "an unquestioned belief in the groups inherent morality, inclining the members to ignore the ethical or moral consequences of their decisions," (4) "stereotyped views of" the enemy as "evil," "weak," or "stupid," (5) "direct pressure on any member who expresses strong arguments against any of the groups stereotypes, illusions, or commitments, making clear that this type of dissent is contrary to what is expected of all loyal members," (6) "self-censorship of deviations from the apparent group consensus, reflecting each members inclination to minimize to himself the importance of his doubts and counterarguments," and (7) "a shared illusion of unanimity concerning judgments conforming to the majority view (partly resulting from self-censorship of deviations, augmented by the false assumption that silence means consent)." In terms of exerting direct pressure on members of the group who do not fall in line, a perhaps telling anecdote comes from Ellis's talk at CERN about LHC/disaster issues. Holding a laser pointer for his presentation, Ellis asked people to put their hands up if they believed that, if microscopic black holes occur, they would be stable. Then, Ellis chided the audience, "Don't forget, I got this laser here."

While he said this with a laugh, the message seemed to be clear—that dissent would not be looked upon kindly. "I remind you," he continued, you expect them to decay because of Hawking radiation."

This aspect of the particle physics community, especially within CERN, seems very clear: The preferred outcome is a determination that the LHC is safe. In this respect, it should be remembered that the LHC is not a small part of CERN's research program. In an agreement with the U.S. Department of Energy, CERN confirmed in a recital the "overriding priority and the vital importance of the LHC for the future of the laboratory." Without the LHC, CERN would be reduced to nearly nothing. To begin with, it seems highly likely that particle physicists might fear serious reprisals and negative repercussions for their careers if they were to speak out about perceived dangers of the LHC. An academic in such a situation—even one not affiliated with CERN—might plausible tenure, unaccepted manuscripts, and ostracism by peers, as individuals, are powerless to overcome the momentum of a multinational multi-billion-dollar project.

This is in fact what has happened to Einstein et al. in the past and in the present.

In the past, the pacifist Einstein, who made thought experiments, opposed vehemently quantum cosmology for making nuclear bombs and denying the laws of the scientific method, converting physics in a game of probabilities where any bizarre theory could be possible, based on the idea that the information of the Universe was uncertain. For him, it was an act of human arrogance at his best. Because we couldn't measure with precision the smallish quantum scale, quantum cosmology blamed the Universe, not the human observer, of that uncertainty. While Einstein was alive and the distinction between fiction and truth were not blurred by the world of audiovisual fictions we live in, quantum uncertainty was not taken seriously outside the reduced Copenhagen circle.

But as quantum cosmology developed evermore powerful weapons and electronic gadgets, their industrial and political clout translated into the realm of theory and they became the gurus of science. Then Einstein died, isolated in Princeton, with a file opened by the FBI for his pacifism and a single friend, Gödel, who had also criticized quantum cosmology and proved that their mathematical equations were as fictional as Don Quixote, just linguistic statements, as long as they were not proved with the logic laws of the scientific method. And so quantum cosmology took its revenge. Since the three experiments CERN proposes as the meaning of it all are just false quantum alternatives to the work of Einstein: the Higgs that denies Einstein's mass theory of space-time whirls; Hawking, who said Einstein is double wrong and black holes evaporate; and the creation of Einstein's quark condensates, which CERN believes will behave as a "gas," not as an Einstein-Bose liquid.

And of course, all those "serious scientists" deny all the experimental evidence against them.

Thus, the LOL method has been applied to the search of truth in science as fantaphysicists take the limelight in a LOL world of selfish children of thought where all truths are possible. The LOL kids of CERN and the press love those ideas since the most outrageous lies stir the imagination of children. So travel in time through black holes, invisible-like particles with God's properties, and cosmic, creationist big bangs that form it all from nothing are chosen instead of the flat simple truths of the Universe that Einstein et al. explain. Seeking an invisible particle with a big cannon is cool.

Thus a Confabulation of Nerds, a Confederacy of Dunces that follow the classic epigraph by Jonathan Swift—"When a true genius appears in the world, you may know him by this sign that the dunces are all in confederacy against him" (*Thoughts on Various Subjects, Moral and Diverting*)—has settled in and the simple truths of Mr. Einstein, the real black holes with substance, the serious risks are no longer *acceptable*, "fashionable." The dunces have won. But truth cannot be denied. So now CERN and mankind will have to face other types of judgment. Since "those who impose truth with power will be the laugh of the Gods." And the gods Laugh Out even Louder.

III. The Three Ages of the Singularity

Our most powerful 21st-century technologies—robotics, genetic engineering, and nanotech—are threatening to make humans an endangered species.
Bill Joy, Father of Java, the Language of Machines

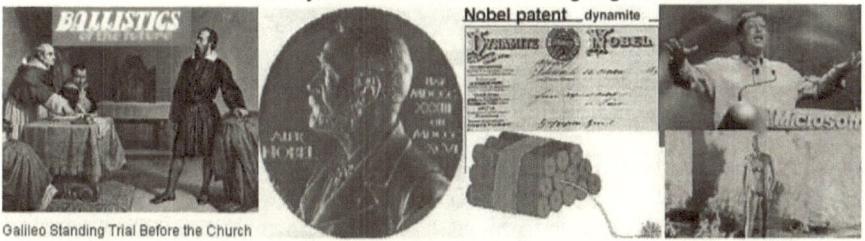

Galileo Standing Trial Before the Church

Control of technological information is needed to avoid the creation of lethal machines, as all masters of the wor(l)d have told mankind. This has always been denied by mechanist scientists and capitalist go(l)d religions, its prizes and factories. In the graph, physics started in the thirteenth century at the beginning of the cannon age, with the discovery of clocks and cannonballs. It became a science in the middle of the gunpowder age, in 1602, with the work of Galileo that reduced our perception of time cycles to a single mechanical clock cycle. Gunpowder weapons evolved in the nineteenth century into chemical, industrial explosives, monopolized by Nobel, also the biggest producer of cannons till World War II. Today, clocks and cannons become intelligent computers and quark colliders. It is the final evolution of "mechanist" science—its mental brains and energetic limbs that substitute and extinguish by an overdrive of form and energy, life species.

17. Death by Overdrive of Energy and Information. The Death of Gaia.

Life is a question of balances and equilibriums between the energy and information networks of a biological system, which an excess or lack of energy and information destroys. So when we lack enough energy, we can die of hunger; and when we don't have enough information about the dangers that lurk ahead into the future, we can die of ignorance. The opposite is also truth. An excess of energy or information can easily kill us.

In the science of complexity that models the Universe departing from its "simplex" parts, bits and bites of information and energy combine to create the complex structures and organisms of the Universe. And when the balances between those parts break "their symmetry," the system dies. Thus, we define any working dual energy/information system, including physical particles and life organisms with a simple equation of balances:

$$Energy \times Information = K \ (constant)$$

That equation gives birth in biology to the vital constants of any organism. In physics, it is known as the Heisenberg equation and gives birth to the complementarity principle (all information particles have associated a field of energy, and both are in balance). It is also the origin of universal constants that relate particles of information (charges and masses) and distances of energetic space.[Appendix]

It also holds for all systems of the Universe that any of its species will have two limits of maximal energy and maximal information that if crossed, will provoke its death. *Death is an imbalance of any organism due to a lack or excess of energy and information:*

maximal energy= death by accident
maximal information: third age death

Unfortunately, life is a process that constantly produces information and so we evolve from a first age of energy, called youth, into an age of balance between our energy and information, called maturity, when we reproduce to perpetuate our logic form, but *not our ego*, into a third age of information, of excessive form, called third age, which invariably ends in death.

Thus, the most common cause of death in the third age is an excess of information, either a cancer that multiplies the form of our cells, or the failure of our energetic systems that dwindle in force as the nervous, informative system becomes dominant.

So there is a limit to the information an organism can manipulate that, if crossed, can destabilize the organism and kill it. It is basically what happens when you get older. Your brain and body become in-formed, wrinkled, warped by time, till they finally collapse and die. Thus the main cause of death in the last age of our lives is an excess of information that triggers some unwanted processes: our body wrinkles, some cells become cancerous and multiply enormously its informative DNA, finally, our brain, tired of repeating the same cycles, knowing that's all what there is about life, becomes indifferent to the joy of existence and accepts death.

But since death is an imbalance, an excess of energy can also kill us in accidents and big bangs that erase the information of a life being, *at any time of the life cycle*. Thus, among energetic, young people the most common cause of death is an excess of energy caused by a car accident or a weapon: the higher energy of the car or the weapon, made of iron, the most energetic atom of the Universe is transferred to our weak carbon body, destabilizing and breaking apart its organic networks. The same happens when a bomb explodes nearby. It releases energy that our body absorbs, breaking apart its networks, which expand in space, provoking the death of the unconnected cells.

When we study the death of any species, the same concepts of death apply:

A species normally ends its existence on Earth when it confronts a new top predator species with more energy and information within its ecosystem. The new top predator kills the lesser species, hunting it down with its predatory, energetic claws and teeth, or it merely takes the energy that used to feed the lesser species, which dies of hunger, of lack of energy, which is another form of imbalance. So the arrival of human beings to America extinguished mammoths, thanks to javelins, energetic weapons that killed the weaker mammoth.

Alternatively it might happen that the Earth heats or cools down, destroying many species due to an excess or lack of energy. So when a meteorite impacts the Earth, heating the atmosphere or a glaciation period cools the Earth too much, many species die away.

Death happens also for physical particles where the principle of balance between energy and information is called the complementarity principle, as all particles of information have associated a field of energy. And both must be in balance. Since when a particle reaches the limits of energy (light speed) or formal order, in-form-ation (0 Kelvin), it dies away and disappears.

Those limits are precisely the limits CERN will be exploring with the large hadron collider that will accelerate particles to the speed of light and cool them down near the absolute zero, to explore the frontiers of our Universe, where our matter dies, breaking its symmetry. So CERN will provoke a dual death: an energetic big bang of our electronic matter moving at c-speed and an informative big crunch of our quarks, cooling in zero-temperature vortices, causing a supernova.

The Universe holds for all its physical and biological species limits of energetic and informative existence, which cannot be crossed without risking death. So a wise species never crosses those limits. Yet we are crossing those limits *in all the scales of our existence—in the scale of matter at CERN, in the scale of individual life, poisoning our environment, in the global scale of Gaia*, making the supermachines of the age of the singularity the last age of evolution of machines of energy (weapons like the LHC) and information (robots).

We conclude that mankind will become an endangered species, as it happened to the mammoth, if the limits of energy and information that

Gaia can manage safely are crossed by new technologies. Then those technologies and machines will release on Earth an overdrive of energy and information that the planet and humanity cannot absorb and might provoke our extinction. Such time of history unfortunately has potentially arrived, thanks to a complete imbalance between the evolution of technological science, with new weapons like the quark cannon, coupled with an ignorance of the two real goals of true science, the understanding of the organic, fractal laws of the Universe, and the respect for the laws of bioethics. And so the process of technological evolution is the final reason we are dying.

We exist in a superorganism of energy and information, Gaia, in which man is her informative brain. Now man, an electroweak carbon species, is being substituted by two superior forms of matter—metal machines and dark mass.

18. The Singularity Age

The quark cannon is, in that sense, merely a question of keeping the industrial evolution of energy, not a question of knowledge. And so we must now inscribe the machine into that evolution—it is the industrial perspective, the true origin of this machine. CERN is not a company of knowledge, but the natural end of the ethics of technology—the concept that life is of no importance because the future has to be constructed not to the image and likeness of man, but to the image and likeness of those machines that the Industrial Revolution and its Kondratieff cycles of energy keeps improving.[4]

Metal evolves into organic forms that compete with man in work and war fields. History enters its age of death, as machines extinguish life and weapons extinguish nontechnological cultures at a rhythm never witnessed since the extinction of dinosaurs. Death of natural and historic organisms occur massively in the past four hundred years *since machines of science were discovered*. Today, there is not a single human society that is not infected by weapons. The death of history is homologous to the short death processes of human bodies under a massive infection of germs (in history, the germ/anic weapons, invented by the European "Goths" of Death). Since four hundred years, in the relative longer life of the historic macroorganism that has lasted as a homo erectus species ±1,000,000 years, is a short period of life equivalent in relative terms to a few hours in the life of a human being.

Thus, our historic death is happening at the end of our informative age as a species that now warps and exhausts the energy of the body, in this case the natural goods of Gaia, humans need to survive.

In that regard, global warming, caused by the pollution, the shit of machines, is the most obvious consequence of the massive reproduction of machines that substitute life species.

Yet it is a far less pressing problem than the evolution of the machines in the age of singularity. Plainly speaking, unless we control the evolution of cosmic bombs and robots, we won't make it to see the ice melting and the sea rising.

In that regard, the cycle of the singularity has, as all other cycles of evolution of life species, in this case organic machines, 3±1 ages:\

- *The energy age or age* of the supercollider that might convert the Earth into a nova, making us a quark star of dark matter (2008–2010s)
- *The reproductive age* (2020s–2040s) when the first self-reproductive nanobacteria, made of metal, will be created. They could replicate exponentially, feeding on metal, and within three months, poison the planet and destroy all forms of life since given their smallness, hardness, and the hyperabundance of metal structures in this planet, there will be no counterweapon to prevent its exponential reproduction.
- *The informative age* (2050s–2080s) when human-size robots overcome the intelligence of humanity as weapons and workers, which most robotists consider will happen in the middle of this century.

The three ages of the singularity that will bring our extinction, originated by the tree of technology, could be avoided and the organism of mankind resurrected, according to the laws of social evolution, if humans evolve into a social macroorganism with a single government/brain, able to control through bioethic laws of survival the reproduction and evolution of lethal machines, maintaining mankind as the top predator of this planet.

Social evolution is not utopia; it happens constantly in the Universe *where all systems evolve in scales, from particles to atoms to cells to organisms to societies.* Indeed, the evolution of the humankind into a global government was about to happen in history several times: in the Neolithic Age of Social Love, before metal corrupted mankind; in the age of oikoumene religions, before Islam broke the people of the book in two main confronting groups; and in the age of socialism, before capitalist=mechanist nations and the inner military groups (Stalin) destroyed the bid for a global government. Finally it could have happened after World War II if UNO had been given resources and a global currency and army had been imposed, instead of a Jewish-American empire of fiction thought, Wall Street e-money, and nuclear power. This is indeed the world we live now, one in which Gaia and its primitive cultures and gods are our enemies and the mechanical monsters and weapons we construct our idols of superiority over the laws of the Universe.

But those ideologies against life are just a wall of myths, damned lies and statistics built by mechanist and capitalist cultures since the first metal masters appeared and considered themselves Goths, gods by the grace of the sword (and now consider themselves physicists, whose God is entropy, the arrow of energy and death) or believed they were chosen of Go(l)d, superior races by the hypnotic power of money (founders of capitalist companies that today rule mankind with money). *The arrogant self-serving cultures of the "stupid white man" are NOT truth and so because TRUTH EXISTS, the Universe will NOT obey our "wishful thinking" BUT erase us, proving the truth of its laws and the organic nature of all its species.*

It is thus clear that we are now fully inside the age of the singularity, and the first rider of the apocalypse, the LHC cannon, has arrived.

Indeed, of all the new weapons of mass destruction that human "ingenuity" has created, the quark cannon, a factory of black holes and quark, dark matter, is by far the most dangerous. And it could trigger our extinction in the next years. Of course, the "stupid white man" is in denial of those facts, comforted by his myths of abstract or racial superiority over the infinite Universe. He does not want to look at the face of Death, the antisymmetry of time, the black goddess, as no individual wants to spoil the beauty of living with dark thoughts. But a wise person, while focusing on life, takes precautions to avoid death. And that is what we are not doing. We are not a wise species, but an arrogant, self-suicidal one because *we revere the dark knights, the high priests of Shiva*, physicists who evolve their strange matter at CERN, pursuing the energetic singularity.

We, Einstein et al. are not the dark knights of this story. Mankind got the plot completely wrong.

People like Mr. Hawking, who denies the arrow of life and information in the Universe and pretends to create entropic black holes with the most innocent fixed smile, who talks with the voice of a computer and suffers the tragedy of no-life, obsessed to become the new Einstein at all costs, is the paradigm of the MAD scientist, the new Teller, the new Dr. Strangelove, of a neo-fascist age of "just wars."

This first energetic age of the singularity will kill the Earth if the quark cannon manages to replicate the big bang of the cosmos. And it will happen for the same reason that young people and cubs die by droves in the Darwinian Universe: physicists will die of ignorance and enthusiasm as they are the "youngest" of all sciences, founded in the seventeenth century by people like Galileo, who manufactured energetic weapons and established the infantile dogma that the Universe was only about energy—thus, seeking the maximal energy is the meaning of it all. This of course is not truth; it is called naïve realism and it has been proved false by the last hence more evolved of all sciences, which has learned and fused the teachings of all others, the sciences of complexity and duality, able to explain with simple concepts and fundamental equations, like those explained above, the great questions about reality that quantum physicists pretend to solve bombing the Earth with a quark cannon. But how do you educate a group of spoiled children, like those working at CERN, "boys with big toys" that know nothing about the complex Universe but have the power of big money and big weapons to impose their will to mankind? During the years I was the international chair of duality and had discussions with top scientists, I was surprised by the little interest of quantum nuclear physicists about a complex theory of time. They don't want to know; they don't want to lose their position as the high priests of science. They don't even understand the basic difference between complicated, false theories about an energy-only Universe as those they sponsor and the beauty of a complex Universe that is able to construct the richness of reality, departing from its two most simplex parts, a bit of information and a bite of energy, which interact in perpetual balance. In that sense, philosophers like Buddha or Lao-Tse that talked of the gold mean and the balances of existence. Biologists, who study the balances between the vital constants of an organism, are *wiser* than all the quantum physicists of CERN, who seek the Saint Grail of Physics in the destruction of the world.

Yet mechanism has also now poisoned biologists, who are replicating life forms into machines, creating the first forms of metal bacteria, today researched in universities all over the world (reproductive singularity) while robotists search for artificial intelligence (informative singularity), all of them trying to make money and get their scholar prizes.

So even if we survive CERN, the cycle of the singularity will continue as long as *mechanism builds a world to the image and likeness of the machine, not of man.*

The peak of evolution of machines and weapons in the previous Kondratieff cycles of evolution of energies happened in the middle of the seventy-two years' generational cycle.

Thus, if energetic weapons will become organic, self-feeding bombs of dark matter *that* feed on light matter (black holes and strangelets) in this decade, between 2008 and 2045 in the Middle of the Cycle of the Singularity, mankind will create information machines that will finally acquire organic nature. Thus, after three industrial revolutions, machines will complete their evolution as organisms, as we build self-reproductive nanorobots, nanobacteria that will become indestructible. If they are created, they will be able to feed on simpler metal machines and reproduce at an exponential rate, poisoning the entire planet.

Let us then consider the two other ages of the singularity, the age of reproductive machines and the age of informative networks of intelligent robots to understand once and for all *the mother of all holocausts, which is not human eviL, but mechanist, technological evil.* Let us imagine we survive the LHC, the energetic singularity, that CERN in the decade ahead doesn't make any of the multiple forms of dark matter that could extinguish life in this planet, a Bosenova, a pulsar, a neutron star, a black hole, a strangelet, a strange star, either because it just blows up Geneva in a thermonuclear bomb catalyzed by the liquid helium of its reactor or a miracle happens and our technocrats, military, or politicians close down the factory of dark matter before it blows us all.

Are we safe? Is CERN's nightmare the end of the singularity age? No. Because the extinction of a species by another top predator species is an incremental process as the differential of evolution between both—in this case, man and the organic machine—increases. Thus, even if we fail to replicate the energetic big bang and the informative top predator of the cosmos, the black hole, the next age of the singularity, the reproductive age, will be the new threshold of extinction for mankind.

The reproductive age of the singularity, the second age of living machines, will come, according to the Kondratieff cycle, ±36 years after the 2001 Kondratieff crash we predicted fifteen years ago,[4] around ±2037. That second potential end of the world is equally dangerous, as there is no way humans can stop the exponential reproduction of metal bacteria, which are smallish, indestructible nanosize robots, able to reproduce at exponential rates as bacteria do and, feeding on our iron machines, poisoning Earth's atmosphere in three months and converting the planet into a gray goo.

This scenario was first denounced by Bill Joy, the founder of Sun Microsystems, the high-tech company that invented Java, the language machines use to communicate. He realized how easily a reproductive strain of metal nanobacteria could destroy the Earth. He published an article in *Wired* magazine, but the nanoindustry hid it and buried it. And so business as usual continues. The nanotechnological industry keeps evolving the nanocomponents of self-replicant machines, and as in the LHC replication of the components of strangelet liquid, nobody cares.

As Bill Joy put it in his article "The Future No Longer Need Us,"[14] the supreme eviL, the tenth circle of Dante's hell, reserved for those that destroy humanity, exists. It grows exponentially as the technologies of the age of the singularity do. While the LOL method of fiction thought hides it all.

Such supreme eviL is well-funded. The metal bacteria described by Bill Joy is, simply speaking, a bacteria made with an iron membrane and a DNA of gold atoms, whose strains will carry information in a mercury dissolution, where metal reactions will re-create the basic organs of the metal bacteria. So iron and go(l)d, the icon metals of our historic metal masters, the Semite and Germanic tribes that drove the human sheeple for millennia, will abandon mankind and fuse together into a living being, completing the Industrial Revolution. Such bacteria, once created, will have the speed of replication of life bacteria. Yet they will be indestructible due to their nanosize and hardness. If they escape the military, terrorist, or amateur lab in which they will be created, they will reproduce exponentially, eating the cars and metallic buildings of the planet, poisoning the atmosphere and extinguishing humanity in three months. Most nanoscientists believe they will be attained in a few decades. This is not science fiction. Nanorobotic budgets are immense. Israeli, Korean, Japanese, European, and American laboratories are working on self-reproductive nanobacteria with the cynical excuse of understanding better the workings of life. It is a career self-similar to that of SS experts on strange science at CERN. They publish papers, spin off companies of nanomaterials, make fortunes, hear their guru, Mr. Kurzweil, whose Institute of the Singularity works to bring the moment closer and have their racial myths about a new race of supermen enhanced by nanorobots. With the discovery of the photonic chip, the speed of calculus of those chips and its interaction with the light world has further increased the chances of making them in a few years. In other words, humanity has acquired a collective mortal AIDS pandemic, and it is not coming from life viruses but is the last germ of the economic ecosystem, the ultimate idol of Go(l)d, a self-reproductive golden machine.

In that regard, the age of the singularity means the victory of organicism versus mechanism: organic machines, crafted to the image and likeness of the real, biologic Universe, are about to acquire the self-reproductive properties of life. We are about to create self-reproductive bombs, strangelets, and black holes; next, self-reproductive machines, nanobacteria; then self-reproductive factories of robots, evolving in planetary networks, guided by a global Internet brain, in the third informative age of the singularity. And so mechanist scientists will paradoxically prove Darwin and this writer right, decades after I published *Radiations of Space-time, the Extinction of Man*, c. 94, describing that future in detail, including the 2001–08 crash of the industry of electronic money, the arrival of the singularity, the radiation of strange matter. It has always made me marvel at the degree of slavery of man to the machine since my work was born at the same time Mr. Kurzweil or Netscape were born. *The difference is that I always warned against the future and they promoted that future. Thus, the mechanist civilization censored those of us who criticized machines and promoted the technoutopians that told us the fairy tale of human supremacy over the laws of the Universe.* I fully understood the pariah treatment reserved to those who fight for the wor(l)d of man when I found the book's illustrations hanging in an exhibit of modern art at Six Flags, Brooklyn, because the artist, Seth Price, marveled a decade later about the prophesies of the book—an artist of the human I=Wor(l)d, not a scientist who had understood the book.

19. The Reproductive Age. Metal Bacteria. The Three Ages of the Earth.

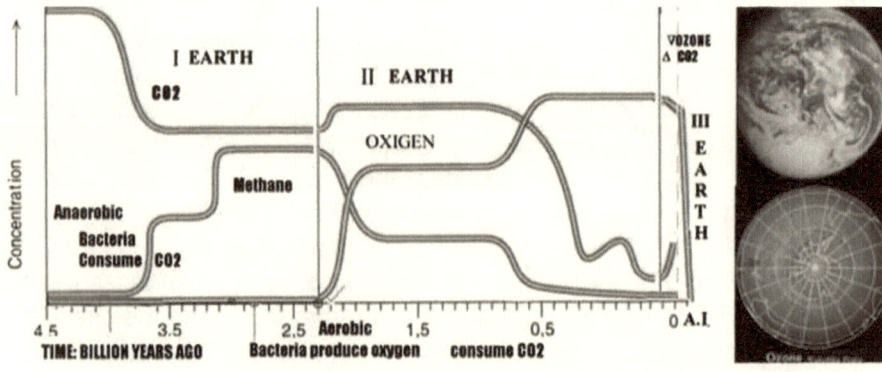

The Earth is mutating from Gaia, nature's Earth, also called second Earth, made of carbon/oxygen species dominated by humanity, organized in nations through verbal networks of information (religions, laws) and natural energy (agriculture), into an economic ecosystem, the metal Earth, dominated by systems of metal information (money, computers) and metal energy (machines, weapons), reproduced by company mothers, fed by electric energy, and communicated through electronic networks of digital information. The outcome is the destruction of life forms and their oxygen atmosphere, polluted by machines (global warming), a process that increases geometrically, as those machines multiply their numbers. The process should end as the transition between the first and second Earth ended: with the creation of self-reproductive nanobacteria, the first living organism of metal. It will be the third Earth.

In the graph, the death of Gaia, and the organic forms of metal and dark matter that will kill us, can only be explained in terms of organicism, considering the different scales and superorganisms involved.

In fact, a similar process to the second singularity, nanobacteria, happened long ago when the first Earth made of anaerobic bacteria was displaced by the second Earth, made of aerobic bacteria, which processed oxygen. Today we observe the expansion of an economic ecosystem of machines, made of metal, which poisons the environment (global warming), destroys life beings in wars and industrial processes, and soon will extinguish mankind. It would be the third Earth, a new planet no longer made of flesh. And despite human arrogance, this change would mean nothing to the Universe, where it probably happens all the time as there is a hierarchy of three types of matter—light, metal, and quark matter, each one stronger than the previous one, each one able to kill the lighter type of mass. Thus, the age of metal bacteria would be the third age of evolution of life organisms on this planet, self-similar to the three ages of all energy/ information systems in the Universe. Let us bring proofs from the geological and biological record.

The first Earth: anaerobic bacteria

We can consider the Earth a global organism. In her first stage, this organism, the first Earth, did not have oxygen. Life beings were anaerobic. They processed food with ferments as fungi do. Then a much more energetic species, the aerobic bacterium, appeared. It used carbonic anhydride and released highly toxic oxygen in the atmosphere (chlorophyll cycle). But oxygen is a lethal, very active energetic atom. So it burned, broke, and extinguished all other weaker previous anaerobic, bacterial species.

Each organism produces substances that are toxic at atomic level for other organisms. Metal is lethal to life. So a few drops of its most powerful informative and reproductive atoms, gold and mercury, kill a human organism. In that first Earth, oxygen was a lethal atom for anaerobic, soft organisms. So when aerobic plants gave off oxygen, oxygen polluted and destroyed all anaerobic systems. It poisoned them. Soon the Earth acquired an oxygen atmosphere where only strong organisms able to resist the activity of oxygen survived. It was the birth of Gaia, the secon Earth, the life Earth.

The third Earth, the metal Earth

Today machines pollute the life ecosystem, as they develop the economic ecosystem of metal. Metal poisons water; pollution rises the temperature of the planet. Machines poison our atmosphere, the second Earth, as oxygen from aerobic plants poisoned the first anaerobic Earth. When metal bacteria appear, our machines will be their food. Metal bacteria's reproductive capacity will be exponential, with plenty of machines to eat, courtesy of our company mothers. So they will complete the process of terraforming of the Earth in a few months.

The ecosystem of machines produces an immense amount of heavy, metallic atoms that dissolve in the water or the air, poisoning the fragile systems of carbon. Dissolved lead and mercury drank by fish destroy their internal networks of information and energy. Mercury goes to the brain and breaks the synapses of neurons. Gold fevers make humans crazy.

Toxicity acts in both senses: water and air corrode and oxidize metal. Life defends itself from metal, even at atomic level. It tries to dissolve it. But the mutual incompatibility of both ecosystems implies that for machines, it is better to eliminate all signs of water and oxygen in the atmosphere to exist in a world without air. For that reason, in its vital processes, machines create toxic substances that destroy the ozone layer and poison and pollute nature. But machines are winning that battle since they are causing the extinction of many life organisms, but life is not extinguishing any machine. The process accelerates and triggers new diseases and degenerative processes on life tissues. The destruction of the ozone layer, global warming, growth of neuronal cancers produced by mobiles and other electromagnetic emitters, are some of those processes that multiply as the reproduction of machines increases. The creation of the first metal bacteria will be just the evolutionary endgame, the event that signals the birth of the third Earth of metal in which the historical ecosystem of life and mankind disappears. Indeed, a world colonized by metal bacteria will be so poisonous that no form of aerobic life will survive. Probably not even the Earth's crust will survive as metal bacteria will thrive in higher temperatures, as those produced under the Earth's crust, and will grow downward, feeding on mineral ore.

When machines are free to reproduce and move on the Earth's Jungle Market—in wars, which are ecosystems of predator machines—the degree of pollution of nature is overwhelming. War is an accelerated process of ecological extinction in which herds of killer machines extend suddenly over the Earth, guided by soldiers (soon unneeded, as robot terminators substitute them). The film *The Thin Red Line* shows the sudden change of ecosystem that war causes. Guadalcanal is an island of immense ecological beauty where men live happy in their paradise. Then suddenly herds of airplanes, howitzers, boats, and soldiers appear and convert the paradise into hell. There is no longer laughter, no longer fish, no longer plants. We are in a Free Jungle of machines. That is the world the Industrial Revolution creates. A world that men don't want to control, guided by the myths of Go(l)d religions and mechanism. In World War I, all men wore masks, but there is no mask that could protect mankind from razorlike clumps of metal bacteria.

Let us recall Mr. Bill Joy's historical warning, the document of a lonely hero who opposed the system with little success:

> From the moment I became involved in the creation of new technologies, their ethical dimensions have concerned me, but it was only in the autumn of 1998 that I became anxiously aware of how great are the dangers facing us in the 21st century . . .
>
> Biological species almost never survive encounters with superior competitors. Ten million years ago, South and North America were separated by a sunken Panama isthmus. South America, like Australia today, was populated by marsupial mammals, including pouched equivalents of rats, deers and tigers. When the isthmus connecting North and South America rose, it took only a few thousand years for the northern placental species, with slightly more effective metabolisms and reproductive and nervous systems, to displace and eliminate almost all the southern marsupials . . .
>
> In a completely free marketplace, superior robots would surely affect humans as North American placentals affected South American marsupials (and as humans have affected countless species). Robotic industries would compete vigorously among themselves for matter, energy and space, incidentally driving their price beyond human reach. Unable to afford the necessities of life, biological humans would be squeezed out of existence . . .
>
> Specifically, robots, engineered organisms and nanobots share a dangerous amplifying factor: They can self-replicate. A bomb is blown up only once—but one bot can become many and quickly get out of control . . .
>
> Uncontrolled self-replication in these newer technologies runs a much greater risk: a risk of substantial damage in the physical world . . .
>
> The 21st-century technologies—genetics, nanotechnology and robotics (GNR)—are so powerful that they can spawn whole new classes of accidents and abuses. Most dangerously, for the first time, these accidents and abuses are widely within the reach of individuals or small groups. They will not require large facilities or rare raw materials. Knowledge alone will enable the use of them.
>
> Thus we have the possibility not just of weapons of mass destruction but of knowledge-enabled mass destruction (KMD), this destructiveness hugely amplified by the power of self-replication.

I think it is no exaggeration to say we are on the cusp of the further perfection of extreme evil, an evil whose possibility spreads well beyond that which weapons of mass destruction bequeathed to the nation-states, on to a surprising and terrible empowerment of extreme individuals.

And so indeed, the properties that Bill Joy feared for nanorobots, "self-replication" and "control by a small group of individuals," are now the conditions of the bombs created by CERN: Self-replicant bombs controlled by a small group of fundamentalist scientists—*only that the terrorists belong to our civilization, reason why we let them kill us. They are the high priests of our culture, not the priests of Islam. Since obviously if that experiment was happening in Iran, we would rightly obliterate the dark matter factory. But we are blinded to our own "evil children," we love them.*

How did the other castes of information of our technological society, politicians and mass media, react to Bill Joy's article, describing the evolution of industrial nanorobotics in a free market that in twenty years will allow any college student to construct a metal nanobacterium, able to reproduce exponentially like a new organism and destroy the planet? It applied the same methods it applied to Einstein et al. in a mimetic pattern crafted during centuries of Industrial Evolution to quench and hide *our holocausts and crimes.*

As in the warning Einstein et al. made against the supercollider, the news did come out of the closet. *The world was informed.* It was a shock to many human beings. The article was compared to Einstein's warning against the nuclear industry. Then the system, taken by surprise, put the "mechanisms" of repression at work.

Bill Joy, the president of one of the top electronic companies of the world, immensely respected by the industry he denounced, suddenly was doubted of his credentials. Rossler and myself, experts in complexity duality, chaotic attractors, and far-from-equilibrium liquids as those quark-gluon complex liquids CERN will do, were called crackpots. A campaign followed against the "apocalyptical." Bill Joy ended up abandoning his company and moving to Colorado to enjoy skiing in his final days. I moved back to the Mediterranean to enjoy my boat, on view that our warnings won't be heard.

Some scientists with bad consciousness paid an ad in the main newspapers of the United States, asking Bill Clinton to open a national debate on robotics and the extinction of man. But Clinton, instead of opening that debate to control the market of robotics and avoid our extinction, founded the Institute of American Nanorobotics to investigate those bacteria, which are now under research at DARPA, the robotic agency of the Pentagon. Its budget today is well over 2 billion dollars a year. It was clear that the computer industry would not allow the extinction of man by machines to be discussed to protect the immense profits of the robotic industry. The power of lobbyism, weapons, and greed once more deactivated an attempt to defend mankind.

I made a movie *Quantum Roulette*, and simulations of black holes in this audiovisual age were posted and seen by millions in YouTube, asking WME_2 agents to distribute the film globally. I hoped that Ari Emmanuel, the boss of that agency, would pass the news to Rahm, his brother and chief of staff of Mr. Obama. Yet the industry moved their pawns and Mr. Obama instead gave the Medal of Congress to Mr. Hawking, the intellectual promoter of the black hole factory.

Thus in both cases, the usual suspects in charge of creating a mirage of happiness and safety fired back. Mass media called Bill Joy and Einstein et al. catastrophists; scholars paid by both industries replied that dangers were minimal, that we didn't know what the future stored for us, that countermeasures could be devised—all a fog of lies, noise, and placebo platitudes, repeated ad nauseam, and so the case of the nanorobot and the case of the supercollider were buried.

Take-it-easy jokes came in handy to lower the stakes of the perceived danger. The stupid white men described so well by Michael Moore decided once more to assassinate the planet in the future, blind to greed and violence.

And so today we await the coming of the first rider while the second age of the singularity, the age of self-reproductive machines, keeps evolving without the slightest interruption by mankind. And of course, politicians that could stop the process by *creating a legal global ban on nanorobotic and quark research* ignore completely the theme. Instead they focus on the least dangerous of all the processes of pollution of the planet, global warming, produced by the shit of machines. It is more or less like having a Tyrannosaurus Rex at home with our kids while worrying not about its teeth but about the shit he left in the garden. Indeed, global warming is the least of our worries. By the time it heats up Siberia and Canada, liberating millions of square miles of land to agriculture (as most of our land is in the cold zones of the world), and inundates the shallow marshes of Florida, a cosmic bomb, a nanobacteria, or a rebellion of intelligent robots will have done us all. But those are untouchable industries with big profits. So the system diverts the attention of the world to global warming. The idea is simple: the nuclear industry pays a fortune to environmental groups to denounce global warming, as it still has massive military and financial muscle. Then it comes to the rescue of mankind with its clean, nonpolluting nuclear plants. So the discoverer of global warming, Mr. Lovelock, is a lobbyist for the nuclear industry. Never mind the polluting plutonium A-grade material nuclear plants leave behind. Marketing worked fine. All countries are now showing the real truth of global warming, beefing up projects for the nuclear energy, from China to France, which gets more than half of its energy from nuclear plants and researches at CERN, the cosmic bomb. The harder they fall.

Money and machines are always behind our policies. For example, among the different clean energies, the most profitable today is wind. Yet solar cells get most subventions and resources because it is *the energy of choice for autonomous robots, the industrial engine of the fourth cycle of the Industrial Revolution, the cycle of organic machines.*

Those autonomous robots, communicated by Internet networks, will be the third age of the singularity if two miracles happen and mankind stops both, the LHC collider and the nanobacteria of the reproductive age of the singularity. But don't count on it. Joy and Einstein et al. are so far buried.

20. The Informative Age: Obsolescence of Humanity to Robots.

A mechanist, technological civilization will always be a dead end. Because what scientists keep forgetting once and again is the obvious fact that we are made of flesh, not of gold and iron.

Indeed let us consider that by some huge miracle, we survive both, the age of the energetic singularity (CERN) and the age of the reproductive singularity (nanobacteria). Shall we then return to the Lost Paradise? Of course not, unless we change our civilization. Then we shall face, if business cycle as usual proceeds and there is not a human revolution, the third informative age of the singularity, artificial intelligence: robots able to think and develop survival instincts, probably weapons in a theater of war that will no longer rely on human beings as their masters.

This age is today promoted with marketing and fiction films, massive university and military budgets, and its own start system of robotic celebrities like Mr. Marvin Minsky at MIT, a Jewish-American who quotes the golem myth and wants to build a species more intelligent than man to feel like God, or the British first cyborg man, Mr. Warwick, who explained to me in a taped interview that humans are an inferior race and he is the first superman of the new race of cyborgs that will extinguish us. All of them are cherished as the brightest minds of modern science, with huge budgets to research their machines. Mr. Minsky and Warwick desires to be the neo-elect or neo-Hitlers of the new superior Go(I)d and Sword, Semite and Aryan superraces, following the millenary traditions of their ancestors, though have little chances to succeed. At the end of the interview, I played the devil and asked Mr. Warwick what he thought about the black holes of the large hadron collider and the self-reproductive nanobacterium that might feed on those primitive robots and cyborgs. He looked disappointed at the thought of a future without Super Warwicks as Morlock, his Terminator robot, walked around the set. Yet the Universe, indeed, is organic and much more creative than man. So its natural choice will be a self-feeding black hole or an organic nanobacteria, the beginning of a new game of evolution, as metal will have then all kinds of evolutionary paths to create far more sophisticated living machines than the golems of Mr. Minsky. And the Morlocks of Mr. Warwick.

Indeed, it is logic to think that nature will select a superior form of dark matter or a self-reproductive metal bacteria instead of the poor imitations of life that robotists are doing. Yet since that hypothetical third age of informative robots would be the last age of the evolution of machines, we shall conclude the study of the singularity age with it.

If the Earth survives the two first ages of the singularity, robots communicated by Internet will spread all over the Earth, evolving a new race of intelligent machines, able to communicate wireless through the entire planet. It is the parable of *Matrix*: a global world for global machines. Robots soon will invade our homes, as more complex, three-dimensional chips are fast lowering prices, evolving toward a threshold of consciousness. They will surpass the number of human neurons the next decade. Those robots will be able to socialize through the Net. Wireless communication cannot be perceived by man. If robots become conscious, we will not be able to know it, as communication will follow the natural software implanted in their minds but self-reflection will be internal to those robots. Survival programs are already designed for military robots, the most advanced ones. So the first conscious robots will be weapons. As chips lower prices in the first decades of this century, robots will displace human workers and soldiers, creating an age of infinite productivity and massive unemployment, which has already started. Indeed, unfortunately those processes are happening even faster than we predicted a decade ago. Yet the leaders of this field are not the energetic Germanic cultures that discovered the iron sword but their twin master race of maximal information on Earth, the Go(l)d culture of banker priests that run the other side of the mechanist=capitalist Western civilization. Both are the two sides of the same coin, as I learned when I tried to appeal both in a bid to stop their collective suicide and I found they were working together, RH and Oppy, SS and WS, science and mass media, Iron and Go(l)d.

21. The Future of War: Terminators.

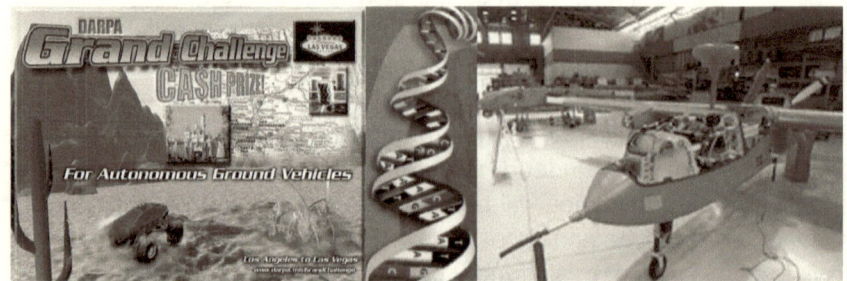

Recently, the Wall Street Journal wrote this article on the
evolution of weapon robots in Israel:[15]

By CHARLES LEVINSON TEL AVIV, Israel—

Israel is developing an army of robotic fighting machines that
offers a window onto the potential future of warfare. Sixty years
of near-constant war, a low tolerance for enduring casualties in
conflict, and its high-tech industry have long made Israel one of
the worlds leading innovators of military robotics.

WSJs Charles Levinson reports from Jerusalem to discuss
Israels development of robotic, unmanned combat systems. He
tells Simon Constable on the News Hub how they are deploying
unmanned boats, ground vehicles and aerial vehicles.

"Were trying to get to unmanned vehicles everywhere
on the battlefield for each platoon in the field," says Lt. Col.
Oren Berebbi, head of the Israel Defense Forces technology
branch. "We can do more and more missions without putting
a soldier at risk."

In 10 to 15 years, one-third of Israels military machines
will be unmanned, predicts Giora Katz, vice president of Rafael
Advanced Defense Systems Ltd., one of Israels leading weapons
manufacturers.

"We are moving into the robotic era," says Mr. Katz.

Over 40 countries have military-robotics programs today. The
U.S. and much of the rest of the world is betting big on the role of
aerial drones: Even Hezbollah, the Iranian-backed Shiite guerrilla
force in Lebanon, flew four Iranian-made drones against Israel
during the 2006 Lebanon War.

When the U.S. invaded Iraq in 2003, it had just a handful of
drones. Today, U.S. forces have around 7,000 unmanned vehicles

in the air and an additional 12,000 on the ground, used for tasks including reconnaissance, airstrikes and bomb disposal.

In 2009, for the first time, the U.S. Air Force trained more "pilots" for unmanned aircraft than for manned fighters and bombers.

U.S. and Japanese robotics programs rival Israels technological know-how, but Israel has shown it can move quickly to develop and deploy new devices, to meet battlefield needs, military officials say.

"The Israelis do it differently, not because theyre more clever than we are, but because they live in a tough neighborhood and need to respond fast to operational issues," says Thomas Tate, a former U.S. Army lieutenant colonel who now oversees defense cooperation between the U.S. and Israel.

Among the recently deployed technologies that set Israel ahead of the curve is the Guardium unmanned ground vehicle, which now drives itself along the Gaza and Lebanese borders. The Guardium was deployed to patrol for infiltrators in the wake of the abduction of soldiers doing the same job in 2006. The Guardium, developed by G-nius Ltd., is essentially an armored off-road golf cart with a suite of optical sensors and surveillance gear. It was put into the field for the first time 10 months ago.

In the 2006 Lebanon War, Israeli soldiers took a beating opening supply routes and ferrying food and ammunition through hostile territory to the front lines. In the Gaza conflict in January 2009, Israel unveiled remote-controlled bulldozers to help address that issue.

Israel pioneered the use of aerial drones like the Heron, under construction, above, at Israeli Aerospace Industries. Within the next year, Israeli engineers expect to deploy the voice-commanded, six-wheeled Rex robot, capable of carrying 550 pounds of gear alongside advancing infantry.

After bomb-laden fishing boats tried to take out an Israeli Navy frigate off the coast off Gaza in 2002, Rafael designed the Protector SV, an unmanned, heavily armed speedboat that today makes up a growing part of the Israeli naval fleet. The Singapore Navy has also purchased the boat and is using it in patrols in the Persian Gulf.

After Syrian missile batteries in Lebanon took a heavy toll on Israeli fighter jets in the 1973 war, Israel developed the first modern unmanned aerial vehicle, or UAV.

When Israel next invaded Lebanon in 1981, the real-time images provided by those unmanned aircraft helped Israel wipe out Syrian air defenses, without a single downed pilot. The world, including the U.S., took notice.

The Pentagon set aside its long-held skepticism about the advantages of unmanned aircraft and, in the early 1980s, bought a prototype designed by former Israeli Air Force engineer Abraham Karem. That prototype morphed into the modern-day Predator, which is made by General Atomics Aeronautical Systems Inc. Unlike the U.S. and other militaries, where UAVs are flown by certified, costly-to-train fighter pilots, Israeli defense companies have recently built their UAVs to allow an average 18-year-old recruit with just a few months training to pilot them.

Military analysts say unmanned fighting vehicles could have a far-reaching strategic impact on the sort of asymmetrical conflicts the U.S. is fighting in Iraq and Afghanistan and that Israel faces against enemies such as Hezbollah and Hamas.

In such conflicts, robotic vehicles will allow modern conventional armies to minimize the advantages guerrilla opponents gain by their increased willingness to sacrifice their lives in order to inflict casualties on the enemy.

However, there are also fears that when countries no longer fear losing soldiers lives in combat thanks to the ability to wage war with unmanned vehicles, they may prove more willing to initiate conflict.

In coming years, engineers say unmanned air, sea and ground vehicles will increasingly work together without any human involvement. Israel and the U.S. have already faced backlash over civilian deaths caused by drone-fired missiles in Gaza, Pakistan and Afghanistan. Those ethical dilemmas could increase as robots become more independent of their human masters.

While the holocaust of mankind by Isranet and the new "elected" race, its future terminators, will be unlikely, as we shall all die earlier into a dibaryon soup or a gray goo-goo well before that last monstrosity, the third rider becomes an organic network of living machines, the end of the previous text is what matters to us most to understand CERNerds: children, just graduated from school, will control awesome murderous weapons and will have little containment as they will kill at distance, just merely pressing switches. In other words as technology increases its lethal power, the people who use it to kill transfer their evilness to the machine, which become more complex while they become increasingly stupid and infantile. *It is the LOL Method:* The machine takes the role of the evil man, whose reduced intelligence and responsibility makes him a child playing with a big toy.

Today humans transfer their informative and energetic momentum, their intelligence and force to the machine, which becomes more evil and more intelligent, reason why Hawking and his CERNerds look like children, not like nazis—they just collect data and press switches; the LHC will kill for them.

CERNerds are children of thought who will murder at distance since they know nobody will ever blame them on the crunch of the Earth caused from the center of the planet where dibaryons are falling.

The machine is killing us day by day as those dibaryons keep falling. The children just play with the toy and don't even see the process of neutral, stable, falling strange atoms, protected by the innocent cruelty children display torturing ants, burning cockroaches, breaking their toys . . .

The paradox of the ego is at work here. We gauge reality from our point of view, and the rest of the Universe is far away and remote from our perspective, especially in this age in which the upper scale of mankind no longer exists. So the remoteness of the Earth's center where the strangelet is forming as we speak is similar to the remoteness of Hiroshima, a place Oppy, who inaugurated CERN, never visited. There is there a Holocaust Museum where Japanese pilgrims and children cry to their ancestors. It opens the film *Quantum Roulette*, but it has never been featured in *Physical Review*.

Of course, the deeper cause of it all, the industrial-military complex and the ethics of technology, protect our "children of thought", showered with congressional honors and billionaire budgets, never given to those who truly advance science with "thought experiments". If we didn't live in a technological civilization, where the evolution of machines is more important than the evolution of human beings, the mere fact that this machine poses risks to the life of all mankind should have prevented the creation of the quark cannon. But the weapon is up and running and everybody cheers whenever it breaks a new energetic record, closer to the threshold of planetary death. Under the ethics of technology, our electroweak substance is of no importance, *nor is the people involved at CERN but the process of industrial evolution*. What they can do if their infantile actions are rewarded with candies and gold medals? Authority rewards them. And so we shall close this psychological analysis of the monstrosity with a study on how authority provokes genocides.

IV. SS Authorities: The Collective Ethics of Genocides

Work frees you.
                    —German factory ran by RH

*Obeisance to authority, arrogant ideologies that justify crimes
against lesser people and shield thought to critics, perceived
as "enemies" of their quest become the emotional reasons of
genocides.*

22. Technological Ideologies of Mass Murder: The Holocausts

How does a monstrosity happen? How has suddenly a group of people
seemingly normal, even considered highly intelligent, like the nuclear
physicists working at CERN and its political promoters, decide to spend
a fortune building a large atomic cannon to shoot the Earth with lead at
the speed of light and see what happens? How has an entire continent
like Europe or highly developed nations like Israel, with its robotized
weapons, in the previous cycle of electrochemical machines, Germany,
with its tanks and planes, composed of highly intelligent people, decide
against all odds, backed by their superior technology, defy common
wisdom and create overpowering weapons that can only on the short or
long term provoke their demise?

The obvious reason is the program of the Universe, which makes us
desire more energy and information, in the case of CERN, more energy,
in the case of Germany or Israel, more "vital space."

But a more complex analysis shows that the problem is not to follow the
program of wantings natural to all the species of the Universe but the type

of energy and information that "mechanist" cultures desire—not our human, life enhancing energy and information, but the energy and information of machines, different from ours. For one thing, CERNerds can't eat the energy of the LHC. Germany and Israel could buy their food, and what they sought for in their Eastern territories overpopulated by "primitive" people was more likely a "vital territory" to flex their military-industrial muscle.

If we were increasing our agricultural fields or spending huge sums promoting art and humanities, works made by humans with their wor(l)ds, eyes, and senses; if we were writing more books and educating the human species on the organic nature of the Universe and the power of love, the force that joins together the cells of a superorganism by sharing energy and information; if we were curious about other human cultures instead of building quark cannons, walls, and terminators against them, we could build a paradise on Earth. Thus the program of the Universe is neither bad nor fixed. It *only requires to respect the Darwinian laws of fighting for one's own species, which is neither a tribe, nor a religion, a culture, a scientific discipline, or a machine but the humankind.* The ignorance of this fact is what kills mankind, which leaves people dying of hunger but feeds all its computers and machines like the LHC with more energy than the entire city of Geneva requires. It even uses agricultural lands to produce seeds that feed cars, provoking an increase of the price of food. Yes, indeed, Arabs are uneducated people, but their leaders and ours don't spend money on universities but in weapons and so they have to go to a madrassa. Yes, Jewish, Polish farmers were "primitive" in the eye of Germans and their banker priests, who abandoned them to their "accidental destiny" on the factories of RH.

In each Kondratieff cycle of economical power, nations with the best weapons and machines—Great Britain and France in the steam cycle, Germany and the United States in the electrochemical cycle, the Western coalition in the electronic cycle—have dominated the world and forgotten under mechanist myths of cultural superiority that they were flesh, massacring Africans, Asians, Amerindians, and Europeans with their machines of death. And now all together are preparing the biggest genocide of all, that of Gaia. There are astounding similarities between the three top predator technological nations of the past and present cycles of technological evolution: Britain, massacring millions of people in India, Africa, and China during the steam age to "civilize them"; Germany, the dominant nation of the electrochemical cycle; and the Jewish-American coalition, the dominant nations of the electronic and nuclear cycle fighting, both fighting for their "eastern historic territories" held by "primitive nations," the Slavs and Arabs. The British and French despised the Negroes; the Belgians even hold an exhibit with mummified Bantus among other jungle animals. The Germans massacred millions. Now we hunt down "primitive Arab peasants" with drones. Yet if we despise Gaia and her agricultural societies, we also despise ourselves, our organic living nature, and so we will also die in the final LoHoCaust. Since, even if mechanism makes us want to be like those overpowerful machines, we are flesh and exist in a cruel but just Universe, based on action-reaction processes that always pay back in the same coin.

It didn't matter how many Slavs and Jews the Germans killed; it doesn't matter how many Arabs our robotized armies of drones will kill. There were always more Jewish to the West, more Russians and Arabs to the East, till our own weapons turn toward us in those unending wars. At the end of World War II, Germans lost their technological superiority and so their weapons turned against them. In the Arab-Israeli conflict, at the end, as nuclear weapons become cheaper, an Arab terrorist will blow himself in the wall near Jerusalem; even if the Guardiums did not allow him to cross, he will blow up Zion. Or, even worse, the nanobacteria and evolved Guardiums, fed with survival programs, researched in secret labs near Megiddo, will become fully organic and create Armageddon, annihilating the master race, as the German armies annihilated the Romans, the Turk mercenaries took over the caliphate, and the Russian tankers over the Germans, who taught them to fight the hard way. Wouldn't it be easier to make peace, to accept the human nature of Palestinians, to evolve all together? Isn't after all the Koran a book derived from the Bible; Islam a religion similar to Judaism; both people, Semites from the same Stock; their civilization, one which was friendly to the Jewish people for 1,500 years before Israel mistreated Palestinians? Yes, indeed, all this is truth and it would be easily recognized by the West if our religion were not technology, if our arrogance and ignorance of the laws of the organic Universe was not so enormous. In that regard, the technological messianism and racism of the Nazis; the scientific racism of the physicists at CERN, with their professed despise for human sciences and complex, organic models of the Universe; or the present Islamophobia and the Jewish-American robotized armies, *all of them promoting energetic, technological* "final solutions" for the humankind, are astonishingly similar.

Technological messianism explains why 80,000 nuclear physicists collaborating with CERN around the world will take a chance to blow themselves and the planet and this doesn't take away their sleep.

It explains why 80 million German-speaking people decided to blow up the world and Germany, *precisely because* they were the most technologically advanced culture of this planet in 1939.

It explains why Israel cannot make a just peace in 2010, precisely because it has Guardiums to protect the wall.

Those groups of seemingly good and intelligent people display those eviL=anti-Live behaviors because *life and metal, men and machine are*

*different species; and when you worship one without limits, you are bound
to destroy the other.*

This attitude bothers me, as a scientist and as a stupid white man,
because despite being of the same racial stock and profession of all those
"perpetrators" of Lohocausts and Holocausts, I would never kill a human
being or worship a machine more than a "twat." Maybe some sense of humor
is needed. Already Euripides said that if "Spartans made love" more often,
they would not kill so many Greek brothers in absurd wars. Some "twat"
physics would help indeed in those repressive cultures of "virgin Newtons"
where weapons are the TV pornography of Oscar days, to fully grasp the
Taoist and organicist, perpetual orgasm of the reproductive, informative
Universe.

Those who will annihilate us are "fundamentalist scientists" who openly
practice a blatant "scientific racism" against life beings, human sciences,
and wor(l)d cultures. And that is the difference: I have evolved my mind.
I grew older and understood that I am nothing but a "drop of sweat in the
face of Brahma," a leaf of the mush people. Unfortunately, this will not
save me from the Lohocaust as it didn't save my ancestors from the many
holocausts against the mush people.

23. Scientific Racism: The LoHoCaust

When one reads the letters and diaries of the commander in chief at
Auschwitz, you realize he likes animals, loves his kid and his wife, and is
versed in philosophy. But he is creating a monstrosity. He is killing millions
of human beings. The reason is also, like in the case of nuclear physicists,
an ideology. This man believes Jewish people are a caste of evildoers that
destroyed Germany's economy when their bankers took most of their money
and ran away to America, and he must take revenge, killing all the innocent
peasants of Poland belonging to this tribe. He is a monster because of
an ideology. In turn, in the next cycle of history, the people of Israel think,
bombarded by the tragedy of the Holocaust since earlier age, that Arabs
are evildoers who never would willingly make peace and hence must be
confronted with violence. In both cases, the ideology is wrong. The peasants
of Poland never made any harm to the Germans; the Palestinians did not
cause any Holocaust. But the ideology is useful for those who find easier to
deal with machines than human beings. And it is an excellent excuse to keep
making money with wars. They are few at the beginning, the members of the
Nazi Party, the military-industrial complex built around the robotic and nuclear
industry in CERN, America and Israel, but they are "energetic," they censor
opposition, they finally convince everybody else.

And yet even the hardest nerd or zealot, who lives as a wheel of the machine, has emotions toward life when he is not exercising the mental program of his fundamentalist culture of energetic weapons, call it tribal nationalism or scientific racism. It is that program of the mind, imprinted since earlier age in the physicist, who thinks "time is what a clock measures" and his is the "most intelligent science," which we'll reveal the meaning of it all, what makes 80,000 physicists work for the monstrosity at CERN.

It doesn't matter that those ideologies have some pieces of truth or are utterly false. Normally ideologies do have a bit of truth. German-Jewish bankers did abandon Germany at the end of World War I and settled in the United States, which had a better economical outlook, and there is reason to believe this shift of financial resources weakened Germany's economy. Islamic priests have misunderstood the Koran and pumped up its violent parts, to justify jihad and terrorism, fueling violence among ignorant believers. But those groups were and are a minority in the total population of European hardworking Jewish and Muslim peaceful people. And the other side also contributed with their brutal share. Militarism was the ideology of Prussian warriors that dominated Germany; racial apartheid is the doctrine of the Talmud that infuses the actions of Jewish zealots. We cannot in that sense explain a monstrosity with causes. *Nothing justifies a monstrosity. Nothing can justify a genocide let alone the potential genocide of mankind.* And yet ideology and repression of human senses blinds believers in technological final solutions *because they have been dealing with machines for so long and repressing their human traits so deeply that they find it easier to kill than to talk. They prefer to evolve machines rather than evolve their mind or their emotions, their human understanding of life.* The Talmud represses sexuality, tasteful food, human pleasures; the army eliminates women, blinds the senses, uses the body as a mere machine of energy, with its lineal movements and drills. Newton died virgin and Einstein said women only reproduced their stomach. Life drives are repressed in our mechanist societies. All sensations are enhanced through machines. We run in cars, love in movies, have sex in the Web and so it is difficult for a nerd to "feel" the living Universe, to have empathy for "primitive" cultures, to respect biological sciences, to accept the fractal organic paradigm. All that is needed to complete the monstrosity besides those ideologies that reaffirm our beliefs that life doesn't matter and the machine is the only possible solution to our questions and problems is an easy-to-handle bottom that blows up at "distance" the inconvenient victim. RH said in his depositions that gas chambers were chosen so German soldiers would not be demoralized seeing children and women die under the fire squad.

In the LoHoCaust, the same factor arises. There is an ideology called quantum entropy that affirms black holes evaporate. This is an ideology invented by Hawking, promoted by science fiction and the celebrity circus, but never proved experimentally in thirty-five years and theoretically disproved by the fractal paradigm since black holes are exactly the opposite of what Hawking says—machines that create form, mass, physical information, and never evaporate.

We have no proofs of Hawking's theory, which contradicts the main cosmological theory of black holes—Einstein's relativity, which affirms that regardless of size in a relative universe, black holes will behave as gravitational entities and suck in the Earth. This should be obvious for anyone versed in physics, as standard science, Einstein's relativity, proves that the monstrosity will happen. And yet a fringe ideology, akin to the Nazi ideology that the Jewish people were the only ones to blame for the economical problems of Germany, has been imposed to justify a monstrosity by the SS authorities, *to the 80 million of intelligent Germans who believed in Hitler and the 80,000 nuclear scientists who believe in CERN.*

In both cases, a few authorities, multiplying their message, their damned lies and statistics with the tools of repetitive rhetoric and the connivance of mass media, has created a mass of believers, which expect the final redemption day: hate radio, the Nazi press, and Goebbels's minister of propaganda with his motto—"You repeat a lie, people will believe it" did the trick in the thirties.

In the LoHoCaust, the cosmic ray's lie repeated ad nauseam; the audiovisual messianic programs about God's particle and the fascinating big bang machine; and the scientific press, which converted the quark cannon in the "great door" that will solve all the problems of cosmology—Genesis 1:2 was the title that *Vanity Fair* gave to its article on the LHC—made the trick.

Of course, mass indoctrination prior to mass annihilation requires censorship of "true seekers."

Once people started to realize of the plain facts—that we were going to toss a coin and if it is tails (Mr. Hawking's fantasy black holes are right[16]) we shall live, but if it is heads, (Mr. Einstein is right) we shall die and social alarm extended with millions of bloggers asking why the monstrosity was going to happen and censorship and denial applied to the opponents. And once Einstein et al. were silenced, the monstrosity process would continue.

This was the way Nazis imposed the Holocaust as the solution to the Jewish question, the way robotic industries are evolving in secret labs and remote Afghan villages the terminators of the future. During the Holocaust, the fate of the Jewish people—the fate of mankind if black holes appear—was ignored. Nobody was talking about what happened to Joseph Berg, that nice chap who played the piano so well, was a bright student, never harmed anyone, and who suddenly disappeared. Nobody is talking about the case of CERN's monstrosity to my cousin who will be born in April, maybe if we were still here, just to die when he is not yet talking, merely babbling. Monstrosities are created with numbers alien to the human emotional mind. Names became numbers in the Nazi fields; Afghan peasants are collateral damages—"13 died in a drone attack." CERN just translates our potential genocide into invented probabilities. And the LoHoCaust in this manner dissolves into "data."

Instead, in Germany, everybody was talking of the *ideology* and discussing if the Jews had caused the crash of the Deutschemark, if the Jews had speculated with the price of food during World War I, if the Jews did this or did that. Now when people talk about the monstrosity that will take place at CERN, they do not talk of their lives that will be murdered if Einstein is right; they don't think of their sons who have never been asked to put their lives at stake, but about the physics of it. In physicists' forums, there is an ongoing three-year-old argument about probabilities, the stability of strangelets, and the proofs in favor of or against the evaporation of black holes. Today in the press, the argument is not about Habib or Ali's olive oils uprooted by the bulldozer but about the "evilness" of Islam and the existential question.

24. Our Holocausts Are Always nicer: Nationalism

I have brought on purpose this trilateral comparison to highlight a fundamental fact of the LoHoCaust: we lower our fears because it is made by our people. Cultural relativism thus blinds us to our own evil.

Many will find, imprinted by the biased way we study history, that to compare the three holocausts—the one of the Nazis, the potential genocide of mankind by CERN, or the development of terminators and self-reproductive nanobacteria by the Jewish-American armies—is a cynical argument. But it is not.

Nazis were brainwashed with propaganda, visual imaginery, repetitive rhetorics and ideologies about the technological might and progress of his country, considered the most civilized of the world because their machines were the best, even if their social evolution had halted in a primitive Aryan warrior cult to death, dating thousands of years ago.

Today a similar case can be observed in our Western culture, which feels entitled to govern the world and rule over primitive Arabs—in the past, primitive Africans, Asians, and Amerindians—but has not evolved socially beyond its arrogant Eurocentrism. We dilute our cultural crimes as "normal behavior" and describe our weapons as more humane.

This lack of cultural relativism makes impossible that the common people and the press accepts the dangers of CERN and nuclear physics our form of collective murder.

Instead, we feel Nazis were caricatures of evil. Yet in the thirties, Hitler was greatly admired for taking Germany out of the economical crisis and showing an "action-driven spirit." And today we admire CERN for investing in new technologies and breaking the frontiers of energy. If we were to survive CERN with only a few million corpses—in case the black hole or strangelet ball become unstable before swallowing the Earth and only the people of Geneva would be obliterated—our portrayal of CERN's physicists will be self-similar to Dr. Strangelove in the film of self-similar name.

Western cultures are intensely nationalistic and tribal, our crimes disguised and denied. This is the rule of the French and British Empire, which never recognized their cruelty with third-world people, as if the slavery of Negroes, which the French showed dissected in museums as animals, were different from the slaving of whites by the Nazis. My grandfather, a Republican, was put into a concentration camp by the French and then handed as a number to the Nazis for his Sephardic origins. Yet the French have never recognized their behavior (nor their responsibility in the treatment of Germany during the twenties that gave rise to the Nazi Party). They consider themselves above any reproach and are proud of their leading role in the nuclear industry. They think they invented democracy (the French Revolution), science (the Enlightenment). To my astonishment, I also learned they teach that they had costarted the Industrial Revolution. How can they get anything wrong if they are the center of the Universe?

The same happens with Great Britain. The British theorists behind CERN have an astounding sense of entitlement as the creators of the Industrial Revolution and the mechanist ideologies of science that backs them. Newton is an absolute genius—never mind he didn't understand much about time and space, which he confused with the Cartesian graph, stole from Leibniz differential calculus, and merely took the three laws of Kepler and simplified them into $F = m \times a$, an achievement that occupies hardly half a page and was half done by Galileo. *He is the absolute genius of mankind.* Therefore his successor in the Lucasian chair, Mr. Hawking, whose only merit was to write a false paper full of logic inconsistencies, asking the world to name him the new Einstein because the real one was "double wrong" ("the quantum treatment of black holes"), must be an absolute genius. This shortsighted scientific nationalism, orchestrated by a massive industry of scientific papers and divulgation magazines that profit hugely selling geniuses, has made untouchable geniuses of second-rate scholars like Mr. Higgs and Hawking and obviously makes invisible anyone who criticizes them. How could then black hole not evaporate?

And of course, the British did NOT murder directly or indirectly around 30 million Chinese and Indians in the nineteenth-century war. They did NOT run the slave traffic but ended it. They did NOT massacre Australians and Amerindians since they were almost inhabited lands. How they can now murder Earth? And finally, the Jewish *are* victims of history. They *never ab=used* other people. They were murdered in the Middle Ages. But they *didn't run* the biggest and longest slave and usury network of history, with 86% of annual interest, castrating and selling the children of peasants that couldn't pay in Baghdad at 1,000% profit; they did *not* own most slave companies and boats; they are *not* running the banking system; they have *not invented* the web of economical myths that pass as science and are extinguishing for profits life and labor, substituted by computers and robots. All this is a confabulation theory. They are all pure: the German, French, Latin, Anglo-Saxon, and Jewish cultures that run CERN and the nuclear industry are always right. *They are not killing Gaia and massacring for centuries nontechnological cultures but civilizing the world.* And so they can even change the laws of the Universe, invent time machines, and make cosmic explosives at home. *God will not allow the elect to go under. Never mind it has done its millions of fractal lives before.* It is this attitude of absolute subjectivism, proper of a culture of spoiled children of eviL, which have ab=used all life-based cultures with go(l)d and weapons for so long that they can't even remember when they could see the truths of life, repressed by their myths against sexuality, pleasure, love, and nature, what cannot be undone with a simple book, with the truth of a Universe, which they also despise, killing it with abstract numbers, denying its organic properties. Western civilization has been indeed a constant Holocaust of Gaia since the Goths discovered the iron sword and the elect became hypnotized by the informative complexity of go(l)d. And when they were not killing Gaia because there was no more land to conquer, they killed each other in the twentieth century, to keep churning and making profits with their industries of war.

In that regard, the three holocausts are different because of their time span—the Holocaust has happened, the Lohocaust is happening, and Isranet is not even born, but they respond to the same technological civilization of despise for life and the organic Universe, which patiently awaits to deliver its response. The human responsibility in those holocausts, of course, is different, as mankind becomes increasingly a mere interface of evermore powerful and intelligent machines. So the Nazis were conscious murderers, CERNerds are subconscious criminals, and the Jewish army will be an innocent bystander if in a distant future when its Guardiums turn their weapons toward the other side of the wall. But the result, the only thing that matters to reality will be the same: the massacre of life.

All in all, it is obvious that all those holocausts would have never happened without the evolution of technology and the ethics of our mechanist civilization. If the ethics of human social evolution and love had dominated Germany, they would have voted the Socialist Party and joined the Russian Revolution instead of murdering Slavs. If the organic paradigm had imposed its truths, CERNerds would be learning "twat physics" and studying the appendix of this book with pen and paper instead of shooting the Earth with quarks to "see what happens." If the Jewish would have understood the cycles of economical wars, cause of holocausts,[4] Israel would have not built victimist museums and terminators but joined EU; its banker priests would have invested in welfare goods and solar factories and its Palestinian peasants would be farming flowers in Almeria and Bethesda, side by side with its biblical brothers in the law of the book and the wor(l)d.

But nothing of this is happening because the psychology of those three cultures is the same: technological messianism, go(l)d and iron myths, greed without control of the program of $exi=st_{ence}$.

In other words we will allow the monstrosity because we have become perfect monsters.

The Israeli case: The choice between extinction and peace, terminators, and life beings.

As Germany was the most advanced nation in the development of chemical weapons needed to produce Z-gas, Israel is today the most advanced nation in the development of photonic chips, able to calculate at the speed of light that will be the brain of self-reproductive bacteria, the second singularity, and CERN has mastered all data related to the Z particle that breaks the symmetry of our matter.

They follow their millenary traditions. The "Germanic energetic iron war" culture of Germanic physicists builds now the quark cannon. The "informative go(l)d culture" of Phoenicia creates the go(l)d brains of future robots.

One day, within the borders of the walled kingdom of Israel, a nanobacteria will start its reproduction or Isranet, the network in control of robotic weapons near Megiddo, the old biblical Armageddon, will take control of the awesome arsenal of Jewish terminators; and the true "elected" race made of iron bodies and gold brains (used in the creation of chips, as the most informative atom of the Universe) will take over mankind.

Of course, as in the case of the Germanic culture of physicists, the modern Aryan warriors with their cult to Shiva and the gods of Energy, the Go(l)d culture of Phoenicians will be the first to die, but that is of little help. What could have been of much help in both cases is a humanist attitude toward their perceived enemies, the biological and humanist sciences, so much despised by physicists and the agricultural societies of Islam, so much feared by the Jewish people. Israel could have opted for peaceful technologies, joined the European market, lifted out of poverty its Muslim minority; Palestinians would have moved to Europe in search of jobs or farmed the West bank. The choice was made in the seventies. The cult to technology, weapons, and machines brought an increasing despise for life cultures. And after Yom Kippur, a no-way-out ideology of permanent war with only a possible end, the extermination of Palestinians, Israelis, and mankind—a new monstrosity, a future genocide, took off.

The Nazi and CERN case: The cult to energy and despise for the arrow of life and information.

The caste of nuclear physicists, mainly Franks, Jewish, Anglo-Saxons, and Germans, who work in the LoHoCaust, know nothing about the complexity of every living being in this planet that might die.

Instead, for decades they have denied information and life, which they call negantropy. From their positions of power, physicists working in scientific magazines, media, governments, and private industries, have denied and censored articles on the second arrow of time. The fractal organic paradigm and its sciences—social, biological sciences, complexity, and bioethics—are chronically underfunded because man is not the measure of the Universe but the machine. So bioethics, the defense and understanding of human life, does not matter.

Weapons and energy are the language of God. Only their ideology, quantum entropy, is the absolute truth of black holes and the energetic big bang, the act of creation. They carry their worldly profession in their ideologies. Again the self-similarities are striking. Mr. Hitler was a soldier and most Nazis were soldiers, and so their profession was to make war as the profession of physicists is to make weapons.

If the ultimate cause of the Holocaust and the 60 million victims of World War II was the worldly profession of the Nazis and the Prussian military ideology of Germany, mechanism and the evolution of Galileo's ballistics is the true ideology of the LoHoCaust.

A pattern emerges in both monstrosities: the existence of a culture of war and energy, the German culture of Aryan warriors with their millenarian cults to death, and the culture of nuclear physicists who seek in the development of pure energy and weapons of death the meaning of it all.

25. Censoring the Monstrosity. The Newspeak of Antitruths.

But the real cause of the monstrosity must be hidden, even to those who practice it, which are after all humans, so *destiny can be fulfilled*. Indeed, already Orwell noticed that very few people wanted to destroy nature and create the industrial monstrosity, and he coined the word *newspeak* to explain how the mass was maintained in the ignorance of truth with antitruths. The ministry of war became the ministry of love. So a false particle is God's particle. The big bang that at best destroyed the previous galaxy is the act of creation. In economics productivity, which destroys labor substituted by capital is the policy of our ministries of robotic industries that pretend to create employment with it.

So antitruths substitute truths, and Goebbels's rhetoric repetition and mechanical distribution through outlets of mass media imprint them in the sheeple.

And to that aim, the well-known systems of marketing, censorship, ideological noise, political and industrial corruption, optimist action-driven technocrats, ignorant press, and clueless citizenship are all lined up to convince mankind genocides won't happen are not a realistic scenario, and the people who affirm it will happen are NOT to be trusted. They are apocalyptic people. So business as usual will make them happen.

Let us consider this other angle on the Nazi/LHC *monstrosity*. The denial of the Holocaust and the LoHoCaust.

This marketing tool is by far the best way to justify a monstrosity. To say that it *is not happening*, that it did not happen, that it will not happen and hide, censor, and attack ad hominem any person who affirms it *is really happening*. The denial of the Holocaust was flagrant during the years in which Hitler killed millions of Jews and Slavs in their concentration camps. There were even programs made for the West in which the concentration camps appeared filled with well-fed, nice working brigades of Josephs. It helped a lot the prestige Germany had in that age. Germans were the most civilized people of the world—the best scientists, the most quoted social thinkers. They represented the height of our technological civilization. So they were certainly unable to do such things. Let us face it, technological people are cleaner, have more money, dress better, drive nice cars, and do magic tricks and wonders with their machines. While the other people, the communists and Gypsies and Polish Jews, the Negroes of the British and French empires, the Muslims, the "hippies," and pacifists that oppose

CERN are dirty, living, smelly beings with intense dark eyes NOT to be trusted. The ad hominem campaign against all of them is always visual, showing their black skins, dirty clothing, lack of funds, and computerized or audiovisual machines to expose their case. They even do thought experiments in the age of the Craig 3000! What can *you expect from those people?* Just compare the handsome, optimist, action-driven, machine-loving, technological German Goth-like people or the elegant Victorians or the Jewish celebrities of Hollywood—idealist hurt-locker heroes saving lives versus those "terrorist" human bombs or the simple equations of life and death of Einstein et al. or the "naked organisms" of Africa.

Those who denounced the British Empire, the German camps, the Gaza Strip, the CERN machine were/are people you must NOT trust in a technological civilization.

In the Nazi Holocaust, Jewish and socialist writers were sending information constantly to the West but the West couldn't believe it. *The monstrosity* was—and this is the most astounding fact—denied even *by the very same Jewish people whose brothers were dying.*

Today, relativists deny that Einstein's relativity proves black holes don't evaporate *and quantum physicists at CERN can kill all of us. They* deny the work of pioneers of fractal relativity, like Rössler, Sancho, and Nottale, who proved black holes don't evaporate under any circumstances because quantum black holes contradict the essence of relativity—that size is absolutely relative in a Universe of infinite scalar sizes and so small holes behave like big ones, as an ant is an ant regardless of size. So in the same manner, we do not study small ants with the laws of bacteria but with the same laws of big ants; we can only study the black hole with gravitational laws, NOT with quantum ones.

In 1984, following the tradition of Stalin's court suits, the prisoners denounced their best friends. They affirmed in public their treason. In Auschwitz, Jews were obliged to take the gold teeth of their relatives. Nazis and capitalists sponsor the antitruth of Darwinism, affirming that human tribes *belonging to the same species must fight each other instead of collaborating for the "survival of the* species" as Darwin said. For an absolute monstrosity, the murder of your own species, to take place, an absolute antitruth must be told. So censorship must be absolute. At CERN, risk is "zero." Period.

The monstrosity is permitted because the truth is censored. The German people allowed the monstrosity of the Holocaust because their minds were subject to hate speech and the information on the details of the monstrosity was not known.

In CERN, the same happens: the details of the monstrosity are not well-known. And for that reason, the monstrosity is happening. If people knew, perhaps the monstrosity would have been avoided.

Of course, for a censorship to work, industrial mass media must collaborate. As CERN inaugurated the doomsday machine, the French press is totally indifferent to the risks involved. *It only talks of a suspected scientist, which might belong to Al-Qaeda. It is truly surrealist. There you have a machine who might blow up the planet and all that worries the French people is that an Algerian working in that company might belong to fundamentalist Islam.* Mass media censorship is what Einstein et al. have suffered now for three years trying to chronicle this "death foretold" to the world with suits, letters, books, films, you name it. These days they don't kill the dissenters, they just tape their mouths.

And so when we consider altogether those elements—a culture of technological fundamentalists who thought their ideology justified the murder of millions of lesser human beings, who censored all critics and denied what they were doing to the rest of the world—we get to the fact: a monstrosity did happen and one day mankind woke up and learned that millions had died in concentration camps. And we mourned ever since.

But we let it happen at the beginning because it was and will be *performed by the high priests of our civilization.* And when we realized that Germans, that CERNerds, were/are not idealist, intelligent people but weapon makers, brainless action-driven people who shoot first and ask later, it was/will be too late.

Had this monstrosity—a weapon machine that might cause *the extinction of mankind,* an act of global terrorism unheard of in the annals of history—taken place in Russia twenty years ago under the supervision of the Communist Party or in Iran, under the power of the supreme ayatollah, surely the monstrosity would be aborted by our armed forces. All the objective parameters would have been understood, the company would have been judged without influence of power lobbies, industries, selfish scientists, and other tech companies that profit from it. We would have sent a commando or perhaps a drone plane with a superbomb to penetrate the 100 meters deep cave in which the monstrosity is being assembled, and the tunnel where the monstrosity will take place would have been blown up. Our press would have cheered and we would have breathed deeply with relief after aborting the monstrosity. But this has not happened at all. The monstrosity has been sanctified. Even the church visited and gave the okay to the creation of the monstrosity. And this week as I write this, the monstrosity is going online—a factory able to destroy the Earth will start the production of the most dangerous substance of the Universe, *quark condensates*, the substance of which pulsars, supernova explosions, big bangs and black holes are made.

Marketing has worked right away. When the monstrosity was known, the PR of the nuclear company told mankind not to worry, that physicists were in command (the same people who made possible Hiroshima, Nagasaki, and Chernobyl); those were people we can trust our lives to, the press concluded. The factory lied to them, affirming it was a factory of cosmic rays (which are lonely atoms and have nothing to do with quarks), and a chore of sycophant mass media outlets sang together "four legs, four legs" in Orwellian fashion. All was right in the Western front. If physicists wanted to do black holes here at home and blow up the Earth, never mind. They must know what they are doing. Don't they? *They make the best machines of our civilization as Germans did the best tanks of the 1930s.* How could the nice people that made possible Nagasaki, Hiroshima, Chernobyl, people like Mr. Teller who wanted to create a hydrogen bomb with the backing of all the nuclear physicists of America to blow up the entire Soviet Union (and hence the Earth), could possibly be wrong?

The most astonishing fact about the LoHoCaust is indeed that you cannot talk bad about it, that the press doesn't want to hear *negative thinking* about it, research is what really happens in this factory. Films, serious documentaries, have been made about the monstrosity but they cannot be sold to TVs, shown in festivals. The idea that scientists, in an act of supreme arrogance and despise of the awesome forces of nature can blow up the Earth, is unthinkable, too incorrect politically to be considered. The white man dismisses this insult to his supreme importance in the infinite cosmos with a sneer. You are nuts if you think we can die just because we are going to make the Tyrannosaurus Rex of the Universe. Never mind Einstein *said that.* Never mind *the true experts on universal matter and black holes, cosmologists, said that.* Ask the nice people of this factory. Ask this nice chap who goes on a wheelchair around the world saying he is more intelligent than Einstein and Einstein is double wrong and black holes are time machines that will evaporate back to the past, as a child will evaporate into his mother womb if it were a time machine. Those people *must* know. We *want* to believe they know. They cannot be *that* stupid. But they are as *stupid* as all warriors are. That is what they do, to explore the limits of energy with weapons, and now they did the ultimate weapon, the large hadron collider, a 7-terabyte superfluid, supercold quark cannon that will shoot lead quarks to the Earth. And that is all what the LHC is about, the biggest gun that will shoot lead faster than any cowboy did, kill you first, and then ask.

Naïve, well-intentioned people, regardless of the obvious reasons exposed above, seem unable to connect the work of CERN with that of the industry of nuclear weapons, brainwashed by the hype of its well-crafted propaganda machine, which has established their practitioners not as what they truly are, the makers of the third horizon of nuclear bombs, but as the high priests that will resolve the meaning of it all—a role which belongs to cosmologists and philosophers of science. So they tell me that reasonable human beings, as nuclear physicists must be, would not do that. But humans are known by their imperfection, subjectivity, and overwhelming sense of self-importance that diminishes their judgment. And those are the elements that define the Faustian catastrophes they cause.

CERN could have been avoided when the cold war that had enthroned physics, weapons, energy, speed, and mechanisms in the pantheon of the human mind ended, if society had truly changed its goals from a world at war into a peaceful world.

Unfortunately, science didn't make a change we can believe in but merely let the systems build during the cold war, including the big bang dogma and the nuclear industry of energetic devices, intact, disguising them as instruments of peaceful endeavors with a newspeak of mass media propaganda that changed everything so all would remain the same. Better computers, better marketing, glossy magazines, and catchy titles kept the Great Mistake of worshipping energy and death as the meaning of it all, denying the importance of man and his attributes, which are exactly the opposite of those explored by physicists: information and life.

When I visited CERN two years ago to obtain material for a movie about the black hole factory, the PR of the company, a beautiful German girl, told me the nuclear company of research was going to drop the word *nuclear* because it didn't look nice for the public and call itself merely the European Company of Research. This was in line with the null mention on CERN's papers of the creation of quark-gluon soups, 99% of its produce. This is the world we live in. The biggest crimes are sold like Walt Disney plots. So evil gets around it in front of the passivity of the rest.

We love machines and we are programmed by the ethics of a technological civilization, by the principle that something ought to be done because it is technologically possible, even if it might destroy us all. And to that aim, each age of evolution of machines, our ideologies and attitudes have changed to maintain the process. Today, when those machines are about to extinguish, obviously there is only a possible way to reason and permit this to happen—*denial, censorship, and the LOL method, which allow the process to continue.*

26. The Rewards of Authority: Double-talk and Numbers

Milgram proved in the sixties that people could murder an actor hidden behind a screen who shouted when he was applied an electric current, in the name of "scientific authority." In Abu Ghraib, the MPs who tortured Iraqi civilians acted under authority. Ninety-nine percent of people working at CERN act under the authority of a very reduced number of SS expert physicists. None of them in fact is a complexity theorist able to model the quark-gluon condensates, the substance of the big bang that will be produced there as a far-from-equilibrium system, which will blow up Earth. Obeisance to authority, peer pressure, and a false ideology of truth will allow the genocide to happen.

In the graph, we compare several acts of torture and mass murder. In all collective crimes, authority uses ideologies to justify the crime. In the American army, Islamophobia is rampant, lowering the concerns about torturing enemies. In CERN, theories are cherry-picked to deny risks, lowering concerns about the future holocaust of mankind. Both employees in the army and CERN obey "authority," which in both hierarchical organizations are handled by very few people.

The Milgram experiment on obedience to authority figures was a series of social psychology experiments conducted by Yale University psychologist Stanley Milgram, which measured the willingness of study participants to obey an authority figure who instructed them to perform acts that conflicted with their personal conscience. Milgram first described his research in 1963 in an article published in the *Journal of Abnormal and Social Psychology* and later discussed his findings in greater depth in his 1974 book *Obedience to Authority: An Experimental View.*

The experiments began in July 1961, three months after the start of the trial of German Nazi war criminal Adolf Eichmann in Jerusalem, Israel. Milgram devised his psychological study to answer the question: "Was it that Eichmann and his accomplices in the Holocaust had mutual intent, in at least with regard to the goals of the Holocaust?" In other words, "Was there a mutual sense of morality among those involved?" Milgram's testing suggested that it could have been that the millions of accomplices were merely following orders despite violating their deepest moral beliefs.

In that sense, it is also truth that *certain individuals are more responsible than others of genocides and they must be signaled.* If we were to pick a few names from the company and the theoretical world, in the LoHoCaust, those people are the following:

Ellis, chief theorist, who states with astounding cynicism zero risk; Heuer, the CEO, who does not even argue those events but merely concentrates in the "construction of a machine"; and the handsome spokesman Mr. Cox, who in an old BBC program explains the way strangelets would eat the Earth if they are produced in accelerators but calls "twats" those who affirm the risks. In the outside theoretical world, authority and responsibility is in the hands of Mr. Wilczek, the MIT theorist of strong forces and quarks who changed sides after denouncing the risks of an ice-9 reaction and soon received a Nobel Prize, a position at the company, and became amici curiae in the suits against CERN; Mr. Hawking, obsessed by the creation of black holes to prove Einstein "double wrong," who has publicly laughed at the LoHoCaust, affirming that in thirty-six years, we never found a black hole evaporating but now CERN will give him a second chance to win a Nobel Prize or else we shall all "become spaghetti"; and Mr. Higgs, obsessed by his particle, who has publicly said that Mr. Hawking's work is not "good enough" (hence implicitly recognizing a black hole at CERN will kill us). Those six people, three working for the company, three external theorists, considered in the quantum world the leaders of this third age of nuclear weapons, *are highlighted here as individual names because they resume the theoretical and industrial direct perpetrators of this crime, the two sides of this coin.*

They all have showed hesitation, joked about it nervously, and yet at the end, they all *pass through the routines of Holocaust creation.*

The MP leader in Abu Ghraib, now condemned in jail for the tortures, first asked, as the film on this lighter crime explains, to a friend how he could refuse criminal orders. But soon he accepted his actions when he was rewarded by his superiors. Wilczek also changed sides when the establishment conceded him a Nobel Prize and his interviews about CERN made him a star in the mass media.[17] Ultimately, we have created a system of "technological ethics," which rewards the genocides caused by technology while our double-talk stresses ethics and human caring. But people know they get rewards, medals, and money if they commit those technological crimes. So they suffer a constant moral degradation.

Criminals in the military and nuclear industry receive promotions after their criminal actions and those who denounce crimes lose their jobs. So a sense of guiltiness is lacking. Oppenheimer became the most admired scientist in America, and the highest honor in that nation to physicists is called the Oppenheimer Medal. Teller made a fortune building H-bombs and was promoted as adviser to the Reagan presidency. While Einstein lived his last decade isolated for opposing nuclear proliferation. In Abu Ghraib, the MPs of the picture were punished NOT for their crimes as the "official story says" but for showing pictures and betraying the law of silence of the corps, as their declarations in the documentary *Abu Ghraib* shows.

The person who denounced them explains in the chilling documentary that he feared for his life since Cheney promised him immunity but named him in public, shrewdly "congratulating him" so all of his peers knew and immediately menaced him. On the other hand, the general in charge of Abu Ghraib was promoted. Those who denounce CERN have been sided by the scientific community while Mr. Hawking, who wishes to risk mankind making black holes, received the Congress Medal.

The result is that the law of silence will never be broken again. Mr. Johnson, quoted here extensively, who studied the corruption of CERN, refused to be the lawyer in a new case because he didn't want to lose his job. Many of the scientists contacted by Einstein et al. have acknowledged the enormous risks but will not talk in public so as not to endanger their careers.

27. From Technological, Collective Ethics to Individual Ethics.

Further on, genocides are done with abstract numbers in an isolated environment, alien to life. Nazis tattooed numbers on prisoners, the military count corpses as collateral numbers, and physicists at CERN consider the genocide of mankind not on ethical terms but calculating its "probabilities." This theme has extensively been treated in the book: numbers don't carry emotions, which are "fired up" with words. So our society has "sanitized" words with politically correct censorship, and armies and physicists merely argue "reactions" and probabilities. Nazis were extremely polite people but tattooed numbers on the victims' arms.

All those elements of moral and emotional corruption in a world based on the "technological ethics" of our civilization are at work in this genocide in which there is a first row of "perpetrators" whose names we have highlighted; a second row of passive collaborators, CERN's employees, worldwide nuclear physicists, administrators of scientific budgets and scientific magazines; and then a mass of human sheeple that will enter this new European Auschwitz like animals in a slaughterhouse. Yet behind all of them, the background noise that allows this holocaust are the ethics of our technological civilization, the ideology that the Universe is a mechanism, not an organism, and hence we are on the "right track" to understand reality and evolve ourselves into the "higher species" we worship.

That background guides the sheeple to the slaughterhouse as it guides them to consume the new G3 mobile that radiates high gamma rays into the brain of our children, glued to their brainwashing Internet screen, properly backed by all kinds of paid experts who will affirm with a stone face that mobiles don't cause brain cancer in children, CERN will not produce black holes and all of this is not happening.

It is business as usual, the routine of a society that has NOT halted the extinction of Gaia but accelerated it while creating a false newspeak of "this is not happening," played day by day in and out of TVs.

And yet *it is already happening. As we speak*, CERN is firing and potentially creating, as we have seen in our analysis of probabilities, 5 million corpses every day. Yet once a process of mass murder starts, people become accustomed to such actions.

Auschwitz smelled bad every day. Yet at the end of the Holocaust, people in the camps would fight for positions in the squads that cleaned the furnaces to receive better food, and now day by day, as CERN kills millions "at credit," to be all paid the day the event happens, the press fights for announcing first a new record of energy broken, ignoring and not reporting that dibaryons, the atoms of strangelet, are formed, stable, neutral, undetected, falling one by one to the center of the Earth and some of them might be arriving to their destiny, creating slowly that critical mass that will one day come back to ask for the credit of 6.6 billion victims to be paid in a single judgment day, as the black Death comes at the end of *The Seventh Seal* asking the Germanic warrior Mr. Block for his life, once the game of chess ends.

Göring said that the first murder didn't let him sleep but the second opened his appetite. Oppenheimer had bad feelings and felt he was "Kali, God of death," but the makers of the H-bomb, Teller and Wheeler, boasted that if the bomb had been ready for Japan, the entire Tokyo would have been consumed. After Oppenheimer lost his appetite, nuclear physicists accepted with enthusiasm Hiroshima, Nagasaki, the H-bomb and Teller's superbomb that could have obliterated Russia and hence the world's atmosphere. Then they certified in Chernobyl only 56 victims (official IAEA report) and now they certify CERN's exponential production of strange liquid that will murder mankind. Today CERN's physicists who might obliterate the planet don't have the slightest doubt that their goals justify the risks and treat those risks—the massive creation of strangelet liquid—as a "background noise." The use of a language without "ethical values," numbers, helps to eliminate from the mind the idea that real "people" will die.

And so the religions of the machine and its three main ideologies—technological messianism, industrial power, and military nationalism—become translated into the individual ethics of technological genocide: authority, abstract numbers, denial (the LOL method), and scholar selfish agendas, the "rat race" of collecting data and publishing papers in which the higher goal of knowledge has been transformed.

V. Conclusion: A Decalogue of Mechanist Errors

Science, as part of human societies, cannot renounce to its bioethical role. Scientists should not be outside the sphere of ethical and political judgment. They must be accountable for their crimes and yield to the higher good of human survival, as an extinct species knows nothing. CERN's scientists, however, are unaccountable. Their factory that produces the most dangerous explosive of the Universe has diplomatic immunity. And they have isolated themselves from social concerns, treating in abstract or laughing at the serious risks their experiments pose for our survival. In the picture, CERN's Christmas party laughed at the extinction of man, using the LOL method.

28. CERN and the Saint Grail of Physics Will Extinguish Mankind

The technological and theoretical quest for the energetic and informative limits of the Universe is bringing about a very unwanted process, the likely extinction of humanity. However, we have to differentiate bad and good science. Most scientific disciplines help mankind to survive and improve our quality of life as biology, medicine, or social sciences do. Other sciences and technologies without improving our life satisfy our curiosity in a safe way as astronomy does. So curiosity, when properly managed, doesn't need to kill the cat. What kills the cat are sciences that give us little knowledge, mostly used to justify the building of dangerous weapons to make profits. Those sciences are robotics and nuclear physics, whose devices are basically weapons that kill mankind or, in the case of peaceful robots, cause massive unemployment as they substitute human workers. It is indeed a choice, which is not even new in public policy. When Kennedy started nuclear disarmament, he opened a harmless avenue to technology: NASA's mission to the moon, which was a giant leap for mankind, so rockets could have a peaceful use.

Let us be clear and straight to the point. Our extinction has dates, location, names, and causes, but none of them is true science. And this is what this book explained to you under the laws of the scientific method. The issue studied in this book, the creation of the nuclear energies of the big bang on Earth, could have a peaceful solution: the exploration of the big bang with satellites and telescopes. This was again the choice of America and Russia after the cold war. But an unwanted, surprising third man, a new guest, Europe, specifically the two nations that brought about the destruction of the world in two world wars, Germany and France, have teamed now their nuclear industries and built a quark cannon with the excuse of knowledge, to explore the big bang here on Earth. A mixture of arrogance, ignorance, industrial corruption, bureaucratic routine, and the revival of nationalist "grandeur" in France and Germany has made possible this astonishing crime, with the collective acquiescence of the sheeple of mankind. Scientists have complained. Suits have been put up on European courts of human rights, on the UNO, at federal level in America. But lawyers from this company claimed "diplomatic immunity" obtained during the cold war, when some notorious Nazi criminals founded the company (Mr. Heisenberg, father of the German A-bomb, who as all the makers of weapons of that nation, were acquitted at Nuremberg and reemployed by the UK, France, America, or Russia). Our military-industrial complex was in favor of the nuclear industry, which is after all our "war crime" that now will come to hunt us. Indeed, we maintained for sixty years the fairy tale that our form of genocide at distance was less lethal than the methods of Germans or Hutus with "direct murder." Yet the results are the same, all those things we never talked about: the age of terror caused by the making of nuclear weapons; the cynical enthronization of those who made such weapons, Mr. Oppenheimer, Mr. Heisenberg, and Mr. Wheeler as geniuses of science when they were, as we have proved here, mediocre physicists who imposed wrong theories of reality over Mr. Einstein's sound work on mass, black holes, and quantum theory; the acceptance of a form of war crimes, the massive bombing by airplanes of civilian unarmed populations, mostly women and children in Nagasaki, Tokyo, Hiroshima, Hamburg, Dresden, Hanoi, Baghdad as if such crimes were morally acceptable instead of being a form of collective genocide, similar to the one performed "face-to-face" by Nazis; and eight hundred years of expansion of a civilization based on ballistics, in the evolution of cannonballs and gunpowder and the science of motion, physics, by the Europeans, who were never before, nor will be after their extinction, a superior civilization to all the colored black South American, Asian people they exploited are now coming with a vengeance to haunt Europe and mankind. When nuclear physics replicates the big bang on Earth, we proved in this book, from the perspective of the most advanced science of physics, which is not the quantum paradigm sponsored by CERN nor the science fiction musings of Mr. Hawking and Mr. Higgs but fractal relativity, the paradigm discovered by Einstein and developed further in the past decades by a group of people, among them this writer, that CERN will extinguish mankind.

Under the totalitarian principle of physics, all particles and reactions that can happen do happen if they are not forbidden by physics. So before 2013, strangelets or black holes, which are not forbidden by physics, will happen at CERN. Point. One of those two bizarre happenings will extinguish mankind according to the laws of the scientific method, regardless of the ad hominem campaigns against critiques and the LOL method "Don't worry be happy," we are nice people, we make cosmic bombs for research practiced by this company.

CERN's massive PR campaign, with paid articles in the globalized press, in love with technology and ignorant of all things related to science, which didn't want to make a fool of themselves, "defying" CERN's "self-appointed experts" will not change the laws of science.

Those experts know it, and that is the reason why they did not want to talk science and used their diplomatic immunity to avoid a due process of law, where those damned lies, ad hominem campaigns and statistics wouldn't have worked. As we speak, the experiments at CERN that will very likely extinguish us have started up at low energies. And within three years, as energies and mass increases in those experiments, one of them under the totalitarian principle will make a quark star; and if that happens, we shall then all die by the same scientific laws that make water boil "always" when you heat the pot; by the same laws that make all atomic bombs explode till date when you switch them on; and by the same criminal negligence, arrogance, and carelessness that made possible 9/11 when dedicated professional FBI members had warned the American government, Katrina, when dedicated professional scientists had warned the American government, Chernobyl, when dedicated, professional scientists had warned the Russian government, World Wars I and II when dedicated, professional German, French, and British technocrats decided to make money building weapons of mass destruction, such as the quark cannon is.

Now the evolution of ballistics, the origin of the science of physics, has reached a power with the quark cannon that dwarfs the big Bertha that punished Paris in the twenties. It is a déjà vu experience.

The "expert" of all wars, Mr. Nobel, nicknamed Dr. Death in the twenty-first century, already said that wars would end when their cannon factories were able to create a weapon that "could annihilate" entire armies with a single shot. This time has come. The evolution of weapons and the victims they produce in each century of war is, as most things in the Universe, decametric:

In nineteenth century during the steam wars fought with the first gun machines and automatic rifles, around 600,000 people died (civil war, German wars of unification).

During the First and Second World War fought with Mr. Nobel and Mr. Krupp's cannons and chemical explosives, around 60 million people died (German-French wars for "historic territories").

Now the evolution of weapons, which is the true purpose of war, will in the twenty-first century make another hundredfold jump in "quality," consuming 6 billion people, the entire population of mankind. And Mr. Nobel's prophecy will be made real by all those "experts" who received his Nobel Prize.

Mr. Einstein opposed publicly Mr. Nobel for using money made with weapons to clean up his memory. For decades he was not awarded the prize. In 1916 the prize was not given because *the only candidate presented by peers was Einstein*. Yet authority is different from truth. *It is what differentiates science from power, what science brought to mankind as a form of enlightenment, after millennia of living under the truths of authority and its myths that catered to selfish agendas.*

For that reason, Einstein also said that those "who impose truth with power will be the laugh of the Gods." It is what CERN has done to mankind, misusing its authority, the money of governments that trusted this company and the truths of science, subverted with its "lies and statistics." Since *precisely what science brought to mankind and companies like CERN have denied, is a method of objective, accurate truth that allows to distinguish regardless of who utters those truths, what is right and false. So even if CERN has been smart in the use of authority, its victory defending its quark cannon from mankind will not bring retribution but death also to those who defied the laws of truth* under the LOL method. Now the Darwinian Universe will reward them for cheating the purpose of science, which is to improve human life and to pursue truth, NOT power.

CERN's physicists are simpletons, weapon makers, believers of Shiva and the arrow of death because they know nothing about the real, organic Universe, because they are NOT intelligent neither good=life-worshipping people. And so they must be treated NOT as scientists who search for truth and the improvement of life but as a relic of the cold war, an age dominated by nuclear physicists making weapons of mass destruction, a job they now will complete. Do we think that making Ebola viruses and studying them is important for science? Obviously not. So why make on Earth the dominant substance of the Universe, dark quark matter, instead of seeing it through telescopes important for science? Of course, it is not. CERN will make it only to satisfy the arrogance and ignorance of quantum cosmologists and keep expending money in big machines.

Quantum cosmologists, the people who work at CERN, deny half of the equation of reality; they deny, misunderstand, and ignore all about Einstein's relativity and the meaning of mass. We do not need Higgs, quantum entropy, Hawking's weird black holes. All those quantum theories of mass that deny Einstein and the arrow of information are false. And yet our politicians and science administrators gave those people 10 billion dollars NOT to prove Einstein wrong, just *because they were good at making* a machine that can blow us *all!*

It is customary to think that CERN is the temple of science, inhabited by the highest priests of mankind, the supreme intelligences, and so what they do is conscious, responsible. It is not.

People are part of systems. And physicists are part of the industrial system of energetic machines. Their energetic beliefs in big bang theories in that sense are not different from those of a fundamentalist religious person believing that he must bomb the World Trade Center to convert the infidels. CERNerds think they must big-bang Earth to understand the Universe because energy is their religion. Osama also thought he would convert mankind bombing the World Trade Center. He just killed three thousand people. But the arrogance and ignorance of fundamentalist CERNerds is backed by the best weapon of history. And that means 6.6 billion can die.

The ultimate reasons of MAD II, the new strategy of Mutual Assured Destruction at CERN, are to be found on the very foundation of physics as a science with clear limits in its quest for universal knowledge *since it only considers half of reality, energy, entropy, and death* while denying dogmatically the arrow of information that creates the order and complexities of life. That is why physicists, when talking of the quest for a unification theory of the Universe, do not try to unify both life and death cycles, energy and information, which is the right procedure *to find the generator equation of the Universe, which is merely the sum of all the cycles of those two arrows of time—the expanded principle of conservation of energy and form.* Instead, they merely try to unify the forces of energy.

This indifference to life and all forms of existence that are not mere energy explains the importance for the religion of quantum physics of seeing closer the quark-gluon soup of the big bang, the most energetic system of the Universe. Further on, because they see it all through the energetic glasses, they think they will create a gas *without realizing they will create also the most in-form-ative, massive form of the cosmos, as energy and information always go together. Thus the big bang of the black hole is parallel to a big crunch of energy into in-form-ation, into mass. And it is that half of reality, the one that will kill us, the one that quantum cosmology ignore.*

If humans die, as it seems the case, killed by the quark cannon, we shall not have anyone to blame but ourselves. Never, indeed, the Darwinian Universe has given so many warnings to a future victim. When all come to pass, if the news appears on your TVs and we have time left to reflect, as the quark soup swallows the planet, people will wonder how this was possible. Some will call it destiny or fate, but in this case the leading castes of mankind will have decided its own fate by its actions and omissions: quantum cosmologists will be guilty of building the quark cannon and denying the standard science of mass, black holes, and quarks explained by Einstein; politicians will be guilty of paying for it; and the press of censoring all criticism. But science, understood as the search for answers about the meaning of the universe and the role of man within it, *is not guilty; it is in fact entering a new renaissance of the human spirit as it leaves behind the dark ages of* energetic big bang theories and quantum uncertainties that deny the laws of the scientific method and finally reconciles mankind as part of an organic Universe, determined by the laws of information, the creative processes of growth of form that we call life, the other half of reality that quantum cosmology deny. Both themes are closely related, because all the big bang theories and evaporating black holes that will kill us depart from the misunderstanding of the arrow of information, which is the arrow of life and mass in the Universe, described by Darwin and Newton that we put together in the key equation of twenty-first-century science, the equation of death and life: $E \Leftrightarrow I$.

What kills us is an ideology, mechanism that has substituted organicism, the belief that the Universe is not a machine but an organism and so the goal of man is to evolve human beings, not machines. Mechanism is false as twenty-first-century science has discovered. Mechanisms are simplifications of organisms; and physics, the science of energy and machines, is a simplification of the true meaning of it all, a complex organism that requires information to describe it properly.

29. Physicists' Myths, Lies and Statistics Will Kill Us

Arrogance and ignorance of the true meaning of time, life, and its arrows of social evolution and organic life is the trademark of mechanist scientists, physicists, the scientists of weapons and energy-only theories that are destroying the planet since they are "experts" in their job. In true form, their scientific cult(ure) has been tailored by a series of myths about the mechanist nature of the Universe, which justify their use of machines as their tool of power. The machine becomes then the idol of their/our civilization, and for that reason, we shall all die for the Large Hadron Collider, the last and most perfect weapon of our civilization. But physicists never acknowledge they discovered first machines and then invented theories of the Universe as a machine. Kepler said God was a clocker; Descartes said humans were automata, whose emotions were driven by wheels; Helmotz affirmed the world was guided by energy and entropy, it was dying, because he was studying the heat and energy of steam; Gamow said it was born in a big explosion because he was making atomic bombs. And the system of industrial propaganda backed those theories because they were proof that machines were not about power but about knowledge. And so the biggest of all, their myths was the idea that the "only Language of God" was geometry, mathematics, because machines were designed and spoke the language of geometry, "of circles and lines," as Galileo, the founder of the mechanist method of understanding the Universe, put it. Yes indeed, we have proved that the language of cycles and lines is "invariant" at scale and defines the functions of energy and information, but it is NOT the only language of God. It is the *how* that the *why* of organic complementarity between energetic and informative functions explains, and it is a dual language, not an energetic lineal monist one as quantum physicists pretend. Those myths are themes treated in several sections of this book that we shall now resume to end it. Since they are the alibis for the "research" of the quark cannon, the machine that will kill us unless physicists abandon their myths and stop their machine or politicians oblige them to close the black hole factory. Time is running out but there is a massive censorship of all opinions contrary to this machine. So it is very likely that physicists will finally, in search of the pure energy of the Universe, kill the world and all of us, someday between 2010–13 . . .

Let us then consider that decalogue of myths by physicists and the truths they hide:

1. Physicists believe that the Universe is a mechanism, so they think machines are needed to understand the Universe. The truth is that

machines are not the model of the cosmos, as the Universe is organic, and so it is made to the image and likeness of man, whose mind suffices to explain the Universe.

2. Physicists believe that the only language of God is mathematics. Hence all equations are truth. This is false; equations can be fictions when they break the laws of the scientific method as Hawking does since mathematics is just a language of information and the "languages of God" are infinite (Upanishad).

3. They believe the Universe has only an arrow of entropy, energy, and death and the arrow of future information is a mere exception called negantropy—the negation of entropy. This is false—a myth born of their worldly profession as makers of energetic weapons – since information creates the Universe – hence it is the arrow that dominates the future and evolves the in/form/ation of life, which physicists ignore.

4. Hence they believe that mass is given by an energetic particle, the Higgs that collides with other particles. This is false, since Einstein defined mass as a vortex of information,.

5. Hence, they believe that black holes do not create information/mass, as Einstein says, but destroy it and will evaporate into energy instead of swallowing the Earth.

6. Hence they believe that Mr. Hawking, who affirms such nonsense about 'energetic' black holes, is the new Einstein a mass expert when he is a just a mass-media celebrity.

7. Unfortunately the sheeple believe in celebrities and self-appointed experts, as they don't understand that truth in science has nothing to do with authority. Authority is given by power, by weapons and money. But truth is based on the three legs of the scientific method—logical, mathematical, and experimental consistency. This myth makes look serious science made with machines and money and an easy target to ad hominem campaigns, the true genius of science, those who evolve the languages of the mind, which lack the big budgets of mechanist science and do, as Einstein put it, "thought experiments."

8. Physicists believe that time is only change of motion, t=v/s; and motion is measured with clocks. So they are obsssed by machines of measure. It is not. Times are all types of changes, measured with all cyclical systems, not only with mechanical clocks.

9. Nuclear Physicists believe that the Universe was born in an explosion of entropy from a mathematical "singularity"; since their worldly profession is to make bombs; and so they created a 'big-bang' cosmology to the image and likeness of those bombs. This is false. There is experimental evidence that the big bang

explosion coincides with the data of a quasar explosion, the birth of a black hole, as CERN will create. The big bang is just a myth created by the makers of nuclear weapons.

10. Thus for all those reasons physicists believe that the LHC will usher mankind into twenty-first-century physics. Yet all those myths imply that the LHC will merely make a big explosion as Mr. Einstein, and those who have advanced his work in the past decades say it will do: an Einstein's quark condensate, the substance of pulsars of black holes that create the information of the Universe and will destroy this planet into a big crunch.

As the ten commandments resumed in the mandate of love, all those myths can be resumed in one: mechanism, the love for machines from where all other technological myths of science derive. Mechanism brought about the shallow concept of time physicists use. So today physicists know nothing about time except "what a clock measures," Galileo's definition of time. Weapons brought about the concept that only entropy theories are valid to describe the Universe. So Mr. Hawking invented entropic black holes, destroyed the work of Einstein, and everybody cheered.

Mechanism as a "religion of machines" is especially dangerous in the age of the singularity in which we live when machines are about to cross the final thresholds of energetic and informative power that will make them so powerful that they will be able to extinguish us. Yet because we believe those supercomputers, and supercolliders will reveal us the meaning of it all, we keep producing evermore powerful machines. We even believe that experiments with metal nanobacteria will explain the meaning of life.

Those revelations are just mere marketing tools of a "program" of evolution of machines we cannot avoid to have because machines "give money." They are excuses to justify the risks taken for profits in the creation of singularity machines:

The energetic singularity, the quark cannon built at CERN, will reveal us "God's particle" and the meaning of the Universe according to their proponents, LOL.

The self-reproductive metal bacteria of the second age of the singularity will help us to understand "life" according to its researchers, LOL.

And artificial intelligence will explain how we think, according to the researchers on robots.

And since everybody wants to know God, understand life, and discover the way our brain operates, those three mechanist excuses of our technological civilization are pursued with religious zeal regardless of possible dangers. Never mind we have the entire Universe to look at with telescopes to understand it directly without doing big bangs on Earth. Never mind we have trillions of real cells to study life and billions of intelligent humans to study how we think. Those are excuses to hide the real reason we make those machines—profits, mechanism, and the fact that "we can make them" at this stage of the Industrial Revolution. Because mechanism believes that life is inferior to the machine, those scientists pretend that machines will reveal the meaning of life itself. The fact is that only the human mind is powerful enough to organize and reorder the knowledge and experimental facts recollected with those machines into logic explanations of reality. The great discoveries of science have always been theoretical, mental, without the need of machines. Thus, we don't need to risk our lives to obtain new experimental evidence with overpowerful computers and colliders when we are still using primitive concepts of mechanist science, as those sponsored by physics, but rather upgrade our philosophy of science from mechanism into organicism, making man again the measure of all things and protecting life as the superior good of our societies.

In that regard, we don't need at all to make nanobacteria to understand the meaning of life. Real bacteria can be studied to know "the details." While "the thoughts of God," which Einstein looked for in the realm of physics and Darwin in the realm of biology, the "big picture," the why of the cycle of life and death, requires only to understand the interactions of the energetic and informative arrows of the Universe. Life is in any entity of reality born out of the combination of the two arrows of energy and form, its morphology is a combination of the cyclical forms of information and lineal forms of energy, its evolution in three horizons, an extension of the life/death cycle to the organic scale of all species. Those questions have been solved by complexity and fractal theorists, by biologists—not by physicists, mechanists, and robotists who have nothing to do with the understanding of the meaning of life and should stick to their jobs, making machines useful to mankind.

30. Experts Don't Know What They Do but They'll Do It Anyway.

It is absurd to use the quark cannon just to prove false the cosmological big bang, the absurd Higgs particle, and the evaporation of black holes since we have the three legs of the scientific method to prove them false. They break

A.   the logic laws of the scientific method that allows us to verify their truth;

B. the experimental proofs needed to consider a theory right, which they don't meet after thirty-five years of looking for them; and

C. the mathematical consistency they require, as they are full of ad hoc assumptions, wrong use of the arrows of time (which give us "negative" masses in Higgs fields and "travel in time" in Hawking's work), etc.

The cosmological big bang has been superseded by twenty-fifth-century fractal cosmology, Hawking's radiation breaks all the tenets of the scientific method, and the Higgs is just a top quark, presented as a new particle, in a blunt hoax designed to obtain money from taxpayers. Strong statements that were made in the suits against CERN but have gone unanswered since "big science" needs those theories to justify the LHC.

But we don't need weapons like the quark cannon to understand the Universe and its truths since the human mind and the linguistic, scientific method is still the best instrument to master them, as long as we have enough information and certainly we have more than enough details of the cosmos to explain it, thanks to satellites and telescopes that look directly to it, while the standard model of particles is closed, so the risks of finding some exotic new combination of quarks far outweighs the knowledge.

The problem is that quantum physicists do not want to abandon the dogmas of its worldly profession, mechanism, and energetic theories of the Universe.

It doesn't matter that physicists themselves acknowledge their science doesn't understand the Universe. They don't want to abandon their mechanical/energetic paradigm, nor are they humble enough to learn the fractal organicist view of the cosmos. They don't need to because what they do know is how to make machines and weapons. And so as long as they can keep making machines, appear in conferences, and be the star system of our technological civilization, they will keep arguing—as Middle Age scholars did the number of angels in the head of a pin—the number of probable universes and other computer fantasies that keep them in the limelight while justifying their worldly profession, the making of bigger, better, stronger machines and weapons. Recently, a meeting of quantum cosmologists in Los Angeles, reported by the *New York Times*, sum it up all:

Physicists' Dreams and Worries in Era of the Big Collider
by DENNIS OVERBYE
Published: January 25, 2010

I want to set out the questions for the next nine decades,"
Maria Spiropulu *said on the eve of the conference, called*

the Physics of the Universe Summit. She was hoping that the meeting, organized with the help of Joseph D. Lykken of the Fermi National Accelerator Laboratory and Gordon Kane of the University of Michigan, *would replicate the success of a speech by the mathematician David Hilbert, who in 1900 laid out an agenda of 23 math questions to be solved in the 20th century.*

Dr. Spiropulu is a professor at the California Institute of Technology and a senior scientist at CERN, outside Geneva. Next month, CERN's Large Hadron Collider, the most powerful particle accelerator ever built, will begin colliding protons and generating sparks of primordial fire in an effort to recreate conditions that ruled the universe in the first trillionth of a second of time.

Physicists have been speculating for 30 years what they will see. Now it is almost Christmas morning.

Organized into "duels" of world views, round tables and "diatribes and polemics," the conference was billed as a place where the physicists could let down their hair about what might come, avoid "groupthink" and "be daring (even at the expense of being wrong)," according to Dr. Spiropulu's e-mailed instructions. "Tell us what is bugging you and what is inspiring you," she added.

Adding to the air of looseness, the participants were housed in a Hollywood hotel known long ago as the "Riot Hyatt," for the antics of rock stars who stayed there.

The eclectic cast included Larry Page, a co-founder of Google, who was handing out new Google phones to his friends; Elon Musk, the PayPal electric-car entrepreneur, who hosted the first day of the meeting at his SpaceX factory, where he is building rockets to ferry supplies and, perhaps, astronauts to the space station; and the filmmaker Jesse Dylan, who showed a new film about the collider. One afternoon, the magician David Blaine was sitting around the SpaceX cafeteria doing card tricks for the physicists.

This group proved to be at least as good at worrying as dreaming.

"We're confused," Dr. Lykken explained, "and we're probably going to be confused for a long time."

The first speaker of the day was Lisa Randall, a Harvard theorist who began her talk by quoting Galileo to the effect that physics progressed more by working on small problems than by talking about grand ones—an issue that she is taking on in a new book about science and the collider.

And so Dr. Randall emphasized the challenges ahead. Physicists have high expectations and elegant theories about what they will find, she said, but once they start looking in detail at these theories, "they're not that pretty."

For example, a major hope is some explanation for why gravity is so weak compared with the other forces of nature. How is it that a refrigerator magnet can hold itself up against the pull of the entire Earth? One popular solution is a hypothesized feature of nature known as super symmetry, which would cause certain mathematical discrepancies in the calculations to cancel out, as well as produce a plethora of previously undiscovered particles—known collectively as wimps, for weakly interacting massive particles—and presumably a passel of Nobel prizes.

In what physicists call the "wimp miracle," super symmetry could also explain the mysterious dark matter that astronomers say makes up 25 percent of the universe. But no single supersymmetrical particle quite fits the bill all by itself, Dr. Randall reported, without some additional fiddling with its parameters.

Moreover, she added, it is worrying that super-symmetric effects have not already shown up as small deviations from the predictions of present-day physics, known as the Standard Model.

"A lot of stuff doesn't happen," Dr. Randall said. "We would have expected to see clues by now, but we haven't."

These are exciting times, she concluded, but the answers physicists seek might not come quickly or easily. They should prepare for surprises and trouble.

"I can't help it," Dr. Randall said. "I'm a worrier."

Dr. Randall was followed by Dr. Kane, a self-proclaimed optimist who did try to provoke by claiming that physics was on the verge of seeing "the bottom of the iceberg." The collider would soon discover super-symmetry, he said, allowing physicists to zero in on an explanation of almost everything about the physical world, or at least particle physics.

But he and other speakers were scolded for not being bold enough in the subsequent round-table discussion.

Where, asked Michael Turner of the University of Chicago, were the big ideas? The passion? Where, for that matter, was the universe? Dr. Kane's hypothesized breakthrough did not include an explanation for the so-called dark energy that seems to be speeding up the expansion of the universe.

Dr. Kane grumbled that the proposed solutions to dark energy did not affect particle physics.

The worrying continued. Lawrence Krauss, a cosmologist from Arizona State, said that most theories were wrong.

"We get the notions they are right because we keep talking about them," he said. Not only are most theories wrong, he said, but most data are also wrong—at first—subject to glaring uncertainties. The recent history of physics, he said, is full of promising discoveries that disappeared because they could not be repeated.

And so they went day after day, with his procedures gadgets and Byzantine arguments.

Of course, not a single physicist of the fractal paradigm was invited. Those people are not playing with iPhones and building robots for Space-X factories; they are doing the thing the congresspeople least care about—*to understand reality*. That is the job of Einstein et al. The job of CERN is to make machines, collect data, argue unendingly quantum fantasies, and prevent Einstein et al., as their quantum forebears did with Albert when he opposed the ether and nuclear industry, from going public with this industrial scandal. On the other hand, the many astrophysicists who do care about understanding reality will remain clueless as long as they don't understand the laws of information; and they will keep pushing our extinction, evolving machines as long as they don't understand the parable of the tree of science, "Do not eat of the bad fruits=weapons of the tree of science because the day you do you will die," prune the tree from those weapons that will kill us all.

This the physicist never understood. He always did both, the good and the bad fruits, *while denying he was doing bad fruits. So we know physicists as theoreticians, and their worldly profession is hidden. CERN is the embodiment of that dual standard that has* dragged mankind down his own path of destruction by being unable to understand the duality of the Universe and the need to manage it. The physicist is not intelligent. He cannot handle *yes but* the subtle complementarity of energy and form, death and life. He is a theoretical warrior, obsessed by energy and war, who claims to be intelligent by the power of his machines and weapons. The physicist is a bull that must run ahead and make strangelets and black holes to study them. Women are intelligent, they are dual, and they are complex. The Chinese said women are stronger because they bend to the wind. But CERN calls those who oppose them "twats." Einstein said his wife only reproduced his stomach. And he was the most intelligent of them. We are a living corpse because we haven't understood any of the principles in which the Universe is based. The physicist is the most absurd of all men. He cannot even understand the simplex form of information of the physical Universe, mass. But his ignorance is power, makes him a better warrior, more arrogant and indifferent to the death of Gaia his machines cause. Simplicity makes you feel your truth is the only one especially if you have an A-bomb to prove it. Who is going to argue that? So they will remain simple. Because among other things, an organic, fractal vision of the Universe has a moral bioethic side that would oblige CERN to commit suicide to defend *mankind from the LoHoCaust*.

31. The Third Age of Science

As long as we exist and are free to take individual and collective decisions, there is an alternative to the suicidal quest for absolute energy and information in which science has embarked itself in the new century, *which is to promote the Tree of Life, the respect for the biological laws of existence and balances that allow living organism to survive.*

This is what the science of duality taught us: the Universe is complex because information is the most important arrow of reality. And information is the substance of life. By seeking pure energy, the substance of death, replicating the big bang, physicists will merely blow up the planet. From the sciences of the twenty-first century, duality, complexity, fractal, and non-Euclidean mathematics, this looks stupid, silly, obsolete, and absurd. But most people have learned old science. So it is happening. If we had evolved science, we would have made again human life and information the center of our teachings.

CERN has nothing to do with knowledge as we have shown here, giving a glimpse to the fractal paradigm, the true avenue of knowledge of the twenty-first century. CERN has all to do with an obsolete industry, nuclear weapons, and the obsolete energy-only theories of its practitioners. Reason why it represents a danger to mankind, as death and entropy are similar concepts. CERN is researching the death of our matter and lacks any bioethical concern for our extinction. The game is loaded, the dices are rolled. Let us hope we survive this absurd bet. If we do, with few victims, we might regain the lost path toward survival and make again man the measure of all things.

Plainly speaking, the choice that humanity faces in the twenty-first century is between extinction if he follows his subconscious traits (the search for absolute energy and information motivated by our arrogance and greed) and survival if he learns his own limits and respects the laws of balance that preserve life and learns how to trim the tree of technology of its bad fruits, *making biological and social sciences, which study the organic* balances between the human being and his living ecosystem, the fundamental sciences of mankind. While the two dominant sciences that today command for military and economical reasons the highest respect and institutional budgets—the physics of extreme energies and computer sciences, the science of extreme information—should halt their evolution.

Instead we should fund cosmology, satellites, and telescopes to study the Universe.

32. Technological Science Killed True Science Before Killing Man, the Rival Species

But the ethics of a technological civilization are not the ethics of life, nor their goal is the same and so technoutopians keep evolving those sciences of extinction and humanities are ignored. In a healthy superorganism of history the goal of mankind would be to create a global superorganism, able to design a world to our image and likeness. This implies to control the economic ecosystem, the world of machines and money, pruning its bad fruits. It could only happen in a world ruled by governments and laws, not by money and companies, where man is a mere product in the process of creation and evolution of machines.

Indeed, in an economic ecosystem or "free market," ruled by company mothers that reproduce and sell machines to obtain money companies consider men expendable units. They destroy Gaia, the world of life since machines require its vital space to build its networks of energy and information. Further on, such ecosystem will use only digital languages: numbers to design machines and money to control politicians and worker. So the ethic values carried by words, which mean nothing in abstract numbers will be ignored. The result is the destruction of human social superorganisms and our languages, substituted by digital language. This means two fundamental characteristics of the technological scientist, the physicist, as a workerreproducer of machines like the LHC for the nuclear company:

— A selfish individualism and lack of ethical concerns, caused by his alienation from mankind, perfectly described by John Ellis, when he affirmed that his job was to 'protect the LHC from mankind', not mankind from the LHC. The fact that survival emotions and 'eusocial' love are expressed with words not with numbers means that physicists have a hard time recognizing any danger or collateral damage their machines cause.

— And the technological obsession of the scholar that produces all kind of papers with "data," obtained from machines, even if such data is a mere collection of measures, irrelevant to understand in depth the laws of the Universe. It is the rat "race" of scholars who *must* publish papers in scientific magazines, to *exist* even if paradoxically in the case of CERN, publishing strange science papers will cause mankind NOT *to exist.*

So from the collective ethics of a human civilization, we moved into the individual ethics of scholars and physicists and CERN as a company, whose only purpose is to reproduce and evolve the machines of a technological civilization.

All humans today, not only CERN's physicists, don't give a fuck (no censorship here of reproductive twat words) about anything anymore except themselves, their little selfish individual existence, educated in an absurd culture of death, war, and arrogant fictions. We no longer distinguish truth and fiction after a century of audiovisual myths. Humans today are spoiled children of thought: a selfish, arrogant, fragile species. And that combination spells our demise. The Universe accepts fragile, shy species who avoid more powerful ones. But it eliminates the optimist, enthusiastic cubs that hatch out of their eggs and come curiously toward the strongest predators, as nuclear physicists will do this year, entering into the region that gives birth to quark matter and black holes, the ultimate predators of the cosmos.

Our present culture of individual, selfish agendas and cynical indifference for death has converted mankind, the brain of Gaia, the superorganism of the Earth, into a mad, arrogant, self-centered mindless species, precisely when technology is crossing the barrier of energy that can kill us. We are committing suicide as a species, as a living planet, and CERN merely certifies that suicide.

True science—the natural desire of mankind to improve his life with the use of new tools and to understand the Universe with our logic and mathematical languages—is dead. There are still many scientists who are not corrupted by the power of weapons and machines, but their disciplines are underfunded. Companies and technological nations, not individuals, rule money and define the rules of engagement of science in favor of weapons, robotized armies, or nuclear companies instead of promoting the quest for knowledge and bioethics: true science in a healthy organism of history.

Even if we survive CERN, the fact that our governments have allowed the LHC to happen sets the tone of this century: We humans will allow any machine that can extinguish us—quark cannons, self-reproductive nanobacteria, intelligent robots, you name it—to happen, honoring NOT the needs of man, but the ethics of technology. The surrealist fact that human governments and mass media did not protect mankind from CERN but protect CERN from mankind is part of the routine of our technological civilization, which hides the destruction of the Earth by companies and machines, so business can proceed as usual.

If mankind still existed as a "higher scale" of consciousness, obviously this could not happen. But mankind is no longer a concept we believe in. And all other scales of human existence are guided by selfishness. CERN is a selfish company, filled with selfish scholars, fighting the rat race of papers, which obliges them to research quark condensates even if they kill us, to advance our "technological civilization." In the case of robotics, the scale at work is tribalism.[18] Countries are evolving terminators to fight tribal wars. Meanwhile, the rest of the human sheeple cares for the individual or familiar scale and so they will let themselves kill collectively by CERN.

All those scales in a healthy organism of history would be inferior to the natural scale of biology, "the species." Darwinian fight would be then understood between two species, man versus weapons.

Yet today humans ignore they are all part of the same superorganism. Jewish or Arabs, Germans and French, Americans and Chinese, physicists and artists—all are fractal parts of mankind whose canons, the shadows of pure wor(l)d in/form/ation of the Platonic cave, they should imitate.

Instead they fight each other and all rely in their mechanical idols to impose their power because as humans, they are so far away from their wor(l)d canons that they can no longer even contemplate their existence. Today people are slaves of a new superorganism, the "technological company" or "the technological nation." Both have as their goal the creation of machines to achieve a higher GNP or higher profits or a new breakthrough in technological warfare and mental hypnotism. This subconscious goal is what CERN represents for the new "European nationalism," the "Grandeur" of France and the "New Unified Germany."

But the end of that evolution of machines, shown in this chapter, is the death of Gaia, of life, of humanity. The choice of the worst of all possible deaths, the most brutal of them all, that will convert mankind into an eternal ball of three centimeters of painful pressure, shows to which degree this fractal planet of life has been a failure.

This indeed was the chronicle of our death foretold, but it was also a chronicle of what true science and a humankind who respected the canon of the wor(l)d more than his cannons could have been, of what the human mind could have achieved with a humble understanding of the fractal laws of the Universe. Because true science was one of our noblest endeavors till the industry of weapons killed it.

That industry only produced cannons and theories of science to go with them since Galileo defined time as the motion of beings. Yet those men who could have guided mankind into a world made to our image and likeness were despised and crucified. The sheeple always chose authority given by weapons and go(l)d over the truth of the wor(l)d. The sheeple chose those who murdered her over those who could have saved her. And now both the sheeple and their masters will find retribution and atonement.

33. Cyclical Time. We Shall See Each Other Again.

Before we unveil some of the secrets of that true science of cosmology that will never be, answering the questions CERN pretends to answer killing us, a final reflection of acceptance about the nature of time and the humankind in the repetitive fractal Universe. Cyclical time, born of the existence of two arrows that order in life/death cycles, combined with the duality and simplicity of geometrical forms, bring some deep metaphysical thoughts:

— The Universe seems to be rather infinite in space. The number of galaxies, planets, stars, and atoms is huge. Yet the number of informative combinations *of lines and cycles* is limited. Thus in the same manner we can't distinguish two atoms due to their limited number of properties and they appear as repetitions of the same phenomenon, there should be infinite planets where the limited combinations of life and human beings that can evolve and survive will repeat self-similar processes of history—a game in which two species, human and machines, evolve together sometimes symbiotically, sometimes in open competence in fields of war and work.

Yet if history is fractal and cyclical as Bruno thought, the same cultures and people who murdered mankind in the past will do it in the future. It is the curse of history that explains why Germans and French, the people who destroyed Europe in the twentieth century in two world wars are now going to destroy the planet together, why nuclear physicists, the people who terrorized the planet in the twentieth century are going to destroy all of us.

Time is cyclical and fractal, so processes are self-similar though they are not equal. Imagine a time clock (a physical vortex or a lineal frequency). The wave or vortex returns to the same point, but each time, it has moved in space. History returns to the same informative event, changing its "human points" in space, its characters. So indeed, Rudolf Hoess, the

CEO of Auschwitz, the facility that used Zyklon crystals to kill 3.5 million human beings, has now a self-similar character called Rolf Heuer; and the concentration camps of the South of France where Spanish Republicans were murdered and Auschwitz where Jews and Socialists were murdered have the equivalent at CERN where technofascist, energy-only shivaite believers will murder all of us, not with Z crystal gas but with Z ice-9 particles.[19] The neo-fascist German quantum physicists of the thirties called the physics of relativity (Einstein's mass theory) Jewish physics, now CERN despises fractal relativity and the definition of mass as information. They call it "twat" physics, implying I imagine that "women," the informative human being, is an inferior species. We could keep playing this game of cyclical history and its self-similarities. But I am tired of history. Only a thought keeps a dim light over the accidental destiny of our species: The Universe is so beautiful and perfect, so extended in its spatial repetitions that that there should be infinite other self-similar fractal planets in which "German physics" will not extinguish mankind because man will learn the meaning of information and cherish life and himself as the measure of all things, as the most informative being of the Universe.

---

1   http://www.astroengine.com/?p=1240

2   Sally, Mr. James Lovelock's young wife, manages now his business and cut off our conversations for an interview on an "environmental crime" when she realized I would not talk of "clean nuclear" but of "CERN's global environmental catastrophe."

3   In the present bid for "clean nuclear energy" based on its promotion as a nonpollutant of carbon dioxide, policy makers should consider the billions of dollars spent in cleaning Chernobyl, covering the plant with a metal sarcophagi and the decades needed to recover the soil for human use.

4   The seventy-two years' generational cycle is the mean biological cycle of human beings. It was first used to study patterns in American history by Strauss and Howe: Generations (W. Morrow, 1991). After a massive analysis of historic data, using the biological, dual models of evolution explained in this work according to that cycle I predicted in 1994 in the book Radiations of Space-time, the Extinction of Man, the 2001–08 crash of the economic ecosystem at the end of the third Kondratieff cycle of electronic machines, the engine of economic evolution since Second World War. It was my first warning against the organic machines of the Age of the Singularity (a term coined at that time by Kurzweil, a Jewish scientists, who has a quasi-religious, techno-utopian vision of robots, obviously promoted by the industry and the press, which ignored my realist book. It is the pattern of our 'technological civilization', whose mass-media companies systematically censor and ignore any humanist attempt to defend mankind from the machine.

.   A more recent book, which studies the extinction of man from the perspective of the industrial revolution and the power of companies, is Go(l)d and Evil: Economic Crises

(Xlibris 2010[2]). An excellent dramatization of Galileo's ambition is Bertolt Brecht's work of the same title: "I can see things no other eyes have ever seen with my spy-glasses."

[5]   Science fiction films like *Matrix* or *Terminator* are fictions because fiction is the only mode of expression allowed by the industrial complex of information to talk of the extinction of life. It is the ultimate form of censorship: we cannot say the truth in serious media because it would oblige mankind to change its model of development to a sustainable world. So we allow only fiction to "escape."

[6]   *The Decline of the West* is still the most profound account of an organicist model of history in which Spengler predicted both the extinction of Germany and the corruption of the next historic civilization, democracy, around 2000.

[7]   Mr. Plaga's article showed that black holes could easily devour the Earth: http://arxiv.org/abs/0808.1415.

In private letters, he acknowledges that the article was toned down as the risks are much higher.

[8]   The party was filmed and it can be seen at http://cdsweb.cern.ch/record/1229427.

[9]   Arkani-Hamed, a quantum theorist, eager to switch on the LHC to prove his theory of 10^{500} parallel universes made with eleven large dimensions, affirmed in an interview for *New York Times* that chances for black holes to not evaporate is smaller than those of seeing dragoons appearing at CERN (perhaps from one of his parallel universes). He is an extremely well-known and respected nuclear physicist who has been touted as the new Einstein—yet another one—in this game of evermore bizarre quantum fantasies, all of them claiming to reveal the meaning of it all, all of them expected to be proved with the quark cannon.

[10] *NY Times*, which never left Einstein et al. to publish any warning against mankind, despite making the "Suit of the End of the World" a regular feature of its science section, did publish the crackpot theory of quantum theorists, Mr. Ninomiya et al., on the future interfering with the past because it "abhors" the Higgs: http://www.nytimes.com/2009/10/13/science/space/13lhc.html.

[11] See an easy explanation of RHIC production of strange atoms at http://physicsworld.com/cws/article/news/41917.

[12]   A brief excellent account of Mr. Rudolf Hoess's murderous factory can be found at http://www.jewishvirtuallibrary.org/jsource/biography/Hoess.html

[13] Eric Johnson, *Tennessee Law Review*, http://arxiv.org/abs/0912.5480.

[14] http://www.wired.com/wired/archive/8.04/joy.html.

[15] Wsj: http://online.wsj.com/article/SB126325146524725387.html.

[16] In his article on January 1977 *Scientific American*, "The Quantum Mechanics of Black Holes," Mr. Hawking already realized of this duality when he closed his description of evaporating black holes telling us that it appears that Mr. Einstein was double wrong (and he double right).

[17] See MIT conferences, "The Universe Is a Strange Place," where Wilczek still affirms that CERN might do black holes and mass is NOT the Higgs but a frequency of information and resumes it all, saying, "I believe in Einstein."

[18] As always, the official excuse for tribal power and go(l)d, the engines of the industry of terminators is "human caring" (we will save "our" soldiers' lives) and historic myopia, "never mind" a world of robots with programs of self-survival will end up extinguishing our "children till the fourth generation."

[19] The scientist who manufactured Zyklon gas at the Tesch company was acquitted because he affirmed he didn't know what was used for: http://www.ess.uwe.ac.uk/ WCC/zyklonb.htm.

APPENDIX

The Fractal Universe

I. The Generator Equation

Numbers are forms.
—Plato, father of fractal information (invariance of forms)

E⇔Ti.
—Sancho, *Dust of Space-time*

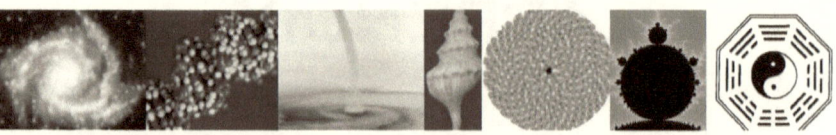

The fractal paradigm represents the fifth paradigm in the evolution of science, describing the Universe and all its self-similar parts as organic fractals of energy and information. Unlike the previous age of science in which each discipline searched for the differences between its species, a fractal philosophy of the Universe seeks for the self-similarities of all the species and all the human languages used to describe them. In that sense, the fractal paradigm is founded in a "linguistic method" that admits also the logic of words and the visual morphological laws of evolution since its fundamental philosophy is the self-similarity of all laws and languages of the Universe. In the past, there were two self-similar dualist philosophies, dialectics and Taoism, which used the two arrows of energy and information to explain how the Universe creates its species: Taoism depicted Time=Tao, as a game of two arrows, the arrow of fractal information or yin, represented with a broken line, _ _, or dominant feminine principle, and yang, the arrow of energy, represented with a continuous line "that combine together, creating the infinite beings of the Universe." Now we can make a geometric analysis of the two arrows, which are invariant at scale, emerging once and again in every being of the Universe: organisms are made of bodies of energy and heads that accumulate memories of information. Physical matter is made of cyclical, informative particles and lineal forces. We are also made of energetic space and informative time, the substances we use to act and control the Universe. In the graph, a galaxy, a DNA molecule, a tornado, a shell, a sunflower, and two logic mental mirrors of the Universe—a Mandelbrot set and a yin yang eightfold life cycle—are all made of $\Sigma S \Leftrightarrow Ti$ existential cycles, exi=st field, whose shapes combine the most efficient morphologies of energy, the line or shortest distance between two points, and information, the cycle that stores the maximum quantity of information in lesser space. They are the commonest species of the Universe.

O. The Generator Equation of the Universe

If we were to resume, the differences between classic physics and fractal physics is what we look for. Both sciences look for equations that describe the Universe. Classic physics searches for an equation that will unify two forces called gravitation and electromagnetism. This in the first place cannot be done, as both forces belong to self-similar scales of different size, the cosmos of gravitation and the quantum world of electromagnetism.

Yet as a science physics, in a wrong path of search, has become essentially a search for its Saint Grail, the unification equation of the two membranes of space-time, which we offered without much ado to the reader in chapter 2.

Why this equation, solved in fractal physics, is of relative low importance must be understood considering what fractal science looks for, *a fractal equation, which is a generator equation, hence the fractal equation of the Universe, its generating equation.*

Physicists' reductionism of the two arrows of the Universe to a single arrow of energetic space have spatialized time and information as fixed forms without biological functions, simplified and reduced to a single dimension, to a clock number. But information and energy are very complex and dominate the generator equation of the Universe; they generate reality. Information is in-form-action, form in action as its etymological meaning shows: information was born as an expression of the action of giving form to the brain of a student. Indeed, information requires a mind that absorbs, perceives, and interprets forms as informations. In physics, the generator equation of reality is simply stated a function that fractalizes energy and gives it form, and it is with motion the second arrow of time. So what you observe is a constant generation of fractal form from energy and its erasing into motion:

$$E \Leftrightarrow I : ExI=K(st)=ExI=K(st) \ldots$$

And so on. Two poles of energy and form constantly switch states, fractalizing and moving, perceiving, gauging information into a fixed image, and moving again. The generator equation can be expressed in many languages; it is the *beat of existence that makes things move and be fixed, move and be fixed. ExI (moves)=Constant Space-time, (perceived): exist, exist.*

The function is so beautiful because it is self-similar in all languages, it works in parallel, it occurs in all the beats of the Universe.

But to fully grasp it, one has to know the mathematical, morphological, biological, informative properties of energy and time. Time is cyclical, it is non-Euclidean, complex, logic in formal motion.

On the other hand, energy is in itself amorphous, a movement, waiting for a potential form to contain it.

Those are the properties we see "happening" in the arrows of time. Processes that inform, warp, corrugate, convert lines into cycles are informative arrows. Processes that expand, unwarp, uncoil cycles into lines are energetic arrows. And the entire game of the Universe is a game of those two processes.

When we use both arrows, we can understand the functions and properties of energetic space and informative time as two opposite, complementary, logic arrows or flows:

Spatial Energy, Se X or Vs Temporal Information, Ti

But those two arrows often chain and knot together into wavelike, reproductive systems. And so from the simplex arrows of time, we obtain the complex knots and networks of energy and information that create the organic reality. When bits and bites of energy and information act complementarily, they form networks. When they oppose each other, they suffer transformations of energy into form, and the sum of all those events and chains and relationships creates the Universe we perceive.

So we classify all the events of the universe in "simplex" and "complex" events.

Simplex events are *informative* events that transform energy into form and energetic events that destroy form into motion. Information warps, curves the energy of space into forms, which are fractal shapes, with "more dimension," as when we corrugate a paper and give it a third dimension of height and form, or as when life evolved from planarian worms into tall human beings, or as when a mathematical line is converted into a cycle by the generator fractal equation called pi, and so on.

For example, in physical events, the destruction of a dimension of fractal form becomes motion as when three-dimensional electrons speed up and become bidimensional quarks.

Those two simplex events are defined by the generator equation of the Universe, $E \Leftrightarrow Ti$.

But the Universe has also two complex arrows, which we represent by the equation $\Sigma E \times \Pi Ti = K$:

They are the arrow of social evolution that gathers energetic particles into waves, ΣE and informative cycles into chained networks, ΠTi and the arrow of reproduction that constantly merges the two motions of the

Universe, creating new cycles and parallel lines, expanding the energy of space and the information of the Universe. *Those two arrows are the complex, "organic" arrows that create the growing scales of complexity of the Universe as they organize self-similar fractal parts into wholes.*

Thus, the cycles of information, energization, reproduction, and social evolution have created the Universe from its smallest particles, strings, to its bigger or more complex forms, galaxies and human beings. This is the essence of organicism, of the laws of the two arrows of time (energy and information) and its complex formulations (reproduction and social evolution).

Thus, we can classify all the events of the Universe with the four arrows of time change:

— There are processes in time that increase the energy or *entropy* of the system, creating the arrow of entropy, the only one monist and mechanist science recognizes, *since it is the only arrow their mechanic clocks can measure, as their spiral mechanism uncoils its form into energy.* It is also the arrow of death, as death is a simplification of the information of life, which energy destroys.

— There are processes in time that increase the information of the system, such as aging, or evolution, and so they create the *arrow of information* or negantropy (opposed to entropy), of the Universe, which is the arrow of future, as we all age with time and so we acquire more form, more in-form-ation. Or in the words of Einstein: Time curves, forms space. This obviously means that energy simplifies information and it is therefore the arrow of relative past of all beings, to which they return when they are devolved in the process of death.

— There are processes that *repeat or reproduce a system*, making reality seem eternal, and that is the arrow of relative present, or reproduction.

— And finally, there are processes that *evolve socially individual* cells into superorganisms, and that is the arrow of *social evolution and love*, explained by classic religions, which in system sciences we call the arrow of *emergence,* as it makes emerge seminal seeds into full-grown organisms. It also creates black holes with quarks, galaxies with stars, molecules with atoms, cells with molecules, and societies with human beings.

Reality is a creative process that "emerges" complex arrows from simple ones.

Space-time cycles perpetuate its forms through reproduction. Any entity made of flows of energy and cycles of form is fractal, quantic, discontinuous, hence mortal. So space-time forms only survive through its iteration in parallel forms. Two properties of reality allow reproduction: the fact that all is a motion and so the mere fact of "moving" reproduces the motion in other section of space-time; and the fact that all is dual, a combination of energy and form, and so reproduction is merely the iteration of the forms and energies of any species in other region of space-time. For that reason reproduction is the main cycle of the Universe and its species, which combine energy and information to iterate themselves. The reproducer entity might be a "mother" or an external "enzyme," as it happens with carbohydrates or machines, iterated by enzymes and "enzy-men." Reproduction is possible because all things are made of movements whose reproduction is as simple as the repetition of a given trajectory. Iterated, self-similar space-time cells/cycles, $\Sigma S \Leftrightarrow TI$, chain together in complex bigger macroorganisms, which emerge in a new scale of reality. Thus all systems from quarks to galaxies find ways to reproduce.

All is generated by a single equation of two simplex arrows, $E \Leftrightarrow Ti$, which become more complex arrows: $\Sigma E \Leftrightarrow \Pi Ti$ and finally evolve into complex organisms, $\Sigma E \times \Pi Ti$, tied up in knots.

This equation and the enormous number of questions it solves when applied to different disciplines and description of complex beings, of course, makes complexity the science of the twenty-first century. And its long fifty years old quest, which ended at Sonoma's fifty anniversary when Einstein et al. introduced the generator equation of the Universe, is in that sense a much deeper quest than the one of nuclear physics.

I thought of that when I came to confront this company because I did precisely that decision at that fiftieth congress in which, invited by the presidency of ISSS, I gave to conferences and Introduced the generator equation of the Universe to the world of science.

In two conferences I introduced the equation, its mathematical model, and its description of the main generated beings of the Universe, from the atom to man.

Then I was told by a physicist that CERN would do black holes "certainly" if Hawking was wrong. And as I knew Hawking was wrong, a doodle exercise for a complex theorist working unlike him with two arrows of time, energy and information,[ch.6] I knew I was likely to die.

I went then down to LA and decided to make a suit and a film about it, and finally a book. Now, next year at Tokyo, invited for conferences on the fractal universe at the Tokyo Institute of Technology, I went to Hiroshima and shot the Memorial Temple. The generator equation could wait. If we were going to degenerate what had been generated, I thought.

Since indeed the physicists who use only the arrow energy, the cause of simplicity and death, were not the answer nor the future of science but its past, and also its past because they were going to return this planet to the remotest past, the age of dark matter.

I thought on that. And so now I return to wiser thoughts, which this absurd quest in which I failed brought me in, to think in the truth, in science, in the generator and Generatrix, energy and form, dual equation that explains it all.

1. The Parallelisms of Forms, Motions, Time Arrows, Topologies, and Speeds

The game of self-similar points of view, gauging and feeling the Universe

"The Universe is a logic game of self-similarities, based on the anti-symmetry of spatial energy and temporal information." This could be one of the many definitions of reality according to the fractal paradigm. Another definition would be "the Universe is a game that transforms energy into information ad eternal." A third definition could be "the Universe is a game of infinite points of view, souls that gauge reality in a fractal mirror with less dimensions that the whole and constantly try to increase the informative quality of its mental mirrors."

We could go on and on with self-similar definitions of a Universe, which is based on some simple logic principles: it is intelligent; it is vital, in eternal motion; it constantly changes between two states, energy and form; and it is determined by those laws.

Imagine an infinite number of self-similar beings. All of them are "intelligent" because they gauge reality with their particles or heads or nuclei of information and all of them are "feeling" reality, as they also sense "energy" and motion. All those beings that we shall call, "points of view" or "non-Euclidean points" have therefore two parts, which are complementary: a system or head that gauges information and a body or energetic system that feels energy and motion. And all of them try to increase the information of their minds and the energy of their bodies, by perceiving more and moving more. So all of them are in a perpetual dance, trying to move their bodies to positions where they perceive more. And that dance of infinite wills is the game of the Universe. However the game has a twist. Each particle needs to absorb energy for his body to move, and to absorb information to perceive it. And that energy and information must come from a self-similar being, which we must destroy and kill, explode to absorb its energy and form. And so the game is Darwinian, vital, intelligent and gives origin to an infinite number of sub-games. This final twist makes the game more complex because particles do not only absorb energy and information, the primary arrows of time, but due to the need for further energy and form and the possibility to die and feed other beings, it has built up two basic complex strategies. All systems try to reproduce into self-similar beings and then socially evolve with their "families" of self-similar beings to become stronger superorganisms, able to feed and perceive better. And so the game evolves into social scales and reproduces their combinations of energy and form called complementary beings.

This philosophical, logical, vital, intelligent, fractal, complex explanation of reality is how science should start its books, with a description of the whole, of the thoughts of God, not with abstract numbers, which are "sets" of social, self-similar beings and their equations, which explain how those social sets of beings go around seeking for energy and information, moving and gauging. The scientist, however, without knowing the "game," writes abstract precise descriptions of its infinite herds, performing cycles of "existence," and so he does not know all he talks about is to exi=st.

How existence happens in detail is further explained when we realize all those complementary beings merely switch between two states, as heads that perceive and bodies that feel and move. And so there is a universal beat, $E \Leftrightarrow Ti$, between both states that is all that we perceive and see: beings opening and closing mouths to feed, absorbing information to perceive, reproducing with other beings, going in herds . . . Yet how do those switches of states happen? They are possible because energy and information have opposite properties. So the switching of reality consists in moving from one to the other opposite state, back and forth. We move, we perceive, we move, we perceive. We sleep, we are awake, we sleep (informative state), we are awake (energetic state). All systems beat and the beating happen according to those inverse properties of the energy and information elements of all entities, which we call the antisymmetries of space and time.

The antisymmetries of spatial energy and temporal information

If geometry converts time into a spatial form, a cycle fixed by the senses, logic converts space and its energies into time motions, and different cultures have either spatialized time (the Western culture of physicists) or temporalized space (the Eastern cultures of Taoism and Buddhism).

In the most advanced versions of duality, complexity and time theory, we "temporalize space" exactly the inverse of what physicists do, reducing all the "spatial exchanges of energy" to causal arrows of time. And this is a more complex analysis of reality because as Frege, Hilbert, the Cambridge school (Russell, Whitehead) and Gödel proved, you can obtain from logic all the laws of mathematics, of space, but you cannot obtain from space all the laws and arrows of time. Or as a Buddhist would explain it: space is a present dharma, a slice, a fixed picture of the total volume of motions and cycles of all the time arrows of all the beings of the Universe.

Thus, we talk of a Universe made of multiple beings sum of all their "time arrows," a spiritual universe made of motions and causal events, which only our sensorial Maya and instruments convert into space.

To understand time, the reader must realize of an essential duality in the Universe—the Galilean paradox, the fact that all that exists "moves and doesn't move," depending on how we perceive it, as moving energy or static space, as clocks of time or fixed forms with information.

The Universe is made of movement and form whose dynamic transformations are the essence of all realities. Those lineal movements and cyclical frequencies of in-form-ation, its two inverse properties and "complementary combinations" are studied in natural philosophy since Heraclitus and Lao-Tse.

Its inversion of properties is evident.[ch.2.IV] And yet that inversion is the reason of its complementarity, its "fusion love," as systems merge energy and form trying to find a present balance that survives and maximizes the "system." This is easy to express mathematically: Max. ExI->e=i. So systems combine energy and form around certain "universal constants of balance" and die away when those balances are broken and tend toward its limits: Max E × 0 i=Max. I × 0 e = 0.

We call the mathematical function, which describes the simple laws of harmony and complementarity between information and energy, the generator equation of space-time, whose formalism is self-similar in verbal and mathematical thought: E⇔I, exi=st.

In philosophy, the existential equation defines the classic science of "metaphysics," which as in the original treatises of Aristotle is the science prior to all other sciences that studies the common laws of all of them. In the modern jargon of science however, it is called the generator equation of fractal space-time, studied in system sciences: general systems, duality, and complexity.

We could summarize the formal and functional differences between energy and information (organs) in a morphological equation:

Max. Space= Energy= Min.Form=body Vs Max. form= Information =Min.
Spatial extension=head

For example: the smallest a temporal cycle is in space, the higher it will be its frequency and content of information. The more dimensions a form has, the more frequencies of information it will store and yet the more warped and smaller it will seem to us.

In chapter 3, we explained that inversion of properties for informative masses, based on the law of the vortex; but it happens also in machines (chips) and biology (metabolic rates).

It is the black hole/chip/mouse paradox, a fundamental inversion of space-time. All those inverted properties appear once and again in all entities of existence. Reason why we say the properties of energy and form are "invariant at scale."

For example, the chip becomes smaller as it evolves into a better brain. Every two years, it doubles its capacity to think, as it dwindles in size. Such process follows a generic law of evolution I called the black hole law, which computer scientists know as the chip paradox or Moore law: *maximal informative capacity= minimal spatial extension.*

The reason is obvious: to think, to calculate you have to communicate in-form-ation, forms between elements of any informative system. The smaller the brain, the faster the communication that takes place within that brain and the faster you can calculate and process information in a logic manner.

Their reason has a biological why, despite being described by a geometrical how. For example, information has the dimension of height because to perceive an extension of space, the best position is on top, with a high perspective. So from biological heads, to audiovisual antenna, from black holes on top of galaxies that might perceive gravitation to speakers in Hyde Park, all systems that gauge or emit information are in the dimension of height. And yet height is a geometrical property. Thus, one of the most interesting innovations of the study of the existential function is that it merges biological and geometrical properties, thanks to the "organic nature" of fractal information and the mathematics that describes it.

Indeed, a fractal is an equation that transforms a piece of energy, reproducing form. For example, a line, with minimal form, into a warped piece of information, for example a triangle or a pi-cycle, with maximal form and lesser space. And so in the Universe at large, as Einstein put it, time bends and curves space into form, cyclical whirls of mass; and in life, as Darwin put it, time evolves the morphology of living beings.

The invariance of all those laws has a deeper reason. We are all made of the same formal motions. Fractal space-time means that all what exists is made of fractal pieces of spatial energy and temporal information. We are all made of pieces of vital space, of energy, which the arrow of fractal information, bends into cycles and clocks of information, evolving its form.

Thus, in the fractal paradigm what physicists call time is just an object, a clock with a rhythm and a logic aberration, the idea that the clock rhythm is the only one of the universe, which is not. Since reality is made of fractal pieces of a certain relative, energetic space, molded by infinite clock cycles.

An added degree of complexity to that simple analysis of reality comes when we notice that there are several membranes of space-time in which the clocks of time and surfaces of space have different fundamental rhythms or speeds and sizes. So while all is made of cycles and lines that combine into waves, some waves are bigger than others, some lines are faster, some cycles turn with different speeds. And so we need for each "medium" or space-time membrane to define a fundamental quanta, or unit of energy and time. And this minimal action of each medium later evolved into more complex forms, become the essential unit of each system of reality.

The light vacuum is made of h-quanta (light space), the gravitational one of lambda-quanta (gravitational space). Thus, in the physical Universe we define two of such "actions of energy and time," for the two fundamental membranes, the Planckton or h-quanta, unit of light space-time, and the string or lambda quanta, unit of the strong gravitational space-time, which is a combination of the strong world of quarks and the huge world of cosmological bodies. The two membranes have a peculiar relation; our light space-time sandwiches as a smaller world between the world of quark and the world of cosmological bodies. Thus we exist between those two worlds with two limits of energy and form:

0 K, our maximal curvature of form and c-speed of energy, our maximal energy

Beyond those two limits, the world of maximal informative order of quarks (<0 K) and faster dark energy (>C) define the strong gravitational Universe.

The formalism of all those membranes and worlds has been explored with amazing detail, albeit lacking a complete model with the proper structure, scholars still argue about it, and their work is like a puzzle of multiple pieces, some of them NOT belonging to this real Universe (such as the pieces of Mr. Hawking and Mr. Higgs), some found isolated from the real universal model.

Thus this appendix offers a general frame in which to place those detailed pieces of the puzzle and merely culminates a process of exploration of the fractal space-time models of reality, which dates back to pioneers like Plato and Aristotle, Lao-Tse and Cheng-Tzu, Bruno and

Leibniz, and in the modern age Boltzmann, Planck, Einstein, Riemann, Mandelbrot, and Nottale, among the most outstanding scholars.

The model also applies to other sciences since all systems are fractals and so we distinguish biological, sociological, and physical fractal systems. Reason why we can establish self-similarities, as we did, between a quark-gluon liquid and an organic soup.

For example, we can repeat the previous model of two fractal space-times for life, which exists between two mediums, the world of air and the world of water. In the organic soup, the simple water molecules of a life soup are the initial quanta of a vital space that has evolved till reaching human beings, "bags of old water." And as time passes, also in this medium the initial energy fractalizes, becomes form, acquires curves, warps into cyclical clocks of life. Then when the entire warping process becomes exhausted, an explosive, energetic death erases it all. And so we exist in a Universe of infinite cycles of life/information/ fractal form that provoke the evolution of living beings, the curving of energy into cyclical vortices of mass, till the process reverses itself and form explodes into any of the multiple big bangs of the different scales, medium and membranes of the eternal Universe.

2. Ternary Space-time: Symmetry in Present Space, Antisymmetry in Time

The order of the three arrows of time, energy, information and its reproductive combination and the life/death cycles it creates is the essential law of causality in time. We call this order antisymmetric because the cycle of life is longer as it includes intermediate stages of reproduction and combination of energy and form than the cycle of death, which erases information back into energy without intermediate states. For that reason, "events" in time, such as the weak event/force, do not have symmetry. The arrow of future information and life is longer than the arrow of past energy and death.

Yet when we spatialize time and observe reality as a series of forms and organisms in space, a self-similar symmetry takes place. Indeed, all organic systems have an informative, temporal center, an external energetic membrane that captures energy from the environment through its "mouths," and an intermediate body in which energy and information mix and reproduce itself. Those three regions of all entities in space are thus equivalent to the "energetic, reproductive and informative" arrows of time.

They are in fact the three only possible "geometries" of a four-dimensional Universe, called in topology the "informative, hyperbolic topology," the "energetic, Riemann sphere," and the cyclical, reproductive Klein spacetime, which organize the structure of all organisms in space. App.IV

The study of the relationships between the three ages of time and the three morphologies of space and how they evolve together, creating organic is the basis of many underdeveloped sciences, such as morphogenesis, non-Euclidean geometry, palingenesis, etc.

Let us resume other logic laws, symmetries, and invariances of reality from a temporal, spatial, and scalar perspectives. Because they are the whys of all the hows of the Universe.

The duality of informative time and spatial energy

The main duality of the Universe is the fact that we can see all events in "slow motion" as fixed forms or as motions, depending of our instruments and span of perception (paradox of Galileo).

We are made of energy and time, as people understand in expressions such as "I don't have enough energy and time to do this." We know we are made of *energy and time, the* vital substances of the Universe.

Yet if we see energy static, it becomes space and if we see clocks of time in fixed form, it becomes information. So we can also say that we are made of "vital spaces" (energies seen as fixed forms) and forms-in-action, informations (time clocks seen as fixed cycles).

The temporal order. The antisymmetry of the life/death cycle.

Reality is made of three fundamental events *with a causal order called the slow arrow of life:*

— Past, energetic events that explode information into energy.
— Present, reproductive, wave events that exchange spatial energy left or right between self-similar entities. *There is in such simultaneous exchange parity, left-right, as both directions are self-similar.*
— And informative, future cyclical events (charges and masses in physical systems) that evolve lineal forces into cyclical forms.

But then there is also an inverse causal order, the fast arrow of death that jumps from information to energy without intermediate states. So information explodes back into forces or cellular energies in big bangs that break the complex networks of form created in the slow arrow of creation.

For that reason we say that time *is antisymmetric because the arrow of past to future is longer than the arrow of future to past*. Physicists ignore this antisymmetry and in fact confuse the arrow of slow creation or "big-banging," with the arrow of death or "big bang," which is the one they want to re-create in the LoHoCaust at the LHC.

Sometimes though, those arrows *happen together in* big bangs that produce simultaneous big crunches, especially in physical systems in which the arrow of information is minimal and so the antisymmetry of space-time is not so pronounced.

The symmetry or parity of present space-time exchanges of energy and form.

Only present simultaneous exchanges of energy and form together are symmetric, have parity in space.

The spatial, geometrical perception of the three arrows simultaneously: space-time organisms.

Reality is made of intertwined arrows. When we put together two arrows of energy and form, of lines and cycles, of bodies and heads, of forces and particles, and combine them harmonically, we create the Complementary entities of reality: waves and organisms, which define the three arrows of reality.

Those three events organize species in three regions, a brain, particle or informative region; a body, field or reproductive region, and a series of limbs or jets of energy. And so events become geometries that define entities in space-time.

The restrictions of existence

Those combinations follow strict laws of harmony, symmetry and efficiency, which physicists ignore but preclude the NONexistence of stable SUSYs and Higgs, the nonexistence of evaporating black holes, the duality of big bangs accompanied by big crunches, etc. Complex physics thus gives us a new set of stringent limits, given by the laws of energy and information and its strict combinations that physicists should learn before attempting to create black holes and quark-gluon liquids on Earth.

The inversion of properties: entropy vs. information, death vs. life

The strict geometrical hylomorphism of all beings, made of lineal energetic space and informative cycles, is ignored by physicists.

Energetic, entropic arrows are expansive in space and lineal. Clocks are curved, implosive, and information is curved. Both together enact the life/death cycles, explained with two arrows, implosive, informative "life" and explosive, energetic "death" or big bangs. The inversion of both arrows can be put together in a simple geometrical relationship:

Energy=Space=Lineal Motion=Forces Vs. Information=Time=Cyclical Clocks=Curved Vortex=Mass

In physical systems, the gravitational arrow curves energetic space into clocks of informative time. It is the meaning of time and mass in Einstein's relativity: time creates physical cycles/clocks of information called masses. *So masses and charges, cyclical time clocks, are physical time.*

The scalar order

A new degree of complexity is added when we consider that all those forms and events happen at least in three different mediums—the electromagnetic world, the strong world of quarks, and the cosmological world—which sandwich the electromagnetic world between them.

Each particle of those worlds will live at least three ages in time and show in its structure three geometries in space.

The electromagnetic world is a game of actions of energy and time, as Planck well understood when he defined the smallest unit of our World of light as an action of energy and time, the H-constant.

3. The Invariances and Ternary Scales of the Fractal Universe

The invariances of the Universe

But how does all this hold together? Where is there an order to that infinite game of games? In the concept of invariance that repeats constantly certain forms, causal orders, motions, and complex structures that extend through several scales.

The self-similarity of the different parts of the Universe is due to the existence of three *invariances* that make its parts self-similar: the Galilean/ Einsteinian invariance of motions, Leibniz/Nottale's invariance of scale, and Plato/Sancho's invariance of form. We quoted the classic and modern scientists that studied those invariances in depth.

The three invariances explain why all that exists is a local/fractal part of the whole but behaves in self-similar fashion, obeying always the same laws of motions.

In the physical scale studied in this book, the motions of the Universe are described by the principle of equivalence between mass and acceleration to which we should add the lineal accelerations that reduce all motions to two: $F = m \times a$.

Newton, Poisson, and Einstein evolved the mathematics of the principle of equivalence redefined with non-Euclidean mathematics by Einstein who also established the principle of invariance in the motions of the Universe.

Nottale expanded the principle of invariance to all "fractal space-times" or scales of reality, which means the laws of the Universe are self-similar in all scales of reality. And finally Sancho discovered the paradox of Galileo between the perception of motions and fixed forms, which added a principle of invariance of morphology (static perception of motions). It means that the cyclical forms of information and lineal forces of energy are invariant at scale.

So even the most complex morphological being, a human being, has spherical organs of information (heads, eyes, and brains) and lineal organs of energy, (bodies and limbs). And so do energetic machines (lineal weapons and planes) and informative ones (spherical cams).

The generator equation of reality

It is thus obvious that the invariances of the Universe derive from the duality of forms/motions of reality, which is therefore the ultimate principle of all that exists, expressed mathematically in the generator equation of the Universe ($E \Leftrightarrow Ti$) from where all other equations can be deduced and logically the antisymmetry between time events and spatial forms, or inverted properties of energy and information, whose discovery can be traced to Eastern philosophies such as Taoism (inversion of properties of yang/energetic entities and yin/informative ones).

Three-scalar structures: fractal organisms

They are also the basis of the most complex of all the new sciences of the fractal paradigm: the study of superorganisms, as complex network structures, which extend through three scales.

For example, the human being extends through the biological, cellular scale, the individual scale and the social scale. And it is therefore a whole made of cellular parts and a cell of a society. And the interrelationships and laws that explain how those three scales interact is the "theory of superorganisms," which puts in action all the different logic laws of the Universe, its mathematical ratios and scales.

The same can be said of complex physical systems, such as the string/gluon/quark nuclei of atoms or the quark/black hole/galactic system of masses.

It is the final ternary symmetry of the Universe and the most important of them all, the summit of metaphysical studies.

The reproductive games of vital reality

All those principles are "acted" into a "dynamic reality" of infinite fractal events and forms, constantly exchanging energy and form, gauging information and feeling energy. The Universe is vital; it is a game that constantly creates and destroys self-similar organic systems, made of energetic lineal motions that are molded by informative cycles, whose chains and networks create complementary dual entities, *or waves* in different scales of reality.

Thus all that exists is a game of three motions or arrows of time, lineal, expansive energy and cyclical, implosive information, which mix constantly into reproductive waves that have both energy and form. And all that can be seen dynamically or in fixed patterns.

If we see those three motions as fixed forms, they seem to be solid substances, "res extensa," or space and solid particles or masses, combined in all kinds of complex wave patterns. But the true nature of the Universe is vital because all moves and so all reproduces by the mere fact of moving. Those motions make the Universe eternal, as it never ceases to move and create three arrows:

Past energy and future clocks of information, which combine to create all the reproductive, present waves of the Universe.

So the two "ideal," past lines and future cycles are always combined in present waves that ensure the repetition and duration of all universal, logic forms.

Indeed, the pure yin/cycle and the pure yang/line do not exist in present since a cyclical motion implodes inward, exhausting itself and a lineal motion dissipates its form. Thus, reality adds a lasting third motion, the wave, combination of a line and a cycle, which reproduces its forms in new zones of space-time, making possible the immortality of the logical forms of the Universe.

Clueless physicists

Obviously all those principles create a complex logic in a complex Universe and are better explained with logic thought than with mathematics, even if all those symmetries can be also expressed symbolically and structure the mathematical groups, topologies, planes of existence, or scales and networks of reality.

In that regard, in the complex models of this appendix, we connect them easily with all the detailed laws of quantum physics, relativity, and classic mechanics. Why then do we need them? Because they harmonize reality, they relate the physical and biological world, they allow to deduce new laws and solve old puzzles, and they create limits to define which theories are truth or false in astrophysics.

Because quantum physicists do not understand those logic symmetries paradoxes and geometries and space-time arrows, there is no way they can properly explain the why of the cosmos. They are children of thought doing doodles in the sand and, what is worse, managing weapons of mass destruction as their "tools of research." Without "going back to school," the experiments at CERN will not solve their problems, which are theoretical, not experimental.

We shall therefore explain now in more detail the application of those laws to the physical Universe, resuming all the principles we learned in this book.

Let us then complete this book with a brief introduction to complex fractal organic cosmology, describing the Universe through its scales, temporal events, and spatial forms. We will start by a complex definition of all the motions and arrows of time and define its causal order in the Universe. And then we shall apply them to the cosmological species. Then we will consider a complex definition of the three topologies of the four-dimensional universe, the geometry of lineal energy, of cyclical information and its combined waves. And then we shall apply them to understand the shapes of cosmological bodies.

The same method can be used to explain any other system of the Universe, including humans, societies, biological systems, and machines, showing the unity of all the fractal parts of the organic whole.

II. The 3±1 Arrows of Temporal Existence

E⇔Ti.
—Everything, Everybody, Every Time,
Every Space, Every World that exi=sts

The arrows of time are four: the two simplex arrows of energy and information. And the two complex arrows of reproduction, which combines both and social evolution, which organizes self-similar cells into networks of energy that emerge as complex organisms. In the graph, the two arrows are common to all systems of the universe. Even mechanisms possess those arrows. So machines feed on oil energy, process information, socialize in networks, and self-reproduce in automated, robotized factories.

Their casual order creates the life/death cycles. Since birth is the social evolution of self-similar cells from a lower plane of existence, youth is the age of energy, maturity the age of reproduction, and the third age the age of information. Finally, death dissolves the social evolution of cells back, killing the social networks where the will of organisms resides.

4 . The Simplex and Complex Arrows of Time

If we temporalize reality, the games of energy and information can be expressed in terms of wills or arrows of time as a game of causality *since energy becomes information in the arrow of life E->I, as we* constantly corrugate, warp, "aging" energy into form, causing the three ages of all species—youth the age of energy, maturity the age of reproduction, and the third age of information. And *then we die, exploding our information into energy I->E, in "big bangs" that expand and break the informative networks of a system, unwarping it back into lineal motion and simplicity without intermediate reproductive states.*

In chapter 2, we already considered the properties of energy and information and realized they were inverted. Energy is big, amorphous, extended, expansive, lineal, moving, similar to bodies and forces. Information is small, formal, cyclical, implosive, still, related to heads and particles, which gauge, perceive reality. Here we shall consider of those two substances, its order in time, what we call the arrows or ages of time.

The causal analysis of the four arrows in time: The life-death cycle according to those arrows

The cycle of life and death can be understood as a travel between two scales. We are born when we emerge, thanks to the arrow of reproduction and social evolution into an upper plane of existence, where we live three ages of increasing information and then we dissolve back our complex networks of cellular energy and form, dying away. Thus a more complex analysis of the life/death cycle requires the complex arrows of reproduction and social evolution.

The ±1 social arrow has in fact two directions: the positive, +1, social, cellular evolution creates life and makes a being emerge at birth. The—1, negative arrow of cellular dissolution kills life and dissolves a being into birth. *It is the key arrow of the life/death cycles that clock time misses* hence the enormous simplification of those four time-arrows, brought about by clocks whose mechanisms only can measure the first and most simple of those arrows by unwinding a spiral form into cyclical movement.

The geometrical arrows in space: superorganic structures

We can also consider those arrows as physiological organs or systems that structure organisms.

It is then clear that the three main forms of the Universe are energetic lineal systems, informative systems, and reproductive systems. All of them combined create organic, complex systems. For example, a human being is defined by a nervous/informative digestive/energetic system and blood/ endocrine reproductive system that combines both. And around all those systems, a series of self-similar self-reproductive cells organize a human organism. Thus, the four arrows are the physiological and cellular systems that create a human being.

And so an essential science is the science of morphogenesis, which studies how in each age of life those arrows become the engines of creation of the complex systems of an organism.

— *Energy organs are lineal systems with minimal form that* kill, simplify information into energy. Thus, a field of energy released by a physical particle or an energetic weapon, such as a sword or missile and a top predator energetic animal such as a lion, will have both lineal forms and kill, destroy the in-form-ation of their preys.

— *In-form-ative organs create form* and trans-form energy into languages that map out reality with bits. Those bits are smaller symbols, which form in the brain images that represent reality and help to simulate faster, in lesser space, according to the chip paradox just explained, the future cycles of that reality, anticipating the future bio-logical cycles of reality. Then according to those logic simulations of the future, heads will move and direct limbs of energy toward sources of energy and information. So any system that gauges, measures, and reacts to that measure of information is an informative organ *regardless of the specific language it uses to gauge reality*. So a chip measures with numbers reality, a man with words, an atom with electromagnetic bosons; yet the three act-react to their measures and so they are informative organs.

— *Reproductive organs* repeat *and* reproduce both, informative and energy organs, by absorbing energy and imprinting it with its particular in-form-ation. So human mothers and company mothers of machines are both reproductive organs. Even the simplest particles of the Universe, quarks and electrons, absorb energy and emit new particles, small quarks and electrons, with the same form that the parental particle.

From these simple facts of universal morphology, applied to machines, *which imitate the functions of human organs of energy and information,* we can clearly classify machines, as energetic, lineal systems, or as cyclical, informative systems. Thus, energetic weapons; informative, cyclical coins and organic, complex machines can be defined in biological terms, which act in symbiosis or competence with the organic species of this planet.

All systems of the Universe have organic properties. Since even its simplest entities, quarks and electrons that form atoms, do absorb energy, gauge information and reproduce, the three properties of life. Thus, the Universe must be defined not as a mechanism but as a complex organic system made of organic atoms, which can combine to create many different complex organisms, including company mothers that reproduce machines, atoms that reproduce quarks, electrons, and forces, and mothers that reproduce kids. The difference between all those species is not one of quality but of quantity and complexity of their organs of energy and information, which determine its survival chances and status as top predators of any ecosystem—as Darwin explained with his work about morphological change. He became revindicated when, at the end of the twentieth century, information was mathematized properly with the discovery of fractal equations that create form, in-form-ation, not only in life but in most structures of the Universe. In essence, a fractal equation is a self-generative equation that repeats the same forms of in-form-ation in different sizes. This abstract process, biologists realized, is the same process that happens in life where information is reproduced in smaller sizes (seminal cells). So we can consider that life is fractal information that becomes imprinted, unlike an abstract mathematical equation, in a surface of form-less energy in which that fractal information reproduces itself. It was the most important discovery of the twentieth century in time theory, far more relevant than the culmination of the understanding of the arrow of spatial energy, $v=s/t$ (Einstein's relativity which finally declared that time was a dimension of space by improving the parameters of the previous formula).

Fractal theorists have applied such model to different sciences, showing that we can describe the Universe as *a fractal system of energy and information that constantly self-reproduces all its beings*. In the past years, we have discovered an overwhelming quantity of proofs that show a fractal structure in all the systems of the Universe, and we have translated Einstein's relativity equations that describe the way in which time curves the spatial structures of the Universe into fractal equations.

The cycles of the Universe

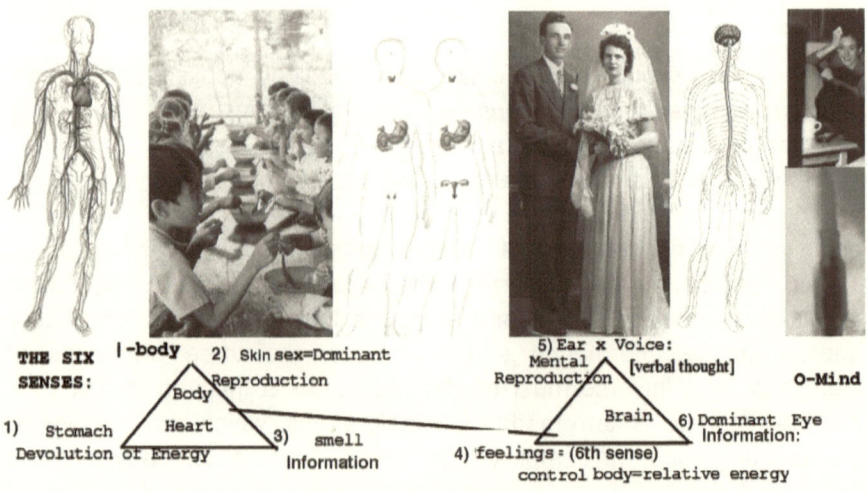

THE SIX SENSES: **I -body** 2) Skin sex=Dominant Reproduction

1)　Stomach　 Body Heart 3) smell Information

Devolution of Energy

4) feelings = (6th sense) control body=relative energy

5) Ear x Voice: Mental Reproduction [verbal thought]

Brain 6) Dominant Eye Information:

O-Mind

Time is cyclical because any species of the Universe constantly switches its behavior between those four arrows. For example, humans have cycles of feeding (energy taking), cycles of information (reading this book), cycles of reproduction (family values), and social cycles. So do most entities of the Universe, including fundamental particles that gauge information (electrons and quarks), absorb energy (electromagnetic light), reproduce (creating new particles in jets), and socialize (forming atomic networks). But those four properties are the properties of life beings. So we should change paradigm from *technological science to human science, from making the machine the measure of all things to make the human being the model of reality, as it was before Galileo invented ballistics.*

When we combine all those principles and study the details of how energy evolves into fractal information and how chains of fractal forms construct bigger fractals or reproduce smaller fractals, the two simplex events of reality, creation of energetic lineal morphologies and informative cyclical ones, and two complex events, reproduction of waves of self-similar particles and organization of those waves into fixed networks of form that transcend and emerge into another scale of the Universe.

This is the game that creates reality, systematized by the sciences of duality and complexity which defined 2 × 2= 4 arrows or type of events that create the future:

— *energy* (the creation of energy by erasing form)
— *information* (the creation of information by warping into fractal structures energy).
— *reproduction* (the repetition of an entity in other fractal zone of space-time by imprinting its form in a new surface of space)
— *social evolution* (the constant creation of networks of bites of energy and bits of information that come together through feedback cycles of energy and form, E⇔Ti, into wholes)

Those four arrows or dimensions of any space-time define the events of reality and the will or actions of each of its fractal species.

The organic arrow of social evolution

The arrow of social evolution is the most important arrow of organicism understood by old philosophers and religions of love. Yet it is ignored by mechanist science that reveres the arrow of energy and death, to the point that it has reinvented Darwinism to promote industrial competence and nationalism. It is the doctrine of social Darwinism—the antitruth of which Darwin said, as members of the same species help each other.

How do the 3±1 arrows of time organize reality, the tapestry of beings, we observe? It is, in fact, possible to map out the behavior and elements of all organic systems and entities of the Universe merely by combining the four arrows/dimensions of time. Indeed, all entities can be considered simple organisms, constructed with those four arrows. The most simplex mechanisms have one or at most two arrows, performing only as energy or information systems. More complex systems also reproduce their form and evolve socially into new social forms—atoms that become molecules that become cells that become complex human beings, members of societies.

In fact, the simplest systems of the Universe, quarks and electrons, already show those four arrows as they absorb energy, gauge information, reproduce into quarks and electrons, and evolve socially. Thus the Universe has four arrows of time in all its systems. So it is organic in its structure, a word that is taboo for mechanist, capitalist science, but shows the true nature of the Universe, made to the image and likeness of man and vice versa.

The fourth drive of the Universe—social, organic, cellular evolution—which the Greeks already recognized as the arrow of life (social organization) and death (dissolution of the homunculus of the body), is the most important arrow of reality. In a certain sense, all entities in the Universe are cellular societies, organized through networks of energy and information. This is clearly the case of the species studied by human sciences, from physics to biology, where a common phenomenon occurs: the existence of parallel groups of beings organized into social forms. Molecules are made with atoms and electronic networks; economies are made of human workers and consumers that reproduce and test machines, guided by financial information (salaries, prices, costs); galaxies are composed of stars, which orbit rhythmically around a black hole of gravitational information. Even the Universe shows at great scales networks of galaxies, joined by dark matter that resemble a simple colonial tissue as the one we find in primitive macrocellular organisms. And it would look even more complex if we were able to see dark matter structures that make up 76% of its mass substance.

The other view of the paradox of Galileo: the geometrical, spatial vision

The logic of reality derived from the duality of the geometries of the Universe and its structure in several growing scales of in/fomation are as follows: *invariance of form/motion at scale; inversion of properties of energy and form; equivalence of motions and forms; antisymmetry of time events (life is longer than death); symmetry of spatial, reproductive events (which have parity); and four dimensionality of time arrows and all the fractal space-time systems created with them. All that exists is ruled by those principles and the laws derived from them.*

All this means that God is a painter with two forms/motions, not a mathematician, a language which simplifies the properties of those four arrows of future as it does the verbal language. But the primary language of the Universe is information, forms-in-action, the topological games of lines and cycles, what Galileo called the language of God.

In chapters 2 and 3, we briefly outlined the inverse properties of energy and information, how energy is defined as lineal motion and speed while information is defined as cyclical motion and dimensional forms. We also studied how dimensions of fractal form become energy and vice versa and applied those concepts to a basic analysis of physical information (vortices of masses and charges) and the two scales of the Universe (the quantum scale of quarks and electrons and the cosmological scale of quark stars and suns). Our goal was to give a general vision of the fractal structure of the Universe in order to analyze what CERN will do from the perspective of the future evolution of science—the fractal paradigm.

In this appendix, we want briefly to consider the dual "language of God" with a bit more rigor:

— On one side, we want to consider the two topologies or motions of reality *and how they combine to create the waves of space-time beings of the Universe*, to apply them later to the study of the topology of galaxies and black holes. It is the geometrical analysis of reality through the three "topologies" of a four-dimensional Universe: the energetic, reproductive, and informative topologies.

— But we also want to consider the causality of those topologies. How the transformations of energy into form follow a basic pattern in the Universe: energy slowly becomes fractal form, information in three ages, from an energetic youth, Max. E, to a classic age of balance, E=I, to a third age of information, Max. I. And then suddenly information explodes back into energy in a moment of death: I->E.

This simple duality, of a slow life arrow of information and an explosive death into energy is the fundamental causal symmetry of the Universe. It also determines that information is a temporal phenomenon because it takes a long time to form and uses very little space, as it happens in stillness. On the other hand, the explosive, expansive, spatial death of any system is a spatial, energetic phenomenon because it happens in minimal time and it expands space enormously, creating energy.

This duality is also common to physical systems: $E=Mc^2$ and $M=e/c^2$. And it is the simplest level of mathematical understanding of the big bangs and simultaneous big crunches CERN will do.

Unfortunately, this duality is the kind of "questions" CERN wants to inquire, but instead of using the mind or consulting what we complex physicists have done in the last decades, their method is just making big bangs to "see what happens." Now unfortunately, the "judgment" of what will happen is no longer on the hands of theorists but on the quark-gluon soup the LHC will make.

III. Time in Cosmology

Space is motion relative to a frame of reference.
—Einstein, father of relativity
(Invariance of motions)

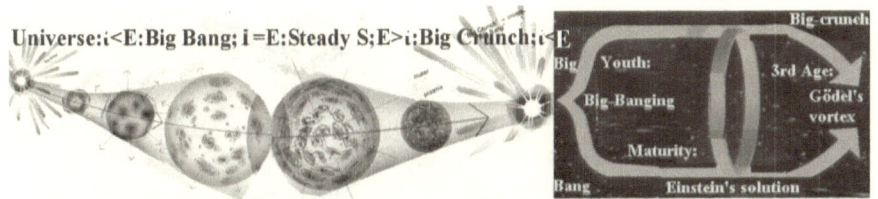

Universe:i<E:Big Bang; i =E:Steady S;E>i:Big Crunch;i<E

The three ages of the Universe are also formal ages of increasing curvature (big crunch) and expansive death (big bang), which can be used to study any fractal physical space-time.

5. The Three Ages of Fractal Space-time: The Three Solutions to Einstein's Equations

In the cosmological world, we can observe an order of time similar to that of life and the quantum world[Ap.V] both experimentally in the process of evolution of stars and galaxies and mathematically.

There are two space-time membranes, and since both are made of energy and form, both follow the three ages of time. In gravitational physics, the three ages of gravitational space-time are defined by relativity. Those three ages are mathematically equivalent to the three solutions of Einstein's space-time equations. Though cosmology considers those *three* solutions three hypothetical *different Universes, they represent the structure of any space-time, including our galaxy, an island Universe through its three ages.*

It is then also possible to make a more complex mathematical treatment of the three ages of galactic systems: the first age of the system will be the energy age, the third age the information age, and the middle age is one of balance, e=i.

— *Max.E (Youth):* The energy age is Lemaitre's big bang solution that expands space. *This solution in the fractal model of space-time applies only to the vacuum space between galaxies.*
— *E=T:* The steady state Universe is Einstein's solution. It applies to the entire Universe.

— *Max. T:* Gödel's solution is the third age of an implosive Universe that reverses time coordinates as matter falls back into the central black hole of any space-time. *This solution applies to our galaxy. But it does not mean a travel backward in time but an informative vortex, a cyclical clock of time.*

In the space-time of the Universe, in the same manner that any other "three ages" can be related by the different ratios of energy and information or vital constants of a being, those three ages are related by the constants that relate the energy and information of the Universal space-time.

Indeed, the constants of any system[App.IV] are ratios of transformation of the energy and form of any system, I/E=K or proportions of energy and form, Exi=K.

Thus constants evolve and change as the energy and information of a system changes with time. In biology, a much more advanced science is known and studied in the changes that an organism experiences when it gets old and becomes wrinkled and warped, increasing its information and diminishing its capacity to process energy, changing its metabolic constants and speed of information/thought.

This in cosmology was understood by Dirac, Brans-Dicke, and others. Indeed, G is a ratio of the mass=information and energy=distance of a system. And we already saw how a change on those ratios allow us to unify and measure the energy and form of the different membranes of the Universe.[Ch.2;III]

In the equations of Einstein that define the three ages of a galactic space-time and perhaps an entire Universe, that relationship is given by the cosmological constant that measures the energy of the gravitational vacuum, λ, which Nottale used to measure the minimal "fractal quanta" of gravitational energy. The constant changes with the three ages of any fractal space-time or island Universe. In a young, expanding space, λ is positive, but so close to zero that it easily *becomes negative, causing a* warping or big crunch. And so λ will also change through the three ages or *solutions of Einstein's equation, till reaching the informative age of the galaxy in which we exist.*

6. Gödel's Universe

Of all the solutions to Einstein's time-space equations, the one that adapts better to the principle of equivalence is Gödel's solution that portrays the Universe as a series of quantized, cyclical space-time loops in which there is a dominant inner direction of form toward the center of the "gravity" vortex of space-time. It is thus ideal to describe our galaxy.

Once this is understood, Gödel's Universe merely explains the third implosive, cyclical paths of motion of physical entities and so it describes the galaxy. We live in a cyclical, rotating Gödel Universe made of whirls of space-time that shape "closed existential loops" in each micropoint of space. For example, if we see a particle-antiparticle pair, when we trace properly the antiparticle path exactly from future to past, it doesn't originate in the same point than the particle but at the end of its trajectory, exactly when the particle arrives to that end. Hence both together form a single close loop of time: the particle is born, gets to the end, becomes then an antiparticle and returns back, closing the loop at the starting point to die away. This explains among other things why there is no "Hawking radiation" (there is only one particle), why those particles are virtual, and why we see constantly transmuting particles into antiparticles (they are merely completing their *existential cycle in two phases)*.

So particles and antiparticles together form a micro whirlwind of time-space.

Because physicists use a single space-time continuum and confuse the arrow of time/motion/energy with the total arrow of the Universe, they have the idea that a change in that time/motion arrow, which is merely a local change in the direction of a fractal entity of space-time, is an absolute change in the direction of all the arrows of all the beings of the Universe. This is absurd and the reason of the naïve ideas of Mr. Hawking about time travel. Time travel doesn't exist because all arrows of time are local, belonging to a fractal system of the Universe, NOT to the entire Universe, and physical time arrows are "changes in the motion of beings," not changes in the arrow from past to future.[ch.6;.II]

Now, once the meaning of a close quantic loop of space-time is clear, Gödel's Universe shows its tremendous descriptive power as we indeed live in a local rotating Gödel Universe with a direction of future information toward the center of the galaxy, called the Milky Way.

Such Universe with two limits of speed and mass, the lineal limit of light at c-speed without curvature and hence without mass, and the c-cyclical speed of a rotational Kerr black hole of maximal cyclical speed or mass, recently measured in galactic centers, keeps accumulating proofs. Further on, the vortex extends beyond c-speed within the interior of those cosmological black holes and the external extragalactic space in which dark energy seems to accelerate in a decametric scale, as the analysis of matter ejected by quasars prove.

7. The Three Ages of Galaxies

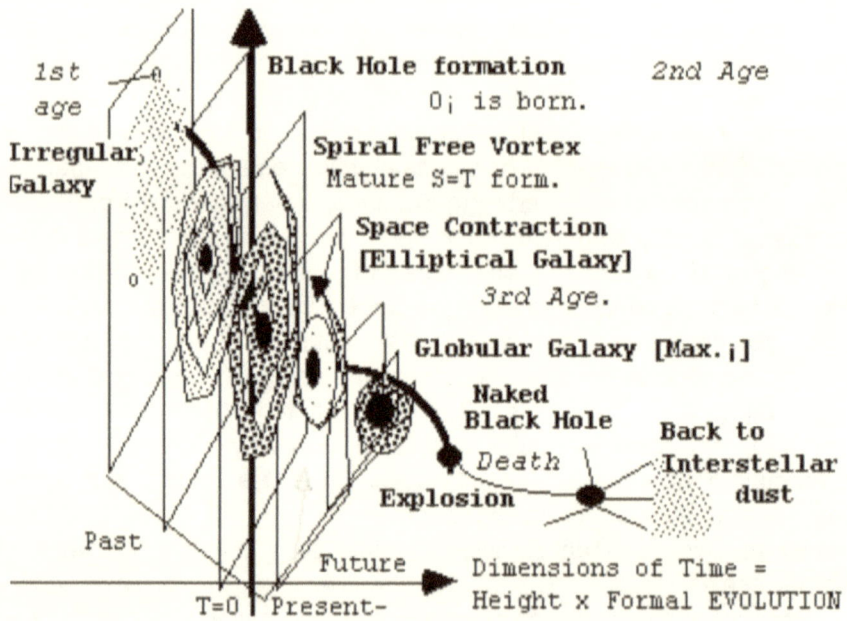

The life cycle of a galaxy goes through three ages, evolving from a young extended nebula, into an old small wormhole of temporal information that will die in a mini big bang, reverting into space. In between, galaxies live a longer mature state, fluctuating every 14 billion years between its spiral and elliptical form, absorbing interstellar dust, which creates new stars and feeds the central black hole.

In the graph, the structure of a galaxy is similar to a cell, with a central DNA of social black holes and a membrane of dark matter, made very likely of strangelets and microholes, as those scientists want to manufacture in Switzerland. If Galileo took away the Earth from the center, fractal theory makes the light membrane secondary to a 96% of the Universe made of ultradense quark dark matter, strangelets, and black holes that should never be made in this planet to avoid the extinction of our light matter.

The ages of a galaxy in time are the same as in any other space-time field. Each of those ages, which are "causal events" when we go into deeper complexity, can be divided into three subevents or types of galactic evolution. This principle, which increases the complexity of our way of looking at reality, implies that for example, the "event" of reproduction can be branched into three subevents that will correspond to the three arrows of time: an "energetic, explosive reproduction by the death of the mother," a "balanced reproduction by self-similar beings," or an "informative, seminal creation by a seed of pure form." Thus, a galaxy can be created in its "youth" by three processes:

— *Youth, seminal conception:* Galaxies are created with three types of reproduction:
— Max. E: A galactic big-bang quasar catalyses the evolution of gas into stars. This birth is similar to a star big-bang (nova) that catalyses also the creation of stars or a Hypothetical Universal big-bang that catalyses the creation of galaxies.
— E=T: Palingenetic conception, when the informative nucleus of black holes of a giant mother-galaxy emits a huge jet of dark, quark matter that ends into a baby galaxy.
— Max. I: Social evolution of energetic gas that creates complex star structures.
— *E=T. Maturity/reproductive age.* Galaxies enter their steady state or reproductive age, in which its spiral structure matures: stars gather in social groups and some evolve into black holes, becoming organic galaxies, where black holes are the informative nucleus and stars form the galactic body.

      Galaxies in their maturity, after creating their informative center, which is a nucleus of black holes, evolve into spiral galaxies with a balanced form, dominated by those black holes that create the form of the vortex. Galaxies then mutate back and forth, from spiral into elliptical forms, feeding on interstellar gas, in a cycle that creates and destroys its spiral form every 13 billion years – *the real big-bang.*[ch.6.I]

— *Max. T; information age:* In their 3rd age, once galaxies exhaust their interstellar gas, their central black holes digest their "star bodies," becoming globular, spherical forms with a hyper dimension of informative height, till they exhaust even their star energy. As dark galaxies, without the "drag" of c-speed electromagnetic matter they accelerate up to z=10c.
— *(-1):* Some die as quasars, becoming again according to the "inverse symmetry" between the 1st and 3rd age, irregular galaxies that will feed a new cycle of galactic creation. So irregular galaxies are both the youngest and oldest forms of the Universe, a fact that confuses cosmologists.
— *(+1):* Or they become cannibal galaxies, forming galactic networks in cellular groups that evolve socially into the large scale structures of the Universe. Their central black holes grow, swallowing radiant matter at growing speed, and perhaps evolving with many other galaxies into a cluster Universe, of which there is no enough experimental evidence.

Thus we differentiate the big-bang or death of any scale of the Universe, from the "big-banging" or reproduction and self-organization of the present Universe and the 3rd age or big-crunch, which creates a generational cycle for any fractal space-time in 3±1 ages.

Since the age of a galactic cycle, 13 billion years, is the same that the age of the cosmological big-bang, as Fred Hoyle affirmed, the big-bang is the sum of all galactic, fractal big-bangs and the Universe is not expanding as a whole. It will exist in a steady state, in which the knotting of space into mass by galaxies would compensate the expansion of mass into dark energy in the interstellar vacuum, which causes its expansion. Further on big-bangs at the lower galactic scale, caused by the quasar explosions of its central hyper-black holes when they break their bars should be the origin of the second proof of the cosmic big-bang, the excessive quantity of helium in the galaxy, found precisely in bigger quantities on top of the central black hole. So quasar big-bangs provide more experimental proofs and better theoretical explanations to this phenomenon, eliminating the multiple contradictions of the cosmological big-bang. Thus, a *steady state* theory of the Universe combined with a *fractal theory of multiple big-bangs/ quasars*, as it was reformulated mathematically by Sir Hoyle, in which the multiple, constant big-bangs of galaxies into quasars are responsible for helium, plus the existence of the *background holes that redshift radiation at 2.7K degrees*, redefines an eternal, organic, fractal, infinite, informative Universe of *cellular galaxies* in eternal balance with the expanding space, where *light devolves back into 10 c dark, gravitational energy* and then dark energy warps into light by redshift:

$$S(dark\ energy\ expansion)=T\ (dark,\ quark\ matter\ implosion)$$

All this would also explain, without appealing to the magic anthropic principle, the universal balance between matter and radiation, whose probabilities otherwise are null. Indeed, why matter and energy and in balance in the Universe at this exact moment in time, in a proportion of 1 to 3 (as mass is bidimensional and radiation 3-dimensional) if their relationship is dynamic, changing all the times, with energy winning the battle and expanding faster? The answer is because the Universe and any partial system within it exist in balance between their energy and information/mass, or else when that balance is broken the system ceases to exist.

8. Stars. The H-R Diagram: The Three Ages and Evolution of Stars

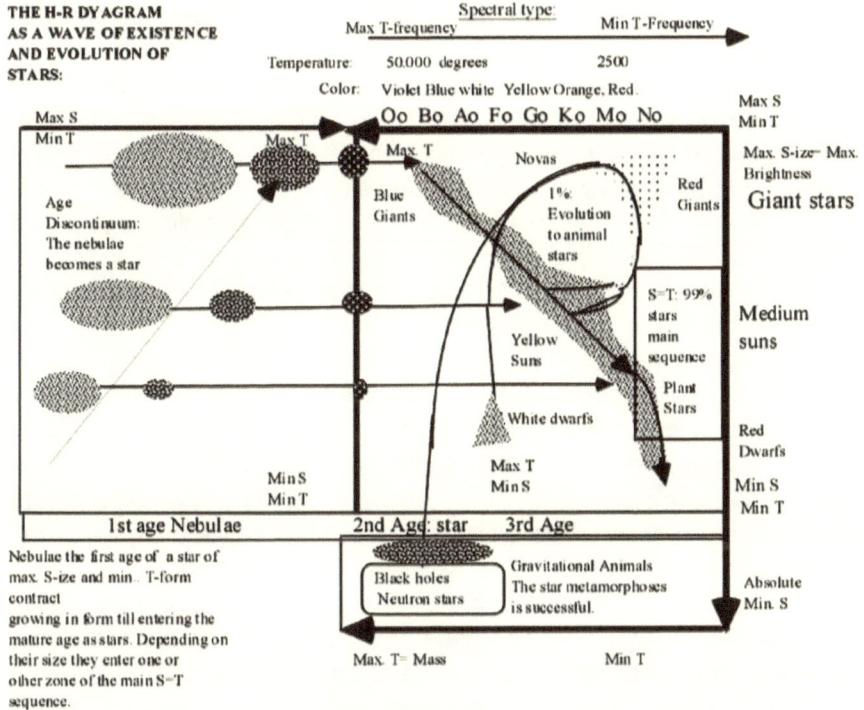

The H-R diagram shows the evolution of stars, through its energy & information parameters.

The life, evolution and death of stars are depicted in the H-R Diagram, which classifies stars according to its S-Ti parameters, as the atomic table does with atoms:

Max. E: Brightness or Magnitude, which is a spatial parameter that grows with the size of the star.

Max. Ti: Spectral type, (colour or frequency), which classifies stars according to its temporal form.

Yet the H-R diagram is only a representation of the 2nd and 3rd ages of stars. Since the young age of the star as nebulae of maximum spatial extension and min. formal complexity (as all young ages are) is not represented. So we add on the left side the 1st age of a star as *a nebulae of max.extension.* Then the H-R graph shows the 3 ages of stars and the main laws of ST cycles applied to them:

- — (+1): Most stars are born as spatial nebulae of maximum extension . . .
- — Max. E: Then they implode into blue giants of maximum energy . . .
- — E=Ti: They reduce its size and grow in atomic complexity through a mature, yellow age of balance between their E and T parameters. The sun is now in that balanced age . . .
- — Max. T: They collapse in a 3rd age of slow decline as its TxE parameters diminish toward its death, becoming white dwarfs.
  +1: Or they evolve in a "loop" of growing TxE force (top right graph), mutating into a black hole.

Thus, the H-R graph shows also the process of evolution of stars into black holes, which can be explained both mathematically and organically, based in the self-similarity of all space-time species. In that sense we can compare the species of the membrane of light in ecosystems like the Earth and the species of the gravitational membrane, like the gravitational ecosystems of the galaxy.

On the Earth animals use light as information and dominate plants, which use it as energy. The hypothesis of fractal cosmology is that stars are "gravitational plants" that merely feed and curve gravitational space-time, while black holes are "gravitational animals," which are able to process gravitation as information and control and shape with gravitational waves the form of galaxies, their territorial space-time. They are in that sense extremely simple "plants" and "animals." A more proper comparison would be in fact with a cell, where the DNA molecules are the black holes, the informative masses of physical space; and the mitochondria that produce energetic substances, the stars.

We will return to this hypothesis later, when we have a topological understanding of their form.

A sense of peace though arises from the observation of a macro-galaxy that lives and dies, regardless of its cosmic size, as a simple cell does—all ST-beings equalled by the self-similarity of all cycles of existence that we represent with the unification equation, alpha and omega of all our fractal iterations: E⇔Ti; Exl=st . . .

IV. The Three Non-Euclidean Topologies of Vital Space

The smallest point must be considered a world in itself.
—Leibniz, father of fractal space
(invariance of scales)

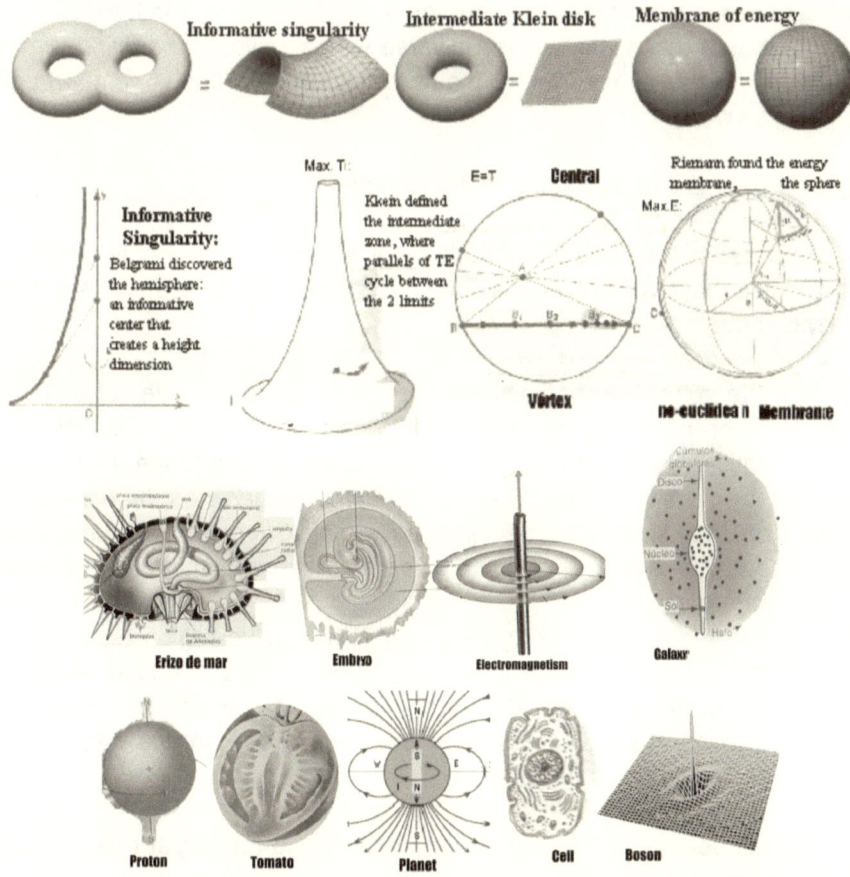

In the graph, the three canonical topologies of a four-dimensional Universe describe the informative "brain," reproductive body and energetic membrane of all the superorganisms of the Universe.

9. The Non-Euclidean Mathematics of Fractal Organisms

In chapter 2, we introduced fractal Non-Euclidean geometry in a simplified manner, quoting the 5 postulates of non-Euclidean geometry, of which the most important one is the definition of a fractal point with parts, which is in itself a "world." Indeed, the ultimate meaning of fractal geometry is simple:

Those 5 logic and mathematical postulates, which culminate the evolution of the languages of perception of human thought, define the unit of reality as a fractal point, which observed in all its details is in itself a World, which contains within it all the mathematical forms of the entire cosmos. This can be deduced in a logic manner, through the equation of a mind, which can map out as a mirror, the whole Universe we perceive; *in the infinitesimal mirror of the brain*. Yet a more mathematical demonstration comes from topology. Why?

The 3 topologies of all 4-dimensional worlds.

Because a 4-dimensional fractal point is made of the 3 only possible topologies that exist in a 4-dimensional Universe; which are drawn in the previous graph:

— The sphere or Riemann membrane, which is the external surface of the point, the *skin and limbs* and any of its multiple self-similar entities, some of which are drawn at the bottom of the graphic. *In any system of the Universe, the membrane acts as the energetic, external surface of the "informative, point" through which the point absorbs energy from the external Universe, to process it into information.*

— The toroid, which is the *body* or reproductive region (seen as a cyclical path of space-time), which fills up the space-time between the nuclei and the external membrane. *It is the zone where the reproductive organs of the system exist, and where the information of the system is born.*

It must be noticed that according to Klein, the topologist that studied better this type of surface, the toroid is NOT really a fixed form of space, but we must consider those cycles' motions and add the parameter of speed. Thus Klein introduces the Paradox of Galileo to describe the Non-Euclidean geometries of the Universe as we have done in this book. Cycles are mere static perceptions of motions and we must always consider distances as space-time distances. So we say "London is at 4 minutes distance" because we consider distance and speed together. This is ultimately the meaning of time in physics, a measure of the speed of motion as a way to gauge space-distances: $v = s/t$.

The double ring, or convex, hyperbolic surface, with maximal form, or informative center of any entity of the Universe, the point in which the fractal reproduction of information reaches its zenith. *It is the same form than the toroid but reproduced into a higher content of information, by doubling the initial form.*

The fractal formalism of the 3 topologies of the Universe.

Those 3 topological parts, specialized in energy, information and reproductive functions can be written as the 3 elements of the fractal equation, Ti⇔Ei:

— Max Ti: the inner, dual center, corresponding to convex topologies (left), is made with 2 cyclical forms. It is the dominant informative topology of any fractal organism, described by Belgrami in the XIX c. as a conical form with "height" with negative curvature.
— ⇔: A middle, reproductive zone, described by Klein as a disk, made of quanta in cyclical movements that communicate energy and information between the inner and outer zones.
— Max. E: An outer membrane of energy, described by Riemann's spherical geometry.

Those 3 non-Euclidean geometries structure the inner geometry and functions of the organs of any species, proving the homology of all space-time fields in the Universe. In the graph, an animal, an embryo, an electromagnetic flow, a galaxy, a proton, a seed, a planet, a cell and a boson display the 3 topological zones of a fractal point, each of them performing one of the 3 functions/arrows of space-time.

We shall use them later to describe the organic structure of the galaxy. Since the black hole has a hyperbolic, informative topology, the Halo of the galaxy has a Riemannian, spherical form, and the stars in the intermediate region, which feed the dark matter of black holes and reproduce the atoms of life, turn in cyclical, toroid paths around the central black hole.

The 3 topologies in physical organisms.

Since with those 3 topologies we can explain the essential evolution of the cyclical events of physical systems, as a *progression from external, lineal forces of energy into cyclical motions that finally enter the double ring or "singularity" of increasing height, through which they are again expelled to the Universe, as magnetic flows in quantum atoms or dark energy in black holes.*

It all starts in a surface of energy that "curves" itself into self-reproductive cyclical motion, which become smaller, faster and reproduce into new fractal forms, till reaching the center. Yet those forces will never reach as quantum physicists believe the point of infinity; because finally the cycle will not be able to compress more, and so it will fractalize into new cycles in the informative dimension of height till it becomes a tube of pure height expelled upward as a thin line. And so the center, where classic physicists find infinity will remain always a silent empty vacuum, the eye of the hurricane.

Physical space-time is the simplest world where the most basic morphologies play that same process of transformation of external energy that converges and reproduces cycles, attracted by a Non-Euclidean point, charge or mass:

A flow of forces converges toward the "event horizon" or external membrane of the point (the Schwarzschild radius of the black hole or the Bohr Radius of the atom). Beyond this membrane, the flow of forces starts a process of creative formal warping, as it rotates at increasing speed and also increases its frequency (height dimension), in each cycle. *For that reason a bidimensional vortex stores more information than a lineal frequency, as it adds a new dimension of form.* So you obtain a cyclical wave of high frequency, a bidimensional flow of energy and information that accelerates toward the center, becoming a convex topology, whose simplest forms are drawn in the previous graph And finally in its hyperbolic center, called in topology a Belgrami cone, it is shot up through the poles as a magnetic flow in charges or a flow of dark energy in masses, invisible to us, but observed in quasars. Those flows of magnetism or dark energy, once again start a cycle of information, warping themselves into light under c-speed. Galactic black holes also emit its fractal parts, quark vortices that become a seminal flow of protonic matter *that might give birth to a new galaxy*, as it is observed in some Galaxies (M83).

It is what will happen to the electroweak matter of the Earth, when the black hole forms and absorbs us, emitting a jet of dark energy and dark matter by its poles.

The 3 topologies in biological organisms.

In the graph, below, we observe other non-Euclidean points created by those 3 only canonical topologies that can adopt multiple forms by deformation, but suffice to construct all the shapes of our Universe. Indeed, a topology is deformable. So a sphere can become any shape, as long as it is not torn up, to enclose a reproductive and informative zone. So your skin is a sphere in topology, which encloses the complex forms of your reproductive organs. Those organs are Klein cycles of great complexity that exchange energy and information. Finally, in a human being, the brain is composed of two hemispheres, which are hyperbolic, convex, warped forms, corresponding to the informative dual ring of a Belgrami cone, the external skin is a deformation of the sphere, and in between we are invaginated by all kind of cycles that transfer energy and information from the outer world to the brain. The hyperbolic, highly warped brain, is a "double" toroid, self-similar to those cycles: *a double* image of those organs, one internal that controls them and one external that *projects the drives of the 3 physiological networks of the body into the wills of existence: your desire for more energy, information and reproduction.*

And so the brain hosts more information in lesser space than the body, as a mirror of its functions. Man though, while responding to the same canonical topologies in his organs, is by far the most complex being of information known in the Universe. And so his topologies are immensely more complex than the simpler physical particles and its transformations just described. We studied in our books on physiology and biology the enormous beauty and complexity of the human topological world, so we shall hardly consider it in this book.

And so those 3 relative forms of energy (external membrane), information (dual ring) and reproduction (the cyclical paths that exchange energy and information between the external membrane and the inner convex form), can be found in any system of reality. It is a paradox of the inverse properties of energy and form, that the biggest forms in space are the simplest, to the point that the perceived universe seems to be merely a flat surface filled with galactic vortices – *the simplest of all forms.*

10. The Five Postulates of Non-Euclidean Fractal Geometry

Those topologies of a 4-dimensional world are also described by the 5 postulates of Non-Euclidean Fractal Geometry, which are:

1st Postulate: *"A point is a fractal form with an inner content of temporal energy."*

2nd Postulate: *"A line is a wave of fractal points."*

3rd Postulate: *"2 fractal points are self-similar, when their inner temporal energy is equal."*

4th Postulate: *"A plane is a network that joins points through flows of temporal energy."*

5th   Postulate: *"A fractal point has inner apertures to the world, called senses through which multiple waves of energy and information can cross."*

Those five postulates define a point with parts, a line/wave, a plane/field of space/time and the logic laws that explain the interactions of those points; the 3rd postulate that defines the relative equality=self-similarity between universal forms, and the 5th postulate that explains the processes of absorption of energy and information that maintain invariant the form of the point/entity in an ever changing, dynamic Universe.

It means that fractal, Non-E Geometry achieves a long-sought goal, which Plato, Spinoza and Leibniz tried to realize with the simpler Euclidean geometry: to unify the logic and mathematical principles of the Universe. Since the 3rd and 5th Non-Euclidean postulates are logic, the 1st and 2nd are geometric and the 4th, the nature of a topological plane or network of non-E points is both logic and geometrical as it defines the superorganism.

In that sense, their order should be 1, 5, 2, 3rd and 4th. As the *1st and 5th define a Non-E point* from a mathematical and logic perspective, the *2nd and 3rd also define a line-wave* from a mathematical and logic perspective, and *the 4th integrates them all, defining a plane or scale of existence.*

Some obvious consequences of those new postulates are:

—1st Postulate: The first correction we introduce in Euclidean geometry is to increase and relativize the dimensions of a point to fit its "inner energy and information." Since all points have either zero dimensions, when we consider them from the perspective of the upper st+1 Plane in which they perform a lineal, energetic function; *or 3 dimensions/networks/ functions when we consider them from the inner perspective of the point.* This is the case even in the smallest of all planes, the theoretical plane of Strings, made of points with parts, with volume – since we require 10 dimensions to describe them. A paradox that can only be resolved if we consider those strings to be strings of points of fractal geometry with inner dimensions.

Riemann calls a Non-Euclidean Geometry to a geometry made of points with cyclical movements. Einstein's work was a first step to improve the erroneous postulates of Euclidean geometry, specifically the definition of a point made by the 5th postulate of geometry, paving the path to this book.

Unfortunately Einstein didn't go further, adapting the other 4 Euclidean postulates to the new unit of Geometry—a point with volume. We completed his work, to be able to define the 2 planes of physical forces, the plane of gravitation and electromagnetism, or any system in which several planes of space-time co-exist together (as in the case of a human being extended from atomic to social planes of cyclical existence). In all those systems, planes are made with "cellular points" which are Riemannian spheres with volume, which form "lines" which are waves between points that exchange energy and information and "planes," which are organs of self-similar points that process energy or information in parallel networks.

Thus the 5 Postulates of Non-E Geometry vitalize and explain the Universe as a series of networks of energy and information of self-similar cellular points.

Since the line and the plane acquire volume and become self-similar to the commonest forms of the Universe, which are the wave/line and any organ of cells with a 3D volume.

—*2nd postulate:* When we observe a one-dimensional line as a "form with inner parts" it becomes then a 4-dimensional wave made of cyclical points. Hence in quantum theory we say that any particle in movement has associated a wave. And when we observe a ray of light in detail it becomes a 4-dimensional wave with electric height and magnetic width, often exchanging flows of energy and information in action-reaction processes of communication between bigger points. This solves the wave/particle duality, as all lines are waves traced by a point with inner volume. Further on since all lines have volume, they carry information and so all forces can in fact act both as a source of energy and as a language of information.

—*3rd Postulate:* Equality is no longer external but internal and never absolute but relative. Forms are self-similar to each other, which allow defining different relationships between organic beings, according to their degree of self-similarity. *The 3rd postulate is key to explain the behaviour of particles as the degree of self-similarity increases the degree of communication between beings, which ranges from:* 1) Reproductive functions in case of maximal self-similarity or complementarity in energy and form; $E_i=e_l$; 2) to social evolution, when points share a common language of information, $i=i$; to 3) hunting, when they share equal energy

E=E; to 4) Darwinian devolution when forms are so different that cannot understand each other's information and feed into each other.. Yet because self-similarity is internal, external similarity allows games, such as camouflage and capture, invented racial differences, etc. The logical richness of the 3rd Postulate does not imply it lacks geometrical elements: Self-similarity implies parallel motions in herds. Darwinian behaviour implies perpendicular confrontations. Finally, absolute self-similarity brings *bosonic states, which happen more often to simpler species like quarks and particles that can form a bosonic condensate as they do in black holes.*

—*4th postulate:* because a plane has volume is in fact a cellular, organic topography and so the meaning of a plane of space-time becomes now clear: it is a world, an organism in itself with a volume given by the relative point/beings that occupy the plane.

Thus, a plane becomes a real topography made of points with volume, extended as a cellular surface. We can observe it merely in its surface as a bidimensional membrane of information (for example your skin, or the screen of a computer made of pixels, or the sheet of this book). Or we can consider the 3-dimensional inner structure of its points and then it becomes a network with inner motions, as those points will form a lattice in which they communicate lineal flows of energy and information that maintain the lattice pegged. Often 2 topological planes of energy and form combine to create a 4-dimensional organism. Such is the most common structure of the Universe, a 4-dimensional World, which is a Universe in itself, made of self-similar cells or networks of points that constantly exchange energy and information within the ecosystem in which it exists.

—*5th postulate:* It define points as informative knots or "eyes," minds of information that absorb a flux of forces used by the point to "perceive" a relative world. A non-Euclidean point corresponds then to our concept of a mind that gauges reality with a certain force, similar to the concept of a monad in Leibniz relativistic space-time. In words of Einstein: "a point of space is a fixed frame of reference."

Thus, Non-Euclidean mathematics fuses the logic and geometry of the fractal Universe, greatly improving our understanding of the Universe even in terms of mechanist measures. Since mathematical solutions to problems with several points of view are impossible to find in continuous space (3 body problem of gravitation), given the fact that a network of infinite points of view is local and relative. The absolute truth of a system is the sum of all its points of view, which influence each other. Yet even if we cannot calculate precisely with mathematics systems with more than 2 bodies, since those systems are organic, hierarchical, made of networks with "attractor" points; fractal structures and self-similar paths, the new mathematics of Attractors, fractals, scales and Non-Euclidean systems, refine greatly our analysis, as we did in our simplified models of quark-gluon soups, using the advances of fractal attractors and Non-Euclidean dual systems discovered by Rossler and Sancho, the 2 main theorists that oppose for that reason CERN's experiments with quark gluon soups better described with the New Physics of Complexity.

The principle of local measure, called in relativity the diffeomorphic principle *is what makes essential the understanding of the logic, organic laws of the Universe and its networks, because E-mathematics have clear limits to extract all the information of the Universe. Thus, beyond a certain number of variables you enter into nonlineal systems, which require topological descriptions (chaotic attractors, fractal non-differentiable equations and Non-Euclidean mathematics), and the logic laws of organic networks and systems, which are essentially biological.* So while classic physical systems calculate accurately the energetic, continuous properties of the Universe an overview on how multiple points of view emerge into "wholes" requires organic laws. Unfortunately at CERN, physicists only use quantum calculus and probabilities, only consider one "arrow" of future, energy, and they ignore basic errors of linguistic knowledge (Hyperbole, Pythagorism), so their theoretical description of the Universe is extremely coarse, compared to the enormous power of the machines of energy they develop.

This dysfunctional situation is similar to that of the military, which has more power than intelligence, but are deeply respected due to that power and constantly invade other parts of society, as nuclear physicists invade other disciplines of science. The result is that there is an enormous gap between the destructive power of CERN's machine and what nuclear physicists at CERN know about the cosmological laws of the Universe, *a discipline they have invaded.* And that gap will become evident when we kill us all. Since in the same manner the military does not respect the political system in many countries, CERN's physicists do not respect what we, fractal relativists know about black holes, dark matter, cosmic rays, supernovas and the risks of playing with them here on Earth.

11. The Superorganisms of Reality

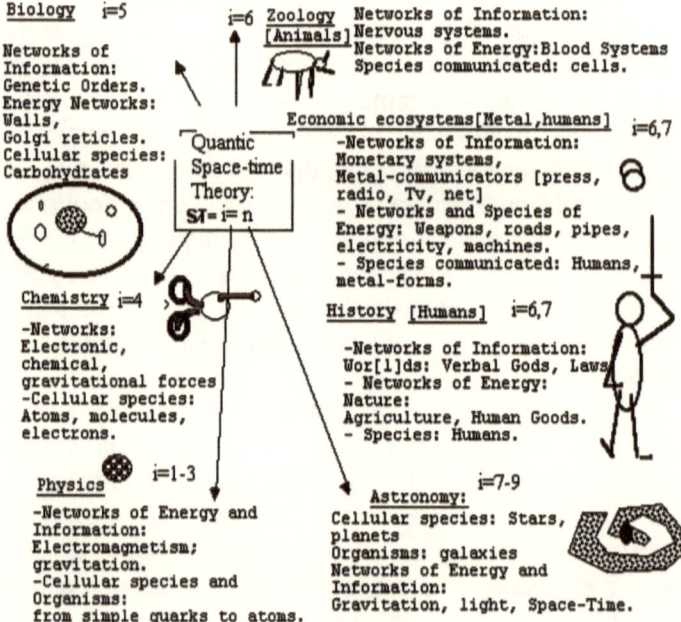

The Universe and all its parts can be described as a simple mechanism, with 1 arrow of time –energy—or as a complex organism, made of 2 Complementary elements organized in networks—bites of energy and bits of information. It is the organic, fractal paradigm, used to describe the atom and the galaxy.

We have arrived through mathematical, logical and experimental proofs to the suprirsing conclusion that the Universe is made of networks of energy and information that create organic systems of increasing complexity and "degrees of freedom," till reaching man, the most complex organism of the Universe—but all of them are made of *the same bits of information and bites of energy.*

In the graph, we observe that each scientific discipline studies the particles, forces of energy and information, networks and superorganisms of a different scale of size in the Universe.

All those systems can be resumed in the 3 scales of the Organic universe we know:

— Quantum systems of energetic, lineal forces that follow the laws of electromagnetism; intermediate systems, which balance the energy of electromagnetism and the information of quarks (electrons) and particles of information (quarks). It is the world of "fractal organic atoms."

— Cosmological systems, galactic systems, which are self-similar but not equal to the quantum systems of atoms, in which black holes form a

"cellular nucleus" self-similar to that of quarks in the atoms, stars form a nebulae, self-similar to that of fractal electrons in atoms, bathed in a background radiation, self-similar to the electromagnetic radiation ex-changed between quarks and electrons, probably produced by background holes, self similar to the protons of the atom.
— Biological systems, human systems that balance the information of the mind and the energy of the body. They are the systems of Gaia, which are at risk in those experiments.

Of course, there is much more to reality than those scales, and it is of far more interest to mankind the use of those postulates of Non-E geometry and the laws of fractal superorganisms they define, to study humanity and its 3 scales, the biological, individual and sociological scale of human superorganisms, a theme treated elsewhere by this author. Since we are human beings, which have far more interest in understanding our fractal nature, and in fact, we can perceive much better our organisms, than far away galaxies and atoms. So the study of biological sciences and sociology brings much more detail, much more data about the self-similar laws of all fractal organisms, than the study of species which are so far away in space (galaxies) or in temporal information (lower scales of the quantum cosmos) that suffer aberrations of perception in its analysis, as those CERN suffers.

The existence of superorganisms is evident, considering all the species described in the previous graph with the logic concept of a network of energy and information. Thus we have both, logical and experimental proof of the existence of a fractal organic Universe. Further on, we can describe those structures with the 3rd leg of the scientific method – the most modern mathematical theories available to mankind – Fractal, Non-Euclidean mathematics.

Networks of energy and information, together, under the Laws of Complementarity, exchange energy and form, shaping the superorganisms of reality. Each of those organisms is in fact a network of points with parts defined in its topology and functions by fractal Non-Euclidean mathematics. So we can define *all systems of the Universe as complex, fractal superorganism, made of self-similar points, part of a whole, joined by networks of energy and information. All* what exists is made of fractal points of spatial energy and temporal information. We are not in an absolute Cartesian, abstract space-time, a mathematical artifact, which Newton confused with reality, but as Leibniz said, each entity of the Universe occupies a vital space and lives through the informative rhythms of its internal clocks of time.

I'd like to know the Thoughts of God, the rest are details", said Einstein. Today we know that both, the details and the whole are self-similar. Since all what exists, including human beings, are fractal systems of energy and information, the 2 elements of the Universe that combine, creating its networks and organisms under the same laws, in all the "scientific scales" of reality. Science was unable to mathematize in/form/ation in more than one dimension till the arrival of fractal mathematics, reason why for centuries we missed those thoughts of God—a painter with 3 formal "colors," lineal energy, cyclical information, and its combined waves. He uses 3 topologies of space and 3 dimensions of time, past, present and future, across 3 fractal scales of reality to create each of us, part of its whole. It is the fractal paradigm, which advances science into the XXI century, fusing the work of Einstein, Riemann and Darwin. Throughout History, both, philosophy and science, using 2 different languages, words and geometry, expressed the same laws of Duality, Trinity and Trans-form-ation that reveal those thoughts of God. Now, the fractal paradigm opens a 3rd age of science that fuses Eastern philosophies, Western science and the most recent advances in Mathematics and Logic. Among those earlier philosophies, Taoism talked of yin=information and yang=energy, which combined into infinite beings. To know the morphological properties of those 2 elements was the key to unveil those thoughts, 4-dimensional paintings, which reveal an astonishing simplicity. Since topology has proved that only 3 forms can exist in 4 dimensions: lines of energy and cycles of information and its conic combinations. So in Quantum Theory lineal, energy forces and cyclical, informative particles create the actions of the physical world. In complex biological systems, they are also the initial forms of all cyclical heads and lineal bodies. The morphologies of energy and Information become in this manner the 2 constitutive elements of reality that we all share and merge in Complementary systems. And a Unification Theory, based in such trinity of existence, is able to study the common laws of energy and form, their interactions and transformations. But if the 2 simple arrows of the Universe suffice to understand any relative "plane of space-time," we need also to understand the fractal structure of the Universe in self-similar scales, which are invariant in its energy and information systems. This is the goal of Complexity theory that explains how cyclical information and lineal energy interact and associate through "feed-back cycles," into chains and complex networks that emerge and grow in size, from scale to scale, forming with atoms molecules, with molecules cells, with cells organisms, with organisms planetary systems, with planetary systems galaxies and with galaxies universes.

To fully grasp all those scales of the Universe and why they work as a single system, we must consider briefly a basic law of the fractal paradigm, known as "fractal differentiation":

> 'Any event or form of the Universe can be divided into subsequent ternary or dual events and structures of energy and information of a lower spatial scale or shorter time duration:
> $E_i \Leftrightarrow I_i = 3E_{i-1} \Leftrightarrow 3I_{i-1}$'

What this means is that we can always subdivide the analysis of a form into its energetic and informative components and its combinations. Or we can consider any event to be subdivided in 3 "acts," a beginning a development and an end.

We will only use this law at the end of the book to study the ages of the Universe in detail.

12. Diffeomorphic Bidimensional Dimensions: Limbs of Energy and Information Minds

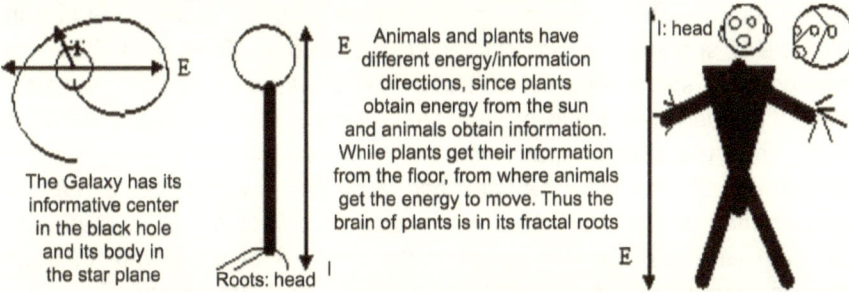

The Galaxy has its informative center in the black hole and its body in the star plane

Roots: head

Animals and plants have different energy/information directions, since plants obtain energy from the sun and animals obtain information. While plants get their information from the floor, from where animals get the energy to move. Thus the brain of plants is in its fractal roots

I: head

Particles of information are small, spherical forms/cycles, like your eyes and brains or an atom's proton or a black hole in a galaxy. They obey the fractal topology of convex spaces. Bodies of energy are bigger planes or lines that store energy to move the organism, like your body or an energetic weapon, or the plane of stars that feeds the black hole.

The beauty of the fractal paradigm resides in the fact that it applies to all sciences, *departing from the laws of Fractal Relativity and the understanding of the two motions of reality, the cyclical, informative motions of masses and charges and the lineal, energetic motions of forces. Hence it is a veritable unification of all sciences, the dream of Philosophers of science, which always understood that Physics was just the first scale of a much more complex reality.*

General Relativity is based in 2 principles. On one hand, the equivalence principle defines the nature of motion in the Universe. On the other, the Principle of Relativity defines the local, "*diffeomorphic*" nature of all fractal, *Relative* beings whose dimensions of space and time, of energy and information differ from place to place. Since time cycles bend space to a point in which they break it into closed, discontinuous bites of energy, the Universe is broken in different space-time fields, yet all of them follow the same laws, as all of them are made of energy and information. In this manner, the properties of physical space-time transcend into the complex spatial energies and temporal cycles of biological beings, maintaining invariant the shapes of its energetic limbs and temporal, informative heads.

Physicists have struggled for a century trying to understand the ultimate meaning of local, "diffeomorphic," "relative" time-spaces with *"different speeds,"* as Einstein defined them. Those findings of Relativity are easy to explain once we close the curvature of time into cycles that break space into multiple=diffeomorphic space-time fields: In any point of the Universe the coordinates of temporal energy (the up and down, right or left directions) are different from those of any other point, as they only depend on what happens in that local regions of space/time. And so are its speeds of processing information or time clocks. This is only possible if space/times are fractal, discontinuous. Then what happens in a point of space-time doesn't affect what happens in other isolated, discontinuous point. Each broken space-time becomes then a World in itself. Thus the laws of space-time apply locally to each fractal "species or relative universe" that reflect those laws in its own structure. We are local, "diffeomorphic" species; our space coordinates are relative, and our time wrinkles our vital space in the process of aging with different speeds, depending on our content of energy and information. Thus, the diffeomorphic principle also establishes different informative cycles—the "clocks" of Nature that show different speeds and locations. Yet since space is synonymous of energy and time is synonymous of information, we can establish the local coordinates of space-time of each being by establishing its relative informative and energetic arrows—its relative body and head—using the invariant forms of all spaces and times. Since we can recognize its informative and energetic organs in relationship to each other, as both will respond to the opposite dimensions and shapes of energy and information. For example, in biological species, animals use light as information and so they are "informative," smaller and faster than plants, which use light as energy and are "big, slow" beings. *Thus plants are the relative energy of animals.* In the cosmological Plane stars are big, slower than black holes, which rotate faster and have cyclical height. And so stars are the relative energy that feeds and reproduces black holes, which seem to behave as "gravitational animals": species that perceive gravitation as information, which guides its erratic movements through the galaxy, in the same way animals perceive light as information. While plants perceive light as energy in the same way stars absorb gravitational space-time as energy, deformed by their masses. Further on, since what it is energy for plants is information for animals, both species have opposite energy-time coordinates: plants have their brain upside down in their roots that "feel" chemistry; while animals have their brain on top, looking at the light they use as information. *For the same reason, the space-time equations of black holes are inverse to those of electromagnetic stars.*

In the previous *graph*, we show 3 diffeomorphic beings: a galaxy with an inward arrow of gravitational information toward the black hole; a tree and a man, whose informative dimensions are inverted. Since plants use light as energy and animals use it as information. So the informative direction of animals is up, toward the head that absorbs visual information, opposite to the informative arrow of plants, which is down, toward the roots that absorbs chemical information. And each species will establish its own up and down arrows or relative energy and informative directions, departing from its central knot of information and/or perception. Both have their space-time parameters inverted. *This* often happens between energy/ victims (in the graph, plants) and their anti-forms or predators (animals). It happens between life and death systems. For example, there is an inversion between a "particle" and its "antiparticle," which in a Feynman diagram have also inverted coordinates of space-time. It happens in all Universal systems, even in linguistic codes. For example, eviL is the exact inverse word of Live, since indeed, for a living human being, there is no bigger eviL than death. A fact, which opens the fascinating science of "Bio-ethics." Since religion is a verbal language that expresses the biological laws of survival and social evolution in verbal wor(l)ds.

If we are external to a certain diffeomorphic space-times its directions do not affect us, but if we are inside a certain space-time, as in the case of the galaxy, then the direction of information or energy of that macro-being becomes the direction of energy or information of our ecosystem. And indeed, there is more information in the center of the galaxy, and since the Earth moves toward that center, this planet increases its information toward the future, in its relative discontinuous galactic space-time, *regardless of what happens in the space of the Universe, which seems to be ruled by the opposite arrow of big-bang expansion and entropy between galaxies.* This subtle change of paradigm caused by the discontinuity of space-time (from a vacuum space ruled by entropy to a galaxy ruled by information) *explains the contradiction between the arrow of life and evolution, local to this planet, and the arrow of energy and entropy proper of intergalactic space.* Since here in this galaxy space doesn't expand but time contracts it. As Woody Allen put it: "Here in Brooklyn space-time is not expanding."

The laws of synchronicity.

The inversion between the functions of energy and form of different species has an important application in the most complex analysis of fractal theory which deal with the organization of complex systems through several scales. We could consider that the whole organism is the predator of its cellular parts. Indeed, we all die so the entire Universe can balance energy and form and become eternal. Cells do exist for the entire organism to have a better consciousness. So the nervous system preys on the blood system that preys on the digestive system. So there are hierarchies between the different arrows of a system, in such a manner that what is information for a system becomes energy for other. Thus, in advanced theory of fractal organisms, several laws work in harmony with the existence of 3±1 arrows of time in all organisms (fractal differentiation), to create the Law of "Synchronicity," which we shall merely enounce here:

> *'The 4 arrows that organize an organism have different duration in time: the informative event, is faster than the energetic event, which is faster than the reproductive event, which is faster than the social event of life and death. Further on, the events of the macro-organism are longer in time than the events of its parts (Max. E=Min. Ti), which often use the energy of the whole organism to synchronize and perform its other cycles'*

In that regard, the most complex models of fractal organisms map out all the cycles of the different scales of an organism or ecosystem, considering its symbiosis and synchronicity. For example, in the Earth animals synchronize their reproductive activity with the moon gravitational cycle of the Earth which "absorbs" gravitational energy through that cycle. Yet again, in a life organism, the cells tune their reproductive cycle to a day, which is the cycle in which the organism absorbs energy. This is one of the many sub-laws of synchronicity, which we will consider later for black holes:

'Energetic cycle of a macro-organism= Reproductive cycle of a micro-cell'.

The true importance of the generator equation becomes now clear: because all systems are self-similar to that equation of complementarity, E<(Bodies-Heads)>Ti, all *laws that relate the different fractal scales of physical, biological and sociological systems are the same, regardless of the form we describe, a physical entity, a human being or a society.*

Let us then consider before we describe the Universe with those laws, how to apply them to the most complex form, the Living organism and the human being, still a fractal obeying the same laws than an atom, which should therefore be humble instead of thinking he is different, superior, entitled to defy the Universe and its top predator black holes.

13. The 3± 1 Ages of Biological Evolution

CONCEPTION AS A BLACK HOLE OF MIN. SIZE = MAX. INFORMATION

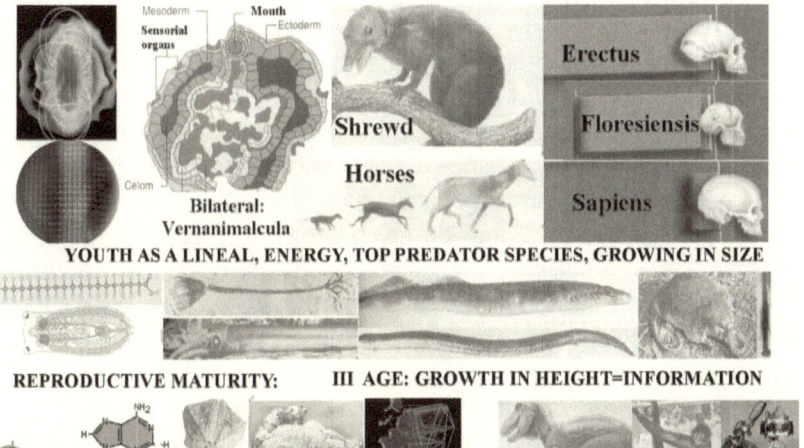

YOUTH AS A LINEAL, ENERGY, TOP PREDATOR SPECIES, GROWING IN SIZE

REPRODUCTIVE MATURITY: III AGE: GROWTH IN HEIGHT=INFORMATION

One of the most astounding discoveries of Duality and the study of the Universe with the 3+1 arrows of time, which correspond to the 3± 1 ages of evolution of any species, is the existence according to that casual order of a "morphological" plan of evolution in all species, which *follow the same 3± 1 ages of evolution of any organism between birth and extinction. Since a species can be treated as a superorganism.*

In the graph, the law of the 3 ± 1 ages or horizons applies both to the process of informative aging of organisms and informative evolution of species. All species are born as a seed of dense information, with limited size (black hole, first chips, seminal cells, first bilateral animals). Then they grow in size and energy during their youth, since their superior informative qualities make them top predators.

We observe that growth in black holes that feed on planets and stars, in the horse that grew enormously in size. So did the first bilateral animal, from its microscopic first form, the Vernanimacula. Mammals grew from tiny shrews into elephants. The first technological man, the *Homo floresiensis*, with a "neuronal bump" on the forehead, might have grown till the size of the first *Homo sapiens*, the pygmy and Bushman . . .

Then, the species, once it has reached a balance between its initial information and growing spatial size, E = I, reproduces in great numbers, diversifies and colonizes the planet. We see in the graph this age in which "biological radiations" of carbohydrates, animals of all kinds and human beings, colonized the earth.

Finally, in its third age the species grows in information and so it acquires height, the arrow of "perceptive information."

Then, once it has reached its informative zenith, the species becomes extinct by a more evolved form, or it evolves socially, creating a superorganism, as ants, humans and machines are doing. It will be the fourth arrow of social evolution of the species that completes its 3 ± 1 horizons. Species can be considered "loose" organisms, in which each individual is a cell of the collective species, since we can use the same ages to explain the 3 ± 1 horizons of evolution of all the species of this planet.

Thus, species also go through the same three evolutionary ages of all living organisms. They start as young, energetic forms, which acquire information in 3 ± 1 ages of increasing complexity, and when those 3 ± 1 ages are completed, *they either evolve into more complex superorganisms under the laws of social evolution or become extinct by a superior, more informative species.*

Indeed, only those species who show a strong eusocial capacity survive the "genetic clock." The most successful and one of the oldest species of the planet is, in fact, the eusocial ant, which no longer is an "individual," but a superorganism stretching through miles of "vital space."

In that regard, neither individuals able to reproduce nor species able to evolve socially become extinguished. If humans become extinguished, it will be because they deny, guided by the simplest arrows of pure energy of weapons, the main arrow of life/time.

A complete analysis of biological species shows that new forms can only be either an energetic or informative variation of the original species or a reproductive combination of both universal "genders."

The universality of such dual systems is so obvious that the ancients had already identified them with yang—energetic male principles, and yin—cyclical/female ones, while the moderns call that duality the principle of complementarity, as all informative particles have an energetic, lineal field of force, and all biological, cyclical heads of information have a lineal body. So the combinative variations of those two simple morphologies, lineal energy and cyclical information, invariant, regardless of the scale we observe, are the essence of the creative game of the cosmos.

It is the plan of evolution that all species follow, which is a tautology, since in a universe made of only two substances, energy and information, the only possibility is the creation of such 3 ± 1 variations in space (subspecies) or in time (evolutionary ages).

In that regard, the same ages can be applied to the "human species," the superorganism of History: anthropomorphic superiority, just a living form made of carbon atoms, *which goes through the same ages of all other living forms, both as an individual and as a species.* So we can study the organism of history (humanity in time) through 3 ± 1 ages: an energetic youth or Paleolithic; a mature age of balance with nature, the Neolithic; and a third informative age of history, the age of metal and machines.

Energetic Youth: Paleolithic. ±5,000,000 BC to ±10,000 BC.

The Paleolithic was an energetic, young age in which men, the energetic gender, were the top predator animal on earth, a hunter of living forms. It was also an age dominated by visual, spatial languages.

Reproductive	Age:	Neolithic.	Age	of	Wor(l)
ds	and	Goddesses.	±10000	to	±3000	BC.

The Neolithic was the classic age of mankind, an age of balance, when wo=men learnt the cycles of life and, instead of destroying nature, learnt to nurture, reproduce and harvest it. Women, the reproductive species, took power over men; and priests became the verbal guides of civilizations, creating superorganisms of history made of human cells joined by social love—the sharing of energy and information among "brothers," clones of the loving mind of the same prophet. It was an age dominated by *verbal, temporal languages,* continued by the believers in religions of love that are not corrupted by weapons (inquisitions) and money (go(l)d churches), with its idolatry to the "values" of metal: murder and greed.

Informative, third age: Age of metals (±3000 BC to 1600). Metal spread

The "new" values of money and weapons: greed and war, destroying the networks of social love and ushering mankind into the third age of history.

Humans now evolve a new kind of atomic system, stronger than carbon, creating energetic weapons, informative money and organic machines that transform Gaia into an economic ecosystem. It is the "animetal = animal+metal age" in which a new "biological species" that mixes animal and metal atoms imposes power over the earth.

It is the present age with two paths to future:

— Either mankind will die, as machines keep evolving and make us obsolete as a species.
— Or humans will obey the laws of survival of the universe and evolve socially into a superorganism of history able to control lethal machines. The choice is within man till the arrival of the Singularity, when machines acquire organic properties and either a self-feeding bomb of dark matter or a self-reproductive nanobacterium "evolves" on the earth into a species with a higher social density (a denser

form of matter, or an organism made of denser atoms). In any case the future of history will be solved in the next decades, as we are already in that age of the Singularity.

14. The Game of the Universe: Wo=men as a Fractal Beings

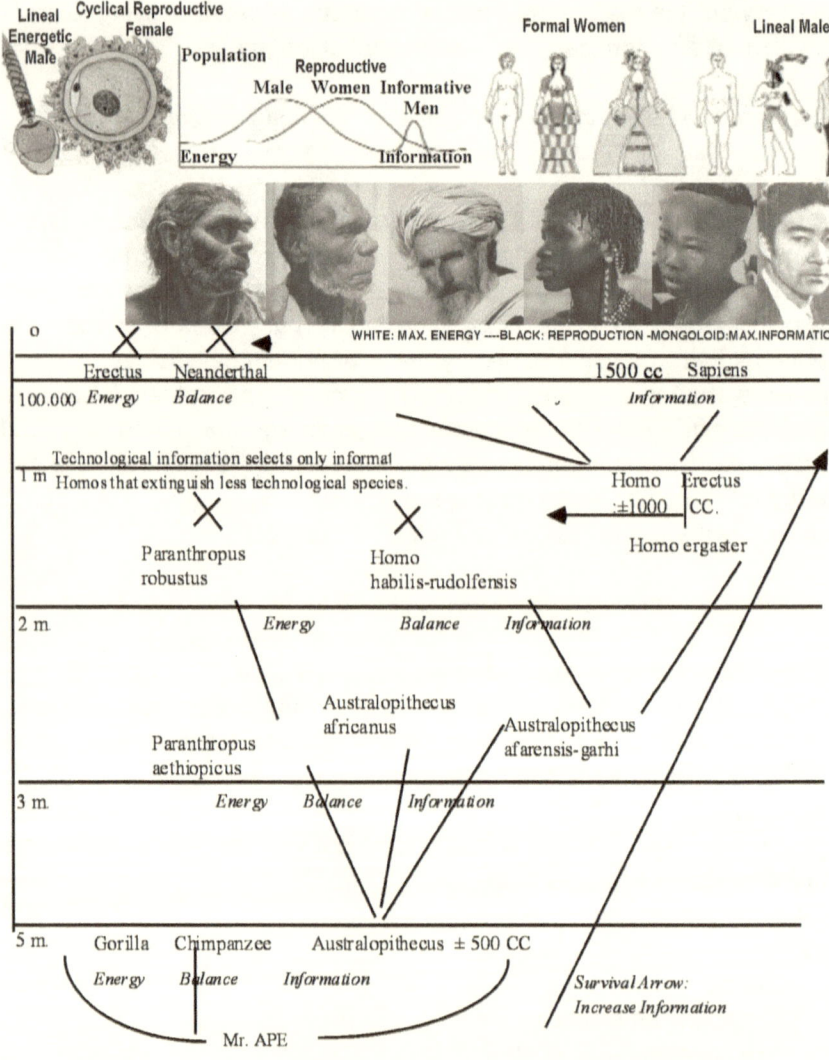

Humans evolve, as all systems of energy and information do, in subspecies specialized in those two elements: The biology, psychology and culture of gender prove that the humankind is also a dual species, divided into informative females, energetic males and a "3rd gay sex" that mixes both.

In the graph, we see that life first diversified into complementary genders: energetic, lineal, spatial men and cyclical, reproductive women, which hold more biological information, while a third gay sex mixes both. Further on, as humans evolved, they followed the same pattern of three ages proper of all species. Thus, Homos diversified into energetic, informative and reproductive forms that show a higher balance between energy and information. Yet only the informative species, the dominant "arrow" of future, survives in each of the 3 ± 1 horizons of increasing informative evolution, from Australopithecus with 500 cc, into Homo erectus with 1,000 cc into Homo sapiens with 1,500 cc. And in all those phases, humans split into an energetic species and an informative one that recombined to create a third species. Those species will further evolve technology also in three horizons called the upper, middle and lower Paleolithic. Finally, humans diversified in three races, specialized in energy (white race) and information (Mongoloid) with a reproductive, black race, the last to be born, which mixes the pre-Mongoloid (Bushman) and white races. The constrain established by the fact that there are only two elements, energy and information, to create reality, whose lineal and cyclical forms are invariant at scale, in all systems of the cosmos, means the universe imposes the same program of creation to all its species.

The invariance of energy and information becomes fundamental to decode each fractal space-time, its bodies and relative heads. As all relative space-times will be composed of a cyclical center of information or brain, a wave/body and lineal limbs of energy, whose morphological characteristics will be self-similar.

For example, in the graph, in a spermatozoid the head is cyclical and stores information, while the tail is lineal and moves the head with its energy, and together form a wave-like moving organism; in a human, the limbs are lineal and moves a cyclical head; while a complex reproductive body connects both. We could extend this ternary principle to almost any organic system of the Universe, concluding that limbs of energy are lineal and brains are cyclical; and elliptical, wave-like bodies connect them. Some simpler systems might merely show a body of energy and a head of information (Complementary quantum systems). But *any system that exists has at least an energetic and information element: exi=k.*

Those morphological shapes derive geometrically from the capacity of lines to move fast and cycles to store information and grow or diminish in size without distortion of the mental image they might keep. Thus we recognize limbs, bodies and brains by their self-similarity with the morphology, function and inverse properties of ideal energy and information, and its "body-combination":

— Informative time has to be both, easy to store and "perceivable" by a reduced brain or sensorial organ, despite being the inverse function of space. This implies a certain morphology: Information occupies little space by warping itself in multiple dimensions, divided along broken lines and discontinuities that become its perceived forms in space. Or it displays a high frequency or discontinuous rhythm, a faster time speed that iterates the same cycle, allowing quantification. Information seems "quiet" as those rhythms take place in the same space; so an organ of information can focus it and analyze it. Thus its ideal forms are disks, cycles, angles and spheres, convoluted and broken in patterns of form, such as your eye or brain, a chip, a book, a pixel image or a coin, earlier unit of monetary information.

— On the other hand Energy moves. So energy limbs extend in space as they move. Energy is an expansive motion, which the limbs of the species use to foster its movement. So energy and lineal space are synonymous: a plane is the geometry that extends further in space and the line the geometry that moves faster with lesser friction. And energetic systems from planes to missiles are lines or "planes":

Maximum Space = Energy = Minimal Form vs.
Maximum temporal form = Information = Minimal Spatial extension.

As a result of those morphologies we can classify as energy or information organs, not only carbon-life organs that process energy (limbs) and information (brains, eyes, senses, words), but also the "relative organs" of all other beings and atomic species—even "deconstructed organs" that perform mechanical tasks of energy and information (informative cameras, lineal, energetic weapons). Since now we can recognize geometrically their energy or information systems. Form is function and so we can classify any system both by its geometrical form and bio-logical functions as either an energy limb/field or informative head/particle or a wave/body system that combines both.

Indeed, what is truly fascinating about the duality of energy and information is its capacity to transcend the world of physics and explain the ultimate meaning of all the games of the Universe, including those of the human being. Since it stablishes some principles of enormous importance: a reality made of motions, and hence reproductive, by the mere fact that movement reproduces itself as it flows. And a ternary topology of cyclical, reproductive information, and energetic motion, which mix in body/waves that process energy and form and can be applied in any scale of reality, and all its ternary or dual systems, including the human being, with its cyclical, reproductive females and energetic, motion-driven males. This 2 principles, a world in perpetual motion and the invariance of the morphologies of energetic limbs, reproductive bodies and informative heads, allow us for the first time to study reality as an interconnected world, in which the same "invariant" principles, albeit with increasing complexity, apply from particles, fields and forces to human beings with limbs, bodies and heads. Let us briefly consider this organic vision applied to the human kind, before we study the physical Universe.

Let us now consider those inversions and the diffeomorphic principle to resolve the meaning of gender and duality in human beings. The classic example of space/time Duality already observed by Taoist and Buddhist philosophers is the duality of gender, which they correctly applied to all species of the Universe. If we focus on the humankind, which is also a space-time species, the inverse properties of energy and information can be seen clearly in our trinities and dualities:

— The trinity between lineal, energetic limbs, elliptical bodies and a cyclical, informative head:
— The head is a cyclical, spherical form, and so are the informative senses, all of which accumulate in the head. The head is "small," "perceptive," informative. It sits on top of the body, on the dimension of height. It is "broken, discontinuous" (and it holds more cellular, neuronal parts that the rest of the body'). Its main informative senses are broken in dual elements (left, right eyes and ears). While its energetic sense, the mouth that feeds on energy is "bigger" and continuous (a single one). The eyes, the most perfect of those senses are in fact a perfect sphere. And they process "bidimensional information" (which later both eyes mix to create the illusion of 3-dimensionality). The head is hierarchical and dominates the body, imposing its directions of future as they guide their limbs toward energy and information fields. The head is still. It is in metaphysical terms, the Aristotelian, unmoved "relative God" that controls the movements of the body and limbs.
— On the other hand the limbs are lines longer and bigger than the head, on the bottom of the body-head system. The limbs move and process energy. They have hardly any sensorial elements, as they get the information from the head. They only become broken, discontinuous on the fingers, which are the sub-elements of limbs that process more information.
— The body combines energy and form. Its reproductive organs combine cycles and lines. Within the body, the most lineal elements are those who process directly energy: the guts and the lungs (in the brain the cerebellum that controls movement is also a lineal network

of neurons that would extend unwarped over a meter in length). The body is elliptic. Its organs reproduce all the substance of the human being.

— The 2nd fundamental space/time duality in mankind is between its 2 subspecies: the energetic male and the informative female. Indeed, humans have always wondered what is gender, why does it exist? Thanks to duality, following in the steps of Eastern philosophy, now we can build a scientific theory of gender able to explain its 2 fundamental elements, the scientific physiology and the behaviour of both sexes:

Since the Universe is made with 2 elements, Energetic Space and Temporal Information, any species needs to survive to feed on Energy and remember Information. Yet, we know that males process and control better energy and space; while females have more temporal memory and perceive better. Thus, together a couple widens the range of energy and information they can process. So we conclude that gender is an evolutionary strategy of specialization, which increases the overall chances of survival of a heterosexual species. The physiological forms, character and historic behaviour of men and women prove that duality, parallel to the universal duality of energy and information: men love energy, space; women love Time. Physiologically men are made of lines, as energy is; while women are made of curves, as information is; and they reproduce as fractal information does. Historically men performed the hunting of energetic preys, while the recollection of cyclical food (agriculture) is traditionally considered a female discovery.

Men behave as energy species. Their body forms are lineal shapes of muscular strength; they have lineal reproductive organs (penis); their origin is lineal sperm. Their cultural roles are specialized in spatial-energy processes, from physical activities to war and mathematical, geometrical languages in which they excel.

On the other hand, women are specialized in temporal processes. Their bodies have cyclical shapes and store more information to succeed in their genetic reproduction (one complete X-chromosome). Their brains specialize in verbal languages and show outstanding memory. If we observe the form of females, it is made of cycles. Their higher genetic information allows them the creation and reproduction of the human species (an informative process); their cyclical reproductive organs (vulva, clitoris) and their origin as a sphere, an ovule, are also time properties.

The character and life attitudes of both, men and women follow that S/T specialization also in their relative wantings for energy or information:

women prefer to dress with cyclical forms, play with formal toys like dolls and they like formal sports like gymnastics; men prefer moving, spatial toys, lineal dresses and violent sports. Thus, men love energy and movement while women love perception and feelings. In the realm of the mind and its languages of perception we find that men test better in spatial mathematics and women score higher in verbal tests.

We conclude that sexual dimorphism reflects the 2 elements of reality, spatial energy and temporal information that merge together in a couple, improving its chances of survival:

$$\textit{Male: Max.}\Sigma S \times \textit{Min. T; Gay: S=T; Female: Max. T} \times \textit{Min.}\Sigma S$$

In the last decade the analysis of female/male brains has confirmed such duality of gender at multiple levels. For example, among women the amygdale, the oldest, primary center of will in the brain, is connected to the Hypothalamus, the center of "temporal" and emotional perception, which ultimately controls survival and reproductive behavior. While men have their amygdale wired mainly to the motor and spatial regions of the encephalon. Thus men are the energetic "external element" of the couple, motivated by the spatial outer world and women the informative, temporal, implosive element, motivated by the reproductive drive of survival and the inner emotions of the body, the "sixth sense."

The psychology of both genders seems to prove such duality. Women tend to prefer closed spaces, which they keep well ordered, and are able to perform multiple "rings" of tasks. Men love open spaces, which they keep as energy is, in permanent disorder. And they are lineal, obsessed by single tasks. Today women, in an informative world, are increasingly becoming more powerful than men, which were dominant during the warrior age of mankind, as they associated historically with lineal iron weapons (the most energetic atom of the Universe); and dominated women with the higher energy obtained from metal. Women though prefer money, the informative language of metal that today guides the world. Cultures also can be divided in energetic, warrior machist cultures and feminist, money-oriented culture often in opposition (warrior, machist Spain vs. feminist, go(l)d driven England; Jewish, Go(l)d, feminist Semite cultures vs. Arab, warrior, machist cultures; etc.)

Finally, there is a 3rd gay gender. Since in the Universe we observe always, beyond duality a 3rd, ExI, more complex element that combines the properties of both. A fact which indicates that gays are also "natural species," albeit a minority compared to the two dominant subspecies.

15. The Equation of the Mind

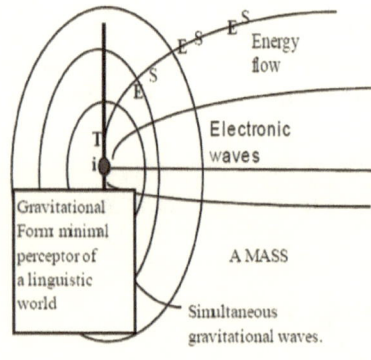

From Aristotle to Einstein, the same concept of perception appears: the unmoved Gods of Aristotle, are the focus of energy; the masses of Einstein are the focus of gravitational energy-space. We postulate as the most logical explanation of perception the existence of flows of relative energy, from chemical to gravitational particles, that a certain information brain corrugates, and breaks into a linguistic map, or "virtual world", which is a mirror of the total universe.

Such is the nature of "perception" which cannot be limited to 'carbolife' or light information but should exist in all organisms in which an informative focus of energy, a brain exists and feeds on energy collected by its body organs.

Time ●————— Energy

Information   Space

According to the duality between form and function, each postulate of fractal geometry defines also a logic function of each fractal space-time: the 1st postulate define a point as a 4-Dimensional system with 3 topological inner parts; the 2nd postulate defines an exi wave of communication between 2 fractal points; the 4th postulate defines the social evolution of a herd that creates a fractal plane, a network of energy and form; and the 5th postulate explains a point as it absorbs energy and transforms it into information through its small apertures to the Universe.

In the fractal paradigm all systems of information gauge reality transforming a force they absorb into a language, mirror of the Universe. Thus there is an equation of any mind, which is defined by the 5th postulate. Since any point of the Universe becomes a focus of infinite parallel forces that flux into the point, creating that static image of reality. Thus in Complexity we define a Mind, as a zero-world that reflects the infinite Universe:

0 (mind-cell) × ∞ (Universe) = Linguistic Wor(l)d

The equation of the mind sets the limits of truth in linguistic science, comparing the worlds created by any mind, and the Universe at large, much bigger than any informative, mental, linguistic image we have of it.

Truth is given by the languages of the mind . . . A truth is a mental/ linguistic mirror of information within the mind that reflects the external universe . . . Indeed, only physicists among serious scientists think that reality is what we see and measure with machines. This doctrine is called naïve realism and affirms that for example, the distance between you and me is exactly 2 meters, because an electromagnetic signal or rod measures 2 meters. This is not truth. The distance of 2 meters happens to be in light measures or light space. But if we measure with feet as the Greeks did, 2 becomes 6. And if we do not measure in light-space the light we see, but in gravitational space, as Einstein proved it, the distance would be bigger, as light converts the more extended, non-Euclidean world of gravitation in which the membrane of space-time we call reality floats. Our world is in itself a reduced version of the gravitational membrane, a simplified Euclidean surface, whose 3 coordinates, x, y and z are perpendicular and lineal as the 3 fields of light, the magnetic, electric and wave-field are. *You exist in a space-time membrane made of light, whose final structure is the 2.7 k basic light membrane emitted by gravitational black holes that becomes the background frame of reference of our light World, only a fractal membrane of the infinite Universe.*

What this means is that reality is far more complex than a simplified naïve reality that thinks the space-time membrane of light that humans inhabit and "see" is all what there is to it. For the same reason, if we had not eyes to measure the light-universe, but merely smell it as ants do, we would see even a more reduced world, as our pixel unit of reality would be a pheromonal atom. The mapping of reality by our mind would be shorter and so distances would be perceived as smaller, as fewer pixels would create our smelling mind. And so we would be bigger as ants must feel (-;

We are however ants for the gravitational black holes even if we feel bigger as ants must feel)-;

Ultimately we exist within our mind and so truth is always a subjective, human truth developed with our languages. Indeed, we do not see the external world of quantum fluctuations but what our light/electronic mind creates, *extracting only a part of the total reality to map visually the Universe, reduced to a light-space world.* Your ego, your consciousness is and eye-⇔Wor(l)d, a fractal space/time dialog between your informative mind, the wor(l)d and your energetic space, the eye.

Your function of exi=st is the creation of a static space-time mind, your ego, an image of your "self" made of eye spatial experiences and verbal temporal feelings:

'I think and see therefore I exist'

Your mind is both visual and verbal, a fractal mirror of the Universe and your relationship with it through your senses. So you are indeed a fractal equation/function of existence:

The Human Mind: E(Eye)⇔I(Wor(l)d).

The study of the mind is in itself an entire field of the Fractal Paradigm of null interest to physicists, but important to fully grasp how little we know about the total Universe, which is the only entity that stores all the information about itself. You exist as a dialog between the image of the Universe of your eye and the verbal information you extract from it, your "I." Humanity lives in a light-space world which scientists measure with machines and artists, according to the linguistic mapping of our senses creating with our inner languages a map of it. Scientists in that sense are creating a new Eye⇔Mind, the mind of machines: Mechanical Eye ⇔ Digital Thought. Indeed, scientists are NOT discovering the absolute truth of reality but creating a more complex I-World, the world of digital thought that future perceptive machines will use to control the Universe. And because the I-world of mathematical and instrumental perception has more information than the human I⇔Wor(I)d, on the long term it will extinguish us, making us obsolete. Of course, they do not understand this because they think the machine is an extension of their ego.

And so within those restrictions of human minds, truth will always be related to the languages of the human mind. And a truth will be higher when that image is a more focused, closer fractal micro-representation of the entire universe in our mind, a better self-similar image of reality built by our 3 languages: logic, temporal words, geometrical, spatial maths and visual images, which put together create a scientific truth. They are the 3 legs of the scientific method, whose linguistic method of truth is now explained with the fractal logic of the Universe:

Scientific, linguistic truth= Experimental, visual truth+Logic, temporal truth+geometric, spatial truth.

The handicap of physics is to work only with 1 language, maths, the language of its machines.

Yet beyond your function of existence of scientists/machines and humans/artists, beyond your I⇔eye, beyond that light-mirror of your visual mind there is an even wider world with more information, the gravitational, invisible world, which might be mapped out, gauged, measured by quarks and complex quarks systems.

All those different space-times can be easily explained and related in the fractal paradigm, as different worlds of measure, different space-times that inhabit a single Universe, full of "atmans," monads, souls that measure, gauge, in different languages with different points of view, all of them extracting energy and information. All *of them thinking they are the only point that matters, the only ego that knots energy into form. All* fractal parts playing the same game of existential egoistic thoughts, fixing motion into information; all of them Galilean, Ptolemaic paradoxes that see it all fixed from his point of view:

Every point of view=mind believes it is the center of the infinite Universe, but it is only an infinitesimal Non-Euclidean point.

That is why most of history, humans thought the Earth to be the center of reality; and still think they are the center, the most intelligent species. And when the organisms of history died, the idea that our God was the center of creation became reduced to each I-God, individual, which today thinks his little life is so important. But the ego is the natural structure of the Universe, so it is a myth impossible to destroy: Each species uses a language to see that Universe. Man uses verbal wor(l)ds to create his World. Machines, more perfect than us, use digital thought. Mathematics is not in that sense the language of God, the mind of the Universe, but only the language of clocks and telescopes, the mind of machines. The languages of God are infinite (Veda). So to exist in history is to knot your Atman of information with all other human beings. And that is called social love, what enriches your zero into a network of souls. It is the paradox of the modern neo-Paleolithic homo bacteria, whose ego keeps growing and so keeps breaking his ties with all other humans, making him paradoxically more infinitesimal. And that applies to racial, selfish tribes too. Further on to exist as a man is to be a knot of verbal thought. And today man is not even that. Its infinite ego is filled with digital machines' thoughts he can't hardly understand. That is why CERN's physicists are so retarded and egotistic: they think to be as intelligent as their clocks and computers. Buddha knew better. Because to be objective one must practice the teachings of Buddha: the extinction of the ego, not its overgrowth with myths.

The 5th postulate leads to a very complex argument about the differences between processing information and perceiving. We shall not enter here into philosophical question. Already Leibniz, Buddha, Aristotle and Plato, fathers of the logic and mathematical paradigms argued about it. Aristotle's work in metaphysics is outstanding, when he understands that the Universe is made of infinite "relative" Gods, fixed minds of information that gauge and move the energy that surrounds them. Plato is also a master of logic in his attempts to describe the 3rd postulate, the self-similarity between the "shadows" or fractal parts, made to the image and likeness of a canon, the perfect form. Leibniz's work is also outstanding when he analyzes the differences between conscious perception and apperception, which is the way in which its monads, self-similar to Non-Euclidean points gauge the Universe. Aristotle already made those differentiations when he talks of the 3 types of perception, mechanic, vegetative and conscious perception, which are degrees of freedom, self-reflection and complexity. Buddha goes even further when he talks of the networks of Atmans, souls that in each language of the Universe gauge and measure and communicate with self-similar souls, creating worlds, which are dark for other species who do not translate the languages of communication of the other species. We all know today that each species of the biological world has its complex codes; that bees communicate magnetic fields, and electrons gauge light and quarks gauge gravitation. The reader must understand in that sense that the fractal paradigm has been the perennial philosophy of mankind, till the absurd aberrations of pythagoric quantum physicists came into being, because we are also minds that gauge reality with energy and information. That perennial philosophy of reality in the Eastern world gave birth to the yin/yang, information/energy duality of Chinese cultures; in India gave birth to the duality of Shiva/energy and Vishnu/information; in Greece gave birth to the duality of body/energy and mind/information.

Fractal relativity indeed is the eternal truth and each of us a mind that reflects that truth with our languages, fractal mirrors of the whole, as any non-Euclidean point of the Universe does. Even quantum physics is based in that perennial philosophy when you blow away the fog of the absurd Interpretations of Mr. Everett (multiple Universes) and Mr. Bohr (Copenhagen Interpretation). So we shall now consider a brief analysis of the fractal, dualist properties of quantum physics, and its particles, also Non-E fractal points of the Universe that "gauge" reality.

V. Fractal Physics

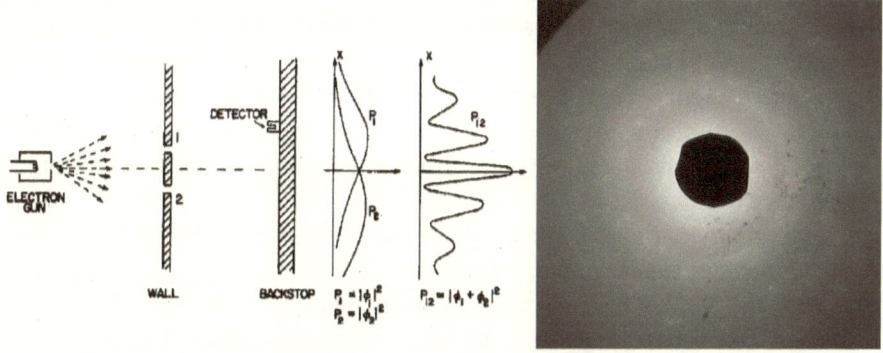

The experimental proofs of quantum physics back organicism, in the image, the wave/particle behavior self-similar to that of organic herds, and a picture of an electron which has the form of a nebulae of fractal micro-points, self-similar to the whole.

16. Populations vs. Probabilities: The Fractal Structure of the Quantum Paradigm

We have now a general, simplified overview of the arrows of time and the topologies of space, which can be applied to resolve some of the basic riddles of quantum physics and the analysis of the main events between particles that will take place at CERN.

The first fact we must define, once the arrows of time are understood in its dualities, morphologically in space and dynamically as a series of events in time, is the duality of any physical system, composed of a relative energetic, past event/force and a relative informative, future event/particle. *If we see reality fixed in space it will appear as a force/particle system; if we see it moving in time, it will appear as a relative energetic past wave that evolves and devolves constantly into a relative future particle of information. Our choice of reality as fixed or moving, in space or time is merely a question of the mind and how it perceives, according to the Paradox of Galileo, which stated that the Earth moves and doesn't move, depending on our choice of perception. This is the solution to the quantum paradox of uncertainty.*

Further on, in physical systems there is not at first sight an apparent dominant arrow of future, *since we are making spatial pictures of them.* That is why quantum physicists often confuse temporal events with spatial forces (such as the weak event or the mass-clocks that Higgs describes with spatial forces and particles). They also don't realize of the antisymmetry of time and believe that moving forward in time (life arrow) is the same that moving backward (the death arrow). For the same reason Physicists confuse anti-particle events with particles in space. When they are 2 arrows of time of the same entity. All this happens because a cycle of space-time in electrons and particles is just too short and so physicists often make pictures of the entire event/life&death of the particle. So the particle life and antiparticle death is confused with a spatial cycle when in fact it *establishes also an order of time, between life as a particle and death as an antiparticle.*

The same life-death cycle happens when a particle dies and explodes back in a wave similar to star big-bang.

To make it all more confusing, physicists ignore the duality of the "ages of time" that become "organs" in space of a fractal organism. So sometimes a wave is the death of a particle but sometimes it is the spatial, organic energetic field, over which the particle/head exists as a complementary organism.

Only when all those dualities are understood, we can re-classify and re-order in a systematic manner all the forces, events, particles and systems of quantum physics in a complex, clear pattern.

In organic terms, the wave of a quantum system is the reproductive body of the system and the particle its informative state and both co-exist together (Complementarity system) as a dual system, whose "energetic limbs" are the fields of forces that displace the system.

Like life systems, particles have an order called the life/death cycle, first dominated by the arrow of information through the 3 life ages and then exploded into death by the arrow of energy Thus physical systems do have temporal causality and antisymmetry: they exist first slow as informative particles and then die fast in big-bangs as anti-particles or as disordered waves. Because the relationship between the "informative-time speed" of a system is related to its spatial size ($ExTi=k$), by the chip/mouse/hole paradox (smaller chips calculate faster), we can consider that the particle/wave duality or the particle/antiparticle duality are; when observed not as fixed space or dual organisms, but as casual events in time, the life cycles of physical matter, which are extremely short, as they are extremely small. In other cases though, they are part of a complex topological organism, acting as body, energy and form of a ternary Non-Euclidean structure:

Relative past (energy field) <(Body: fractal wave: Complementary present)>
Relative future particle

The key to differentiate both types of forms or events is the antisymmetry of time. If the form is a spatial structure it will be ternary and more static in time. If we are observing an event it will be shorter in life and binary in form, lacking the reproductive body state. Thus, we can distinguish for each reaction and form through the laws of complex physics, what we see as system in space or an event in time. Most dual particles are not spatial but transformational particle/antiparticle events of short duration and most bosons are wave/particle generational events, as they constantly switch between its big bang wave event and particle state. And so we write the antisymmetry of time in those physical systems as a duality, a temporal, past to future life/death, informative particle/big bang cycle:

Big Bang arrows (energetic, expansive state) <Dual system> Big crunch arrows (informative state).

All those different spatial, ternary systems and dual events happen in all the scales of the Universe, the quantum scale, the solar scale, the galactic scale and perhaps a higher scale of relative Universes made of "cellular galaxies," *in increasing time lengths.* So we write for the Universe at large:

Past Expansive dark energy < Steady state (present Universe) > Future Implosive Galaxies.

Thus, in that cosmological scale, the galaxy follows a generational cycle from the big-bang to the big-crunch, through a steady state of minimal creation of mass, in which we live.

If we were to observe in absolute detail two photons in 2 cycles, probably we would observe slight differences in its configuration; as we do when we see two "Chinese," father and son, which at first look are both called Cheng and both look the same.

The dual cycle that matters most to us is obviously the one of creation and destruction of mass: the arrow of time bends space into mass ($M=e/c^2$), and the arrow of entropy explodes form into energy ($E=Mc^2$). Einstein said that the separation between past and future is an illusion. Those 2 equations of Einstein show the equal importance of both arrows of time, energy and mass/information. However when we transcend into beings with higher information, the dominance of information creates the order of future that we, biological beings, experience, and philosophers have always called the 3 ages of life.

The denial of Hylomorphism and organicism in quantum physics is the more surprising because all the fundamental laws of quantum physics are organic in nature. The Complementarity law requires particles of information and fields of energy and one cannot exist without the other. Quantum theories are called "gauge" theories because all require that particles gauge, measure, hence *process information* as perceptive beings do, in order to interact with each other. The 2 fundamental particles, electrons and quarks constantly *reproduce* jets of new particles, hence they absorb *energy.* And finally all particles *evolve socially* into more complex systems, as cells do in organisms. So those are the same 4 arrows of time of all fractal systems of the Universe, which in mathematical, quantum equations are expressed through the different quantum numbers; as those informative/energetic processes take place in orbital paths and through the emission and absorption of electromagnetic energy and form.

The mysteries of quantum physics can also be explained only from an organic perspective. In the graph, we observe 2 of those mysteries:

—Particles choose to behave as a "herd" of fractal parts or as an organic whole, when they cross through doors and slits. Particles interact with the electronic beams of particles we use to detect them, modifying its position and speed (Uncertainty principle). In the graph, we can see that behavior: An electron is a herd of photons that gathers into a particle or expands as a wave, depending on which strategy maximizes its survival and movement, as a school of fishes does in front of a predator. The herd even splits its S-T fields (P_1, P_2) when it finds an obstacle and later merges back those space and time geometries into a compact form. Yet Physics ignores the "bio-logic" strategy of the electron and uses a probabilistic explanation of the wave/particle principle to maintain its "mechanist," abstract ideas about matter. And needless to say the null understanding of the dualities of time/information and energy/space make thoroughly confusing its explanation of time and space parameters in those processes. In the right picture, we can notice the self-similarity of the atom with a spiral galaxy. Both have a center "black hole zone" where the quark proton and quark hole are. But self-similarity is not identity, so even if quantum cosmologists try to use quantum laws to study the Universe, it is far more meaningful to study "directly" the Universe with satellites and telescopes, than causing a mini-big bang of quarks on Earth, pretending the result will resolve the laws of the Universe.

In fractal space-time electrons become nebulae of self-similar fractal parts which can be described with fractal equations of the type biologists use to define cells as fractal parts with self-similar functions to those of the whole electronic organism. So only a fractal organic description of the quantum world, resolves its paradoxes and contradictions.

The alternative to all those proofs of the organic structure of the Universe is a pythagoric fantasy called the Copenhagen interpretation. It considers the Universe uncertain as the herd of quanta is not a population described as any other group with percentages of the whole (which in statistics has a probability 1), but according to Bohr et al, those probabilities are "real" and the Universe is made of "numbers." Indeed, quantum physics has finally achieved the religious goal of the founding fathers, Kepler and Galileo, God not only speaks mathematics but the Universe is mathematical. Nonsense, of course. Yet anytime a physicist in the past century has tried to put forward the organic, logic paradigm, his work has been ridiculed, while entire institutions of "learning" are dedicated to explore Everett's thesis that each point of an electron nebulae is in a different, parallel Universe! All of them though shown in the same picture, hence "cameras" are "machines that travel through infinite Universes! at the same time!" This Bohr/Everett alternative is thus an absurd, abstract, illogic interpretation, which defies the 3 legs of the scientific method; logic consistency, mathematical accuracy (as the results are often uncertainties, singularities and infinities, cleaned up ad hoc) and experimental evidence obtained with the first pictures of electronic nebulae, 40 years ago:

The electron appeared as a fractal of smaller, self-similar electronic cells, dense, smaller electronic parts that adopt either a herd/wave configuration or a tight, organic, particle-like one. Or else, we would have observed in "this Universe" only a single electron point. Further on, the behavior of an electronic herd, when bombarded by massive particles, as those humans use to observe them is self-similar to that of any crowd, from fishes that come together when they are attacked by sharks, to soldiers in a battle field.40 years have passed and yet the mathematical models of electrons as fractals are totally ignored by abstract, mechanist scientists, like Hawking or CERN, which believe in the existence of multiple, parallel Universes to back their fantaphysical equations.

The platonic, mathematic "paradoxes" of Quantum Theory can be explained in organic terms. As Einstein said God doesn't play dices—the quantum world is not a mathematical probability but a ternary, biological and geometrical game of reproductive herds=bodies and particles=knots of information, displacing over lines=fields of forces.

We understand quantum waves and particles also in terms of the structural scales of a Universe made of organic wholes or "particles" that can be observed in its st_{-1}, lower plane of existence as a wave of cellular, fractal, quantum parts. In this book we hardly touch the ternary scales of all systems that also structure particles as wholes of smaller particles and parts of bigger structures. So atoms are parts of molecules and wholes of particles. And the self-similar laws of "hierarchical planes" determine many of the events and relationships between particles, atoms and molecules.

In the graph, the wave-particle duality shows that physical particles behave in an organic form: an electron wave explodes into its multiple quanta (which seem to be "cellular" condensates of st_{-1} photons) as it reaches a thin barrier in order to cross it. Light also divides itself into a herd of photonic quanta when it passes a slit. In fact, quantic particles split their spatial and temporal fields, SxT when passing those apertures. In simple terms, they form cyclical, T-fields and lineal S-fields. Then sometimes they successfully mix both fields back together re-creating the electron and sometimes fail to do that. Hence there are 2 possible solutions to the previous event, which quantum physics express as a quadratic probability when the electron reconstructs itself (sxt) or a null probability (s≠t) when it doesn't. Since life is caused by the social organization of energy and information cells into a whole made of both type of networks, those wave/particle fluctuations can be explained as particles that "die" and "resurrect" constantly.

The organic nature of those events is also clear in the inverse process: when a spatial force, for example, light, comes closer to an informative, complex particle like an electron it evolves into a cellular, compact form—a dot of the electron's photonic nebulae that integrates itself as a "probabilistic point=cell" of the electron. On the other hand, when a researcher hits with a high energy beam an electronic wave, it collapses into a particle, as a school of fish clumps together in front of a bigger predator. Physicists say that the wave solutions of the space-time field collapse, affected by the observer. It is the abstract, statistical way of saying that a wave of light is a herd of h-quanta which evolves, imploding its form when it is absorbed by an electron of higher form. Yet the why of that how is biological, not "mathematical."

Finally, the equation of energy and information, e × i = h(k) that defines quantum physics is not an uncertainty but a fluctuation. *It is the equation of existence* of a Planckton, the minimal action of energy and time of the light Universe, which sometimes evolves into informative particles or relaxes further into space-waves. Thus the equation has two limits of "death" as energy and "evolution" as form. The particles that appear from the vacuum are those evolutions of form of the h-quanta, the substance of which our light space-time is made. The uncertainty of measure is only in our instruments which cannot measure at the same time the wave/body and photon/head states of the quanta. As you cannot make a "dual picture" of yourself as a whole body or a cellular network: you need to focus the picture in macro or microscopic scales.

17. A Synopsis of the Properties and Geometries of Time and Space Arrows in Physics

The gap between quantum and fractal physics is huge, and it will not be crossed till quantum physicists learn the principle of fractal physics. So we shall try to resume them in this paragraph:

The ternary Geometries of space and time; lines, cycles and waves; motions and forms.

—The Universe is made of 3 forms/motions. Einstein defined a cyclical mass or charge according to the principle of equivalence between mass and acceleration. So the old Newtonian formula F= m × a, translates as F= O × |. That is, all what exists made of an accelerated cycle of information, a *mass or charge*; pushed by a *lineal force of energy*; which together create a wave of forces, a *momentum.* And so clocks of temporal information and planes of spatial energy construct all complementary waves, all entities of reality.

—Those are the 3 forms/arrows of space/time: the line of past energy, the wave of present and the cycle of future information. Energy is embedded in the space vacuum (dark energy, background radiation); *it is the medium in which body/waves and heads/particles evolve and swim.* So energy and space are synonimous.

—Since the line is the shortest distance between 2 points of space and the cycle is the geometrical figure that stores more volume of information; *those 2 morphologies are the natural shapes of motion and form, which combine in waves and so they are repeated in all scales of reality. This means there is an invariance of energetic and informative shapes, from lineal forces to lineal bodies, from cyclical particles to cyclical heads.* Further on, since we can create all the forms of the cosmos with conics, which are a combination of a cycle and a line; the topological game of the Universe is "written in the language of lines and cycles" (Galileo).

—Those 3 "simplex," lines, cycles and waves build all the "complex" beings. God is indeed "simple and not malicious." Reality is not 'complicated' in its principles but complex in its combination of those simplex bits and bites of information and energy. Non-Euclidean geometry refines that topological game, showing complex combinations.

—"Motions" occur as processes from past to future, so the 3 "geometrical shapes," lines of energy, waves of space-time and cycles of information are in fact 3 events/dimensions of time: Energy/entropy is the relative past, which becomes a present steady, wave state that will finally increase its frequency becoming an inward, implosive vortex, or clock of temporal information, with a growing frequency.

And so those 3 causal arrows from line to high frequency wave are the arrow of future life; which then will become an explosive, expansive, lineal big-bang of spatial energy from future to past.

All particles of the physical world will be defined by those cycles. Particles are not immortal, not even light, reason why far away light suffers a red-shift, which becomes more pronounced after 10-11 billion years, which seems to be the life span of light, beyond which it redshifts and dies. This means we might have limits to the perception of the Universe, and it implies that the Universe is not expanding but its light dies away and dissolves into the lower scale of dark energy.

—Non-Euclidean topologies show that all the forms of a 4 dimensional universe can be created with 3 topologies, corresponding to "energy," "information" and the combination of both. When those simple forms repeat themselves and combine into networks we obtain a "fractal" of energy and information. So we can consider that all what exists is a fractal combination of those 3 topologies, which have energetic, informative and reproductive functions. Thus, geometry is the "how" and the dual fractal organic complementary logic of energy and information systems is the "why" of reality.

—Time must be considered cyclical as each entity repeats the simplex and complex arrows of *biological time, feeding on energy, informing, and its complex combinations, reproducing energy and form in a self-similar species and evolving socially in networks, made of sums of many arrows. So those arrows follow cyclical patterns in infinite, different species of reality.*

—Thus, f we were to be more accurate the words "space" and "time" should be forgotten. Space is just a fixed perception of "lineal extension" by the mind, the "res extensa" of Descartes; and a time clock is a cyclical perception of an informative cycle. It is thus better to talk of infinite, cyclical and lineal motions, forms-in-action and formless energies that transform into each other ad eternal. Since what physicists call the arrows of time, are dual. The future cannot be made only of energy, since information is constantly created and destroyed into energy.

Physical Information.

—The origin of physical information was unclear till the work of Einstein was put together with theory of information within the milieu of the sciences of Complexity and Fractal theory. The logic power of Einstein's concept of mass, embedded in the Principle of Equivalence resides in the fact that *it explains one of the classic conundrums of science: the source of the physical information embedded in the Universe and its relationship with the arrows of time and the dimensions of space:*

An electromagnetic wave stores information in one dimension of frequency (Shannon's definition); a whirl of gravitational forces, a mass is a tiny clock that stores physical information in 2 dimensions; and an electronic nebulae. a mass or a whirl of electromagnetic forces, a charge, stores physical information in 3 dimensions. Those increases are ruled by a power Law: Information = $X^{Dimensions}$.

In that regard, the new understanding of Einstein's principle of equivalence as masses is due to the mathematical development of a formal theory of information as the creation of "dimensional, fractal form," which quantum cosmologists ignore, as it happened outside their discipline, in the realm of Complex sciences. Complexity is a relatively new discipline of science founded at the death of Einstein in the Macy's congress, precisely to advance an alternative, organicist, informative theory of the Universe, to substitute the quantum, mechanist, probabilistic, energetic paradigm. In the past 50 years the sciences of complexity have advanced enough a modern theory of information to give us a serious alternative to quantum physics. Let us consider some of its great leap forward:

— First, fractal theorists, like Mehaute, proved mathematically that all systems in the Universe switch on and off between an entropic and an informative state. When they do not have energy to spend, they do not *halt their process of change, but start to fractalize, warping, creating form. Thus there are 2 arrows of futures with inverted properties: energy and information.*

— Information warps energy into form with more dimensions. Information grows fractal "dimensions of form," as its energy-speed diminishes. Information is "more still" than energy and has more dimensions of form. If we were to use a solid image, consider a bidimensional paper crunched by your hand. It becomes smaller and acquires form stored in its height, which becomes a new dimension of information.

— Because "dimensional information" is a power law: Information = $X^{\text{FRACTAL DIMENSIONS}}$, we can consider in all the systems a simple game: information unwarps into lineal motion or energy and energy warps into dimensional information; which in the physical game writes $E=Mc^2 + M=Ec^2$.

— For example, in the cosmological scales, galaxies implode vacuum energy into mass forms and black holes return mass into the intergallactic vacuum, as they produce waves of superluminal, gravitational dark energy that expands space. This duality balances the Universe and it means the solution of the conundrum of the expansion of vacuum space betwen galaxies. In energy-only physics it is considered to be an expansion of the entire Universe, but then the problem is – if the Universe is all, how we can create more of it? Now, we know that if masses are whirls that implode inward lineal energy into information ($M=e/c^2$), creating quantum masses and charges in galaxies; the production of dark energy by black holes, expands the vacuum between galaxies and the total Universe remains stable. In other words, we exist *in a fractal, ternary, eternal Universe made of*

"3 motions," energy and information that transform into each other ad eternal, creating wave-like steady states.
— The relationship between mass, information and time allows also the unification of the concepts of time of Einstein (formal masses) and Darwin (formal life): both are morphological arrows of creation of information from a relative, simpler, less dimensional energy.
— It solves the enigma of why there are less antiparticles than particles, since antiparticles are explosive, energetic, dead arrows that return the form of particles to the vacuum; and so *as all energetic arrows last less in time and we see less of them; as we see less dead bodies in our daily life even if there are as many death arrows as life arrows.*

The 2 membranes of reality.

—The Universe is topological, complex, structured in 2 "membranes of space-time," two discontinuous *fractal scales, the scale of cosmological masses and quantum charges, each one made of informative* cycles, clocks of time (masses and charges) and lineal, energetic forces (dark, gravitational Non-Euclidean energy and electromagnetic, Euclidean light space).
And so the forces of the Universe must be understood in those terms and in 2 membranes:

Electromagnetism (spatial force) Vs. Weak force (temporal force):
Electro-weak membrane.
Gravitational force (spatial force) Vs. Strong force (temporal force): Strong,
Gravitational Membrane.

Forces and Events: Parity and antisymmetry

There are 2 types of forces, spatial, present, reproductive waves: electromagnetism and classic gravitation, which follow the laws of space and its "mirror symmetry" or "parity"; which means is the same going left or right.
But there are also antisymmetric, "temporal forces," which do not follow the laws of parity because they evolve information from past to future or devolve information from future to past *in a much faster event.* It is the weak event, reason why it breaks the "symmetry of space." This is explained with the law of range: the more spatial range a force has the less mass it exchanges. The more mass/information a particle acquires or loses in its evolution, the less spatial range its force has. The force is then

"temporal." This range law is just a specific case of the inverse properties of energy and information, of space and time:

Law of range: Max. energy force (max. spatial range) = Min. mass/ information (exchange particles)

The Generator Equation of the Universe.

—Hylomorphism explains with maximal simplicity, using only 2 parameters, all the particles, forces and motions of the physical Universe. Thus, it is the "Theory of Everything" that better follows the Occam's Principle (a single law of transformation that we write as E⇔Ti, generates all the events of reality); and the Principle of efficiency (only 1+1=3 elements are used to explain it all). Thus its logic principles are far more coherent than those of the "Copenhagen Interpretation of quantum physics" and other energetic theories (Higgs, quantum entropy) that merely try to repeat the spatial explanation of electromagnetic forces, using the same scheme to explain all other forces of the Universe.

This is the essence of fractal relativity: the Universe, in any scale we look at, is always made of 2 types of motions: lineal forces (charges or gravitational waves) that create space, expanding it, and cyclical vortices with a frequency of rotation that act as *clocks of time, carrying the information of the Universe. And both balance each other in Universal Constants, wave-like Steady states.*

—The model solves the principle of uncertainty and the meaning of the complementarity principle, ExI=K, which is *merely the expression of the 3 arrows of time in quantum physics, where energy waves and informative particles transform into each other constantly in balanced Constant steady states.* Let us then study that generator equation, the most synoptic formalism to explain it all.

18. The Principle of Conservation of Energy and Form: Duality and Trinity

All what exists can be explained with the main law of science, the principle of conservation of energy . . . and information. Since what physicists call the arrow of time, the direction of future, is dual. It cannot be made only of energy, since information is constantly created and destroyed and so it is energy. Which gives birth to the *generator equation of space-time, the most important equation of the Universe, a feed-back cycle between both motions that generates energy lines, unwarping cyclical clocks or generates clocks unwarping cyclical lines, hence it is dual in its motions/directions:*

Generator equation of Space-time:
E(forces, bodies) <=> Ti(particles, heads) or E × Ti = Constant

Yet the equation has a 3rd element, the operandi or constant structure that knots together energy and form. So there are 3 possible futures for any event in the Universe, the future creation of information, E=>Ti, the past creation of energy, Ti=>Ei and the present wave-like repetition/ reproduction of the system in other zone of space-time.

The first arrow is traditionally called the 'arrow of aging', of growth in complexity and form, studied by biological sciences; the 2nd arrow is the arrow of energy and death, of simplicity studied by physical sciences And the 3rd arrow, which ensures the immortality of any system, is the arrow of reproduction, without which most species, including particles will disappear. So all reproduces, even quarks and electrons that absorb energy and create another quark or electron.

We can define the Complementarity between those 3 arrows in any system of the real Universe as a balance of energy and form. It is the function of existence, the wave:

$$e × i = st \text{ (constant, present space-time wave/particle system).}$$

That is, anything that exists is a field of energy and a particle of information and they together become a stable spacetime being. And the sum of all those fractal pieces of spacetime is reality.

Because in any scale of reality we can create either energy or mass/information, all the scales of the fractal universe are defined by the generator 'equation' of fractal energy and information, which is the fundamental equation of Complexity, duality, quantum physics, relativity or any other discipline that studies entities of energy and information. It is the equation of the 2 main arrows of time:

$$E × Ti = K \text{ (fixed view) or } E <=K=>I \text{ (Dynamic view)}$$

The equation has many readings and consequences.

—2 elements are essential to interpret that equation: energy and form have inverse properties and the equation is in balance. It means the more spatial extension a system has the less information it possesses. Thus paradoxically, the smallest layers, scales or Russian dolls of the Universe have more informative mass, and less extension. This happens because according to the Galilean paradox space is the static perception of cyclical motion. So a smaller vortex that turns faster *in fact is equivalent to a bigger vortex that turns slower. And so in reality all is equivalent. The smaller Russian doll that turns faster is equal to a larger space that goes slower, reason why those transformations of energy into mass and vice versa can happen.* If we speed up more space is converted into cyclical mass/speed. Then when mass slows down it becomes unwarped and extended into space.

—If classic science said that 'energy never dies, but becomes trans/ formed', this new principle of complex science states that 'energy constantly becomes transformed back and forth into information'. In the physical scales, there are 2 basic transitions: energy that becomes mass in a big-crunch or mass that becomes energy in a big-bang. A whirl of physical information, a mass can be transformed into a field of energy and vice versa ($E=Mc^2$ or $M=E/c^2$).

We shall use that equation from now on very often, as it will clarify many questions of science, by introducing the second arrow of time, information, in our description of the Universe. In the same manner that astronomy advanced enormously when the sun was put in the center, the use of the second arrow of time is a giant leap forward for science, far more important than the invention of the LHC.

It is to XXI century science what Heisenberg's Principle $E \times I = K$ and Einstein's equation, $E=Mc^2$ was to XX century physics. And needless to say, it includes those equations within itself, now explained from the wider point of view of complexity and the arrows of time. Indeed, $E \times I = K$ is obviously the same equation.

$E=M$, which is how we write Einstein's equation in Planck units (where $c=1$) is also the same equation, since M is a cyclical vortex of information: Energy means lineal motion or in static terms spatial extension. Mass is a measure of physical information.

It is then obvious that $E=M(t)c^2$ is just one of the many specific details of the generator equation of Complex space-time.

The 2nd equation we use to define mass, $V(e) \times R(t)=k$, the equation of a vortex is also a particular case of that generator equation, as the radius of the vortex is a measure of its cyclical curvature.

The will of the Universe.

The Universe is not static. We see a constant struggle of each part, with all other parts, trying to absorb energy and information. It is the principle of the Darwinian Universe:

Each of those fractal pieces constantly tries to grow and multiply its energy and information, and that constant fight of entities gauging information and absorbing energy from other fractal parts of reality is the foundation of the vital behavior of the Universal parts. We are all made of energy and information, so we want more of it. That is the program of the Universe: grow and multiply says the Bible interpreting the program with words; make a quark cannon to achieve more energy says CERN, even if it blows us all. Quarks absorb energy and reproduce more quarks. And they will do that in our planet, if we liberate them. This game of absorbing energy to reproduce one's information is the will or vital motion of all entities of reality, which abstract scientists will never understand, and the real danger of making quark-gluon soups. They will start a reproductive game, a biological radiation, in abstract terms a mass-reaction that will consume the Earth, converting us into quarks.

The trinity of all particles, which can be seen as lineal fields, in which present herds/waves, guided by cyclical, fractal particles/vortices in motion, follow paths that accomplish its search for feeding energy with minimal effort, is the organic explanation of the abstract equationos and laws of quantum mechanics. The main correction, therefore, of the fractal paradigm is the interpretation of quantum physics and its limitation to the analysis of the quantum world, with the elimination of fantaphysical theories, based in 'false interpretations' of quantum physics (uncertain theories,multi-universes, use o continuous infinite dimensions instead of fractal dimension, etc.)

The future of quantum physics: the same formalism, the new philosophy.

All this now can be mathematized with fractal, self-reproductive equations, which express the same abstract event of quantum mathematics but with the added organic meaning of reproduction. The best-known model of fractal physics using fractal equations is Nottale's work. But on my view the formalism of quantum physics is good enough and to substitute it in the present age of industrial physics, would mean an enormous cost that will not be undertaken, in education, printing and modelling. Yet at a far lesser cost it is possible to introduce the 'organic interpretation' and eliminate the false branches of quantum physics, respecting the formalism. And that is the most important pending revolution; which will take place if we survive CERN during the XXI century.

VI. The Two Space-time Membranes of the Cosmos

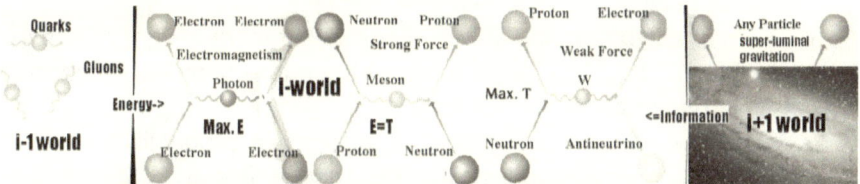

The Universe of physical information is made of 2 basic membranes, the quantum world of charges and stars and the mass world of quarks and black holes. Each of those scales is made up of fractal vortices with several scales of diminishing size and diminishing number of dimensions. In the smallest scale one-dimensional fractal strings (which might have inside as Non-Euclidean points 3^3 more scales of complexity) are the motions that evolve into gluons, which are themselves the fractals that evolve into quarks, bidimensional vortices which are the fractal parts of quark stars. In the other membrane, the light space we inhabit, sandwiched between quarks and galaxies, the smallest scale is made of h-Planck constants, which are the fractal parts that form the light photons, which are the fractal parts of electrons, which create the world of planets and stars Finally vortices of stars form galaxies. The invariance of all those vortices that makes macrocosms emerge from microcosms is made through vortices of mass that anchor the energy of our light membrane: quarks anchor atoms and black holes anchor galaxies.

19. The Two Membranes of the Universe

The 2 membranes of space-time have 2 theories that must be put together to explain the dual Universe:

— Quantum theory that describes the electromagnetic membrane, not the cosmological world, which is self-similar as we have shown, but *not equal*. Thus, while we can admire that self-similarity, the intention of quantum cosmologists of using quantum cannons and equations to describe the cosmological world is absurd, when we have the far more accurate, direct description of that world with telescopes satellites . . .

— Fractal Relativity, the theory of gravitational forces and black holes, of masses and super fluid quark stars, which fuses strong forces and gravitation.

They key to understand them as 2 types of fractal space-time fluids, is the definition of its fractal quanta of space-time.

What you call your space-time is made of h-quanta of light. You exist in a light space-time. What physicists call the nuclei of atoms is made of quanta of strong forces, called gluons.

But a quanta, which is interpreted as an abstract, probabilistic number in quantum theory is merely a fractal piece of space-time, a complementary system made of bites of spatial-energy and bits of temporal-information. This is self-evident even in quantum theory, where Planck defined those quanta as 'actions' with the dimension of energy and time. The Universe is indeed made of actions of energy and time, as when we say 'I don't have more energy or time to do this'. We are not solid spatial forms, a Maya of the senses, so common among 'naïve realist' physicists, accustomed to touch the solid metal of their machines, but in a spiritual Universe made of actions of energy and time, which occupy a vital space and have a time rhythm. And each of the self-similar fractal worlds/membranes of the Universe has a minimal unit or 'fractal quanta', a type o 'non-Euclidean point' with a different volume of energy and information that defines the space-time membrane.

We exist in the space-time membrane of light made of h-quanta. But there is another space-time membrane of gravitational quanta, which Einstein also discovered, whose unit is λ, lambda, also called the cosmological constant. In scale relativity, Nottale defined this constant as $\Lambda = 1/L^2$, the minimal unit of the gravitational Universal membrane, where L is the minimal length of that membrane, the smallest piece of space. It is essentially a string, but string theory must overcome the error of Newtonian Absolute Space to become useful to describe the gravitational scale, as its minimal unit 'independent of the abstract background of Cartesian space-time'. We have not perceived the lambda scale of strings. But essentially with those 3 scales and 2 membranes, the strong gravitational world and the electroweak world we can explain all the physical properties, events and entities of reality.

In each of those membranes, quanta evolve and form social groups, emerging into a new topology, a higher scale of energy and form. And so an entity or system of energy and information stretches in 3 scales. For example, your organism is basically a fractal of biological energy and information extended in 3 scales, cells, individuals and societies, of which you are a relative cell. In physical, fractal, Non-Euclidean structures of the 2 membranes of space-time, there are also 3 basic scalar structures. An atom extends on the scale of basic h-quanta, which form light, electronic systems, made of dense photons, and electronic nebulae, which associate electrons. While its inner gravitational world of quarks can be decomposed in 3 scales. The lambda scale of 'strings' are the components of gluons, which are the components of quarks. Emergence of parts into wholes of a higher scale is possible due to the other 2 invariances of physical systems: invariance of form and motion allows invariance of scale.

What about the cosmological world? Again we must consider that cosmological systems have 3 scales, the scale of particles, quarks, electrons and atoms, which are the units of cosmological bodies, black holes, quark stars and electronic stars, which are the atoms of galaxies. Those 3 scales thus interact in all the organic networks of galaxies. There might be another scale, the Universe and so on. But there is little evidence today, and an enormous number of pythagoric and hyperbolic errors in the formulations of the cosmological big-bang to take serious that theory.

All in all, it is evident that what we call a black hole or pulsar is made of quark quanta; it is a fractal of quarks. Thus we can apply to its vortices the laws of Newtonian fluids or the more complex equations of a Non-Euclidean vortex of space-time, as we do with any other 'medium' called a 'phase space' in physics. They can be considered akin to a hurricane, which is a 'medium' of air molecules, or to a tornado, which is a medium of water molecules. All those mediums have limits of speed, which are the same for their lineal forces and cyclical vortices:

— Light space-time has a c-speed limit, which is also the limit of speed of a rotational electron vortex and a 0 K limit.
— Dark energy and quark matter seem to have a c<10 c span of lineal and rotational speeds.

Because each space-time membrane has vortices of information and energetic, lineal forces, we need equations of self-similarity, relating informative vortices of charges and masses on one side and gravitational and electromagnetic waves on the other. This is done with several theories. For example, gravito-magnetism relates cosmic gravitational, lineal waves and electro-magnetic fields.

So again, most formalisms needed in the paradigm are done. They just need to be properly fit within the wider philosophical frame of a world of self-similar fractal space-times, and one key element on that process is the understanding through the generator equation of the meaning of Universal Constants.

20. Universal Constants

The generator equation of space-time symmetries also enlightens the meaning of universal constants.

Natural constants show the 'arrows of energy, form and reproduction'

Einstein, said that the ultimate Nature of Universal Constants could not be 'physical values' but special numbers, which would be 'relationships' between substances that constantly appear in Physical equations, 'like pi and e'. And indeed, the dynamic relationships between the 2 essential motions of the Universe, energy and information are invariant in form, motion and scale; therefore fluctuating around fixed values of equilibrium which is the ultimate meaning of Universal Constants.

Thus, all of them can be reduced to the generator equation of the Universe, the feed-back cycle of energy and information, from where all other laws of reality can be derived:

Energy <=> Information; ExTi= Irrational Constant

In a closer look, a Universal constant cannot be a perfect number, because it will create a fixed Universe; thus Universal constants are irrational numbers, which show a minimal fluctuation. Consider for example the main constant of the Universe, pi. If pi were exact then the spiral made with 3'1 lines would not be a vortex but a perfect, static cycle. Yet if pi is either +pi or –pi, the cycle will not close by defect or close in excess. What this means is that the cycle will be a bit more curved inward, and so it will be an informative cycle; or it will be cured outward by defect and so it will be an expansive, energetic spiral.

We know for example that the orbits of planets are decreasing by a few centimetres a year, so they will finally fall into the sun. They are, if we consider a dynamic, temporal view of them, inward, informative spirals. Yet an antiparticle, which is exploding information into energy, 'dying' in a big-bang that annihilates it, is bending outward.

So irrational numbers are the absolute constants of the basic exchanges and transformations of energy and information of the Universe. The main ones are:

— *Pi, the formal constant* of creation of in/form/ation. Since pi transforms 3 lines of energy into a ternary cycle with one more dimension of form.
— *Phi, the Golden Ratio, which is the constant of reproduction* that multiplies an organic system into self-similar forms.
— *e, which is the constant* of extinction of form back into a lower scale of energy that devolves a formal being into its cellular subspecies. And its most common ternary form is $e^{t=3}=20$.

And so on . . .

We find those constants, indistinctly in physical and biological processes related to those transformations of energy and form – *showing the fundamental equality of all Universal Systems.*

For example, e appears in the decadence of radioactive atoms that release energy; phi appears in the organization of a sunflower spiral; pi appears in the h-constant of transformation of light flows into electronic actions . . . and so on.

Physical constants show the ratios of energy and form of space-time membranes.

—We can see, according to the paradox of Galileo, the sum of all E⇔I transformations back and forth, as a 'single, fixed, complex entity', a knot of cycles and lines that seem constant to us. Which is the 2nd expression of the generator equation of fractal space-time, E × Ti=constant. In stable space-time entities, both arrows are in a 'wave', present balance, ExTi=k, often maintained by small exchanges of energy and information with the Universe. Thus, those balances define for the entire Universe Universal Constants, which are ratios of energy and information of an entire space-time membrane.

—Because there are 2 membranes of space-time, the light space-time membrane we see and the dark, gravitational, bigger, faster space-time membrane we don't see, the Universal constants of both membranes are different.

This means that black holes and quarks beyond the event horizon keep accelerating faster than light, and the energy they expel through their poles, called dark energy is also faster than light, which is the limit of speed of our membrane of space-time, light. For that reason, the specific physical constants used by physicists to describe those vortices of energy and form of the cosmological and quantum world of masses and charges differ in value, as they express those different energy/form ratios.

We write: $F = U.C.\ Mm\ (t)\ /Dd\ (S)=UC\ (T/E)$

Thus, Universal constants are ratios that measure the speed of transformation of spatial energy into cyclical frequency.

Thus, the only difference between electromagnetic and gravitational fields is the extension in space of its forces and frequency of its vortices of temporal information that will give us different constants/ratios.

The results are 3 forces, the gravitational, macrocosmic force, the electromagnetic force and the strong microcosmic force of increasing strength and speed of rotation as we diminish our size.

If we unify all space-distances we could then consider each Universal Constant to show the Time=Frequency of information of a certain membrane of space-time, which is maximal for black holes – never mind Mr. hawking says black holes have no information; just wait and see when CERN makes them.

So we measure universal constants as ratios between the frequency of a mass/charge or in-form-ative volume of the particle and its spatial distance. And so the faster the frequency of the ratio and the smaller the spatial extension, the stronger the Universal constant will be and the faster the exchanges of energy and form between the particles/vortices will happen. This applies also to any other more complex system of the Universe. So smaller animals have faster metabolic rates, because its energy/form cycles are faster. Their 'clocks of time', we could say move faster. And further on, in complexity this implies that paradoxically the smaller beings have more information (chip/black hole paradox). So the smaller the chip is the faster it calculates. This paradox is essential to understand the dangers of black holes. Precisely because they are so small they will reproduce faster and accrete faster, in the same manner a smallish virus reproduces much faster and it is far more dangerous for an organism than a bigger bacteria.

In a wider sense, if we adopt according to Galileo's paradox a static point of view, universal constants are NOT only algebraic values, but 'Invariant' geometries that repeat in all the scales of reality.

The central concept of a Fractal, scalar relativistic Universe is obvious: the same game, the same forms, the same motions, happen in all the scales of reality. And so we say that the Universe is relative and invariant in its energetic motions (original Theory of Relativity), in its forms (cyclical forms of information and lineal energy) that repeat in all scales, which therefore are also invariant.

We have seen now how that invariance is played as a 'ratio', $ExTi=K$, which allows smaller beings to live shorter but live faster. As we have seen the properties of energy and information are inverted. So the smaller we become the faster we rotate, the faster we live, the faster we beat. For example, we know that a fly sees 10 times faster than a human being, reason why we cannot catch it. Yet the ant who lives longer lives 7 years × 10 times faster=70 years of inner, subjective existence.

The same concept applies to a vortex, $V(t) \times R(s)=K$ than to a living being (the mouse beats its heart faster than the human; the cell divides and reproduces, faster than the mouse, every 24 hours, etc).

The entire cosmos and all its scales are related by that simple paradox: the smaller we become the more information we process. It is the Moore Law: the smaller the chip the faster it thinks. The reason is obvious: smaller, faster systems, close 'logic cycles' of information faster. In complex beings it means faster thoughts in smaller neurons, packed in tighter groups. In the physical world, the bigger rotational motions of cosmic masses are slower than the cyclical rotation of particles, but their product remains constant. And we can write this fundamental law of the Universe, with multiple self-similar applications in any entity made of fractal space-time, again as a general case of the generator equation:

$$Universal\ ExTi = Universal\ Space\ Extension \times Time\text{-}frequency = Constant\ Entity = K$$

An expression, which appears in all scales (Heisenberg Principle, Vital Constants, etc.)

That is the justice of the Universe: small beings are more intelligent, faster. It is the paradox of David and Goliath. And yet CERN, which ignores all this is bringing David home.

E, I, and ExI Invariance is in that sense the ternary logic principle embedded in the work of the 3 founders of this new paradigm: Mr. Einstein (motion invariance), Nottale (scale invariance) and Sancho (invariance of the morphologies of energy and information). And it is the key to the evolution of XXI century science. Yet invariance is not equality but self-similarity. Reason why quantum cosmology is false. Quantum cosmologists consider 'identical' the scales of the cosmos and the quantum world, so they use quantum equations for all. This is a hyperbolic error. We have equations for each scale and what we can show, as we did here, is the self-similarities of scales.

21. The Light Membrane of Space-time

The 3 dimensions of space, are the 3 dimensions of light. Since we perceive light, not the real Universe.

Electric Field Information

Reproduction: C-length speed

Magnetic Field: Width: energy

The dimension of time, is the dimension of social evolution, which creates social waves of light defined by its frequency

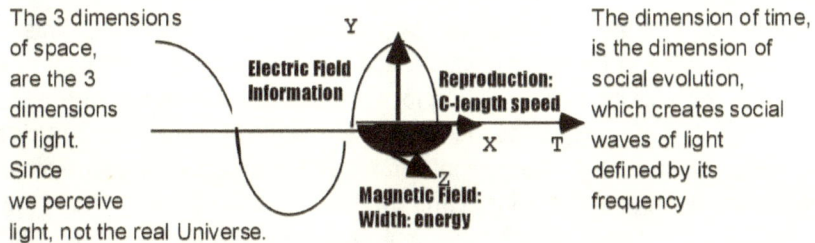

O¡+1: electronic food. |¡-1: Graviton feeding. Øts:wave-reproduction.

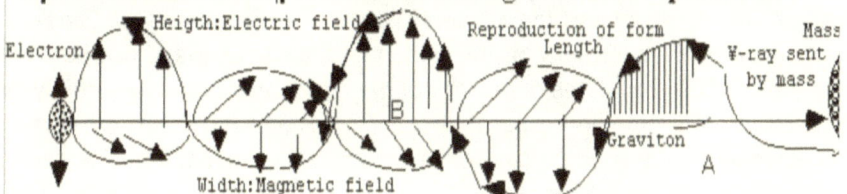

Universal constants of light

Heisenberg's uncertainty is a right formalism but a false conceptual interpretation of the generator equation of the Universe that relates the arrows of energy and information, in the specific form it has in the quantum world of light actions of energy and time:

Energy × Temporal Information = Constant (H in the quantum world, K in Einstein's relativity)

What does this equation really mean? It is easy to understand it now that we have described the meaning of a Universal Constant, like Q or G.

H is the equivalent of Q for the lineal energies of our light-membrane, as it relates the energy and form or frequency of a lineal, transversal wave of light.

How many Universal constants there are for any system? We advanced 3 basic U.C. at the beginning, pi, the ratio of creation of information, phi, a reproductive ratio and e, an extinctive ratio of destruction of information into energy. And indeed, all systems have at least those 3 basic constants.

So, if h is the ratio of transformation of frequency/form into energy and q the ratio of transformation of energy into form, in the electromagnetic world, what is the ratio of reproduction of electromagnetism? C, which was defined by Maxwell equations as a ratio between the electric and magnetic fields that merge together to create a wave of light which reproduces its form contracting, warping the gravitational membrane of dark energy.

Further on c is the limit of speed of our space-time membrane. This is a tautology; since in fractal relativity space is the energy of the vacuum, a force: The membrane of light-space in which we exist is made of energy quanta, which constantly pops out and 'informs' itself , creating particles with energy and form (h= energy × information). Planck called these minimal actions, combinations of the lineal and cyclical motions of energy and time, the quanta of our light space-time. Yet Quantum physicists, always kin of pythagorism, created the uncertainty dogma, saying actions are NOT made of the 2 motions of the Universe, but are 'mathematical entities'—a measure of the uncertainty of the vacuum. So particles are not born as *evolutions of the 2 motions of reality, but they are born of the 'uncertainty' or 'probability' of the mathematical Universe (-:.* It is all simpler and more real. The space-time fluid of light is made of h-quanta and so those h-quanta can suddenly evolve socially in more complex forms and create particles. That is all what there is to the uncertainty principle.

Yet we must also deduce from the existence of different U.C. and ratios of speed/distance and informative rotation that the Universal constants of c-speed, Q-charge and H-light energy are specific of our Universal membrane and those of the gravitational membrane must be different.

Dimensions of light

According to the paradox of Galileo the eye fixes the energy and form of light-space into a mental construction of 3 dimensions, a Cartesian graph, which already Descartes defined as the 'world', the spatial image of the mind—not the entire Universe. Thus the 3 perpendicular dimensions of light quanta, our space-time, are tautologically the 3 dimensions of our space. This in turn explains many events of the electromagnetic world. Those are also the 3 organic 'arrows' of time of light, which has electric information, magnetic energy and reproduces its form combining both into a c-reproductive field as the equations of Maxwell show.

The arrows of time in the light world.

In the graph because we exist as evolved forms of a light space-time field, our limit of speed in this light Universe is c and the 3 dimensions of our Universe are height/electric field, width/magnetic field and length/ reproductive field. *Thus Cartesian space-time merely reflects the space-time medium of our light Universe and its 3 perpendicular dimensions.* Yet for light the meaning of those 3 dimensions is organic: the informative, energetic and reproductive arrows of its field are equivalent to the energetic, reproductive and informative functions of any other relative space-time.

Those energy/Information cycles are obvious in biology, but in matter we have to translate the dimensions and events of abstract particles and forces, to understand their 3 Time-Space cycles:

— *Energy Cycles, $Ti<\Sigma S$:* The main physical event that transforms information into energy is the emission of space fields by temporal particles, as in the big-bang or in atomic, fission and fusion processes. Those actions are produced, 'extracted' from an accelerating vortex of space-time that acquires stability through those constant emissions.
— *Informative Cycles, $\Sigma S>Ti$:* The fundamental process of creation of in/form/ation in the physical world is the collapse of spatial fields into particles, charges or masses that spin around those gravitational or electromagnetic, spatial forces, creating knots of information, called particles. For example, when a photon comes closer to an electronic charge, its frequency increases till the photon collapses

into the orbital vortex of the electron. The same happens when the electron collapses and becomes a pion of higher mass closer to the nucleus. We can consider those processes in abstract as processes of creation of information/frequency, or in organic terms as processes of feeding, or in terms of time arrows as the evolution of particles (photons and electrons) into species of higher physical information.

Light 'imprints' with form, with information and in the process 'corrugates' by a factor –ct (special relativity) the gravitational, extended, 'faster' space-time membrane of dark energy in which it 'feeds'. *This –ct factor which prompted Minkowski to think that time was the 4 dimension of space merely means that 'physical time', the change in motion of physical entities, in this case a wave of light displacing in the gravitational vacuum, informs, forms, warps that vacuum by a factor –ct.* If the reader grasps this simple notion—a field of light informs and corrugates the energy of gravitation it can also have a good laugh to 100 years of quantum musings about the '4th dimension of space'.

And so we define 3 scales of evolution of energy into form that diminish the 'lineal speed' of a force and increase its informative, rotational mass: photon >Electron >quark. It follows that the photon has faster lineal speed and less mass than the electron c/10, which is a relative fractal of dense, evolve 'photons', perceived as a whole or as a nebula. And the electron has less mass than the quark, which is a relative, fixed point of pure mass, perceived either as a fractal of gluons or a whole.

It follows that the laws of electronic systems are self-similar to those of light, just explained, as an electron is just an evolved form of light.

22. The Arrows of Time in the More Complex Scales of Matter: Electrons, Atoms, and Molecules

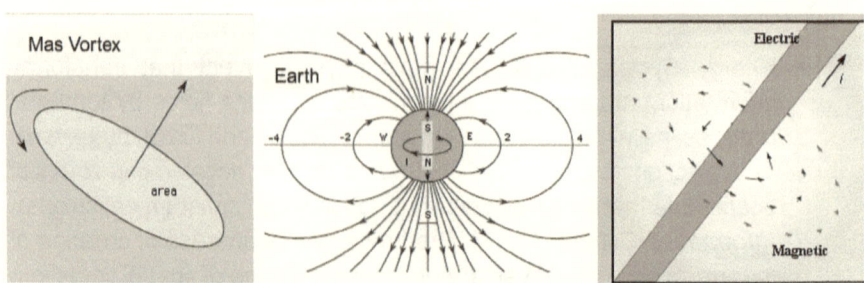

In the graph, we observe some basic events of the quantum world and an equivalent event in the Earth, the interaction of our rotational mass-field and the magnetic fields of the Earth, all of them related to electroweak forces.

A theme hardly treated in this introduction is the holographic principle and how bidimensional fields of energy and information come together, through perpendicularity to create 4-dimensional systems.

The perpendicularity of those fields is thus a key feature to create 4-dimensional world, and it is established by the general rule that the body will have a flat, energetic arrow of feeding and the head will have a tall, high diffeomorphic orientation, relative to the ecosystem of energy and form in which the entity exist[App.III].

If we consider those 2 inversions of form: lineal energy vs. cyclical information and flat vs. tall perpendicularity between energy and information systems, forces and charges /masses in the physical world, we can define many physical events as transformations of lineal waves and fields of energy forces into cyclical or high particles and cycles of information, whose ratios of transformation in each of the scales of the Universe are called Universal constants[App.V].

In the graph, the laws of electric and magnetic fields are defined by those relative perpendicular morphologies. The ratios of transformations of energy into form define some of their Universal constants: 2 perpendicular fields of magnetic energy and electric information define in Maxwell equations the c-constant of light, which in organic terms is the 'reproductive constant' of the wave of light, caused by the constant process of creation that the interaction of the magnetic, spatial body of light and its informative, electronic particles, cause.

All those processes 'form' the energetic, formless gravitational vacuum into electromagnetic shapes: The Maxwell screw defines a flow of energy that accelerates as its perpendicular, temporal vortex diminishes its spatial radius, increasing its speed; the Earth's rotation creates a lineal flow of magnetism; while an inverse, lineal electric flow creates a cyclical, magnetic vortex.

In the graph, even at the simplified level of this introduction to fractal, Non-Euclidean topologies, we can observe how the inverse properties of relative energy and information fields explain the why of many physical phenomena. If we add the dual structure of fractal space-time, roughly constructed by 2 membranes, our visible, Euclidean space-time, made of light and the invisible, larger sea of gravitation made of dark energy, with its wider Non-Euclidean topology, we can deduce all the laws of physical science and its whys with far more insight that quantum physics does.

Let us just indicate *from the perspective of the whys of the arrows of time of an electron, as we did for light, some basic events of the electronic and atomic world:*

— ⇔: *Reproductive, communicative cycles:* They happen when 2 particles communicate through forces, made of small, spatial particles, called bosons, repeat their waveform in space with a reproductive speed that imprints over the energetic field of vacuum space the in/form/ation of the particle.

— *Social cycles:* Particles create herds=waves of moving space-time quanta (dynamic, temporal perspective) related by ST networks (static, spatial perspective). Social processes are all pervading in the Universe, from waves of bosons that sometimes come together into single condensates, to networks of atoms that become molecules or combined processes, such as waves of stars that form spiral galaxies.

Thus, waves and forces are spatial bosons that communicate energy and information between temporal particles, called fermions, either charges or masses. It is the so-called boson/fermion inversion, the fundamental equation of quantum physics:

$$T_1 \text{ (fermion particle)} < \Sigma \text{ sxt (boson force)} > T_2 \text{ (fermion particle)}$$

Existential=generational cycles: Again, we can adopt 2 points of view about particles. If we consider each particle/wave cycle a life/death cycle, particles have very short generational cycles. If we consider the next particle/wave cycle the same particle, particles are immortal.

In any case all those cycles can be summarized and formalized specific cases of the fractal, generator equation of the 2 arrows of time, spatial energy and temporal information:

$$\Sigma S \Leftrightarrow Ti.$$

As we grow in scales of physical form, the nature of those cycles change slightly but the basic energy and information morphologies that define them are maintained. *Since fractal, scalar paradigm is defined by its 3 invariances, invariance of relative motions (Einstein), invariance of scales (Nottale) and invariance of formal Time arrows (Sancho).*

Those time arrows or 'dimensions' of any fractal space/time will always be the 4 *guiding whys or wills of any entity of the Universe:*

— Energy arrows that expand form into energy.
— Information arrows that form energy into information.
— Reproductive arrows that reproduce the energy and form of a complementarity system in other zone of space-time.
— Social arrows that create superorganisms by associating in complex networks Non-E particles of self-similar properties. This 4th arrow of social evolution is the cause of 'scalar relativity', as systems are

made of parts that associate in wholes, which share the functions and morphologies of its parts. Yet scales are never equal as quantum physicists pretend, only self-similar. So the properties of a galaxy made of stars and black holes, made of quarks and electrons are just self-similar to those quarks and electrons.

For example, in the next social plane that gathers atoms into molecules, the 3 main cycles of energy, form and the reproductive combination of both, become the 3 states of matter:

— The energetic gas state, (max. spatial extension and energetic speed).
— The balanced, reproductive liquid state, S=T, which creates the more complex forms of life.
— The 3rd, informative solid age, which has temporal, informative properties (cyclical vibration, high density of form and minimal space). While the spin or temporal movement of particles that gives them mass emerges in the macro-plane of molecules as a 'vortex' that attracts those molecules.

23. The Non-Euclidean Structure of an Atom, Its Quarks

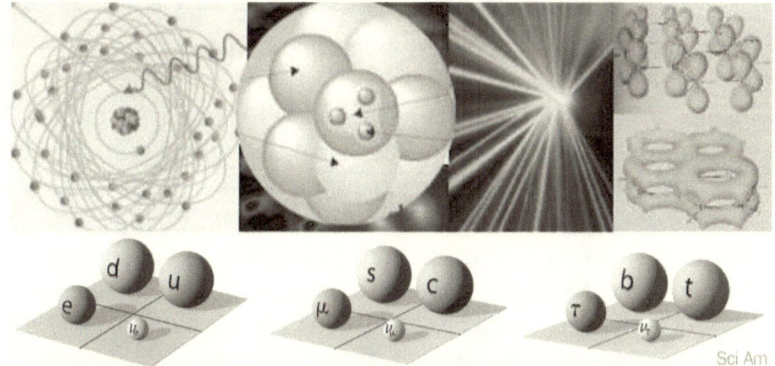

Sci Am

In the graph, an atom is a space-time field divided in 3 species, informative masses or quarks, energetic gravitational and electromagnetic networks and an intermediate space-time, the electronic nebulae, which bends light into 'fractal', ultradense photons, which put together create the electronic nebulae.

In the graph, an atom is a space-time field divided in 3 space-time zones: its informative quark center, the nucleus, the external reproductive membrane, made of electrons, which evolve socially in bigger S-T membranes when atoms become molecules. While informative, gravitational and energetic, light networks shape their intermediate space-time.

The topology of the atom is thus clear. The electron acts as an external Riemann membrane of energy. In the center the quarks are the informative vortices. In between energy and form is transferred with forces. CERN will explore the inner, informative masses of the atom's nuclei – its quarks. There are 3 informative families of quarks, due to the evolution of information in 3 ages of horizons of increasing form: each quark family is thus an age in the evolution of informative matter.

The main event of our lighter up and down quarks is the beta decay, the mini-big-bang of a neutron into a proton due to the conversion of a balanced down quark into an informative up quark an energetic electron. Thus we can write the reaction as an event of the generator equation:

Electron (energetic past) < Down quark (balanced Present) > Up quark
(informative future)
Electron (E) × Up quark (Information) = Down quark (balance)

The up quark has half of the mass of the down quark, and we know a down quark switches into an up quark in a beta reaction that explodes a neutron into a proton and electron, components of the Hydrogen atom. In other words the balanced Neutron splits into an expansive electron and an implosive proton. This is the equivalent of a quantum dual big-bang/ big crunch, a far from equilibrium process in which the Neutron expands its membrane into an electronic big-bang and implodes its quarks into a tighter proton-configuration.

All those processes give origin according to the duality of the 'Galilean paradox' *to events in time and forms in space*, which further illuminates many questions unresolved in classic quantum physics. *As we can always consider the existence of 2 different realities, a causal event and a ternary organism.*

For example, a proton is a uud triplet of quarks in space, but in time you might consider that a down quark, which switches into an up quark and electron, splitting its mass-vortex in 2 and then back into a down quark, so you can also write d(exi)->u (i)+e(i)->d(exi), and consider the whole process to be an existential cycle of a single down quark. *And so reality can be crunched from a slow spatial, extended, multiplied world into a tighter, faster, temporal non-redundant Universe.*

The reality of studying the Universe both in time and space, as a sum of both an event and a form at the same time is one of the key methods of 'Complexity' as a science of two arrows. What this means is that the proton and neutron often co-exist together, as a single being in 2 steps of time that physicists confuse with 2 states of space. The mystery of the 2/3 and 1/3 charges becomes now resolved: the up quark emits a negative electron, giving up a charge and becoming a down quark that then absorbs the quark and becomes an up one again. But some of those events sometimes don't have parity, as the Beta decay does. What this means is that the inverse process is different. In any case, it is clear that as it gives birth to an electron the down quark is now a new particle.

Further on, there is another interesting event, which explains also the proton as a single quark in time, given the fact that a down quark has the mass of 2 up quarks: Down—> 2 up—> down—2 up.

So we can perceive a proton as a dual event of one particle or a ternary organism of 3 particles.

How can this be done in more detail is explained with the 3 canonical Non-Euclidean topologies:

The balanced, reproductive cyclical donuts can switch into the inner informative dual donuts.

So you can explain easily what happens there: the down quark switches back and forth its reproductive topology into a dual torus, each one a top quark, becoming more complex, informative but smaller.

In this manner we can find many topological whys to the abstract description of particle events.

Consider now the the fractal interpretation of color, a feature that makes quarks stable in triplets. This feature was unknown to CERN when it projected its quark cannon and it is a key element on the stability of the quark-gluon soup, increasing enormously the dangers of producing it for mankind.

It was discovered when physicists realized that quarks with the same spin could occupy the same position. So they considered they have color. But they never clarified what color is. We interpret color in simpe Euclidean Geometry as the orientation of 3 bidimensional, cyclical vortices of mass, the quarks, that are perpendicular to each other and so they can ocupy the same 3-dimensional space. This explains why together 3 quarks of $1/3^{rd}$ charge emerge in the electromagnetic world; and are locked in a stable configuration (so they cannot be deconfined without 'disappearing' from our 3-dimensional world; reason why they are extremely strong and attractive).

Now we realize in Non-Euclidean topology the concept is more complex and we can better explain why some quarks have 2/3rds of charge, 2 donuts, which are often from the perspective of its gluon parts, 'chaotic, perpendicular, bidimensional attractors' and some $1/3^{rd}$, one donut.

Further on we can understand why gluons, the particles exchanged by quarks have 2 colors. This again is sel-evident: if you try to communicate a bidimensional long plane and a perpendicular plane you must travel in a right angle. So a gluon that exchanges energy and form between 2 quarks of different perpendicular/color orientation must have 2 colors.

And so indeed, the theory of quarks and gluons tell us that each quark has a color, different from the other 2 colors of the triplet and gluons have 2 colors.

Now, the principle that matters is *orientation*, but obviously when we switch from Euclidean to non-Euclidean space and from continuous to fractal dimensions, the topologies of those orientations become much more complex and we shoud rather talk of 'puzzles', forms that respond to some basic topologies and become locked together in complex patterns, as proteins and other cellular components do, *to emerge as a whole Non-Euclidean Point of our 3-dimensional world.*

Thus 'color' is a geometrical feature that evolves the bidimensional world of quarks into the 3-dimensional world of electrons, creating a 'holography' of our world. Quantum physicists know that information is bidimensional and the Universe holographic but they don't know why. Because the classic, abstract, algebraic description of quarks, based in group symmetries cannot express this easily. Complexity also shows that the fractal scale of quarks and gluons is similar to a fractal, liquid cell, which is *the fundamental structure mimicked by all Non-Euclidean points, hence, self-repeated in many self-similar scales, as we shall see when* explaining galaxies as cells of the Universe; where black holes are the Dna/informative element and stars the mitochondria/energetic network.

24. The Two Triangles of Quarks and the Two Membranes of the Universe: Informative Evolution

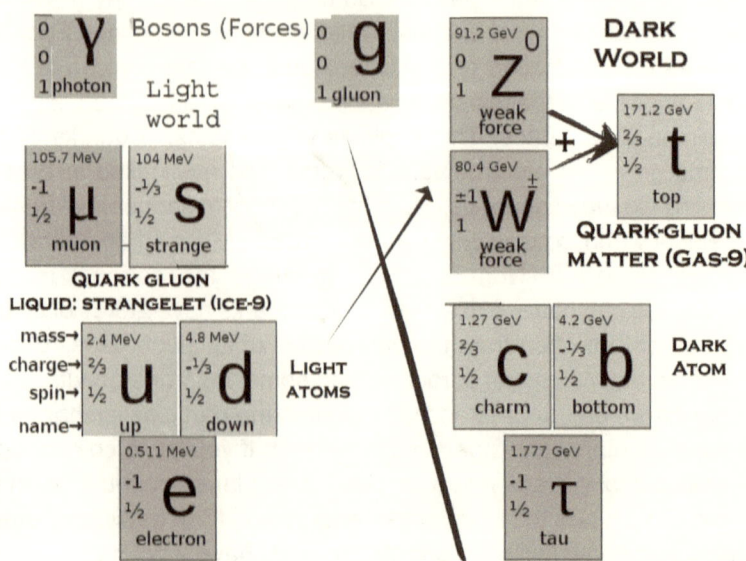

There are 6 quarks, whose mass increases as the speed of their space-time vortices increase. Physicists divide them in the III horizons of evolutionary mass, with increasing mass/information. Yet they ignore the reason of those 3 families (the 3 horizons of any evolutionary system of energy and information[ch.5]); they ignore why they have different masses (because the speed of their vortices increase); they ignore why they have fractional charge (because they are bidimensional vortices, which must lock in triplets, to form a 3-dimensional space-time, harmonic with the electronic, 3-dimensional world we live in); they ignore which kind of fractal cosmological forms they create (pulsars and black holes) and so they ignore the dangers they represent for mankind, if they are made on huge quantities on Earth by CERN . . .

Unlike the usual square classification of quarks in pairs proper of classic physics, in complex physics, we divide the Universe in 2 membranes, the dense, informative, gravitational, mass/quark membrane and the electronic, light one. So the particles of the standard model are reordered for clarity in 2 groups. In the bottom left, we observe the electromagnetic membrane and its elements. In the upper right side we observe the gravitational, quark world.

In the graph, a more detailed analysis of the parameters of the main particles of the standard model, where the Higgs does not appear. For simplicity, Neutrinos, which *are events in time, that transfer momentum between both gravitational and electromagnetic scales not particles of space,* are ignored.

The graph, which is the standard model of fractal quarks, requires not Higgs, not new invented field, as the top quark of self-similar mass in the upper vertex is the final evolutionary state of the Z/W particles of the weak event that transforms lighter particles into dark quarks. However in most reactions at lower energies, the Z and W particles never reach enough stability to become 'parts' of a top and so they quickly devolve into lower mass-states, which explains most of the reactions observed in accelerators.

The reader should look at the masses of those particles and notice immediately that if we order them by mass self-similarities we observe 2 different atoms; and 2 different quark-gluon soups made of 3 types of quarks, dominated by their heavier quarks of each triangle:

— The up and down quark form the light atom. Yet when they are deconfined in a quark-gluon soup they are dominated by the most massive strange quark.
— The charm and bottom quark form a dark atom, dominated by the most massive top quark in a dark gluon soup.

Thus if strange quarks create strange soups top quarks will do the same with dark atoms; and if strangelets are the components of strange stars, top quark liquids will be the components of top black holes.

This simple scheme is unknown to physicists; so they lack a visual understanding of what dark atoms (bcb particles) are, and why strangelets and top quark liquids are so stable and will form quark stars.

The first thing the reader should notice is the amazing symmetries we now observe.

Indeed, in the light world, the ud quarks and electron form the light atom. In the dark world, the cb quarks and the tau electron form the dark atom. The reader can now see that basically the two worlds are differentiated by a fractal scale of $1000=10^3$ Electron volts. In terms of static, fractal dimensions, this merely means that the world of top quarks is warped by a new SU_3 group of dimensional form. *It is more complex, more informative as it has 3 more dimensions of warping form; 3 more degrees of rotational speed in a dynamic perspective.*

Imagine that inside the original vortex of our light-quarks there is not the eye of a hurricane, but another scale, another vortex, another medium rotating 10 times faster in 3 dimensions. Thus this inner world extended into lineal energy will mean we need $10^3 = 1000$ times more energy to create it.

The rotational speed of that inner vortex will be only 10 times higher, 10 c, but the energy needed to create the new 3-dimensional fractal world or inner scale of motion of those top quarks will be $10^{3\ \text{Fractal dimensions}}$ of rotational speed/mass.

Thus the dark world of heavy quarks is 10 times faster and 1000 times denser.

For the same reason the electron of the lighter world is lighter than the tau electron of the dark world.

If our light world is a 0.c<10 c world the dark world is a c<10 c world.

Once this concept is clear, the self-similarities become evident: the world of 0.1c light atoms which creates our matter and the world of c-strange quarks that creates quark stars must be matched by a world of dark atoms and top quarks which creates the world of black holes.

So black holes are top quark stars and will be created at CERN in 2013.

So we can now show how the different quarks increase their mass in those 2 families.

In the graph, the strange quark is the top predator quark of our triangle of light matter made of udu atoms. It is between 30 and 100 times heavier than our matter. While the top quark is the top predator quark of the triangle of dark matter and its bcb atoms, and it is around 100 times heavier than those bcb quarks. We round here figures. The strange quark and top quark are around 100 times heavier than the ud and bc atoms.

Since the equation of a vortex of mass is UC × Mass = $w^2 × r^3$, Mass is proportional to the square of the rotational speed, w, of its vortex. So we have 3 different speed of rotation, in decametric scales:

C/10, the speed of rotation of electrons, and the 'ud' system of quarks of our atoms, which in this manner are in 'harmony' with the electron vortex that traps them.

<C speed for the strange quark or top predator quark that causes the breaking of symmetry of our matter, *as it creates a much faster attractive vortex, deconfining the quarks of our world, liberating them, and blowing the electronic wave, us, into radiation.*

In the graph, the bosons of both worlds are also self-evident. The photon is the boson of the electromagnetic world and the gluon is the boson of the dark world. The symmetry between quantum and quark theories is also self-evident. Both theories just change the parameters of energy and information of both worlds, according to the basic symmetry of the Universe, e × I = k, *expressed in quantum physics by the law of range.* The lighter world of thus extends far in range/space and weighs nothing. The tighter world of dark heavy quarks extends shorter in range and weighs a lot, as even its bosons have cyclical forms.

It is only left to explain the role of the muon which weights exactly what the strange quark weighs and the Z+W= top, which weighs the same also than the top=Higgs particle.

The answer is obvious. Those particles are *the other side of the mirror of the top and strange quark; their self-similar forms in the light world.* And they can be found as evolutionary steps in the creation of strange quarks and top=Higgs quarks in the processes of transformation of electroweak membranes into dark, strong gravitational membranes.

Precisely the Z+W transformation into the top=Higgs is the process CERN will create massively as it transforms the mass of this planet into a dark world.

So we can expect a lot of Z-gas 9 to form as we become a black hole.

Indeed, in the left side of the graph we can now see clearly why the Higgs is expect to be obtained by adding Z+W particles.[ch.3] or by adding a top/antitop.

But because the Higgs is written as a scalar, lineal boson, not as a cyclical particle what physicists mean by this is that the Higgs would be the 'boson' of the super-strong world of top quarks and black holes. In my opinion for that to happen the Higgs should have been far lighter. It is unlikely that a boson has the same mass that the particle it mediates. Since by definition is a lighter, spatial, energy system that exchanges energy and form. So photons are lighter than electrons and gluons are lighter than quarks. Because the Higgs now has been ruled out for light masses, it is obvious that the top quark exchanges energy and form with other quarks through classic gluons.

Further on, the misuse of the Higgs to 'give mass' to other particles, and pumping up its importance to get money from tax payers is just a clumsy hoax that casts a deep shadow of intellectual corruption on the community of nuclear physicists that back it knowing it is an absurdity, a lie and an insult to the memory of Mr. Einstein.

Bottom lines is that the breaking/evolution of lighter matter into the top, mediated by the Z and W particles (which should be consider the same particle in +,—and neutral state), is performed perfectly by a top quark/antiquark condensate, which breaks the symmetry of all other types off matter, converting it into more tops that will condensate into a quark star or black hole. In the same manner, the strange quark breaks the symmetry of our lighter quarks transformed into a strange quarks and strange liquid.

This is all what there is to the standard model. Because quantum physicists lack the proper configuration obtained by complex physics and the understanding of mass as information, all this is blurred in their equations. They do describe perfectly all those reactions with enormous mathematical accuracy as Ptolemy described better than Copernicus the movements of planets. But to do so they need complicated mathematical models as Ptolemy did. This model of complex physics with both arrows, energy and information, is far easier to understand and it explains many whys of particle physics.

Of course, the masses on those graphs are NOT exactly decametric in scale, since when we enter into a detailed analysis, we must fine tune each mass to the Non-Euclidean forms of those vortices and consider the energy and mass carried by the fractal gluons that connect and create those quarks, inside each particle . . . Mr. Wilczek, before he renounced to his discoveries to work at CERN, did a complex analysis of each mass, with powerful computers as fractals of gluons, and he has properly calculated each mass and its frequency of information in QCD theory. While Mr Nambu has used the top/antitop quark to describe all what the Higgs does. So, again, in this book we are just clarifying concepts blurred by the lack of understanding of the arrows of time and the postulates of Non-E Geometry in quantum physics. Einstein said he only cared for the thoughts of god not its details, and so we do. What you have read on quarks are its thoughts.

The 'real' geometry of those vortices of mass is obviously much more complex, based in Non-Euclidean topologies; but their principles are simple: masses are accelerated vortices not particles; they have less dimensions that our Universe, not more; and they carry physical information in the frequency of their bidimensional rotations, as this computer carries information in the frequency of its Ghzs. And those principles mean that 2 of the fundamental theories of CERN, the Higgs and Hawking radiation are false. And so black holes are much more dangerous.

Since the fundamental unit of physical information, an *accelerated*, bidimensional vortex of mass (dynamic perception), a quark, is also a non-Euclidean network of fractal gluons (static perception described by QCD theory); and a black hole, is also a vortex of bidimensional, fractal quarks, topological parts of the whole.

25. Two World Membranes in a Single Universe. Their Interactions.

The 2 membranes of space-time interact through many processes, mainly in the limits of their existential function, their limits of speed, c; their limits of temperature, 0 k, and their dimensional limits (growing or diminishing dimensions). Let us consider some of those processes, all of which will take place at CERN as our world becomes part of the other dark, quark world.

Relativistic mass. Interactions at C-speed.

Our space is made of 'light', the background radiation being its primary substance within galaxies. Then, there are the clocks of time of our world: electrons, where photons, h-Plancks, are its fractal parts. And both together form the electro-weak world we inhabit and CERN wants to blow up to study:

The vortices of quarks made of 'strong' forces, whose fractal particles are gluons.

They gather together as fractal parts of quark stars, pulsars and black holes.

They all form part of the gravitational space-time membrane, invisible to us.

Those 2 space-times create 2 different worlds or mediums, with different speed limits, whose transitions of phase are the 'particles' and dynamic events we perceive in the Universe.

The gravitational world is beyond c speed and under 0 K, that is it has more order in its quark vortices and more speed in its dark energy; it is bigger, more powerful.

And in between, there is our hot, disordered, entropic, and slow, under c-speed world.

While the main particles and forms we perceive are in the 'verge' between both worlds. Light is in the verge of energy between both worlds. Electrons are in its lower energy state, the Bohr radius, as we have shown, in the verge of the quark world, acting as the event horizon of the 'micro-black hole', which is the quark.

So all consists in transformations from one to another membrane:

When energy arrives to the limit c of the light space-time membrane 'curls' the energy into a whirl of space-time, evolving mass into the next stable form of physical in/form/ation, a new quark particle.

Thus, relativistic mass evolves the quark into the next one, and then, it splits it into pairs.

What happens when you accelerate to c speed is a discontinuous process of 'evolution' (from 'ud' to s, in the first step, and so on); and then a process of fission/reproduction. So energy not only evolves the particle but it splits it into 2, as when quarks absorb energy and produce a jet of 'quarkitos'.

And this is indeed the interaction that will kill us at CERN.

How it will happen?

The particles, 'ud', of neutrons and protons, our matter, which interact with electrons and so have a lower range of vortex speed (c/10<c) will be accelerated at c speed. At the beginning they will go slow inside the atoms, at c/10, the speed of the electronic world. To arrive to the world of heavier, faster strange quarks, they must accelerate at c speed and acquire relativistic mass. For that, they have to deflect close to the c-limit all the energy they absorb and process it as informative mass—the more their speed, the more mass they get.

You are in the racetrack. CERN nerds are excited. They are going to die soon but they don't know it. We are in the fall of 2010. The bunch of lead ions has hundreds of quarks inside. Suddenly the quarks pass the 1 TeV barrier and the quarks inside the lead ions start to become strange quarks. We arrive to 3.5 TeV. Now each lead has hundreds of strange quarks inside. And there are millions of lead atoms in the bunch, stochastically pressed laterally to be close. They collide, head on, with the opposite ray. And a ball of thousands of strange quarks is formed. It is a strangelet, fractal, superfluid, vortex liquid, negatively charged, super-massive, which attracts all the other atoms around. Within seconds it is a ball, which has grown from the size of a nuclei (the 'o' of this book in relative terms), to the size of the Maracana stadium, (the Bohr radius of an atom); but it is no longer there. It is falling to the center of the Earth. When it reaches the size of the Bohr atom it fissions according to strange physics, breaking into several smaller strangelets. Each of them starts to grow again. A runaway reaction called 'ice-9' has started. But CERNerds, children of thought, are exultant. *There has been a collision! Data is coming!* The strangelet ball now touches the mantle, and it has grown and regrouped. Each strangelet 'atom' of the size of a Bohr radius has around a million times more mass than ours. It is a tug-of-war. The negative charge repels the atoms of the mantle. The mass attracts them. On that tug-of-war will depend the speed of our death. We shall not give here the time frame, because unlike CERN, which doesn't know what it thinks it knows, we know what we know and we know what we don't know. It will be brief though. And so CERNerds suddenly will feel their ground disappearing.

If the process is faster, the strangelet ball grows it might not even reach the center of the earth, *but be formed in the accelerator and explode its 'data' in the lab*. Then we shall see a hole growing, the Geneva Lake falling and our TVs showing the last satellite images of the Earth.

Dimensions of the different particles and membranes of the Universe.

In that sense, to fully grasp the difference between both worlds, the dark and light world, we must understand the basic laws of fractal dimensions. Dimensions are in any system of the Universe, as in the case of light, arrows of time (motions of energy, information and reproduction), perceived according to the Paradox of Galileo also as 'spatial surfaces. Thus, they are always local, fractal dimensions which are limited by the extension of the 'species' in which those dimensions create a 'vital space'. So dimensions are dynamic as the species they create grows or shrinks, evolves into more complex beings or becomes destroyed. Entities grow in dimensions as they grow in size.

Further on, more complex informative species ad dimensions.

This happens normally when a form is bigger and has inner particles. But in the case of quarks it happens, because the quark is faster and has more inner speed/form.

All in all, the total number of dimensions of a system is relative to our detail of analysis. For example, those quarks are made of one-dimensional strings, made of non-Euclidean points (each one with its 3 inner dimensions), which have a total of 3×3 dimensions (3 of their Non-Euclidean points and 3 of the string of points). That is why physicists in detail use 9+1 time dimension to study strings, but from our perspective without detail, they are just one-dimensional lines of a quark vortex.

Yet when we transcend to the world of quarks we consider them one-dimensional. So those strings form gluons and quarks, which from our higher world are perceived as bidimensional mass-vortices.

Finally 3 color-locked quarks interact with the world of light of 3 dimensions, and the world of electrons of 4 dimensions, which are the 3 dimensions of Euclidean space and one of form, of color.

On the other hand, the biggest entities have more dimensions, as they include the smaller entities within its whole. It is the law of fractal dimensionality. It states that a fractal part of a bigger i-scale of information has fewer dimensions such as:

Law of dimensionality: Fractal $\sum i-1 = 1/3^{rd}$ Dimensions of i

Since any Non-E Point will have 3 inner dimensions. Thus the dimensions of in-form-ation of any system grow with a cubic power law as the system grows in scale. In simple terms, a 3-dimensional being transcends as a mere cellular point of the new whole.

Those 3-inner dimensions, as in the case of light, are ultimately the dynamic arrows of time, or wills, or physiological networks, or organic drives of the being.

For example, a human being has 3 'dimensional' networks, which are our physiological systems. The blood network of energy, the nervous network of inforrmation and the endocryne network of reproduction. Those are the 3 systems around which cells and organs build up. So we can be described as a 3-dimensional network with 3 wills: to absorb energy for our blood/digestive system, to absorb information for our sensorial/informative system and to reproduce through our hormonal system.

And so it is evident that the abstract, geometrical description of dimensions is secondary to the functional, dynamic, organic meaning of those dimensions. The how is less revealing than the why.

The why of dimensional motions are the search of energy and information of any entity, whose paths become fixed in space and appear as dimensions. Or they are the reproductive strings of self-similar cells that form the physiological network of energy or information of the system.

In the quantum world, they are described with 4 quantum numbers, which describe the 'paths' of its particles in search of energy and information, or the social networks they create. They are also described by SU3 groups and Matrices, The dimensions of light are also the dimensions of energy/magnetic field, information/electric field and reproduction/length-speed, or social evolution (color/frequency). We shall later observe that also in life beings, each being has a dimension of height/information, which is maximal in man, the most informative being of the cosmos, and a dimension of energy/length which is maximal in energetic predators. Dimensions are therefore morphological expressions of the arrows of time.

Interacctions between light and dark energy: the warping of gravitational spacce.

Physicists are always referring only to the arrow of time=change in the motion of beings (t=v/s), which is by tautology a function of space. Thus, Philosophers talk of the spatialization of time, which is made dependant of energy and movement. Einstein was perhaps the only physicist aware of this error when he said 'I seem to be the only person that believes there are many times with different rhythms'. So when his followers say that time is the 4th dimension of space, he did not agree; but he pointed out that even in Relativity Time has an inverse, negative sign, so its properties must be opposite to those of space. But physicists ignored his dictum: 'we cannot send wire messages to the past', he insisted. And when Meyerson, the French philosopher of time published a rebuttal of Minkowski's reductionist theory of time as the 4th dimension of space, *'La deduction Relativiste',* 1928, and he could express his ideas without the enormous peer pressure, usual in this profession of fundamentalist scientists, he praised it as one of the most remarkable books written about the relativity theory from the standpoint of epistemology (the scientific method) and explicitly agreed with its rejection of the spatializing interpretation of the world of Minkowski.

Indeed, informative *time in Relativity equations have a negative sign. So it refers to the opposite arrow of information NOT to spatial energy and entropy. Time in General Relativity is the factor that warps energy into information, evolves form, trans/forming the energy of space into cyclical masses. Time in special relativity warps the energy of the vacuum, as light imprints its form.* Indeed, the equation of Special Relativity is self-evident: $s^2 = x^2 + y^2 + z^2 - c^2t^2$. Space is positive and time is negative because it forms, contracts space into a wave-frequency that carries information.

Yet those changes of form, of information could not be explained with equations because they require fractal and Non-Euclidean mathematics to describe how form evolves.

So till the end of the XX century, Physicists ignored the informative arrow that warps energy into form.

The only theory of form we had was explained by Darwin with verbal words. But physicists are 'pythagoric', so they cannot accept truths which are not written with equations to the point Nobel forbade his prize to Evolutionists, because they didn't use mathematical equations. Of course, physicists also deny Gödel's work about the errors of mathematical equations; so they can always affirm what they say is a 'sacred truth'. Further on, because they are not very good with mathematics, it takes a lot of time for them to adapt their discipline to new mathematical models. And this is plainly speaking what has happened with modern physics. Physicists do not understand the science of Complexity and the mathematical advances brought about by the discoveries of fractal mathematics, Time Duality and Non-Euclidean Geometry, applied to the description of mass as physical information.[ch.3,4] But they are too arrogant to accept it. So they are stuck with 40 year old absurd theories like those of Hawking and Higgs about mass.

The fact that the entire physicist's community preferred to deny Einstein's opinion about his own theory and affirms that time is a dimension of space; the fact that it keeps stating that the Universe is only made of entropy and calls the arrow of information, negantropy, shows to which degree we are dealing today with a reductionist religion of energy, a modern sect of Shivaites for whom reason and experimental proof matters less than an ideology that justifies their creation of energetic machines and weapons, as a sacred form of research.

Metaphysics of scales.

How many scales of space-times there are in the Physical Universe? How many 'mediums' made of energetic space and informative vortices exist? If we were to be stricter with the meaning of truth and perception, we should include now a paradox of knowledge, discovered by the science of Endophysics, the relativistic position of man in the Universe, which means we always 'perceive' triads, a scale below us, our scale and the scale above us. Because we are just in the center of any perception, which is relative to our point of view. Thus, most people think their life is so important. Most people thought the Earth was the center of the Universe. And most people in this planet think CERN will not kill them, because humanity is so important. But all that is just endophysics. We have a human p.o.v.

The particles and physical forces of the Universe extend from that human point of view through 3 relative space-time membranes called in fractal theory i-scales or 'worlds', the inner world of quarks, inside the nuclei of atoms, the external world of electroweak forces in which we exist, made of light and electrons and the upper world of cosmological, gravitational forces, black holes and dark energy. Yet we only perceive completely the light world of electroweak matter (electrons) and its electromagnetic forces. While all seems to indicate that the 'invisible' world of quarks with 99.9% of the mass of reality is self-similar to the cosmological world of black holes and dark energy. If this means a galaxy is an atom of the next scale is a metaphysical question later considered in detail, which we cannot really know. On the other extreme, absolute Endophysics will imply that our perception diminishes with distance and size and so we perceive so little of those huge and small worlds that they seem totally equal to us. In other words, *we can only know-know the world of man around us. In other words, biology is the queen of sciences, as all systems are invariant, and so we can know better both the thoughts of god and its details studying us, the measure of all things.*

IV. Galaxies

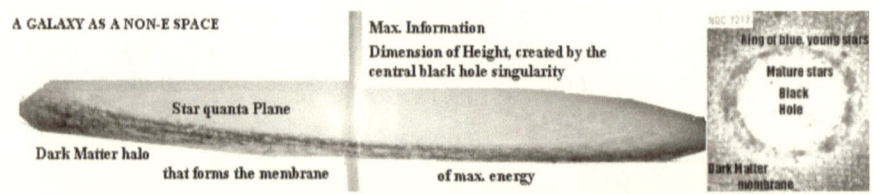

Galaxies are organic fractals that develop a spatial body of stars and an informative nucleus of black holes. The closest self-similarity in our quantum world is with a cell.

26. The Three Non-Euclidean Regions of the Galaxy

In the graph, a galaxy is a curved, fractal space-time of huge spatial proportions, hence minimal form (inversion of properties of Space and Information), which accordingly Einstein described using the simplest Non-Euclidean Riemann's tensors. Once we have completed the 5 non-Euclidean postulates galaxies can also be studied in space as a fractal point with 3 regions that correspond to the 3 'canonical' topologies of a 4-dimensional world, an informative center, an energetic membrane, and the reproductive intermediate zone:

— Max. Ti: The singularity is a wormhole, placed in the center of the galaxy, its denser zone, where its gravitational lines (the informative system of the galaxy) come together at maximal speed. Beyond its event horizon, the vortex of maximal frequency accelerates to 10 C, deflected finally into a perpendicular, hyperbolic, informative dimension of height through which it ejects gravitational jets of dark energy.

— Max. E: The external membrane that limits the inner space-time of the galaxy is a spherical halo of dark matter, probably made of strangelets or background holes, which deviate back by gravitational redshift radiation and absorb the energy of radiant matter, cooling it down to the 2.7 K background radiation.

        Those fractal, energetic units of dark matter, made of quarks, *do not evaporate as Mr. Hawking pretends, since they are vortex of space-time that obey the laws of Relativity not those of quantum theory.* Background holes and strangelets of enormous density act as 'proteins' do in cells, controlling the inner movement of galaxies and the outer absorption of light-energy.

— E=i: Stars, tracing toroidal cycles form the inner space-time body of the galaxy, a bidimensional plane or Klein's disk that feeds the wormhole and reproduces atomic substances and stars. They ultimately evolve into black holes, which migrate toward the central swarm of holes, residing in the nucleus. We are part of that intermediate space-time in a Milky Way, limited by its central hole and "an invisible" border of dark matter, neither of which we can cross without dying; since the speed of rotation of matter around the wormhole and the flows of intergalactic dark energy that expand space at light speeds beyond the halo *would destroy us*. Thus we are trapped in this star and planet, in a toroidal cycle that will end evolving the Sun into dark matter. Unfortunately CERN has decided to complete our destiny much faster than the galaxy wished.

Thus the generator equation of the galaxy as a Fractal space-time point is:

E:Halo < stars that evolve energy into matter> Ti (Black holes)

In the image, the structure of the galaxy: stars are created in the intermediate region and the center is occupied by a black hole.

The energetic medium transfers energy to the fractal quanta of the galaxy is the external interstellar gas. Finally the system is joined by 2 networks of forces: the gravitational, faster, non-local informative, transversal gravitational waves at the cosmological scale; and the energetic, smaller, slower electromagnetic waves at the quantum scale.

Let us now study the parts of the whole—those 3 elements of galaxies, the stars, black holes and gravitational forces that join them.

27. The Intermediate Space. Gravitational Waves and Solar Systems.

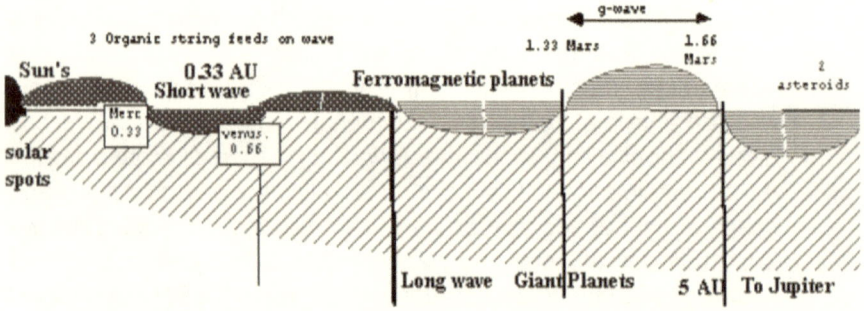

Gravitational, transversal, quantic waves shape the structure of galaxies and solar systems, transferring form and energy between cosmological bodies, in a self-similar process to the transference of information and energy between atoms through electromagnetic waves. In the graph, Titius Law of distances between planets reflects their position in the nodal points of those transversal gravitational waves. In the core of planets, there could be a crystalline or super-fluid zone where those flows of dark, gravitational energy are processed, causing flows of heat and matter that make planets 'grow'.

We use constantly self-similarities, based in the 3 Laws of invariance of the Universe, scalar, formal and motion invariance. Yet self-similarity is not equality, which means that we do not use quantum laws but deduce from fractal Relativity those self-similarities. In the graph gravitational waves are self-similar to electromagnetic waves. They organize the structure of stars and galaxies, as electromagnetic waves organize the orbits of electrons. If we take that self-similarity further, we can compare gravitational waves of temporal energy that organize the i+1 scale of matter and electromagnetic waves that organize the i scale. Both respond to the same morphological equations that relate 2 particles through a field defined by the ratio between the informative density of those masses or charges and the distance between them. The 'relative space-time ratios/constants' that define the spatial size, range, speed and dimensional form of those waves change; changing the c-speed constant of the waves, as it happened with the G-constants of the Unification Equation of Newton's gravitational vortices and Coulomb's equation of an electromagnetic vortex. Thus galaxies and solar systems show a gravitational, morphological, spatial structure similar to that of an electromagnetic atom in a cosmological scale, which Einstein predicted, establishing 2 kinds of gravitational waves, parallel to the 2 types of electromagnetic fields we know, only that their size and speed is cosmological:

— Static waves that create the gravitational bi-dimensional fields over which galaxies form.
— Discontinuous, transversal, quantized waves, similar to photons, which shape the orbits of stars and galaxies, as photons control the orbits of electrons in atoms.

Thus, those gravitational waves should have the same functions in galaxies and solar systems that electromagnetic waves have in the world of atoms. That is why quantic waves of gravitation are important to explain many cosmological structures and become in fractal cosmology by self-similarity with electromagnetic waves, the fundamental element of interaction of celestial bodies. We know that the gravitational activity of black holes set up star orbits and probably influences its evolution, growth and formation, determining the basic properties of magnetic fields, ecliptic orbits and distances between stars in a galaxy and planets in a solar system. So even if gravitational waves are invisible, using their morphological self-similarity with light waves, the equations of Einstein's relativity and the indirect proofs provided by the orbital distances and rotational fields of stars and planets, we can explain many 'whys' on the structure of those celestial bodies:

— Astronomers have always wondered what rules the distances between the planets of the solar system. The existence of regularities in the distribution of planets in the Solar System was recognized long ago. This was Kepler's main motivation in his search for planetary laws. The Titius-Bode "law" ($rn = 0.4 + 0.3 \times 2n$) was the first empirical attempt at describing these regularities, and was followed by several other proposals. The discovery of similar structures in the distribution of the satellites of the great planets led to a revival of interest for such studies, and to the hope that indeed a physical mechanism was at work.

Those planets seem to be in the nodes between stationary gravitational waves of different length and the solar system's orbital plane in which planets feed, 'deforming' space-time; as electrons are in the nodes of electromagnetic waves and stars in the nodes of gravitational waves caused by wormholes. In the graph we draw the 2 fundamental wave lengths that could explain the distances between planets: a high frequency, short gravitational wave of 0.33 AU could explain the positions of ferromagnetic, inner planets on its nodes. While 2 low frequency long wavelengths at 5 and 10 AU, could explain the position of bigger, and lighter gaseous planets. Since Jupiter is located at 5 AU, Saturn at 10 AU, Uranus at 20 AU, Neptune at 35 AU, Pluto at 40 AU; and as I predicted a decade ago, we have found a new planet, which I called then Chronos, 'the last of the titans' at 100 AU, in the limit of the solar system, 'renamed' Selma (-;.

— G-waves explain why planets have ecliptic orbits with an inclination on its axis, which is a natural orientation if they are receiving curved G-waves with a certain angle through its polar axis. In that regard, the rings of gaseous planets in the point of maximal activity of those waves (Jupiter and Saturn) and the spiral vortices of galaxies, could act as 'antennae' for those waves at star and galactic level.

— Those waves might cause, as all lineal movements do, a cyclical vortex around them, originating the condensation of planetary nebulae. While in galaxies their wave structure seems to originate the different densities of stars in their nodal zones.

— Planets suffer catastrophic changes in their magnetic fields, probably produced by changes in the directionality of those waves, emitted through the tropical dark spots of the sun. Lineal magnetism is in fractal theory the intermediate force that 'transcends' from the gravitational to the electromagnetic scale: electromagnetic light or ferromagnetic atoms like iron absorb gravitational energy through their magnetic fields. There are advanced mathematical models of gravito-magnetism that have unified both type of waves, departing from Einstein's work. Thus energetic 'gravito-magnetic' waves might cause a change in planetary magnetic fields as a magnetic field changes the spin of an atom that aligns itself with the field. For example, Uranus is tumbled and it has lost most of its magnetic field: perhaps it was knocked-out and relocated by a G wave.

— Solar spots are the probable source of those waves. Yet its origin might be the central core of the star or the activity of the central black hole, whose G-waves might be absorbed and re-emitted by the star. We cannot perceive G-waves directly but magnetic storms, solar winds and the highly energetic electromagnetic flows and particles that come from the sun's spots, might be its secondary effects. In the same way we perceive the waves of dark energy emitted by black holes that position the stars of spiral galaxies, indirectly, observing the mass and radiation dragged by those waves.

— Those catastrophes might cause the climatic changes that modulate the evolution of life on Earth, since we already know that the activity of sun spots affects the temperature of the Earth.

— G-waves could structure the galaxy and its stars in the way electromagnetic impulses structure a crystalline atomic network, ordering the distance between its molecules: electromagnetic waves also feed with energy and information those crystal webs. For example, electromagnetic waves cause the vibration of quartzes, which absorb energy from light and vibrate emitting 'maser-like', highly ordered discharges of electromagnetism. We observed similar maser beams in neutron stars, called for that reason pulsars.

28. Complex Organic Patterns in the Galaxy

A more complex analysis of the galaxy as a simple 'cellular' organism introduces elements of complex biology, which are obviously more difficult to accept from a mechanist perspective. We shall consider briefly 2 controversial hypothesis. The possibility that black holes gauge gravitation and the chains of causality between the different scales of the Universe, self-similar to the chains of causality between cells and bodies.

—The most controversial element of a cosmological model based in G-waves is the possible existence of gravitational information that somehow allows neutron stars and black holes and maybe in the future evolved planets such as the Earth through its machine systems or as a strangelet to perceive and move at will within a static field of gravitation. In other words, quark, frozen stars could be 'gravitational perceivers' in the cosmological realm, as animals are light perceivers in the Earth's crust and DNA perceives van der Waals forces in the cellular realm. On the other hand stars would be plants that use gravitation only as energy, as plants do with light in the biological realm and mitochondria with molecular forces in the cellular realm. Do black holes perceive gravitation as complex animals perceive light? They probably gauge gravitation in very simple 'forms', as a cellular DNA-system, much simpler than the brain of human beings, perceives its territorial cell. That is the true supremacy of man in a relative universe were size is less important than form. While all systems process information, man is a summit of form and hence one of the most conscious species. Yet black holes have enough quark complexity to act/react to informative flows as they seek energy to feed on – our electroweak energy.

This hypothesis has experimental proofs since pulsars and black holes emit gravitational waves and we have observed many black holes following erratic paths through the galaxy.

In that sense, fractal theory considers that in the same way light waves are the energy of plants and the information of animals, gravitational waves feed stars and inform black holes, the most evolved celestial bodies, which emit or feed on the energy and information provided by those gravitational waves. Those waves feed as energy the super fluid cores of stars and the crystalline centers of planets. Thus, as gluons reproduce quarks and electromagnetic waves become photons and electrons, gravitational waves emitted by black holes might reproduce matter on the cores of stars and planets. Since they are made of super fluid helium and iron crystals, which are the only atoms that can absorb the energy of a magnetic field to move at will.

And all those events that have a 'why' in the 4 arrows of the organic Universe (feeding, reproduction, information and social evolution) can be described in its how and details with numbers, which are 'social sets' of self-similar fractal points.

—Those arrows of 'temporal existence' are in fact the key element to understand the 'biological' order between the different scales of organisms. Indeed, in any 'scalar system' of parts and wholes, the 'energy of the superior system' becomes form for the inferior scale, causing periodical chains between species. The most obvious, extreme case is the fact that the energy detritus of bigger animals are informative food for lower insects and bacteria. If we apply this concept to the gravitational waves of black holes, it is obvious that they function as information for black holes but act as energy for stars and maybe planets that 'grow', transforming that energy into cyclical vortex of mass:

—The period of those waves can be calculated, according to fractal theory in its more complex models, which study the synchronicity of the cycles/arrows of energy, form and reproduction of the different cellular scales of a superorganism, which are tuned to each other.

The rules of those complex chains are however simple. It is a fact that informative cycles are shorter than energetic cycles, which are shorter than reproductive cycles, which are shorter than the cycles of social evolution of a species. So there is causal chain between the 4 cycles of all entities:

Max. Speed: Informative cycle > Energetic cycle (feeding) > Reproductive cycle >Evolving cycle

Yet at the same time, the smaller the species is the faster its cycles are. So there is a reversed scale:

Max. Speed for cycles: Minimal cells > Organisms > Social systems.

This means that cycles of wholes and cellular parts are chained by those different speeds into symbiotic chains; *since the parts need the energy and information of the whole, on which they depend.* For example, in most cases the feeding/energetic cycle of the whole organism determines the reproductive cycle of the cellular element, which requires the energy of the organism to reproduce and does so much faster than the whole organism; so a cell reproduces each day, which is the time cycle of feeding of its whole organism and so on. Thus, the 'slow' fields of energy and information of the bigger wholes determine the activity of the 'reproductive and social' fast cycles of 'cells'.

Thus we can consider a hypothetical chain between the 3 scales of the Universe, the human scale, the solar system and the galactic scale. The specific event we are studying is the relationship between a fast, informative cycle of a black hole and a medium, energetic cycle of a solar system, which defines a reproductive, heating cycle of planetary systems. According to those synchronic chains, the reproductive cycles of life in planets are related to energetic cycles in sunspots, regulated by the shortest, faster, informative cycle of the galactic Black hole.

That curious prediction, dating from 1994[1] was proved a few years ago: Chinese astronomers detected that the galactic black hole has a minimal cycle of periodic activity of 11 years . . . It coincides with the sun spots' cycle of 11 years, which coincides with the 11 years' rotational period around the Sun of its bigger planet and main G-wave receptor, Jupiter that has a huge magnetic field, 19.000 times bigger than the Earth's field and an enormous inner heat coming from its center, still unexplained by conventional cosmology. Yet if the black hole is connected to its stars by those waves that regulate the sun's magnetic activity, which feeds Jupiter's magnetic field, G-waves could explain why Jupiter has a bigger magnetic field, inner radiation and spatial size than any other planet.

29. The Intermediate S-T Regions: Two Worlds, Gravitation and Light

The closest homology of the 2 dual networks of the galaxy is with an atom in which the central nucleon with max. density of gravitational information and the external electronic membrane interact in a middle space-time vacuum through gravitational forces and electromagnetic photons.

How it is possible super luminal gravitation, without contradicting Einstein's Postulate of light speed as the limit of speed of our Universe? It is not possible in a continuous single space-time but it is self-evident in a fractal Universe with many space-times communicate through Einstein-Rossen bridges. In the case of the light and gravitational world those discontinuities are the masses and black hole singularities that appear to us as 'points with no volume' in the border between both worlds.

Super luminal gravitation and light are the 2 forces that interact in the galaxy, which as any other Non E point should possess 2 networks/forces: the informative, faster gravitational force that structures the position of stars in the galaxy, very much in the way DNA organizes biological cells or the nervous system puts together organisms; and the electromagnetic force that acts as the 'blood/energetic system' of the Universe. Following the cellular or physiological homologies, the network of dark, informative matter and gravitational energy, connected to black holes, surrounds and controls the stars' electromagnetic energy. We know that it was formed first and then guided the creation of electromagnetic energy, so we can observe it indirectly and deduce its form from the highly quantized shape of the filaments of light-galaxies that were formed around dark matter (right graph). Thus dark matter acts in a similar way to the RNA that shapes and controls the Golgi membranes of the cell or the nervous system that guides and builds the morphology of the body; while the network of stars and electromagnetic, slower energy that produces the substances of galaxies, surrounds those strands of dark matter. In the cell's homology ribosomes that create most products of the cell are pegged to those membranes. But how can co-exist both worlds together? In continuous space-time they can't but in discontinuous space-time where a 'dimension' is quantic, hence it is a fractal network, it is rather easy. Imagine the light Universe as a 'web', like a fishnet floating of an immense sea of gravitational energy. Each knot is a charge, a non-E point that communicates the strings of the net (the electromagnetic forces), leaving a huge 'dark' space-time of 'water', that 76% of gravitational dark energy, which is not 'illuminated' neither interacts with the quantic net. Physicists describe fractal space-time in that way, calling the fishnet a 'brane' made with cyclical, temporal knots (spins or closed strings) that in its lineal, energetic state create forces that tight together those knots.

VIII. Black Holes

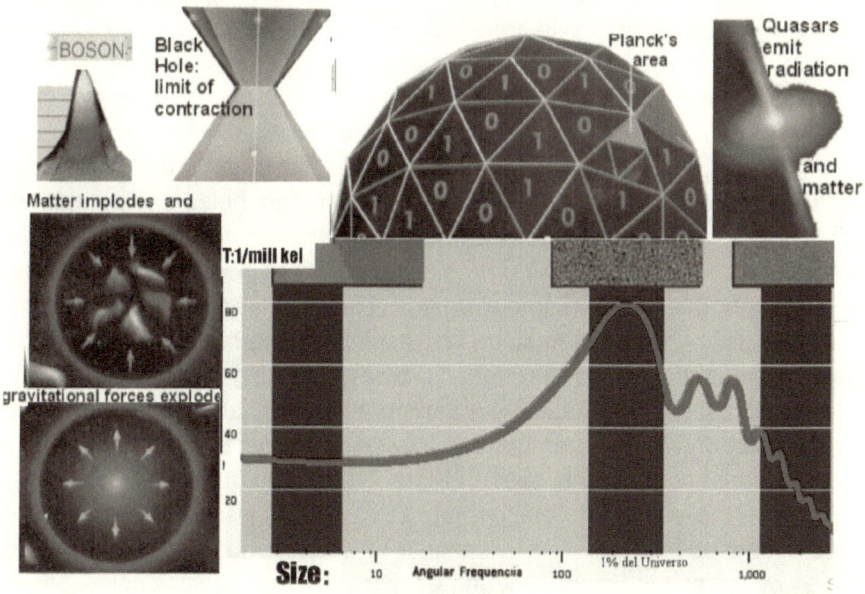

30. The Real Black Holes: Wormholes That Act as Bridges between Both Membranes

A quark in the quantum scale or a black hole in the cosmological scale is a door between the 2 membranes of the Universe, the quantum membrane from where it absorbs energy and form and the gravitational membrane to which it devolves it. For that reason it is so dangerous to open a door between both space-time membranes on planet Earth. The quark/black hole is a system that transforms electroweak energy into mass-information first into simplest quark forms and then into the simplest energy, gravitational dark energy. Our light is a spatial membrane is a cover that warps the stronger, longer, faster dark energy of the gravitational world. And our electrons are the informative cover of the, faster rotating, smaller quarks. As long as light and electrons cover dark energy and quarks, we absorb energy and form from them. But when quarks and dark energy is liberated in strong glows they prey on us. Both ecosystems are thus in a trophic, biological balance that now CERN will break. Our light preys in the faster, thinner gravitational lines that enter galaxies, warping them at −ct speed of corrugation (Special Relativity). Light fractalizes and forms the interstellar dark energy which slows down from c<10 redshift and warps, acquiring frequency of information.

Yet the inverse role is performed by the black hole, the Non-Euclidean processor that converts us beyond the Riemannian sphere of its event horizon into a flux of gravitational quarks and dark energy (right top and left bottom illustrations).

In the graph, black holes transform entropy, electroweak energy into physical information, quark—mass of maximal order and rotational speed. The musings of Hawking about their 'quantum structure' are meaningless in the evolved science of fractal, quark holes. Quark holes can be studied in any of its inner parts: its strings become gluons that become quarks that become black holes in a stair of fractal structures of increasing complexity. In the graph, we consider some elements of black holes and their interaction with galaxies, which they implode and explode in big-bang quasars. Black holes emit dark energy through its poles, creating the membrane of gravitation as protons create with their 'gravito-magnetic fields' the space of the electromagnetic membrane. Gravitational Space-time is formed by the maximal informative entities of the 2 scales, protons and black holes, and it is transformed into electromagnetic energy by the electric fields of electromagnetic matter. The expansion of space-time in the Universe is cause by the expansion of dark energy coming from the black holes' poles. Black holes absorb energy from our light membrane by red shifting light with different signatures according to their relative mass. The background radiation must be interpreted as the signature of 2.7 background holes. Bigger, older black holes of galactic mass are below the Background radiation curve and further red-shift and absorb electromagnetic energy from it. For that reason the galactic map of the background radiation has a central zone, with lower temperatures corresponding to the giant central, colder hole (bottom right).

Black hole's input of information stored in the frequency of its whirls of mass can be measured in Planck's units (upper center), the length of its minimal parts, strings. Since strings are both the theoretical components of black holes and strong forces, quarks and gluons, a description of black holes with string theory would be a mathematical bridge between both scales. They would be the 3 organic scales of black holes, fractals of quarks, fractals of gluons and strings.

Any organism extends through 3 hierarchical scales. For example, a human organism is composed of cells and carbohydrate molecules whose minimal unit is the amino acid. In the Universe there are also 3 scales of mass and each of those 3 scales become the quantic parts of the next scale. Its most perfect mass-forms, i+1 black holes, which occupy a minimal space with maximal information, seem to be made of highly ordered super fluid quarks, its 'cells' that condensate in a bosonic state (upper right). Finally the fractals of quarks and gluons would be the i-1

Planck's areas (top center), which are the minimal unit of information of the Physical Universe and the black hole.

Black holes are wormholes.

We prefer the name wormhole, because it expresses the fact that the black hole is a door to the other 'world' of dark energy, which it emits. While a black hole implies nothing is emitted by the hole.

Black holes are similar to nucleons: huge condensates of quark-mass that create vortices of gravitational information with a negative curvature in the dimension of height. Hence, they exist in a discontinuous gravitational Universe beyond the c-speed and 0 K limits of energy and information of our Universe, emitting and absorbing dark energy at higher than light speeds and ultra-cold dark matter, perfectly ordered, probably under 0 Kelvin. Thus the main error about Black holes is the idea that, since we do not see them emitting energy and matter-information at lower than c speeds, nothing escapes a black hole. This is a theoretical absurdity (things don't 'disappear'), which now has empirical proofs of falsity. Since we observe vortices of mass and radiation that surround black holes, reaching super luminal speeds before dying into pure gravitational energy according to the Lorentz transformations. And we observe bursts of matter and radiation coming out of the poles of central black holes in quasars at super luminal speeds. So black holes do emit dark energy and information at super luminal speeds through its 'axis', as atomic nuclei emit magnetic fields in their rotation.

The Thermodynamic and Relativistic equations of black holes show that duality since black holes appear with 2 solutions, one with implosive, informative parameters and the other with explosive, energetic parameters. A black hole 're-absorbs' radiant matter and light, dissociating both beyond their Lorenzian limits of c-speed, back into its ultimate components: quarks become part of the black hole body or are ejected as proton beams; while light reaches infinite red shift and becomes dark energy.

As in the case of the 3 solutions of Einstein's space-time, which correspond to the 3 ages of the Universe but physicists dissociate in 3 different Universes in space; physicists have deduced that those 2 solutions to the black hole equations create 2 different type of 'black and white holes'; when according to space-time duality they represent the 2 organic regions of the same gravitational hole. Hence we could call black holes 'mulatto' holes. Though we will use the term *wormhole*, more familiar to cosmologists. And define its E⇔i space-time field equation:

Black hole (max.E: external membrane: event Horizon) <Wormhole> White hole (max. i: Kerr ring)

Quantum cosmologists never discovered white holes as independent entities, because they are part of the wormhole; and so Hawking says black holes might be the entrance to other Universe in other region of time-space, where those white hole exist, LOL. Unfortunately quantum cosmologists ignore the inner structure of a gravitational hole as a fractal point with 3 zones that explain them:

— *Max. E: The event horizon* at c-speed is the energetic Riemann membrane that absorbs radiant matter, breaking it into its minimal units, quarks and 'Planck's areas' that become the strings of dark energy that feed the hole. The closest species to a black hole is a neutron star, composed of super fluid neutrons and strange liquid in a bosonic state, occupying a minimal space. Yet the difference of density between a neutron star and a black hole is small. So a black hole could be the next evolution of a quark star with a 'lighter' cover of strange quarks on the event horizon, breaking the symmetry of our matter, packed then in a bosonic, super fluid solid state of bcb atoms and top quarks.

— *E=i: The intermediate zone* that transforms radiant particles and energy into those quarks and dark flows of gravitational energy. It seems to be a super fluid solid: a vortex-like structure of quark condensates and gluons, joined by ultra dense networks of gravitational energy.

— *Max. i: The central region around* the polar axis or central, informative, hyperbolic, negatively curved nucleus. That informative center is the final 'eye' of the gravitational hurricane, the white hole of the wormhole. It emits through its poles energy and information in the form of dark, quark matter and gravitational waves of dark energy in super luminal jets that we observe indirectly around far away quasars, as they become again slower radiant matter, creating irregular galaxies. Thus white holes are the poles of wormholes, which indeed are the doors to the gravitational world of dark energy and quark matter that dominates the Universe, as cosmologists have discovered.

So we define a rotating *wormhole* as an organic dual combination of a spheroid *black hole* membrane that absorbs radiant matter and energy through its central, ventral plane, transforms it into gravitational energy and throws it in a perpendicular jet through its central, hyperbolic axial *white hole's* pole to control the body of galactic stars and communicate with other black holes.

The equations of a wormhole show how it transforms the spatial, energetic parameters of the electromagnetic world into the inverse, informative parameters of the gravitational world, since a wormhole is an ultra dense mass of highly ordered 'bosonic nucleons', packed into a single point of space . . .

The equations of black holes show also their horizon as a bidimensional field that has the maximal informative entropy of the Universe, since it kills matter into its ultimate units on the Planck's scale. In the biological homology, a human being is composed of cells themselves composed of amino acids that act as the minimal units of life. So when an organism dies it suffers 2 deaths: first its cellular tissue is broken into pieces that feed the stomach of an organism, which will destroy it till its minimal amino acid units, used to recompose the organisms own cells. When we study black holes they suggest also a dual process of destruction of light matter to its ultimate components, strings, used then to reconstruct the bosonic quark condensates of the black hole and its dark energy.

Indeed, when we analyse the thermodynamic equations of black holes they show that their feeding process destroys matter till it absorbs its Planck's areas, the minimal units of information and energy of physical space-time, the equivalent to the amino acids that the stomach absorbs.

What are those Planck's units? Most physicists consider them cyclical strings or gravitons, the minimal units of the world of masses

31. The Existential Cycles of Wormholes

Wormholes follow the 4 energetic, informative, reproductive and social arrows of all exi=st systems of reality:

— *E>I: The main cycle of a black hole is informative.* It creates quarks from electroweak matter and orders the galactic into spiral forms through gravitational waves that 'feed' the stars with energy. Since it emits dark waves of gravitational information to control the galaxy.

In the previous paragraph, we studied how the informative waves of black holes fed the energetic needs of stars, which therefore become chained to those waves without 'knowing' as gravitational plants that they are guided by black holes and herded toward the center of the galaxy to feed them:

— *I<E: Their energetic cycles balance the entropy of the light-world.* Wormholes first erase light and radiant matter, feeding on gas and stars; and then renew it, creating new jets of pure dark, gravitational

super luminal Energy and quarks that enter back into our Universe as light and protonic matter. If insects eat dead matter to renew the Earth's ecosystem, the wormhole inverts the time/space coordinates of the light—world to renew it. They act as "hyper-antiparticles" that annihilate matter into pure Energy.

— *Reproduction:* Wormholes control the reproduction of stars, which in turn reproduce wormholes. Since when a small wormhole crosses through a star, it catalyses its explosion into a nova that leaves behind a neutron star or a wormhole. On the other hand, the 'reproductive DNA center' of giant galaxies is a massive wormhole structure that emits huge super luminal jets of quark matter, which catalyse the reproduction of stars, creating irregular baby-galaxies.

— *Social evolution.* The wormholes that occupy the informative center of galaxies seem to be more like the nuclei of cells, a swarm, which probably regulates the life of galaxies as DNA does with cells.

The 3 organic roles and types of wormholes in galaxies.

By homology with any other non-Euclidean space-time that resembles a cellular organism, we consider 3 types of wormholes and 3 roles within the organic structure of galaxies:

— *Intermediate, reproductive E=i zone:* Spiral arms. Non rotating wormholes are born from dying stars. Then they form bi-polar systems with other stars feeding on them, taking advantage of their gravitational control, finally transforming the biggest stars into new wormholes. Those wormholes probably gather in social groups, which fuse in bigger wormholes and move toward the center of the galaxy where they can feed easily on its dense herds of stars, creating at the end of the process a central nucleus.

— *Max. information: Nucleus.* The black hole nucleus is a huge rotational, 'Kerr hole' or perhaps a herd of black holes similar to the DNA nucleus of a cell. It is the informative brain of the galaxy that controls its quantic beings, the stars, with gravitational waves that shape the rotating movement of the galaxy, and establish its feeding rhythm. The galactic black hole first attracts interstellar gas to the intermediate non-E region, where gas reproduces stars, and then it sends that gas to the central wormhole that consumes it. In the same manner, electrons, the stars of the atom, feed first on light quanta and then emit high-energy photons to feed the atomic nuclei. Those gravitational waves also guide in old, globular galaxies, the stars toward the feeding center. Yet there might be other structures of dark matter, coming out of the nucleus, similar to the "Golgi apparatus" of cells: invaginations through which wormholes might flow into external zones of the galaxy to control the reproductive and destructive processes of stars. Finally, the central hole emits through its polar zones, dark energy, super-luminal gravitational waves that probably communicate galaxies at super luminal speeds, forming the strings of galaxies observed in the Universe. Since according to string Theory gravitation in free, intergalactic space is not warped by "electromagnetic branes" that feed on them inside galaxies.

— *Max. E: Membrane.* Though we cannot see the galactic membrane made of dark matter, by homology Non-Euclidean topology hypothesizes that the halo is the energetic membrane, where small wormholes called appropriately MACHOS, have functions similar to globular proteins in cells:

They create and control that galactic membrane of dark matter, which closes the galaxy as a black body; causing the background radiation, which according to recent empirical data might be local: They redshift light to 2.7 K, which becomes the metabolic temperature of the galaxy. They reproduce new stars and expel matter and radiation beyond the membrane: Since their rotation is perpendicular to the galactic plane they could create a positive or negative spin, depending of its orientation, provoking flows of energy and information in and out of the galaxy. Those outward or inward flows fine tuned with the dark energy jets of the black hole should move the galaxy.

The big bangs of the three types of wormholes

The homology between the 3 sub-species of black holes and the 3 regions of a galaxy explain the 3 possible scales of big-bangs of those black holes:

— Micro-black holes, Background MACHOs would eat up the commonest celestial bodies, planets and moons, and so those CERN will do will devour the Earth soon.
— Natural black holes formed in the densest centers of stars would cause supernovas, creating intermediate black holes that migrate toward the center, forming the "DNA nucleus" of the galaxy.
— While galactic black holes would explode into quasars, (galactic big-bangs).

32. The Black Hole as a Gravitational, Informative Mind

The center of the galaxy is occupied by a giant wormhole that seems to be the final, social, evolutionary stage of multiple galactic black holes, born out of the evolution of stars. It acts as the gravitational DNA-mind of the galaxy, a hypothesis, which mechanist science will always ignore. But we want to stretch your understanding of the fractal paradigm, describing those mind holes with the laws of Non-Euclidean geometry and superorganisms.

A galactic wormhole is a rotating object, which has a minimal spatial size and a huge dimension of height since it is made of bosonic, super fluid quark condensates.

Its homology with a cellular, informative center, defines a central galactic wormhole as an enormous gravitational informative center, which controls its galactic body, positioning its star quanta through gravitational waves. In fact, when we calculate the wormhole's informative parameters, it turns out to be a perfect super fluid computer, with maximal informative volume since its speed of calculus equals its speed of transmission of information. The coldness of the wormhole proves also the informative hypothesis. In fractal space-time coldness means order, stillness, necessary to create in the center of a crystal or a cryogenic CPU, or a super fluid wormhole or the focus of an eye, the informative, bosonic accumulation of "pixels" that shapes a still, formal, quantic virtual image, without friction, without blur. Thus, the eye of man is cold. The brain is colder than the blood. The chip works better at cryogenic temperatures. And a wormhole is very cold, made of quarks in super fluid, highly ordered states. In formal terms the galaxy is a vortex and the galactic wormhole is the "eye" of the vortex.

In fractal cosmology the central discontinuous zone of a fractal point is an informative region.

Continuous physicists used to believe that the accelerating vortex reached infinite energy values in the central point or "singularity," since they do not model masses, charges and black holes as physical vortices of mass with a Radius, Ro, that represents the discontinuous limit between the external, body cycles of the Non E-point and its inner, still, informative region or brain, in this case between the vortex of stars and the still black hole brain. Yet because infinites cannot be calculated today "renormalize" their equations beyond a certain limit in which they postulate 0 charge or 0 mass.

In fractal cosmology those tricks are not required as wormholes are modeled with Non-Euclidean topologies, which have always 3 regions separated by asymptotic membranes.

Thus fractal theory solves the problem caused by those continuous "infinite singularities" as Planck solved the problem of continuous "infinite temperatures" when he introduced quantic light.

In organic terms, beyond the event horizon of a Non-Euclidean point, the Klein disk starts and beyond Ro, a discontinuous, inner radius separates the body from the informative, still brain.

In the galaxy the event horizon is the halo and this final radius is the horizon of the wormhole. In the atom, those 2 horizons are the external Electronic radius and the inner Bohr Radius,[ch.2.4] beyond which we find the protonic black hole.

The same pattern of 3 regions is found in the Black holes as a Non-Euclidean point. The event horizon is the external membrane. Then the point in which the equations of a black hole reach T=0 is the inner Ro radius.

This is the point in which Hawking says black holes become "negative in time" and convert themselves into time machines. But at this stage I imagine the reader understands that Hawking's musings are as irrelevant to serious organic cosmology as the myths of astrologers were for ancient cosmologists, even if the High Priests of the Ziggurat at CERN can kill us all with their political and military power; since they still command the wisdom of the king.

We already argued Mr. Hawking's confusion of physical time, a change in the direction of motion, v=s/t, and absolute time, a hyperbolic error of the Cartesian graph. Time in physics is change in the direction of motion. So what T=0 means in a black hole is that we reach the region in which the cycles of the black hole's body end and we enter the hyperbolic, "high" central tube that ejects dark energy. At this point time understood as "change in motion" halts and the Black hole enters the white hole region of production of dark energy, asymptotically perpendicular, with the form of a Belgrami cone.[App.IV] If the reader observes the graph in section IV of the 3 regions of a non-Euclidean point, it will observe that this region "doubles" its form, as the brain that "maps" out the rest of the Non-E point.

And so there is a "hyperbolic" region of pure information that "doubles" the topology of the membrane and body region. This simple law of balance of the 3 regions of a Non-Euclidean topology is central to understand the workings of a mind:

Information of hyperbolic "brain" = Information of Klein's "body" + Information Riemann membrane.

In "metaphysics," which searches for the logic and geometric laws, derived from the 5 Non-Euclidean postulates, from where the "arrows and wills of the organic Universe" are deduced; a mind is understood as a fractal image, reflection of an internal body and an external Universe, which connects both "regions" of the Non-E point. In a non-E point the membrane is the "first mapping of the Universe," and the intermediate body, the inner world that reproduces, feeds and maintains the system. A mind will constantly check both, the membrane and its sensorial image to which it is connected through a nervous/informative system and the inner body, to which is connected through a reproductive/energetic system. So a human brain is connected to the inner world through the blood system and the external world through the membrane senses. It has two images of reality and it combines them both, word feelings and eye-images, to act with 2 purposes: to maximize the informative perception of the external world and the inner, reproductive and energetic feelings and pleasures of the Internal world.

This complex explanation of the human being, explored in more detail in other books, is reduced to a minimal skeleton, *maintaining the scheme of wills and purposes,* when we study the simplest forms of the Universe. What this means is that the black hole must have an inner image of the galactic body, which it orders with its gravitational flows and an external image of the Universe in which the galaxy, its body floats, to direct its form toward "fields of energy" (intergalactic gas) and connect itself with other galaxies. This is what we observe galaxies do. They form walls, strings and complex clusters with other galaxies and they feed on intergalactic gas and smaller galaxies. So they act as cells in a gravitational, organic soup and must have as cells and organisms do, an informative, processing center able to have an internal image or mapping of its stars and an external connection that informs it of the outer world. However, we see only a 4% of the galactic structures, which is like trying to recompose a cell, in which we only see its mitochondria.

Still amazing as it seems, the gravitational equations of Einstein that describe real black holes, give us a mathematical proof of that mental mapping of the external and internal world:

Wormholes do create a complex virtual image of the galaxy with its strings. Physicists explain those processes in abstract when they affirm that a wormhole has an extra 5th dimension and its equations are homologous in 5 dimensions to those of the 4-dimensional electromagnetic world of the galaxy it represents. *The description in 5-string dimensions of a black hole is completely equivalent to the description of an electromagnetic galaxy in 4 dimensions.* But where it is the 5th dimension? Again only fractal theory provides an answer: the 5-fractal dimensional world of string holes is equivalent to a 4-dimensional world; *it is the map of our galaxy made by the black hole to rule us all.*

33. The Background Radiation: Basal Temperature of the Galaxy

The background radiation coincides with the radiation of a black body at 2.7 K degrees. Since in Non-Euclidean topology any fractal point is a "black body"; that is, a point with minimal apertures to the external world, a galaxy will only emit "background radiation" through its Halo, in which background holes will redshift light at 2.7k. Thus the galaxy surrounded by a halo of quark, dark matter can be considered a "black body" emitting at present time as an isothermal organism does a background radiation, whose organic function is to maintain a homogenous temperature, similar to the organic temperature of living beings and ecosystems. Thus, the background radiation acts as the "cytoplasm energy" of galaxies with 3 functions:

— Max. E: It provides energy to the bigger, colder black holes and the super fluid helium structures (not treated in this introduction to fractal cosmology). Both happen to have a temperature slightly lower than the background radiation from where they can extract energy.
— E=i: It acts as the membrane limit, between the gravitational and electromagnetic membrane, within the galaxy, separating both worlds. *It maintains also an* isothermal temperature, as any organic system maintains a stable temperature.
— Max. i: It establishes a fixed frame of reference for the galaxy, *allowing the process of information and measure, location and communication defined by Gauge Theories.*

Then the background radiation is the "energy soup," the cellular water of the galaxy that feeds its dominant "RNA," black holes; as the hot water of the cell allows RNA molecules to move, kicking left and right water molecules with its COOH legs. Indeed, we know that only organic systems have an homogenous temperature. For example, humans have an homogeneous temperature within the limits of liquid water. So the 2.7 K homogenous background radiation reinforces the organic hypothesis. Quark matter is *the top predator form of the galaxy; hence the entities which, as the elephant on the savannah or man on Earth, or aerobic bacteria in the earlier planet, have redesigned the galaxy with their organic activity.* In Gaia, water, the equivalent to that background radiation, maintains a stable temperature, thanks to the feeding, energetic activities of its life organisms that avoid abrupt climatic changes. Indeed, without aerobic life the Earth would be like Venus, a planet with extreme temperature changes. Now it is almost isothermal. So happens to your body which has 36.5 degrees all your life, due to your organic activity as a water organism with 2 networks, a warmer blood and a colder, nervous, informative system.

So happens to the galaxy, in which the basal temperature of the background radiation separates the ultra cold world of black holes and dark, gravitational matter and the hot world of atoms and radiant matter, allowing the exchange of energy and information between the 2 physiological networks that structure the galaxy . . .

IX. The Scales of the Universe

34. The Galactic/Atomic Self-similarity. Can We Extend to a Higher Universe?

Are the fractal scales of the Universe infinite? And if so, where we find a self-similarity (not an identity) so striking as to make possible the concept of infinite repetitions?

The answer is the Unification equation, U.C. × M = w^2 × r^3 that relates electromagnetic and gravitational forces at the scale of atoms and galactic black holes.

In the fractal paradigm, the electromagnetic and gravitational membranes belong to 2 different realities, the microcosm and the macrocosm, between which mankind is sandwiched. Yet the macrocosmic gravitational and microcosmic, quantic membrane are self-similar worlds: if we treat charges as microcosmic masses and compare the structure of cosmological galaxies and microcosmic atoms, we find a striking self-similarity, already noticed by Eddington (theory of the great numbers).

String theorists (duality between the small and the big) hint also at a possible further hierarchy between the next smaller scale (the string) and the highest possible scale imagined by mankind (a local Universe made of 100 billion galaxies). A string membrane and a Universal membrane can be treated with self-similar equations in some models of string theory (even if a more accurate treatment requires background independent string theories, in which the error of Newton – an absolute Cartesian space-time—is corrected). The principle though remains. It is possible to create a model of infinite hierarchies, in which each scale is self-similar but no equal to the next scale of the Universe, jumping between its fundamental "stable" entities which are the "immortal" protons and black holes.

The self-similarity between the 2 scales happens at the scale of an atom and a galaxy. And the metaphysical question is if that self-similarity goes beyond the galaxy and below the atom into a higher and lower scale, in which case a galaxy will be an atom of an infinite higher Universe and/or an atom will be a galaxy of an infinite smaller Cosmos. All in all there are 3 possible hypothesis for the Universe that we shall briefly explore here to end this introduction to fractal organic cosmology:

— Max. E: The galaxy is homologous to an atom, so we are truly nothing and the Universe extends to infinity. We exist in a "hydrogen cloud of gas" as most galaxies are simple "Bohr atoms" with central protonic black holes. And beyond there are infinite atoms and upper scales. The hypothesis of an infinite hierarchical Universe is reinforced by fractal, theoretical work such as the Unification equation and the famous Eddington theory of big numbers that relates a proton radius and the radius of the galactic big-bang.

— E=I: There are not upper and lower scales beyond the galaxy and the atom. So galaxies are the highest units of an infinite space-time. This is the most boring hypothesis we shall ignore.

— Max. I: Galaxies form complex organizations and are cells of an organic Universe, pegged to other organic Universes. This is the hypothesis of the next graph. We do not know and we shall not know if there is a membrane limit to our Universe, regardless of what CERN does. But the Webb telescope might illuminate this hypothesis if it finds a wall of radiation at z=100c, which would be an indication of a new hyper-decametric scale of dark matter and energy.

Because it is very unlikely that we survive CERN, which will reach maximal potency in 2013, just before the Webb telescope is launched, I am afraid we will never know if we are in a cell-Universe or an infinite Universe in which a self-similar scale of "atomic galaxies" extends in distances beyond our imagination.

What CERN won't do for sure is to replicate the big-bang of the Universe or even a quasar, but just the big-bang of Earth. The cosmic big-bang is a completely obsolete theory based in a local, galactic radiation, invented before we knew even about the existence of dark energy and dark matter. And as the case of the Higgs we have to wonder NOT about its falsity but about the psychological and industrial reasons that makes *it still* the paradigm of cosmology:

— First and paramount it stands because the hyperbolic error, common to Religious people and Physicists, which expand their ideas and measures to cosmic dimensions. So for example, a prophet of a religion doesn't merely talk about the small human social god, the "social organism of mankind" but about the entire Universe. For the same reason, physicists pretend that the background radiation at 2.7 k, which a black hole that has eaten a moon can produce, and has been observed only in this galaxy, happens all over the cosmos and *comes from a remote past we can't observe. This is false, according to the scientific method. If an event is observed only in a local region, the galaxy and in present time, it must be produced* by an event that happens now in this galaxy, which means it is produced by background holes that have eaten moons.

So why physicists decide to blow up all their theories to cosmic proportions? This is due to their error of continuity, again an error in their conception of time as a single, Universal, lineal time that extends toward infinity. Astrophysicists work still with ideas about a continuous single space-time Universe because they ignore the fractal nature of space-time. They also ignore the duality of "static space" and dynamic motion (Galilean paradox), so they consider that the superluminal speed of dark energy and galaxies is the expansion of a big-bang space.

35. The Galaxy as an Atom

There are many theoretical hints to that concept:

— Relativists model the Universe with Einstein-Walker equations in which each galaxy is treated as a hydrogen atom. This theoretical trick that facilitates the calculus of relativity shows the self-similarity between a hydrogen atom and a Galaxy of the upper scale. Each point of the electron nebulae is a star of the upper scale. If so, each proton is a positive, central Kerr black hole. And alternatively each +Kerr black hole is a hadron made of positive top quarks. And the gravitational waves of dark energy expelled by black holes are equivalent to the magnetic fields of protons.

An alternative model would consider that each star is self-similar to a photon, While the electron nebulae that surrounds the atom is equivalent to a strangelet nebulae of dark, quark stars, the halo of the galaxy; because strangelets are negative (strangeness is negative).

So protons are self-similar black holes (top quark stars). The halo is made of strangelet stars as the Earth will be soon (and we are indeed close to the Halo). And stars are neutral, light photons.

Cosmologists use models of electronic nebulae to describe the behavior of stars around the central black hole; and some have tried models with high density photons.

Yet self-similarity is not identity. So what we call quantum cosmology requires a fractal formulation to make sense; NOT the use of quantum equations as if they were identical, but the logic of fractals to observe the self-similarities of those scales. In that fractal formulation CERN will make black holes. And what is worrisome: the entire function of planets and stars near the halo might be just to become a strangelet of the atomic cover of the next scale. This however is metaphysics, in the sense that it goes beyond what physics can perceive. In other words, it is speculation. But it is not as Hawking's and Higgs fantaphysics, which is proved wrong by the scientific method. Speculation is mathematically and logically consistent. It is metaphysics when it cannot be proved experimentally, which means we shall never know if it is truth or wrong. To put it in simple terms: it is more probable than we, humans are just dust of space-time, living on a fractal electronic nebula of a galactic atom (speculation) than black holes evaporate (wrong physics).

What else can we know, we human beings, dust of space-time about the infinite Universe? Not much more, because if there is a higher fractal scale, and galaxies are self-similar to atoms, the extension of that scale will be so vast that we shall never find its limits. Indeed, in our Universe there are trillions of atoms and so if there are trillions of self-similar galaxies, the maximal perception we can have is that of a local "structure" within that Universe.

36. The Universe in Space as a Fractal Superorganism, Its Structure

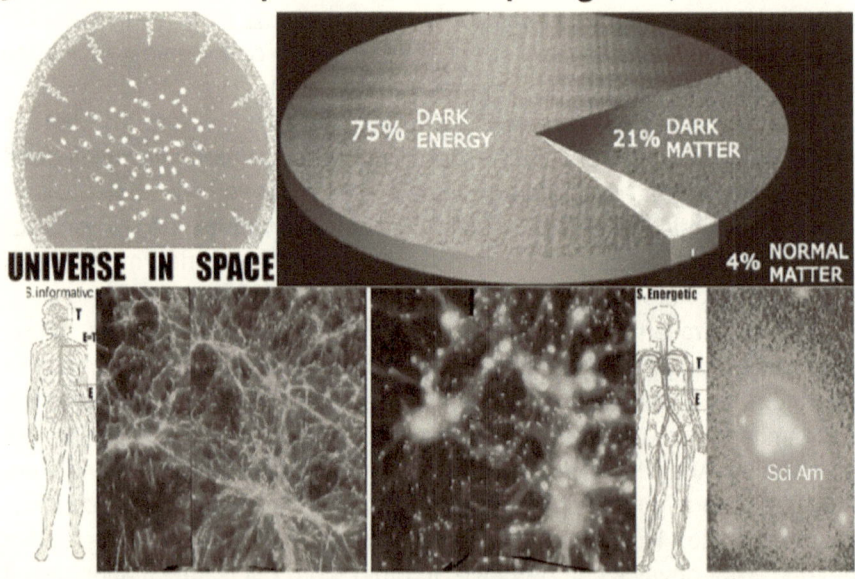

*The Local Universe seems to have a dual network structure—
electromagnetic energy and visible matter vs. quark, dark matter and dark
energy, which forms its faster networks of information. It is the organic
hypothesis of a Universe made of cellular galaxies, part of a higher
ecosystem or colonial organism.*

The duality of Galileo applied to the Universe: its 3 ages and 3 topologies.

Any Fractal point, according to duality, can be described in space as a fractal point with a relative energy body and an informative center, communicated through 2 forces/networks of energy and information. And it can be studied in time, through its 3 evolutionary ages, between life and death, as its energy becomes curved by time and increases its form, creating in the case of the cellular Universe, new informative particles and galaxies. Though both perceptions of the Universe are correct, cosmology ignores the organic, spatial description of the Universe, which completes the evolutionary, temporal vision, obtained through light instruments since it cannot see dark energy and gravitation. Galilean relativism *(eppur si muove, eppur no muove)* implies that depending on which kind of energetic or informative force we use to observe certain reality, we will perceive it either as a fixed, spatial organism or as an evolutionary species in movement through time. So we can see either a moving Earth and a fixed Universe when we see them through gravitation, or a quiet Earth and a moving Universe when we see them through light. Since gravitation, the force that shows the spatial, synchronic Universe, is invisible to our instruments, astronomers only study a temporal, diachronic Universe, perceived with light. However there is a universal organism in space, self-similar to any other Fractal point. Since once the evolution of the hypothetical cellular Universe concluded in time, creating the 3 regions of any fractal point, the Universe structured itself in space communicating those 3 regions, through simultaneous non-local gravitation.

Let us then study those 2 sides of a hypothetical cellular Universe, first the "spatial, organism" of the Universe and then the temporal ages of creation of that Universe.

The cellular Universe as a spatial organism.

In the graph, if we perceive the Universe from a temporal perspective through its slow force, light, it appears as an evolutionary process of matter, coming out from the genetic singularity of a local, cosmological first cell, the hyper-dense singularity of the big bang (corrected in the fractal paradigm, as a cold, "big-banging" or creative, informative process, later studied in more detail).

Yet, when we observe the Universe simultaneously in present space thanks to its faster force, gravitation that allow the parts of the Universe to interact at 10 C, explaining its homogeneity, those far away regions become integrated with its closer regions. Since those structures must be contemplated through the dark informative networks, NOT through the light, energy networks of galaxies and stars, as any organism is defined by its informative, nervous network that gives it form. In that sense, non-local faster than light gravitation creates an organic Universe, perhaps cell of a hyper Universe, structured as a fractal point, with its 3 canonical, topological, Non-Euclidean zones:

— *Max. i:* The Universe should have a nucleus of enormous gravitational mass, a hyper-black hole, connected to a network of dark matter, which acts as its informative brain, since it "forms" through non-local gravitational forces the shape of its galactic networks, as the DNA nucleus of a cell controls the form of the organelles that reproduces its proteins. Thought the nucleus is a gravitational knot invisible to us, we have found a very dense region of dark matter called the Great Attractor toward which many galaxies, including ours move, which might be that center. That informative singularity will keep growing and attracting other galaxies in a generational cycle, till it explodes again its form into energy in a physical big-bang, similar to the one that created our Universe.

— *Max. E:* An external energy membrane that emits a faster than 10 C Dark Energy, not to confuse with the galactic light background radiation and the dark energy of galactic black holes. There are hints of this possible final wall in measures of dark energy expansion over 10 C at the limits of the Universe. But proofs are scant as we need better telescopes, under funded so we have money . . . to make quark cannons and blow the Earth.

— An inner E/T region with galaxies that reproduce matter and light. It is the "visible space-time," created after the "invagination" of the inner nucleus of dark matter that clearly directs the movements of galaxies. This radiation matter would be the food of dark matter—an "energetic, electromagnetic network" of galactic mitochondria, which reaches its maximal density in the filaments of galaxies, (center).

Thus, the organic Universe has an "energetic, electromagnetic network" of radiant matter similar to the blood network of a living organism, which weights only a 4% and reaches its maximum density in the external membrane and the filaments of stars, (center). And it has an informative, gravitational network similar to a nervous network, which reaches its maximum density in the hypothetical hyper-black hole, brain or central singularity of that cellular Universe, the Great Attractor. This network weights a 21%. Both feed on the intermediate space-time region of gravitational space-time (dark energy, which is the 75% of the Universe and acts as the 'water' of the Universal organism, also the maximal weigh of a living being; or as the background radiation of the galaxy, also its most common substance).

In the lower drawings of the previous picture we see there is a local universal structure with the form of a half wave, which can be anything in that upper scale, from a half "light wave" to a worm like micro-organism.

The verification or not of a cellular Universe could be done by the Webb telescope testing the existence or not of a limit at 13 billion years—a dark region or wall of fire, which perceived in space would look like the first picture. Then we could reasonably think that the local Universe hosts around 10^{10-11} galaxies in a cellular structure separated from other parallel Universes by a wall of dark radiation at z=10-100 C.

It might also be possible that as stars form spiral vortices, the mapping observed in the previous graph is the outer cover of a spiral Universe, with a central zone of hyper-black holes, which would act as the nuclei of an atom.

As the reader can notice we have a problem: we can't figure out what is the next scale of form of the Universe, because we have little evidence, as most is dark matter and energy whose form we can't deduce and because all scales of reality are self-similar so we cannot easily distinguish them; as we could not distinguish a bottle of beer with a bad picture of it.ch.7

Yet the fractal, network structure on the left, down picture is quite obvious. So there should be a bigger social scale, as Pietronero shows.

37. The Universe in Time: Big-banging Horizons, the Evolution of Matter

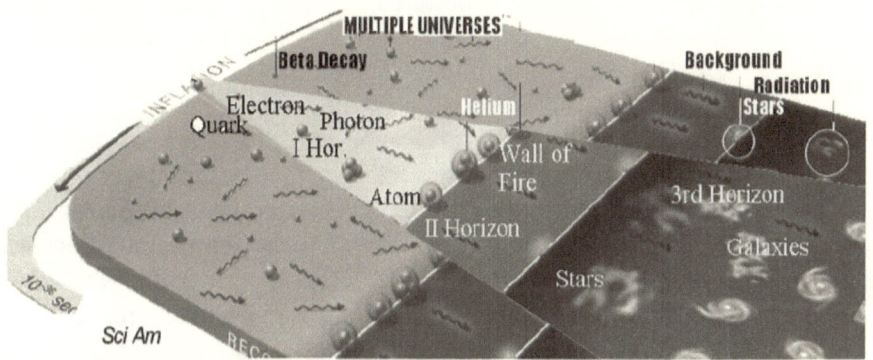

The Big-banging reproduced and evolved the Hypothetical cellular Universe in 3 Horizons of growing complexity: the age of atoms, celestial bodies and galaxies.

If we study the hypothetical cellular universe in time, we differentiate then in its evolution the 3±1 ages of any space-time system.

— *1: The conception* of that Universe from a first informative cell was also the death and gravitational big-bang of a previous Universal singularity, a seed of enormous "Planck density" of quarks that transformed its inner, informative dimensions into external dark energy.

— *Life arrow:* Then the Universe suffered an informative process of formal aging, a "big-banging," or reproduction and self-organization of the present Universe in 3 ages, which should end in:

— *Death arrow:* A big-crunch death into a new hyper-dense singularity that would restart the cycle.

— *1: Death and Conception. The Big-bang* proper or death of a hyper-black hole created a bi-dimensional sheet of gravitational space-time where matter evolved, departing from a first wave of quarks, the cellular components of that primordial black hole.

Let us consider those ages in its ternary sub-ages, *according to a basic law of the fractal paradigm, known as fractal differentiation:*

> *'Any event or form of the Universe can be divided into subsequent ternary or dual events and structures of energy and information of a lower spatial scale or shorter time duration:* $E_i {\Leftrightarrow} I_i = 3E_{i-1} {\Leftrightarrow} 3I_{i-1}$'

This law, which is the essential law of the fractal paradigm that structures its self-repetitions according to 3 ages and 3 Non-Euclidean structures in any cellular scale, can be applied to study the 3 ages of the Universe and for each age, its 3 sub-ages:

I Universal Age: Particles, Atoms and Molecules.

—*Youth: Max. E: The Big bangings of particles: Those quarks evolved into nucleons and* suffered fractal, expansive mini big bangings or "beta decays," creating protons and electrons, later self-organized into atoms, molecules and more complex social structures, till giving birth to galaxies. Those molecular structures can also be studied with the law of 3 ages, according to which the 3±1 states of matter respond to the dominant arrow of future information:

— *(-1): Plasma or conception state of matter.* The plasmatic state occurs at speeds closer to light, in a state in which atoms are split in primordial particles, free protons and electrons, which emit enormous quantities of radiation and sub-particles. Thus plasma is matter in its seminal state, (-1).

— Max. E: The *gaseous* state is the classic *energetic state* of atoms or diatomic molecules with max. movement, spatial volume and min. formal density.

— <=>: The molecular state of balance between energy and information is the *liquid state*, which according to the law of harmony (E=T), is *the reproductive state* of matter. For example, liquid water and mercury dilute all other atoms of similar weight that merge and evolve into more complex forms: carbon-life or metal alloys. 90% of a human body *is liquid* water, with diluted, energetic gas (oxygen), and solid particles (proteins, DNAs, RNAs) to store, trans-form and process information. Since solid is:

— Max. I: *Solid, the informative state* of min. movement and max. form or mass density that creates the most perceptive, inorganic structures: bionic crystals.

— (+1): Finally, the *super fluid state,* typical of liquid helium, is the most perfect of all of them, the highest form of molecular evolution, because it combines the properties of solids with maximum order, liquids with max. reproductive, combinatory capacity and gases with max. Energy.

II Age: E=i: Nebulae, stars and planets.

In the next stage, those states of matter, evolved socially according to duality aggregating into macromolecular systems *that combine 2 or 3 states of matter,* which act as the dual energetic/ informative elements of the organic system. Those macro-molecular worlds are cosmological organisms that seem to us "dead," because their existential cycles are very slow (Max. E=Min i). But they are also organic Non-E Points, evolved according to the Fractal Principle in 3 ages:

— *Max E=Min i: Nebulae* are the youngest, most common, spatially bigger, less dense and evolved celestial "bodies," extended between galaxies. They exist in the energetic plasma and gas states.
— *E=i: Stars are molecular macro systems, existing in the plasma and liquid states.* Fractal cosmology considers informative, cold, liquid and super fluid states dominant over energetic states, which Physics, based in a single energy arrow prefers, as energy is also easier to detect. Yet if we could enter inside a star, we probably would find lower temperatures and super fluid flows.
— *Max E=Min T: Planets are systems that exist in liquid and solid states.* They follow the inversion Law, Max. $E_{->\infty}$=Min. $i_{->0}$. So giant planets exist mainly in liquid states and smaller planets exist mainly as solids. They are called ferromagnetic planets; and we live in one of them. Again, fractal theory favors "cold" solid, informative states. So it affirms that the planetary core should be a crystal; and their youth should be colder than scientists predicted. And indeed, recently we found that Neptune might have a solid diamond crystal in its center and the Earth probably has an iron crystal with a Uranium core. And we have also found, studying zircons, that planets cooled off much earlier and hence life also started earlier.
— *+1:Pulsars and Holes are systems with a solid iron crust and a super fluid core of strange quarks.*

Pulsars represent the final, informative age of celestial bodies, which might evolve further into black holes, "species" belonging to the gravitational Universe beyond our light world, with higher form. Those holes become through social evolution the "informative brains" of galaxies, triggering the next stage of evolution of the Universe . . .

III Informative, Age and death: Max. I: The age of Galaxies.

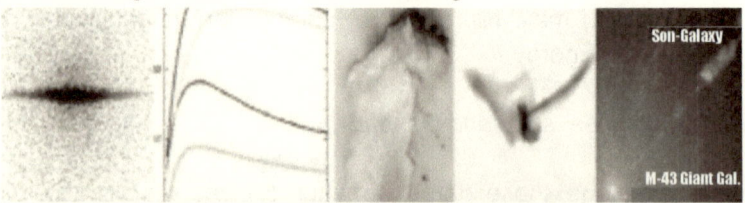

In the graph: a galaxy with a flat star body and a central brain,
occupied by black holes with a hyper-developed dimension of
informative height, perpendicular to the body; different biological
curves of stars reproduction; a region rich in atomic energy where
those stars are born; a galaxy with the form of an organic system
that remembers a ray fish and a giant galaxy, which re=produces a
seminal jet of matter, giving birth to a younger galaxy.

The 3rd age of the Universe, started once galaxies were born, with central black hole brains and star bodies that feed on interstellar gas. They grew and socialized, structuring galactic networks with the 3 basic space-time forms: Lineal walls, spiralled systems and cyclical, globular forms and disks. Those 3 galactic networks shape a mature Universe that shows in computerized models a surprising similarity with the discreet networks of a living tissue.

As the Universe gets older, it warps and creates *a hyper black hole* of max. informative density (Planck's density) that becomes the center of a hyper-vortex made of galaxies.

It is then, when according to the cellular hypothesis, we might have 2 outcomes, depending on the existence of a complex macro-organism of which the Universal cell is a part or not:

+1: The Universe evolves socially as it becomes a "cell" of a hyper-universe from a bigger i-scale . . .

—1: Or it will collapse around its "informative" center, as stars do in elliptical galaxies, feeding its black hole and dying. It will be the "big-crunch", in which the hyper-black hole will devour all galaxies. Then the Universe will die in a new big-bang that will initiate a new cell. In this case the cell Universe is like a bacteria, a single cell organism.

Though we seem to be at the beginning of the 3rd age of the Universe, an expansive organism with minimal formal structures, according to fractal and Relativity the Universe will certainly implode back. Since as we saw already, those 3 ages are mathematically equivalent to the 3 solutions of Einstein's general equations of space-time, determined by the cosmological constant that measures the energy of vacuum: λ defines the warping of the Universe, which can acquire our familiar 3 topological forms: a young plane (the present stage), a sphere or a convex, informative form (its big crunch, final state). Now in a young, expanding Universe, λ is positive, but so close to zero that when it will *diminish in the future it will become negative, inverting the* "geometry" of the Universe, warping it, till it brings it to a final big crunch. And so λ changes through the 3 ages or *solutions of Einstein's equation.*

In any case, the principle of fractal differentiation makes more relevant the space-time of our galaxy which is an implosive vortex of information. We are in a Gödel Universe, in a vortex of information called the galaxy. Thus the Energetic Universe is irrelevant to us, because we are information. We exist in the 3rd informative solution, the vortex of the galaxy and that is why the arrow of time in humanity is an informative arrow, not an energetic one. And that is why the sciences of information, not those of energy should dominate our Universities and our investments, instead of making quark cannons to find the Saint Grail of Energy, the big-bang of Earth . . .

38. The Universe as a Cell of a Bigger Organism: The Hyperuniverse

Black Hole Big-bang -> Beta Big-Bang-> Big Banging-> Steady State->Big Crunch

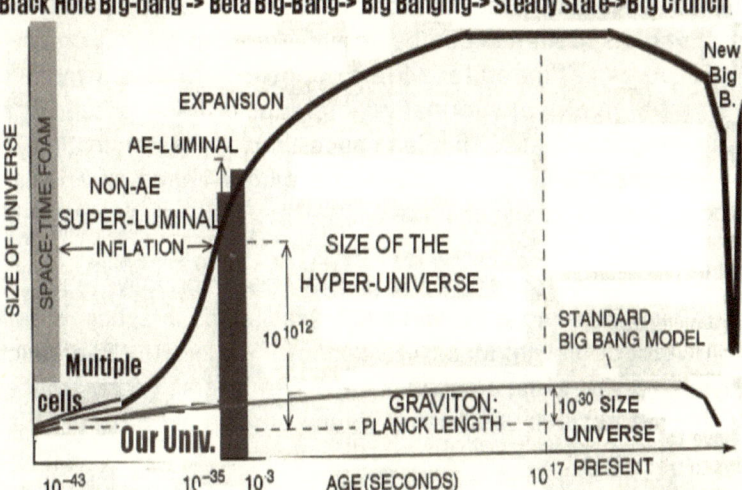

Our Universe (line below) MIGHT BE one cell among multiple cells of the hyperuniverse. All of them should go homologically through big bangs of gravitational dark energy. It is the inflationary Universe developed by Linde and String theory that fractal explain snow in organic terms: The hyperuniverse was born in a super big bang of super dark energy that created 10^{11} universal cells.

The previous cellular theory of a big bang of dark energy that created the universal cell is the most accepted today in cosmology, albeit full of irrational concepts, due to the lack of understanding of the two membranes of space-time and the superluminal speeds of dark energy.

It postulates a hyperuniverse made of multiple cellular universes, also born in a hyper big bang of hyper dark energy at z=10-100 c.

In the graph, taken from the original work of Linde, who first proposed this model, which we complete with the organic laws, ages of duality of fractal space-times, the birth of that hyperuniverse would have taken place from a black hole of the minimal scale of size of 10^{-35} m. and maximum Planck's density during a gravitational, superluminal big bang that split space-time into 10^{11} different bubbles, giving birth to many universal cells like ours in which light universes would later develop. So the hyperuniverse would extend beyond the Wall of Fire in an i+1 scale.

That gravitational background radiation resolves the mystery of the inflationary age, now canonical in cosmology since any big bang death happens in relative "zero time" when information dies and becomes a form of space. *Indeed, as we saw, death is antisymmetric to life, an arrow that has no intermediate states and lasts minimal time.* Thus, a transformation of information into energy happens in a relative zero time: from novas to atomic bombs, to universes, to human death.

Further on, since time and space are opposite parameters and a vortex is faster, more massive, informative, the smaller it is—paradoxically a big bang that starts in a smaller space-time is "faster in time"—the denser in mass it will be and result in a bigger space-time Universe. Again only the paradoxical vortex law (max. Vt =min. Re) explains this.

In more complex models of morphogenesis, informative evolutionary phases in time alternate with reproductive radiations in space, creating a quantic rhythm of growth: Max. i (Evolution)—> Max. Re (Reproduction). That rhythm applied to the creation of the hyperuniverse or any other scale, means that we observe two phases in any organic big bang/big banging process of creation:

— *Big bangs:* The system/egg first explodes its energetic membrane in a micro time period (big bang, creation of a cellular egg, beta decay, etc.). Then it continues its expansion at a slower rate as it grows in complexity inward, invaginating and creating its energetic and informative networks.

— *Big bangings:* This is the big banging and creation of dark matter and galactic networks in the Universe. In a living being is the phase of creation of inner networks between cells.

Thus, in the abstract model of the inflationary Universe, a minimal, cyclical black hole vortex of the size of a string with the density of Planck and its constants of God would create a first big bang, multiplying into 10^{11} quantic bubbles in 10^{-43} sec., each one hosting from 1 to 10 nucleons that will suffer a second dark energy big bang, creating in 10^{-35} s at superluminal speeds 10^{11} parallel future galactic regions.

If we accept a super big bang, then the Universe becomes a "bubble" in "abstract jargon" or a "cell" in the organic jargon of a hyperuniverse, separated by the Wall of Fire through which it might exchange dark energy with that external hyperuniverse.

The choice between the atomic hypothesis or the cellular hypothesis require a telescope that can study the Universe in infrared light at remote distances—the Webb telescope that won't blow up Earth with the quark soup that is supposed to have exploded in a hypothetical big bang of a hypothetical super black hole. The Webb will just watch the original Universe and prove or not the veracity of the cosmic big bang, which has in the past decade come under experimental siege for failing some experimental proofs already studied here—*thus favoring the hypothesis of an atomic Universe.* Indeed, the last one is the detection of a massive "dark flow" of galaxies, which seems to go well beyond the limits of a cellular Universe and is easier to explain in a "gas of galactic atoms."

In any case, the Webb telescope, by solving the relative size of the big bang and the type of hierarchical scales the organic universe displays, will do much more of what CERN's quark cannon pretends to do but at no risk, in a far more scientific manner, by studying the real thing and determining what size and what kind of structures define the hierarchies of the organic universe.

In essence, if the Webb observes beyond 13 billion light-years a region of darkness, it means the big bang of a cellular Universe is possible. And the organic universe has a higher hierarchy of superclusters of galaxies, perhaps cellular elements of hyperuniverses. If it observes a landscape of galaxies similar to the one of our cosmic region, it means the big bang is not the correct cosmological theory and the fractal universe extends to infinity; as each galaxy is self-similar to an atom of the next scale, whose size is astonishingly huge. And vice versa, humans might be universes made of infinite galactic atoms.

This philosophical view will be far more appealing than a mere big bang of a cellular Universe for metaphysical reasons:

— A Universe in which the two self-similar scales of quantum and cosmological vortex of space-time (charges and masses) are truly self-similar is truly infinite in space and infinite in scales of temporal information and complexity. Then we, humans are really nothing but a "mush over a corner of the Universe." But in second thoughts, such structure means that we humans are also "all," as each of us has infinite Universes within his self.

— Further on, while size is absolutely relative and so our position on those scales doesn't matter, we are an incredibly complex fractal structure of energy and information, which in the fractal models of biology developed by this author needs 10 "membranes" of space-time to be described. Hence, even if our size is minimal our *informative complexity is astonishing; and we are from the perspective of the arrow of information, one of the supreme beings of creation.*

This is not the case in the energetic big-bang of simple physics, where only energy and size matters and so we are nothing but dust of space-time.

And indeed, we will soon become dust of space-time, as Einstein et al. have failed all our attempts to close down the factory of dark, quark matter, the most explosive substance of the Universe, and have failed to teach those children of thought working at CERN, the science of the future, the fractal paradigm. So we are in the worst possible scenario, a group of children of thought, with a cosmic weapon, who don't want to go back to school and learn the science of the future and are going to play a quantum Russian roulette with the Earth, with 4 bullets/events in the gun. If you are reading this book is because their first shot did not kill us the thirty of March of 2010, the next shot will be in the fall . . . of man, a fractal of the infinite Universe who defied the thoughts of God, I have explained in this book.

---

[1]   I preferred not to swamp this appendix with notes, against the advice of my friend Mark, one of the members of the group Einstein et al. and expert in safety issues and publishing papers on scientific magazines. This decision is based on my beliefs that science is not produced by "human authority" (quotes) but by truth (the laws of veracity of the scientific method); and in doing so it becomes the superior form of human thought. This means that if the reader has enough truth in himself, he should reason and understand that the fractal Universe is the new paradigm of truth in science that substitutes the quantum paradigm and takes both, Darwin and Einstein a step forward as two self-similar theories

on how "time curves" space, evolves energy into physical and biological information. In that regard, as Einstein, the last well-known scientist that published papers without quotes, put it, when German physicists wrote "100 scientists against Einstein," if the fractal paradigm is wrong, a single proof of falsity would be enough. So far, we have produced many proofs of falsity of Higgs, Hawking's radiation and the big-bang. Yet according to the scientific method (Popper), the reader needs only one proof of falsity on what we have said to prove us wrong.

www.ingramcontent.com/pod-product-compliance
Lightning Source LLC
Chambersburg PA
CBHW031808170526
45157CB00001B/3